AF877678

Longitude West from Greenwich
Longitude West from Paris
This Lake, the limits of which are very imperfectly known from the Journals of Father Escalante, is perhaps the Teguayo Lake, from the borders of which, according to some Historians, the Azteques removed to the River Gila.
LAKE TIMPANOGOS
Sierra de Timpanogos
Yamparicas Indians
Sierra Verde
Savannahs which extend to the East of Sierra Verde
Rio de S. Clemente
Rio de S. Buenaventura
Principal sources of the R. Colorado
Raguapui Indᵗ
Valle Salado
The Western limits of this Salt Lake are unknown
S. Bernardo
Yutas-Tabeguachis Indians
Asuncion
Rio del Sacramento
Rio de Napestle
R. Dolores
It is not known at New ... by what name the R. N... is known in Louisiana ... May it be the Arkansa...
Llanos Playa de Nuestra Señora de la Luz
S. Rustico
Valle S. Jose
Ind. Quacaros
Country seen by Father Antonio Velez y Escalante
S. Maria de las Nieves
Rio Nabajoa
Apaches Llaneros
Here is a chain of Mountains extending to 110° West Lon. from Paris. It is crossed by a river called by the Missionaries Rio de San Felipe, the borders of which are inhabited by the Nochi Indians.
El Rastrillo
S. Gertrudis
Puerto de Bucarelli
Moqui Indᵗ independant since the year 1680
Indᵗ of Nabajoa
R. Nabajoa
Oraybe
Cosninas Indians
Jumbuicrariri Indians
Chemeguaba Indians
Mosanais
Jongoapi
Gualpi
El Moqui
Unknown Country between the Jaquesila and the R. Gila
It is believed at New ... the River, which rises ... East of the Village ... receives the R. Mora ... which in Louisiana is ... Red River. This last ... Oxen and Black River ... the Missisipi below F...
Albuquerque
All this Country, from the Mission of Zuñi to the Cobaji Indians has been visited by Father Pedro Font in 1775.
Yamaya Indᵗ
Yabipias Indians with long beards
APACHERIA
Rio Colorado
Jalchedum Indᵗ
Tejua Indᵗ
Jeniguech Indians
Cocomaricopas Indᵗ
Jumas Indians
Rio Gila
Las Juntas
Cajuenches Indᵗ
Cucapa Indians
Ruins of Casas grandes Second abode of the Azteques from whence they passed by Tarahumara to Huicolhuacan
Valley of Florida
Apaches Gilenos Indians
Pres. de Tubson
S. Xavier de Bac
Pres. de S. Cruz
Unknown Country between the R. Pecos and the sources of the R. Colorado
Popagos Indians
PROVINCIA
Pres. de Tubac
Tumacacori
Terrenate
Buenavista
Sarie
Tubutama
Suanca
Cocospera
Apaches Chiricagui
S. Bernardino
Missions de la Pimeria Alta
GULF OF CALIFO...
Seris Indians
Pres. del Altar
S. Ignatio
Bacanuchi
Cucurpe
Pres. de Bacuachi
Arispe
S. Lorenzo
Caborca
Nueva Andalusia
DE
S. Ildefonso de Ceneguilla
Nacosari
Basochuca
Pres. de Bavispe
Casas grandes Third abode of the Azteques
Ojo Caliente
Las Boquillas
Pres. del Carizal
Los Patos
El Paso
Al Ojito
L. de Candelaria
Apaches Mescaleros
R. del Norte or Las Juntas
Tuape
Opodepe
La Tarahumara
Banamitza
Hucpaca
Sonora
Babiacora
El Populo
Pres. de Horcasitas
S. Antonio
Concepcion
Nacori
Quisuani
S. Jose de Gracia
Hiquane
El Aguage
SONORA
S. Marcial
Hostimuri
S. Francisco Xavier
S. Antonio de la Huerta
S. Jose de Pimas
Cocomarachic
Matachic
Pres. de S. Buenaventura
L. de S. Martin
Agua nueva
Pres. de S. Carlos de Cerro gordo
S. Jose del Potrero
S. Bernardino
Tonachi
Cerro de Sᵗᵃ Clara
I. de la Tortuga
Mⁿ de S. Gertrudis
BOLSON

QE 142 N489 1984 c.2

RECEIVED
MAR 10 1986
LIBRARY

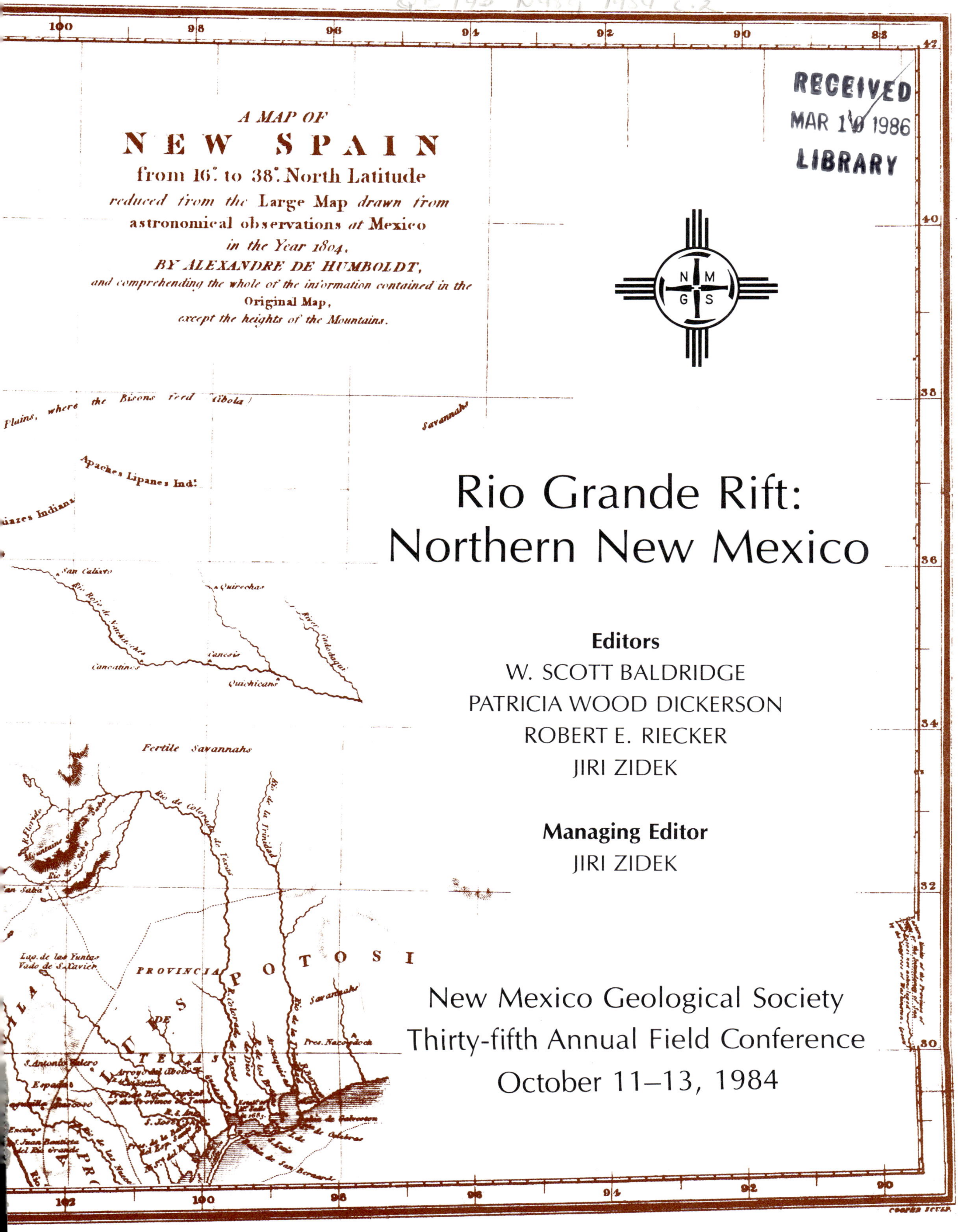

Rio Grande Rift: Northern New Mexico

Editors

W. SCOTT BALDRIDGE
PATRICIA WOOD DICKERSON
ROBERT E. RIECKER
JIRI ZIDEK

Managing Editor

JIRI ZIDEK

New Mexico Geological Society
Thirty-fifth Annual Field Conference
October 11–13, 1984

CONTENTS

ARTICLES

Rift Processes

Tectonics, Structure, and Geophysics

Stratigraphy, Sedimentology, and Paleontology

Volcanic Geology

Precambrian Geology

Economic Geology

Geomorphology and Quaternary Geology

Hydrogeology and Water Geochemistry

Archaeology and History

FIELD-CONFERENCE SCHEDULE AND ROAD LOGS

NMGS COMMITTEES AND CREDITS

EXECUTIVE COMMITTEE

Jeffrey A. Grambling, *President* .... University of New Mexico
Jerry E. Mueller, *Vice President* .... New Mexico State University
Kay S. Hatton, *Treasurer* .... New Mexico Energy and Minerals Department
David W. Love, *Secretary* .... New Mexico Bureau of Mines and Mineral Resources
Russell W. Jentgen, *Past President* .... U.S. Bureau of Land Management

CONFERENCE

Robert E. Riecker, *General Chairman* .... Los Alamos National Laboratory

REGISTRATION

Protocol Office, Los Alamos National Laboratory

TECHNICAL ASSISTANCE

Schlumberger Well Services .... Beverages en route
Welex, a division of Halliburton Services .... Mobile sound equipment
Los Alamos National Laboratory .... Photography, drafting, typing, editing
New Mexico Bureau of Mines and Mineral Resources .... Drafting, typing, editing

CREDITS

Title page: Modified by J. Paskiewicz from a map of New Spain (1810) by A. von Humboldt; by permission of Barker Center, University of Texas at Austin
Satellite photo: LANDSAT false-color composite image of part of Rio Grande rift in northern New Mexico obtained on 19 September 1975 (frame no. E-2240-17022)
Printer: University of New Mexico Printing Plant

To order New Mexico Geological Society publications,
write to: *NMGS, Campus Station, Socorro, NM 87801, U.S.A.*

COPYRIGHT © 1984 by the New Mexico Geological Society, Inc.

The articles and road logs in this guidebook were prepared for the 35th annual field conference of the New Mexico Geological Society, held in Taos, New Mexico, on October 10–13, 1984. No part of this publication may be reproduced, stored in a retrieval system, or transmitted, in any form or by any means, electronic, mechanical, photocopying, recording, or otherwise, without the prior *written* permission of the New Mexico Geological Society, Inc.

PRESIDENT'S MESSAGE

Welcome to the Thirty-fifth Field Conference of the New Mexico Geological Society, co-sponsored this year by the Los Alamos National Laboratory. Another guidebook has made its way to press. Special thanks are due to Scott Baldridge, Pat Dickerson, Bob Riecker, Jiri Zidek, Kim Manley, and Steve Blodgett for their superhuman efforts in finishing the book, Jim Bones for contributing the frontispiece, and Nancy Young for permitting us to reproduce her art on the guidebook cover.

The Society suffered a near-disaster last year when Jon Callender resigned as managing editor. For those of you who don't know Jon, he talks fast, persuasively, and has very long arms which help him pull manuscripts out of recalcitrant authors. Fortunately, we were able to enlist an equally qualified replacement. Jiri Zidek, our new managing editor, is the founder and past editor of the Journal of Vertebrate Paleontology and the current chief editor–geologist for the New Mexico Bureau of Mines and Mineral Resources. This guidebook could not have been completed without Jiri's dedication. Jiri is quieter than Jon, but should instill the same terror into the hearts of future guidebook authors.

Seeing my name under the President's Message may come as a surprise to some of you. Dave Norman was duly elected to this office by the membership, but, fortunately for him and unfortunately for the Society, was awarded a sabbatical leave for 1984–85. His departure resulted in my promotion to President, and we have been fortunate enough to have Jerry Mueller volunteer to serve as Vice President for the remainder of 1984.

Enjoy yourselves in Taos, it is a fine time of the year in New Mexico's northern mountains. I hope to see you again at the 1985 conference in east-central New Mexico, to be coordinated by Spencer Lucas of the University of New Mexico.

Jeffrey A. Grambling, President

PREFACE

The theme of the 1984 New Mexico Geological Society field conference is the Rio Grande rift. The rift is a fitting subject for this first topical conference of the Society: It has become recognized in the last decade or so, and especially since the 1978 International Symposium in Santa Fe, as a major late Cenozoic extensional feature of the continental lithosphere. As such, the rift has become the focus of an increasing amount of interdisciplinary research directed toward developing a broader understanding of the dynamics and thermal history of the lithosphere.

Since it is impossible for a single field conference to cover the entire rift, this year's guidebook focuses only on its northern New Mexico segment. In northern New Mexico, the rift is tectonically and morphologically well defined; hence, descriptions and discussions of this region are relevant both to the rift and to rifting processes generally.

Although the field conference is topical, we have nonetheless encouraged all papers of relevance to northern New Mexico and southern Colorado, for two reasons. First, we felt it was important to continue the Society tradition of soliciting all work related to the region. And secondly, while the rift is a modern tectonic feature, many of its characteristics and perhaps even its existence are determined by geological events certainly as old as late Paleozoic and perhaps even as old as Precambrian. Hence, work which may not seem rift-related to us at this time may eventually turn out to be relevant.

The Rio Grande rift, largely because of its control on drainage, has strongly influenced Indian and Spanish settlement and trade patterns. In recognition of this intimate relationship between the geology and the culture and history of northern New Mexico, this year's guidebook includes an integrated group of papers on the archaeology of northern New Mexico. Three of these papers, by Peckham, Schroeder, and Cordell, were originally presented as part of a lecture series at the College of Santa Fe entitled *Man on the Rio Grande: 10,000 Years of Human History,* organized by John Ware and Myra Ellen Jenkins.

We wish to thank K. Manley for putting together this year's road logs and coordinating the field trips; J. M. Casey, J. A. Grambling, P. W. Lipman, S. J. May, and J. C. Reed, Jr. for their contributions to the road logs; S. Blodgett for aid with road logs, technical editing, and logistics; J. Ware for coordination and assistance in editing of the archaeological papers; G. Ponder for editorial assistance with the initial manuscripts; D. Witt of the Harwood Foundation of the University of New Mexico and R. Richardson of the Colorado Railroad Museum, Golden, Colorado, for historical photographs; A. Bowman for contemporary photographs; D. Carlton and T. Ellis for locating the map on title page; J. Tubb, A. Garcia, and J. Paskiewicz for assistance with illustrations; B. Hahn, P. O'Rourke, B. Reecer, and N. Warnes for word-processing support; L. Woodwell for help with logistical arrangements; and the many reviewers of manuscripts for their time and effort. We are especially grateful to N. Young for the cover print and J. Bones for the frontispiece and end plate. Finally, we acknowledge the support of the Los Alamos National Laboratory and the New Mexico Bureau of Mines and Mineral Resources.

W. Scott Baldridge, Los Alamos National Laboratory
Patricia Wood Dickerson, CONOCO Inc., Midland, Texas
Robert E. Riecker, Los Alamos National Laboratory
Jiri Zidek, New Mexico Bureau of Mines & Mineral Resources

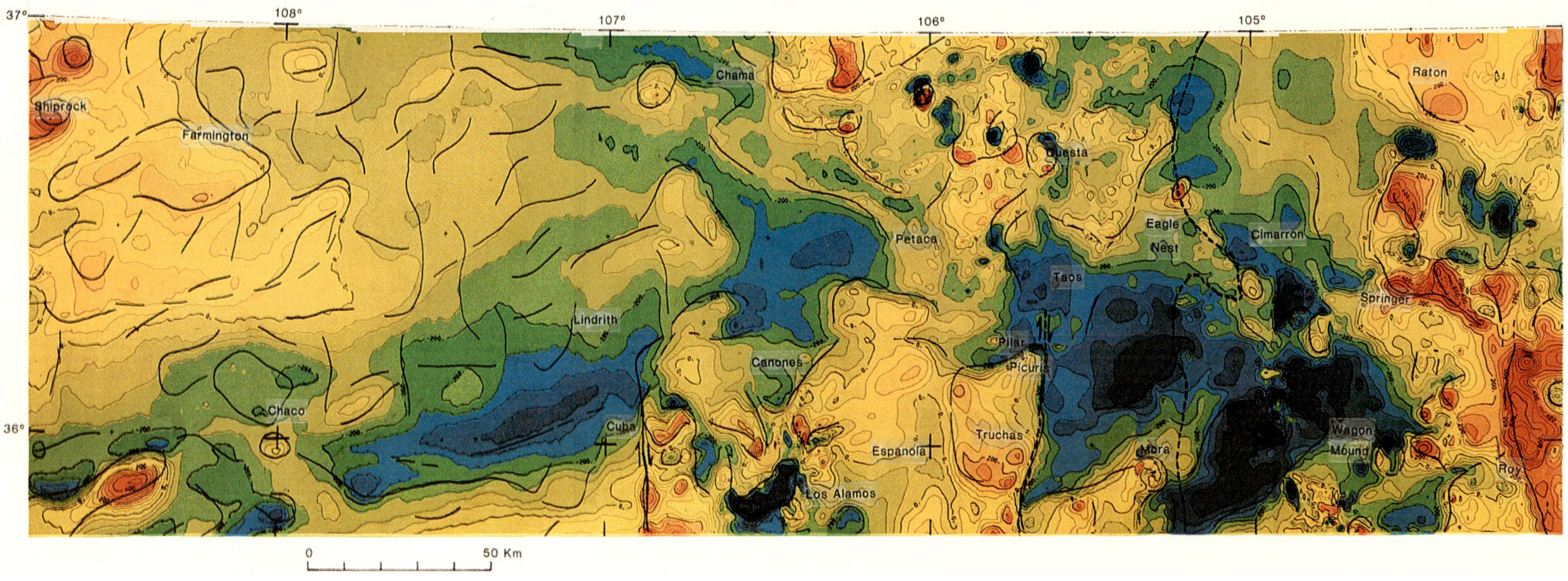

Cordell and Keller, this guidebook, p. 21

FIGURE 1. Residual total-magnetic intensity; contour and color-zone intervals are 50 gammas. Red is high and blue is low. Data from Cordell (1984). Black lines delineate magnetization boundaries and geological features discussed in text. Latitude 37° is the New Mexico–Colorado border.

Cordell and Keller, this guidebook, p. 21

FIGURE 2. Residual gravity map derived from terrain-corrected Bouguer gravity anomaly (Cordell, Keller, and Hildenbrand, 1982), minus second-order orthogonal polynomial surface. Reduction density is 2.67 gm/cm^3. Contour interval is 2 mgals and color-zone interval is 4 mgals, relative to an arbitrary datum. Red is high and blue is low. Red lines indicate magnetization boundaries, from Figure 1. Narrow black lines indicate density boundaries, heavy black lines with dumbbells indicate density boundaries along major faults of Rio Grande rift, as determined by gravity data. Short-dashed lines indicate geological features discussed in the text. Exposures of Precambrian rocks indicated by lined pattern. Latitude 37° is the New Mexico–Colorado border.

105°
106°
107°
108°
38°
37°
36°
Goodpasture
Walsenburg
Trinidad
Raton
Springer
Roy
Wagon Mound
Weston
Cimarron
Westcliffe
Eagle Nest
Mora
Fort Garland
San Luis
Mineral Hot Springs
Alamosa
Questa
Taos
Pilar
Picuris
Truchas
Petaca
Monte Vista
Espanola
Los Alamos
Canones
Chama
Cuba
Lindrith
Dolores
Durango
Chaco
Farmington
Cortez
Shiprock
0
60 Km

STRATIGRAPHIC NOMENCLATURE

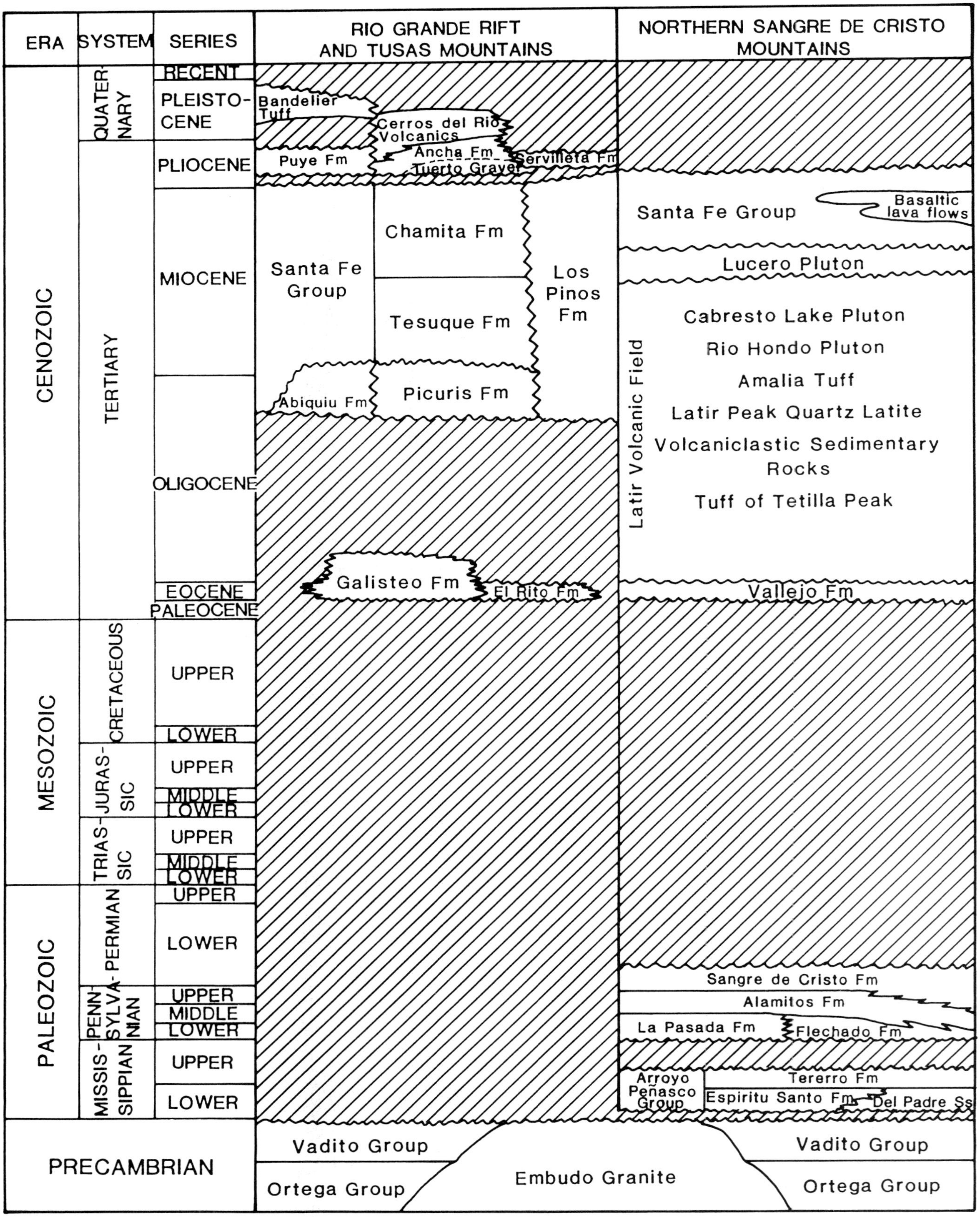

RIO GRANDE RIFT: PROBLEMS AND PERSPECTIVES

W. SCOTT BALDRIDGE[1], KENNETH H. OLSEN[1], and JONATHAN F. CALLENDER[2]
[1]Earth and Space Sciences Division, Los Alamos National Laboratory, Los Alamos, New Mexico 87545; [2]New Mexico Museum of Natural History, Albuquerque, New Mexico 87194

INTRODUCTION

Our intent in this paper is to present new ideas concerning the Rio Grande rift which we hope are at least provocative, if not correct, and to emphasize important unanswered questions which will be fruitful areas for continued research. Our approach is to evaluate, relate, and interpret existing data derived from a number of disciplines, rather than primarily to present new data. We hope thereby to achieve a synthesis of sorts—a new and broad perspective on the Rio Grande rift. This perspective should help guide future observations by directing attention toward the more fundamental questions regarding rifting processes. In this paper we use the Basaltic Volcanism Study Project (1981, p. 838) definition of rifts as elongate depressions overlying places where the entire thickness of the lithosphere has ruptured in extension.

Among the topics and ideas that we address are: (1) the regional extent of the Rio Grande rift, (2) the structure of the crust and upper mantle, (3) whether the evidence for an "axial dike" (i.e., a composite mafic intrusion) in the lower crust is compelling, (4) the nature of faulting and extension in the crust, and (5) the structural and magmatic development of the rift. These issues provide important constraints on the nature of thermal and tectonic processes involved in formation of the Rio Grande rift.

The Rio Grande rift is a region where the lithosphere is being permanently altered through thinning and (probably less importantly) intrusion of mafic magmas. The rift is the culmination of a long and complex geologic history. Initiation of rifting probably resulted from plate-boundary interactions along the west coast of North America (e.g., Lucas and Ingersoll, 1981). However, the presence of a major thermal event associated with the immediately preceding subduction regime may have been a necessary condition for rifting.

DESCRIPTION AND REGIONAL EXTENT

The Rio Grande rift extends as a well-defined series of asymmetrical grabens from Leadville, Colorado, to Presidio, Texas, and Chihuahua, Mexico, a distance of more than 1,000 km (Fig. 1). North of Socorro the rift is a distinctive morpho-tectonic feature. The main rift grabens have undergone vertical structural offsets of as much as 6 km (e.g., near Bernalillo). South of Socorro the rift is not physiographically distinctive, yet can be distinguished from the adjacent Basin and Range province by a variety of geologic and geophysical signatures (Seager and Morgan, 1979).

Over much of its length the rift is part of a broad region of "rift-like" late Cenozoic extensional deformation, i.e., a region characterized by large crustal blocks separated by steeply dipping normal faults. In central New Mexico, west of Albuquerque and Socorro, this region is more than 200 km in width, extending southwestward across the physiographic Colorado Plateau to Springerville, Arizona, and perhaps farther (Baldridge and others, 1983). A broadly linear, northeast-trending array of late Cenozoic volcanic fields, commonly referred to as the Jemez lineament, separates this extended terrain (transition zone) along the southeastern margin of the plateau from the less deformed "core" to the northwest. The Jemez lineament must correspond to a major boundary or zone of weakness in the lithosphere. However, it is not an expression of a fault or fracture zone and does not correspond to any single, simple structure in the upper crust (Baldridge and others, 1983). Farther south, the extended region encompasses the entire Mogollon–Datil volcanic field. In southern New Mexico and northern Chihuahua, the rift is not physiographically distinguishable from the Basin and Range province extending across southern Arizona and southern New Mexico.

Although the style (i.e., extensional deformation) and timing of structural deformation of the entire extended region are similar to that of the main rift grabens, the magnitude of deformation is much less. Vertical offset on normal faults of the Colorado Plateau transition zone probably does not exceed a few hundred meters. In the Mogollon–Datil region of southwestern New Mexico, narrow, deep grabens exist near Datil, Reserve, and Silver City, but these are separated from the larger, main grabens of the rift by a region 100 km or so wide, in which extensional deformation formed only shallow grabens.

This recognition of a broad, "rift-like" region is fully in accord with definitions of the rift based on geophysical and topographic criteria (e.g., Cordell, 1978; Ander, 1981). Whether the term "Rio Grande rift" is applied to the entire region (e.g., Cordell, 1978), or is restricted to the deeper, physiographically distinguishable main grabens, as we prefer, is largely a semantic argument. The important point is that the continuity of structural style and timing across this broad region clearly indicates a continuity in underlying process. That is, formation of the Rio Grande rift and Colorado Plateau transition zone and breakup of the Mogollon–Datil volcanic field all resulted from, and were part of, the very widespread Basin and Range deformational event.

LITHOSPHERIC STRUCTURE

The fact that deformation associated with the Rio Grande rift affected such a large region suggests that it resulted from an event that perturbed the entire lithosphere. The structure of the lithosphere beneath the rift is anomalous in many respects (Fig. 2):

(1) A moderate amount of crustal thinning has taken place beneath the axis of the rift. At 35°N latitude, depth to Moho is 33 km beneath the Albuquerque–Belen Basin compared to about 45 km under the Arizona–New Mexico border and to 50 km under the Great Plains (Olsen and others, 1979). North–south interface dips are small (0–2°) so that uplifted Moho beneath the rift exists from the southern border of New Mexico northward to Colorado. Discontinuous reflection segments on COCORP lines within the rift basin have two-way travel times that correspond well to a depth of 33 km (Brown and others, 1979).

(2) The sub-Moho compressional velocity (P_n velocity) beneath the rift axis is 7.6–7.7 km/s, significantly lower than velocity of "normal" mantle (8.0–8.2 km/s) beneath the Great Plains. Gravity modeling suggests that a low-density layer, which we presume correlates with this low P_n material, persists westward from the rift beneath the Colorado Plateau at least as far as northeastern Arizona, where P_n has been directly measured at 7.8 km/s along the reversed Chinle–Hanksville line. Black and Braile (1982) argue that such low P_n values are primarily due to high temperatures in the mantle. Velocities of 7.6–7.9 km/s therefore strongly suggest that the asthenosphere is in direct contact with the base of the crust beneath the rift and southeastern Colorado Plateau, without any intervening normal mantle.

The very broad (long-wavelength) gravity low extending across the entire state (about 500 km) at the latitude of Albuquerque (Fig. 2) is related in part to the shape of the low-velocity zone in the upper mantle (the asthenosphere). Other east–west gravity profiles between latitudes 32° and 38° also show this asthenospheric diapir, which thus forms a ridge-like Moho upwarp approximately parallel to the surface trace of the rift (Cordell, 1978; Ramberg and others, 1978). Both heat-flow and electrical-conductivity measurements support this gravity picture of an asthenospheric upwarp. The axis of the asthenospheric low does not coincide with the axis of the rift, but instead is displaced about 200 km westward.

(3) The compressional velocity in the lower crust beneath the rift (6.4–6.5 km/s) is substantially less than lower-crust velocities (6.7–6.8 km/s) under the flanks both to the east and west. Lower-crust velocities on the order of 6.5 km/s are rather uncommon in continental North America, existing mainly in the Great Basin section of the Basin and Range province and in the Rio Grande rift (Prodehl, 1979; L. Braile and others, written comm. 1984). We interpret such subnormal lower-crust velocities to be mainly due to high crustal temperatures and to suggest that no major compositional difference exists between the lower

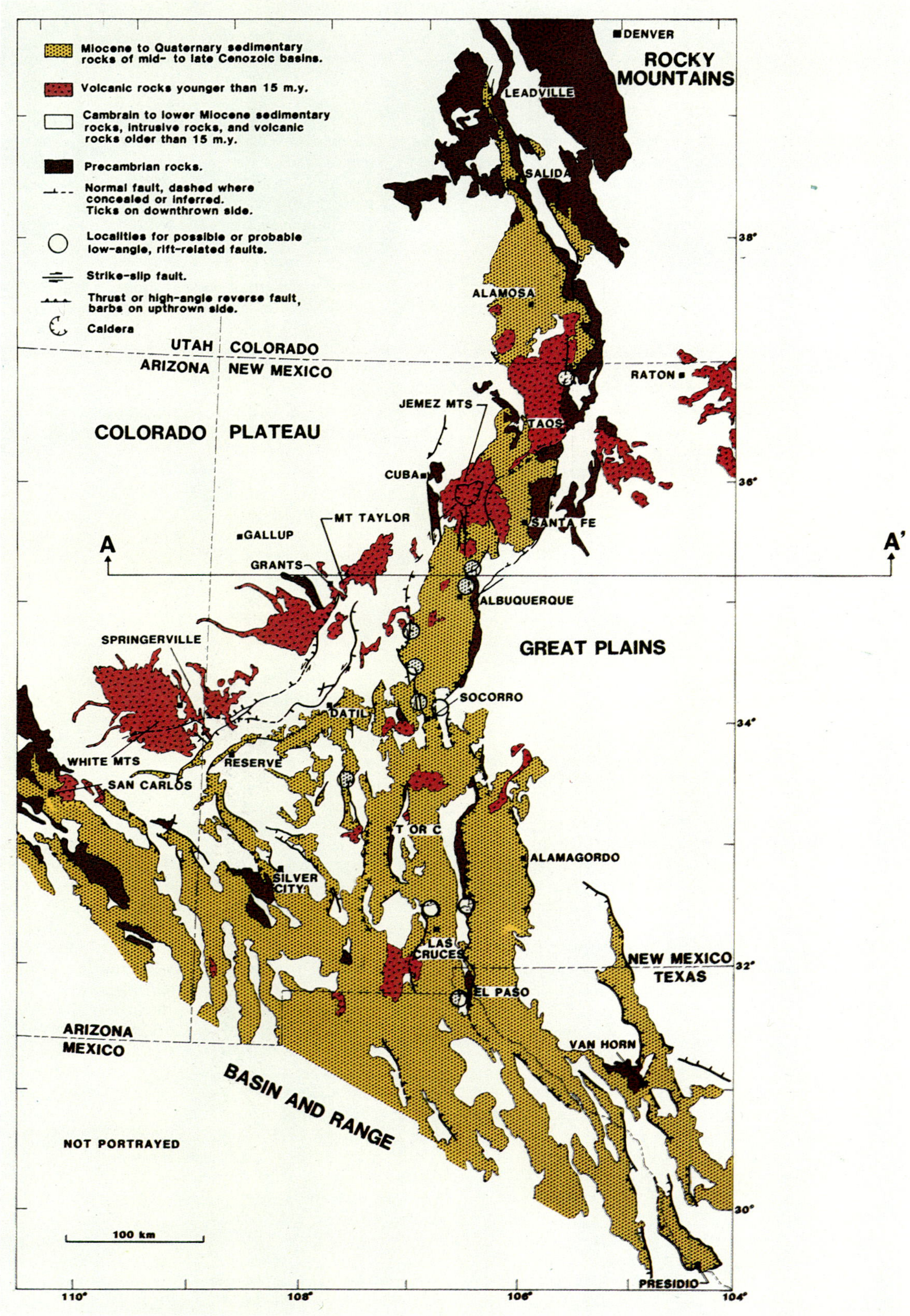

FIGURE 1. Generalized tectonic map of the Rio Grande rift. After Woodward and others (1978), Tweto (1978), NMGS Map (1982), and Baldridge and others (1983). A–A′ shows location of cross section in Figure 2.

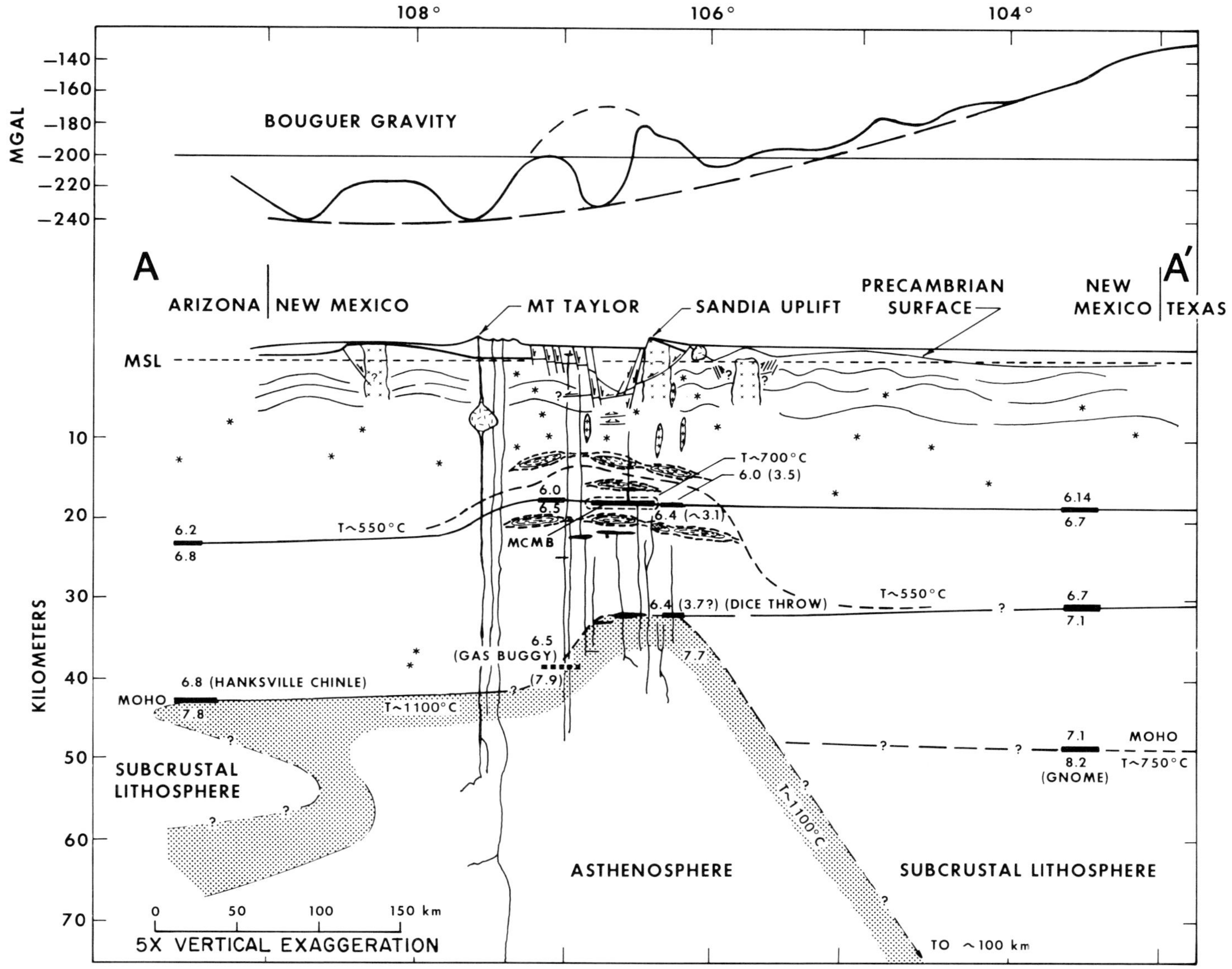

FIGURE 2. Cross section through northern Albuquerque–Belen Basin. See Figure 1 for location. Locations of intracrustal boundaries, interpreted from travel times, and record-sections of correlated phases are plotted as short, heavy lines at the longitudes where seismic profiles intersect the cross section. Numbers are P-wave velocities in km/s (numbers in parentheses are S-wave velocities). Profiles are identified by locations of the shotpoints (Chinle–Hanksville: Roller, 1965) or by code names of the main source explosions (DICE THROW: Olsen and others, 1979; GASBUGGY: Toppozada and Sanford, 1976; GNOME: Stewart and Pakiser, 1962). The Chinle, GASBUGGY, and GNOME profiles were obtained at a relatively early stage of U.S. crustal-profiling efforts, when station spacing was relatively coarse (10–50 km) and true amplitude/waveform data were inadequate to permit more than estimates of possible velocity gradients or fine structure in the principal crustal layers. DICE THROW average station spacing was about 3 km, allowing better gradient estimates using modern synthetic seismogram-modeling techniques. Generalized distribution of earthquake hypocenters in the upper crustal layer is shown by asterisks, as are the specific deep crustal events of 1976/1977 (Sanford and others, 1979). A Bouguer gravity-anomaly profile at the latitude of the cross section is shown at the top (Cordell and others, 1982). The midcrustal magma body (MCMB) at Socorro has been projected into the cross section. Locations of basaltic dikes, though schematic, are based partly on heat-flow data (e.g., Clarkson and Reiter, this guidebook). Lensoidal "megaboudins" in the middle crust represent region of discontinuous ductile flow (see text).

crust beneath the rift and that adjacent to it. The intermediate-wavelength gravity high (Fig. 2) is often interpreted to indicate mafic igneous rocks intruded into the lower crust (an "axial dike" or, more properly, composite batholith). However, in the Rio Grande rift this intermediate-wavelength gravity signature can as well result from the equivalent replacement of lower-density crustal rock with higher-density mantle material because of the thinning of the lower crust. If there were a large intrusion of mafic material, such as a dike swarm, in the lower crust, we would expect a very considerable increase in density and hence a corresponding increase in seismic velocity (to 7.3–7.7 km/s) of the lower crust. In fact, the seismic results show a *decrease* in lower-crust seismic velocity compared to values in the flanks. Therefore, we do not consider the gravity high to be a persuasive argument for a large, lower-crust mafic-dike system; we believe crustal thinning can account for almost the entire effect.

The GNOME profile data (Stewart and Pakiser, 1962) also indicate a third (7.1 km/s), deeper layer beneath the usual 6.8 km/s lower-crust layer along the eastern New Mexico/west Texas border area. Deep, higher-velocity layers are commonly observed in craton and platform areas in other parts of the world and possibly reflect the growth of continental crust through such mechanisms as underplating. Another refraction line running eastward from GNOME (Romney and others, 1962) into central Texas was too sparsely instrumented to clearly show how far eastward this third crustal layer may extend.

(4) Unusually strong wide-angle reflections at precritical distances from refraction profiles in the southern Albuquerque Basin imply the presence of a thin, tabular, low-rigidity (i.e., abnormally low S velocities) layer at midcrustal depth (about 20 km) in the vicinity of the Socorro "midcrustal magma body" (Olsen and others, 1982). The refraction observations suggest that this anomalous sill-like layer probably surrounds, but is more extensive than, the partial melt or magma body between Socorro and Belen that has been extensively documented by Sanford and coworkers (Reinhart and others, 1979; Sanford and others, 1977). We tentatively correlate this feature with the boundary between the upper and lower crustal layers (Conrad discontinuity).

NATURE OF THE MOHO

Beneath the axis of the Rio Grande rift, existing seismic-refraction data (mainly from the DICE THROW profile) indicate that the Moho is a sharp transition at the resolution of 1.0–1.5 km provided by the dominant 1–10-Hz frequencies characteristic of regional earthquake and explosion sources. On the other hand, the more restricted-range COCORP-type deep-reflection experiments, where dominant frequencies are in the 10–80-Hz range, reveal a more complex, sandwich-like structure in the same region. Complex lamellae show considerable lateral variation, but the overall trend follows essentially the lower resolution "boundary" determined over much greater distances by refraction/wide-angle-reflection experiments. Thus, the few-hundred-meter-resolution reflection data indicate that the sub-rift Moho is a complicated structure which, in addition to representing a major petrological and compositional change with depth, has had a complex metamorphic and horizontal-stress history. The relatively small amplitudes of the P_n phase observed by refraction techniques along the Rio Grande rift axis indicate that the P-wave velocity in the asthenosphere actually decreases slightly with depth. Such decrease suggests that the sub-Moho temperature gradient is small and that nearly isothermal conditions apply in the top few kilometers of the asthenospheric upwarp.

Because refraction data for the adjacent Colorado Plateau and Great Plains are relatively sparse, with correspondingly more uncertain variations in amplitude vs. distance behavior of the relevant seismic phases used in analysis, we cannot describe details of the Moho transition and sub-Moho properties with the same confidence. However, by analogy with data from plateaus and shields in other parts of the world, we expect some significant differences from what has been inferred for the central part of the rift (Mueller, 1977).

NATURE OF THE CRUST

Our ideas concerning the nature of the upper crust and the mechanisms of crustal extension are greatly influenced by COCORP seismic-reflection profiling in the Socorro area and southern part of the Albuquerque–Belen Basin (Brown and others, 1979, 1980), and by observations of the Precambrian basement in New Mexico. Basically, we think that the upper and middle crust, and possibly the lower crust, are dominated by horizontal structures. COCORP data show a horizontally stratified upper crust, with local, transparent volumes which Brown and others (1979) suggest may be granitic plutons (e.g., line 2A). Below about 10 km are numerous discontinuous, subhorizontal reflection segments varying in length, dip, density, and amplitude. Brown and others (1979) suggest this type of signature would be expected from a highly deformed metamorphic terrain with a dominant subhorizontal fabric.

This interpretation fits well with our ideas concerning the structure of the upper crust, clues to which are present in the basement blocks surrounding the rift. The lithology and, to a lesser extent, the structure of the Precambrian basement of northern New Mexico are relatively well understood. Three major lithologic groups of Proterozoic age are exposed in the uplifted blocks of the central rift basin: volcanic and volcaniclastic rocks, quartzites and associated mature clastic sediments, and silicic plutons. With local exception, the volcanic sequence is oldest and is overlain structurally and perhaps stratigraphically by the quartzite sequence. Both sequences are intruded by younger batholithic to hypabyssal plutonic rocks. The sedimentary and volcanic rocks are folded into large (kilometers in wavelength) overturned folds, the axial-plane foliations of which yield the seemingly pervasive structural fabric of the basement. This fabric strikes northwest to northeast, with local easterly trends, and dips at a high angle. However, the folds commonly plunge less than 50°, suggesting that stratigraphic layering, although complexly folded, actually dips at a rather low angle. The presence of a low-angle lithologic boundary is also suggested by regional aeromagnetic patterns which show simple, broad, east-trending bands of contrasting magnetic intensity corresponding to outcrops of quartzite and metavolcanics (Zeitz, 1982).

Because of the high-angle fabric so obvious in outcrop, the basement appears (although incorrectly) to be characterized by a mosaic of angular blocks separated by high-angle zones of structural weakness. Late Miocene to Pliocene high-angle rift faults tend to follow basement fabric and reinforce this view. However, structural analysis of the basement suggests that the Proterozoic deformation produced large-scale horizontal transport of the lithologic units, giving rise to nappes and overthrust sheets that were subsequently folded into the high-angle structures. Thus, there was and is a pervasive low-angle fabric in the basement.

In extrapolating to the structure of the crust, the significance of Precambrian structure and lithology is the competence contrast produced by the juxtaposition of quartzite and volcanic terranes. At high levels in the crust, the quartzite is relatively brittle and contrasts strongly with the more incompetent metavolcanics and schists that surround it. Decollement and shear along stratigraphic boundaries are normal features of these terranes. Projecting downward into the crust, it is likely that strong layer contrasts will be found in the upper 5–10 km, and that horizontal fabrics will dominate where ductility contrasts along lithologic and structural boundaries are well developed. Perhaps the high-angle fabric exposed in the uplifted blocks is a manifestation of folding in an upper plate above a delamination surface of Proterozoic age. As previously noted, seismic-reflection data support the suggestion of low-angle lithologic contrasts in the deeper parts of the upper crust.

Strong evidence of a dominant horizontal structure in the middle crust within the rift, corresponding essentially to the Conrad discontinuity, comes collectively from: (1) unusual S-to-S-wave reflections from the horizontal interface from microearthquakes that overlie the reflector; (2) strong P-to-P-wave reflections from near-vertical-incidence COCORP profiling; and (3) unusually strong wide-angle reflections on refraction profiles that can best be explained in terms of an anomalously low S-wave velocity in the material just beneath the reflector interface (Olsen and others, 1982). Combined evidence from these various seismic techniques, plus supporting evidence from electrical, geodetic-leveling, and heat-flow measurements, have led to a widely accepted interpretation that this anomaly is due to a thin, sill-like magma body or partially molten layer at midcrustal depths (18–20 km) beneath the axis of the rift. This midcrustal magma body is most strikingly manifested beneath a 1,700 km^2 "central" area extending from Socorro northward almost to Belen (Sanford and others, 1977; Rinehart and others, 1979). Within this region, Brocher (1981) has summarized the evidence indicating the total thickness of this magma body is of the order of only 1 km and consists of a highly complex, laminated series of sills of various degrees of fluidity and lateral "interfingering." Beyond the 1,700 km^2 central region, the wide-angle data (Olsen and others, 1979, 1982) imply that the horizontal tabular geometry continues outward, but that the seismic-velocity anomaly is confined almost entirely to subnormal S-wave velocities. Velocities for P waves are almost indistinguishable from those in surrounding rocks at these depth ranges. This low-rigidity aureole suggests a metamorphic-reaction zone surrounding the central midcrustal-sill complex. A surprising and not well understood property of the magma body is its very horizontal upper surface, which is "flat" within the limits of resolution (± 0.5 km) of the seismic techniques.

Similar, but sparser and therefore less conclusive, geophysical observations suggest the midcrustal anomalous layer may extend intermittently along the rift and/or into the flank regions. For example, Hermance and Pedersen (1980) argue that magnetotelluric electrical-conductivity data favor a contiguous set of tabular units with a common generic origin along the rift. Although the midcrustal layer is undoubtedly not a continuous lithologic or petrologic unit throughout and involves only a small volume of magma compared to volumes of magmatic rocks in other rifts (e.g., the Kenya rift), the layer undoubtedly plays a very important but little understood role in the style of Rio Grande rift volcanism. Thus, we have projected some of the features of the Socorro magma body into our cross section (Fig. 2) in order to emphasize our perception of its importance, even though the section is about 100 km north of the best data.

In the middle crust, but still within the seismic upper crustal layer (Fig. 2), is a transition from brittle to ductile behavior. The best evidence for the depth of this transition in the Rio Grande rift region is provided by the distribution of earthquake hypocenters. Although the precision of measurements is lower for hypocentral depths than it is for horizontal coordinates (epicenters), almost all regional earthquakes recorded since 1960 are from depths less than 20 km and most are probably shallower than 10–15 km (Sanford and others, 1979, 1981; Olsen, 1979). At strain rates applicable to crustal extension processes for the rift and Basin and Range provinces, the level of the brittle–ductile transition zone is probably at depths where temperatures are in the range 350–500°C. In some local regions of high heat flow, such as beneath the young (1.1–1.4 m.y. old) calderas of the Jemez Mountains, the brittle–ductile transition comes within only a few kilometers of the surface, creating large aseismic areas. Focal mechanisms of the shallow-crust earthquakes are consistent with E–W tensional forces. The only exceptions to the generally shallow depths of regional earthquakes were two moderate (magnitude ~4) events which occurred in 1977 near the base of the crust northwest of Mount Taylor. It is speculated (Sanford and others, 1979) that these deep events may be related to magmatic processes of the Jemez lineament/Mount Taylor system.

Curie temperatures for crustal materials (~550°C) are very similar to the higher temperature range for the brittle–ductile transformation; therefore, buried structures inferred by aeromagnetic data all exist within the brittle domain.

The nature of the lower crust is considerably more uncertain, not only because of limits of seismic resolution, but also because there are not enough exposed sections of the lower continental crust world-wide to give a representative picture of the range of lower crustal conditions. However, there now seems to be a consensus (Smithson and Brown, 1977; Smithson and others, 1981; Kay and Kay, 1981; Bott, 1982) that the lower continental crust consists predominantly of granulite-facies metamorphic rocks having an intermediate bulk composition similar to andesite or diorite. Presence of water (and other volatiles) in the deep crust is of fundamental importance in that it can greatly affect seismic and electrical-conductivity properties. However, definitive laboratory and field studies are so few that while these effects are currently being actively debated (Shankland and Ander, 1983), they remain unresolved. There is little doubt that the dominantly metamorphic nature of the lower crust implies a highly complex structural grain that we at present can describe only in general terms (Smithson, 1978). While we recognize these complexities in both the lower and upper crust, we have refrained from trying to cartoon these in Figure 2 because we do not want to have these misunderstood as specific, proposed features in the Rio Grande rift.

As pointed out above, seismic-refraction data show a lateral variation in average P-wave velocity of the lower crust from about 6.7–6.8 km/s beneath the flanks to 6.4–6.5 km/s beneath the rift axis (Fig. 2). In addition to chemical composition, several other factors, such as pressure, temperature, fluids, and anisotropy due to preferential alignment of certain minerals (usually olivine) or major cracks, can affect observed elastic-wave velocities. P-wave velocities increase with increasing pressures, especially at shallow depths (upper crust) where cracks and voids are closed. For compositions and pressures characteristic of the lower crust, laboratory measurements (Christensen, 1979) indicate the pressure derivatives of wave velocities are small enough so that P and S velocities should not be *increased* by more than about 0.1 and 0.05 km/s, respectively, throughout the 20- to 40-km depth range. Temperature coefficients are negative, however, and Christensen's (1979) measurement of various granulites showed that $-(\delta V/\delta T)_p$ is in the range of $0.5–0.7 \times 10^{-3}$ km/s/°C at approximate lower-crust pressure conditions. Interpretations of the differences in the geotherms between the axial region of the rift and the flanks beneath the Great Plains and (more certainly) the Colorado Plateau indicate that the flank–rift temperature difference in the lower crust may be about 300°C at ~20-km depths and more than 400°C at the 30–35-km level. Thus, granulite temperature coefficients in combination with high sub-rift temperatures are sufficient to decrease lower-crust P velocities from the observed values of ~6.7 km/s in the flanks to 6.4–6.5 km/s beneath the rift *without requiring any substantial changes in chemical composition* between the two regions. This result leads us to conclude that there is no persuasive evidence for a *massive* mafic-intrusive complex or dike swarm ("axial dike") in the lower crust beneath the axis of the Rio Grande rift, as has sometimes been postulated (from gravity data only) for the Kenya rift (Olsen, 1983). Certainly, lithospheric thinning beneath the central rift will generate some (predominately vertical) magma feeder conduits and local dikes that must be the main source of high average temperatures in the lower crust (e.g., Clarkson and Reiter, this guidebook). However, the volumes and masses of mafic material intruded into the lower crust must be less than a few (possibly ten) percent. Otherwise, both the lower-crust density and averaged seismic velocities would be *increased* substantially beneath the rift contrary to observations. The composition of the lower crust beneath the axis of the Rio Grande rift is a problem for further modeling studies.

The process by which large amounts (50–200%) of crustal extension are produced has been concisely described by Hamilton (1982, 1983). Hamilton's model identifies three crustal layers, each with distinctive mechanical behavior: brittle fracturing and rotation in the upper crust, discontinuous ductile flow in the middle crust, and laminar ductile flow in the lower crust. The brittle–ductile transition is near the top of the middle crustal layer. A significant feature of this model is the lenticular, transposed nature of the middle crust (see Callender, 1983, fig. 2), with lenses of more competent rocks ("megaboudins") interspersed with less competent material along ductile-shear zones. We think this hypothesis adequately explains the crustal evolution of the Rio Grande rift, and have cartooned this crustal structure in Figure 2.

As noted by Eaton (1982), the development of this type of crustal structure is strongly temperature and strain-rate dependent. In the Rio Grande rift, significant crustal extension is confined to late Oligocene/early Miocene (see next section), implying that the appropriate combination of high (convective?) heat flow, rapid strain rate, and resulting upward mass transport must have existed only at this time. Subsequently, the rift must have experienced a decrease in strain rate and upward mass transport (both plastic flow and magmatic upwelling) in a manner similar to that described by Eaton (1982) for the Basin and Range province.

Transport paths of magma through a crust altered by significant extension is presumably complicated. In the upper crust, brittle fracturing to depths of 10 km would allow relatively steep, near-surface conduits to develop, and would result in the aligned volcanic centers which are commonly observed in the Pliocene and Quaternary volcanic fields of the rift. However, middle crustal structure would be lensoidal and horizontally stratified, and the contact between middle and upper crust would probably be characterized by a low-angle detachment surface. This crustal environment might yield magma "pillows" and stratiform, sill-like intrusions in the middle crust, such as the Socorro magma body. It would also produce strong seismic reflectors near the brittle–ductile transition.

STRUCTURAL DEVELOPMENT

We must keep in mind that the picture of the Rio Grande rift presented to this point is simply a "snapshot" of an evolutionary process. Features of all ages, including those inherited from the pre-rift history, are superimposed. To understand the rifting process, we must understand structural and magmatic developments of the rift as a function of time.

The geometry of rift basins and Miocene to Pliocene high-angle faults in the Rio Grande rift has been characterized by numerous authors (e.g., Chapin, 1971; Chapin and Seager, 1975; Cordell, 1978; Seager and Morgan, 1979). Until recently, it was generally accepted that extension occurred dominantly on high-angle normal faults (e.g., Chapin, 1971; Woodward and others, 1975; Brown and others, 1980). These faults were thought by some to listrically shallow near the brittle–ductile transition (e.g., Woodward, 1977). Commercial seismic-reflection profiling in the rift (L. Russell, pers. comm. 1981), reinterpretation of the COCORP deep seismic-reflection profile (Cape and others, 1982), and recent structural analyses of rift-bounding faults (e.g., Seager, 1981; Chamberlin, 1983a; Lipman, 1983; Rhoades and Callender, 1983) now suggest this model is incomplete.

Instead, there appears to be a significant component of low-angle faulting in the Rio Grande rift. It is unclear whether low-angle faulting in the rift involves domino-style movement (Morton and Black, 1975; Chamberlin, 1983a), listric faulting (Cape and others, 1982), regional crustal shear (Wernicke and Burchfiel, 1982; Bird, 1979), detachment faulting (Frost and Martin, 1982; Callender and Zilinski, 1976; Rhoades and Callender, 1983), or some combination of the above models.

Figure 3 presents published examples of late Oligocene to Miocene low-angle faults in the Rio Grande rift. The north end of the Sandia uplift (Fig. 3A) plunges approximately 20°N. In down-plunge projection, the map of Kelley (1977) shows the San Francisco–Placitas fault zone to be a listric fault dipping west (L. Russell, pers. comm. 1981). Seismic-reflection profiling in the Albuquerque Basin west of the Sandia uplift supports this interpretation (L. Russell, pers. comm. 1982). In addition, recent work in the Placitas area confirms low-angle, northwest-dipping faults and shears within the Paleozoic section and at or near the Phanerozoic–Precambrian contact (J. Callender, unpubl. data). Rhoades and Callender (1983) described similar low-angle structures along the west side of the Sandia uplift, 10–20 km south of Placitas. In all cases, late Miocene to Pliocene high-angle normal faults truncate the low-angle structures.

Figure 3B shows the low-angle Carrizo fault of late Oligocene/early Miocene age (Callender and Zilinski, 1976) along the western edge of the Albuquerque Basin. The Carrizo fault is intruded by 27-m.y.-old intermediate and mafic hypabyssal rocks, suggesting the fault zone acted as a conduit for local magmatism. Similar structures have been described in the Joyita Hills on the eastern edge of the Socorro Basin, 60 km southeast (Rosen, 1983; Osburn and Eggleston, 1983; Smith, 1983). These low-angle faults are confined to the Phanerozoic section and are commonly associated with large ductility contrasts between stratigraphic units. The low-angle faults in the Lucero uplift and Joyita Hills are truncated by late Miocene to Pliocene high-angle normal faults.

One of the most carefully studied areas in the Rio Grande rift that contains numerous low-angle normal faults is the Lemitar Range near Socorro (Fig. 3C). Chamberlin's (1983a, fig. 5) restoration of this cross section (Fig. 3C) indicates nearly 200% of domino-style extension from late Oligocene to late Miocene time, which was contemporaneous with periods of silicic volcanism from 31 to 20 m.y. ago and from 12 to 7 m.y. ago. Late Miocene to Pliocene high-angle faulting and block-tilting are superimposed on this earlier deformation.

The Jeter fault in the Ladron Mountains (Fig. 3D), one of the first low-angle faults described in the Rio Grande rift (Black, 1964), has been characterized as a thrust fault (Black, 1964), low-angle normal fault (Kelley, 1977), domino-style fault (Chamberlin, 1983b), or detachment zone (T. Shackleford, pers. comm. 1982). It shows many of the characteristics of a regional upper-plate detachment surface (K. Gillespie-Nimick, unpubl. data), including intense brittle deformation in both sedimentary and Precambrian crystalline rocks. It is truncated by late Miocene to Pliocene high-angle normal faults and is commonly the locus for metallic mineralization.

Similar low-angle structures exist in the southern part of the Rio Grande rift (Fig. 3E–H). With the exception of the Franklin Mountains (Fig. 3H), Tertiary low-angle faults are clearly cut by late Miocene to Pliocene high-angle normal faults. In the Franklin Mountains, the age of low-angle faulting is uncertain, but comparison with the carefully studied Organ Mountains to the north (Seager, 1981) suggests that these faults are similar in style and age to late Oligocene to early Miocene low-angle faults in the Organ Mountains.

Additional low-angle structures have been described in the northern Rio Grande rift (e.g., Spiegel and Baldwin, 1963; Lipman, 1983; Muehlberger, 1979). As studies of intrabasin and basin-margin uplifts continue, undoubtedly many more such features will be found.

Thus, late Oligocene to early Miocene low-angle faults may be common features of the Rio Grande rift. They predate the high-angle normal faults of late Miocene to Pliocene age and suggest a major shift in structural style and mode of crustal extension with time. The low-angle structures overlap in time with the later phase of intense intermediate to silicic volcanic activity in southwestern New Mexico (Datil–Mogollon volcanic field), which culminated in early Miocene time. Abnormal crustal heat flow may have allowed significant high-level crustal stretching along low-angle structures. This may have been accomplished by the rise of the brittle–ductile transition in the upper crust. Convective-heat transfer, by massive intrusion of silicic magmas into the continental crust in the middle Tertiary (Elston, 1984), may have provided the thermal input to allow this stretching.

Hence, the structural development of the rift apparently occurred during two discrete time intervals: (1) an early phase characterized by low-angle faults, which began about 30 m.y. ago and lasted an estimated 10–12 m.y.; and (2) a later phase, dominated by high-angle normal faults, which extended from 9 or 10 m.y. ago to about 3 m.y. ago (e.g., Chapin and Seager, 1975; Baldridge and others, 1980; Seager and others, 1984). These phases of deformation are similar to extensional events in the Basin and Range province, where low-angle faults are typical of the early phases of crustal extension, in contrast to later high-angle deformation in the same area (Eaton, 1982; Frost and Martin, 1982). The similarities in style of deformation between the rift and the Basin and Range illustrate that both regions are manifestations of the same event.

Furthermore, possible widespread low-angle faulting early in the formation of the rift has major implications for the total amount of extension. All models of low-angle faulting demand a significant amount of crustal stretching. Extension of more than 200% is possible with this geometry (Chamberlin, 1983a), whereas high-angle normal faulting yields extension of only 5–30% (Woodward, 1977).

VOLCANISM

Most of the volcanism associated with the Rio Grande rift, with some important exceptions such as the Jemez volcanic field (Fig. 1), is basaltic and occurred less than 5 m.y. ago. [Volcanism in the Jemez field began more than 13 m.y. ago and includes enormous volumes of intermediate to silicic magmas (Smith and others, 1970; Gardner and Goff, this guidebook)]. These basalts consist of a variety of nepheline- and hypersthene-normative compositions which are generally similar to other late Cenozoic basaltic rocks of the western U.S. However, the most voluminous of the late Miocene to Recent basaltic volcanic fields, the Taos Plateau field, consists dominantly of a distinctive, low-alkali tho-

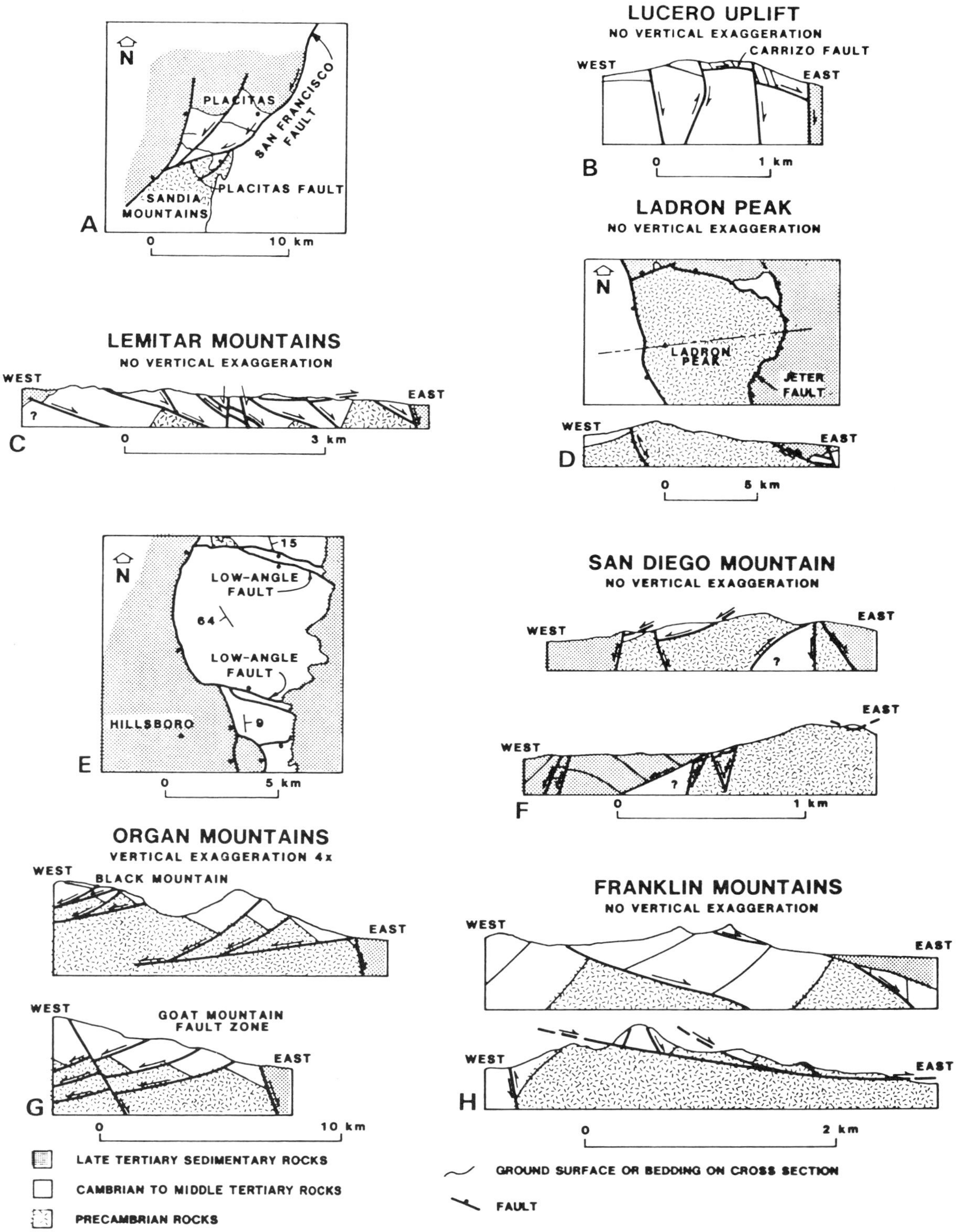

FIGURE 3. Low-angle Neogene faults in the New Mexico part of the Rio Grande rift. See text for explanation. **A,** Map of northern Sandia uplift, after Kelley (1977); **B,** cross section through northern Lucero uplift, after Callender and Zilinski (1976); **C** cross section through Lemitar Mountains, after Chamberlin (1983a); **D,** map and cross section of Ladron Peak area, after Black (1964) and Chamberlin (1983b); **E,** map of the Hillsboro area, after Kuellmer (1956); **F,** cross sections through San Diego Mountain (Tonuco uplift), after Seager and others (1971); **G,** cross sections through east side of Organ Mountains (4× vertical exaggeration), after Seager (1981); **H,** cross sections through northern Franklin Mountains, after Harbour (1972).

leiitic magma (Servilleta Basalt), with lesser volumes of more alkaline (but still hypersthene-normative) magmas. Minor quantities of intermediate to silicic magmas represent melts formed by crystal fractionation processes and by melting of the lower crust. The Taos field apparently reflected a major thermal source which generated large quantities of magmas at or near the top of the mantle, and began to involve the crust in a major way (Lipman and Mehnert, 1979; Basaltic Volcanism Study Project, 1981, pp. 108–131; Moorbath and others, 1983; Dungan and others, this guidebook).

In the central and southern rift, Servilleta-type magmas were not erupted (except for minor quantities near Mt. Taylor) and lavas are not dominantly of any single composition. Neither do their compositions correlate in any unique way with their tectonic setting. (Perhaps because so few volcanic rocks occur, their compositional complexity is the more obvious.) In the Cerros del Rio volcanic field near Santa Fe, the next most voluminous late Cenozoic basaltic field, a variety of compositions were erupted: high-alkali tholeiite, basaltic andesite, alkali-olivine basalt, and basanite (Baldridge, 1979). Farther south, both tholeiitic and alkaline magmas were erupted (Aoki and Kudo, 1976; Kelley and Kudo, 1978; Baldridge and others, 1982). East and west of the main rift grabens (primarily along the Jemez lineament) volcanic fields include more alkaline compositions (nephelinites, basanites, mugearites) than occur within the rift, but also range in composition to include high-alkali tholeiitic rocks that characterize much of the central rift. Minor volumes of Servilleta-type magmas occur near Mt. Taylor (Fig. 1) (e.g., Baldridge and Perry, 1983).

Hence, no single composition is dominant along the rift; neither is there any evolution of composition with time nor unique correlation of composition with tectonic setting. In the Taos area, Lipman and Mehnert (1975) contrasted the Servilleta basalts to basalts outside the axial graben and proposed that the basalt compositions delineated the asthenospheric upwarp. That is, tholeiitic basalts in the axial graben were derived from very shallow depths (near the base of the crust) at the top of the upwarp, whereas more alkaline basalts outside the grabens were derived from the flanks of the uplift at much greater depths. We do not disagree with this model for the Taos area; certainly, our lithospheric profile (Fig. 2) is compatible with it. We emphasize, however, that the Servilleta basalts are compositionally distinctive compared to most other basalts in the rift (and perhaps unique compared to other basalts of the Southwest?) and thus indicate that the Taos Plateau volcanic field represents a unique and localized heat source. Hence we do not generalize the petrogenetic model of Lipman and Mehnert (1975) to the rift as a whole. We think instead that the situation is much more complex. The compositions of volcanic rocks in the rift as a whole seem to indicate that basalts were generated at various levels in the asthenospheric upwelling, implying an unintegrated heat source with only local melting. The fact that volcanic fields outside of the axial grabens include some compositions more alkaline than those inside suggests that at least some magmas from deeper in the asthenosphere are erupted to the surface.

Most workers agree that, in addition to the post-5-m.y. pulse of volcanism, an early pulse also occurred (about 30–18 m.y. ago), coinciding generally with early structural development of the rift. These two magmatic events were separated by a mid-Miocene lull in volcanism (Chapin and Seager, 1975; Baldridge and others, 1980; Seager and others, 1984). We also agree with this picture of two separate magmatic pulses, but emphasize that this earlier event—as a rift-wide phenomenon involving eruption of true basalts—was very minor (Fig. 4). Most of the mafic volcanism during this interval occurred in southwestern and southern New Mexico and involved basaltic andesites (e.g., Chapin and Seager, 1975; Seager and others, 1984). In this region, the enormous amount of basaltic-andesite volcanism was related not to rifting, but to (and was part of) the Mogollon–Datil volcanic field (erupted about 43–20 m.y. ago) of dominantly intermediate to silicic magmas (Chapin and Seager, 1975; Elston and others, 1976). This does not mean that rifting was not an immediate cause of at least some of this volcanism. The extensional tectonic environment that was part of the rifting event may have allowed magmas to escape to the surface, whereas in the pre-rift environment magmas were not able to ascend. [Alternatively, these magmas may have been able to reach the surface only after the upper-crust batholiths had solidified so that they could be brittly fractured (Eichelberger and Gooley, 1977.)] It does mean, however, that the heat source which generated these magmas was part of the previous (subduction-related?) event (e.g., Seager and others, 1984) and not of rifting. This volcanic (and thermal) event obviously must have affected the lithosphere and crust, introducing compositional changes (Elston, 1984) and perhaps, through convective-heat transfer (e.g., Spohn and Schubert, 1983), initiating the rise of the brittle–ductile transition in the crust. The distinction between a prior major thermal event and one specifically associated with rifting is important in recognizing thermal constraints on the rifting process.

Much of the large volume of mafic lavas erupted 30–18 m.y. ago in southern and southwestern New Mexico is basaltic andesite related to the more silicic, calc-alkaline magmas. However, the compositions of magmatic rocks emplaced during this interval north of Socorro range greatly, with no obvious temporal or spatial pattern (Fig. 4). For example, olivine nephelinites erupted southwest and north of Santa Fe 19–25 m.y. ago (Baldridge and others, 1980) were derived from depths greater than 50 km, possibly as great as 90 km (Green, 1970a, b). Yet in the same general area, quartz-normative tholeiitic basalts, derived from much shallower levels in the mantle, were erupted approximately contemporaneously. This lack of a temporal pattern in magma composition confounds attempts to construct simple models of diapir growth beneath the Rio Grande rift, as was done for instance by Wendlandt and Morgan (1982) for the East African rift. Instead, the data seem to require an unintegrated heat source beneath the rift with only highly localized melting.

A striking feature of magmatism associated with the Rio Grande rift is the very small volume of volcanic rocks of any age in the rift. The largest fields within the rift grabens (Jemez and Taos Plateau volcanic fields) are part of the northeast-trending Jemez lineament (Fig. 1). Volcanic rocks are sparse in the rest of the rift. An essential question is, then, whether the low volume of volcanic rocks in the rift is due to the fact that magmas were never generated, or that they were never able to penetrate the crust and erupt on the surface. Based on evidence (largely seismic) presented above, we see no reason to think that large quantities of rifting-related magma were ever generated. If this conclusion is correct, it has enormous consequences for the magnitude of the thermal event accompanying rifting. For example, it would imply that no major heat source is associated with the rifting event and that little or no deep upwelling occurred beneath the rift. This conclusion is compatible with the diverse compositions of volcanic rocks in the rift.

LITHOSPHERIC PROCESSES

The goal of our studies is an understanding of the processes of rifting: what drives the lithospheric system, and what are the effects of rifting on the crust and mantle. Much discussion of rifting is cast into the simplified concepts of "active" and "passive" (e.g., Basaltic Volcanism Study Project, pp. 842–844). Certainly, our interpretation of the volcanism associated with the Rio Grande rift does not lead us to think that it is in any sense "active" (Fig. 5a). There are neither large volumes of volcanic rocks (of any age) in the rift, nor are there convincing arguments for large quantities of magmatic rocks at depth. Thus, though significant crustal thinning has occurred during rifting, this thinning has apparently been accomplished with little addition of heat into the system during the rifting process. A significant component of the modern, relatively high heat flow in the Rio Grande rift may be related to conductive-heat transfer from the middle Tertiary thermal episode, with the delay time dependent on crustal thickness and thermal time constant.

Pre-rift, subduction-related(?) volcanism, however, may have substantially changed the mechanical and lithologic character of the lithosphere and crust, providing an important component of convective-heat transfer into the crust and uplifting isotherms and the brittle–ductile transition. Rifting, particularly when low-angle faulting and large amounts of local crustal extension are involved in the upper crust, probably requires significant heat flow into the crust. Thus formation of the Rio Grande rift may have resulted from the (fortuitous?) combination of a thermally weakened lithosphere with a superimposed tensional-stress

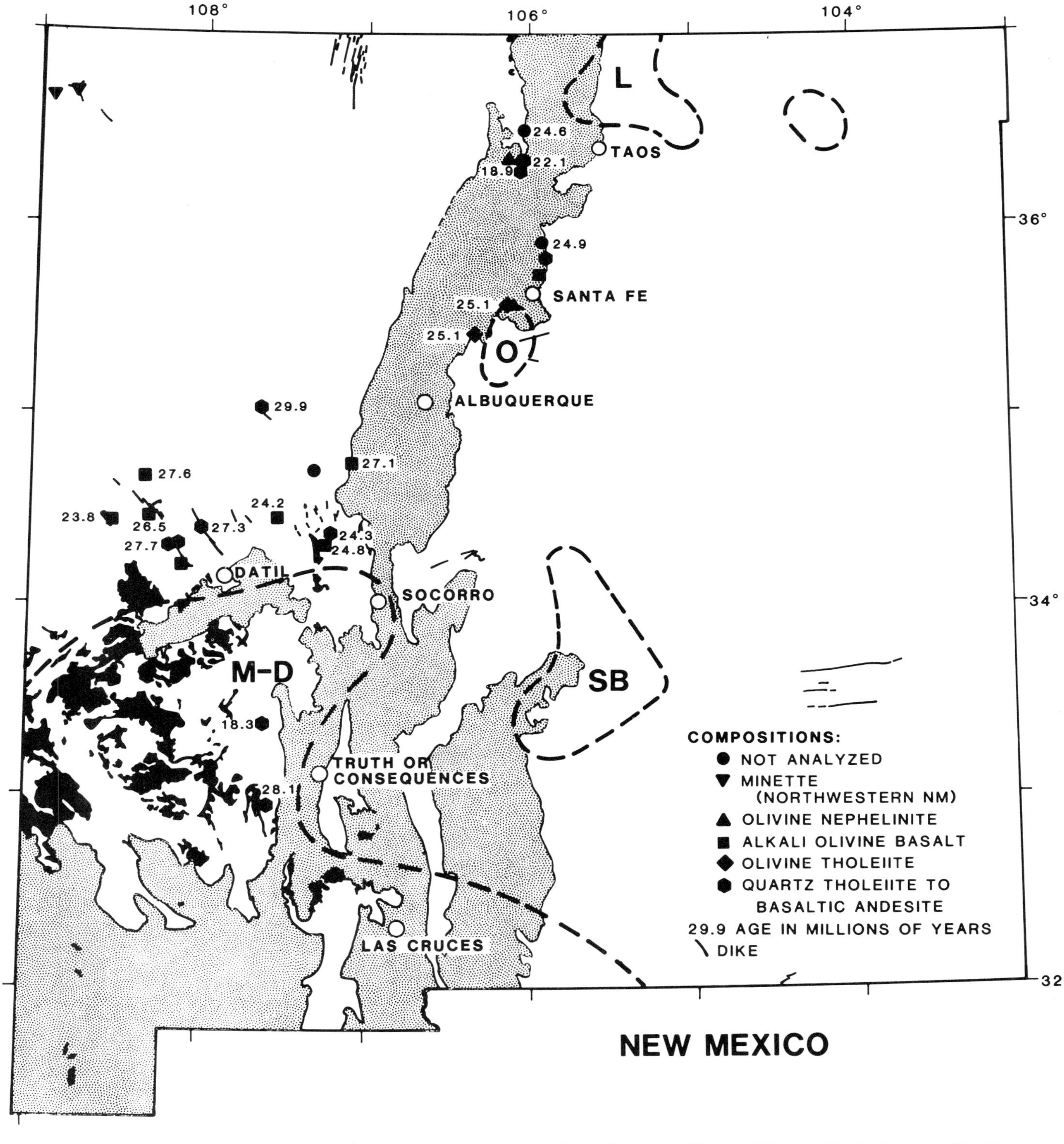

FIGURE 4. Basaltic rocks inferred or known to have been emplaced between about 30 and 18 m.y. ago. Diagram differs considerably in degree of detail represented: North of about 34° virtually every data point and outcrop are plotted; south of this latitude, outcrops are generalized and most data points are not plotted. Much data exists, however, indicating that these rocks are generally basaltic andesites. In the Socorro and Las Cruces areas, large volumes are buried beneath rift sediments. Broad, dashed lines show areas of Oligocene to Miocene (>43–20 m.y.), intermediate to silicic volcanism: M–D – Mogollon–Datil volcanic field, SB – Sierra Blanca, O – Ortiz, L – Latir (including "early rift volcanic field" of Lipman and Mehnert, 1979). Data from Sun and Baldwin (1958), Callender and Zilinski (1976), Massingill (1977), Laughlin and others (1979, 1983), Lipman and Mehnert (1979), Baldridge and others (1980), Kautz and others (1981), NMGS Map (1982), Lipman (1983), Seager and others (1984), and Vaniman (unpubl. data).

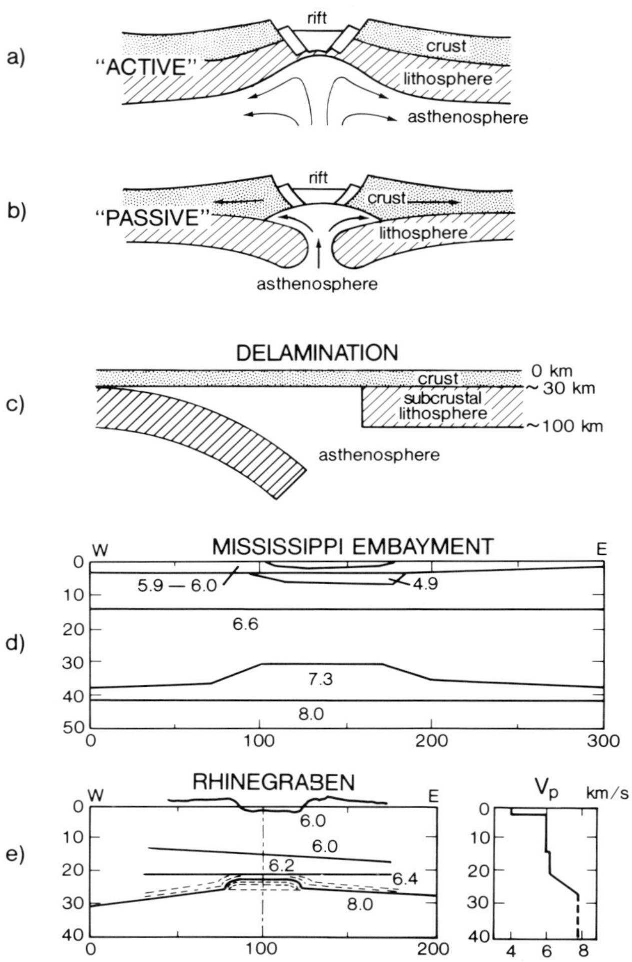

FIGURE 5. Schematic cross sections of the Mississippi embayment and Rhinegraben, and generalized models showing aspects of the rifting process.

regime. Middle Tertiary convective heat flow may have been critical in softening the crust and allowing it to yield.

Our lithospheric profile (Fig. 2) contains information that may pertain to mechanisms of rifting. One particular model for purely passive rifting (Turcotte and Emerman, 1983) postulates an asthenospheric "mushroom" geometry (Fig. 5b), in which asthenosphere is floored by subcrustal lithosphere with a probable compressional seismic velocity of about 8.0 km/s. As noted below, the axial seismic-refraction lines along the Rio Grande rift do not seem to support the mushroom or floored "pillow" picture, perhaps because existing refraction lines coincidentally overlie the "broken window" in the subcrustal lithosphere. Additional profiles in the near flanks might possibly reveal such structures. We feel, however, the existing deep seismic and gravity interpretations of the Rio Grande rift are much more in accordance with the broadly domed picture of Figure 5a than with the mushroom concept. Preliminary interpretation of teleseismic delays by Davis and others (this guidebook) suggests the asthenospheric low-velocity "upwarp" anomaly broadens with depth and may extend to depths of 200 km or deeper beneath the rift in northern New Mexico and southern Colorado. There is, however, an important difference from the model of Figure 5a in that the Rio Grande system is very asymmetric with upper-mantle/asthenospheric anomalies extending westward under the Colorado Plateau (Ander, 1981).

Thus, the concepts of active and passive rifting are obviously highly oversimplified, and probably neither is directly relevant to the Rio Grande rift. A most important feature of the rift is that it is superimposed on a region which underwent a long and complex history of deformation and heating prior to rifting. In order to understand the rifting process, it is imperative to sort out phenomena uniquely associated with rifting (i.e., with crustal extension and lithospheric thinning) from those inherited from previous events. We do not see convincing evidence for major, active, deep lithospheric involvement during rifting. Instead, the discontinuous nature of rifting and the fact that the rift is part of a very broad region of deformation suggest that plate processes exert a dominant control on the formation and evolution of the Rio Grande rift. As we have stated above, however, this does not imply that the major thermal and magmatic event preceding rifting did not influence, or perhaps even control, the initiation of rifting.

The significance of the Rio Grande rift to the evolution of the continental lithosphere is illustrated by contrasting it to other rifts. Based on gravity modeling and several axial and crossing seismic profiles, Mooney and others (1983) derived a crustal model for the northern Mississippi embayment (Fig. 5c). This area is the site of a buried late Precambrian rift that was reactivated in the Mesozoic. Mooney and others (1983) note the presence of the 7.2-km/s "domed" layer, which implies the lower crust in the rift vicinity has been altered by injection of mantle material. They suggest such a "high-to-intermediate"-velocity intrusion or lower-crust alteration may be a ubiquitous feature of continental rifts. A somewhat similar "pillow" feature has been inferred beneath the central region of the Rhinegraben (Fig. 5d), except there the rift pillow takes the form of a 5-km-thick velocity-gradient zone between 6.4 and 8.0 km/s rather than a more uniform 7.2–7.4-km/s layer. In the flanks outside the Rhinegraben, the Moho returns to a first-order discontinuity (Prodehl and others, 1976).

Although there is evidence for a 7.1-km/s deep crustal layer under the Great Plains (Fig. 2), axial-profile data within the Rio Grande rift and even to the west (DICE THROW, GASBUGGY, and Chinle–Hanksville) do not support the interpretation of such a floored, high–intermediate-velocity layer above an approximately 8-km/s mantle. We have made synthetic seismograms for floored models and compared them to DICE THROW data where P_n (at a velocity of 7.7 km/s) can be clearly observed *as a first arrival* between distances of 150 and 350 km. If there were an 8-km/s "floor" at depths of 5–15 km beneath the observed (33-km depth) 6.4–7.7-km/s transition, 8-km/s apparent-velocity headwaves and wide-angle reflections from the floor layer would be very strong and would severely mask the observed 7.7-km/s P_n phase and associated wide-angle reflections.

At present, the origin of the deep, approximately 7.1-km/s layer that appears to occur in some rifts is not well understood. This layer may represent lower crust that has been densified by intrusion of basaltic material, such as in magmatically active rifts (e.g., the East African rift). Its presence in the fossil Mississippi embayment might also suggest that such velocities are attained after significant cooling of the lower crust. The Rhinegraben may represent an intermediate evolutionary stage between a fossil system and a currently dynamic system such as the Rio Grande rift.

Bird's (1979) suggestion of continental delamination (Fig. 5c) may possibly help account for the asymmetry of the Rio Grande rift/Colorado Plateau system. The lower continental (subcrustal) lithosphere is in unstable mechanical equilibrium because it is denser than the underlying asthenosphere. If some process, such as fault breakage throughout the subcrustal lithosphere or plume erosion in the vicinity of the present rift, provided a line of weakness and an elongated conduit connecting asthenosphere with the base of the crust, the denser subcrustal-lithosphere boundary layer could peel away and sink into the chemically equivalent but hotter and less viscous asthenosphere. Observed moderately low P_n velocities of 7.8 km/s extending from the rift westward under the Colorado Plateau (Fig. 2) strongly suggest the existence of hotter asthenospheric material in contact with base of the crust, perhaps as far westward as the Arizona–New Mexico border. Bird (1979) suggests several possible geological and petrological consequences of this model that need to be explored in greater detail in connection with rift evolution.

CONCLUSIONS

Rifting is a long-term process involving changes in the entire lithosphere. Manifestations in the crust of the rifting process are a relatively surficial part of a deeper process. Geophysical studies yield information on the deeper crust and mantle, whereas geological and geochemical studies can provide an historical record of rifting. It is important in all of these studies to focus on the whole process of rifting.

The evolution of the Rio Grande rift does not fit either simple active or passive models. For example, no compelling evidence exists for a major thermal event associated with rifting. Yet heat, inherited from the immediately preceding deformation regime, was a critical factor in, and may have been a necessary condition for, rifting.

Even though volcanism and extension associated with the Rio Grande rift are relatively minor, the lithosphere beneath the rift is being permanently altered. For example, compositional changes in the lithosphere probably lead to a density increase. The effect of rifting will probably not be continental splitting. Rather, we suspect that the lithospheric anomaly associated with the Rio Grande rift will ultimately lead to major basin formation.

ACKNOWLEDGMENTS

We thank J. Aldrich, C. Chapin, L. Braile, G. Eaton, B. Larsen, and F. Perry for their constructive comments on this manuscript. Preparation of this manuscript was supported by the Office of Basic Energy Sciences of the U.S. Department of Energy and by National Science Foundation grant EAR-8018506.

REFERENCES

Ander, M. E., 1981, Geophysical study of the crust and upper mantle beneath the central Rio Grande rift and adjacent Great Plains and Colorado Plateau: Los Alamos National Laboratory, Report LA-8676-T, 218 pp.

Aoki, K., and Kudo, A. M., 1976, Major-element variations of late Cenozoic basalts of New Mexico: New Mexico Geological Society, Special Publication 5, pp. 82–88.

Baldridge, W. S., 1979, Petrology and petrogenesis of Plio-Pleistocene basaltic rocks from the central Rio Grande rift, New Mexico, and their relation to rift structure; *in* Riecker, R. E. (ed.), Rio Grande rift: tectonics and magmatism: American Geophysical Union, Washington, D.C., pp. 323–353.

———, and Perry, F. V., 1983, Spatial and temporal patterns of late Cenozoic basaltic volcanism of the central Rio Grande rift area (abstract): EOS (American Geophysical Union, Transactions), v. 64, p. 753.

———, Bartov, Y., and Kron, A., 1983, Geologic map of the Rio Grande rift and southeastern Colorado Plateau, New Mexico and Arizona; *supplement to* Riecker, R. E. (ed.), Rio Grande rift: tectonics and magmatism: American Geophysical Union, Washington, D.C.

———, Perry, F. V., and Gladney, E. S., 1982, Petrology and geochemistry of the Cat Hills volcanic field, central Rio Grande rift, New Mexico: Geological Society of America, Bulletin, v. 93, pp. 635–643.

———, Damon, P. E., Shafiqullah, M., and Bridwell, R. J., 1980, Evolution of the central Rio Grande rift, New Mexico: new potassium-argon ages: Earth and Planetary Science Letters, v. 51, pp. 309–321.

Basaltic Volcanism Study Project, 1981, Basaltic volcanism on the terrestrial planets: Pergamon Press, New York, 1286 pp.

Bird, P., 1979, Continental delamination and the Colorado Plateau: Journal of Geophysical Research, v. 84, pp. 7561–7571.

Black, B. A., 1964, The geology of the northern and eastern parts of the Ladron Mountains, Socorro County, New Mexico: M.S. thesis, University of New Mexico, Albuquerque, 117 pp.

Black, P. R., and Braile, L. W., 1982, P_n velocity and cooling of the continental lithosphere: Journal of Geophysical Research, v. 87, pp. 10557–10568.

Bott, M. H. P., 1982, The interior of the Earth (2nd edition): Elsevier, New York, 404 pp.

Brocher, T. M., 1981, Geometry and physical properties of the Socorro, New Mexico, magma bodies: Journal of Geophysical Research, v. 86, pp. 9420–9432.

Brown, L. D., Chapin, C. E., Sanford, A. R., and others, 1980, Deep structure of the Rio Grande rift from seismic reflection profiling: Journal of Geophysical Research, v.85, pp. 4773–4800.

———, Krumhansl, P. A., Chapin, C. E., and others, 1979, COCORP seismic reflection studies of the Rio Grande rift; *in* Riecker, R. E. (ed.), Rio Grande rift: tectonics and magmatism: American Geophysical Union, Washington, D.C., pp. 169–184.

Callender, J., 1983, Transposition structures in Precambrian rocks of New Mexico: New Mexico Geological Society, Guidebook 34, pp. 143–146.

———, and Zilinski, R. E., Jr., 1976, Kinematics of Tertiary and Quaternary deformation along the eastern edge of the Lucero uplift, central New Mexico: New Mexico Geological Society, Special Publication 6, pp. 53–61.

Cape, C. D., McGeary, S., and Thompson, G. A., 1983, Cenozoic normal faulting and the shallow structure of the Rio Grande rift near Socorro, New Mexico: Geological Society of America, Bulletin, v. 94, pp. 3–14.

Chamberlin, R. M., 1983a, Cenozoic domino-style crustal extension in the Lemitar Mountains, New Mexico: a summary: New Mexico Geological Society, Guidebook 34, pp. 111–118.

———, 1983b, Stop 4 discussion in Second Day Road Log: New Mexico Geological Society, Guidebook 34, pp. 51–53.

Chapin, C. E., 1971, The Rio Grande rift, pt. I: modifications and additions: New Mexico Geological Society, Guidebook 22, pp. 191–201.

———, and Seager, W. R., 1975, Evolution of the Rio Grande rift in the Socorro and Las Cruces areas: New Mexico Geological Society, Guidebook 26, pp. 297–321.

Christensen, N. I., 1979, Compressional wave velocities in rocks at high temperatures and pressures, critical thermal gradients, and crustal low velocity zones: Journal of Geophysical Research, v. 84, pp. 6849–6857.

Cook, F. A., McCullar, D. B., Decker, E. R., and others, 1979, Crustal structure and evolution of the southern Rio Grande rift; *in* Riecker, R. E. (ed.), Rio Grande rift: tectonics and magmatism: American Geophysical Union, Washington, D.C., pp. 195–208.

Cordell, L., 1976, Aeromagnetic and gravity studies of the Rio Grande graben in New Mexico between Belen and Pilar: New Mexico Geological Society, Special Publication 6, pp. 62–70.

———, 1978, Regional geophysical setting of the Rio Grande rift: Geological Society of America, Bulletin, v. 89, pp. 1073–1090.

———, Keller, G. R., and Hildenbrand, T. G., 1982, Bouguer gravity map of the Rio Grande rift, Colorado, New Mexico, and Texas: U.S. Geological Survey, Map GP-949.

Eaton, G. P., 1982, The Basin and Range province: origin and tectonic significance: Annual Review of Earth and Planetary Sciences, v. 10, pp. 409–440.

Eichelberger, J. C., and Gooley, R., 1977, Evolution of silicic magma chambers and their relationship to basaltic volcanism; *in* Heacock, J. C. (ed.), The earth's crust—its nature and physical properties: American Geophysical Union, Geophysical Monograph 20, pp. 57–77.

Elston, W. E., 1984, Subduction of young oceanic lithosphere and extensional orogeny in southwestern North America during mid-Tertiary time: Tectonics, v. 3, pp. 229–250.

———, Rhodes, R. C., Coney, P. J., and others, 1976, Progress report on the Mogollon Plateau volcanic field, southwestern New Mexico, no. 3—surface expression of a pluton: New Mexico Geological Society, Special Publication 5, pp. 3–28.

Frost, E. G., and Martin, D. L. (eds.), 1982, Mesozoic–Cenozoic tectonic evolution of the Colorado River region, California, Arizona, and Nevada: Cordilleran Publishers, San Diego, 608 pp.

Green, D. H., 1970a, A review of experimental evidence on the origin of basaltic and nephelinitic magmas: Physics of the Earth and Planetary Interiors, v. 3, pp. 221–235.

———, 1970b, The origin of basaltic and nephelinitic magmas: Leicester Literary and Philosophical Society, Transactions, v. 64, pp. 28–54.

Hamilton, W., 1982, Structural evolution of the Big Maria Mountains, northeastern California; *in* Frost, E. G., and Martin, D. L. (eds.), Mesozoic–Cenozoic tectonic evolution of the Colorado River region, California, Arizona, and Nevada: Cordilleran Publishers, San Diego, pp. 1–27.

———, 1983, Mode of extension of continental crust: New Mexico Geological Society, Guidebook 34, p. 9.

Hermance, J. F., and Pedersen, J., 1980, Deep structure of the Rio Grande rift: A magnetotelluric interpretation: Journal of Geophysical Research, v. 85, pp. 3899–3912.

Kautz, P. F., Ingersoll, R. V., Baldridge, W. S., and others, 1981, Geology of the Espinoso Formation (Oligocene), north-central New Mexico: Geological Society of America, Bulletin, pt. II, v. 92, pp. 2318–2400.

Kay, R. W., and Kay, S. M., 1981, The nature of the lower continental crust: inferences from geophysics, surface geology, and crustal xenoliths: Reviews of Geophysics and Space Physics, v. 198, pp. 271–298.

Kelley, V. C., 1977, Geology of Albuquerque Basin, New Mexico: New Mexico Bureau of Mines and Mineral Resources, Memoir 33, 59 pp.

———, and Kudo, A. M., 1978, Volcanoes and related basalts of Albuquerque Basin, New Mexico: New Mexico Bureau of Mines and Mineral Resources, Circular 156, 30 pp.

Kuellmer, F. J., 1956, Geologic map of Hillsborough Peak 30-minute quadrangle: State Bureau of Mines and Mineral Resources, Socorro, New Mexico, Geologic Map 1, scale 1:126,720.

Laughlin, A. W., Aldrich, M. J., and Vaniman, D. T., 1983, Tectonic implications of mid-Tertiary dikes in west-central New Mexico: Geology, v. 11, pp. 45–48.

———, Brookins, D. G., Damon, P. E., and others, 1979, Late Cenozoic volcanism of the central Jemez zone, Arizona–New Mexico: Isochron/West, no. 25, pp. 5–8.

Lipman, P. W., 1983, The Miocene Questa caldera, northern New Mexico: relation to batholith emplacement and associated molybdenum mineralization; *in* The genesis of Rocky Mountain ore deposits: changes with time and tectonics: Denver region Exploration Geologists Symposium, Proceedings, pp. 133–148.

———, and Mehnert, H. H., 1975, Late Cenozoic basaltic volcanism and development of the Rio Grande depression in the southern Rocky Mountains: Geological Society of America, Memoir 144, pp. 119–154.

———, and ———, 1979, The Taos Plateau volcanic field, northern Rio Grande rift, New Mexico; *in* Riecker, R. E. (ed.), Rio Grande rift: tectonics and magmatism: American Geophysical Union, Washington, D.C., pp. 289–311.

Lucas, S. G., and Ingersoll, R. V., 1981, Cenozoic continental deposits of New Mexico: an overview: Geological Society of America, Bulletin, pt. I, v. 92, pp. 917–932.

Massingill, G. L., 1977, Geology of Riley–Puertocito area, southeastern margin of the Colorado Plateau, Socorro County, New Mexico: New Mexico Bureau of Mines and Mineral Resources, Open-file Report 107.

Mooney, W. D,. Andrews, M. C., Ginzburg, A., and others, 1983, Crustal structure of the northern Mississippi embayment and a comparison with other continental rift zones: Tectonophysics, v. 94, pp. 327–348.

Moorbath, S., Dungan, M. A., Harmon, R. S., and others, 1983, Pb-, O-, and Sr-isotopic evidence for mantle and lower crustal involvement in the genesis of intermediate lavas of the Taos Plateau volcanic field (TPVF) (abstract): EOS (American Geophysical Union, Transactions), v. 64, p. 753.

Morton, W. H., and Black, R., 1975, Crustal attenuation in Afar; *in* Pilger, A., and Rosler, A. (eds.), Afar depression of Ethiopia: Inter-Union Commission on Geodynamics, Scientific Report 14, pp. 55–65.

Muehlberger, W. R., 1979, The Embudo fault between Pilar and Arroyo Hondo, New Mexico: an active intracontinental transform fault: New Mexico Geological Society, Guidebook 30, pp. 77–82.

Mueller, S., 1977, A new model of the continental crust; *in* Heacock, J. C. (ed.), The earth's crust—its nature and physical properties: American Geophysical Union, Geophysical Monograph 20, pp. 289–317.

NMGS Map, 1982, New Mexico Highway Geologic Map: New Mexico Geological Society.

Olsen, K. H., 1979, The seismicity of north-central New Mexico with particular reference to the Cerrillos earthquake of May 28, 1918: New Mexico Geological Society, Guidebook 30, pp. 65–75.

———, 1983, The role of seismic refraction data for studies of the origin and evolution of continental rifts: Tectonophysics, v. 94, pp. 349–370.

———, Cash, D. J., and Stewart, J. N., 1982, Mapping the northern and eastern extent of the Socorro midcrustal magma body by wide-angle seismic reflections: New Mexico Geological Society, Guidebook 33, pp. 179–185.

———, Keller, G. R., and Stewart, J. N., 1979, Crustal structure along the Rio Grande rift from seismic refraction profiles; *in* Riecker, R. E. (ed.), Rio Grande rift: tectonics and magmatism: American Geophysical Union, Washington, D.C., pp. 127–144.

Osburn, G. R., and Eggleston, T. L., 1983, Geologic map and cross sections of small area of southern Joyita Hills, Fig. 1-22.4a in First Day Road Log: New Mexico Geological Society, Guidebook 34, p. 10.

Prodehl, C., 1979, Crustal structure of the western United States: U.S. Geological Society, Professional Paper 1034.

———, Ansorge, J., Edel, J. B., and others, 1976, Explosion seismology research in the central and southern Rhinegraben—a case history; *in* Giese, P., Prodehl, C., and Stein, A. (eds.), Explosion seismology in central Europe: Springer-Verlag, Berlin, pp. 313–328.

Ramberg, I. B., Cook, F. A., and Smithson, S. B., 1978, Structure of the Rio Grande rift in southern New Mexico and west Texas based on gravity interpretation: Geological Society of America, Bulletin, v. 89, pp. 107–123.

Rhoades, S. M., and Callender, J. F., 1983, Evidence for low-angle faulting along western margin of the Sandia uplift, New Mexico rift, New Mexico (abstract): Geological Society of America, Abstracts with Programs, v. 15, p. 322.

Rinehart, E. J., Sanford, A. R., and Ward, R. M., 1979, Geographic extent and shape of an extensive magma body at midcrustal depths in the Rio Grande rift near Socorro, New Mexico; *in* Riecker, R. E. (ed.), Rio Grande rift: tectonics and magmatism: American Geophysical Union, Washington, D.C., pp. 237–249.

Roller, J. C., 1965, Crustal structure in the eastern Colorado Plateau province from seismic refraction measurements: Seismological Society of America, Bulletin, v. 55, pp. 107–119.

Romney, C., Brooks, B. G., Mansfield, R. H., and others, 1962, Travel times and amplitudes of principal body phases recorded from GNOME: Seismological Society of America, Bulletin, v. 52, pp. 1057–1074.

Rosen, S., 1983, Geologic map and cross section, Fig. 1-15.2b in First Day Road Log: New Mexico Geological Society, Guidebook 34, p. 8.

Sanford, A. R., Olsen, K. H., and Jaksha, L. H., 1979, Seismicity of the Rio Grande rift; *in* Riecker, R. E. (ed.), Rio Grande rift: tectonics and magmatism: American Geophysical Union, Washington, D.C., pp. 145–168.

———, ———, and ———, 1981, Earthquakes in New Mexico, 1849–1977: New Mexico Bureau of Mines and Mineral Resources, Circular 171, 20 pp.

———, Mott, R. P., Schuleski, P. J., and others, 1977, Geophysical evidence for a magma body in the crust in the vicinity of Socorro, New Mexico; *in* Heacock, J. G. (ed.), The earth's crust: its nature and physical properties: American Geophysical Union, Geophysical Monograph 20, pp. 385–403.

Seager, W. R., 1981, Geology of Organ Mountains and southern San Andres Mountains, New Mexico: New Mexico Bureau of Mines and Mineral Resources, Memoir 36, 97 pp.

———, and Morgan, P., 1979, Rio Grande rift in southern New Mexico, west Texas, and northern Chihuahua; *in* Riecker, R. E. (ed.), Rio Grande rift: tectonics and magmatism: American Geophysical Union, Washington, D.C., pp. 87–106.

———, Shafiqullah, M., Hawley, J. W., and others, 1984, New K–Ar dates from basalts and the evolution of the southern Rio Grande rift: Geological Society of America, Bulletin, v. 95, pp. 87–99.

Shankland, T. J., and Ander, M. E., 1983, Electrical conductivity, temperatures, and fluids in the lower crust: Journal of Geophysical Research, v. 88, pp. 9475–9484.

Smith, C. T., 1983, Structural problems along the east side of the Socorro constriction, Rio Grande rift: New Mexico Geological Society, Guidebook 34, pp. 103–109.

Smith, R. L., Bailey, R. A., and Ross, C. S., 1970, Geologic map of the Jemez Mountains: U.S. Geological Survey, Map I-571.

Smithson, S. B., 1978, Modeling continental crust: structural and chemical constraints: Geophysical Research Letters, v. 5, pp. 749–752.

———, and Brown, S. K., 1977, A model for lower continental crust: Earth and Planetary Science Letters, v. 35, pp. 134–144.

———, Johnson, R. A., and Wong, Y. K., 1981, Mean crustal velocity: a critical parameter for interpreting crustal structure and crustal growth: Earth and Planetary Science Letters, v. 53, pp. 323–332.

Spiegel, Z., and Baldwin, B., 1963, Geology and water resources of the Santa Fe area, New Mexico: U.S. Geological Survey, Water-supply Paper 1525, 258 pp.

Spohn, T., and Schubert, G., 1983, Convective thinning of the lithosphere: a mechanism for rifting and mid-plate volcanism on Earth, Venus, and Mars: Tectonophysics, v. 94, pp. 67–90.

Stewart, S. W., and Pakiser, L. C., 1962, Crustal structure in eastern New Mexico interpreted from the GNOME explosion: Seismological Society of America, Bulletin, v. 52, pp. 1017–1030.

Sun, M., and Baldwin, B., 1958, Volcanic rocks of the Cienega area, Santa Fe County, New Mexico: New Mexico Bureau of Mines and Mineral Resources, Bulletin 54, 80 pp.

Toppozada, T. R., and Sanford, A. R., 1976, Crustal structure in central New Mexico interpreted from the GASBUGGY explosion: Seismological Society of America, Bulletin 66, pp. 877–886.

Turcotte, D. L., and Emerman, S. H., 1983, Mechanisms of active and passive rifting: Tectonophysics, v. 94, pp. 39–50.

Tweto, O., 1978, Tectonic map of the Rio Grande rift system in Colorado; *in* Hawley, J. W. (ed.), Guidebook to Rio Grande rift in New Mexico and Colorado: New Mexico Bureau of Mines and Mineral Resources, Circular 163, Sheet 1.

Wendlandt, R. F., and Morgan, P., 1982, Lithospheric thinning associated with rifting in East Africa: Nature, v. 298, pp. 734–736.

Wernicke, B., and Burchfiel, B. C., 1982, Modes of extensional tectonics: Journal of Structural Geology, v. 4, pp. 105–115.

Woodward, L. A., 1977, Rate of crustal extension across the Rio Grande rift near Albuquerque, New Mexico: Geology, v. 5, pp. 269–272.

———, Callender, J. F., and Zilinski, R. E., 1975, Tectonic map of the Rio Grande rift, New Mexico: Geological Society of America, Map MC-11.

———, Callender, J. F., Seager, W. R., and others, 1978, Tectonic map of Rio Grande rift region in New Mexico, Chihuahua, and Texas; *in* Hawley, J. W. (ed.), Guidebook to Rio Grande rift in New Mexico and Colorado: New Mexico Bureau of Mines and Mineral Resources, Circular 163, Sheet 2.

Zietz, I., 1982, Composite magnetic anomaly map of the United States, pt. A: conterminous United States: U.S. Geological Survey Map, scale 1:2,500,000.

FACTORS CONTROLLING THE PHASES AND STYLES OF EXTENSION IN THE NORTHERN RIO GRANDE RIFT

PAUL MORGAN[1] and MATTHEW P. GOLOMBEK[2]
[1]Department of Geosciences, Purdue University, West Lafayette, Indiana 47907; [2]M.S. 183-501, Jet Propulsion Laboratory, 4800 Oak Grove Drive, Pasadena, California 91109

INTRODUCTION

The Rio Grande rift is the latest development in a complex series of tectonic and magmatic events in Colorado and New Mexico that began in the Late Cretaceous with Laramide orogeny and have continued to the present. Numerous authors have documented all or part of this complex geologic history in northern New Mexico (e.g., Baldridge and others, 1980; Baltz, 1978; Chapin, 1979; Chapin and Cather, 1981; Golombek and others, 1983; Lipman, 1981; Stearns, 1953a). Following a generalization and simplification of this series of events in south-central and southwestern New Mexico by Seager and others (1984) and Morgan and Seager (1983 and unpubl.), we identify four basic Cenozoic tectonic–geologic events in northern New Mexico (Fig. 1): (1) Laramide (Cretaceous–Eocene) compressional deformation, essentially amagmatic, producing uplifts and compressional and strike–slip-related sedimentary basins (Baltz, 1978; Chapin and Cather, 1981; Stearns, 1953a); (2) major Oligocene–early Miocene volcanism (Bachman and Mehnert, 1978; Baldridge and others, 1980; Kautz and others, 1981; Lipman, 1981; Lipman and Mehnert, 1979; Osburn and Chapin, 1983; Stearns, 1953a, b; Tweto, 1979); (3) early-phase (late Oligocene–early Miocene) extension with locally large strain and formation of broad basins (Baldridge and others, 1981; Baltz, 1978; Chamberlin, 1983; Lipman, 1981; Manley and Mehnert, 1981; Rhoades and Callender, 1983; Stearns, 1953a); and, following a middle Miocene lull in volcanic activity (Chapin and Seager, 1975), (4) late-phase (approximately 13 m.y. to present with most extension from 12 to 5 m.y.), extensional block faulting with associated volcanism (Golombek and others, 1983; Gardner, 1983). Major volcanic and modern tectonic activities continue to the present (Aubele, 1979; Bachman and Mehnert, 1978; Doell and others, 1968; Kudo, 1982; Lipman and Mehnert, 1979). Recently, Morgan and Seager (1983 and unpubl.) used the tectonic and magmatic history of south-central and southwestern New Mexico to study the evolution of geotherm and changing styles of extensional tectonism in the southern Rio Grande rift. In the present contribution we attempt to identify the factors which controlled the different tectonic styles of the two phases of Cenozoic extension in northern New Mexico.

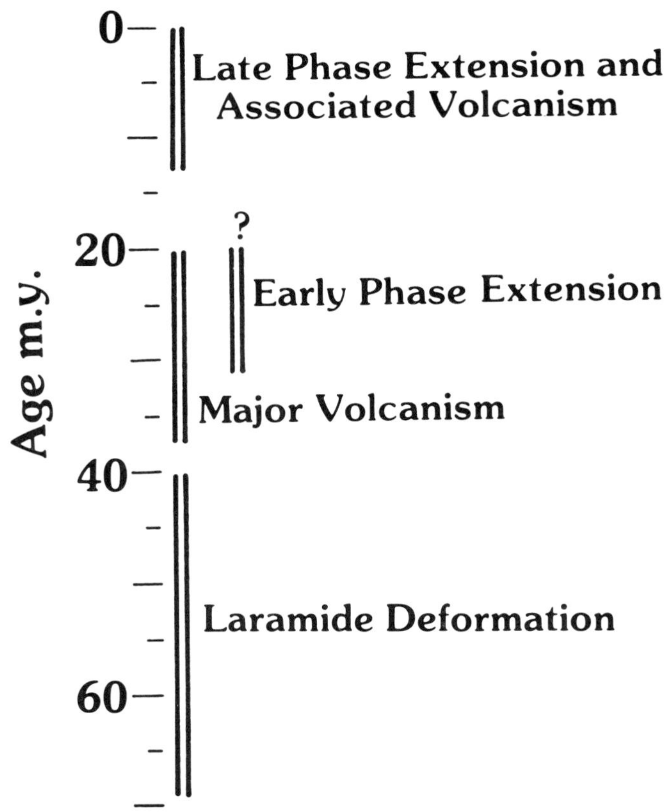

FIGURE 1. Major Cenozoic geologic–tectonic events in northern New Mexico. Bars indicate the approximate duration of each event.

EARLY EXTENSION AND BROAD BASINS

The early (late Oligocene–early Miocene) phase of extension in the area of the Rio Grande rift started during the later stages of major Oligocene–early Miocene volcanism and was similar in timing, orientation, and style to an early phase of extension in the Basin and Range province (e.g., Zoback and others, 1981; Eaton, 1982). The axis of extension during this phase was oriented southwest to west–southwest, oblique to the extension axes of the modern Rio Grande rift and Basin and Range. Age data suggest that extension was not synchronous in all areas, but commonly was temporally and spatially associated with major magmatic events (e.g., Lemitar Mountains, 31–28 m.y., Chamberlin, 1983; Questa, 23 m.y., Lipman, 1981). Palinspastic reconstructions suggest that very large strains (greater than 100%) were locally associated with this extension (e.g., Chamberlin, 1983; Lipman, 1981). Unfortunately, in many areas this event is masked by the later extension event, and although it has only been well documented in a few areas, it is likely that extension associated with this event was more widespread.

In some areas relatively broad and shallow basins were formed by the early extension event. A good example is the Popotosa Basin in the Socorro–Magdalena area (Chapin and Seager, 1975; Chamberlin, 1983). Another example may be the area now occupied by the Española Basin, where several authors have documented the development of a broad basin around 25 m.y. ago (e.g., Baldridge and others, 1980; Baltz, 1978; Manley and Mehnert, 1981; Stearns, 1953a), although it cannot be demonstrated at present that extension was directly responsible for the formation of this latter basin. By 26 m.y. ago, shallow basins had formed along the trend of the Rio Grande rift from southern New Mexico to the Wyoming border (C. E. Chapin, written comm. 1984).

We suggest that most, if not all, of these basins were formed by extension (see below and McKenzie, 1978), primarily on the evidence that formation of these basins was roughly contemporaneous with early-phase extension. If this hypothesis is correct, it is possible that many of the basins have not been identified as extensional either because the style of faulting associated with large strains is only just being recognized in some areas of the rift (e.g., Rhoades and Callender, 1983), or because faulting is not exposed. Basins formed by high-strain extension of the lithosphere have three components of subsidence: (1) a tectonic component caused by thinning of the buoyant crust during extension; (2) a thermal component caused by cooling, thickening, and contraction of the lithosphere after convective upwarp of the geotherm during extension; and (3) a loading component caused by sediments infilling the depression caused by (1) and (2) (McKenzie, 1978). It is only where pre-basin basement or early basin sediments faulted during extension are exposed by uplift that the extensional origins of the basins may be manifest. Late basin sediments, deposited during the thermal phase of subsidence, are often more laterally extensive than early basin sediments due to flexural widening of the basin with loading, and do not record the extensional origin of the basin.

Thus, while we recognize that many of the late Oligocene–early Miocene broad basins may have formed by processes unrelated to extension, we consider it possible that they may be dominantly extensional in origin, but that features diagnostic of this extension have not been identified. For our discussion of factors controlling extensional style, the percentage of the broad basins formed by high-strain extension is unimportant, although this percentage is a measure of the regional significance of early-phase extension.

LATE EXTENSION AND NARROW GRABENS

The second major phase of Cenozoic extension was responsible for the formation of the modern Rio Grande rift, and corresponds in many respects to Basin and Range extension (e.g., Zoback and others, 1981; Eaton, 1982; Golombek and others, 1983). The extension axis during this event was oriented east–west to east–southeast, and both the local and total strain were much smaller than in the earlier event. Estimates of extension for central and northern New Mexico are of the order of 10 km, decreasing to the north (Brocher, 1981; Cordell, 1982; Golombek, 1981; Golombek and others, 1983; Woodward, 1977), which suggests strains of about 10%. In many areas there is no significant association of volcanism with extension, in fact most volcanism following the middle Miocene lull has occurred in the waning phases of tectonic activity (5 m.y. to present; e.g., Baldridge and others, 1980). Although there is some debate as to the form of the faults at depth (e.g., Brown and others, 1980; Jurdy and Brocher, 1980; Brocher, 1981; Cape and others, 1983), there is little doubt that the prevailing form of faulting during the late extension event is block faulting, with the vertical component dominant.

Late Miocene–Pliocene extension in the Rio Grande rift formed relatively narrow and deep, elongate fault-bounded basins in contrast to the broad, relatively shallow basins formed contemporaneously with late Oligocene–early Miocene extension. These basins in north-central New Mexico are commonly asymmetric grabens (e.g., Baltz, 1978), a characteristic of many continental rifts (e.g., Ramberg and Morgan, 1984). Although the ages of rift faulting cannot be tightly constrained in many areas, the block-faulted style of the rift suggests that most of the Neogene faults with major vertical displacement were formed during the second extension event. Vertical displacement of the Precambrian surface between the Albuquerque–Belen Basin and the northern part of the Sandia uplift could be of the order of 6.4 km (Kelley, 1977, p. 42; Baltz, 1978), and the thickness of Neogene fill in the eastern side of the Albuquerque Basin exceeds 2 km (e.g., Birch, 1982). Major vertical displacements (greater than 1 km) associated with late-phase extension are also indicated for other basins of the northern Rio Grande rift. Thus, the late Miocene–Pliocene extension in the Rio Grande rift can be characterized as a minor extension event with major vertical movements, in strong contrast to the late Oligocene–early Miocene event with major extension and minor vertical movements.

EXTENSION AND BASIN FORMATION

We seek to understand the factors which caused the two extension events in the area of the Rio Grande rift to extend the same upper crust in two contrasting tectonic styles. We consider two hypothetical extensional mechanisms of extensional basin formation which are illustrated in Figure 2. In the first mechanism (Fig. 2a) a section of crust is assumed to extend plastically with an increase in length from x to x′. An extension factor B is given by B = x′/x. If the crust is incompressible and no mass is added or lost during extension, the extension will be accompanied by a uniform decrease in crustal thickness from L to L′, and it is easily shown that L/L′ = B. Subsidence of the extended crust will occur because the crust is thinned during extension. Assuming all crustal

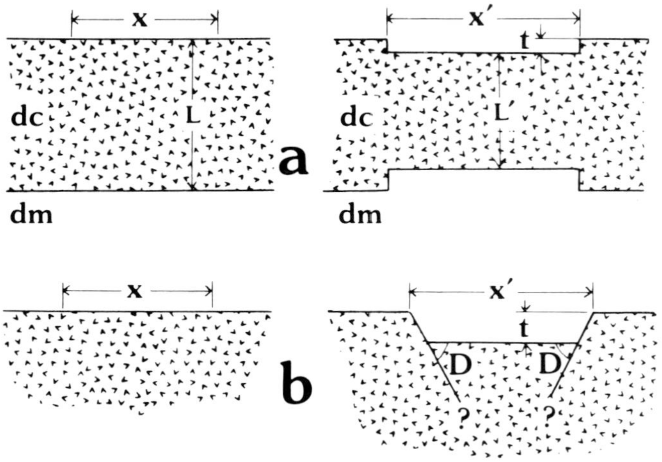

FIGURE 2. Models of unextended (left) and extended (right) crust. **a,** Plastic crustal extension with subsidence resulting from isostatic compensation of crustal thinning. **b,** Brittle fault-block extension with relative subsidence controlled completely by movement on planar normal faults which penetrate to an unspecified depth in the crust. Not to scale. Symbols are explained in the text.

columns to be isostatically compensated, and ignoring any density variations in the mantle, it can be shown that the downwarp, t, is given by

$$\begin{aligned} t &= (L-L')(dm-dc)/dm \\ &= L(1-1/B)(dm-dc)/dm \end{aligned}$$

where dm and dc are the mean densities of the uppermost mantle and crust, respectively (see McKenzie, 1978, for a more complete treatment of this problem). This mechanism is relatively inefficient at producing subsidence: Assuming crustal and mantle densities of 2.8 and 3.25 Mg m^{-3}, respectively, and an extended crustal thickness of 40 km, 10% extension produces only 500 m downwarp; for 50 and 100% extension the downwarp increases to 1.8 km and 2.8 km, respectively. Thinning and subsidence may be reduced by the addition of magmas to the crust during extension (e.g., Lachenbruch and Sass, 1978). We suggest that the style of late Oligocene–early Miocene extension in the area of the Rio Grande rift, with large strains producing broad, relatively shallow basins, is approximated by this extension/subsidence mechanism.

Extension is assumed to be localized on graben-bounding faults in the second extension mechanism; for simplicity, the effects of crustal thinning by extension and the nature of the lower termination of the faults are not discussed at this point (Fig. 2b). More detailed analyses of this problem have been given elsewhere by Vening Meinesz (1950), Bott (1976), Bott and Mithen (1983), and others. To a first approximation, vertical movement, t, on these faults can be calculated simply from the extension across the faults, x′–x, and the dip, D, of the faults, and is given by

$$2t = (x'-x)\tan D$$

for a symmetric graben. For an extension of 10 km (similar to that reported for north-central New Mexico), vertical displacements of 8.7 to 13.7 km are calculated for faults dipping from 60 to 70°. For an asymmetric graben most vertical and horizontal movements occur on one side of the graben. This is a very efficient mechanism for producing deep basins with little extension, and we suggest that late Miocene–Pliocene Rio Grande rift extension is approximated by this mechanism. Total vertical displacements by this mechanism are limited by isostatic and energetic considerations not discussed here; however, in a more complete analysis of this problem, Bott (1976) and Bott and Mithen (1983) have calculated that with a reasonable stress field, sediment-filled grabens on the order of 5 km deep can be produced by this mechanism. Larger vertical displacements may occur if uplift of horst blocks is also considered. The net regional effect of this extension mechanism is to cause regional subsidence because, as with the uniform extension mechanism, the buoyant crust is thinned. However, major horst and graben topography may be produced, rejuvenating the topography.

For simplicity, in the analyses of both extension/subsidence mechanisms we have ignored the effects of sediment loading in the depressions. This loading increases downwarp and basin depth. Complementary erosion of adjacent uplifts will result in further structural uplift of these ranges. It is not possible to explain the difference in extension/subsidence styles in the two extension events from the effects of sediment loading, because these effects are isostatic and modify, but do not change, the relative subsidence from the two mechanisms. We have chosen extreme mechanisms for extension/subsidence, and it is probable that neither extension event is completely represented by these extremes. By concentrating on these extremes, however, we attempt to determine the major factors controlling extensional style in the two events.

FACTORS CONTROLLING EXTENSIONAL STYLE

The basic difference between the two extreme extension/subsidence mechanisms presented above (Fig. 2) is the effective mechanical behavior of the crust. In the uniform-extension mechanism (Fig. 2a), the crust behaves as a plastic material; in the graben-forming mechanism (Fig. 2b), movement on faults indicates control by brittle deformation in at least the upper crust. We therefore seek the factors which control the mechanical strength and failure mechanism of the crust under extension.

Mechanisms of rock deformation are complex, but can be grouped into two basic types, brittle failure and ductile creep. Brittle failure applies to all types of shear motion confined to a discrete zone (i.e., a fault); ductile creep describes distributed solid-state deformation. If the brittle-failure strength of a rock is less than its ductile-creep strength, as stress is increased it will fracture. Conversely, if its ductile strength is less than its brittle strength, it will deform plastically as stress increases. Laboratory and field experiments have shown that the brittle-yield strength of the lithosphere is relatively independent of composition, temperature, and strain rate, and that it can be reasonably approximated as two linear functions of depth, one for compression and one for extension (e.g., Byerlee, 1978; Jamison and Cook, 1980; Lynch, 1983; McGarr, 1980; Stetsky, 1978). Ductile-creep strength depends upon a number of parameters, the primary ones being temperature, composition, and strain rate (e.g., Carter, 1976; Goetze and Brace, 1972; Lynch, 1983; Weertman and Weertman, 1975). In general, the ductile strength increases with increasing mafic-mineral content and strain rate, but decreases with increasing temperature (e.g., Smith and Bruhn, 1984).

In many areas of the Rio Grande rift the second extension event is superimposed on earlier extension; thus, to explain contrasting styles of extension in the same area at different times, a time-dependent parameter controlling crustal strength is required—temperature or strain rate. Strain rate is very difficult to constrain as at best we can only bracket the time period during which extension occurred. Ductile strength increases with increasing strain rate, causing the brittle–ductile transition to become deeper (e.g., see later, Fig. 4). Thus, if strain rate was the controlling parameter, the crust would be expected to deform in a more brittle style in the high-strain (high-strain rate?) extension event than in the later low-strain (low-strain rate?) event. However, as the converse was true, temperature was the more likely parameter controlling extensional style.

EVOLUTION OF THE GEOTHERM AND LITHOSPHERIC STRENGTH

Heat-flow data from northern New Mexico and southern Colorado (Edwards and others, 1978; Reiter and Mansure, 1983; Reiter and others, 1979; Clarkson and Reiter, this guidebook; Decker and others, this guidebook) provide a reasonable constraint on variations in the modern geotherm across the northern Rio Grande rift. Heat flow increases from about 70 mW m^{-2} in the southern San Juan Basin on the eastern margin of the Colorado Plateau to about 120 mW m^{-2} in the rift, then decreases to about 50 mW m^{-2} in the Great Plains to the east (Clarkson and Reiter, this guidebook). Higher heat-flow values are commonly associated with young volcanic fields on both sides of the rift. Geotherms predicted from these heat-flow values are shown in Figure 3. It is interesting to note that the modern geotherm estimated for the rift gives unrealistically high temperatures in the lower crust, suggesting that the thermal gradient decreases with depth and that there has been a significant component of heat transfer by magma intrusion into the lower crust during the past few m.y. associated with the young volcanic fields in this region (see also Lachenbruch and Sass, 1978; Reiter and Clarkson, 1983; Decker and others, this guidebook).

At present no data constrain the history of the geotherm in northern New Mexico, as the present geothermal activity masks earlier thermal events. However, the major Oligocene–early Miocene volcanic activity in north-central New Mexico makes it reasonable to assume that the

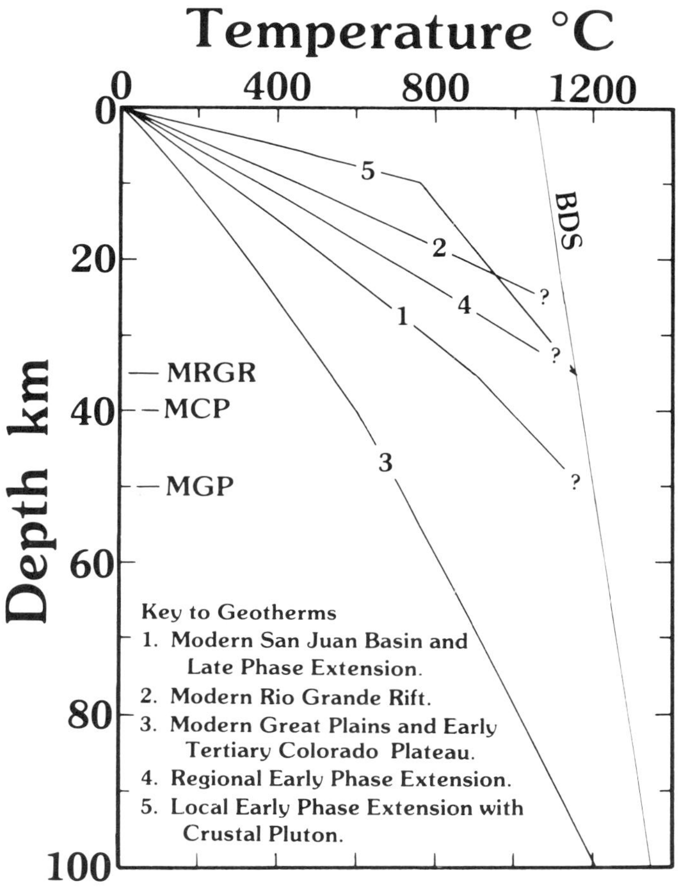

FIGURE 3. Assumed geotherms for northern New Mexico. Modern San Juan Basin, Rio Grande rift, and Great Plains geotherms (curves 1, 2, and 3) are based on surface heat-flow values of 70, 120, and 50 mW m^{-2}, respectively. Early-phase extension geotherms (curves 4 and 5) are based on the implication that crustal melting occurred during the later phases of Oligocene–early Miocene volcanism. Curve 1 (San Juan Basin) also represents the late-phase extension geotherm based on the calculation that surface heat flow would decrease from about 95 mW m^{-2} (curve 4) to 70 mW m^{-2} (curve 2) during 7 m.y. of cooling during the middle Miocene magmatic lull. Curve 3 (Great Plains) also represents the early Tertiary Colorado Plateau on the assumption that the plateau lithosphere was cool at this time. See text for further details of the assumptions for the geotherms. All curves are schematic especially at higher temperatures and do not accurately represent the transient effects of intrusions. However, the strength curves generated from these geotherms are not sensitive to the geotherms at high temperatures, and for the purposes of comparison of strength curves it is the relative levels of the geotherms that are of greatest importance. BDS is the basalt dry solidus (from Lachenbruch and Sass, 1978). MRGR, MCP, and MGP are modern Moho depths for the Rio Grande rift, Colorado Plateau, and Great Plains, respectively (from Cordell, 1978).

crust was heated during that event. Strontium-isotope data from volcanic rocks of this event in the southern Rio Grande rift (Seager and others, 1984) indicate crustal contamination of the magmas during the latter part of this event and suggest possible crustal melting. Further evidence for crustal melting is given by the abundance of mid-Tertiary silicic, alkalic, and rhyolitic volcanic rocks and associated granitic intrusions and calderas throughout the region. We therefore present two speculative geotherms for the rift area during the latter stages of Oligocene–early Miocene volcanism and the early phase of extension. The first geotherm is constrained to intersect the basalt dry solidus at the Moho (35 km), a lower thermal limit to the effects of intrusion of mantle-derived melts (curve 4, Fig. 3). The second geotherm has the same constraint at the Moho, but it is also constrained to intersect a temperature of 760° at 10 km (the tonalite solidus; Wyllie, 1977) to simulate the effect of a major upper-crust pluton emplacement (curve 5, Fig. 3).

To estimate the geotherm at the time of late-phase extension, we deduce from the middle Miocene lull in volcanic activity (Chapin and Seager, 1975; Baldridge and others, 1980) that the lithosphere was probably cooling during this interval. Assuming 7 m.y. cooling of the geotherm (see Vitorello and Pollack, 1980) represented by curve 4 in Figure 3, we estimate a reasonable decrease in surface heat flow from about 95 to 70 mW m^{-2}. As a result, the geotherm at the start of Rio Grande rift extension would be very similar to the modern southern San Juan Basin geotherm (curve 1, Fig. 3). A slightly higher geotherm would be predicted by cooling from the hot geotherm simulating crustal pluton emplacement (curve 5, Fig. 3). However, since the pluton emplacement occurred as local events scattered through the major volcanic phase, we consider the hot geotherm without the effect of a crustal pluton (curve 4, Fig. 3) more representative of the regional geotherm at the start of cooling. Thus, we have estimates of the geotherm at the start of both extension events.

Lithospheric-strength curves for northern New Mexico for the geotherms shown in Figure 3 have been calculated using approximations to extensional brittle-failure and ductile-creep data given by Lynch (1983) (Figs. 4, 5). For these calculations we made the common assumption that the upper part of the crust was dominated by a silicic rheology and the lower part by a mafic rheology. For our purposes of comparison of strength curves, the details of the assumed upper–lower crust division are unimportant. Figure 4 shows the strength curves calculated for the low heat-flow Great Plains and shows the basic features of the strength curves. Ignoring any shallow compositional variations, the strength increases with depth from the surface in a brittle zone in silicic upper crust until a temperature of about 250°C is reached when deformation changes to ductile creep and the strength rapidly decreases with depth. At the silicic–mafic interface strength rapidly increases again in a second zone of brittle strength and continues to increase further until a temperature of about 400°C is reached; then the strength rapidly decreases again with depth in a zone of ductile deformation. This pattern is repeated below the Moho with a brittle–ductile transition in the ultramafic rocks occurring at about 850°C. These transition temperatures and the heavy curve in Figure 4 are appropriate to a strain rate of 10^{-15} s^{-1} (approximately 3%/m.y.). Also shown in Figure 4 are the ductile-strength curves for strain rates of 10^{-14} and 10^{-16} s^{-1} which probably encompass the likely strain rates of Cenozoic extension in New Mexico. With these different strain rates the details of the strength curve change, but the basic form is essentially the same. All further comparisons are made using strength curves generated assuming a strain rate of 10^{-15} s^{-1}.

Strength curves generated using the two hypothetical early extension-phase geotherms (curves 4 and 5, Fig. 3) are shown in Figure 5a. High temperatures in the lower crust and upper mantle predicted by these geotherms result in no significant strength in the mafic and ultramafic layers, and strength is concentrated into the silicic upper crust. The only significant difference between the two curves is the depth of the brittle–ductile transition, a shallower transition being predicted for the hotter, pluton-heated crust. The strength curve predicted for the late extension-phase geotherm (curve 1, Fig. 3) is shown in Figure 5b. Cooler temperatures predicted by this geotherm result in an increase in the brittle–ductile transition depth, a small zone of mafic ductile strength in the lower crust, and a significant zone of ultramafic ductile strength in the uppermost mantle. Obviously, just as the geotherm for this period was transitional between the hot Oligocene–early Miocene geotherm and the modern Great Plains geotherm, its predicted strength curve is also transitional between the curves shown in Figures 5a and 4. The curve in Figure 5b is probably also representative of the modern southern San Juan Basin lithosphere.

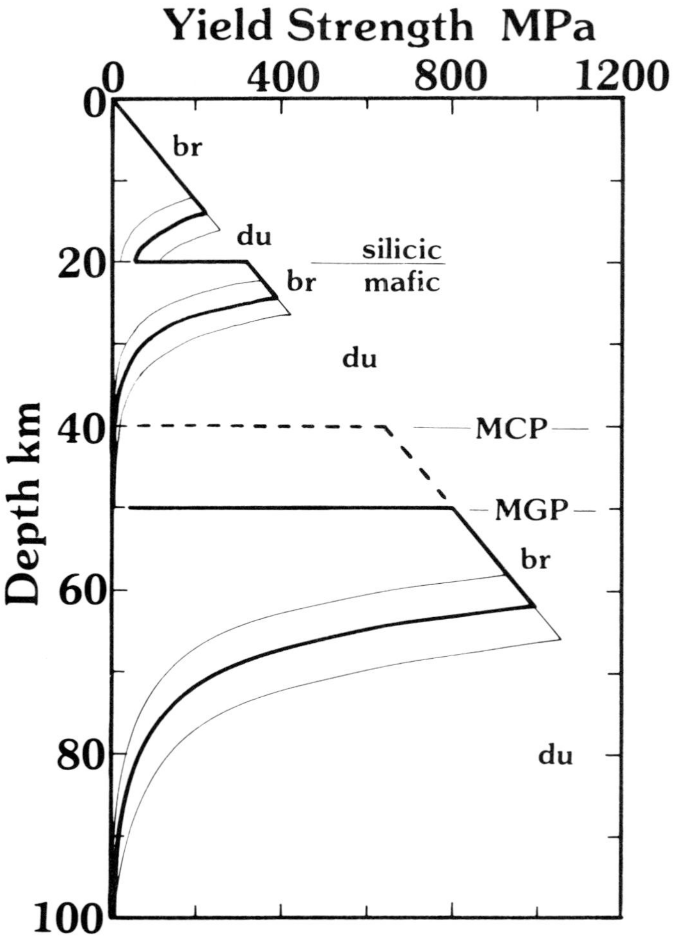

FIGURE 4. Calculated lithospheric-strength curve for the Great Plains (heavy curve) assuming a 20-km silicic upper crust and 30-km mafic lower crust and a strain rate of 10^{-15} s^{-1} (approximately 3% $m.y.^{-1}$). Layers of brittle and ductile strength are indicated by **br** and **du**, respectively. Modifications to the ductile sections of the strength curve assuming strain rates of 10^{-16} and 10^{-14} s^{-1} are shown by the fine lines above and below the heavy lines, respectively. Dashed lines show additional strength predicted for the early Tertiary Colorado Plateau relative to the modern Great Plains strength curve (assuming the same geotherm for both) by virtue of its thinner crust prior to Cenozoic heating and uplift. Depths to Great Plains and Colorado Plateau Mohos are indicated by MGP and MCP, respectively (from Cordell, 1978).

The first obvious result of a comparison of the strength curves in Figures 4 and 5 is that extension occurred where the crust (lithosphere) was weakest. It has been suggested that the lithosphere of the Colorado Plateau has been heating up during the latter half of the Cenozoic (e.g., Thompson and Zoback, 1979; Bodell and Chapman, 1982). Thus, during the early extension phase, and possibly during late-phase extension, the strength curve of the Plateau may have been similar to the modern Great Plains curve (Fig. 4), but is now weakening due to late Cenozoic heating.

The strength curves predicted for the crust during the first stage of extension strongly suggest that extension of the crust would have been dominated by ductile deformation. Brittle failure is predicted in the uppermost crust, but a decollement at a shallow depth, probably close

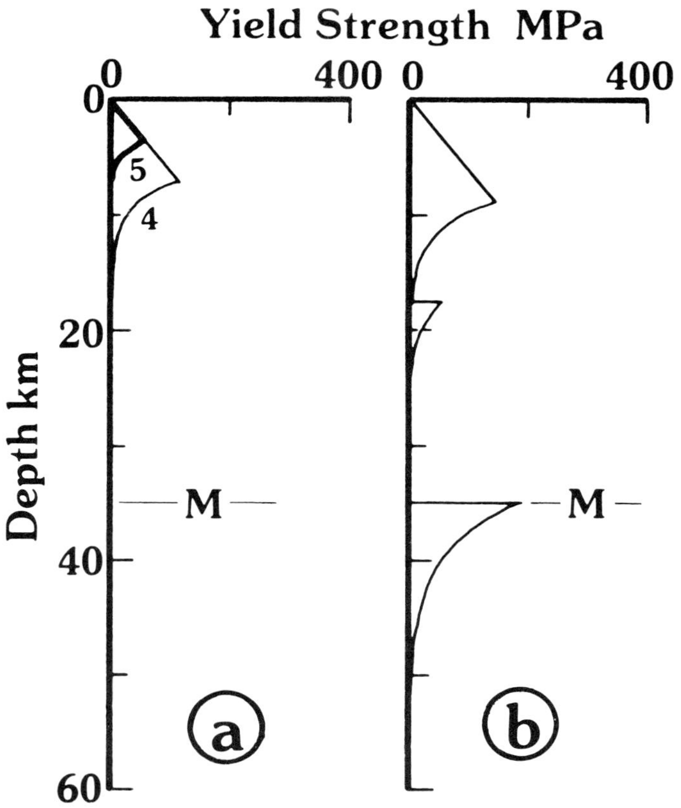

FIGURE 5. Calculated strength curves for the Rio Grande rift lithosphere during the two Cenozoic extension events assuming a strain rate of 10^{-15} s^{-1} and a 17.5-km silicic upper crust overlying a 17.5-km mafic lower crust. **a,** Strength during early-phase extension: curve 4 predicted for normal hot crust during this event (predicted from geotherm 4, Fig. 3); curve 5 predicted for hot crust with upper crustal pluton (predicted from geotherm 5, Fig. 3). **b,** Strength during Rio Grande rift extension (predicted from geotherm 1, Fig. 3). Moho depth is indicated by M (from Cordell, 1978).

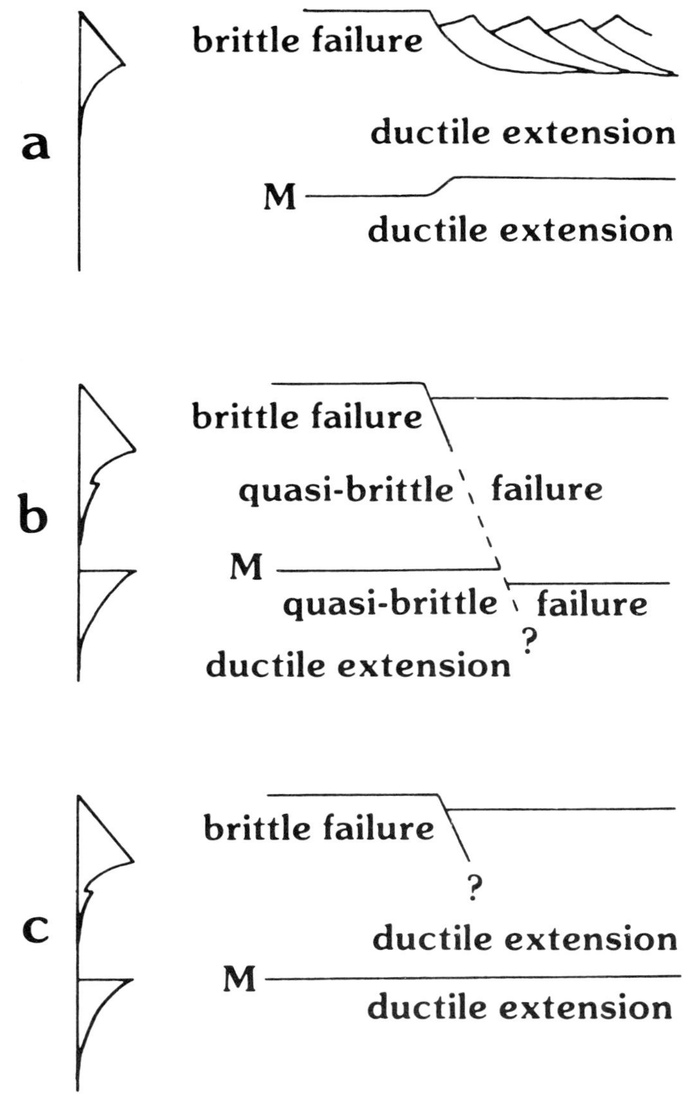

FIGURE 6. Relationships between strength curves (left) and styles of extensional tectonics (right). **a,** Early-phase extension. **b** and **c,** Late-phase extension. Not to scale. See text for explanation. Upwarp of the Moho (M) is shown schematically in a. Only the direct effects of faulting on the Moho are shown in b and c; crustal thinning associated with extension is not shown in b and c.

to the brittle–ductile transition (about 7 km deep or less), would allow strong rotation of locally pervasive brittle fault blocks either on listric normal faults (e.g., Eaton, 1982) or domino-style faults (e.g., Chamberlin, 1983). This style of extension and the controlling strength curve are shown in Figure 6a. Large rotations during this event were made possible by high local strain, which in turn may have been localized by very weak (hot) columns of lithosphere.

Deformation style during the late-phase extension is more difficult to predict from the form of the strength curve. The only predicted brittle–ductile transition is still in the upper crust, but strength in the uppermost mantle could have limited strain and prevented the formation of a shallow decollement. We suggest that perhaps faulting extended right through the crust into the upper mantle with quasi-brittle failure in the lower crust or upper mantle (Fig. 6b), in a manner similar to that described by Vening Meinesz (1950) for graben formation. Alternatively, faulting was truncated by ductile extension in the lower crust as suggested by Bott (1976). However, ductile strength in the uppermost mantle prevented any significant intracrustal decoupling, which in turn prevented any significant rotation of upper crustal blocks in this event (Fig. 6c). Ductile flow in the lower crust and upper mantle after faulting would probably result in both fault-controlled mechanisms, giving a final crustal section similar to that shown in Figure 6c.

DISCUSSION AND CONCLUDING REMARKS

In our simplified analysis of extensional style associated with Rio Grande rift development it has been necessary to generalize geologic events and geotherm evolution. There are obviously many examples of local deformation which do not concur with our simplified models, but we hope that we have identified the basic factor, temperature change, which has controlled the change in extensional style between early and late phases of extension. Local departures from our models may be due to spatial variations in the geotherm, or local compositional or preexisting structural controls of upper-crust faulting. We have no direct evidence to support our speculation on the evolution of the geotherm during the latter half of the Cenozoic, but our models are consistent with the volcanic history of the region. A somewhat better constraint is available in southwestern New Mexico where heat-flow data in regions affected by Oligocene volcanism and early-phase extension, but not affected by young (less than 10 m.y.) volcanism, are in reasonable agreement with the cooling geotherm model (Morgan and Seager, 1983). We hope that new evidence of the paleogeotherm in north-central New Mexico may be gathered to support or refute our models.

Our models of extension associated with the Rio Grande rift are independent of, and ignore, the forces that cause the extension. We simply assume that an extensional-stress field was active only during the two phases of extension. Stress fields locally generated by topography and lithospheric inhomogeneities (e.g., Bott, 1981), and additional weakening of the lithosphere by heating of the crust during plutonic activity may explain the variation in ages of early extension along the rift and the close association of early-phase extension and volcanism in some areas (e.g., Chamberlin, 1983; Lipman, 1981). Considerable regional control of early-phase extension is evident, however, from the similar timing and alignment of extension axes of this phase throughout the western U.S. (Zoback and others, 1981).

If our analysis of geotherm evolution in north-central New Mexico is correct, our models also have important implications for evolution of regional topography. Deposition of the marine Mancos Shale over much of the region indicates a regional surface altitude close to the sea level during Late Cretaceous prior to Laramide deformation. Regional altitudes are now in the range of 1.5–2.5 km, with local elevations in excess of 2.5 km (Cordell, 1978, fig. 2). Epeirogenic uplifts of up to 3 km can be explained in terms of lithospheric heating, thinning, and expansion (Morgan, 1983), and this has been suggested as the dominant mechanism of Cenozoic uplift of the Colorado Plateau (Bodell and Chapman, 1982). Tectonic or magmatic thickening of the crust also results in surface uplift by adding more buoyant low-density material to the lithosphere.

Uplift during Laramide deformation was probably primarily a result of crustal thickening by folding and overthrusting. The lack of any major magmatic activity associated with this event suggests that thermal uplift either did not occur or was minor. Heating of the lithosphere during major Oligocene–early Miocene volcanic activity requires thermal uplift during this event, possibly augmented, at least locally, by magmatic crustal thickening. Evidence for this phase of uplift is given by Axelrod and Bailey (1976) and Elston (1984). Both extension events suggest regional crustal thinning and thus subsidence unless thinning was compensated by the addition of magmas to the crust (e.g., Lachenbruch and Sass, 1978). There is no evidence to indicate regional subsidence during extension at present. Block-faulting during late-phase extension resulted in both local uplift and subsidence and the generation of major topography.

Several authors have suggested significant recent (less than 10 m.y.) uplift of the Rio Grande rift (e.g., Axelrod and Bailey, 1976; Scott, 1975; Seager and others, 1984; Taylor, 1975). We estimate that reheating of the lithosphere to the modern geotherm after the middle Miocene magmatic lull (heating from curve 1 to curve 2 in Fig. 3) could cause about 500 m of uplift, based upon the thermal-uplift models of Morgan (1983). If a significantly more recent uplift has occurred, we suggest that magmatic thickening of the crust may be responsible. Axelrod and Bailey (1976) suggest 1200 m of recent uplift, which would require approximately 5 km of crustal thickening in addition to lithospheric reheating, assuming mean crustal and mantle densities of 2.8 and 3.25 Mg m^{-3}, respectively. A modern magma body or bodies in the Socorro area of the Rio Grande rift (e.g., Sanford, 1983) may be a current example of this magmatic crustal thickening. Transient models of heat-flow data in southern New Mexico (Cook and others, 1978) and in Colorado (Decker and others, this guidebook) also suggest significant recent crustal intrusions. However, contemporaneous uplift of the Colorado Plateau (e.g., Lucchitta, 1979) and the western Great Plains (e.g., see Axelrod and Bailey, 1976) suggests that recent uplift of the Rio Grande rift may not be a local event, and clearly this problem requires further study.

Northern New Mexico shares its two-phase Cenozoic extension history with other areas of the Rio Grande rift and much of the Basin and Range province (e.g., Eaton, 1982; Zoback and others, 1981; Golombek and others, 1983). Possible association of a changing geotherm with changing extensional style has also been suggested for some of these other areas (e.g., Lucchitta and Suneson, 1982; Morgan and Seager, 1983). If this association proves to be correct, we hope that we are one step closer to understanding the complex tectonic history of the western U.S.

ACKNOWLEDGMENTS

Many of the concepts developed in this study have been suggested by previous students of Basin and Range and Rio Grande rift geology; in particular, we have benefited from discussions with Bill Seager and Ivo Lucchitta. We are very grateful to Chuck Chapin and Art Lachenbruch for thoughtful and constructive reviews of a draft of this manuscript. P.M. made his contribution to this study while at the Lunar and Planetary Institute, which is operated by the Universities Space Research Association under contract no. NASW-3389 from the National Aeronautics and Space Administration. Work by M.P.G. on this study was carried out at the Jet Propulsion Laboratory, California Institute of Technology, under contract no. NAS7-100 from the National Aeronautics and Space Administration. This is LPI contribution no. 532.

REFERENCES

Aubele, J. C., 1979, The Cerros del Rio volcanic field: New Mexico Geological Society, Guidebook 30, pp. 243–252.

Axelrod, D. I., and Bailey, H. P., 1976, Tertiary vegetation, climate, and altitude of the Rio Grande depression, New Mexico–Colorado: Paleobiology, v. 2, pp. 235–254.

Bachman, G. O., and Mehnert, H. H., 1978, New K–Ar dates and the late Pliocene to Holocene geomorphic history of the central Rio Grande region New Mexico: Geological Society of America, Bulletin, v. 89, pp. 283–292.

Baldridge, W. S., Damon, P. E., Shafiqullah, M., and others, 1980, Evolution of the central Rio Grande rift, New Mexico: new potassium–argon ages: Earth and Planetary Science Letters, v. 51, pp. 309–321.

Baltz, E. H., 1978, Resume of Rio Grande depression in north-central New Mexico: New Mexico Bureau of Mines and Mineral Resources, Circular 163, pp. 210–228.

Birch, F. S., 1982, Gravity models of the Albuquerque Basin, Rio Grande rift, New Mexico: Geophysics, v. 47, pp. 1185–1197.

Bodell, J. M., and Chapman, D. S., 1982, Heat flow in the north-central Colorado Plateau: Journal of Geophysical Research, v. 87, pp. 2869–2884.

Bott, M. H. P., 1976, Formation of sedimentary basins of graben type by extension of the continental crust: Tectonophysics, v. 36, pp. 77–86.

———, 1981, Crustal doming and the mechanism of continental rifting: Tectonophysics, v. 73, pp. 1–8.

———, and Mithen, D. P., 1983, Mechanism of graben formation—the wedge subsidence hypothesis: Tectonophysics, v. 94, pp. 11–22.

Brocher, T. M., 1981, Shallow velocity structure of the Rio Grande rift north of Socorro, New Mexico: Journal of Geophysical Research, v. 86, pp. 4960–4970.

Brown, L. D., Chapin, C. E., Sanford, A. R., and others, 1980, Deep structure of the Rio Grande rift from seismic reflection profiling: Journal of Geophysical Research, v. 85, pp. 4773–4800.

Byerlee, J. D., 1978, Friction of rocks: Pure and Applied Geophysics, v. 116, pp. 615–626.

Cape, C. D., McGeary, S., and Thompson, G. A., 1983, Cenozoic normal faulting and the shallow structure of the Rio Grande rift near Socorro, New Mexico: Geological Society of America, Bulletin, v. 94, pp. 3–14.

Carter, N. L., 1976, Steady state flow of rocks: Reviews of Geophysics and Space Physics, v. 14, pp. 301–360.

Chamberlin, R. M., 1983, Cenozoic domino-style crustal extension in the Lemitar Mountains, New Mexico: a summary: New Mexico Geological Society, Guidebook 34, pp. 111–118.

Chapin, C. E., 1979, Evolution of the Rio Grande rift—a summary; *in* Riecker, R. E. (ed.), Rio Grande rift—tectonics and magmatism: American Geophysical Union, Washington, D.C., pp. 1–5.

———, and Cather, S. M., 1981, Eocene tectonics and sedimentation in the Colorado Plateau–Rocky Mountain area: Arizona Geological Society Digest, v. XIV, pp. 173–198.

———, and Seager, W. R., 1975, Evolution of the Rio Grande rift in the Socorro and Las Cruces areas: New Mexico Geological Society, Guidebook 26, pp.297–321.

Cook, F. A., Decker, E. R., and Smithson, S. B., 1978, Preliminary transient heat flow model of the Rio Grande rift in southern New Mexico: Earth and Planetary Science Letters, v. 40, pp. 316–326.

Cordell, L., 1978, Regional geophysical setting of the Rio Grande rift: Geological Society of America, Bulletin, v. 89, pp. 1073–1090.

———, 1982, Extension in the Rio Grande rift: Journal of Geophysical Research, v. 87, pp. 8561–8569.

Doell, R. R., Dalrymple, G. B., Smith, R. L., and others, 1968, Paleomagnetism, potassium–argon ages, and geology of rhyolites and associated rocks of the Valles caldera, New Mexico: Geological Society of America, Memoir 116, pp. 211–248.

Eaton, G. P., 1982, The Basin and Range province: origin and tectonic significance: Annual Reviews of Earth and Planetary Science, v. 10, pp. 409–440.

Edwards, C. L., Reiter, M., Shearer, C., and others, 1978, Terrestrial heat flow and crustal radioactivity in northeastern New Mexico and southeastern Colorado: Geological Society of America, Bulletin, v. 89, pp. 1341–1350.

Elston, W. E., 1984, Subduction of young oceanic lithosphere and extensional orogeny in southwestern North America during mid-Tertiary time: Tectonics, v. 3, pp. 229–250.

Gardner, J. N., 1983, Tectonic controls of petrogenesis, Jemez volcanic field, New Mexico (abstract): Geological Society of America, Abstracts with Programs, v. 15, p. 380.

Goetze, C., and Brace, W. F., 1972, Laboratory observations of high-temperature rheology of rocks. Tectonophysics, v. 13, pp. 583–600.

Golombek, M. P., 1981, Geometry and rate of extension across the Pajarito fault zone, Española Basin, Rio Grande rift, northern New Mexico: Geology, v. 9, pp. 21–24.

———, McGill, G. E., and Brown, L., 1983, Tectonic and geologic evolution of the Española Basin, Rio Grande rift: structure, rate of extension, and relation to the state of stress in the western United States: Tectonophysics, v. 94, pp. 483–507.

Jamison, D. B., and Cook, N. G. W., 1980, Note on measured values for the state of stress in the Earth's crust: Journal of Geophysical Research, v. 85, pp. 1833–1838.

Jurdy, D. M., and Brocher, T. M., 1980, Shallow velocity model of the Rio Grande rift near Socorro, New Mexico: Geology, v. 8, pp. 185–189.

Kautz, P. F., Ingersoll, R. V., Baldridge, W. S., and others, 1981, Geology of the Espinaso Formation (Oligocene), north-central New Mexico: summary: Geological Society of America, Bulletin, pt. I, v. 92, pp. 980–983.

Kelley, V. C., 1977, Geology of the Albuquerque Basin: New Mexico Bureau of Mines and Mineral Resources, Memoir 33, 59 pp.

Kudo, A. M., 1982, Rift volcanics of the Albuquerque Basin: overview with some new data: New Mexico Geological Society, Guidebook 33, pp. 285–289.

Lachenbruch, A. H., and Sass, J. H., 1978, Models of an extending lithosphere and heat flow in the Basin and Range province; *in* Smith, R. B., and Eaton, G. P. (eds.), Cenozoic tectonics and regional geophysics of the Western Cordillera: Geological Society of America, Memoir 152, pp. 209–250.

Lipman, P. W., 1981, Volcano-tectonic setting of Tertiary ore deposits, southern Rocky Mountains: Arizona Geological Society Digest, v. XIV, pp. 199–213.

———, and Mehnert, H. H., 1979, The Taos volcanic field, northern Rio Grande rift, New Mexico; *in* Riecker, R. E. (ed.), Rio Grande rift—tectonics and magmatism: American Geophysical Union, Washington, D.C., pp. 289–311.

Lucchitta, I., 1979, Late Cenozoic uplift of the southwestern Colorado Plateau and adjacent lower Colorado River region: Tectonophysics, v. 61, pp. 63–95.

———, and Suneson, N., 1982, Relationships of core complexes, Cenozoic extensional tectonism, and the stable craton near the Colorado Plateau margin in western Arizona (abstract): Geological Society of America, Abstracts with Programs, v. 14, p. 182.

Lynch, H. D., 1983, Numerical models of the formation of continental rifts by processes of lithospheric necking: Ph.D. dissertation, New Mexico State University, Las Cruces, 290 pp.

Manley, K., and Mehnert, H. H., 1981, New K–Ar ages of Miocene and Pliocene volcanic rocks in northwestern Española Basin and their relationships to the history of the Rio Grande rift: Isochron/West, no. 30, pp. 5–8.

McGarr, A., 1980, Some constraints on levels of shear stress in the crust from observations and theory: Journal of Geophysical Research, v. 85, pp. 6231–6238.

McKenzie, D., 1978, Some remarks on the development of sedimentary basins: Earth and Planetary Science Letters, v. 40, pp. 25–32.

Morgan, P., 1983, Constraints on rift thermal processes from heat flow and uplift: Tectonophysics, v. 94, pp. 277–298.

———, and Seager, W. R., 1983, Thermal, mechanical and tectonic evolution of the southern Rio Grande rift (abstract): Geological Society of America, Abstracts with Programs, v. 15, p. 320.

Osburn, G. R., and Chapin, C. E., 1983, Ash-flow tuffs and cauldrons in the northeast Mogollon–Datil volcanic field: a summary: New Mexico Geological Society, Guidebook 34, pp. 197–204.

Ramberg, I. B., and Morgan, P., 1984, Physical characteristics and evolutionary trends of continental rifts: *in* Proceedings of the 27th International Geological Congress: VNU Science Press BV, Utrecht, in press.

Reiter, M., and Clarkson, G., 1983, Geothermal studies in the San Juan Basin and the Four Corners area of the Colorado Plateau II. Steady-state models of the thermal source of the San Juan volcanic field: Tectonophysics, v. 91, pp. 253–269.

———, and Mansure, A. J., 1983, Geothermal studies in the San Juan Basin and the Four Corners area of the Colorado Plateau I. Terrestrial heat flow measurements: Tectonophysics, v. 91, pp. 233–251.

———, ———, and Shearer, C., 1979, Geothermal characteristics of the Rio Grande rift within the southern Rocky Mountain complex; *in* Riecker, R. E. (ed.), Rio Grande rift—tectonics and magmatism: American Geophysical Union, Washington, D.C., pp. 253–267.

Rhoades, S. M., and Callender, J. F., 1983, Evidence for low-angle faulting along western margin of the Sandia uplift, Rio Grande rift, New Mexico (abstract): Geological Society of America, Abstracts with Programs, v. 15, p. 322.

Sanford, A., 1983, Magma bodies in the Rio Grande rift in central New Mexico: New Mexico Geological Society, Guidebook 34, pp.123–125.

Scott, G. R., 1975, Cenozoic surfaces and deposits in the southern Rocky Mountains; *in* Curtis, B. F. (ed.), Cenozoic history of the southern Rocky Mountains: Geological Society of America, Memoir 144, pp. 227–248.

Seager, W. R., Shafiqullah, M., Hawley, J. W., and others, 1984, New K–Ar dates from basalts and the evolution of the southern Rio Grande rift: Geological Society of America, v. 95, pp. 87–99.

Smith, R. B., and Bruhn, R. L., 1984, Intraplate extensional tectonics of the eastern Basin–Range: inferences on structural style from seismic data, regional tectonics and thermal–mechanical models of brittle/ductile deformation: Journal of Geophysical Research, v. 89, in press.

Stearns, C. E., 1953a, Tertiary geology of the Galisteo–Tonque area, New Mexico: Geological Society of America, Bulletin, v. 64, pp. 459–508.

———, 1953b, Early Tertiary volcanism in the Galisteo–Tonque area, north-central New Mexico: American Journal of Science, v. 251, pp. 415–452.

Stetsky, R. M., 1978, Rock friction—the effect of confining pressure, temperature, and pore pressure: Pure and Applied Geophysics, v. 116, pp. 690–704.

Taylor, B. B., 1975, Neogene tectonism in south-central Colorado; *in* Curtis, B. F. (ed.), Cenozoic history of the southern Rocky Mountains: Geological Society of America, Memoir 144, pp. 211–226.

Thompson, G. A., and Zoback, M. L., 1979, Regional geophysics of the Colorado Plateau: Tectonophysics, v. 61, pp. 149–181.

Tweto, O., 1979, The Rio Grande rift system in Colorado: *in* Riecker, R. E. (ed.), Rio Grande rift—tectonics and magmatism: American Geophysical Union, Washington, D.C., pp. 35–56.

Vening Meinesz, F. A., 1950, Les grabens africains, résultat de compression ou de tension dans la croûte terrestre?: Institut Royale Coloniale Belge, Bulletin, v. 21, pp. 539–552.

Vitorello, I., and Pollack, H. N., 1980, On the variation of continental heat flow with age, and the thermal evolution of the continents: Journal of Geophysical Research, v. 85, pp. 983–995.

Weertman, J., and Weertman, J. R., 1975, High temperature creep of rock, and mantle viscosity: Annual Reviews of Earth and Planetary Science, v. 3, pp. 293–315.

Woodward, L. A., 1977, Rate of crustal extension across the Rio Grande rift near Albuquerque, New Mexico: Geology, v. 5, pp. 269–272.

Wyllie, P. J., 1977, Crustal anatexis: an experimental review: Tectonophysics, v. 43, pp. 41–71.

Zoback, M. L., Anderson, R. E., and Thompson, G. A., 1981, Cainozoic evolution of the state of stress and style of tectonism of the Basin and Range province of the western United States: Royal Society of London, Philosophical Transactions (A), v. 300, pp. 407–434.

Taos Plaza, northwest corner, August 1933. Photo by Charles Lord.

REGIONAL STRUCTURAL TRENDS INFERRED FROM GRAVITY AND AEROMAGNETIC DATA IN THE NEW MEXICO–COLORADO BORDER REGION

LINDRITH CORDELL[1] and G. R. KELLER[2]
[1]U.S. Geological Survey, Federal Center, MS-964, Denver, Colorado 80225; [2]University of Texas at El Paso, El Paso, Texas 79968

INTRODUCTION

Long-term efforts to provide aeromagnetic and gravity data for the Rio Grande rift have recently led to the publication of an aeromagnetic map of New Mexico (Cordell, 1984) and gravity maps of New Mexico (Keller and Cordell, 1984) and of the Rio Grande rift covering large parts of Colorado, New Mexico, and Texas (Cordell, Keller, and Hildenbrand, 1982). These maps provide both a synoptic perspective for the study of very large structures and a readily accessible data base for New Mexico Geological Society field excursions and other topical studies. Here, we present a residual Bouguer gravity map centered on the Española–Taos–San Luis Basins section of the Rio Grande rift, and a corresponding aeromagnetic map for the New Mexico sector (digital data from Colorado are not yet available). Preliminary interpretations of these data suggest the presence of some remarkable structures which may record successive Laramide, Eocene, mid-Tertiary, and Neogene deformations. *The maps (Figs. 1, 2) appear in the front matter of this guidebook.*

AEROMAGNETIC DATA

The residual total-magnetic-intensity data shown in Figure 1 are taken directly from Cordell (1984) where data reduction and compilation procedures are described. In the study area, the aeromagnetic data, responsive primarily to bodies of crystalline rock, reflect either Cenozoic igneous rocks or lithologic boundaries within Precambrian crystalline basement. Black lines in Figure 1 delineate boundaries between rocks of differing magnetization, and were drawn by means of a computer-based method described by Cordell and Grauch (1982, 1984). To an approximation, these lines delineate geologic contacts between rocks having differing percentages of magnetite. The magnetization-boundary lines are helpful in interpreting magnetic-contour data because, in these latitudes, magnetic anomalies tend to be skewed southward of their causative bodies, as is apparent from comparing magnetization boundaries with associated anomalies in Figure 1.

GRAVITY DATA

The gravity data used to produce Figure 2 are essentially those used by Cordell, Keller, and Hildenbrand (1982). Major sources of data were the Defense Mapping Agency, Cordell (1976a), and Davis (1979). Reduction procedures are discussed in Cordell, Keller, and Hildenbrand (1982). A reduction density of 2.67 gm/cc was used to produce the Bouguer-anomaly values. Terrain corrections for the region extending from 0.875 km to 166.7 km were calculated using the technique of Plouff (1977). In the study area, gravity data, responsive to variations in rock density, reflect low-density graben or sedimentary basin fill (gravity lows), basement-cored uplifts (gravity highs), and various intrabasement features giving rise to either lows or highs.

Deep crustal and upper-mantle structures, and their associated isostatic effects, produce a broad increase in regional gravity values from west to east across the study area, which obscures the more local anomalies of interest here. In order to remove this regional anomaly, a second-order polynomial surface was fitted to our gravity data in southern Colorado and northern New Mexico using the technique of Lance (1982). The residual-anomaly values, that is the Bouguer-anomaly values minus corresponding values on the second-order polynomial surface, are shown in Figure 2. Black lines delineating contacts between rocks having differing densities were computer-derived by a procedure similar to that used to determine the magnetization boundaries (Cordell, 1979; Cordell and Grauch, 1982, 1984).

DISCUSSION OF GEOPHYSICAL FEATURES BY INFERRED GEOLOGIC AGE

It is not a common practice to discuss potential field-geophysical features chronologically, because there is no time quality intrinsic to these data and because potential field effects of sources of different ages can be superimposed. However, we do so here because a chronological ordering provides the most straightforward presentation of our data and interpretations. The reasons for this are that these geophysical data seem exceptionally clearly correlative with major structural features, and a sense of the evolution of superimposed tectonic events in this region seems to be the key to understanding its present structural style. However, lest the reader unfamiliar with geophysical methods be misled by an historical format, we emphasize that these data are ambiguous, as are all potential field data. Alternative interpretations are possible, and the age and source depth of the anomalies may not be as we have supposed.

We believe from regional considerations that the northeast magnetic grain of the San Juan Basin (southwest quarter of Figure 2; see Cordell and Grauch, 1984), and perhaps the arcuate gravity high extending from Westcliffe through Walsenburg (Kleinkopf and others, 1970) reflect Precambrian structural trends. Geophysical expressions of other suspected Precambrian, Cambrian, and Pennsylvanian structures are difficult to relate to geologic structures formed at a particular time. Examples include the northwest-trending gravity high beneath the San Juan Basin (Lindrith to Farmington); the northeast-trending gravity and magnetic features in the Mora–Wagon Mound area; and the gravity lows related to sedimentary basin fill or, possibly, to unsuspected Precambrian felsic batholiths in the Mora and Weston areas. From Laramide (Late Cretaceous–Paleocene) time on, however, the geological story comes more sharply into focus, and we can try to see the development of the geophysical anomalies in that context.

Laramide structure in the region is characterized by west-to-east compression and thrusting sharply bounded on the east by a hogback monocline, thrust-sheet limit, or similar abrupt structural boundary. The trace of the Rocky Mountain front is delineated in Figure 2 by a short-dashed black line extending from the north border of Figure 2 near Goodpasture to the south border near Mora. This boundary in the New Mexico–Colorado border area is generally interpreted to be a zone of west-dipping, high-angle reverse faults (Grose, 1972; Tweto, 1975), and positive gravity anomalies over basement-cored uplifts in the Sangre de Cristo Mountains north of Fort Garland are consistent with this interpretation. However, south of Fort Garland, despite extensive areas of outcropping basement rocks, the gravity anomalies are transverse to the mapped Rocky Mountain front, a fact which is *not* consistent with the presence of thick basement blocks bounded by rooted, steeply dipping faults. Together, the gravity and geologic data suggest the existence of an extensive allochthonous terrane south of Fort Garland and east of Picuris (Pecos–Picuris fault) in southernmost Colorado and northern New Mexico, with sedimentary basins in the Mora and Weston areas (Raton Basin) extending westward beneath this terrane. Eastward-thrusted Precambrian rocks have been mapped in a few places east of Taos by Clark and Read (1972) and Reed and others (1983). Alternatively, the gravity patterns could be ascribed to combined effects of variable thickness of Pennsylvanian and Permian strata and density variations within basement rocks, although this seems less likely given the data presently at hand.

In early Eocene time the region containing the Colorado Plateau moved north relative to the High Plains (Chapin and Cather, 1981, and cf. Hamilton, 1981). In New Mexico, displacement of 20 to 100+ km is thought to be distributed within a north–south zone of right-lateral shear-faulting about 100 km wide. The north–south-trending, partially right-slip Nacimiento fault (Baltz, 1967) near Cuba, New Mexico (in Fig. 2), and the Pecos–Picuris fault (Miller and others, 1963) are elements of this deformation. However, Miller and others interpreted strike–slip on the Pecos–Picuris fault as being primarily of Precambrian age. The Nacimiento falt is clearly delineated by both gravity and magnetic data, but the Pecos–Picuris fault is most clearly delineated by magnetic trends. Probably, more such faults are obscured by younger Neogene and Quaternary block-faulting and graben fill within the rift where any long, straight, north–south-trending Neogene normal fault can be suspected to be a reactivated early Eocene right-slip fault. The Nacimiento fault dies out north of Lindrith. It is difficult to accommodate much of strike–slip offset in Colorado along the Nacimiento fault trend because it crosses continuous gravity and structural positives that extend northwestward from the northern boundary of the Española Basin. Precambrian fold axes are offset 38 km in a right-lateral sense along the Pecos–Picuris fault (Miller and others, 1963). We think this fault does *not* connect northward along the west side of the Taos–Questa Basin to the east side of the Alamosa (Colorado) gravity high and structural horst (Sutherland, 1972) primarily because the gravity data show the west side of the Taos–Questa graben to be zigzagged. If the Pecos–Picuris fault does extend northward into the Taos–Questa graben, it seems more likely that the Eocene fault controlled the *east* side of the Neogene Taos–Questa graben and that it dies out near the state border.

The aeromagnetic data of Figure 1 show a group of striking, blue-colored magnetic lows east and west of the rift, for which a palinspastic restoration, if valid, indicates a net 130 km offset along an early Eocene right-slip shear zone. The source of the blue anomalies is probably a thick sequence of Proterozoic quartzites and associated meta-supracrustal rocks which are common in northern New Mexico (Cordell and Grauch, 1984). East of the rift, the geologically mapped 38 km of right-lateral offset on the Pecos–Picuris fault is also clearly evidenced by an offset of the blue anomaly in the aeromagnetic data. A magnetization boundary almost exactly coincides with the mapped Pecos–Picuris fault, and the northeast-trending blue anomaly west of the fault at Pilar matches with the northeast-trending blue anomaly east of the fault headed north of Mora, allowing for the 38 km offset. A similar northeast-trending blue anomaly in the vicinity of Wagon Mound suggests another north–south-trending right-slip fault, similar to the Pecos–Picuris fault, in the vicinity of Mora. Restoring the offset on the Pecos–Picuris and inferred Mora faults, one can visualize a long, linear, northeast-trending blue anomaly east of the rift, similar to the long, northeast-trending blue anomaly to the west (Fig. 1), but seemingly displaced from it, again in a right-lateral sense. The sum of all the perceived displacements is about 130 km. Conceivably, the repetition of blue anomalies east of the rift could be explained as ~30 km wave-length synformal infolds of quartzite and other weakly magnetic rocks in an isoclinal Precambrian terrane. However, this alternative interpretation would seem to be inconsistent with the mapped strike–slip displacement of Precambrian rocks along the Pecos–Picuris fault. Eocene deformation, manifested as north–south right slip in northern New Mexico, shows a component of folding in southern Colorado, which is consistent with an Euler pole in the Texas panhandle (Hamilton, 1981). Taken together, gravity and geologic data show a scalloped pattern along the east edge of the basement-cored uplifts, with southeast-plunging synclines in the Canyon City area (north of the north border of Figure 2 in the vicinity of Goodpasture), up the Wet Mountains Valley at Westcliffe, and, in light of preceding discussion, possibly up the San Luis Valley.

During Oligocene time the region was subjected to subduction-related magmatism and back-arc extension to the west–southwest (Eaton, 1980; Lipman, 1981). A salient, possibly Oligocene-age geophysical feature of the region is the gravity low associated with the San Juan Mountains batholith, marked by the large blue spot in the northwest quarter of Figure 2. Density boundaries show this feature to be about 70 km long and 25 km wide, and possibly contained within a larger gravity low about 160 km long and 70 km wide. To the extent that the gravity anomaly and gradients delineate an enormous volume of material of initially low viscosity hydraulically into the crust, these data require a mid-Tertiary episode during which the horizontal component of least principal stress (the extension direction) was north–south. This configuration is contrary to expectation and we have no explanation for it except to speculate that it could reflect a relaxation of the earlier, Eocene north–south compression.

Neogene and Quaternary-age geophysical anomalies are manifested gravity lows over low-density graben fill along the axis of the rift, and local gravity and magnetic anomalies over intrusive rocks. A broad gravity positive is present everywhere along the axis of the rift and provides a measure of crustal thinning and extension (Cordell, 1982). However, this anomaly decreases in width and magnitude northward, and in the region of Figure 2 becomes difficult to distinguish from gravity highs over basement uplifts. The very broad gravity low associated with the structure of the lithosphere–asthenosphere boundary (Ander, 1980) is largely removed from the data of Figure 2 in the course of removing the 2nd-degree polynomial "regional" field.

As elsewhere along the rift (Cordell, 1976b), the graben borders as defined by geophysical data follow an overall linear trend. However, in detail these borders zigzag complexly along segments typically 5–25 km in length, aligned in many cases with Precambrian structural grain. The fault at Pilar represents an east–northeast-trending, transform-like offset of the Española and Taos–Questa grabens (Muehlberger, 1979). In the Taos–Questa area the *west* border of the graben is approximately along the Rio Grande, resulting in a much narrower graben than one would expect driving across the valley. A similarly narrow graben continues north–northwest to Mineral Hot Springs and beyond. The San Luis Valley section of the rift is discussed elsewhere in this guidebook by Keller and others. Here again, the graben is considerably narrower than the topographic valley.

REFERENCES

Ander, M. E., 1980, Geophysical study of the crust and upper mantle beneath the central Rio Grande rift and adjacent Great Plains and Colorado Plateau: Ph.D. dissertation, University of New Mexico, Albuquerque, 218 pp.; and Los Alamos National Laboratory publication LA-8676-T.

Baltz, E. H., 1967, Stratigraphy and regional tectonic implications of part of Upper Cretaceous and Tertiary rocks, east-central San Juan Basin, New Mexico: U.S. Geological Survey, Professional Paper 552, 101 pp.

Chapin, C. E., and Cather, S. M., 1981, Eocene tectonics and sedimentation in the Colorado Plateau–Rocky Mountain area; *in* Dickinson, W. R., and Payne, W. D. (eds.), Relations of tectonics to ore deposits in the southern Cordillera: Arizona Geological Society Digest, v. 14, pp. 173–198.

Clark, K. F., and Read, C. B., 1972, Geology and ore deposits of the Eagle Nest area, New Mexico: New Mexico Bureau of Mines and Mineral Resources, Bulletin 94, 152 pp.

Cordell, L., 1976a, Preliminary complete Bouguer anomaly gravity map of the Taos Basin section of the Rio Grande graben: U.S. Geological Survey Open-file map, scale 1:250,000.

———, 1976b, Aeromagnetic and gravity studies of the Rio Grande graben in New Mexico between Belen and Pilar: New Mexico Geological Society, Special Publication 6, pp. 62–70.

———, 1979, Gravimetric expression of graben faulting in Santa Fe country and the Española Basin, New Mexico: New Mexico Geological Society, Guidebook 30, pp. 59–64.

———, 1982, Extension of the Rio Grande rift: Journal of Geophysical Research, v. 87, pp. 8561–8569.

———, 1984, Composite residual total intensity aeromagnetic map of New Mexico: National Oceanic and Atmospheric Administration, National Geophysical Data Center, scale 1:500,000.

———, and Grauch, V. J. S., 1982, Mapping basement magnetization zones from aeromagnetic data in the San Juan Basin, New Mexico: extended abstract; *in* Society of Exploration Geophysicists, 1982 Annual Meeting, Abstracts with Programs, pp. 246–247.

———, and ———, 1984, Mapping basement magnetization zones from aeromagnetic data in the San Juan Basin, New Mexico: Geophysics, special issue on potential field methods, in press.

———, Keller, G. R., and Hildenbrand, T. G., 1982, Bouguer gravity map of the Rio Grande rift: U.S. Geological Survey, Geophysical Investigations Map GP-949, scale 1:1,000,000.

Davis, G., 1979, A gravity study of the San Luis Basin, Colorado: M.S. thesis, University of Texas at El Paso, 101 pp.

Eaton, G. P., 1980, Geophysical and geological characteristics of the crust of the Basin and Range province; *in* Continental tectonics: National Academy of Science, Washington, D.C., pp. 96–113.

Grose, L. T., 1972, Tectonics; *in* Mallory, W. W. (ed.), Geologic atlas of the Rocky Mountain region: Rocky Mountain Association of Geologists, pp. 34–44.

Hamilton, W., 1981, Plate-tectonic mechanism of Laramide deformation: University of Wyoming, Contributions to Geology, v. 19, pp. 87–92.

Keller, G. R., and Cordell, L., 1984, Bouguer gravity anomaly map of New Mexico: National Oceanic and Atmospheric Administration, National Geophysical Data Center, scale 1:500,000.

Kleinkopf, M. D., Peterson, D. L., and Johnson, R. B., 1970, Reconnaissance geophysical studies of the Trinidad quadrangle, south-central Colorado; *in* Geological Survey Research 1970: U.S. Geological Survey, Professional Paper 700-B, pp. B78–B85.

Lance, J. O., Jr., 1982, Frequency domain analysis of least-square polynomial surfaces with application to gravity data in the Pedregosa Basin area: Ph.D. dissertation, University of Texas at El Paso, 190 pp.

Lipman, P. W., 1981, Volcano-tectonic setting of Tertiary ore deposits, southern Rocky Mountains; *in* Dickinson, W. R., and Payne, W. D. (eds.), Relations of tectonics to ore deposits in the southern Cordillera: Arizona Geological Society Digest, v. 14, pp. 199–213.

Miller, J. P., Montgomery, A., and Sutherland, P. K., 1963, Geology of part of the southern Sangre de Cristo Mountains, New Mexico: New Mexico Bureau of Mines and Mineral Resources, Memoir 1, 106 pp.

Muehlberger, W. R., 1979, The Embudo fault between Pilar and Arroyo Hondo, New Mexico: an active intracontinental transform fault: New Mexico Geological Society, Guidebook 30, pp. 77–82.

Plouff, D., 1977, Preliminary documentation for a Fortran program to complete gravity terrain corrections based on topography digitized on a geographic grid: U.S. Geological Survey, Open-file Report 77-535, 45 pp.

———, and Pakiser, L. C., 1972, Gravity study of the San Juan Mountains, Colorado: U.S. Geological Survey, Professional Paper 800-B, pp. B1P3–B190.

Reed, J. C., Lipman, P. W., and Robertson, J. R., 1983, Geologic map of the Latir Peak and Wheeler Peak Wildernesses and Columbine–Hondo Wilderness Study Area, Taos County, New Mexico: U.S. Geological Survey, Map MF-1570-B, scale 1:50,000.

Sutherland, P. K., 1972, Pennsylvanian stratigraphy, southern Sangre de Cristo Mountains, New Mexico; *in* Geologic atlas of the Rocky Mountain region: Rocky Mountain Association of Geologists, pp. 139–142.

Tweto, O., 1975, Laramide (Late Cretaceous–early Tertiary) orogeny in the southern Rocky Mountains; *in* Cenozoic history of the Southern Rocky Mountains: Geological Society of America, Memoir 144, pp. 1–44.

Taos Pueblo, ca 1920. Photo by Burt Harwood.

SEISMICITY OF THE RIO GRANDE RIFT IN NORTHERN NEW MEXICO, 1973–1983

DANIEL J. CASH and JOYCE J. WOLFF
Geophysics Group, ESS-3, Los Alamos National Laboratory, Los Alamos, New Mexico 87545

INTRODUCTION

In the years before adequate seismograph coverage, reports of felt and damaging earthquakes in New Mexico came mostly from along the Rio Grande rift because most of the state's population was concentrated along the Rio Grande River, which lies within the rift. Even though the seismograph coverage is now fairly substantial and the population more widely spread, the picture is little changed. The Rio Grande rift and related structures continue to be the sites of most earthquakes, although earthquakes are occasionally felt in almost all areas of New Mexico.

The Los Alamos Seismograph Network (LASN) was begun in late 1973. Current LASN station locations are shown in Figure 1. Tectonic research on the Rio Grande rift was the principal reason for establishing the network. Continuous data of good quality have been acquired for about a decade; close to 2,000 earthquakes and microearthquakes have been recorded, analyzed, and cataloged. The largest event occurred in 1976, 60 km east–northeast of Gallup, and had a magnitude $M_L = 4.6$ ($m_b = 5.2$).

SEISMICITY–STRUCTURE RELATIONS

The seismicity of northern New Mexico based on LASN data (Fig. 2), although concentrated near the Rio Grande rift, is not entirely correlated with rift structures. Some epicenter patterns cut obliquely across the rift, some are associated with volcanism, and some outline pre-rift structure. Several noteworthy correlations between recorded seismic activity and geologic features are discussed below.

Calderas of the Jemez Mountains

Very few events have occurred within the calderas, despite the presence of numerous faults associated with ring fractures and resurgent doming (Cerro Redondo) that followed the formation of the calderas about 1 m.y. ago. This aseismicity may be attributed to high crustal temperatures or to the relief of stress on the surrounding active fault zones. The high heat flow of the caldera region is well known (Edwards et al., 1976). Both Los Alamos National Laboratory and Union Geothermal Company have major geothermal projects on the west side. In areas such as the Rio Grande rift, where basin geology and normal

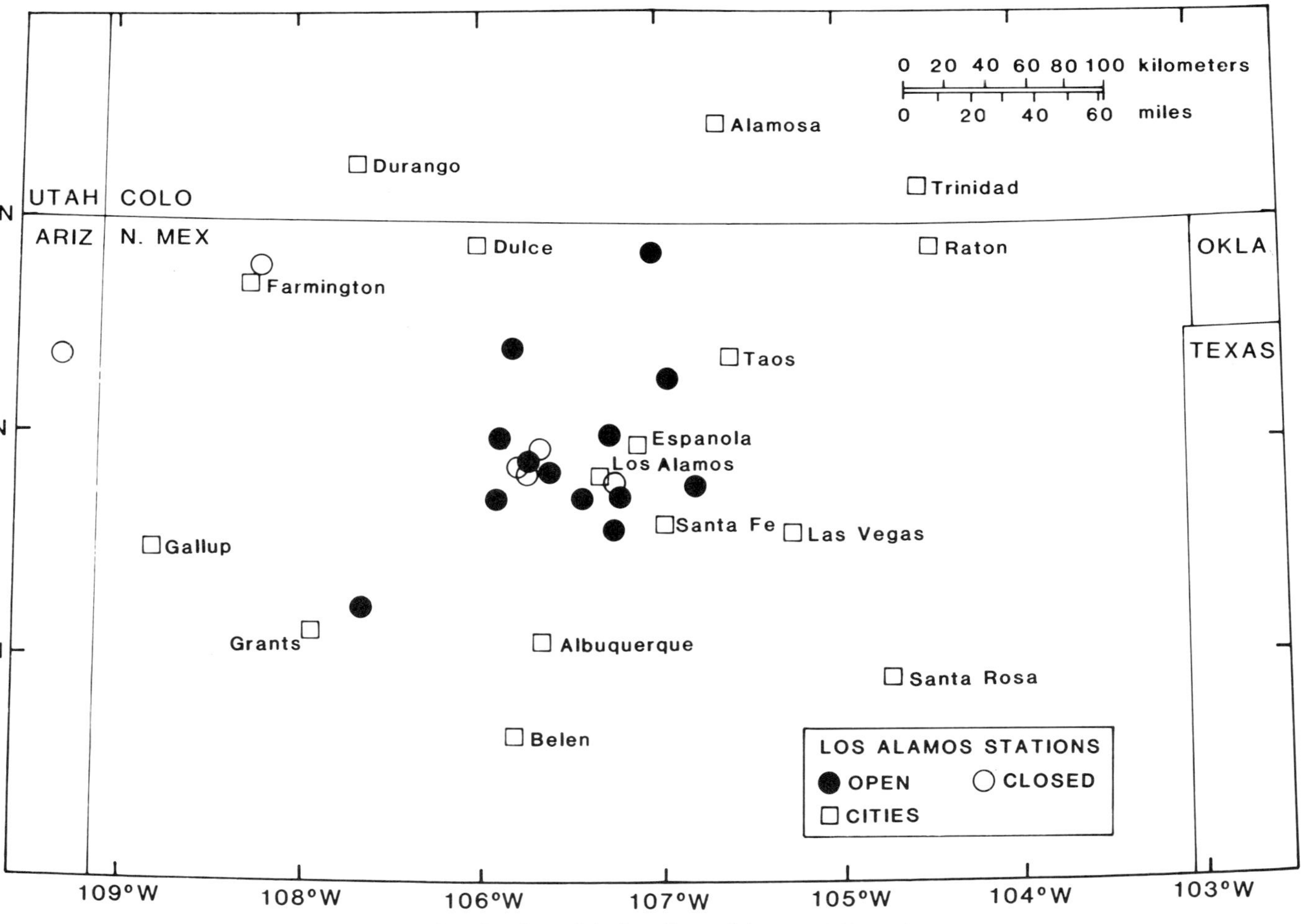

FIGURE 1. Map of stations of the Los Alamos Seismograph Network.

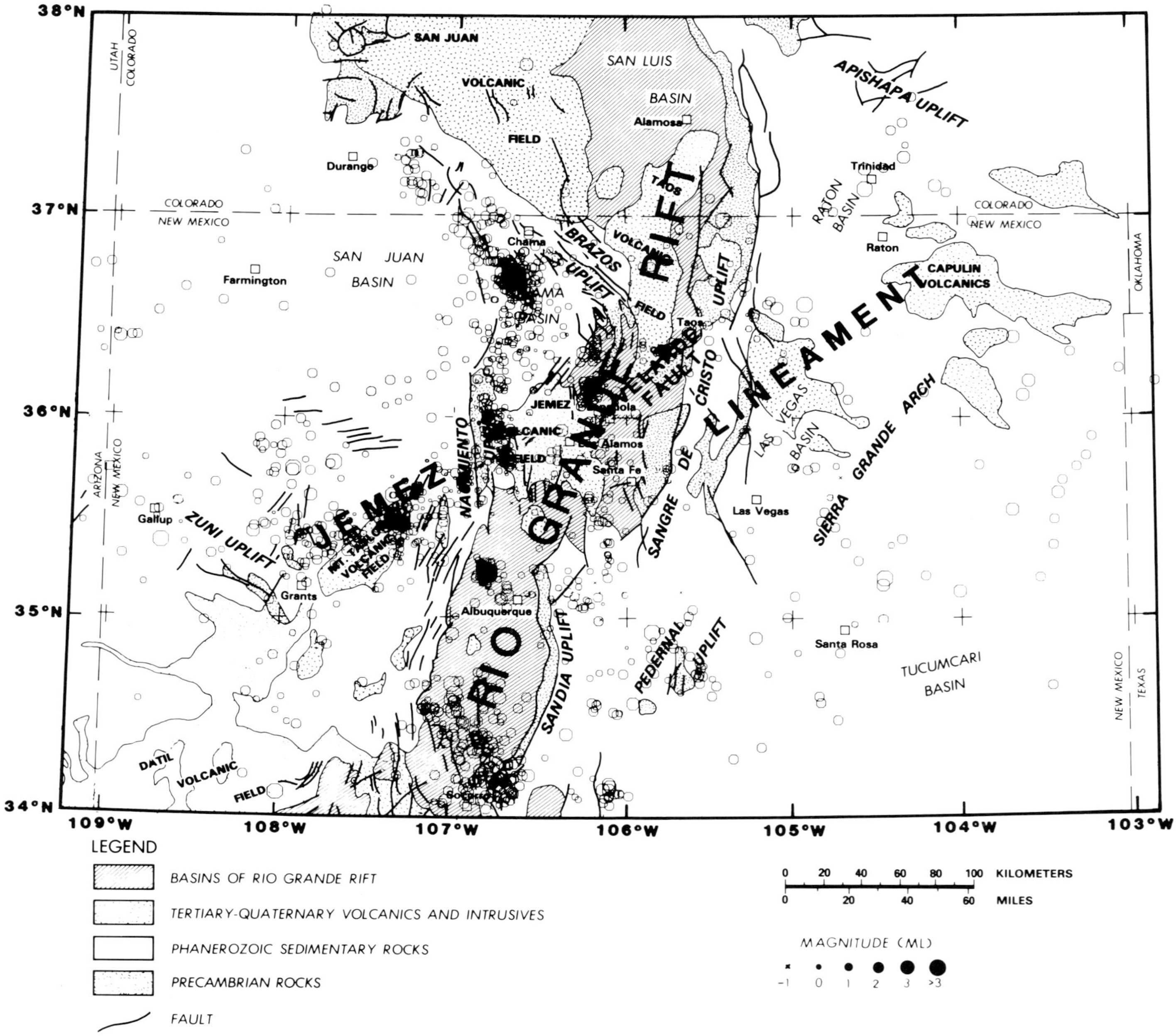

FIGURE 2. Map of earthquake epicenters for northern New Mexico, September 1973–December 1983.

faulting predominate, seismicity tends to be shallow. Where temperatures are high, the brittle nature of the crust at even shallow depths is replaced by a degree of ductility, making the process of brittle fracture or faulting much less probable. Stress that would increase to the breaking point in brittle rock is relieved through creep in the more ductile material. The absence of seismicity may also be attributed to stress being relieved on active fault zones immediately outside the calderas, most notably the north–south zones of activity along the Nacimiento uplift on the west and near the Rio Grande on the east.

San Felipe Fault Zone

Immediately to the south of the Jemez Mountains lies an area that appears seismically similar to the Jemez calderas. The region is nearly aseismic, surrounded by zones of marked seismicity, and contains a high density of normal faults no older than Pleistocene. These faults are particularly visible in the surficial Quaternary–Tertiary basalts of Santa Ana Mesa near the center of the aseismic area (Smith et al., 1970). Little is known about heat flow in this region, so speculation about the cause of aseismicity has little foundation at present.

Nacimiento Uplift

The zone of concentrated epicenters on the west side of the Jemez Mountains lies on the Nacimiento uplift. The uplift was formed in Laramide time by reverse or high-angle thrust faulting upward toward the west; thus the Nacimiento fault dips to the east under the uplift and is the probable source of the seismicity. The tectonic stresses are much different now than during the Laramide time. The Nacimiento fault was originally a result of east–west compressional tectonics. In contrast, the present-day rift structures are more consistent with east–west extensional tectonics. Apparently, the Nacimiento fault is a zone of crustal weakness that has been reactivated by the tectonic processes of rift formation. Fault-plane solutions for Nacimiento events would contribute to answering questions about contemporary tectonics in the Nacimientos, but results of fault-plane analyses have so far produced ambiguous results.

Induced seismicity occurs during hydraulic-fracturing experiments at Los Alamos' Hot Dry Rock (HDR) geothermal project on the west side of the Jemez Mountains. The HDR-project site lies along the east edge of the seismically active region of the Nacimiento uplift. The several hundred microearthquakes that have been induced are not shown in Figure 2 because they are too small to be located by the permanent network, and are too numerous to be shown. A temporary network of nearby seismographs is used to record and research the HDR microquake activity.

Gallina–Archuleta Arch

North of the Nacimiento uplift, the seismic activity closely follows the Gallina–Archuleta arch to the Colorado border. The epicenter of the Dulce earthquake of 1966 lies in this zone. The earthquake had a magnitude $m_b = 5.5$ (Herrmann et al., 1980), a maximum intensity of MM VII on the Modified Mercalli scale, and caused moderate damage.

The concentration of approximately 150 events 30–40 km south–southwest of Chama is the site of El Vado and Heron reservoirs. Nearly two-thirds of the events occurred in 1982, although some seismic activity has occurred there every year. The largest magnitude recorded was $M_L = 2.7$ ($m_b = 3.8$). This may be a case of reservoir-induced seismicity (RIS) in which tectonic stress is being released prematurely by either the weight of the reservoir or the production of abnormally high pore-fluid pressure. In contrast, the nearby Abiquiu reservoir on the north side of the Jemez Mountains does not manifest an abnormal level of seismicity, perhaps because less tectonic stress exists there as evidenced by lower seismic activity in general north of the Jemez Mountains.

Jemez Volcanic Lineament

At the south end of the Nacimiento seismic zone the trend of epicenters turns to the southwest and is coincident with (at least as far as Grants) the Jemez volcanic lineament (Mayo, 1958), a line of extrusive volcanics. Near Grants, basalts as young as 1,000 years (Nichols, 1946) are visible along Interstate 40. The lineament intersects the west side of the Rio Grande rift at the Jemez Mountains. On the east side of the rift, the lineament continues northeastward, but is offset to the south as though some north–south right-lateral displacement has occurred along the rift. Where the lineament meets the southwest edge of the Jemez Mountains, the seismicity lies on the San Diego fault near Jemez Springs. The San Diego fault is colinear with the southwest portion of the lineament, which suggests a correlation between volcanism, faulting, and seismicity.

Velarde Fault

Northeast of the Jemez Mountains seismicity continues northeastward, colinear with the seismic zone to the southwest. Here it follows the trace of the Velarde fault, extending through and to the northeast of Taos. This activity is somewhat to the north of the Jemez volcanic lineament on this side of the rift. There is a concentration of events northwest of Española where the Velarde fault projects to the Pajarito fault zone, and the northeast trend of epicenters crosses another trend oriented north–south. This area of seismicity, which lies about 25 km north of Los Alamos, has been the site of earthquakes felt generally throughout the Los Alamos–Española region, four of them since the town of Los Alamos was established in the early 1940's. One occurred in 1952, two in 1971, and one in 1973. All of these occurred before the establishment of the LASN, and, therefore, the epicenters are not precisely known. The first two may have been very near Los Alamos, while the last two are estimated to have occurred about 25 km north of there. The seismicity along the Velarde fault and that along the San Diego fault are separated by the aseismic region of the Jemez calderas.

Pajarito Fault Zone and the Española Basin

The Pajarito fault, a major vertical offset downthrown to the east, occurs on the east side of the Jemez Mountains just west of Los Alamos. The fault offsets the Bandelier Tuff 100–150 m at Los Alamos (Bailey and Smith, 1978). Despite its relatively recent age and significant displacement, the Pajarito fault was not very active near and south of Los Alamos in the 10-year period of coverage in Figure 2. This observation must be accompanied by caution, because 10 years is an extremely short interval for drawing conclusions about absolute levels of seismicity. North of Los Alamos the Pajarito fault zone becomes quite active. This activity is part of a band of epicenters that runs from an area northeast of Abiquiu almost due southward along the center of the Velarde graben in the Española Basin (Manley, 1979) and on to the Tijeras Canyon east of Albuquerque. From there the seismic trend turns westward through Albuquerque to the volcanoes on the city's West Mesa.

Albuquerque Volcanoes

In 1978–1979, a swarm of approximately 160 small earthquakes occurred near the volcanoes on Albuquerque's West Mesa. A few events still occur there, but 95% occurred within a period of 15 months. The largest recorded magnitude was $M_L = 2.4$ ($m_b = 3.6$). This activity and that along the Jemez volcanic lineament suggest that volcanism-induced stresses are still occurring in these regions. A detailed investigation of the swarm is reported by Jaksha et al. (1978).

GENERAL DISCUSSION

Information about earthquakes that occurred in New Mexico before 1962 is quite sketchy. The first high-magnification seismograph in the state was installed at Albuquerque in 1959. The World-wide Seismograph Station Network (WWSSN) was begun in the early 1960's and included a station in Albuquerque installed in 1961. New Mexico Institute of Mining and Technology installed another one at Socorro in 1962. Before then, the only reports of New Mexico earthquakes were from people who experienced the effects. Such reports are subject to the imagination of the reporter and tend toward exaggeration and misinterpretation of observations. Often earthquakes large enough to be felt go completely unnoticed, or at least unreported, if they occur in an area where the population density is low. Even when reported from low-population regions, estimates of the epicenter locations can be in error by many kilometers.

The seismic history of New Mexico includes several earthquakes that caused moderate damage and frequent events that were felt locally. Earthquakes that have caused moderate damage include those at Socorro in 1906 (Reid, 1911; Sanford, 1963), Cerrillos in 1918 (Olsen, 1979), and Dulce in 1966 (Cash, 1971; Hermance et al., 1980). Earthquakes have been felt in almost all parts of the state, but most often along the Rio Grande River, in part due to the higher population density there (Northrop, 1976, 1982). That no tremor has been severe is not surprising. Very strong shocks usually occur in subduction zones, as in Japan, Alaska, and South America, or on long transform (strike–slip) faults such as found along the San Andreas fault system in California. Severe shocks are much less common in areas of extensional tectonics, where normal dip–slip and short transform faulting predominate. The Rio Grande rift is an example of extensional tectonics in a generally east–west direction, although some structural evidence exists for strike–slip displacement parallel to the rift, i.e., approximately north–south.

Although New Mexico's level of observed seismicity is only about 5–10% that of highly active areas of California, the potential for seismic disaster exists. The LASN was established by the Los Alamos National Laboratory in 1973 to address the seismic-hazard question and to pursue research on the local structure and tectonics. There are currently 15 stations in the network. About the same time, the U.S. Geological Survey (Albuquerque Seismological Laboratory) installed a network of seismographs to the south of the Los Alamos network. These two networks and the New Mexico Institute of Mining and Technology seismographs provide excellent coverage of the Rio Grande rift region of New Mexico from south of Socorro to the Colorado border. The three institutions continue frequent exchange of pertinent seismic data.

REMARKS ON THE DATA

Los Alamos publishes a quarterly catalog containing a list of earthquakes and a seismicity map (see, for example, Cash et al., 1983).

Figure 2 is a plot of all events listed in the catalogs for the period September 1973 through December 1983. We believe that very few non-earthquake events are included. Careful effort has been devoted to locating the sources that are explosions produced by mining and construction activities. Other man-made events, such as sonic booms from supersonic aircraft, are readily identifiable by the character of their seismic signal or by their relative time of arrival across the network.

The accuracy of epicenter locations varies widely, dependent on the number and proximity of seismograph stations. We believe that epicenters of events large enough to be recorded at five or more stations and located within the network are determined correctly within 3 km. Because of this uncertainty, attempts are usually not made to associate a specific event with a specific fault. But where trends of events can be seen to lie on or near a fault or fault zone, the correlation becomes credible and is included in the discussion in this paper. Because of low station density and complex velocity structure, few depths are well determined. Depth calculations are much more susceptible to poor velocity control than epicenters. Events whose depths are reliably determined are usually shallow, i.e., less than 10 km and often between 0.5 and 5.0 km.

Magnitude is determined from the measured duration of each event, defined as the time interval from the arrival of the first P wave until the signal level drops to the level of the background noise. The formula we derived for this purpose is

$$M_L = 2.79 \log T - 3.63$$

where T is the event duration in seconds and log is the common logarithm (base 10). The formula was obtained by correlating the duration of local events with the value of M_L calculated from records of the standard Wood–Anderson torsion seismometers at the WWSSN station, ALQ, in Albuquerque. It is therefore nominally the same magnitude scale as the local magnitude scale derived by C. R. Richter for use in California. Our values of M_L compare very well with those published by the U.S. Geological Survey for earthquakes occurring in New Mexico and Arizona.

REFERENCES

Bailey, R. A., and Smith, R. L., 1978, Volcanic geology of the Jemez Mountains, New Mexico; *in* Hamley, J. W. (compiler), Guidebook to Rio Grande rift in New Mexico and Colorado: New Mexico Bureau of Mines and Mineral Resources, Circular 163.

Cash, D. J., 1971, The Dulce, New Mexico, earthquake of January 23, 1966: location, focal mechanism, magnitude, and source parameters: Ph.D. dissertation, New Mexico Institute of Mining and Technology, 89 pp.

———, Olsen, K. H., Stewart, J. N., and Wolff, J. J., 1983, Earthquake catalog for northern New Mexico, October–December 1982. Los Alamos National Laboratory, LA-9782-PR.

Edwards, C. L., Reiter, M., Shearer, C., and Young, W., 1978, Terrestrial heat flow and crustal radioactivity in northeastern New Mexico and southern Colorado. Geological Society of America, Bulletin, v. 89, pp. 1341–1350.

Herrmann, R. B., Dewey, J. W., and Park, S., 1980, The Dulce, New Mexico, earthquake of 23 January 1966. Seismological Society of America, Bulletin, v. 70, no. 6, p. 2171.

Jaksha, L. H., Locke, J., and Gebhart, J. H., 1981, Microearthquakes near the Albuquerque volcanoes, New Mexico. Geological Society of America, Bulletin, v. 92, pp. 31–36.

Manley, K., 1979, Stratigraphy and structures of the Española Basin, Rio Grande rift, New Mexico; *in* Riecker, R. E. (ed.), Rio Grande rift: tectonics and magmatism: American Geophysical Union.

Mayo, E. B., 1958, Lineament tectonics and some ore districts of the Southwest. AIME, Transactions, pp. 1169–1175.

Nichols, R. L., 1946, McCartys basalt flow, Valencia County, New Mexico: Geological Society of America, Bulletin, v. 57, pp. 1049–1086.

Northrop, S. A., 1976, New Mexico's earthquake history, 1849–1975; *in* Tectonics and mineral resources of southwestern North America: New Mexico Geological Society, Special Publication 6, pp. 77–87.

———, 1982, Earthquakes of Albuquerque country: New Mexico Geological Society, Guidebook 33, pp. 171–178.

Olsen, K. H., 1979, The seismicity of north-central New Mexico with particular reference to the Cerrillos earthquake of May 28, 1918: New Mexico Geological Society, Guidebook 30, pp. 65–75.

Reid, H. F., 1911, Remarkable earthquakes in central New Mexico in 1906 and 1907. Seismological Society of America, Bulletin, v. 1, pp. 10–16.

Sanford, A. R., 1963, Seismic activity near Socorro: New Mexico Geological Society, Guidebook 14, pp. 146–151.

Smith, R. L., Bailey, R. A., and Ross, C. J., 1970, Geologic map of the Jemez Mountains, New Mexico: U.S. Geological Survey, Miscellaneous Geologic Investigations Map I-571.

TELESEISMIC DEEP SOUNDING OF THE VELOCITY STRUCTURE BENEATH THE RIO GRANDE RIFT

P. M. DAVIS[1], E. C. PARKER[1], J. R. EVANS[2], H. M. IYER[2], and K. H. OLSEN[3]
[1]Department of Earth and Space Sciences, University of California at Los Angeles, Los Angeles, California 90024; [2]U.S. Geological Survey, Menlo Park, California 94025;
[3]Institute of Geophysics and Planetary Physics, Los Alamos National Labortory, Los Alamos, New Mexico 87545

INTRODUCTION

A 1,000-km-long linear array of 20 seismic stations between Moab, Utah, and Odessa, Texas, crossing the Rio Grande rift near Santa Fe, recorded 40 useful teleseisms during December 1982 and January 1983. The orientation of this array was chosen to be at an angle of 45° to the strike of the rift because teleseismic arrivals commonly occur along this azimuth. P-wave arrival times show a clear delay of as much as 1.5 sec centered on the rift with measurable delays extending several hundred kilometers to either side of it. This delay pattern correlates well with the broad negative Bouguer gravity anomaly in this area which is in isostatic compensation with the topographic uplift. Variation in crustal velocities is too small to fully explain the delay pattern. Preliminary inversions of the data require a low-velocity, low-density body situated well below the Moho. Constraining it to lie completely within the crust significantly degrades the fit of the model to the data and rules out an Airy type of isostatic compensation. Both a damped-least-squares block inversion and fitting of a bowed "lithosphere–asthenosphere interface" to the data indicate a rift-centered delaying structure at upper-mantle depths (70–200 km), at least 500 km wide. We compare velocity, gravity, and topography across the rift.

The Rio Grande rift, which separates the Great Plains to the east from the Colorado Plateau and Basin and Range province in the west, lies just to the east of the continental divide in a region of broad uplift, high heat flow, and low gravity. Figure 1 (from Cordell, 1978) illustrates the topographic uplift and regional gravity low extending from the Colorado Plateau into the Great Plains. This profile and several similar ones at 1° intervals lying between 33° and 37°N latitude, suggest that a broad, linear low-density zone several hundred kilometers wide is associated with the uplift, and has given rise to the rift, a low-lying block (some 50 km wide). Ander (1980) has presented a wide range of gravity models which fit the observations. We have used the teleseismic P-wave delay technique to provide extra constraints on the models in order to better identify the position and shape of low-density zone.

Spence et al. (1982) reported differential travel times of 0.6 sec between within-rift stations of a northern New Mexico seismic net. This pattern was interpreted as due to a deep-seated, low-velocity body beneath the Valles caldera. They also reported a northeast-trending, low-velocity zone aligned in the same direction as the Jemez lineament. Superimposed on this pattern was up to a 1.8 sec difference between the rift stations and the distant WWSSN station at Lubbock, Texas. Our experiment was designed to investigate this larger, longer-wavelength anomaly, which necessarily required installation of stations over a much longer baselength than that covered by the northern New Mexico net.

THE EXPERIMENTAL METHOD

Geophysical methods to investigate localized velocity heterogeneity near the base of the lithosphere are limited. Surface waves which are generally used to determine lithospheric thickness are difficult to interpret where the structure is not approximately plane-layered. Structure lying beneath the Moho is too deep for economically feasible conventional refraction and reflection studies. Teleseismic deep sounding represents the most practical method. The P-wave-delay technique is well developed and has been used to search for magma chambers and deep structures in diverse tectonic environments. However, this is the first linear transect of a major rift. Descriptions of the technique can be found in papers by Aki (1982), Iyer (1975, 1984), Iyer et al. (1981), Evans (1982), and Sharp et al. (1980). Briefly, the method consists of timing arrivals from distant earthquakes and subtracting from them arrival times for laterally homogeneous earth. Any residuals remaining are attributed to lateral heterogeneity beneath the seismic array, since distant heterogeneities such as those at the source will smear out due to geometric spreading. Angles of incidence and azimuth are calculated from source parameters using a standard earth model. The pattern of residuals has been observed to shift in response to different angles of incidence and azimuth of the incoming waves. Given enough rays at different angles, the data can be inverted to determine velocity at depths of up to about half the array dimension, with a resolution equal to station spacing near the surface but decreasing with depth.

THE EXPERIMENT

Figure 2 shows the 20 recording-station locations along a 1,000-km line for the (joint UCLA, USGS, LANL) 82/83 experiment. A NW–SE orientation was used since teleseismic arrivals lie predominantly along this azimuth so that rays from the NW and SE intersect in regions lying beneath the array. This geometry was important for our initial line to avoid ambiguity in the inversion of the data. Otherwise we must

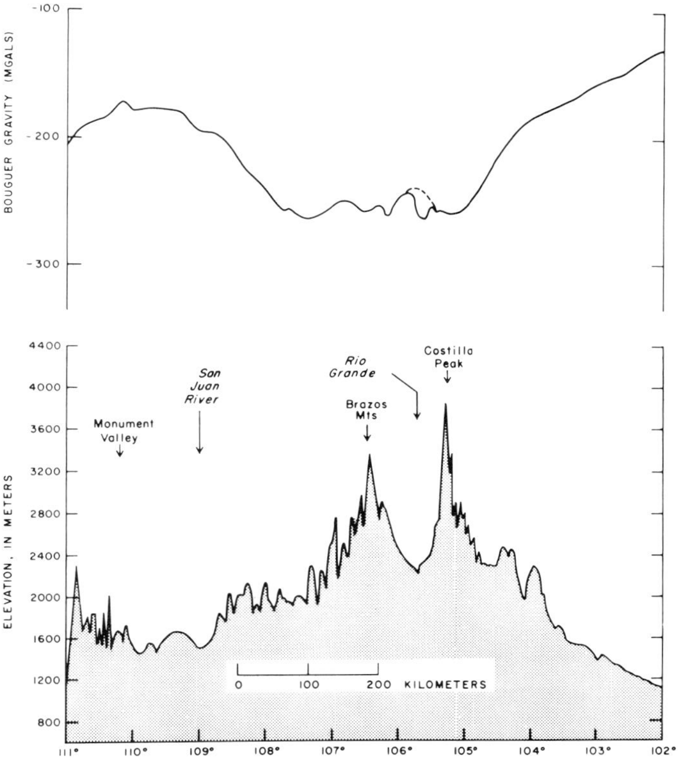

FIGURE 1. E–W gravity profile across the Rio Grande rift at latitude 37°N (from Cordell, 1978).

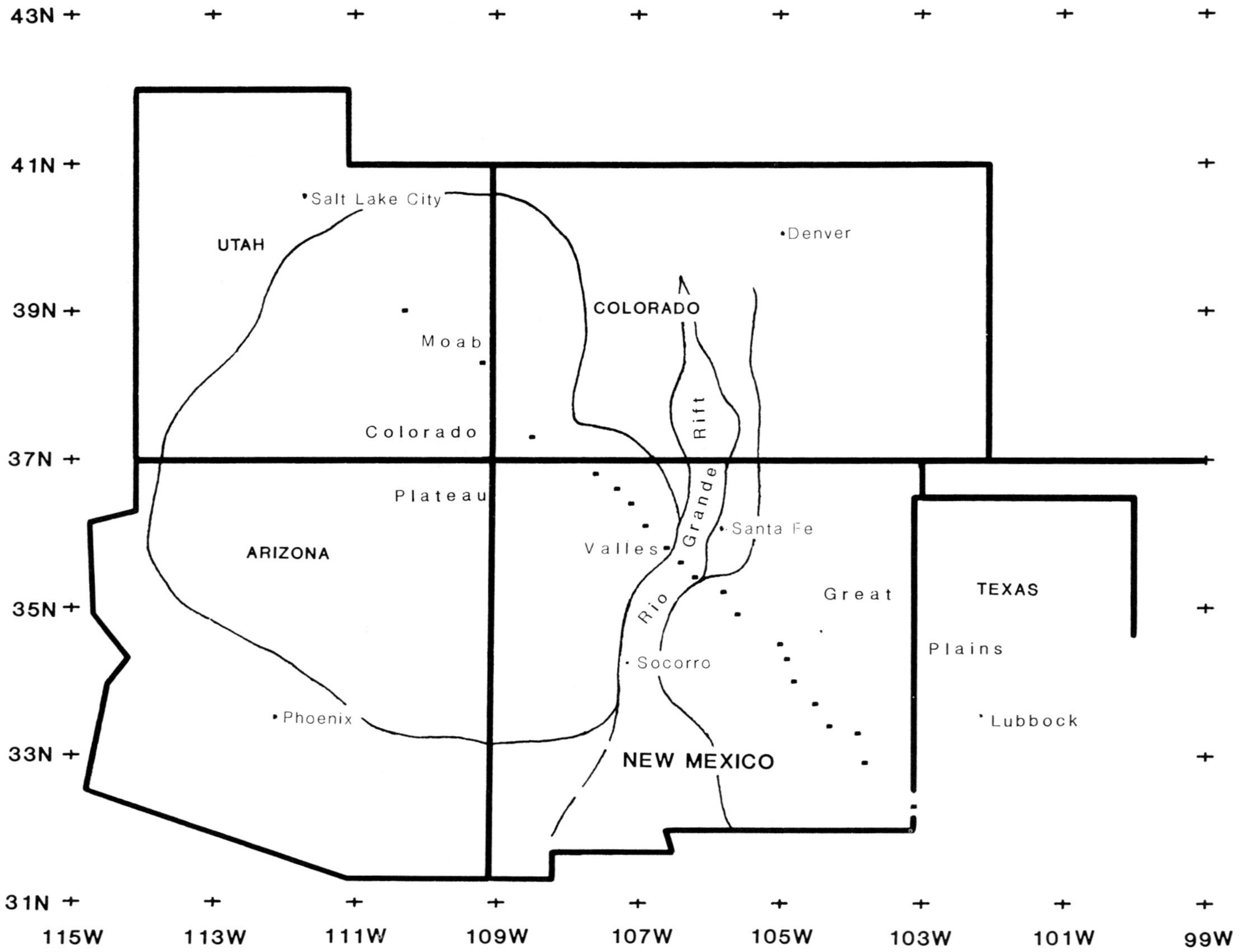

FIGURE 2. Location map of portable seismic recording stations used in this experiment.

assume that the structure is two-dimensional. The array ran from late November 1982 until late January 1983. Field base stations were set up at Moab, Utah, and Santa Fe and Roswell, New Mexico. Field parties, comprised of two people and a 4-W/D vehicle, serviced six instruments over 300-km section of line. This involved synchronizing time, changing tapes and batteries every five days at each site, as well as data playback and instrument repair, resulting in a full schedule. Typical driven distances per vehicle amounted to 10,000 miles.

Figure 3 is a sample of the data for a teleseism from Honshu. The relative P-wave arrival times can be measured to 0.1 sec. Residuals are found by subtracting from the arrival times computed travel times listed in the Herrin tables. Relative residuals are formed for each event by subtracting the mean residual from the raw residuals.

Of the 60 earthquakes recorded, 40 are of sufficient signal-to-noise ratio to be used in an inversion. Twenty-five of these lie along the array azimuth. For graphical purposes we have grouped these 25 into two groups of delta (epicentral distance) greater than or less than 50°. The groups are further subdivided and averaged to give residuals from the NW and SE. These four curves are shown in Figure 4a, b—4a shows averaged teleseismic residuals for delta greater than 50°, steep incidence, and 4b shows these for delta less than 50°, shallow incidence. In both cases the patterns shift in response to the reversal in azimuthal incidence. For the shallower incidence the shift is greater. Modeling this shift with a simple spherical-velocity anomaly places it at a depth of 100 km.

INTERPRETATION

Interpretation of the residual pattern requires additional assumptions to give a unique result. Consider the data plotted in Figure 4a, b. We assume that the regional shifting pattern is mainly due to upwarp of the asthenosphere–lithosphere boundary (which we define as the boundary at the base of the lithosphere separating high P-wave and S-wave velocities from lower values in the low-velocity zone of the asthenosphere); that localized hetergeneity gives rise to the small-scale variations. We can test how to invert for such a boundary using synthetic models. We computed and inverted data from a synthetic model which gave similar residual patterns for the range of incident angles and azimuths of our experimental data set in two ways: (1) downward projection of the data to give the boundary shape directly as outlined below, and (2) the two-dimensional (Aki et al., 1977) damped-least-squares block method as implemented by Ellsworth (1977). The first method gives an exact result; the second smeared the negative velocity perturbation at depth to shallow depths. This occurs because of the limited range of incidence angles, I less than 38°; therefore, the depth resolution is poor—e.g., parts of the residual patterns for I $= \pm 38°$ that overlap are independent of incidence and can equally be placed at

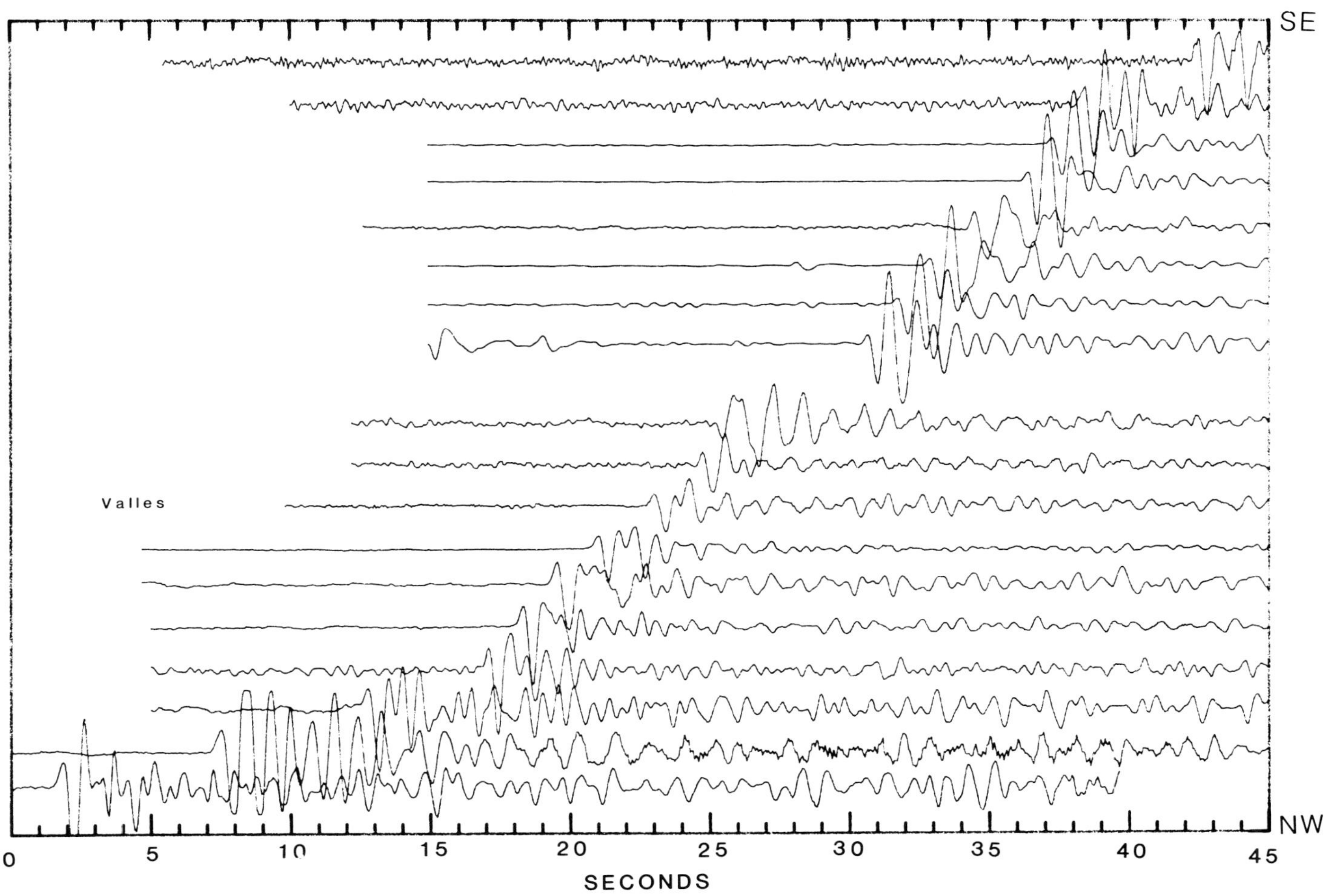

FIGURE 3. Record section of arrivals from Honshu earthquake of 15 January 1983 showing that timing to 0.1 sec is feasible for this signal-to-noise ratio.

either deep or shallow depth in the crust. The least-squares solution favors neither and smears this part of the solution over the whole depth range.

The first method of downward projection is justified on the grounds that it achieves the simplest and smoothest geological model and requires a minimum number of parameters (i.e., 7) in the inversion.

DOWNWARD PROJECTION OF TELESEISMIC RESIDUALS TO SOLVE FOR A VELOCITY INTERFACE

We assume that the long-wavelength teleseismic anomaly found on our profile across the Rio Grande rift is due to a single velocity interface—the lithosphere–asthenosphere transition which upwarps beneath the rift and has a north–south strike, i.e., parallel to the rift. Figure 5 shows our analysis method. If the interface is not too deep and velocity contrasts not too extreme, a straight ray and plane wave approximation for the teleseismic waves is adequate. The raw residuals will then be proportional to the length of the ray in the anomalous velocity zone. The raw residual pattern gives a distorted picture of the interface, i.e., it is projected onto a rotated axis. Since we know the angle of incidence, we can transform the projection and regain the original shape (dotted curves in Figure 5).

The ith ray length h_i in low-velocity material is

$$h_i' = V_1V_2t_i/V_1-V_2$$
$$= A\ t_i \text{ at positions } x_{in}' \text{ on the surface}$$

where t_o is the ith residual, V_1 and V_2 are upper and lower velocities, respectively, and A is a parameter given by the fit. The transformation to give the interface geometry is

$$h_i = h_i' \cos I$$
$$x_i'' = x_i' + h_i' \sin I \qquad (1)$$

where I is the angle of incidence for a given event. If the depth is D, the true lateral position of the anomaly is given by

$$x_i = x_i'' - D \tan I \qquad (2)$$

Relative residuals corresponding to 22 earthquakes (Table 1) arriving from either the NW or SE, i.e., along our 1,000-km line of stations, are shown in Figure 6. Raw residuals for each event have been demeaned to form relative residuals. Transformations (1) and (2) were applied to the data and parameters A and D varied until the sum of squares of residuals to a sixth-order polynomial fit was minimized. The minimum is fairly broad. The range of acceptable models is shown in Table 2. Figure 7 shows the best-fit model. Table 1 lists acceptable-alternative models lying in a range given by the criterion (Jenkins and Watts, 1968) that S is less than

$$S_o\,(1 + m/n{-}m\ f(m,\ n{-}m,\ (1-a)))$$

where n is the number of data, m the number of parameters, f the value of the F-statistic for confidence limit a with m, and n–m degrees of freedom. S is the sum of squares of residual differences for a particular model, and S_o is the sum of squares of residuals for the best-fitting model. If we take V_1 = 8 km/sec, the velocity contrast for our chosen model is 8%.

Values of the parameters used to contrain acceptable models of Table 2 were m = 7, n = 270, a = 95%. The average misfit of the polynomial model to the data was 0.3 sec. The best-fit model has a mean depth of 140 km and A = 80 km/sec.

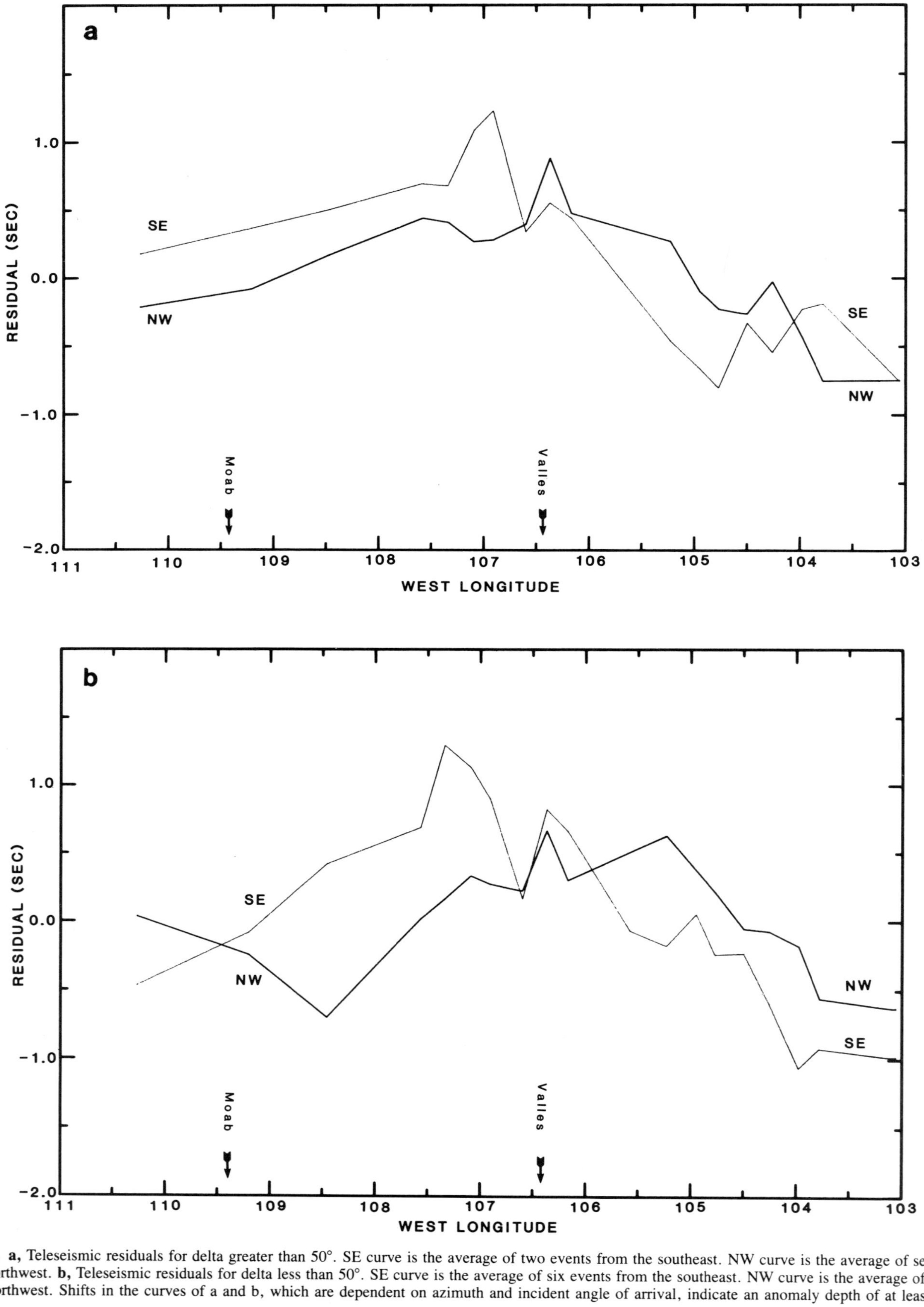

FIGURE 4. **a,** Teleseismic residuals for delta greater than 50°. SE curve is the average of two events from the southeast. NW curve is the average of seven events from the northwest. **b,** Teleseismic residuals for delta less than 50°. SE curve is the average of six events from the southeast. NW curve is the average of ten events from the northwest. Shifts in the curves of a and b, which are dependent on azimuth and incident angle of arrival, indicate an anomaly depth of at least 100 km.

TABLE 1a. Station locations. Each station consists of a 1 Hz, L4C vertical seismometer and recorder.

sta	latitude		longitude		elevation
RG1A	32	19.35	103	3.81	1013.
RG1S	32	19.35	103	3.81	1013.
RG2A	32	57.42	103	46.61	1323.
RG2S	32	57.42	103	46.61	1323.
RG3A	33	18.94	103	58.51	1173.
RG4A	33	28.87	104	15.33	1187.
RG5S	33	43.66	104	30.03	1247.
RG6A	34	2.62	104	46.33	1443.
RG7A	34	17.53	104	57.30	1641.
RG8A	34	30.73	105	13.51	1768.
RG8S	34	30.73	105	13.51	1768.
RG9A	34	52.42	105	35.03	2080.
RG9S	34	52.42	105	35.03	2080.
RGAA	35	13.40	105	49.08	2100.
RGAS	35	13.40	105	49.08	2100.
RGBS	35	24.25	106	10.06	2006.
RGCS	35	37.24	106	22.16	1652.
RGDS	35	48.66	106	36.04	2408.
RGES	36	9.70	106	54.46	2502.
RGFS	36	24.96	107	5.19	2262.
RGGS	36	35.00	107	20.51	2246.
RGHS	36	47.45	107	34.41	2086.
RGJS	37	16.38	108	27.61	2554.
RGLS	38	20.65	109	12.44	2359.
RGMS	39	.34	110	15.94	1311.

TABLE 1b. Hypocentral coordinates of earthquakes used in the inversion.

No.	Event		Time		(UT)	Longitude		Latitude		Depth	Place
01	82	358	03	49	52.7	94.543	W	15.304	N	57	OAXACA
05	82	363	07	02	31.4	139.336	E	33.763	N	19	HONSHU
06	82	363	11	06	25.8	91.603	W	14.405	N	85	GUATMALA
07	82	363	12	17	46.1	139.333	E	33.732	N	26	HONSHU
10	82	365	01	35	35.2	68.046	W	21.386	S	130	CHILE
11	82	365	02	46	50.7	78.962	W	10.743	S	28	PERU
12	82	365	03	47	28.5	68.464	W	20.993	S	118	CHILE
16	83	001	05	31	56.1	69.114	W	16.943	S	172	PERU/BOL
17	83	001	11	18	07.7	147.174	W	61.336	N	55	ALASKA
77	83	006	12	29	48.6	148.120	E	44.543	N	66	KURILS
78	83	009	10	42	47.7	163.125	E	55.163	N	33	KAMCHTKA
79	83	009	21	03	54.2	163.244	E	55.179	N	33	KAMCHTKA
34	83	010	09	17	34.7	68.470	W	22.006	S	121	CHILE
35	83	010	12	32	21.6	63.301	W	27.237	S	558	ARGENTIN
39	83	014	18	20	52.6	154.154	W	55.911	N	33	KODIAK
40	83	015	00	39	34.2	136.040	E	33.268	N	435	HONSHU
41	83	015	00	49	53.6	160.076	E	52.844	N	30	KAMCHTKA
50	83	018	07	44	28.6	88.817	W	12.930	N	70	CENTAMER
80	83	021	10	25	53.5	142.988	E	42.422	N	76	JAPAN
57	83	023	14	29	28.5	77.543	W	6.485	N	9	COLOMBIA
62	83	026	01	05	50.7	66.997	W	23.971	S	186	CHEL/ARG
65	83	027	19	01	35.3	66.960	W	24.105	S	188	CHI/ARGN

TABLE 2. Sum of squares of residuals, SSQ, as a function of $A = V_2V_1/V_1 - V_2$, and depth, D, for the projection model inversion. Acceptable models are denoted by asterisks.

Depth	40	90	140	200	250	300	350
A							
40	25.7	23.6	22.5*	23.0*	25.8	29.5	34.0
80	25.4	23.4	22.3*	23.1	25.3	28.8	33.1
120	25.3	23.5	22.6*	23.0*	25.4	28.6	32.5
160	25.4	23.6	23.1	23.8	25.7	28.6	32.2
200	25.5	24.3	23.8	23.8	26.0	28.6	32.5

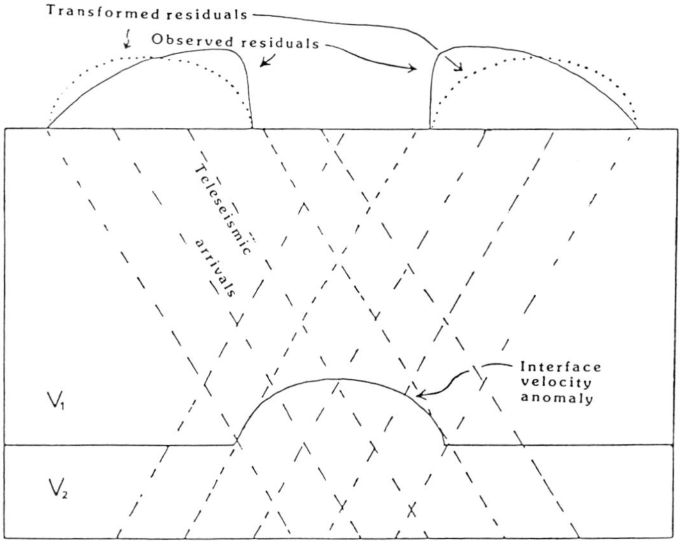

FIGURE 5. Cartoon showing downward-projection method of inverting teleseismic residuals to find the shape of a velocity interface.

DAMPED-LEAST-SQUARES TWO-DIMENSIONAL BLOCK MODEL

Relative residuals were inverted to determine percent velocity perturbations about average layer velocities following the approach of Evans (1982). No earth-flattening transform was used since our final model geometry, 1,000 km × 200 km, was of the same order as those tested by Evans (1982), who found earth-curvature effects to be of second-order importance. The blocks were aligned north–south, roughly parallel to the strike of the rift.

We ran a series of tests to determine the uniqueness and minimum parameterization of the block model and, having determined an optimal model, tested the hypothesis that all the residuals are due to slowness in the crust or upper lithosphere rather than in the deep lithosphere. The hypothesis was rejected at the 99.9% confidence level. The data require a deep, low-velocity zone.

Since our station spacing ranged from 30 to 140 km, we chose a block width of 100 km. Unperturbed crustal and upper-mantle velocities were set at 6.5 and 8.0 km/sec, respectively, with the transition at a depth of 40 km. We then varied the number of layers (from six down to our final value of three), the positions of the interfaces between layers, and the total depth of the model. Since the number of layers determines the number of parameters, for any given model depth we were able to arrive at an optimum number of layers, three by means of the F-test. However, non-uniqueness became a problem when comparing different interface depths. Models with an overall depth of 500 km were found to have a fit negligibly different from those of 200-km depth (average misfit 0.285 sec cf. 0.297 sec). Although the velocity variation was qualitatively similar, its amplitude was proportionately reduced in the deeper model. Also, placing the second interface at 70 km (misfit 0.297 sec) was not significantly different from placing it at

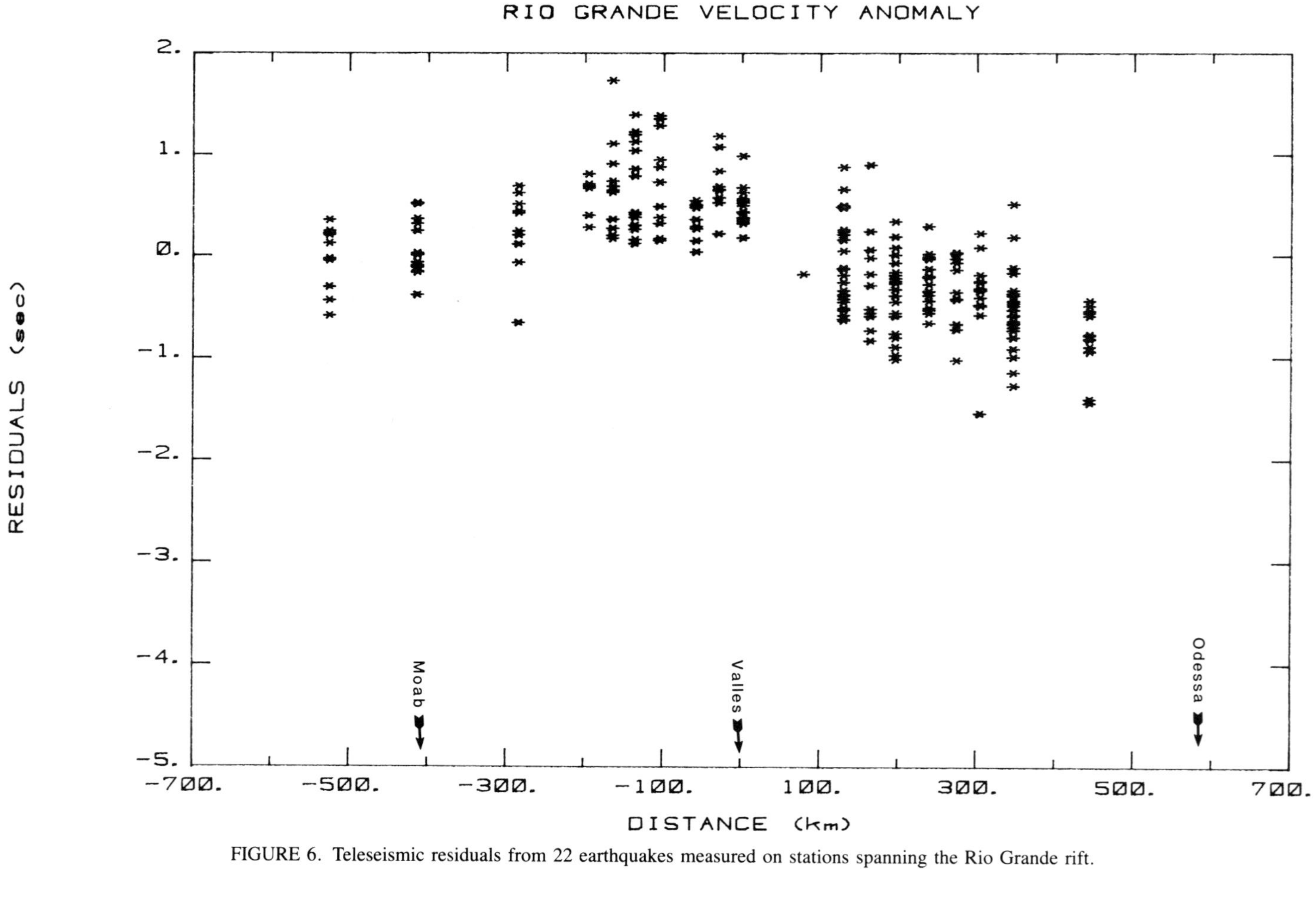

FIGURE 6. Teleseismic residuals from 22 earthquakes measured on stations spanning the Rio Grande rift.

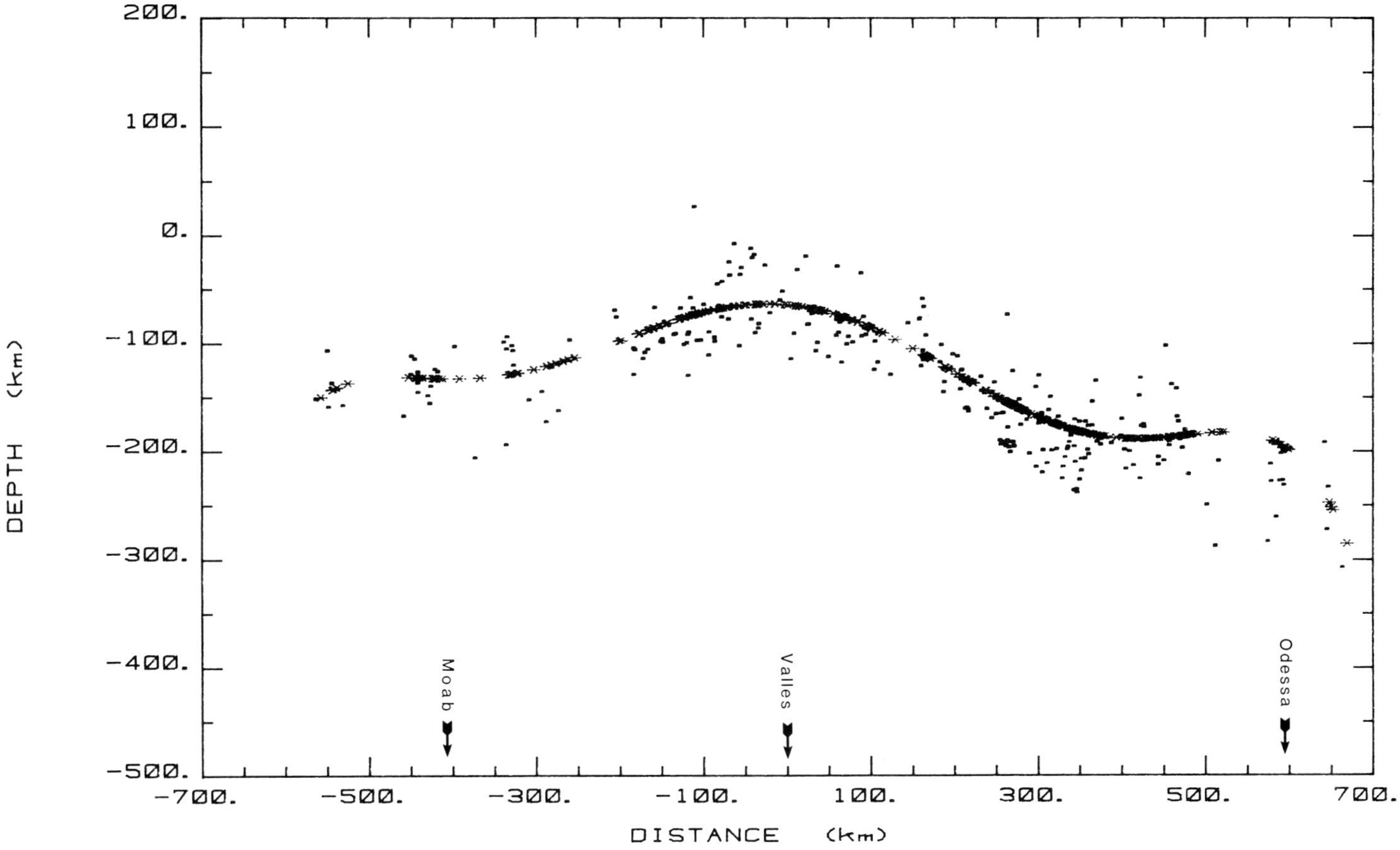

FIGURE 7. Results of downward-projection modeling of the residuals. The best fitting polynomial is in asterisks.

150 km (misfit 0.300 sec). Therefore, the model we present in Figure 8 was chosen from this non-unique set because its interfaces correspond to those of the polynomial model of the previous section, that is a 40-km crust, a 70-km upper and 200-km lower boundary of a low-velocity body. The main feature of this model is a 5% reduction in velocity in the lowest center-most block, beneath the Valles caldera, with coherent negative variation on either side. Smaller variations in the upper two layers are incoherent. Standard errors given by the least-squares solution are all about 0.6%. The diagonal elements of the resolution matrix ranged from 0.55 in the top layer to 0.86 in the lowest layer.

The percent reduction in variance after subtracting model values from the data was 71%. This model is in reasonable agreement with the polynomial model (Fig. 7). The velocity contrast of 6% is slightly less than the 8% found in that case, but this may be due to the smearing effect mentioned earlier. To test the hypothesis that the source of the relative residuals lies no deeper than the crust, we confined the block model to the upper 40 km, keeping the number of blocks constant. The variance reduction was only 50% compared with 71% for the deeper models. Comparison with F-statistic values rejects this hypothesis with greater than 99.9% confidence.

COMPARISON WITH BOUGUER GRAVITY

The Bouguer gravity anomaly calculated for both the polynomial model and the block model fit the measured data for the area well.

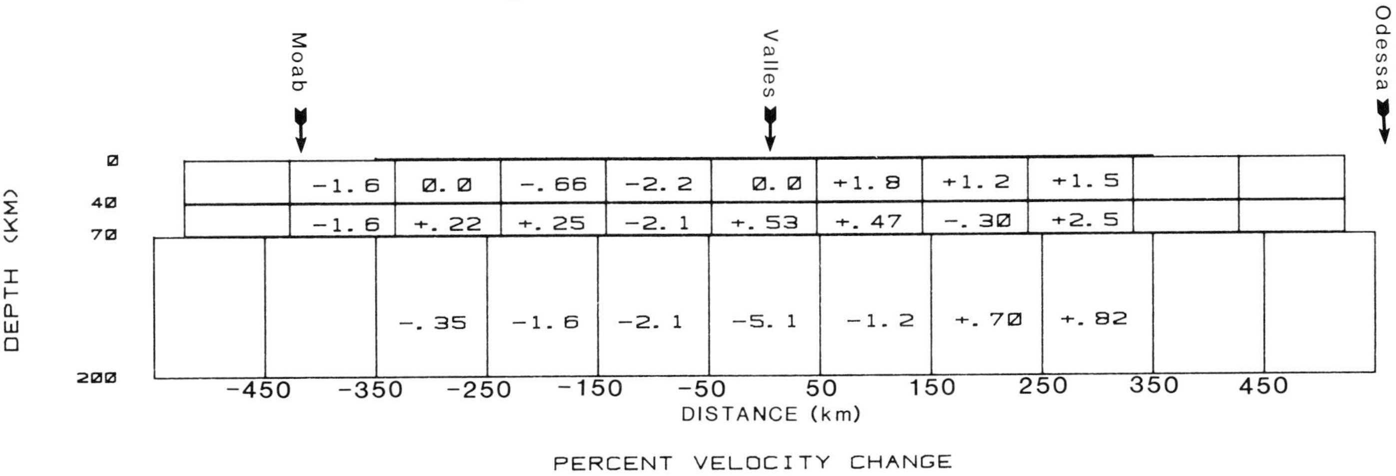

FIGURE 8. Results of two-dimensional block modeling of the residuals. The numbers represent percent change in velocity relative to starting model velocities.

Figure 9 shows the fit for the gravity from the polynomial model for which the adjusted density contrast in the low-velocity zone gave a value of −0.05 gm/cc. This is a much smaller contrast than that calculated from the Nafe–Drake relationship (0.16 gm/cc). However, we are not in a position to pursue the implications here.

The shape of the topography is also somewhat in agreement with topography in isostatic balance with the polynomial model. However, this requires a lower density contrast (−0.03 gm/cc), and we have not yet reconciled these two values.

PREVIOUS LOCAL SEISMIC STUDIES

Previous seismic work has been concentrated within the rift zone. Figure 1 shows a positive gravity anomaly centered on the rift and superimposed on the regional gravity low. The smaller-scale anomaly is thought to be due to a combination of mantle upwarp into the base of the crust and crustal thinning due to tectonic extension (Cordell, 1978). However, broad-scale igneous intrusion of the shallow crust represents another possibility. Support for the mantle-upwelling hypothesis is seen in the seismologically determined crustal thicknesses as summarized by Olsen et al. (1979), in which the depth to the base of the rift crust (33 km) is shown to be 12 and 17 km shallower than for the Colorado Plateau (45 km) and Great Plains (50 km), respectively. Corroborating evidence comes from Keller et al. (1979), who combine surface-wave dispersion measurements and refraction data to infer a similar structure with the rift Moho at a depth of 35 km, as well as from the tentative identification of a Moho reflector at 12 sec on the COCORP line (Brown et al., 1979), which translates to a depth of 35 km. In contrast, Phinney (1964) used the spectral-ratio method to find a Moho depth of 35–40 km on the rift beneath the WWSSN station at Albuquerque. The high value of this result may be a consequence of allowing both depth and velocity to vary during modeling (Olsen et al., 1979).

One of the most striking seismic results seen on the rift in the vicinity of Socorro is the evidence for the existence of a 19-km-deep crustal magma layer (Sanford and Long, 1965; Sanford et al., 1973; Rinehart et al., 1979; Rinehart and Sanford, 1981) confirmed by the COCORP line (Oliver and Kaufman, 1976; Brown et al., 1979). Evidence that it might be a widespread phenomenon beneath the rift is seen in the study of the amplitudes of arrivals from the refraction experiment of Olsen et al. (1979), who infer a layer thickness of a few km.

Other seismological evidence as to the nature of the rifting includes the observation of low P_n velocities of 7.6 km in the upper mantle beneath the rift (Olsen et al., 1979) compared to 7.8 km/sec beneath the Colorado Plateau and 8.2 km/sec beneath the Great Plains. Crustal thinning and low P_n velocity have been observed in other rift zones, e.g., the Rhinegraben (Giese et al., 1976), the East African rift (Long et al., 1973), and the North Sea (Christie and Sclater, 1980). However, a recent study of Murdock and Jacksha (1981) finds a normal P_n velocity of 8.0 km/sec beneath the Rio Grande rift, excluding the possibility of a wide zone of asthenospheric intrusion at the base of the crust.

Tracing rays through these crustal structures gives rise to a long-wavelength lateral variation no greater than 0.2 sec, providing further evidence that the delaying structure lies at depth in the mantle. Smaller-scale variations of 0.5 sec have been seen corresponding to deep sedimentary basins.

REGIONAL SEISMIC STUDIES

Our approach to modeling the residuals has been to build on what is already known of the lithosphere/asthenosphere structures in the region by incorporation of the delay model. Biswas and Knopoff (1974) examined Rayleigh-wave dispersion along a number of paths across the United States and found significant regional variations in S-wave velocities in various subregions of the continent. The south-central part of the continent and the Basin and Range are characterized by a well-

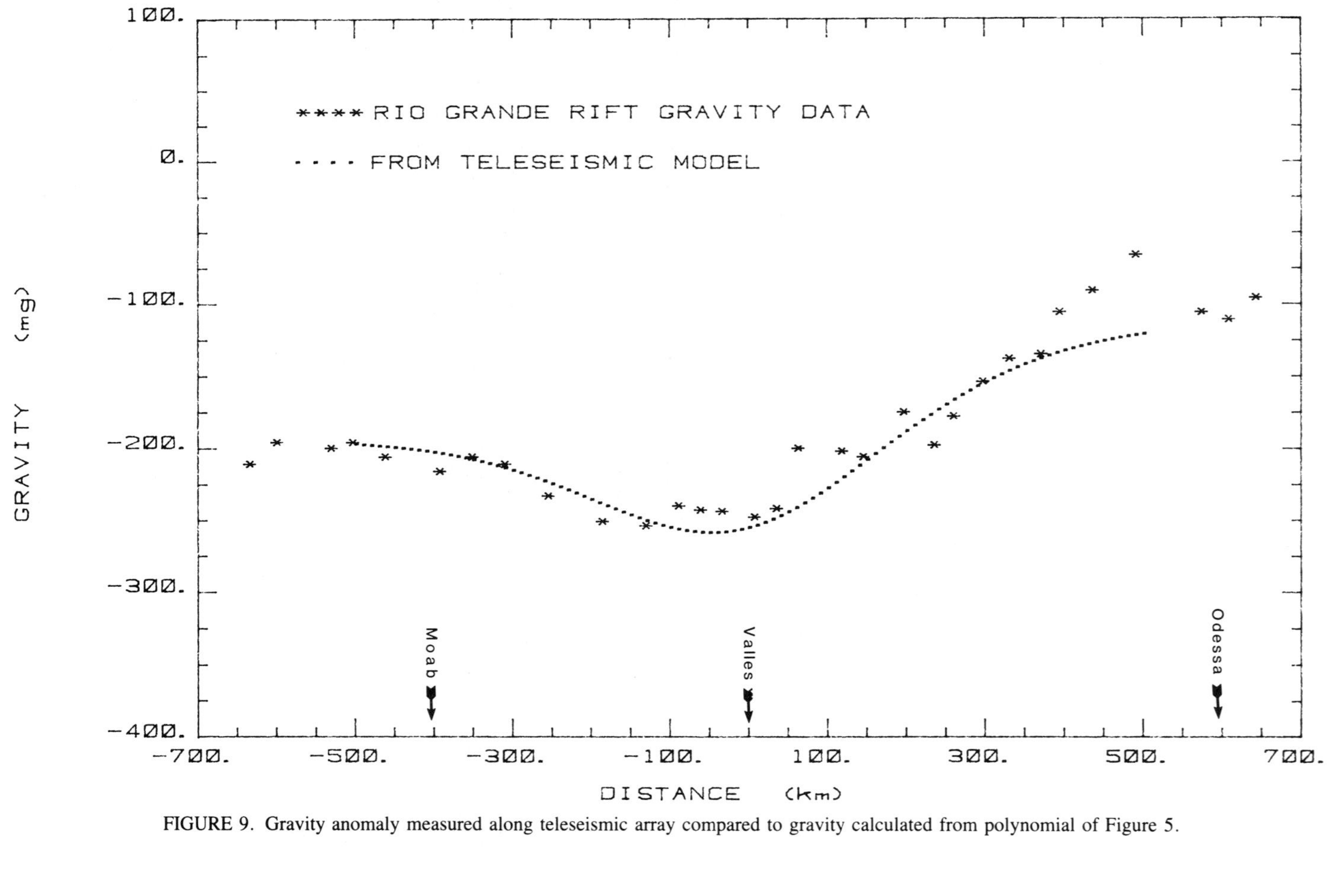

FIGURE 9. Gravity anomaly measured along teleseismic array compared to gravity calculated from polynomial of Figure 5.

developed low-velocity channel (LVZ) for S waves in the upper mantle. The north-central United States has high upper-mantle S-wave velocities without the development of a significant low-velocity channel. They presented a suite of models for a path (GOL–DUG) lying just to the northwest of our array and a path (SHA–LUB) lying to the southeast. They find a lid thickness of 35–95 km underlying a 48-km crust with lid (shear-wave) velocities around 4.6 km/sec and low channel velocities around 4.3 km/sec, i.e., a lithospheric thickness (lid plus crust) in the range of 83–145 km. The maximum lithospheric thickness interfaces to the shallowest polynomial boundary (Fig. 7) in the west. The eastern boundary lies somewhat deeper, 190 km; the western side lies at a depth of 130 km.

Further to the west, in the Basin and Range province, Biswas and Knopoff (1974) found a well-developed low-velocity channel with S-wave velocity in the range of 4.1–4.3 km/sec. More recent studies (Priestley and Brune, 1978; Priestley et al., 1980) of the Great Basin of Nevada and western Utah show 65-km-thick crust and lid (35-km crust and 29-km lid), with the lid-LVZ-velocity contrast 4.5 km/sec to 4.05–4.12 km/sec. The central part of our polynomial fit (Fig. 7), which lies beneath the rift at 75-km depth, is not very different from the Great Basin lithospheric thickness. However, although the polynomial fit of Figure 7 shows a degree of agreement with structures determined from Rayleigh waves, there are some remaining points to be resolved. Where P-wave velocities have been associated with low S-wave velocities (Priestley et al., 1980; Dziewonski et al., 1975; Burdick and Helmberger, 1978) in the LVZ, the percent reduction in P velocity is about half of that in S velocity. Our model reduction in P velocity is 8%, implying an unprecedented reduction of 14% in S velocity. Also, the LVZ has been associated with increased density values (Biswas and Knopoff, 1984; Dziewonski et al., 1975), which is opposite to our conclusion. We can trade off less P-wave-velocity perturbation with increased amplitude of the polynomial variation. However, this would place the flanks unacceptably deep to interface with the surface-wave models. It therefore appears that if we invoke asthenospheric upwarp to explain the gravity, topography, and teleseismic delays, it is anomalous asthenosphere compared to known properties elsewhere. However, this might be expected since it underlies the most negative region of the east–west continental Bouguer gravity profile (latitude 37°) (Fig. 10).

CONCLUSION

Teleseismic residual delays of as much as 1.5 sec, seen on an array of aperture 1,000 km which spanned the region of uplift surrounding the Rio Grande rift, require low velocities to exist deep beneath the crust. Both the block-inversion method and a simple downward projection of the residuals place the low-velocity zone in the depth range 70–100 km, with a velocity contrast of −8%.

The associated Bouguer gravity variation matches the observations well, given a density contrast of −0.05 gm/cc, adjusted to optimize the fit, i.e., a contrast of 1.6%. A large velocity contrast and small density contrast are indicative of a small fraction of partial melt which, while having a small effect on density, will have a disproportionately larger effect on shear modulus and hence P-wave velocity.

Airy-type isostatic support of the regional uplift through a thickened crust is ruled out in favor of a much deeper low-density zone in the region of the lithosphere–asthenosphere transition.

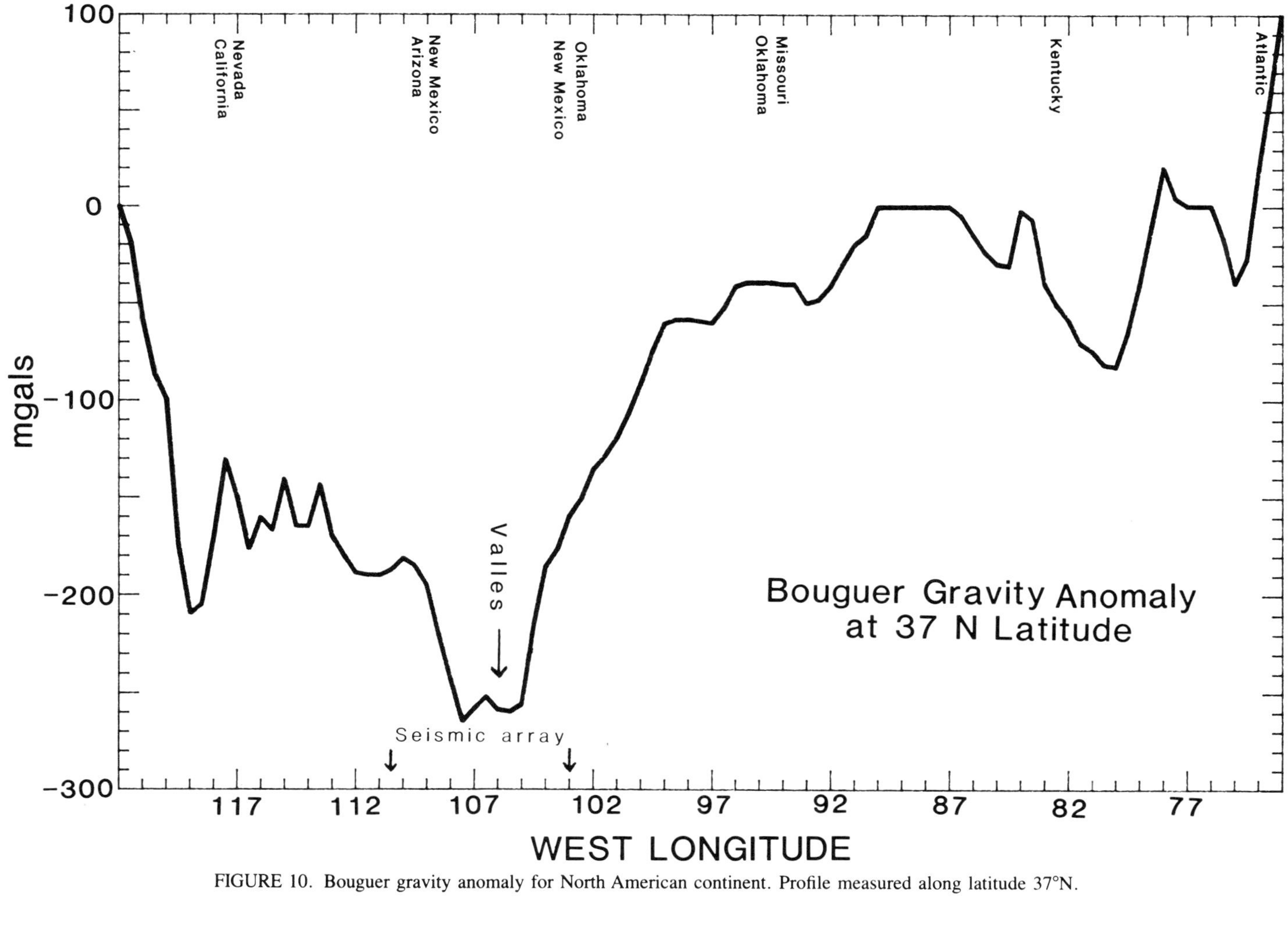

FIGURE 10. Bouguer gravity anomaly for North American continent. Profile measured along latitude 37°N.

ACKNOWLEDGMENT

This work was supported by grants from the Institute of Geophysics and Planetary Physics at Los Alamos and the University of California at Los Angeles.

REFERENCES

Aki, K., 1982, Three-dimensional seismic inhomogeneities in the lithosphere and asthenosphere: evidence for decoupling in the lithosphere and flow in the asthenosphere: Review of Geophysics and Space Physics, 20: 161–170.

———, Christofferson, A., and Huseleye, E. S., 1977, Determination of the three-dimensional seismic structure of the lithosphere: Journal of Geophysical Research, 82: 277–296.

Ander, M. E., 1980, Geophysical study of the crust and upper mantle beneath the central Rio Grande rift and adjacent Great Plains: Ph.D. thesis, University of New Mexico, Albuquerque.

Biswas, N. N., and Knopoff, L., 1974, The structure of the upper mantle under the United States from dispersion of Rayleigh waves: Royal Astronomical Society, Geophysical Journal, 36: 315–450.

Brown, L. D., Krumhansl, P. A., Chapin, C. E., Sanford, A. R., Cook, F. A., Kaufman, S., Oliver, J. E., and Schilt, F. S., 1979, COCORP seismic reflection studies of the Rio Grande rift; *in* Riecker, R. E. (ed.), Rio Grande rift: tectonics and magmatism: American Geophysical Union, Washington, D.C., pp. 169–184.

Burdick, L. J., and Helmberger, D. V., 1978, The upper mantle P-velocity structure of the western United States: Journal of Geophysical Research, 83: 1699–1712.

Christie, P. A. F., and Sclater, J. G., 1980, An extensional origin for the Buchan and Witchgraben in the North Sea: Nature 283: 729–732.

Cordell, L., 1978, Regional geophysical setting of the Rio Grande rift: Geological Society of America, Bulletin, 89: 1073–1090.

Dziewonski, A. M., Hales, A. L., and Lapwood, E. R., 1975, Parametrically simple earth models consistent with geophysical data: Physics of Earth and Planetary Interiors, 10: 12.

Ellsworth, W. A., 1977, Three-dimensional structure of the crust and mantle beneath the island of Hawaii: Ph.D. thesis, Massachusetts Institute of Technology, Cambridge.

Evans, J. R., 1982, Compressional wave velocity structure of the upper 350 km under the Snake River Plain near Rexburg, Idaho: Journal of Geophysics, 87: 2654–4670.

Giese, P. C., Prodehl, C., and Stein, A. (eds.), 1976, Explosion seismology in Central Europe—data and results: Springer-Verlag, New York.

Iyer, H. M., 1975, Anomalous delays of teleseismic P-waves in Yellowstone National Park: Nature, 253: 425–427.

———, 1984, Geophysical evidence for the locations, shapes and sizes, and internal structures of magma chambers beneath regions of Quaternary volcanism: Royal Society of London, in press.

———, Oppenheimer, D. H., and Hitchock, T., 1979, Abnormal P-wave delays in the Geysers–Clear Lake geothermal area, California: Science, 204: 495–497.

Jenkins, G. M., and Watts, D. G., 1968, Spectral analysis and its applications: Holden–Day, San Francisco.

Keller, G. R., Braile, L. W., and Schlue, J. W., 1979, Regional crustal structure of the Rio Grande rift from surface wave dispersion measurements; *in* Riecker, R. E. (ed.), Rio Grande rift: tectonics and magmatism: American Geophysical Union, Washington, D.C.

Long, R. E., Sundaralingam, K., and Maquire, P. K. H., 1973, Crustal structure of the east African rift zone: Tectonophysics, 20: 269–281.

Murdock, J. N., and Jacksha, L. H., 1981, P-wave velocity of the uppermost mantle of the Rio Grande rift region of north-central New Mexico: Journal of Geophysical Research, 86: 7055–7063.

Oliver, J. E., and Kaufman, S., 1976, Profiling the Rio Grande rift: Geotimes, 27: 20–23.

Olsen, K. H., Keller, G. R., and Stewart, J. N., 1979, Crustal structure along the Rio Grande rift from seismic refraction profiles; *in* Riecker, R. E. (ed.), Rio Grande rift: tectonics and magmatism: American Geophysical Union, Washington, D.C.

Phinney, R. A., 1964, Structure of the Earth's crust from spectral behavior of long-period body waves: Journal of Geophysical Research, 69: 2997–3017.

Priestley, K., and Brune, J., 1978, Surface waves and the structure of the Great Basin: Journal of Geophysical Research, 83: 2265–2272.

———, Orcutt, J. A., and Brune, J. N., 1980, Higher-mode surface waves and structure of the Great Basin of Nevada and western Utah: Journal of Geophysical Research, 85: 7166–7174.

Rinehart, E. J., and Sanford, A. R., 1981, Upper crustal structure of the Rio Grande rift near Socorro, New Mexico from inversion of the microearthquake S-wave reflections: Seismological Society of America, Bulletin, 71: 437–450.

———, Sanford, A. R., and Ward, R. M., 1979, Geographic extent and shape of an extensive magma body at midcrustal depths in the Rio Grande rift near Socorro, New Mexico; *in* Riecker, R. E. (ed.), Rio Grande rift: tectonics and magmatism: American Geophysical Union, Washington, D.C., pp. 237–251.

Sanford, A. R., and Long, L. T., 1965, Microearthquake crustal reflections, Socorro, New Mexico: Seismological Society of America, Bulletin, 55: 579–586.

———, Alptekin, O. S., and Toppozada, T. R., 1973, Use of reflection phases on microearthquake seismograms to map an unusual discontinuity beneath the Rio Grande rift: Seismological Society of America, Bulletin, 63: 2021–2034.

Sharp, A. D. L., Davis, P. M., and Gray, F., 1980, A low velocity zone below Mount Etna and magma storage: Nature, 287: 587–591.

Spence, W., Gross, R. S., and Jacksha, L. H., 1982, P-wave velocity structure beneath the Jemez lineament, New Mexico: EOS, American Geophysical Union, Transactions, 63 (45): 1117.

ANALYSIS OF TERRESTRIAL HEAT-FLOW PROFILES ACROSS THE RIO GRANDE RIFT AND SOUTHERN ROCKY MOUNTAINS IN NORTHERN NEW MEXICO

GERRY CLARKSON[1] and MARSHALL REITER
New Mexico Bureau of Mines and Mineral Resources and Geoscience Department,
New Mexico Institute of Mining and Technology, Socorro, New Mexico 87801
([1]Present address: Department of Geology, Western Michigan University, Kalamazoo, Michigan 49008)

INTRODUCTION

The Rio Grande rift within the southern Rocky Mountains complex of northern New Mexico is a tectonically and volcanically active region. The southern Rocky Mountains have been uplifted ~1.1 km since middle Miocene time (Axelrod and Bailey, 1976), with most of the uplift occurring between 7 and 4 m.y ago (Chapin, 1979). Within the Rio Grande rift recent bimodal volcanism has been extensive; the Jemez lineament and Taos Plateau have been especially active in the last 5 m.y. (Chapin, 1979; Lipman and Mehnert, 1979). On the eastern periphery of the southern Rocky Mountains in northeastern New Mexico extensive late Tertiary and Quaternary magmatic activity has occurred in the Raton and Las Vegas Basins and in the Capulin and Clayton volcanic fields (Baltz and Bachman, 1956; Gabelman, 1956; Baltz, 1965; Dane and Bachman, 1965; Johnson, 1968; Lipman and others, 1973).

The tectonic and volcanic activity in the rift is often associated with a thinned crust, upwarped mantle, and extensive subsurface magma bodies (e.g. Cordell, 1978; Baldridge, 1979; Keller and others, 1979; Lipman and Mehnert, 1979; Olsen and others, 1979; Rinehart and others, 1979). Based on heat flow, electrical conductivity, gravity, and the elevation profile across the Rio Grande rift, Cordell (1978) suggested the presence of a symmetrical anomalous crustal and upper mantle structure along the axis of the rift. Bridwell (1976) interpreted gravity anomalies in northern New Mexico along 36°N latitude in terms of an intrusion of low-velocity mantle material into an overlying thinned lithosphere across an area 400 km wide (E–W). Thermal models consistent with the gravity data also suggest that anomalous heat sources occur in the crust from 10 to 40 km depth (Bridwell, 1976). The presence of shallow magma bodies, 15–30 km deep, within the rift has been proposed on geochemical, seismic, heat-flow, and gravity data (Lipman, 1969; Sanford and others, 1973; Decker and Smithson, 1975; Cordell, 1976; Reiter and others, 1978). Cook and others (1979) suggested partial melting, or the intrusion of mantle material, at the base of the crust in the southern Rio Grande rift.

As would be expected in a region of tectonic and volcanic activity, the Rio Grande rift–southern Rocky Mountains region has high heat flow (Decker, 1969; Reiter and others, 1975; Edwards and others, 1978; Reiter and others, 1979). Edwards and others (1978) state that the heat-flow pattern across the southern Rocky Mountains suggests more widely distributed and perhaps deeper heat sources than along the Rio Grande rift itself. High heat flow in the rift can be related to elevated geotherms associated with a thinned crust and to crustal magmatic activity. High heat flow in the surrounding uplifted regions may not be as obviously related to magmatic processes. Heat-flow values in the region could also be elevated by ground-water circulation (Reiter and others, 1979; Swanberg, 1979). High radiogenic-heat production can contribute to elevated heat flow, but this does not seem to be a significant anomalous heat source in the Rio Grande rift (Decker and Smithson, 1975; Edwards and others, 1978).

The purpose of this study is to consider heat-flow data across northern New Mexico, and to discuss possible trends in the data and the subsequent implications concerning heat sources in the Rio Grande rift–southern Rocky Mountains region. Heat-flow data are considered in terms of steady-state, finite-difference, isothermal-step models similar to those models presented by Reiter and Clarkson (1983) and Clarkson (1984). Fundamental parameters in the finite-difference models used to analyze the data are the mean heat flow within regions and the variation in heat flow with distance between regions of different mean heat flows. Isothermal-step models are surely a geometrical and phenomenological simplification of actual subsurface thermal sources; however, such models do allow basic appreciation of subsurface temperature discontinuities which may generate observed heat-flow variations between regions. The time constant for isothermal sources at depths $\lesssim$30 km is relatively short, and, therefore, steady-state models may be reasonable first-order approximations to "relatively deep" subsurface heat sources. For the models to be considered in association with the northern Rio Grande rift and the San Juan volcanic field, heat-flow values and profile characteristics (half widths) reach about 3/4 of steady-state values 10 m.y. after step initiation, and approximate steady-state values after 20 m.y. (Clarkson and Reiter, unpubl.). As such, steady-state models may be generally consistent with the repeated volcanism in the Española Basin (Manley, 1979), and may suggest heat replenishment in the San Juan volcanic field (Reiter and Clarkson, 1983).

DISCUSSION OF DATA

Heat-flow data across northern New Mexico are shown in Figure 1 along with the geologic localities in which they occur. Possible trends in the data are indicated in Figure 2. Using the data, one may consider various aspects of the heat sources associated with the Rio Grande rift and southern Rocky Mountains in northern New Mexico. If the rift is associated with a relatively deep heat source, such as an unwarping in the mantle, one may expect to see a heat-flow anomaly across the rift having a relatively broad half width (although such an anomaly may be obscured by shallow effects related to ground-water circulation or transient heating associated with shallow magmatic activity). Alternatively, shallow thermal sources will produce heat-flow anomalies having a relatively narrow half width (or high spacial variability).

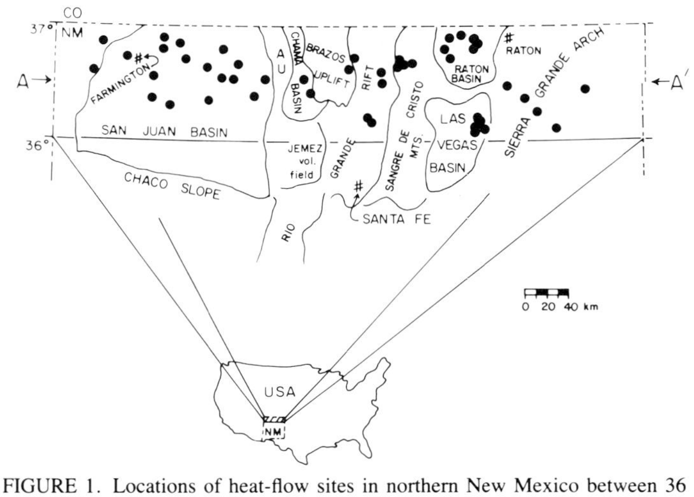

FIGURE 1. Locations of heat-flow sites in northern New Mexico between 36 and 37°N latitude. AA′ defines the projection profile shown in Figure 2. Data from Decker (1969), Sass and others (1971), Reiter and others (1975), Edwards and others (1978), and Reiter and Mansure (1983).

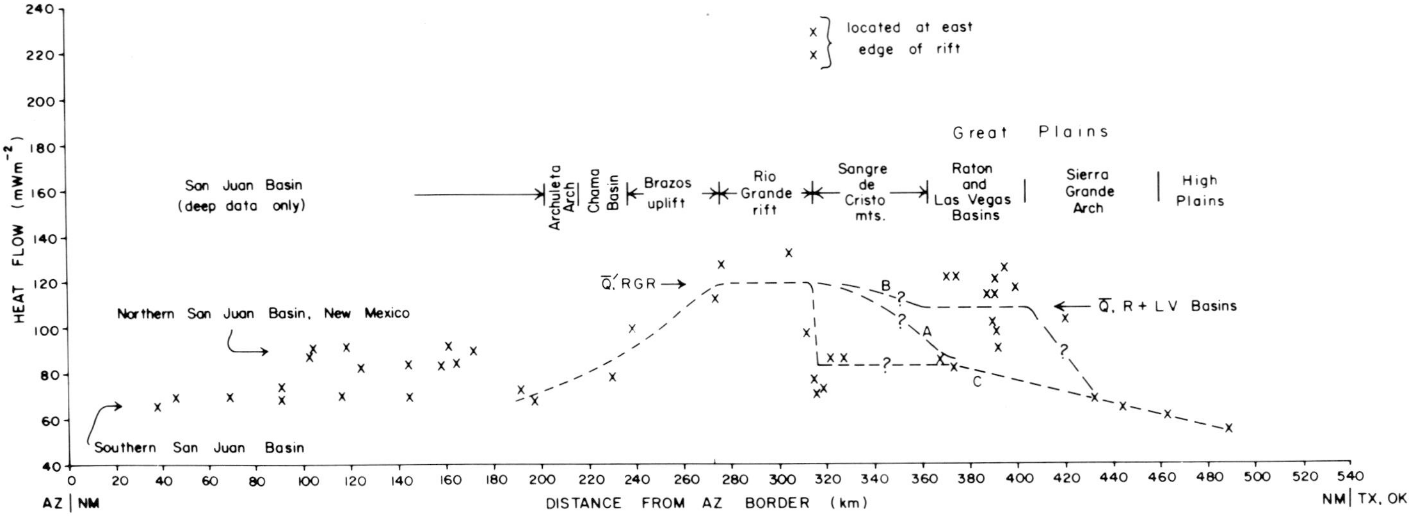

FIGURE 2. Heat-flow data and possible heat-flow trends across northern New Mexico (see Figure 1 for profile projection). Trends A, B, and C, east of the Rio Grande rift, represent various interpretations of the data (see text).

An example of a heat-flow trend in northern New Mexico which is likely to be associated with a relatively deep thermal source is shown in Figure 3a (data and profile locations are shown in Fig. 3b). The data are from heat-flow measurements in the San Juan Mountains and from deep heat-flow measurements in the San Juan Basin. The trend of increasing heat flow toward the San Juan Mountains can be modeled from a steady-state isothermal step rising from ~100 km to ~30 km depth (Fig. 4) beneath the northern San Juan Basin with the shallow isotherm extending underneath the San Juan Mountains (Reiter and Clarkson, 1983). More complicated models could be employed to explain the observed heat-flow trend; however, the isothermal-step model provides a good fit to the experimental data and allows a basic appreciation of a subsurface-temperature discontinuity which would produce the heat-flow increase approaching the San Juan volcanic field. Although ground-water movement can surely abstract heat-flow profiles by advecting heat from one region to another, the trend of increasing heat flow toward the San Juan volcanic field is believed to represent a basic change in the deep thermal state of the region, i.e. greater mantle and/or magmatic heat. Ground-water movement into the San Juan Basin from the northern and northeastern uplifted regions near the San Juan volcanic field (Stone and others, 1983) may influence the half width of the heat-flow profile; however, such water movement would serve to diminish the magnitude of the heat-flow contrast between the two areas by advecting heat from a warmer to cooler location. If the endpoint heat flows and the profile half width in Figure 4 are representative of conductive heat flow, then the temperature discontinuity or anomaly should be relatively deep as shown (i.e. in the lower crust or upper mantle). This suggestion would be consistent with the crustal low-velocity zone predicted by Prodehl and Pakiser (1980) for the San Juan Mountains region.

The isothermal-step model is not directed toward understanding the phenomena producing the subsurface-temperature discontinuity consistent with observed heat flow, and a discussion of the causes of elevated heat flow in major volcanic fields, rifts, and mountains is not the purpose of this study. The step model may, however, allow some limited speculation concerning the causes of upwarped isotherms. The temperature step shown in Figure 4 could be interpreted as a type of plume or advecting mechanism bringing heat from depth to nearer the surface. This basic idea may be consistent with the geochemical data presented by Lipman and others (1978), who suggest subduction processes may ultimately be involved with the production of the San Juan volcanics. The ~1200°C isotherm at ~30 km depth (Fig. 4) is approximately consistent with diffusion models, but is about 200–250°C higher than temperatures predicted from models incorporating convective-heat transfer at depth (Lachenbruch and Sass, 1978). Although it would seem quite reasonable that some type of mass-transport process is occurring in association with the elevated isotherm shown in Figure 4, the essence of the heat-flow contrast between the San Juan Basin and the San Juan volcanic field (and the subsurface temperature distribution) is approximated by the isothermal-step model.

A trend similar to that associated with the San Juan Basin–San Juan Mountains trend may be seen in the heat-flow data along a profile from the Rio Grande rift westward to the San Juan Basin (Fig. 2). This trend also appears to reflect a relatively deep heat source. The average heat flow from three values in the northern Rio Grande rift (excluding the two extremely high values at the eastern edge of the rift) is 118 mWm^{-2}. The heat flow decreases to about 70 mWm^{-2} at the eastern edge of the San Juan Basin, approximately 100 km away from the edge of the Rio Grande rift. The similarity between the San Juan Basin–San Juan Mountains profile and the Rio Grande rift–San Juan Basin profile suggests that the heat-flow data west of the rift in northern New Mexico would also be consistent with a relatively deep heat source (an isothermal step from ~86 to ~32 km depth; Fig. 5). It must be kept in mind, however, that only a few data points have been used to define this trend and that the data outside the San Juan Basin are from relatively shallow wells where the temperature gradients may be perturbed.

Locally high heat-flow values within the northern Rio Grande rift could certainly be induced by ground-water convection, and should be complemented within the hydrologic framework of the northern rift by low heat-flow values. Regionally high heat flow in the northern rift, complemented by convective heat transport, still reflects anomalous sources of heat such as magmas, a shallow asthenosphere, etc. Although the broad half width of the heat-flow profile from the Rio Grande rift into the San Juan Basin may be influenced by ground-water movement into the basin (which would lower heat flows in the uplifts on the western edge of the rift), the suggested isothermal discontinuity near the northern Rio Grande rift (Fig. 5) is generally consistent with other studies (discussed above) proposing magma sources at rather similar depths. If the 1200°C temperature at 30 km depth, as predicted by the diffusion model, is somewhat high, then mass transport of heat may be occurring in the lower crust and upper mantle along the northern Rio Grande rift.

The heat-flow data from the Rio Grande rift east to the Great Plains are characterized by high spatial variability. Such variability must be due to relatively shallow causes. For example, heat-flow values along the eastern edge of the rift vary from 239 to 71 mWm^{-2} (Fig. 2). This variability may well result from ground-water circulation or near-surface magma emplacement along major rift faults (Reiter and others, 1979; Swanberg, 1979). The variability within the Raton and Las Vegas Basins

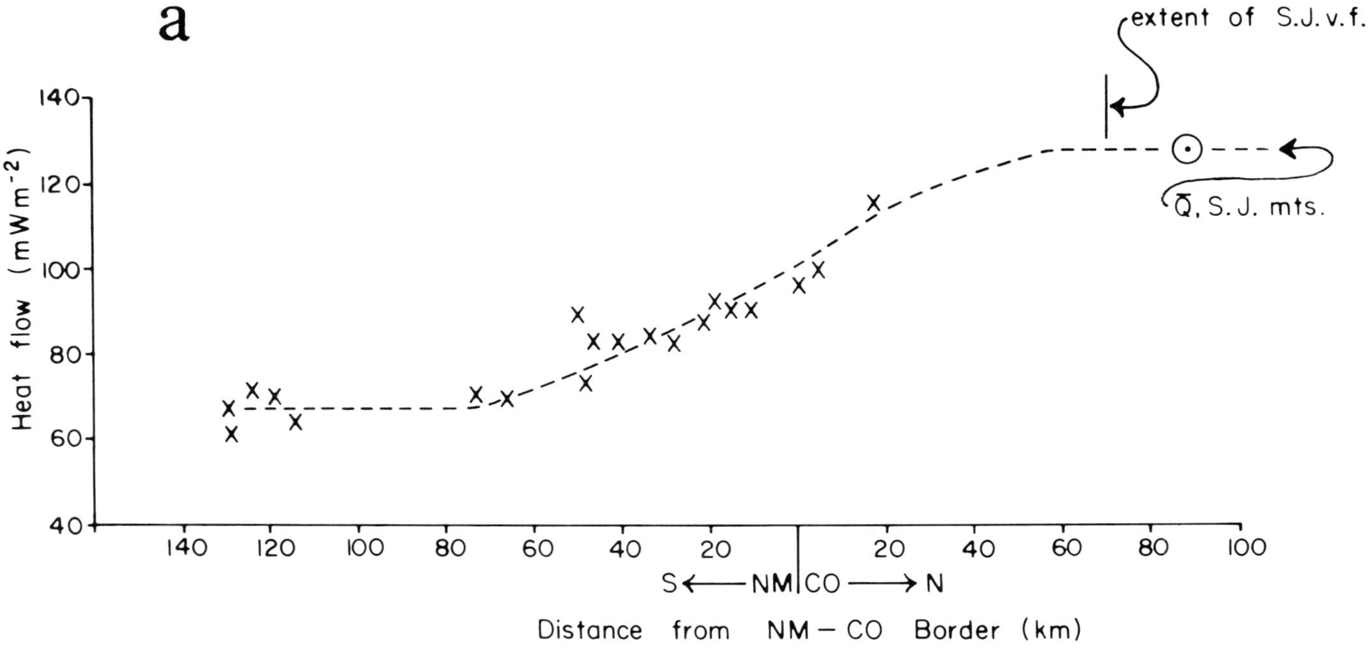

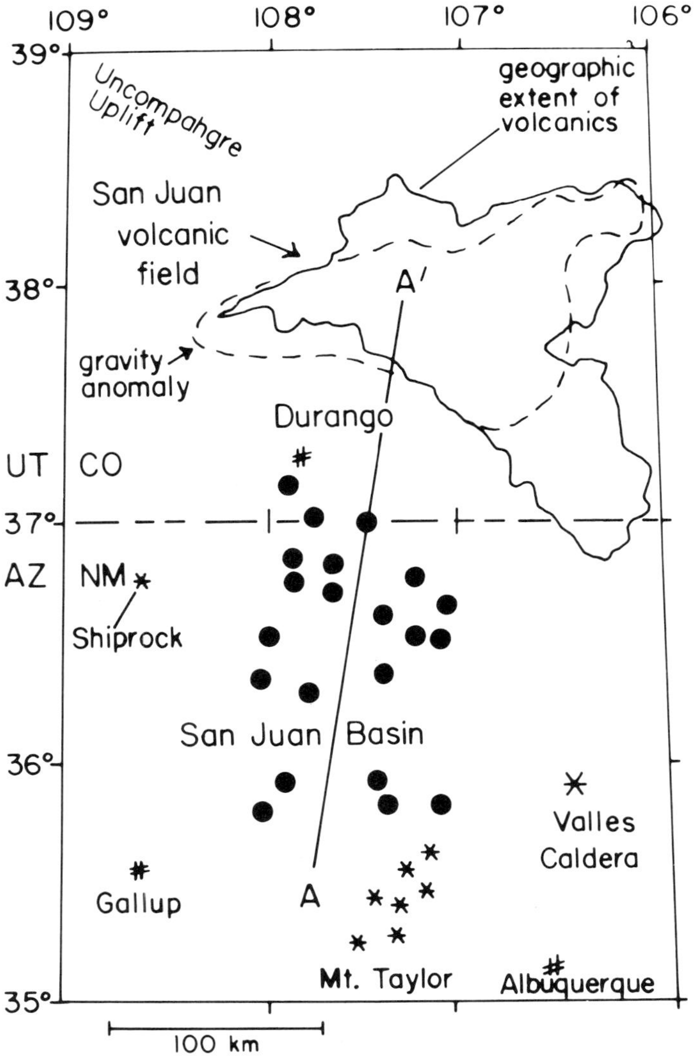

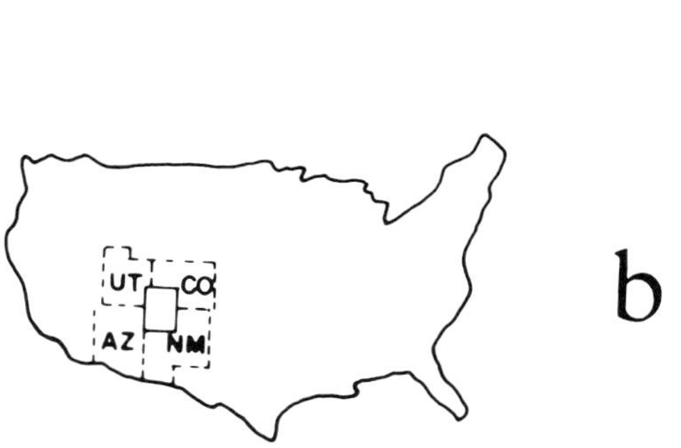

FIGURE 3. **a,** Heat-flow data and theoretical steady-state heat-flow projection resulting from an isothermal step of depth ~100 to ~30 km. Step occurs ~10 km north of Colorado–New Mexico border. **b,** Deep heat-flow data sites (temperature logs >750 m) in the San Juan Basin and projection profile (AA′) for Figure 3a. Data from Sass and others (1971), Reiter and others (1975), and Reiter and Mansure (1983).

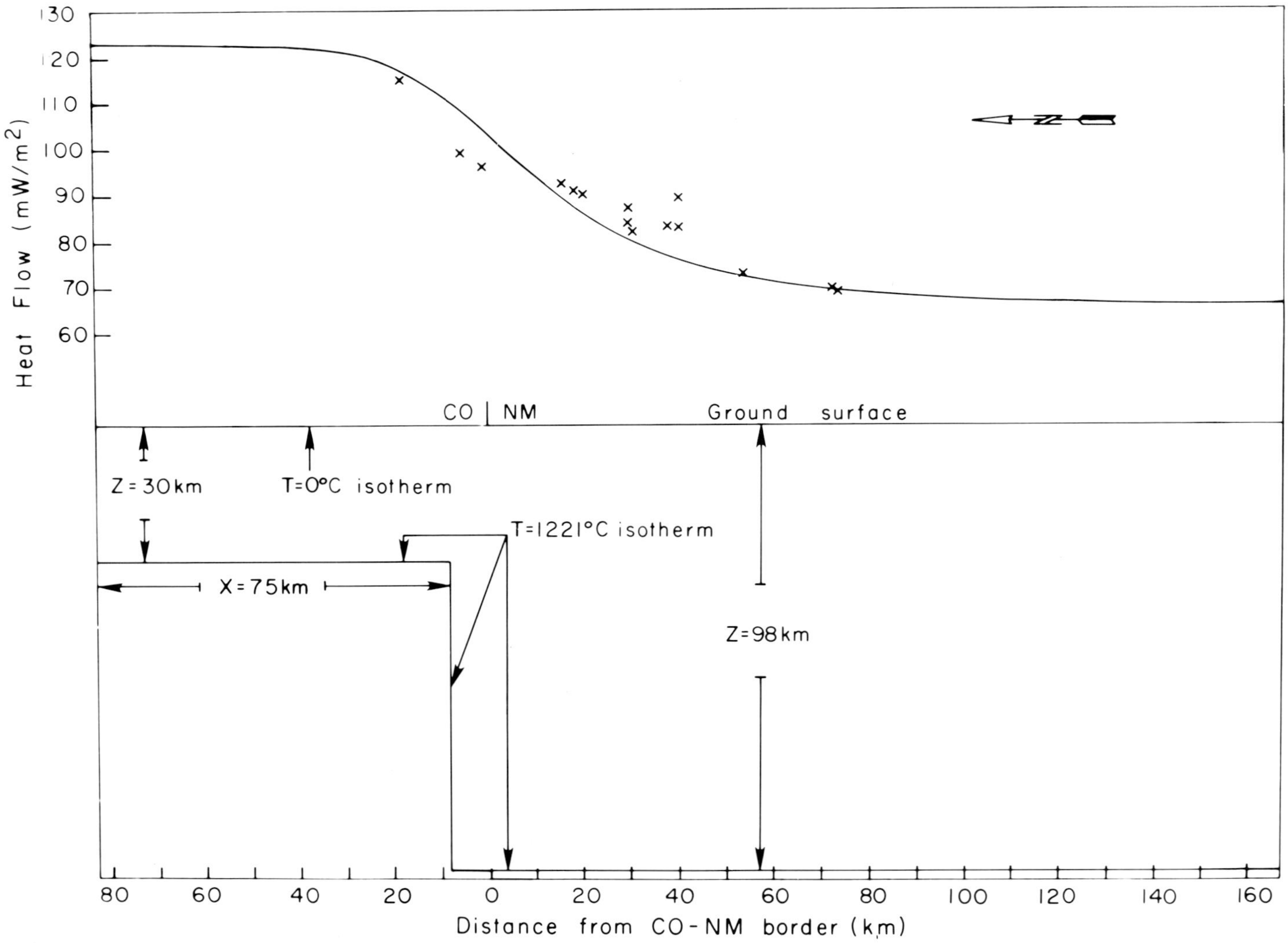

FIGURE 4. Steady-state isothermal-step model for the heat-flow profile approaching the San Juan volcanic field. Experimental data projected onto profile AA′ (Fig. 3b) indicated by x, solid line is the calculated heat flow resulting from step isotherm. Endpoint heat flows determined by averaging data in San Juan volcanic field and southern San Juan Basin. Note comments in text concerning 1,221°C isotherm.

could also be associated with similar shallow phenomena. The Raton and Las Vegas Basins have experienced late Tertiary and Quaternary magmatic activity (Baltz and Bachman, 1956; Gabelman, 1956; Baltz, 1965; Johnson, 1968). In addition, the Las Vegas Basin contains a geothermal area, the Montezuma Hot Springs.

Even though the heat-flow data from northeastern New Mexico are highly variable, it may still be possible to discern a broader trend. Two such possibilities are illustrated in Figure 2. Trend A results by considering the two deepest heat-flow measurements (thermal gradients measured to >0.5 km) in the Raton Basin to be representative of the regional heat flux. With this assumption it is possible to suggest a heat-flow profile which may be similar to the Rio Grande rift–San Juan Basin profile, although the eastern profile eventually approaches a lower value of 50 mWm^{-2} in the High Plains (as opposed to the 70 mWm^{-2} value approached in the San Juan Basin). Taken together, the Rio Grande rift–San Juan Basin and Rio Grande rift–Great Plains heat-flow trends may suggest a fundamental, relatively deep (~30 km) thermal source which underlies the Rio Grande rift and possibly the western part of the Sange de Cristo Mountains and the eastern part of the Brazos uplift (Figs. 1, 2, 5).

A different trend in the heat-flow data across northeastern New Mexico may result if the average of all heat-flow data in the Raton and Las Vegas Basins is representative of the regional heat flux between the southern Rocky Mountains and the High Plains (trend B, Fig. 2). With this assumption a much broader high heat-flow region emerges. This profile would suggest a heat source underlying much of the southern Rocky Mountains, and the Las Vegas and Raton Basins, i.e. the Rio Grande rift, Sangre de Cristo Mountains, and the Raton and Las Vegas Basins. As Edwards and others (1978) stated, the heat-flow pattern across the southern Rocky Mountains suggests more broadly distributed, perhaps deeper sources than along the Rio Grande rift.

CONCLUSIONS

As previously noted by many authors, the Rio Grande rift–southern Rocky Mountains is a region of high heat flow. In particular, very high heat-flow values are found within the rift and within the Raton and Las Vegas Basins. The large variability of values within the rift and the basins suggests that much of the anomalously high heat flow is due to shallow heat sources such as crustal magmatic intrusion and/or ground-water circulation.

Possible trends in the heat-flow data across northern New Mexico suggest some constraints on relatively deep thermal sources; however, it should be noted that the scarcity of data and the potential effects of shallow heat sources and ground-water convection in the region result in ambiguous interpretation of the data. The heat-flow trend between the Rio Grande rift and the San Juan Basin is consistent with an iso-

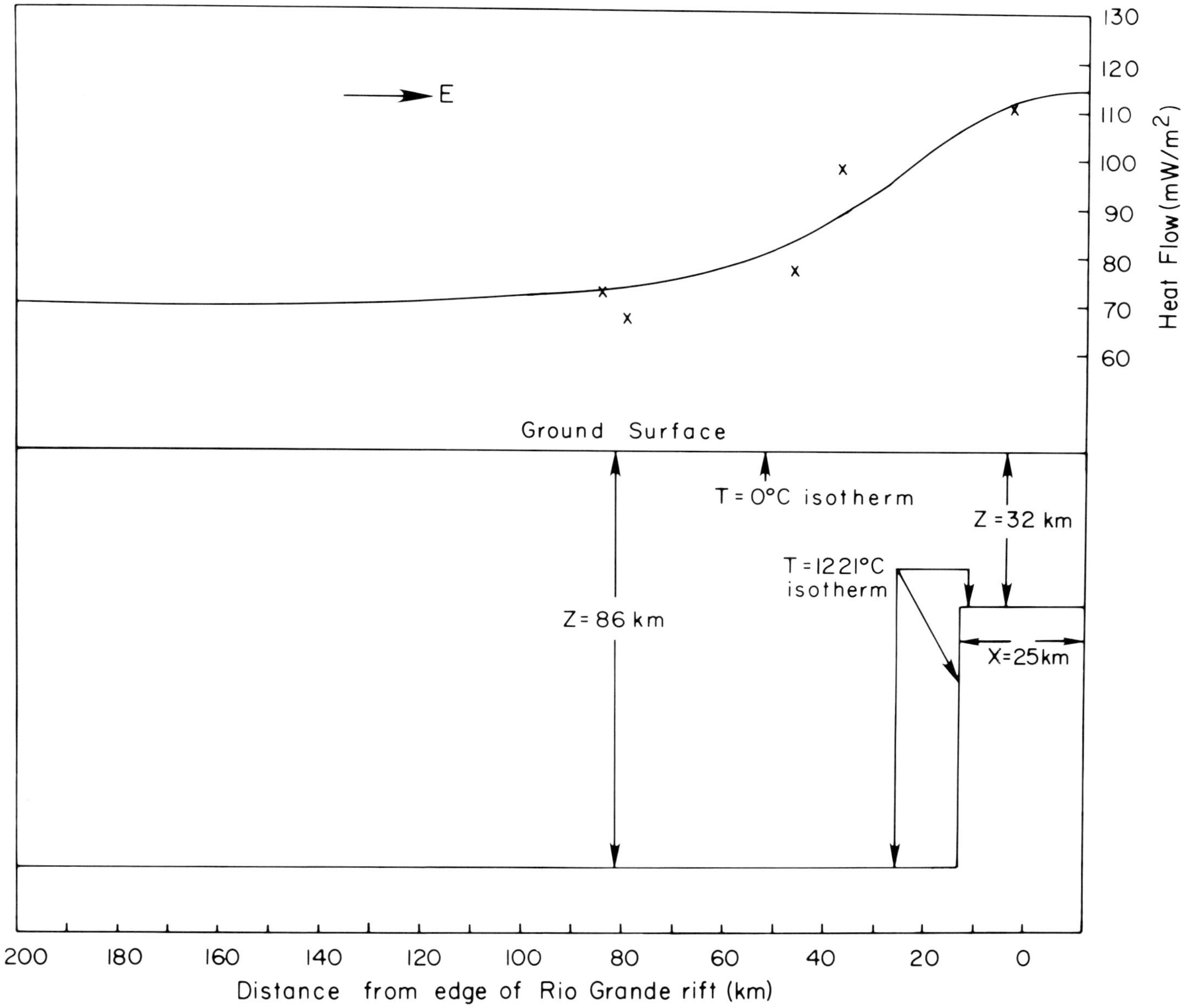

FIGURE 5. Steady-state isothermal-step model for heat-flow profile from northern Rio Grande rift into the San Juan Basin. Experimental data projected onto profile AA′ (Fig. 1) indicated by x, solid line is the calculated heat flow resulting from isothermal step. Rio Grande rift heat flow (endpoint) determined by averaging data (see text). Note comments in text concerning 1,221°C isotherm.

thermal step rising from depths of ~90 km to ~30 km (Fig. 5) beneath the Rio Grande rift and the eastern part of the Brazos uplift. If such a relatively deep heat source underlies the rift, then its effects may also be present east of the rift, e.g. under the Sangre de Cristo Mountains. A profile of the Rio Grande rift–Great Plains heat flow, which is very similar to the Rio Grande rift–San Juan Basin heat-flow profile, is possible (trend A, Fig. 2); however, a clear trend in the heat-flow data east of the rift is not observed, possibly due to the obscuring effects of shallow thermal phenomena. The heat-flow profile from the San Juan Basin across the rift and into the Great Plains (profile A, Fig. 2, east of the rift) would be consistent with a relatively deep heat source underlying the Rio Grande rift and parts of the uplifts on either side, i.e. the Brazos uplift and the Sangre de Cristo Mountains (Fig. 5).

Another possible trend of the heat-flow data in northeastern New Mexico (trend B, Fig. 2) may suggest a more geographically extensive heat source which reaches beneath the region of the Raton and Las Vegas Basins as well as the Rio Grande rift and Sangre de Cristo Mountains. However, the sharp drop-off in the observed heat-flow profile between the Raton and Las Vegas Basins and the High Plains–Sierra Grande arch (~25 km) is inconsistent with a relatively deep steady-state heat source in this region. A gradual increase of heat flow from the High Plains–Sierra Grande arch to the Raton and Las Vegas Basins (trend C, Fig. 2) suggests a thinning of the lithosphere approaching the southern Rocky Mountains. Caner and others (1967) and Reitzel and others (1970) have suggested increased electrical conductivity in the mantle going from the Great Plains to the southern Rockies; this would complement the observed heat-flow trend.

In summary, the observed heat flow across northern New Mexico is consistent with the existence of both relatively shallow (crustal) heat sources within the Rio Grande rift (which may result from magmatic intrusion and/or ground-water circulation) and with a relatively deep heat source underlying the Rio Grande rift and part of the surrounding uplifts (Figs. 2, 5). As such the heat-flow profile in northern New Mexico may reflect the effects associated with both the Rio Grande rift and the southern Rocky Mountains. However, additional heat-flow data will be needed in order to accurately define trends associated with relatively deep heat sources.

ACKNOWLEDGMENTS

Paul Morgan made many helpful suggestions to improve the manuscript.

REFERENCES

Axelrod, D. I., and Bailey, H. P., 1976, Tertiary vegetation, climate, and altitude of the Rio Grande depression, New Mexico–Colorado: Paleobiology, v. 2, pp. 235–254.

Baldridge, W. S., 1979, Petrology and petrogenesis of Plio-Pleistocene basaltic rocks from the central Rio Grande rift, New Mexico, and their relation to rift structure; *in* Riecker, R. E. (ed.), Rio Grande rift: tectonics and magmatism: American Geophysical Union, Washington, D.C., pp. 323–353.

Baltz, E. H., 1965, Stratigraphy and history of Raton Basin and notes on San Luis Basin, Colorado–New Mexico: American Association of Petroleum Geologists, Bulletin, v. 49, pp. 2041–2075.

———, and Bachman, G. O., 1956, Notes on the geology of the southeastern Sangre de Cristo Mountains, New Mexico: New Mexico Geological Society, Guidebook 7, pp. 96–108.

Bridwell, R. J., 1976, Lithospheric thinning and the late Cenozoic thermal and tectonic regime of the northern Rio Grande rift: New Mexico Geological Society, Guidebook 27, pp. 283–292.

Caner, B., Cannon, W. H., and Livingstone, C. E., 1967, Geomagnetic depth sounding and upper mantle structure in the Cordillera region of western North America: Journal of Geophysical Research, v. 72, pp. 6335–6351.

Chapin, C. E., 1979, Evolution of the Rio Grande rift—a summary; *in* Riecker, R. E. (ed.), Rio Grande rift: tectonics and magmatism: American Geophysical Union, Washington, D.C., pp. 1–5.

Clarkson, G. W., 1984, Implications for thermal histories of the San Juan Basin and San Juan Mountains since Late Cretaceous time: Ph.D. thesis, New Mexico Institute of Mining and Technology, Socorro, N.M.

Cook, F. A., McCullar, D. B., Decker, E. R., and Smithson, S. B., 1979, Crustal structure and evolution of the southern Rio Grande rift; *in* Riecker, R. E. (ed.), Rio Grande rift: tectonics and magmatism: American Geophysical Union, Washington, D.C., pp. 195–208.

Cordell, L., 1976, Aeromagnetic and gravity studies of the Rio Grande graben in New Mexico between Belen and Pilar; *in* Woodward, L. A., and Northrop, S. A. (eds.), Tectonics and mineral resources of southwestern North America: New Mexico Geological Society, Special Publication no. 6, pp. 62–70.

———, 1978, Regional geophysical setting of the Rio Grande rift: Geological Society of America, Bulletin, v. 89, pp. 1073–1090.

Dane, C. H., and Bachman, G. O., 1965, Geologic map of New Mexico: U.S. Geological Survey, Washington, D.C.

Decker, E. R., 1969, Heat flow in Colorado and New Mexico: Journal of Geophysical Research, v. 75, pp. 550–559.

———, and Smithson, S. B., 1975, Heat flow and gravity interpretation across the Rio Grande rift in southern New Mexico and west Texas: Journal of Geophysical Research, v. 80, pp. 2542–2552.

Edwards, C. L., Reiter, M., Shearer, C., and others, 1978, Terrestrial heat flow and crustal radioactivity in northeastern New Mexico and southeastern Colorado: Geological Society of America, Bulletin, v. 89, pp. 1341–1350.

Gabelman, J. W., 1956, Tectonic history of the Raton Basin; *in* Guidebook to the geology of the Raton Basin, Colorado: Rocky Mountain Association of Geologists, pp. 35–39.

Johnson, R. B., 1968, Geology of the igneous rocks of the Spanish Peaks region, Colorado: U.S. Geological Survey, Professional Paper 594-G, 47 pp.

Keller, G. R., Braile, L. W., and Schlue, J. W., 1979, Regional crustal structure of the Rio Grande rift from surface wave dispersion measurements; *in* Riecker, R. E. (ed.), Rio Grande rift: tectonics and magmatism: American Geophysical Union, Washington, D.C., pp. 115–126.

Lachenbruch, A. H., and Sass, J. H., 1978, Models of an extending lithosphere and heat flow in the Basin and Range province; *in* Smith, R. B., and Eaton, G. P. (eds.), Cenozoic tectonics and regional geophysics of the western Cordillera: Geological Society of America, Memoir 152, pp. 209–250.

Lipman, P. W., 1969, Alkalic and theoleiitic basaltic volcanism related to the Rio Grande depression, southern Colorado and northern New Mexico: Geological Society of America, Bulletin, v. 80, pp. 1343–1354.

———, and Mehnert, H. H., 1979, The Taos Plateau volcanic field, northern Rio Grande rift, New Mexico: *in* Riecker, R. E. (ed.), Rio Grande rift: tectonics and magmatism: American Geophysical Union, Washington, D.C., pp. 289–311.

———, Bunker, C. M., and Bush, C. A., 1973, Potassium, thorium, and uranium contents of upper Cenozoic basalts of the southern Rocky Mountain region, and their relation to the Rio Grande depression: U.S. Geological Survey, Journal of Research, v. 1, pp. 387–401.

———, Dow, B. R., Hedge, C. E., and others, 1978, Petrologic evolution of the San Juan volcanic field, southwestern Colorado. Pb and Sr isotope evidence: Geological Society of America, Bulletin, v. 89, pp. 59–82.

Manley, K., 1979, Stratigraphy and structure of the Española Basin, Rio Grande rift, New Mexico; *in* Riecker, R. E. (ed.), Rio Grande rift: tectonics and magmatism: American Geophysical Union, Washington, D.C., pp. 71–86.

Olsen, K. H., Keller, G. R., and Stewart, J. N., 1979, Crustal structure along the Rio Grande rift from seismic refraction profiles; *in* Riecker, R. E. (ed.), Rio Grande rift: tectonics and magmatism: American Geophysical Union, Washington, D.C., pp. 127–143.

Prodehl, C., and Pakiser, L. C., 1980, Crustal structures of the southern Rocky Mountains from seismic measurements: Geological Society of America, Bulletin, pt. I, v. 91, pp. 147–155.

Reiter, M., and Clarkson, G., 1983, Geothermal studies in the San Juan Basin and the Four Corners area of the Colorado Plateau, II. Steady-state models of the thermal source of the San Juan volcanic field: Tectonophysics, v. 91, pp. 253–269.

———, and Mansure, A. J., 1983, Geothermal studies in the San Juan Basin and the Four Corners area of the Colorado Plateau, I. Terrestrial heat-flow measurements: Tectonophysics, v. 91, pp. 233–251.

———, ———, and Shearer, C., 1979, Geothermal characteristics of the Rio Grande rift within the southern Rocky Mountain complex; *in* Riecker, R. E. (ed.), Rio Grande rift: tectonics and magmatism: American Geophysical Union, Washington, D.C., pp. 253–267.

———, Shearer, C., and Edwards, C. L., 1978, Geothermal anomalies along the Rio Grande rift in New Mexico: Geology, v. 6, pp. 85–88.

———, Edwards, C. L., Hartman, H., and others, 1975, Terrestrial heat flow along the Rio Grande rift, New Mexico and southern Colorado: Geological Society of America, Bulletin, v. 86, pp. 811–818.

Reitzel, J. S., Gough, D. I., Porath, H., and Anderson, C. W., III, 1970, Geomagnetic deep soundings and upper mantle structure in the western United States: Royal Astronomical Society, Geophysical Journal, v. 19, pp. 213–235.

Rinehart, E. J., Sanford, A. R., and Ward, R. M., 1979, Geographic extent and shape of an extensive magma body at mid-crustal depths in the Rio Grande rift near Socorro, New Mexico; *in* Riecker, R. E. (ed.), Rio Grande rift: tectonics and magmatism: American Geophysical Union, Washington, D.C., pp. 237–249.

Sanford, A. R., Alptekin, O. S., and Toppozada, T. R., 1973, Use of reflection phases on microearthquake seismograms to map an unusual discontinuity beneath the Rio Grande rift: Seismological Society of America, v. 63, pp. 2021–2034.

Sass, J. H., Lachenbruch, A. H., Munroe, R. J., and others, 1971, Heat flow in the western United States: Journal of Geophysical Research, v. 76, pp. 6376–6413.

Stone, W. J., Lyford, F. P., Frenzel, P. F., Mizell, N. H., and Padgett, E., 1983, Hydrogeology and water resources of the San Juan Basin, New Mexico: New Mexico Bureau of Mines and Mineral Resources, Hydrologic Report 6, 70 pp.

Swanberg, C. A., 1979, Chemistry of thermal and nonthermal groundwaters in the Rio Grande rift and adjacent tectonic provinces; *in* Riecker, R. E. (ed.), Rio Grande rift: tectonics and magmatism: American Geophysical Union, Washington, D.C., pp. 279–288.

PRELIMINARY INTERPRETATION OF HEAT FLOW AND RADIOACTIVITY IN THE RIO GRANDE RIFT ZONE IN CENTRAL AND NORTHERN COLORADO

EDWARD R. DECKER[1], GERALD J. BUCHER[2], KENNETH L. BUELOW[2], and H. P. HEASLER[2]
[1]Department of Geological Sciences, University of Maine at Orono, Orono, Maine 04469;
[2]Department of Geology and Geophysics, University of Wyoming, Laramie, Wyoming 82071

INTRODUCTION

Several values of heat flow have been published for Colorado and adjacent states in the past 15–20 years (Spicer, 1964; Roy et al., 1968a; Decker, 1969; Decker and Birch, 1974; Costain and Wright, 1973; Reiter et al., 1975, 1979; Edwards et al., 1978; Bodell and Chapman, 1982; Reiter and Mansure, 1983; Sass et al., 1971b), and, as a result, a more complete picture of regional thermal regimes is beginning to emerge. The most obvious feature of present data for the mountains in Colorado is that most of the values are high. There is evidence also for unusually high flux in parts of the Rio Grande rift within the southern Rocky Mountain complex, but interpretation of the thermal regime in the northern rift system is difficult because the number of control points is small. This paper summarizes new heat-flow and radioactivity data for the rift zone in central and northern Colorado, and presents a preliminary interpretation of the results.

Data acquisition, calculation methods, and correction procedures are not discussed herein because they directly followed well-established philosophies and techniques (e.g., Roy et al., 1968a; Sass et al., 1971a, b; Decker, 1973; Birch, 1950). Instead, summaries of completed results are emphasized. Details of unpublished data used in the present study are in an advanced stage of preparation for publication, and will be submitted to a journal in the second half of 1984.

GEOGRAPHY AND GENERALIZED GEOLOGY

Figure 1 shows geography and generalized geology of Colorado and nearby parts of adjacent states. Most of the ranges in the southern Rocky Mountains are elements of late Paleozoic highlands that experienced uplift and erosion during the Laramide orogeny (Tweto, 1975). The Colorado parks are Laramide structural and sedimentary basins that contain significant thicknesses of sediments, the Uncompahgre highland was uplifted in late Tertiary as well as Laramide time, and the La Plata Mountains were the site of Laramide intrusions into Paleozoic sediments (Tweto, 1975). The current physiography of the region is considered to have resulted from uplift and differential erosion in late Tertiary time. Chapin (1979) suggests that the southern Rockies experienced rapid uplift betwen 7 and 4 m.y. ago, and Larson et al.

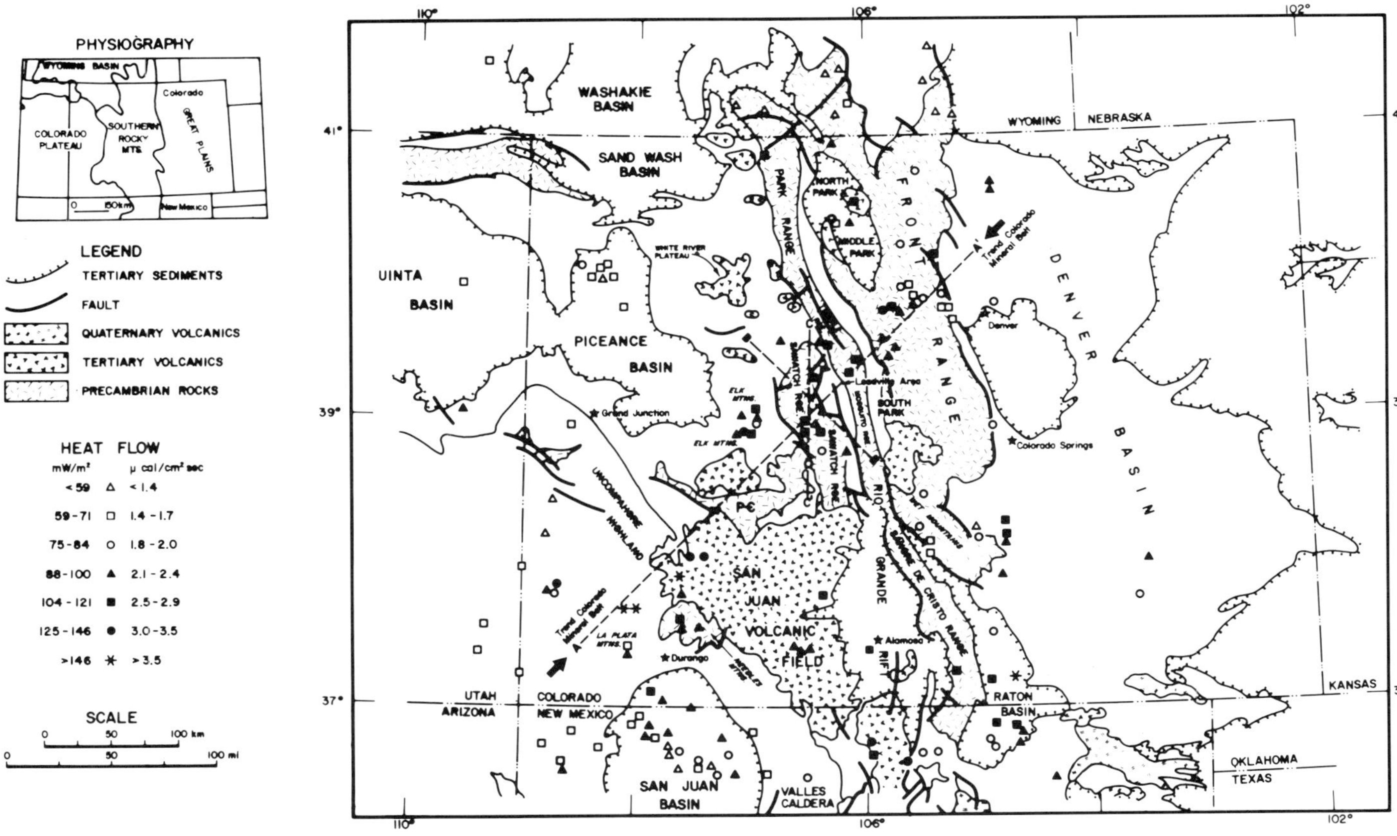

FIGURE 1. Map showing physiographic provinces, geography, generalized geology, and generalized heat flow in Colorado and nearby parts of adjacent states. Heat-flow values after Birch (1950), Spicer (1964), Roy et al. (1968a), Decker (1969), Costain and Wright (1973), Decker and Birch (1974), Sass et al. (1971b), Reiter et al. (1975, 1979), Edwards et al. (1978), Bodell and Chapman (1982), Decker et al. (1980), Buelow (1980), Reiter and Mansure (1983), and Decker et al. (in prep.). Geology after King and Beikman (1974). See Figure 3 for summarized thermal data along lines A–A′ and B–B′, and Figure 5 for summaries of gravity and thermal data near line C–C′.

(1975) find evidence for two periods of rapid uplift of the central and northern Colorado Rockies about 10 m.y. and 1.5 m.y. ago, respectively. Axelrod and Bailey (1976) suggest that late Tertiary uplift of the rift zone approached about 1.1 km.

Sizable masses of Laramide igneous rocks crop out in the environs of the Colorado Mineral Belt (Fig. 1). In Oligocene time, a major volcanic field covered most of the Rockies in southern Colorado and northern New Mexico, and much of the Colorado park areas to the north (Steven, 1975). The youngest (30–25 m.y.) Oligocene volcanism created a large ash-flow field in the San Juan Mountains, and perhaps smaller fields near source areas in the Sawatch Range and the Never Summer Mountains near North Park (Steven, 1975). Miocene and Pliocene extrusive activity in the San Juan Mountains consisted of less voluminous bimodal volcanism (Lipman et al., 1978).

The Rio Grande rift in Colorado developed since latest Oligocene and/or earliest Miocene time (Lipman and Mehnert, 1975; Lipman et al., 1978; Chapin, 1979; Tweto, 1979). Although the main-graben zone ends about 20 km north of Leadville (Chapin, 1979), Tweto (1979) suggests that extension within the rift system reaches northward to areas near the Wyoming border. The rift between Alamosa and Leadville is notable for a near absence of Neogene volcanism in the axial basins (Chapin, 1979). In contrast, Neogene tholeiitic and alkali basalts occur in the environs of the rift south of Alamosa (Lipman and Mehnert, 1975), and Neogene basalts with alkali affinities crop out in the rift system north of Leadville (Larson et al., 1975).

HEAT-FLOW AND RADIOACTIVITY DATA

Generalized heat-flow data in Colorado and nearby parts of bordering states also are plotted in Figure 1. Fourteen of the values in the southern Rockies are from Reiter et al. (1975) and Edwards et al. (1978), and two are from Birch (1950). The other values in the mountains in Colorado and Wyoming were determined by Decker and co-workers (Decker, 1969; Decker and Birch, 1974; Roy et al., 1968a; Decker et al., 1980; Decker et al., in prep.). The new values by Decker and others directly follow basic data for more than 70 holes that were drilled for economic purposes. Ten of these holes are ≤200 m deep. The other holes are deeper than 250 m.

Several studies (Birch et al., 1968; Roy et al., 1968b, 1972; Lachenbruch, 1968) suggest that data on bedrock radiogenic-heat production are required for reliable regional heat-flow interpretations. Therefore, our research in Colorado and Wyoming focused on combined heat-flow (Q) and heat-production (A) studies in crystalline rocks. The relevant results for stations in the Colorado Rockies are shown in Figure 2. Figure 2 also shows Q–A bands for "normal" portions of the Basin and Range thermal province, and the Battle Mountain, Nevada, heat-flow high (after Lachenbruch and Sass, 1977). The data for two sites in the Front Range (the Adams and Moffat tunnels) are after Q values in Birch (1950); radioelement data for these stations are after analyses in Phair and Gottfried (1964) and Lovering and Goddard (1950). The other values of Q are our determinations (Fig. 1); A values for these stations are based on radioelement (U, Th, K) measurements using gamma-ray spectrometry (for instrumentation see Adams, 1964; Decker, 1973).

DISCUSSION

Figures 1 and 2 contain a large body of new data on heat-flow and radioactive-heat production for the Rio Grande rift system in Colorado. We call attention here to some of the implications of the now available data and briefly discuss their significance in regional thermal regimes.

Regional Heat Flow

Heat flows at seven localities in the Colorado Rockies are in the range 62–71 mW/m². The much larger number of values that exceed 71 mW/m² confirms earlier suggestions (Decker, 1969; Roy et al., 1972; Reiter et al., 1975, 1979; Edwards et al., 1978) that most of the southern Rockies in Colorado are characterized by high flux. From the data available, the mean flux in the Front Range is about 92 mW/m². The mean of our values in the vicinity of the rift near Leadville is about 113 mW/m², with a scatter of approximately 4 mW/m².

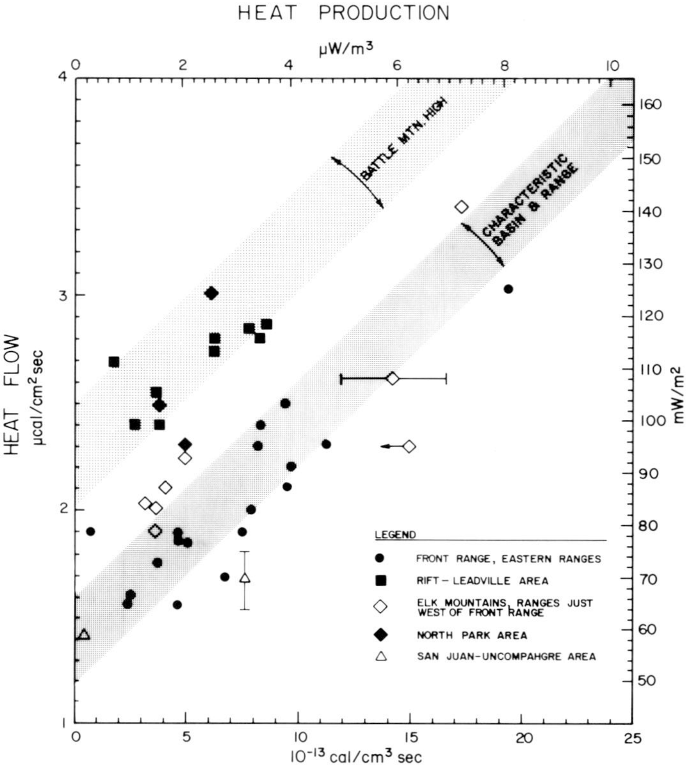

FIGURE 2. Heat flow as a function of near-surface radiogenic-heat production for the Colorado Rockies. See text for references. "Battle Mountain High" and "Characteristic Basin & Range" zones drawn from data in Lachenbruch and Sass (1977).

Heat Flow and Thermal Regime in the Rift near Leadville

The concomitant changes of Q and A in Figure 2 imply that near-surface radioactivity accounts for a substantial fraction of the lateral heat-flow variations in the Colorado Rockies. Additionally, the Q–A data for the Front Range plot near the heat-flow–heat-production band depicted for the Basin and Range thermal province, whereas the data for the rift–Leadville area and the North Park region plot in or near the zone shown for the Battle Mountain area in Nevada. The data therefore suggest that the rift–Leadville and North Park areas are unusual thermal highs that are bordered by a high heat-flow zone in the Front Range to the east.

The Q–A values for the Front Range determine a line with an intercept of about 54–59 mW/m², and that for a line through the rift–Leadville area data is about 88 mW/m². The slopes of the lines are in the range of 7–11 km. The contrasting intercepts imply that the deep flux in the rift near Leadville is unusually high. Thus, profiles of surface and reduced heat flow [flux with the variable crustal component removed (after Roy et al., 1972)] were constructed to further investigate the indicated anomaly. The resulting profile parallel to the Mineral Belt shows that the reduced flux in the rift–Leadville area changes from 84–92 mW/m² to 50–60 mW/m² in the Front Range (Fig. 3b). Similarly, a profile perpendicular to the Mineral Belt indicates a narrow, reduced heat-flow high in the rift zone near Leadville (Fig. 3a). The number of control points west of the Leadville area is small, but reduced values in discordant intrusions in the Elk Mountains (R, P, P1) are 25–21 mW/m² lower than those in the rift zone to the northeast (Fig. 3b). The borders of indicated anomaly near Leadville are only 30–50 km wide (Fig. 3). Therefore, the unusually high reduced flux is likely due to heat sources in the crust.

Crustal-heat sources are also suggested by steady-state geotherms for the rift–Leadville area. For example, calculated equilibrium tempera-

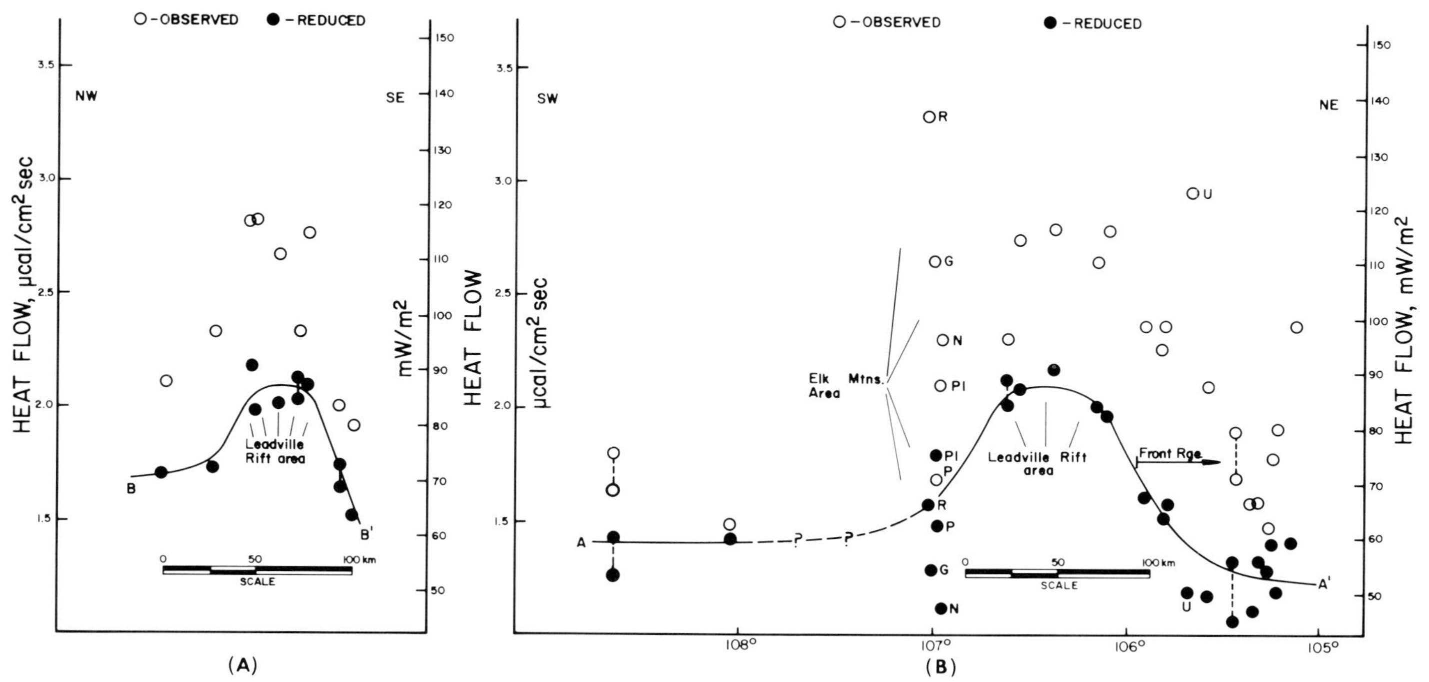

FIGURE 3. Surface and reduced heat-flow profiles parallel (A–A′) and perpendicular (B–B′) to the Colorado Mineral Belt. See Figure 1 for locations of profiles.

tures for the Leadville area exceed the liquidus of dry basalt in the lower 5–10 km of the crust (Fig. 4). Under such conditions there would be widespread melting in the lower crust and upper mantle, and large amounts of basaltic magma could be generated. However, the near absence of synrift volcanics in the Leadville area suggests that the large amounts of basaltic magma implied by the steady-state geotherm probably are not present in the crust. Similarly, P-wave velocities in the upper mantle should be lower than the observed values of 7.9–8.0 km/sec (Prodhel and Pakiser, 1980; Archambeau et al., 1969; Jackson and

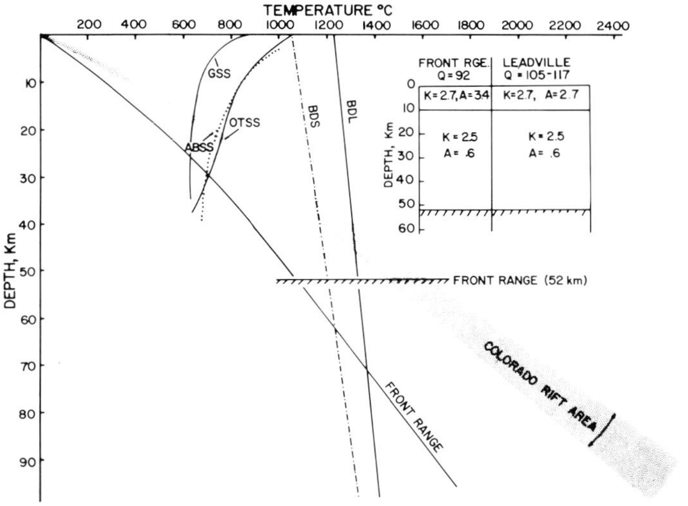

FIGURE 4. Possible steady-state temperature–depth curves for the Front Range and the rift–Leadville area. Crustal thicknesses after Prodhel and Pakiser (1980) indicated by hatching. Mean surface Q and shallow A values are after data summarized in this paper; other model parameters after accepted values used by Roy et al. (1968b), Blackwell (1971), Roy et al. (1972), and Smithson and Decker (1974). Melting curves after Yoder and Tilley (1962), Wyllie (1971), and Lambert and Wyllie (1972): GSS = granodiorite saturated solidus; ABSS = alkali-basalt saturated solidus; OTSS = olivine-tholeiite saturated solidus; BDS = basalt dry solidus; BDL = basalt dry liquidus. Units: heat flow (Q) in mW/m^2; thermal conductivity (K) in W/m°K; radiogenic heat production (A) in $\mu W/m^3$.

Pakiser, 1965), if large volumes of molten material exist near the crust–mantle boundary.

A major Bouguer gravity low is associated with the Mineral Belt–Arkansas Valley area near Leadville (Tweto and Case, 1972; Tweto, 1975, 1979). This anomaly, geologic mapping, and radiometric-age dates indicate that Laramide and Oligocene intrusive rocks are widespread in the underlying upper crust (Pearson et al., 1962; Tweto, 1968, 1975; Cunningham et al., 1977; Tweto and Case, 1972; Isaacson and Smithson, 1976). Additionally, post-Oligocene igneous units crop out as small bodies in the area, thin silicic-felsite and vitrophyre dikes are present in rift-related faults, and numerous rhyolitic pipes with post-Miocene ages(?) occur near Leadville (Tweto, 1979, pers. comm. 1980). Considered together, the Neogene igneous activity, the narrow borders of the thermal anomaly, and the unrealistic inferred equilibrium temperatures for the mantle suggest that the unusual reduced flux in the area arises from non-radiogenic time-dependent heat sources in the crust.

The gravity anomaly and the regional Neogene igneous activity permit speculation that the reduced heat-flow high could be due to temporal conductive effects of young, low-density intrusions in a buried batholith. This is evident from the gravimetrically constrained cooling-intrusion model depicted in Figure 5. From this figure, the surface flux produced by a 5-km-thick intrusion emplaced in the 15–20-km-depth interval between 1 and 2 m.y. ago closely approximates the heat-flow field based on reduced values projected to the gravity profile (line C–C′in Fig. 1). Buried intrusions with 1–2 m.y. ages may in turn be consistent with rapid uplift and erosion in central Colorado in the past 1.5 m.y. (Larson et al., 1975).

The model depicted in Figure 5 is artificial and others involving buried intrusions, anomalous radioactivity not indicated in surface rocks, or non-conductive heat sources could be designed. But the data available are not consistent with unusually high radioactivity below the Leadville area. Clearly, models with cooling late Neogene intrusions could be used to explain the gravity low, the heat-flow anomaly, and the uplift history. Alternatively, convective sources might be employed because Neogene faulting, rifting, and intrusions imply young extension in the area. Lachenbruch (1978) shows that steady convection in the lithosphere in tectonically extending regions (e.g., the Basin and Range) can produce very high flux without producing unrealistically high tem-

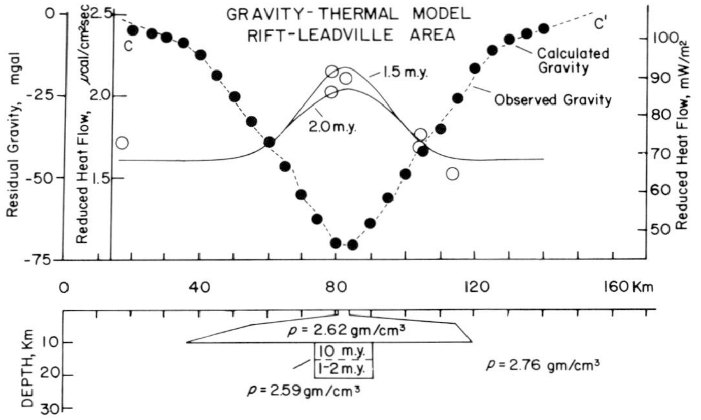

FIGURE 5. Gravimetrically constrained crustal structure—cooling-intrusion model for the rift area near Leadville. Profiles along line C–C′ shown in Figure 1. Gravity after Tweto and Case (1972); two-dimensional modeling used methods of Talwani et al. (1959). Conductive-cooling models with latent heat after Jaeger (1968) and Lachenbruch et al. (1976). Density values from Tweto and Case (1972): Neogene(?) rhyolites = 2.59 gm/cm³; Late Cretaceous–middle Tertiary porphyries = 2.62 gm/c³; Precambrian country rocks = 2.76 gm/cm³. Neogene(?) intrusions implaced in crusts with equilibrium temperatures like those calculated for the Front Range; therefore, the intrusions with latent heat initially were at temperatures of about 700°C (1100°C − 400°C) above those of bordering units in 10–20-km-depth intervals (see Fig. 4).

peratures in the crust or mantle. Although time-dependent, penetrative convection models involving uplift (e.g., Bodell and Chapman, 1982; Morgan, 1983) could be used to explain the Leadville area data, it is desirable to confirm the magnitude of the excess reduced flux and fix the widths of the western borders of the anomaly before presenting detailed interpretations.

Northern Rift Zone

Based on regional geology and a 120 mW/m² heat-flow value in the North Park area, Decker et al. (1980) speculated that very high flux occurs in parts of the mountains in northern Colorado, and implied that the northerly borders of the anomaly were narrow (≲70 km). Reiter et al. (1979) also implied high heat flow for northern Colorado, but their stations (4) in sediments in eastern Colorado and southeastern Wyoming were separated by distances that exceeded 90–100 km.

In Figure 6, available surface and reduced values for northern Colorado and southern Wyoming are projected to a profile along longitude 106°15′W. This plot shows that the surface and reduced heat flows increase from normal to high in a narrow transition zone (≲50 km) near the Colorado–Wyoming border, followed by very high reduced values in the park areas in the Colorado Rockies immediately to the south. The new evidence for unusually high reduced flux in Middle and North Parks is consistent with conjectures by Tweto (1979) and Decker et al. (1980) that the Rio Grande rift zone extends into extreme northern Colorado.

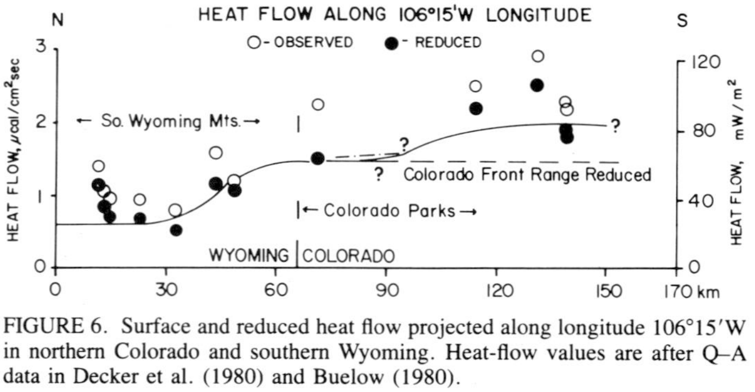

FIGURE 6. Surface and reduced heat flow projected along longitude 106°15′W in northern Colorado and southern Wyoming. Heat-flow values are after Q–A data in Decker et al. (1980) and Buelow (1980).

Buelow (1980) analyzed gravity, flux, and radioactivity data for the park areas in northern Colorado, and suggested that parts of the Bouguer gravity lows and reduced heat-flow highs could be explained by low-density, post-Miocene intrusions in the upper 10–15 km of the crust. Periods of late Cenozoic alkali-basalt volcanism during different time periods (24–20 m.y., 14–9 m.y., 8–10 m.y., ≲1.5 m.y.) in northwestern Colorado (Larson et al., 1975) may be consistent with late Neogene intrusions in the northern park areas, as may be late Miocene (12–7 m.y.) intrusions and extrusions in the Park Range–Elkhead Mountains regions (Tweto, 1979). A low-velocity zone in the upper crust in the Park Range and other mountain areas east of the Front Range (Prodhel and Pakiser, 1980) also may imply low-density, young intrusions and elevated heat flow in the vicinity of Middle and North Parks.

CONCLUDING REMARKS

Combined heat-flow and heat-production data for the Colorado Rockies indicate a large positive heat-flow anomaly in the vicinity of the Rio Grande rift near Leadville, and perhaps in the rift zone in the Middle and North Park areas near the Colorado–Wyoming border. The borders of both anomalies are narrow (≲50 km) and inferred steady-state geotherms are unusually high; therefore, the excess flux in both areas is considered to be due to heat sources in the upper crust. Because negative Bouguer gravity anomalies and Neogene igneous rocks occur along the Mineral Belt near Leadville and in the park areas in northern Colorado, one reasonable hypothesis is that the unusually high heat-flow zones are partly caused by low-density, young intrusions (<5 m.y.) in the upper crust. This view is consistent with periods of rapid uplift and alkalic magmatism in central and northwestern Colorado in late Miocene and Pleistocene times. High reduced flux along the Colorado rift zone also could arise from convective-heat sources in a tectonically extending crust. However, the rift between Alamosa and Leadville is not characterized by extensive synrift volcanics (Chapin, 1979), alkali basalts in northwestern Colorado probably were derived from mantle below a thick crust (Tweto, 1979), and seismic data for the Leadville–northern park areas do not imply pronounced crustal thinning locally (Prodhel and Pakiser, 1980). Thus, questions arise about the likelihood of appreciable extension and lithospheric convection in the rift zone north of Alamosa. These views do not preclude other heat sources below the rift zone. However, we do not have concrete evidence for enriched crustal radioactivity below the thermally anomalous areas, and nothing in our deep measurements of flux unequivocally confirms widespread hydrothermal circulation in the upper crust in the rift zone in central and northern Colorado.

High regional heat flow in the Colorado Front Range implies near-liquidus equilibrium temperatures in the upper mantle (Fig. 4). Below a developing rift zone to the west, therefore, basaltic magmas from the mantle could rise into the lower crust and produce partial melting. Partial melting of parts of the lower crust might in turn produce rhyolitic magmas that could collect together and rise into the upper crust (Eichelberger, 1978; Eichelberger and Gooley, 1977). A two-stage melting process like this may account for extensive bimodal volcanism in rapidly extending parts of the southern Rio Grande rift and the western Great Basin (Chapin, 1979; Eaton, 1979; Christiansen and Lipman, 1972; Lipman et al., 1971, 1972). In contrast, a relatively tight, less-extended rift zone in Colorado north of Alamosa would imply slower ascent of magmas and perhaps encourage solidification within the crust. Furthermore, large Oligocene and older intrusions below the Leadville–northern park areas could have hindered magma ascent. Hence, it does not seem unreasonable that these areas could be underlain by low-density plutons that were trapped in the crust during younger evolution of the rift zone in central and northern Colorado.

ACKNOWLEDGMENTS

Partial financial support for our research was received from U.S. Department of Energy agreement DE-FC07-791D12026, Los Alamos National Laboratory contracts N28-3896G-1 and N28-4934H-1, and U.S. National Science Foundation grants GA-18450, DES 75-15119, EAR-8207036, and RII-8011448. The University of Maine at Orono provided partial financial support for drafting of figures. Paul Morgan provided an excellent, constructive review of a preliminary version of the manuscript.

REFERENCES

Adams, J. A. S., 1964, Laboratory gamma-ray spectrometer for geochemical studies; *in* Adams, J. A. S., and Lowder, W. M. (eds), The natural radiation environment: University of Chicago Press, pp. 485–497.

Archambeau, C. E., Flinn, E. A., and Lambert, D. G., 1969, Fine structure of the upper mantle: Journal of Geophysical Research, v. 74, pp. 5825–5865.

Axelrod, D. I., and Bailey, H. P., 1976, Tertiary vegetation, climate, and altitude of the Rio Grande depression, New Mexico–Colorado: Paleobiology, v. 2, pp. 235–254.

Birch, F., 1950, Flow of heat in the Front Range, Colorado: Geological Society of America, Bulletin, v. 61, pp. 567–630.

———, Roy, R. F., and Decker, E. R., 1968, Heat flow and thermal history in New England and New York; *in* Zen, E., and others (eds.), Studies of Appalachian geology: Interscience, J. Wiley and Sons, New York, pp. 437–451.

Blackwell, D. D., 1971, The thermal structure of the continental crust; *in* Heacock, J. G. (ed.), The structure and physical properties of the Earth's crust: American Geophysical Union, Geophysical Monograph 14, pp. 169–184.

Bodell, J. M., and Chapman, D. S., 1982, Heat flow in the north-central Colorado Plateau: Journal of Geophysical Research, v. 87, pp. 2869–2884.

Buelow, K. L., 1980, Geothermal studies in Wyoming and northern Colorado, with a geophysical model of the southern Rocky Mountains near the Colorado–Wyoming border: M.S. thesis, University of Wyoming, 150 pp.

Chapin, C. E., 1979, Evolution of the Rio Grande rift—a summary; *in* Riecker, R. E. (ed.), Rio Grande rift: tectonics and magmatism: American Geophysical Union, Washington, D.C., pp. 1–5.

Christiansen, R. L., and Lipman, P. W., 1972, Cenozoic volcanism and plate tectonic evolution of the western United States. II—Late Cenozoic: Royal Society of London, Philosophical Transactions (A), v. 271, pp. 249–284.

Costain, J. K., and Wright, P. M., 1973, Heat flow at Spor Mountain, Jordan Valley, Bingham, and LaSal, Utah: Journal of Geophysical Research, v. 78, pp. 8687–8698.

Cunningham, C. G., Naeser, C. W., and Marvin, R. F., 1977, New ages for intrusive rocks in the Colorado mineral belt: U.S. Geological Survey, Open-file Report 77-573, 16 pp.

Decker, E. R., 1969, Heat flow in Colorado and New Mexico: Journal of Geophysical Research, v. 74, pp. 550–559.

———, 1973, Geothermal measurements by the University of Wyoming: University of Wyoming, Contributions to Geology, v. 12, pp. 21–24.

———, and Birch, F., 1974, Basic heat-flow data from Colorado, Minnesota, New Mexico and Texas; *in* Sass, J. H., and Munroe, R. J. (compilers), Basic heat-flow data from the United States: U.S. Geological Survey, Open-file Report 74-9, pp. 5-1-5-60.

———, Baker, K. H., Bucher, G. J., and Heasler, H. P., 1980, Preliminary heat flow and radioactivity studies in Wyoming: Journal of Geophysical Research, v. 85, pp. 311–321.

Eaton, G. P., 1979, A plate-tectonic model for late Cenozoic crustal spreading in the western United States; *in* Riecker, R. E. (ed.), Rio Grande rift: tectonics and magmatism: American Geophysical Union, Washington, D.C., pp. 7–32.

Edwards, C. L., Reiter, M., Shearer, C., and Young, W., 1978, Terrestrial heat flow and crustal radioactivity in northeastern New Mexico and southeastern Colorado: Geological Society of America, Bulletin, v. 89, pp. 1341–1350.

Eichelberger, J. C., 1978, Andesitic volcanism and crustal evolution: Nature, v. 275, pp. 21–27.

———, and Gooley, R., 1977, Evolution of silicic magma chambers and their relationship to basaltic volcanism; *in* Heacock, J. G. and others (eds.), The Earth's crust: American Geophysical Union, Monograph 20, pp. 57–77.

Isaacson, L. B., and Smithson, S. B., 1976, Gravity anomalies and granite emplacement in west-central Colorado: Geological Society of America, Bulletin, v. 87, pp. 22–28.

Jackson, W. H., and Pakiser, L. C., 1965, Seismic study of crustal structure in the southern Rocky Mountains: U.S. Geological Survey, Professional Paper 525-D, pp. 85–92.

Jaeger, J. C., 1968, Cooling and solidification of igneous rocks; *in* Hess, H. H., and Poldervaart, A. (eds.), Basalts (the Poldervaart treatise on rocks of basaltic composition): Interscience, J. Wiley & Sons, New York, pp. 503–536.

King, P. B., and Beikman, H. M. (compilers), 1974, Geologic map of the United States (exclusive of Alaska and Hawaii): U.S. Geological Survey, Reston, Virginia.

Lachenbruch, A. H., 1968, Preliminary geothermal model of the Sierra Nevada: Journal of Geophysical Research, v. 73, pp. 6977–6989.

———, 1978, Heat flow in the Basin and Range province and thermal effects of tectonic extension: Pure and Applied Geophysics, v. 117, pp. 34–50.

———, and Sass, J. H., 1977, Heat flow in the United States; *in* Heacock, J. G. (ed.), The Earth's crust: American Geophysical Union, Monograph 20, pp. 626–675.

———, Sass, J. H., Munroe, R. J., and Moses, T. H., 1976, Geothermal setting and simple heat conduction models for the Long Valley caldera: Journal of Geophysical Research, v. 81, pp. 769–784.

Lambert, I. B., and Wyllie, P. J., 1972, Melting of gabbro (quartz eclogite) with excess water to 35 kilobars, with geological implications: Journal of Geology, v. 80, pp. 693–708.

Larson, E. E., Ozima, M., and Bradley, W. C., 1975, Late Cenozoic basic volcanism in northwestern Colorado and its implications concerning tectonism and the origin of the Colorado River system; *in* Curtis, B. F. (ed.), Cenozoic history of the southern Rocky Mountains: Geological Society of America, Memoir 144, pp. 155–178.

Lipman, P. W., and Mehnert, H. H., 1975, Late Cenozoic basaltic volcanism and development of the Rio Grande depression in the southern Rocky Mountains; *in* Curtis, B. F. (ed.), Cenozoic history of the southern Rocky Mountains: Geological Society of America, Memoir 144, pp. 119–154.

———, Prostka, H. J., and Christiansen, R. L., 1971, Evolving subduction zones in the western United States, as interpreted from igneous rocks: Science, v. 174, pp. 821–825.

———, ———, and ———, 1972, Cenozoic volcanism and plate tectonic evolution of the western United States. I—Early and middle Cenozoic: Royal Society of London, Philosophical Transactions (A), v. 271, pp. 217–248.

———, Doe, B. R., Hedge, C. E., and Steven, T. A., 1978, Petrologic evolution of the San Juan volcanic field, southwestern Colorado: Pb and Sr evidence: Geological Society of America, Bulletin, v. 89, pp. 59–82.

Lovering, T. S., and Goddard, E. N., 1950, Geology and ore deposits of the Front Range, Colorado: U.S. Geological Survey, Professional Paper 223, 319 pp.

Morgan, P., 1983, Constraints on rift thermal processes from heat flow and uplift: Tectonophysics, v. 94, pp. 277–298.

Pearson, R. C., Tweto, O., Stern, T. W., and Thomas, H. H., 1962, Age of Laramide porphyries near Leadville, Colorado: U.S. Geological Survey, Professional Paper 450-C, pp. 78–80.

Phair, G., and Gottfried, D., 1964, The Colorado Front Range, Colorado, U.S.A., as a uranium and thorium province; *in* Adams, J. A. S., and Lowder, W. M. (eds.), The natural radiation environment: University of Chicago Press, pp. 7–38.

Prodhel, C., and Pakiser, L. C., 1980, Crustal structure of the southern Rocky Mountains from seismic measurements: Geological Society of America, Bulletin, pt. 1, v. 91, pp. 147–155.

Reiter, M., and Mansure, A. J., 1983, Geothermal studies in the San Juan Basin and the Four Corners area of the Colorado Plateau, I. Terrestrial heat flow measurements: Tectonophysics, v. 91, pp. 233–251.

———, Mansure, A. J., and Shearer, C., 1979, Geothermal characteristics of the Rio Grande rift within the southern Rocky Mountain complex; *in* Riecker, R. E. (ed.), Rio Grande rift: tectonics and magmatism: American Geophysical Union, Washington, D.C., pp. 253–267.

———, Edwards, C. L., Hartman, H., and Weidman, C., 1975, Terrestrial heat flow along the Rio Grande rift, New Mexico and southern Colorado: Geological Society of America, Bulletin, v. 86, pp. 811–818.

Roy, R. F., Decker, E. R., Blackwell, D. D., and Birch, F., 1968a, Heat flow in the United States: Journal of Geophysical Research, v. 73, pp. 5207–5221.

———, Blackwell, D. D., and Birch, F., 1968b, Heat generation of plutonic rocks and continental heat flow provinces: Earth and Planetary Science Letters, v. 5, no. 1, pp. 1–12.

———, ———, and Decker, E. R., 1972, Continental heat flow; *in* Robertson, E. C. (ed.), The nature of the solid Earth: McGraw-Hill, New York, pp. 506–543.

Sass, J. H., Lachenbruch, A. H., and Munroe, R. J., 1971a, Thermal conductivity of rocks from measurements on fragments and its application to heat flow determinations: Journal of Geophysical Research, v. 76, pp. 3391–3401.

———, ———, ———, Greene, G. W., and Moses, T. H., Jr., 1971b, Heat flow in the western United States: Journal of Geophysical Research, v. 76, pp. 6376–6412.

Smithson, S. B., and Decker, E. R., 1974, A continental crustal model and its geothermal implications: Earth and Planetary Science Letters, v. 22, pp. 215–225.

Spicer, H. C., 1964, Geothermal gradients and heat flow in the Salt Valley anticline, Utah: Bolletino de Geofisica Teorica ed Applicata (Trieste), v. 6, pp. 263–282.

Steven, T. A., 1975, Middle Tertiary volcanic field in the southern Rocky Mountains; *in* Curtis, B. F. (ed.), Cenozoic history of the southern Rocky Mountains: Geological Society of America, Memoir 144, pp. 75–94.

Talwani, M., Worzel, J. C., and Landisman, M., 1959, Rapid gravity computations for two-dimensional bodies with application to the Mendocino submarine fracture zone: Journal of Geophysical Research, v. 68, pp. 49–59.

Tweto, O., 1968, Leadville district, Colorado; *in* Ridge, J. D. (ed.), Ore deposits of the United States, 1933–1967 (Graton–Sales Volume, Vol. 1): American Institute of Mining, Metallurgical, and Petroleum Engineers, Inc., New York, pp. 681–705.

———, 1975, Laramide (Late Cretaceous–early Tertiary) orogeny in the southern Rocky Mountains; *in* Curtis, B. F., (ed), Cenozoic history of the southern Rocky Mountains: Geological Society of America, Memoir 144, pp. 1–44.

———, 1979, The Rio Grande rift system in Colorado; *in* Riecker, R. E. (ed.), Rio Grande rift: tectonics and magmatism: American Geophysical Union, Washington, D.C., pp. 33–56.

———, and Case, J. E., 1972, Gravity and magnetic features as related to geology in the Leadville 30-minute quadrangle, Colorado: U.S. Geological Survey, Professional Paper 726-C, 31 pp.

Wyllie, P. J., 1971, Experimental limits for melting in the Earth's crust and upper mantle; *in* Heacock, J. G. (ed.), The structure and physical properties of the Earth's crust: American Geophysical Union, Geophysical Monograph 14, pp. 279–301.

Yoder, H. S., and Tilley, C. E., 1962, Origin of basalt magmas: an experimental study of natural and synthetic rock systems: Journal of Petrology, v. 3, pp. 342–532.

A GEOPHYSICAL STUDY OF THE SAN LUIS BASIN

G. R. KELLER[1], LINDRITH CORDELL[2], G. H. DAVIS[3], W. J. PEEPLES[1], and G. WHITE[1]
[1]University of Texas at El Paso, El Paso, Texas 79968; [2]U.S. Geological Survey, Federal Center, MS-964, Denver, Colorado 80225;
[3]Heritage Resources, Inc., 1000 Paragon, Midland, Texas 79701

INTRODUCTION

The San Luis Basin (Fig. 1) is an elongate intermontane valley located in south-central Colorado and north-central New Mexico. Stretching approximately 240 km in a north–south direction, it is bordered on the west by the San Juan and Tusas Mountains and on the east by the Sangre de Cristo Mountains. At its northern terminus near Poncha Pass, the valley is only 10–12 km wide, but it widens to approximately 75 km near Alamosa, Colorado. The valley floor, although appearing relatively flat, is built from the coalescence of alluvial fans which slope gently from the mountain flanks toward the center of the valley. Fifteen kilometers south of Alamosa, this sloping terrain is broken by the San Luis Hills, a series of flat-topped, low-lying volcanic hills extending along a north–south trend to the Colorado–New Mexico border. Geographically, the valley has been arbitrarily defined as terminating 15 mi south of the state line (Siebenthal, 1910). However, Upson (1939) noted that the valley is not a geological entity unto itself and instead merges southward into the Taos Plateau which is underlain by a Holocene volcanic field (Lipman and Mehnert, 1979). In this study, we have adopted the structural definition of the San Luis Basin (Kelley, 1956) and extended it to include the Taos Plateau with a southern termination at the Embudo constriction just southwest of Taos.

Since the early studies of Siebenthal (1910), Upson (1939), and Kelley (1956), the San Luis Basin has been recognized as being one of the major structural elements of the Rio Grande rift (Chapin, 1971). However, its subsurface configuration is poorly known, and it is thus a good target for geophysical studies. Early gravity studies (Gaca and Karig, 1965) were hampered by the lack of topographic maps. Cordell (1978) conducted a gravity study of the Taos Plateau region, and Davis (1979) conducted a gravity study of the Colorado portion of the San Luis Basin. These surveys became the major source of gravity data for this study and were merged into a consistent data base which was used to create Bouguer anomaly maps of the Rio Grande rift (Cordell et al., 1982) and New Mexico (Keller and Cordell, 1983). A Bouguer anomaly map of the San Luis Basin area is shown in Figure 2. Zietz and Kirby (1972) published a regional aeromagnetic map of Colorado, but the flight-line spacing was sparse in the San Luis Basin area. An aeromagnetic map of New Mexico has recently been completed (Cordell, 1983) and a portion of this map, which includes the southern San Luis Basin, is presented elsewhere in this volume. Seismic data of sufficient resolution to be of use in this study are very sparse. In fact, the reflection profiles of Davis and Stoughton (1979) are the only publicly available data. Although quite limited in areal extent, these data provide valuable constraints on the subsurface structure in the northernmost San Luis Basin.

Reviews of the geologic setting and history of the San Luis Basin area can be found in Baltz (1965, 1978), Tweto (1979), Lipman and Mehnert (1979), and Woodward and Ingersoll (1979). The brief discussion which follows is based on these studies and documents the main aspect of the geologic history impacting geophysical interpretations, which is the fact that little or no pre-Eocene Phanerozoic rock is present beneath the basin because the region has been high-standing throughout most of the Phanerozoic. During the early Paleozoic, the area was located on the Transcontinental arch. The Uncomphagre–San Luis highland extended through the area in the Pennsylvanian, but its eastern margin may have been located in the eastern portion of the area presently occupied by the San Luis Basin. The entire region was uplifted again during the Laramide orogeny as part of the Sangre de Cristo–Brazos uplift and virtually all pre-existing sediments were eroded away at that time. Some Eocene alluvium was deposited at least locally (Tweto, 1979), and then rocks of the San Juan volcanic field covered the area in the Oligocene (Steven, 1975). Drilling data and outcrops in the Taos Plateau area, San Luis Hills, and Sangre de Cristo Mountains confirm that these volcanic rocks still underlie the basin. The formation of the Rio Grande rift caused block faulting which has preserved these volcanic rocks in the resulting grabens. Thus, for the purpose of geophysical interpretations, the basin can be considered to consist of graben fill overlying Oligocene volcanic rocks which lie on the Precambrian basement.

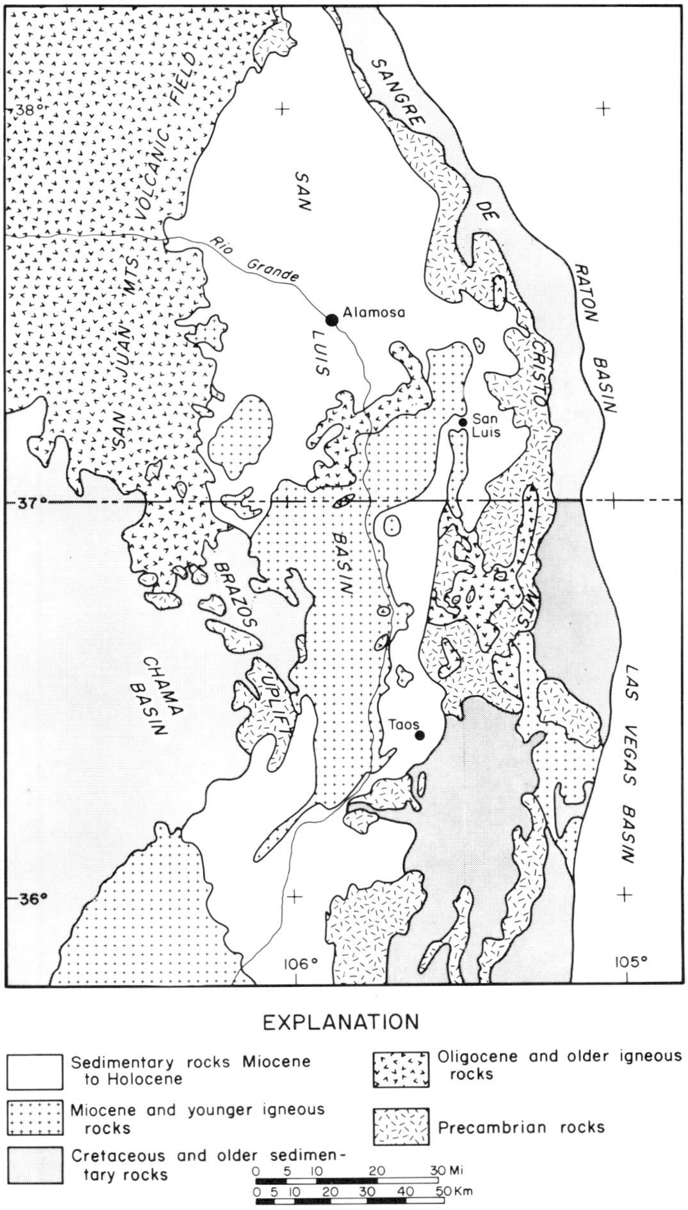

FIGURE 1. Generalized geologic map of the San Luis Basin area.

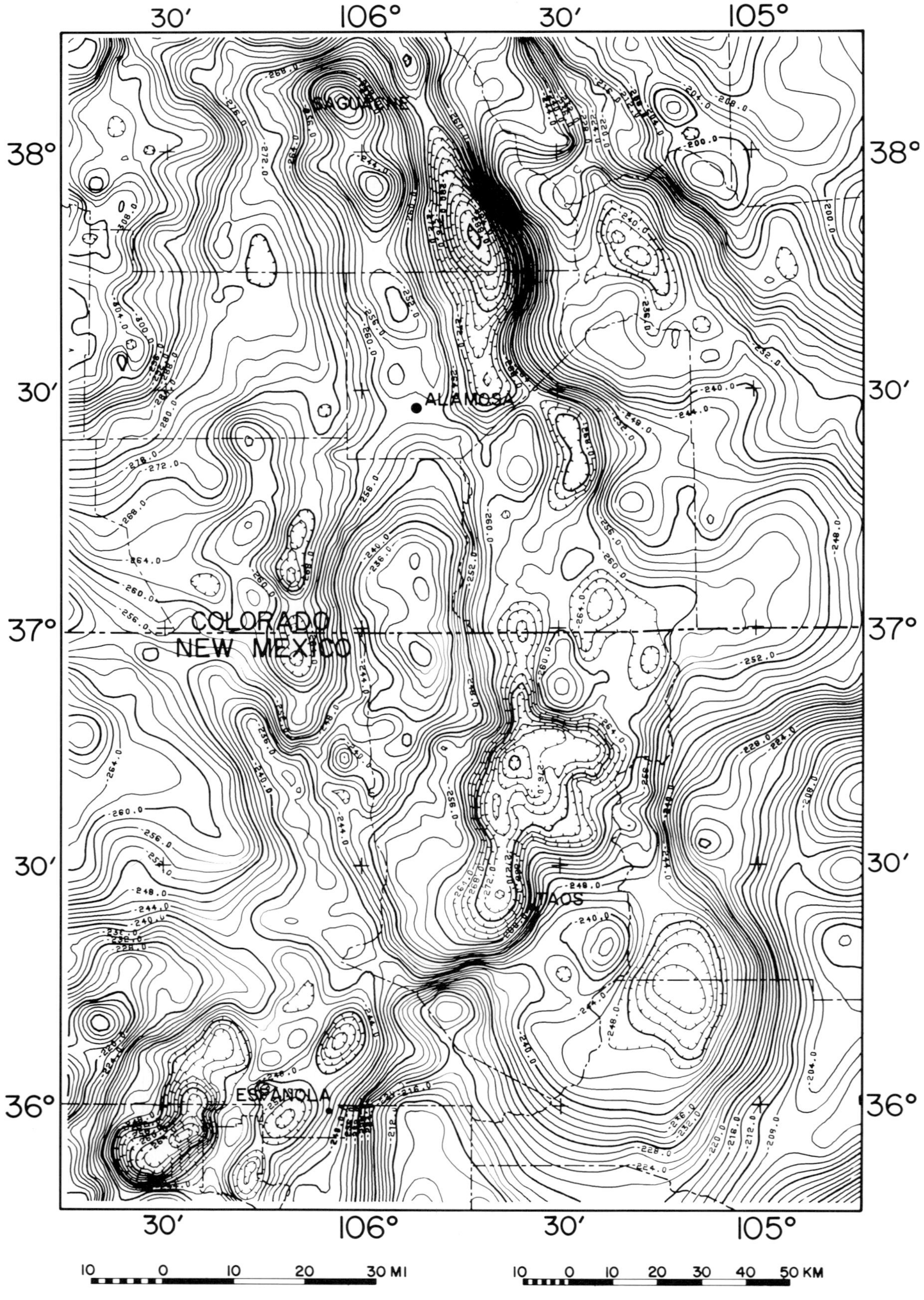

FIGURE 2. Bouguer anomaly gravity map of the San Luis Basin area. Contour interval = 2 mgal. Datum = sea level. Bouguer reduction density = 2.67 gm/cc.

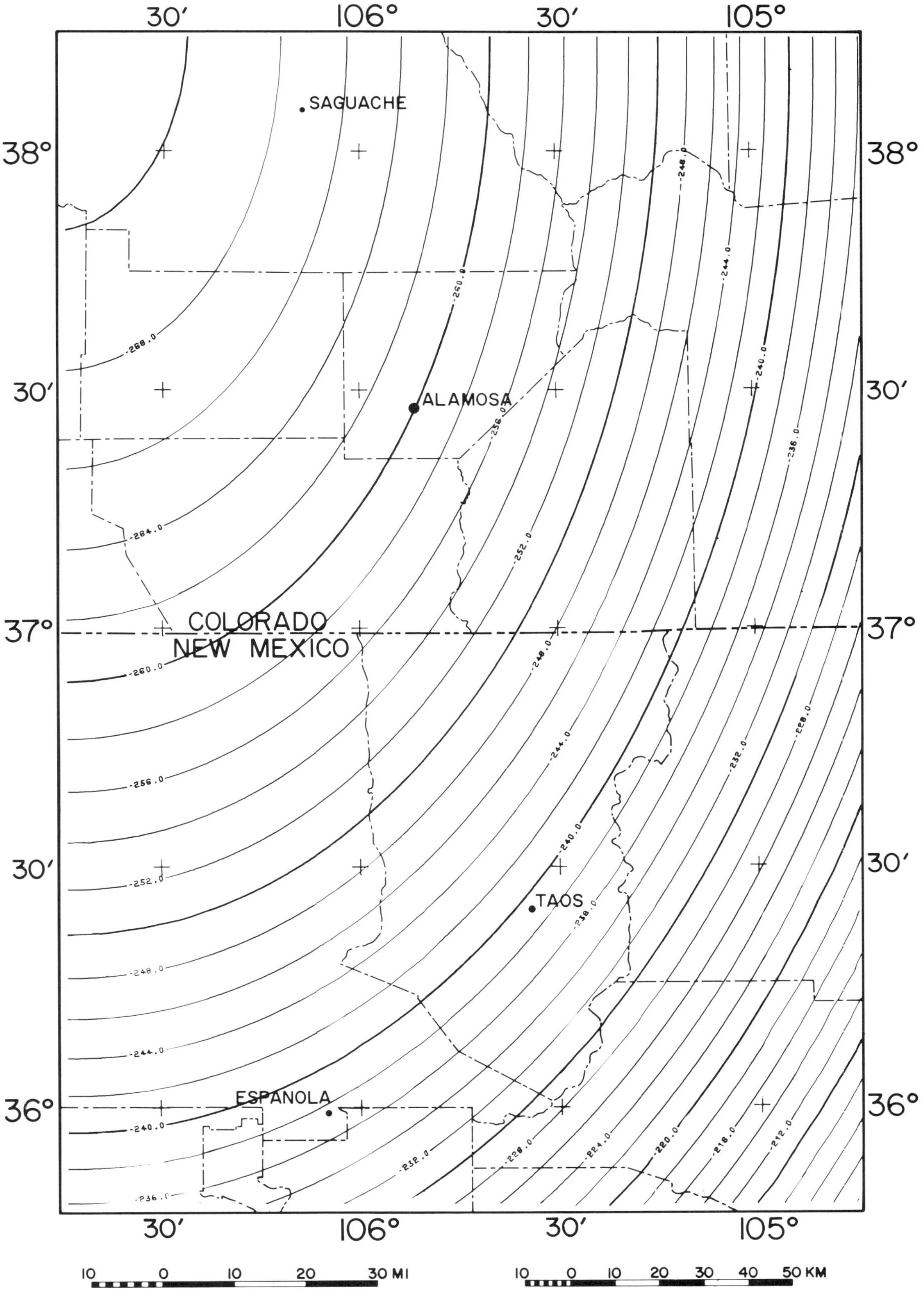

FIGURE 3. Regional gravity trend in the San Luis Basin area as modeled using a second-order polynomial surface. Contour interval = 2 mgal.

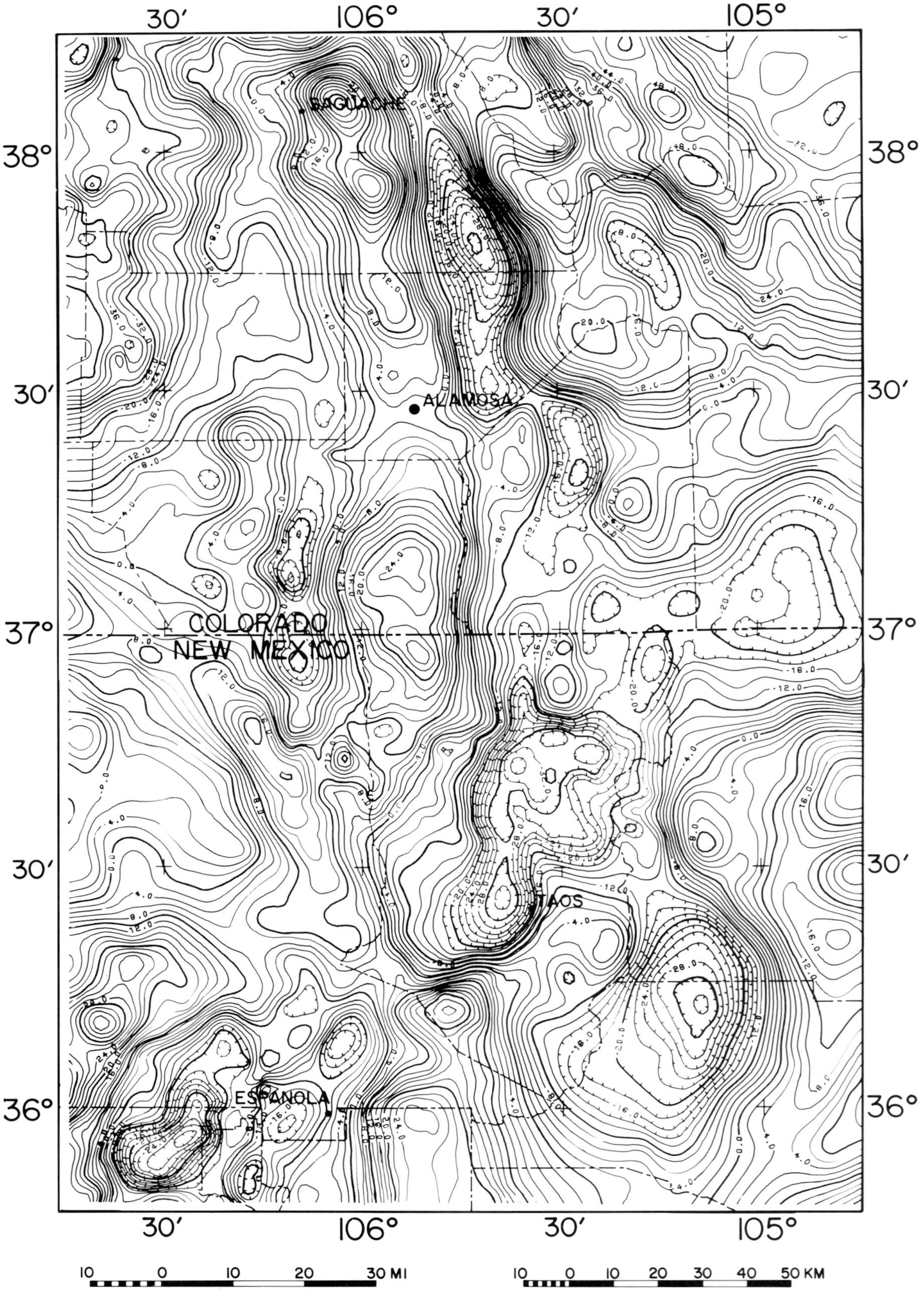

FIGURE 4. Residual gravity anomalies in the San Luis Basin area. These values were obtained by subtracting values on the surface shown in Figure 3 from the Bouguer anomaly values displayed in Figure 2. Contour interval = 2 mgal.

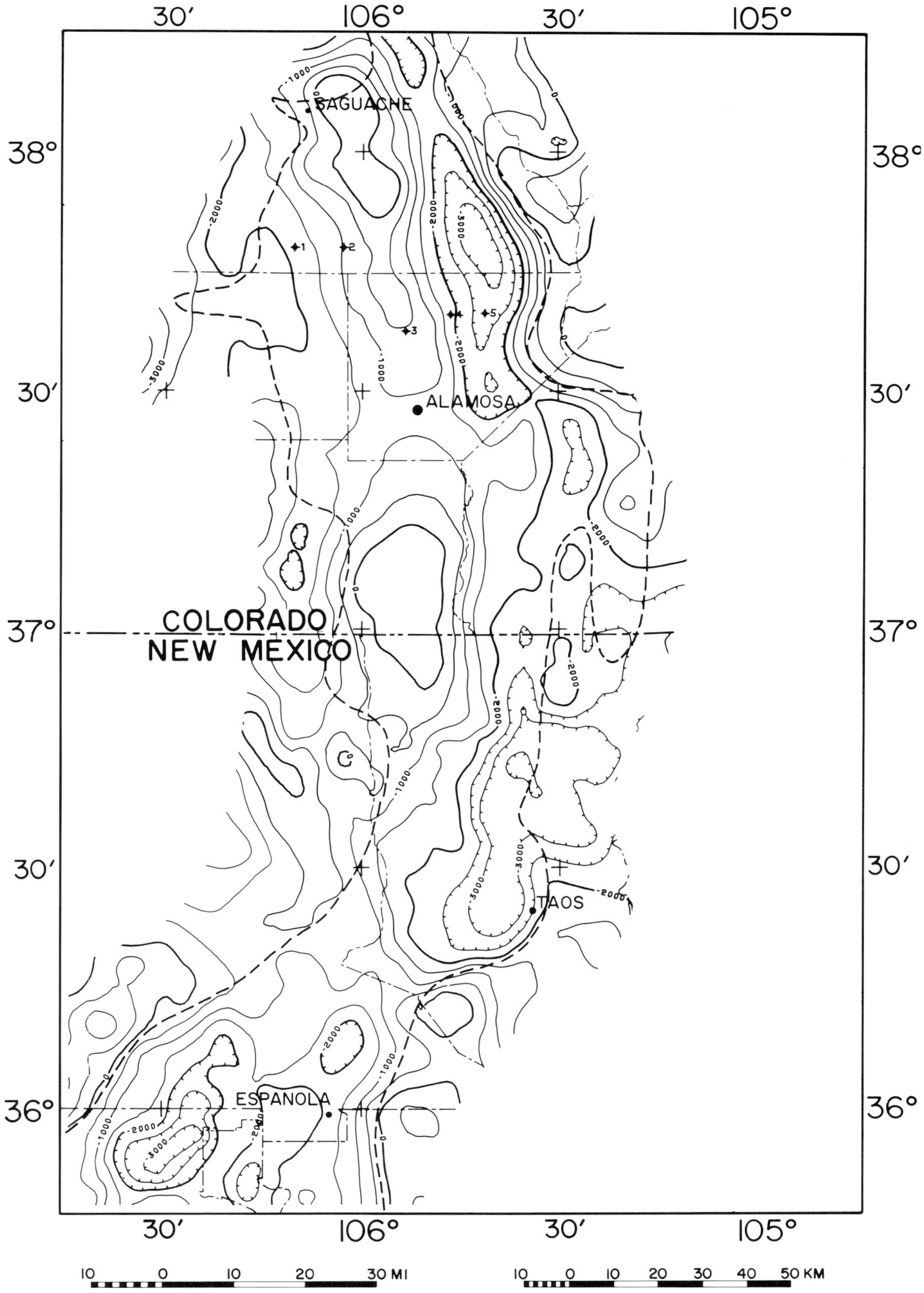

FIGURE 5. Contour map of estimated depth (meters) to the base of the bolson fill. The calculations assumed a density contrast of 0.35 gm/cc between the fill and adjacent rocks. Contour interval = 1,000 m. Wells used for calibration are shown as circles with crosses. Approximate boundary of the basin in dashed line.

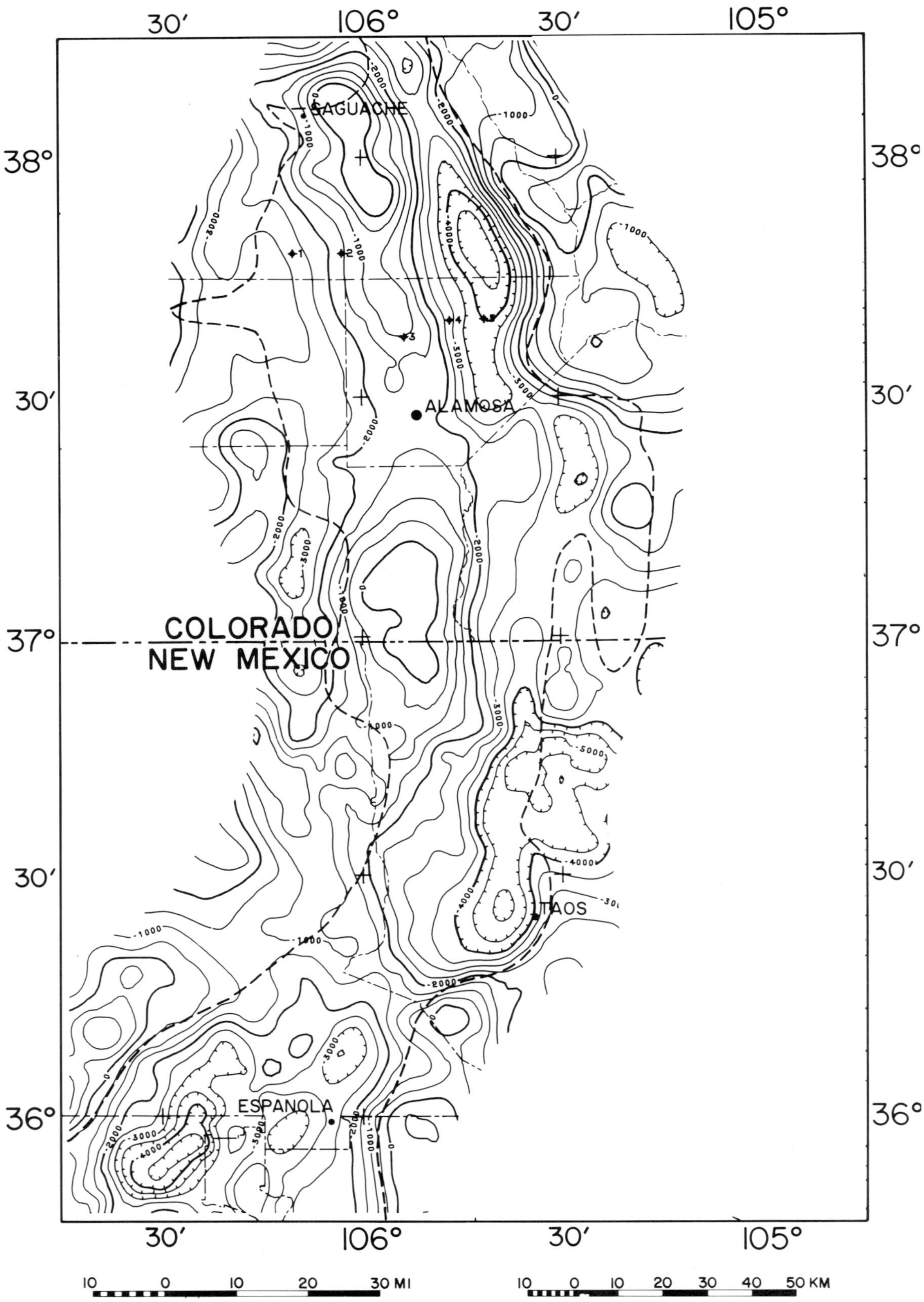

FIGURE 6. Contour map of estimated depth (meters) to the Precambrian basement. The calculations assumed a density contrast of 0.25 gm/cc between the basement and younger rocks. Contour interval = 1,000 m. Wells used for calibration are shown as circles with crosses. Approximate boundary of the basin in dashed line.

GEOPHYSICAL DATA AND ANALYSIS

With the exception of a small addition of data in the Sangre de Cristo Mountains, the data base of gravity readings used in this study is the same as shown on the maps of Cordell et al. (1982) and Keller and Cordell (1983). The reduction parameters and procedures are given in Cordell et al. (1982). A major consideration is that all these data have been terrain-corrected for topography in a zone extending from 0.875 km to 166.7 km from each station. These corrections were calculated using the technique of Plouff (1977). However, the resulting complete Bouguer anomaly values cannot be interpreted quantitatively without further processing because a large regional variation in the gravity field is present and distorts the local anomalies due to features such as basin fill. This regional variation was modeled by fitting a second-order polynomial surface to the data in the area encompassing all of southern Colorado and northern New Mexico using the technique of Lance (1982). Contour values of this surface in the San Luis Basin area are shown in Figure 3. In this area, the regional gravity field can essentially be represented by a linear increase in gravity values from west to east. That this is an appropriate choice of regionals can be verified by looking at individual profiles (e.g., Davis, 1979). This removal of the regional can be considered as a high-pass filter with the residual gravity values (i.e., the differences between the grid of observed values and the corresponding second-order surface values) representing the high-frequency component being passed. A map of these residual gravity values is shown as Figure 4. It is geologically reasonable (but not necessarily valid) to assume that density variations within the basement are small. Thus, the gravity anomalies in San Luis Basin area (Fig. 4) would be at least mostly due to variations in the thickness of the basin fill and underlying volcanic rocks. A linear gravity high delineates a prominent N–S trending horst running up the center of the basin. The San Luis Hills and outcrops of Oligocene volcanic rocks in the Taos Plateau (Lipman and Mehnert, 1979) represent exposures of this horst whose presence is also indicated by drilling data (Tweto, 1979) and seismic-reflection data (Davis and Stoughton, 1979). The deepest portions of the basin generally lie along the western margin of the Sangre de Cristo Mountains (Cordell, 1978; Davis, 1979), and the smaller graben west of the horst dies out to the south as the Taos Plateau is encountered.

In an attempt to quantify the estimates of basin fill, we have applied the computer-modeling techniques of Wen (1983) to the residual anomalies. This technique assumes that the regional has been adequately removed and that a single density contrast between basin fill and basement is a good approximation to reality. The results are shown as contour maps of the thickness of basin fill in Figures 5 and 6.

Five deep oil tests in the San Luis Basin were used to calibrate the choice of density contrasts for the fill-thickness determinations. These wells are located in Figures 5 and 6. A bulk-density log was available for the Amoco Mapco 1-32 state well. This log showed that the density of the fill averaged 1.9 gm/cc for the interval extending from the surface to 570 m, 2.1 gm/cc from 570 m to 1,325 m, and 2.3 gm/cc from 1,325 m to 2,630 m. Oligocene volcanic rocks were encountered at 2,630 m and had an average density of 2.6 gm/cc. The biggest unknowns in the modeling were the density of the basement rocks and the extent of Eocene sedimentary rocks. However, a density contrast of 0.35 gm/cc produced a good fit between calculated fill-thickness values and drilling data. Thus, the map shown as Figure 5 can be considered to display the general distribution of post-Oligocene basin fill. A density contrast of 0.25 gm/cc produced a good fit between the calculated-thickness values and the depths to Precambrian basement from drilling data. Thus, Figure 6 cn be interpreted to indicate the approximate depth to Precambrian basement.

REFERENCES

Baltz, E. H., 1965, Stratigraphy and history of the Raton Basin and notes on San Luis Basin, Colorado–New Mexico: American Association of Petroleum Geologists, Bulletin, v. 49, no. 11, pp. 2041–2075.

———, 1978, Resume of Rio Grande depression in north-central New Mexico: New Mexico Bureau of Mines and Mineral Resources, Circular 163, pp. 210–228.

Chapin, C. E., 1971, The Rio Grande rift, pt. I—modifications and additions: New Mexico Geological Society, Guidebook 22, pp. 191–202.

Cordell, L., 1978, Complete Bouguer anomaly map of the Taos Basin section of Rio Grande rift: U.S. Geological Survey, Open-file Report 78-317.

———, 1983, Aeromagnetic map of New Mexico: NOAA, National Geophysical Data Center, Scientific Map Series, scale 1:1,500,000.

———, Keller, G. R., and Hildenbrand, T. G., 1982, Bouguer gravity map of the Rio Grande rift: U.S. Geological Survey, Geophysical Series Map GP-949, scale 1:1,000,000.

Davis, G. H., 1979, A gravity study of the San Luis Basin, Colorado: M.S. thesis, University of Texas at El Paso, 101 pp.

Davis, T. L., and Stoughton, D., 1979, Interpretation of seismic reflection data from the northern San Luis Valley, south-central Colorado; *in* Riecker, R. E. (ed.), Rio Grande rift: tectonics and magmatism: American Geophysical Union, Washington, D.C., pp. 185–194.

Gaca, J. R., and Karig, D. E., 1965, Gravity survey in the San Luis Valley area, Colorado: U.S. Geological Survey, Open-file Report (unnumbered).

Keller, G. R., and Cordell, L., 1983, Bouguer gravity anomaly map of New Mexico: NOAA, National Geophysical Data Center, Scientific Map Series, scale 1:500,000.

Kelley, V. C., 1956, The Rio Grande depression from Taos to Santa Fe: New Mexico Geological Society, Guidebook 7, pp. 109–114.

Lance, J. O., 1982, Frequency domain analysis of least-squares polynomial surfaces with application to gravity data in the Pedregosa Basin area: Ph.D. dissertation, University of Texas at El Paso, 190 pp.

Lipman, P. W., and Mehnert, H. H., 1979, The Taos Plateau volcanic field, northern Rio Grande rift, New Mexico; *in* Riecker, R. E. (ed.), Rio Grande rift: tectonics and magmatism: American Geophysical Union, Washington, D.C., pp. 289–311.

Plouff, D., 1977, Preliminary documentation for a Fortran program to compute gravity terrain corrections based on topography digitized on a geographic grid: U.S. Geological Survey, Open-file Report 77-535, 45 pp.

Siebenthal, C. E., 1910, Geology and water resources of the San Luis Valley, Colorado: U.S. Geological Survey, Water-supply Paper 240, 128 pp.

Steven, T. A., 1975, Middle Tertiary volcanic field in the southern Rocky Mountains; *in* Cenozoic history of the southern Rocky Mountains: Geological Society of America, Memoir 144, pp. 75–94.

Tweto, O., 1979, The Rio Grande rift system in Colorado; *in* Riecker, R. E. (ed.), Rio Grande rift: tectonics and magmatism: American Geophysical Union, Washington, D.C., pp. 33–56.

Upson, J. E., 1939, Physiographic subdivisions of the San Luis Valley, southern Colorado: Journal of Geology, v. 47, pp. 721–736.

Wen, C. L., 1983, A study of bolson fill thickness in the southern Rio Grande rift, southern New Mexico, West Texas, and northern Chihuahua: M.S. thesis, University of Texas at El Paso, 74 pp.

Woodward, L. A., and Ingersoll, R. V., 1979, Phanerozoic tectonic setting of Santa Fe country: New Mexico Geological Society, Guidebook 30, pp. 51–57.

Zietz, I., and Kirby, J. R., Jr., 1972, Aeromagnetic map of Colorado: U.S. Geological Survey, Geophysical Series Map GP-880, scale 1:1,000,000.

The southbound train at Española in 1941. The locomotive is a K-28 class, outside frame Mikado, built in 1923 and used on the "Chili Line" after 1933 (photo by Bob Richardson).

STRUCTURAL ANOMALIES IN THE ESPAÑOLA BASIN

BRUCE A. BLACK
Black Oil, Inc., P.O. Box 537, Farmington, New Mexico 87499

INTRODUCTION

Over the past decade, the tectonic basins of the Rio Grande rift have been the focus of an expanding interest in oil and gas exploration. This is primarily due to the recognition of the potential for Cretaceous oil and gas production in these areas. The Cretaceous section has good source rocks and reservoir-quality sands in certain parts of these basins, and this section has probably been buried to levels of organic metamorphism which are high enough for the generation of commercial quantities of oil and gas.

The tectonics of formation of the rift basins have provided abundant potential for structural traps. This structural activity, in conjunction with the stratigraphic variation in the Cretaceous package, makes a double-trap potential in those areas where the rifting has not been severe enough to have destroyed the reservoirs. In some areas, the tectonics have worked against economic oil and gas production by dropping the rocks so deeply that any organic matter has been "overcooked." The last decade of exploration (Black, 1984) has added greatly to the understanding of the rift itself, as well as its tectonic style and the stratigraphy of the contained rocks. Primarily as a result of the recent detailed mapping by a host of excellent workers too numerous to list, as well as the recent oil and gas exploration efforts along the rift, some interesting and previously unsuspected structural styles are now being documented.

DISCUSSION

Of particular interest has been the recognition of probable significant Laramide thrust faulting preserved within the Española Basin (Fig. 1) portion of the rift. In 1976 through 1979, TransOcean Oil Company acquired over a hundred miles of seismic reflection data on certain areas of this basin. Although of relatively good to excellent quality, much of the data originally presented more questions than answers. As a result of early misconceptions and preconceived ideas, much of the data was ignored or misinterpreted. For example, large areas of the basin were assumed to be filled primarily with Tertiary basin fill. Because of such assumptions, these areas of the rift were also assumed to be less (or non-) prospective for oil and gas exploration. In retrospect, some of these areas may have good potential.

The exploration drilling in the southern portions of the Española Basin (Santa Fe embayment) has confirmed the presence of significant thrust faulting in the Cretaceous. In this area, seismic data originally

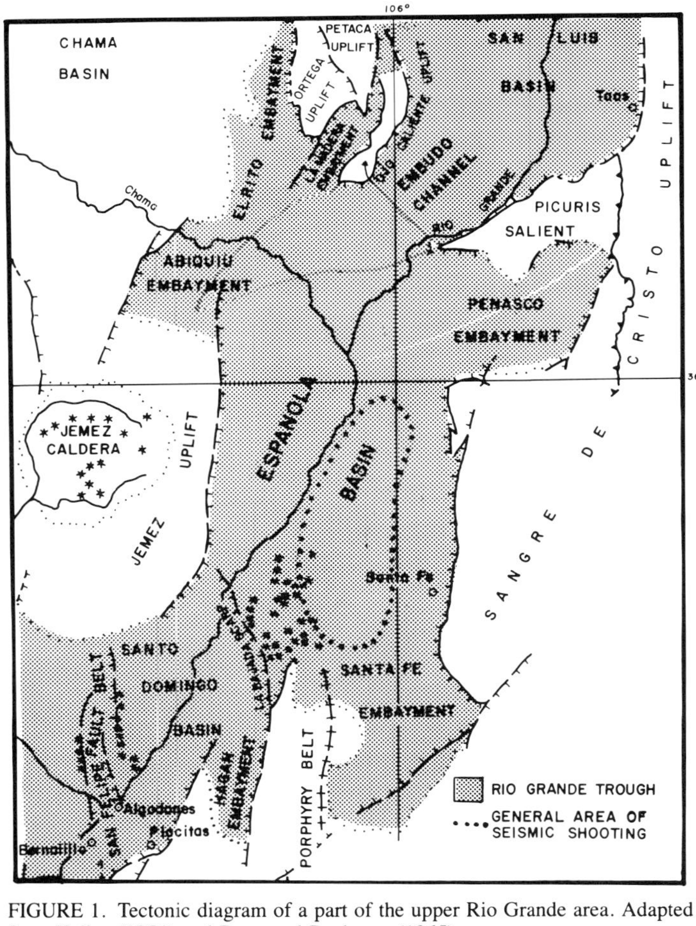

FIGURE 1. Tectonic diagram of a part of the upper Rio Grande area. Adapted from Kelley (1954) and Dane and Bachman (1965).

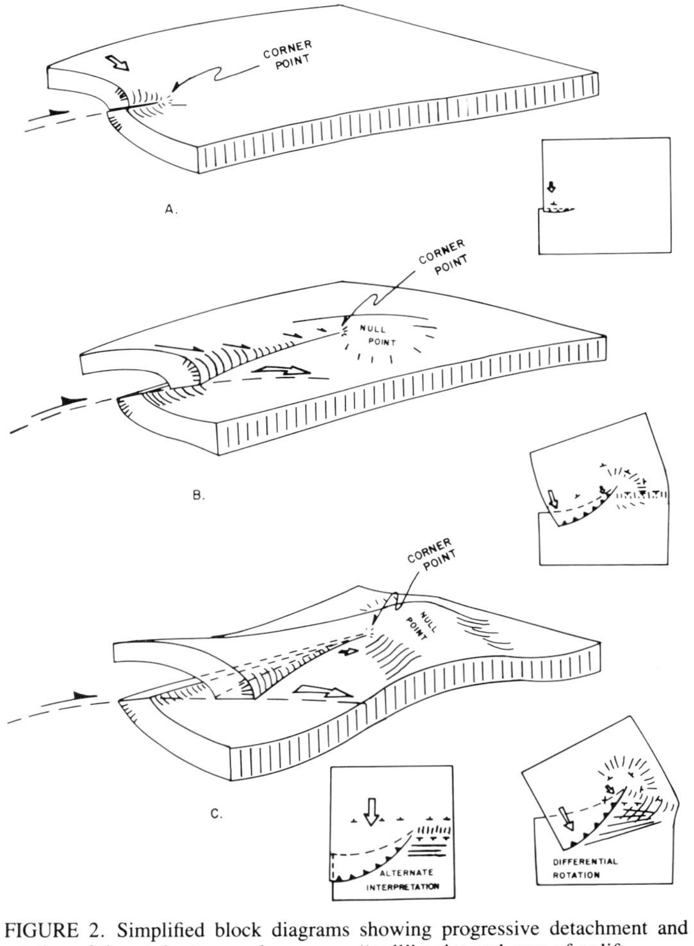

FIGURE 2. Simplified block diagrams showing progressive detachment and rotation of thrust sheet around corner or "null" point and area of uplift.

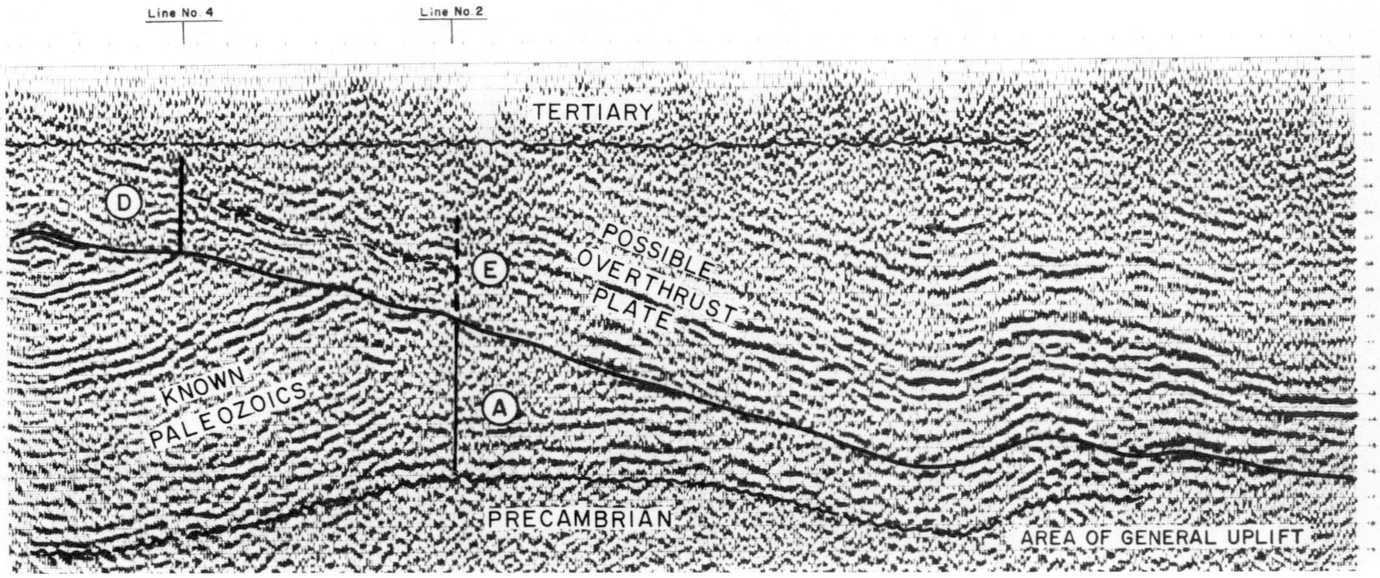

FIGURE 3. Seismic line no. 1.

thought to be "noise" are now recognized as real data. The integration of information on water-wells and core-holes which bottomed in Paleozoic rocks, along with the growing wildcat-well control and a logical reconstruction of the pre-rift geologic history of the basin, strongly suggests that the Laramide thrust faulting (so well demonstrated in the Paleozoic and Precambrian rocks of the Sangre de Cristo Range east of Santa Fe) should also be present and preserved in areas of the rift.

If Laramide thrusting has taken place here, evidence of it should be preserved in the downthrown Paleozoic and Mesozoic sections beneath the Tertiary sands and gravels of the rift grabens. Although thrusting is now confirmed in the Cretaceous section by both seismic and well data, definite evidence of it in the Paleozoic section awaits confirmation by drilling. Seismic data strongly suggest, however, that Paleozoic rocks are involved.

TENTATIVE SEISMIC INTERPRETATIONS

Figure 2 shows progressive detachment and rotation of a thrust sheet around the corner point of a thrust fault.

In the Española Basin, unique documentation of this structural style is suggested in the Paleozoic rocks at the termination of what appear to be thrust faults. In the basin east of Los Alamos, it is possible to run seismic lines across an area where an apparently in-place and unthrusted Paleozoic section becomes detached and is thrust over itself to varying degrees along the fault.

On subsequent crossing and subparallel seismic lines (which form a seismic loop), it is possible to correlate from the lower Paleozoic plate laterally out of the first seismic section onto a perpendicular line which passes into the "corner point" area. We can then correlate parallel to the first seismic line on a third seismic section passing out of the "corner point" onto a fourth section. Correlations on this line can be made back to the original line where one apparently emerges in an upper Paleozoic plate above the original starting place—an obvious mistie. Although the correlations do not tie, they are nonetheless apparently correct, and in this sense the lines in fact do "tie."

Such a geologic puzzle reminds one of the Möbius strip where observers can start on one side of a Möbius loop at any point, and by going once around the loop find themselves on the other side of the loop.

Figures 3 through 6 are portions of seismic lines in the Española Basin. Because of ongoing exploration, the precise location and orientation of these lines cannot be shown. The correlations on the lines are tentative and the conclusions are those of the author only; however, the ramifications of these correlations, if correct, are considerable.

Beginning in known Paleozoic strata on line 1 (Fig. 3) at section A and correlating from this point on a perpendicular seismic line 2 to section B (Fig. 4), we can turn parallel to line 1 on seismic line 3 (Fig. 5) and correlate to section C where line 3 (Fig. 6) crosses line 4. We can then correlate back from section C to the original line 1 along line 4 to section D. If we now correlate back toward point A on line 1, we find ourselves at point E in an "upper" plate—and yet our correlations have seemed reasonable and continuous.

What we may have done, in effect, is to accurately tie our seismic net across low-angle thrust faults and through what at first appeared to be a mistie. In reality, our seismic lines may have been shot across both the thrusts and the "corner" point, and our correlations are actually correct. If so, we are not mistied at all—we are in a sense simply halfway through a seismic Möbius strip—back on the line where we began but on the "other side" (in this case, the top side) of the loop.

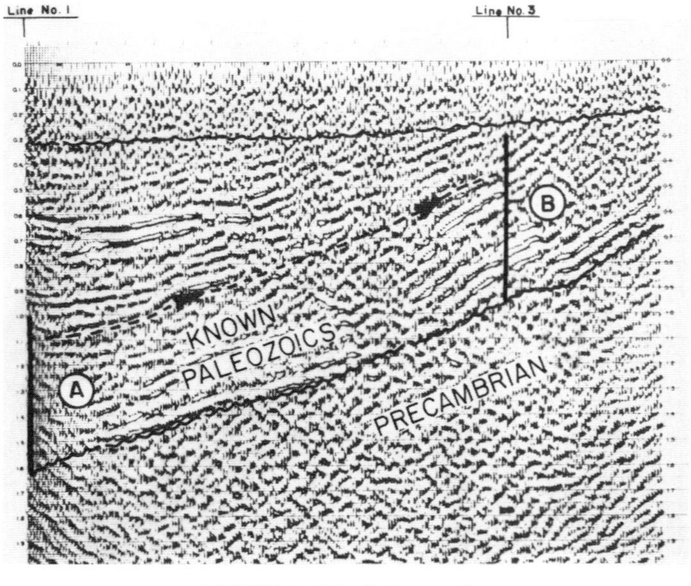

FIGURE 4. Seismic line no. 2.

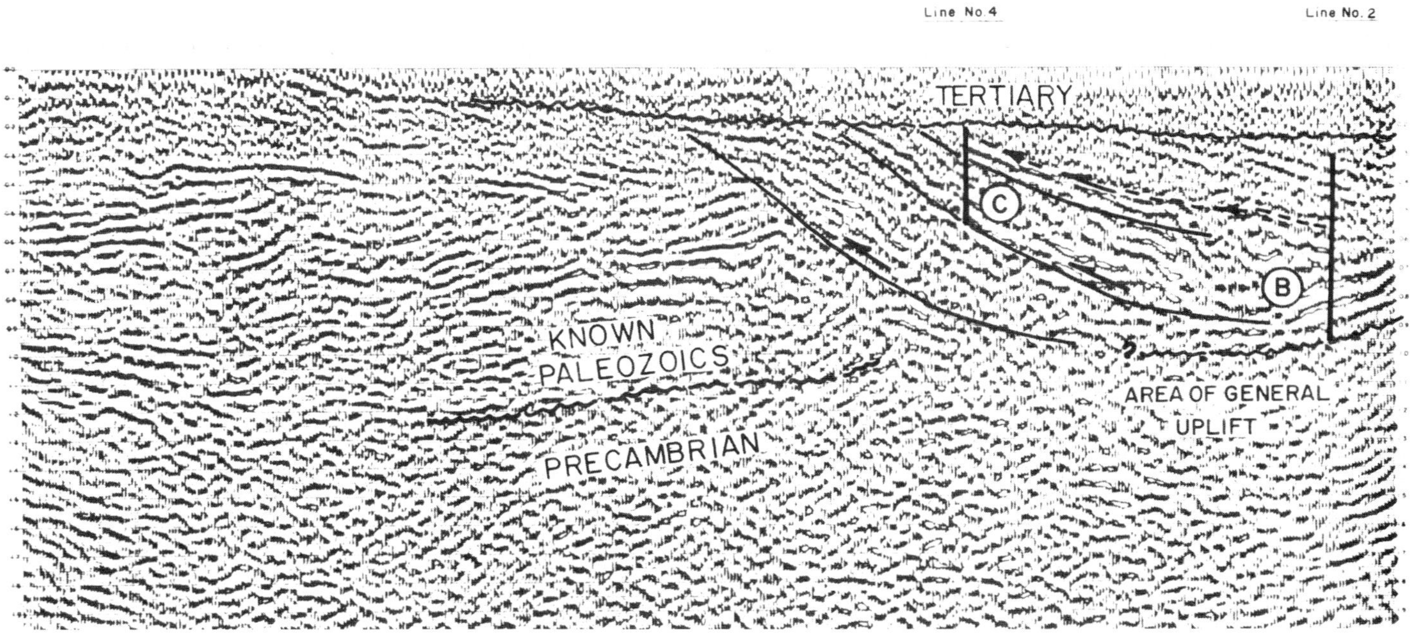

FIGURE 5. Seismic line no. 3.

Figure 7 shows the relative location of the four seismic lines in relation to the thrusting and the "corner point."

Figures 3 through 6 are the actual lines which forced our reconsideration of the sediment package and thus the possible structural configuration in the Paleozoic rocks of the Española Basin. A fifth line (Fig. 8) dramatically shows the nature of the suspected imbricate thrusting as well as part of the corner point, and can also be tied into the seismic net with the same accurate "misties" as demonstrated above. Seismic line 1 in Figure 3 shows the sole thrust well, and Figure 8 shows a splay of imbricate thrusts as the corner point is approached. Note also how the basement and overlying rocks are generally uplifted as one approaches the corner point.

Nowhere within the basins is thrusting observable in outcrop within the Paleozoic section. However, the shallower effects of Laramide thrusting can possibly be seen in outcrop where both Cretaceous and basal Tertiary Galisteo sediments have been folded sharply and thrust-faulted in the area from the town of Galisteo southwestward to at least as far as the old Cash Ranch. Some early workers (Stearns, 1953; Lisenbee, 1976) have reasonably assumed deformation in the Cretaceous in this area had been caused by shale diapirism. Others (Black, 1979; Lisenbee and others, 1979) inferred the tectonic style as being more indicative of compressional tectonics and representing local thrust faulting.

Both seismic and well control now strongly suggest that a major component of early movement was probably due to Laramide compression, although local deformation of the sediments due to compression associated with lateral movements along the San Lazarus, Los Angeles, and Lamy faults may very well be important. This zone which underwent deformation in Precambrian and Paleozoic times (Lisenbee and others, 1979) may have served to focus the effects of Laramide thrusting on the Cretaceous rocks.

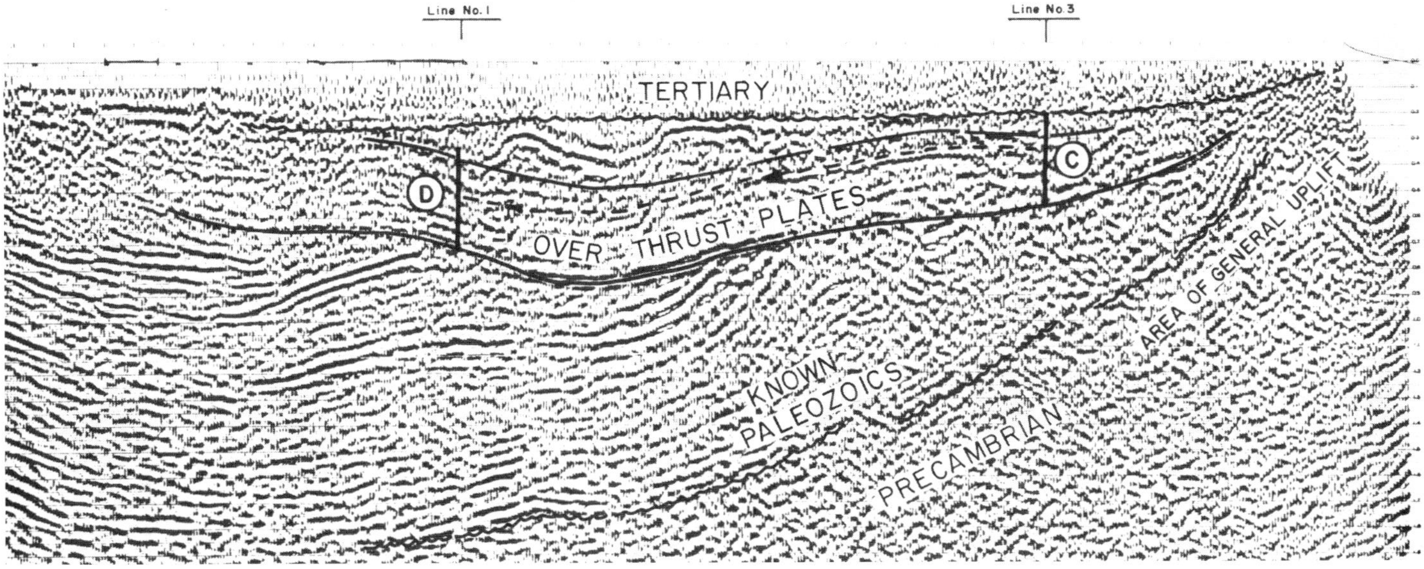

FIGURE 6. Seismic line no. 4.

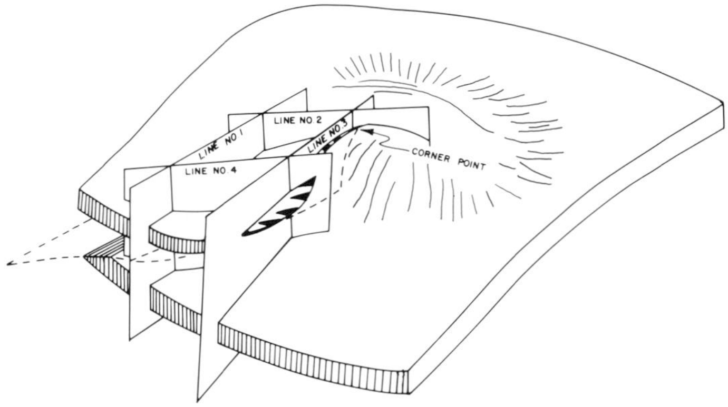

FIGURE 7. Block diagram showing relative location of seismic lines to thrust sheet and corner point.

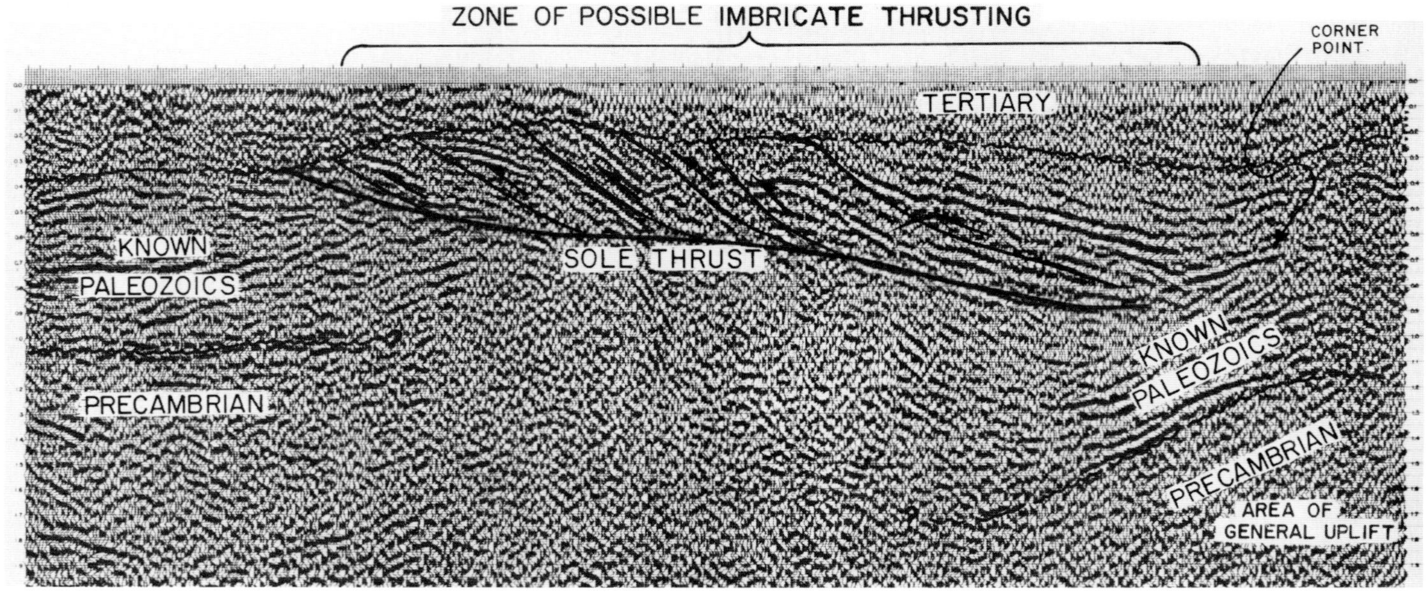

FIGURE 8. Seismic line no. 5. Note that area of general uplift coincides with termination of sole thrusts and imbricate thrusting at the apparent corner point.

CONCLUSIONS

Strata that were originally assumed to be Tertiary basin fill lying on Precambrian rocks may actually consist of imbricately thrust Mesozoic and Paleozoic rocks overlying an undeformed Paleozoic section. The whole Mesozoic and Paleozoic section may have been thrust over itself during Laramide deformation and then covered by a mantle of Tertiary sands and gravels in most of the Española Basin.

ACKNOWLEDGMENT

The author wishes to thank Mobil Texas–New Mexico Producing Company for gracious consent to use portions of the seismic lines shown in this paper. Their generosity and help in making these data available has been invaluable in attempting to explain some of the features of the Española Basin.

REFERENCES

Black, B. A., 1984, Oil and gas exploration in the Rio Grande rift; *in* Oil and gas fields of New Mexico: Four Corners Geological Society, v. III, in press.

Dane, C. H., and Bachman, G. O., 1965, Geologic map of New Mexico: U.S. Geological Survey.

Kelley, V. C., 1954, Tectonic map of a part of the upper Rio Grande area, New Mexico: U.S. Geological Survey, Oil and Gas Investigation Map OM-157.

Lisenbee, A. L., 1976, Shale diapirism and the structural development of Galisteo syncline, Santa Fe County, New Mexico: New Mexico Geological Society, Special Publication 6, pp. 88–94.

———, Woodward, L. A., Connolly, J. R., 1979, Tijeras–Canoncito fault system—a major zone of recurrent movement in north-central New Mexico: New Mexico Geological Society, Guidebook 30, pp. 89–99.

Stearns, C. E., 1953, Tertiary geology of the Galisteo–Tongue area, New Mexico: Geological Society of America, Bulletin, v. 64, pp. 459–508.

BRIEF SUMMARY OF THE TERTIARY GEOLOGIC HISTORY OF THE RIO GRANDE RIFT IN NORTHERN NEW MEXICO

KIM MANLEY
Consulting Geologist, 6133 East 97th Street, Tulsa, Oklahoma 74137

INTRODUCTION

The Rio Grande rift of Colorado and New Mexico can be divided into a series of structural basins. Within one of these basins (the San Luis Valley), from the New Mexico state line southward, is the Taos Plateau (Fig. 1). Adjoining it to the south is the Española Basin. The boundary between is marked by the termination of the Taos Plateau volcanic field and the presence of a basement ridge (Cordell, 1978) crossing the rift from the Picuris Range, through the Precambrian exposure at Cerro Azul, to the La Madera area (Fig. 2).

This paper presents an overview of the Tertiary history of the Taos Plateau, the northernmost Española Basin, and parts of the adjacent Tusas Mountains to the west and the Sangre de Cristo Mountains to the east.

STRUCTURE

In the Taos Plateau, the Rio Grande rift is roughly 35 km wide. The east side of the rift is defined by a major fault system along the base of the Sangre de Cristo Mountains that has a possible offset of 7–8 km (Lipman and Mehnert, 1979). The western boundary is formed by a series of north–northwest-trending normal faults that bound eastward-dipping blocks along the eastern half of the Tusas Mountains (Manley, 1982a, b). Trending parallel to the Sangre de Cristo Mountains, in the eastern half of the rift valley, are an intra-rift horst and graben. The horst brings Tertiary volcanic rocks to the surface. Pliocene basalts show about 500 m of post-emplacement offset along the eastern boundary (Lipman and Mehnert, 1979), and overlying Quaternary deposits have been faulted 17 m in the late Pleistocene (Personius and Machette, this guidebook).

The Española Basin is about 65 km wide. On the northeast it apparently lacks a prominent border fault; the westward-tilted blocks on the east side are bounded by down-to-the-east normal faults (Manley, 1979). The northwestern margin is formed by a major northeast-trending fault zone that places Tertiary rocks against rocks of Permian age. Displacement of 8 m.y. old basalts is about 670 m (Manley and Mehnert, 1981).

Located within the north-central Española Basin is a roughly 8 km wide graben, the Velarde graben. The western side of the graben is formed by a fault passing southwestward from beneath Black Mesa, across the Rio Grande, and along the southeast side of Arroyo de la Presa into the area of Clara Peak where it turns southward to become the Pajarito fault. The eastern fault border is visible from Velarde southwestward until it is lost beneath the Rio Grande floodplain northeast of Española (Manley, 1977). Gravity data indicate a total thickness of 2.5–5 km of basin fill in this area (Cordell, 1979). Pliocene subsidence of the graben is at least 300 m (Manley, 1979).

The boundary between the San Luis Valley and the Española Basin is complex. The two basins have opposite directions of major tilting, and the intra-rift grabens of each are approximately 19 km apart in a right-lateral direction. The basement ridge has a possible 3 km of structural relief (Cordell, 1979). The Velarde graben pinches out northeastward; its bounding faults appear to join each other and merge into the Embudo fault along the northwest flanks of the Picuris Range where reverse and left strike–slip motion are reported (Muehlberger, 1979). Kelley (1978) indicates a reversal of throw on the western side of the graben near the confluence of the Chama and Ojo Caliente Rivers; however, my detailed mapping in the area contradicts this interpretation. I concur with Muehlberger (1979) that the point of reversal probably overlies the basement ridge which separates the basins. Other relationships in this boundary area require further investigation.

STRATIGRAPHY

Tertiary sedimentary and volcanic rocks overlie Precambrian metamorphic rocks to the east and west of the Taos Plateau. The Precambrian rocks, uplifted in the Laramide orogeny, were almost totally stripped of later deposits. Paleozoic and Mesozoic rocks are present only locally

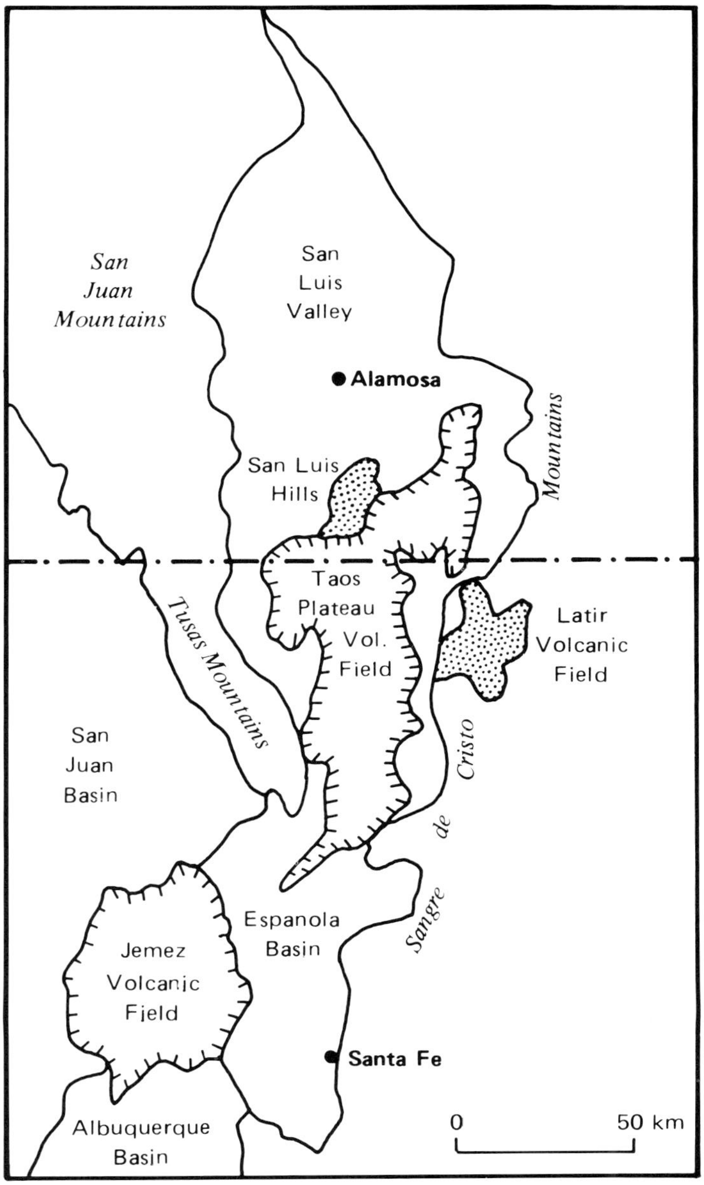

FIGURE 1. The Rio Grande rift in southern Colorado and northern New Mexico.

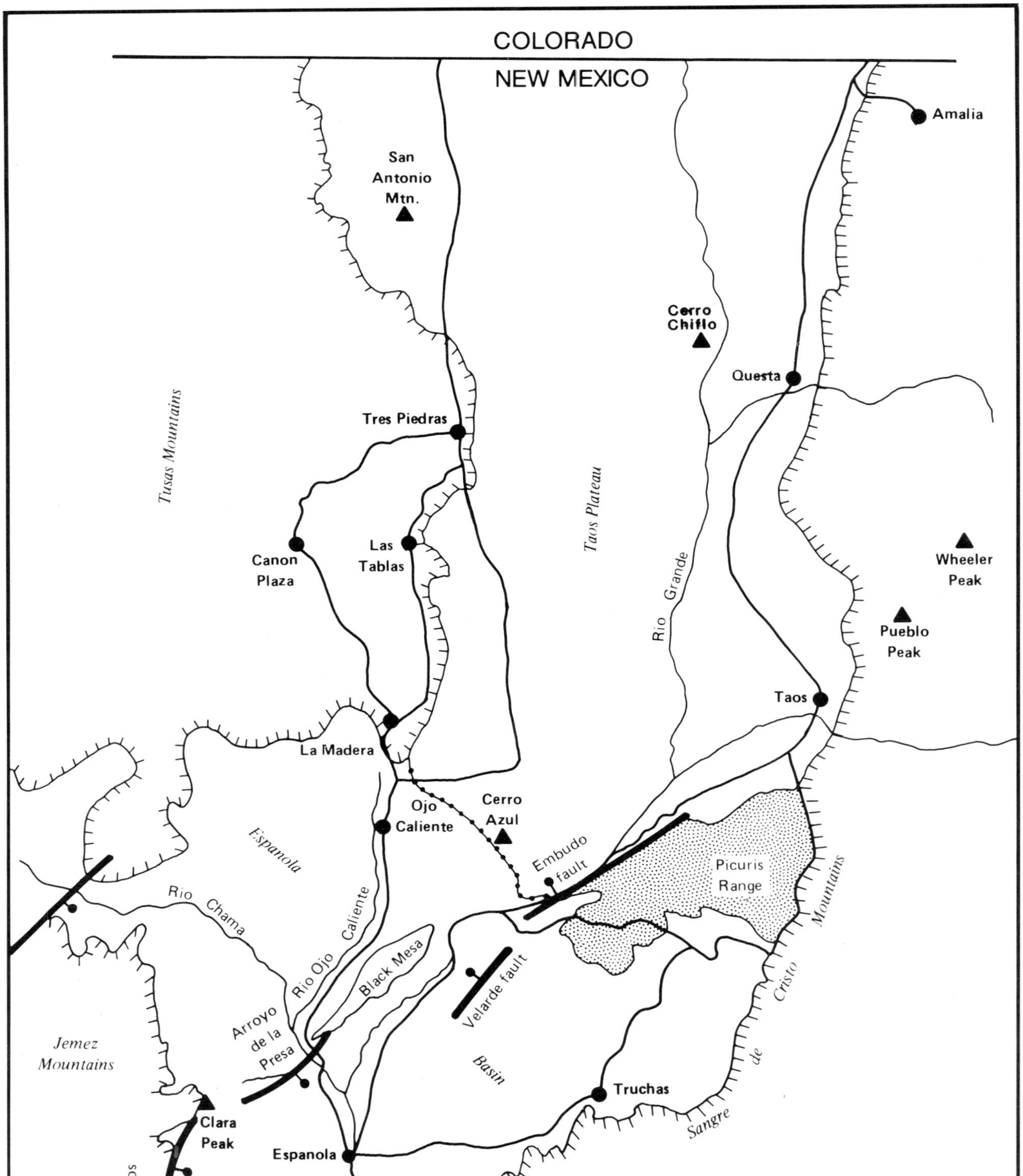

FIGURE 2. Index map for the Rio Grande rift area in north-central New Mexico.

in the westernmost and southernmost Tusas Mountains (Muehlberger, 1968; Bingler, 1968). Mississippian and Pennsylvanian rocks are exposed in the Sangre de Cristo Mountains south of Pueblo Peak (Armstrong, 1967; Dane and Bachman, 1965); structural relationships are imperfectly known.

The earliest Tertiary deposits on both sides of the rift are discontinuous coarse conglomerates and sandstones that were derived from the underlying Precambrian rocks. These units, the Eocene El Rito Formation of the Tusas Mountains and the Eocene(?) Vallejo Formation of the Sangre de Cristo Mountains, filled valleys in areas of high relief and may be locally as much as 250 m thick (Manley and Wobus, 1982a). Both are assigned an Eocene age, although no paleontological evidence is available.

Pre-rift volcanism began in the Tusas Mountains with intermediate-composition breccias that erupted from numerous local centers. In the north these are equivalent to the Conejos Formation, whose age is 35 to 31 m.y. (Lipman and others, 1970). South, near La Madera, these breccias may be as young as 26 m.y. (Manley and Wobus, 1982a). Local eruptions of intermediate-composition lavas and breccias also occurred in the Sangre de Cristo Mountains between 35 and 26 m.y. (Lipman, 1983) and in the San Luis Hills (northern Taos Plateau in southern Colorado) prior to 28 m.y. (Burroughs, 1971).

The volcanic breccias, Eocene sedimentary rocks, and remnant Precambrian hills were soon buried by an influx of volcaniclastic detritus. In the Taos Plateau and the Latir Mountains, the detritus was locally derived. Elsewhere only a small percentage of the material had local sources: (1) the northern and central Tusas Mountains were covered by alluvial fans building southward from the volcanic highlands of the San Juan Mountains; (2) in the central and southern Tusas Mountains and northern Española Basin, detritus was primarily derived from the early Taos Plateau and Latir volcanic fields. The alluvial material comprises the Los Pinos and Abiquiu Formations (Butler, 1971), and a part (Manley, unpubl. data 1980) of the Picuris Formation of Miller (Miller and others, 1963). The Picuris Range did not create a barrier to northern-derived volcaniclastic sediment in the northeastern Española Basin because it was not in existence at this time.

The age of volcaniclastic deposition is fairly well known. The Los Pinos overlies the 30 m.y. old Treasure Mountain Tuff and the 28 m.y. old Masonic Park Tuff (Lipman, 1975) in the northern and central Tusas Mountains (Manley, 1982a, b; Wobus and Manley, 1982b). The termination time for Los Pinos deposition is highly variable: in the northern Tusas Mountains it continued to 5 m.y. (Lipman and Mehnert, 1975); in the northwestern Española Basin the transitional and interfingering upper contact appears to be 16–18 m.y. (Baldridge and others, 1980; May, 1979 and this guidebook; Manley, 1981; Manley and Mehnert, 1981; Ekas and others, this guidebook). On the east of the Taos Plateau, comparable volcaniclastic deposition ranged from about 22 to 16 m.y. (Lipman, 1983).

Sedimentary rocks of the Santa Fe Group were deposited throughout most of the area. Early deposits are volcaniclastic (Galusha and Blick, 1971; May, this guidebook) but in the northern Española Basin are distinguishable from the underlying Los Pinos and Abiquiu Formations because of the difference in composition and/or percentage of volcanic clasts. In the northwestern Española Basin, the oldest Santa Fe Group deposit, the Chama–El Rito Member of the Tesuque Formation, contains locally derived volcanic clasts (May, 1979; Manley, unpubl. data 1980). As vertical tectonism increased in the rift, Precambrian source areas became re-exposed and their contribution to the Santa Fe Group sediments came to dominate the fans spreading westward from the southern and central Sangre de Cristo Mountains. In much of the Española Basin these fans built across the rift, but they barely extended into the Tusas Mountains due to the eastward-tilting of the Taos Plateau. A similar shift from dominantly volcaniclastic deposition to Precambrian clasts is shown in the sedimentary rocks of the northeast edge of the rift (Lipman, 1983).

Volcanism continued in the area contemporaneously with the deposition of the Los Pinos and Abiquiu Formations. Within the Taos Plateau, rhyolite (including ash-flow tuff), rhyodacites, and andesites were erupted roughly 26 to 22 m.y. ago (Lipman and Mehnert, 1979). Similar eruptions occurred in the Sangre de Cristo Mountains and include the Tuff of Tetilla Peak (27 m.y.) and the Amalia Tuff (26 m.y.) (Lipman, 1983). Rhyolite domes (Manley and Wobus, 1982) and the Amalia Tuff, with a K–Ar age of 26 m.y. (Bingler, 1968), are interlayered with the Los Pinos Formation in the southern Tusas Mountains (Butler, 1971; Manley, 1981).

Volumentrically minor basalt flows are interlayered with the Los Pinos Formation; many now cap eastward-tilted mesas in the northern Tusas Mountains. These flows have been variously named in the past (Butler, 1971), but all of them are currently considered to belong to the Hinsdale Formation (Lipman and Mehnert, 1975). In southern Colorado these flows have a K–Ar age range of 26 to 5 m.y. (Lipman and Mehnert, 1975). In the northern Española Basin, K–Ar ages on three flows are 19, 21, and 22 m.y. (Baldridge and others, 1980; Manley and Mehnert, 1981).

The older Santa Fe Group deposits in the northwestern Española Basin contain a few scattered volcanic flows, dikes, and tuff rings. These were sporadically erupted from 13 to 7 m.y. ago (Baldridge and others, 1980; Manley and Mehnert, 1981). Included in this group is a quartz-latitic dome, Cerro Chiflo, in the central part of the Taos Plateau, which has an age of 10 m.y. (Lipman and Mehnert, 1979).

Major basaltic volcanism, beginning within the Rio Grande rift about 5 m.y. ago, constitutes the Taos Plateau volcanic field. This field covers more than 1,500 km^2 and consists of numerous volcanoes, flows, and several domes of compositions ranging from basalt to silicic rhyolite (Lipman and Mehnert, 1979; Dungan and others, this guidebook). The most voluminous rocks are the olivine tholeiites of the Servilleta Basalt whose K–Ar ages range from 4.5 to 2.8 m.y. (Ozima and others, 1967; Manley, 1978), but may prove to contain several older flows (Baldridge and others, 1980). The Servilleta Basalt flows were erupted from shield volcanoes to the west-central part of the Plateau.

TECTONISM

In early Miocene the Rio Grande rift was no more than a slight depression. Sedimentary rocks were deposited, from highlands to the north and northeast, across the locations of the current boundary faults. The northwest boundary faults in the Española Basin were active between 7 and 3 m.y. (Manley and Mehnert, 1981). In the Tusas Mountains the age control is poor; only Miocene Los Pinos and Hinsdale Formations are commonly present. Along the Rio Tusas valley Santa Fe Group deposits are faulted, but their age is unknown (Manley and Wobus, 1982b).

In both the Española Basin and Taos Plateau the central grabens may have formed slightly later than the outer fault boundaries. Creation of the Velarde graben is constrained between 5 and 3 m.y. (Manley, 1979); the Taos graben is less than 4.5 m.y. (Dungan and others, this guidebook). During this interval of tectonism the Picuris Range and the Sangre de Cristo Mountains were uplifted rapidly.

Tectonism and volcanism appear to have slowed by the end of the Pliocene. The stage was set for the Quaternary erosional cycles that led to the creation of the present landscape.

REFERENCES

Armstrong, A. K., 1967, Biostratigraphy and carbonate facies of the Mississippian Arroyo Peñasco Formation, north-central New Mexico: New Mexico Bureau of Mines and Mineral Resources, Memoir 20, 79 pp.

Baldridge, W. S., Damon, P. E., Shafiqullah, M., and Bridwell, R. J., 1980, Evolution of the Rio Grande rift, New Mexico—new potassium–argon ages: Earth and Planetary Science Letters, v. 51, pp. 309–321.

Bingler, E. C., 1968, Geology and mineral resources of Rio Arriba County, New Mexico: New Mexico Bureau of Mines and Mineral Resources, Bulletin 91, 158 pp.

Burroughs, R. L., 1971, Geology of the San Luis Hills, south-central Colorado: New Mexico Geological Society, Guidebook 22, pp. 227–288.

Butler, A. P., Jr., 1971, Tertiary volcanic stratigraphy of the eastern Tusas Mountains, southwest of the San Luis Valley, Colorado–New Mexico: New Mexico Geological Society, Guidebook 22, pp. 289–300.

Cordell, L., 1978, Regional geophysical setting of the Rio Grande rift: Geological Society of America, Bulletin, v. 89, pp. 1073–1090.

———, 1979, Gravimetric expression of graben faulting in Santa Fe country

and the Española Basin, New Mexico: New Mexico Geological Society, Guidebook 30, pp. 59–64.
Dane, C. H., and Bachman, G. O., 1965, Geologic map of New Mexico: U.S. Geological Survey.
Ekas, L. M., Ingersoll, R. V., Baldridge, W. S., and Shafiqullah, M., 1984, The Chama–El Rito Member of the Tesuque Formation, Española Basin, New Mexico: This guidebook.
Galusha, T., and Blick, J. C., 1971, Stratigraphy of the Santa Fe Group, New Mexico: American Museum of Natural History, Bulletin 144, 127 pp.
Kelley, V. C., 1978, Geology of Española Basin, New Mexico: New Mexico Bureau of Mines and Mineral Resources, Geologic Map 48.
Lipman, P. W., 1975, Evolution of the Platoro caldera complex and related volcanic rocks, southeastern San Juan Mountains, Colorado: U.S. Geological Survey, Professional Paper 852, 128 pp.
———, 1983, The Miocene Questa caldera, northern New Mexico—relation to batholith emplacement and associated molybdenum mineralization; *in* The genesis of Rocky Mountain ore deposits—changes with time and tectonics: Society of Exploration Geologists, Denver Region, Symposium Proceedings, pp. 133–148.
———, and Mehnert, H. H., 1975, Late Cenozoic basaltic volcanism and development of the Rio Grande depression in the southern Rocky Mountains: Geological Society of America, Memoir 144, pp. 119–154.
———, and ———, 1979, The Taos Plateau volcanic field, northern Rio Grande rift, New Mexico; *in* Riecker, R. E. (ed.), Rio Grande rift—tectonics and magmatism: American Geophysical Union, Washington, D.C., pp. 289–311.
———, Steven, T. A., and Mehnert, H. H., 1970, Volcanic history of the San Juan Mountains, Colorado, as indicated by potassium–argon dating: Geological Society of America, Bulletin 81, pp. 2329–2352.
Manley, K., 1977, Geologic map of the northeastern Española Basin, New Mexico: U.S. Geological Survey, Miscellaneous Field Studies Map MF-877.
———, 1978, Structure and stratigraphy of the Española Basin, Rio Grande rift, New Mexico: U.S. Geological Survey, Open-file Report 78-667, 24 pp.
———, 1979, Stratigraphy and structure of the Española Basin, Rio Grande rift, New Mexico: *in* Riecker, R. E. (ed.), Rio Grande rift— tectonics and magmatism: American Geophysical Union, Washington, D.C., pp. 71–86.
———, 1981, Redefinition and description of the Los Pinos Formation of north-central New Mexico: Geological Society of America, Bulletin 92, pp. 984–989.
———, 1982a, Geologic map of the Broke Off Mountain quadrangle, Rio Arriba County, New Mexico: U.S. Geological Survey, Miscellaneous Field Studies Map MF-1450.
———, 1982b, Geologic map of the Bighorn Peak quadrangle, Rio Arriba County, New Mexico, and Conejos County, Colorado: U.S. Geological Survey, Miscellaneous Field Studies Map MF-1451.
———, and Mehnert, H. H., 1981, New K–Ar ages for Miocene and Pliocene volcanic rocks in the northwestern Española Basin and their relationships to the history of the Rio Grande rift: Isochron/West 30, pp. 5–8.
———, and Wobus, R. A., 1982a, Reconnaissance geologic map of the Las Tablas quadrangle, Rio Arriba County, New Mexico: U.S. Geological Survey, Miscellaneous Field Studies Map MF-1408.
———, and ———, 1982b, Reconnaissance geologic map of the Mule Canyon quadrangle, New Mexico: U.S. Geological Survey, Miscellaneous Field Studies Map MF-1407.
May, S. J., 1979, Neogene stratigraphy and structure of the Ojo Caliente–Rio Chama area, Española Basin, New Mexico: New Mexico Geological Society, Guidebook 30, pp. 83–88.
Miller, J. P., Montgomery, A., and Sutherland, P. K., 1963, Geology of part of the southern Sangre de Cristo Mountains, New Mexico: New Mexico Bureau of Mines and Mineral Resources, Memoir 11, 106 pp.
Muehlberger, W. R., 1968, Geology of the Brazos Peak quadrangle, New Mexico: New Mexico Bureau of Mines and Mineral Resources, Geologic Map 22.
———, 1979, The Embudo fault between Pilar and Arroyo Hondo, New Mexico—an active intracontinental transform fault: New Mexico Geological Society, Guidebook 30, pp. 77–82.
Ozima, M., Kono, M., Kaneoka, I., and others, 1967, Paleomagnetism and potassium–argon ages of some volcanic rocks from the Rio Grande gorge, New Mexico: Journal of Geophysical Research 72, pp. 2615–2622.
Personius, S. F., and Machette, M. N., 1984, Quaternary faulting in the Taos Plateau region, northern New Mexico: This guidebook.
Wobus, R. A., and Manley, K., 1982, Reconnaissance geologic map of the Burned Mountain quadrangle, Rio Arriba County, New Mexico: U.S. Geological Survey, Miscellaneous Field Studies Map MF-1409.

TECTONIC HISTORY OF THE NORTHERN RIO GRANDE RIFT DERIVED FROM APATITE FISSION-TRACK GEOCHRONOLOGY

SHARI A. KELLEY and IAN J. DUNCAN
Department of Geological Sciences, Southern Methodist University, Dallas, Texas 75275

INTRODUCTION

This study is concerned with the uplift history of the mountain ranges that bound the eastern margin of the Rio Grande rift in northern New Mexico (Fig. 1). The principal objectives of this paper are to present the results of fission-track dating of granitic rocks from three ranges on the eastern margin of the rift (Fig. 1), and to use these dates to infer the approximate timing and the rate of uplift and erosion in each range. These data are then used to place constraints on the paleogeographic evolution of northern New Mexico between the Late Cretaceous and the early Miocene. The paleogeography shows that the ranges on the eastern margin of the early rift were not as continuous along the length of the rift as the ranges observed today.

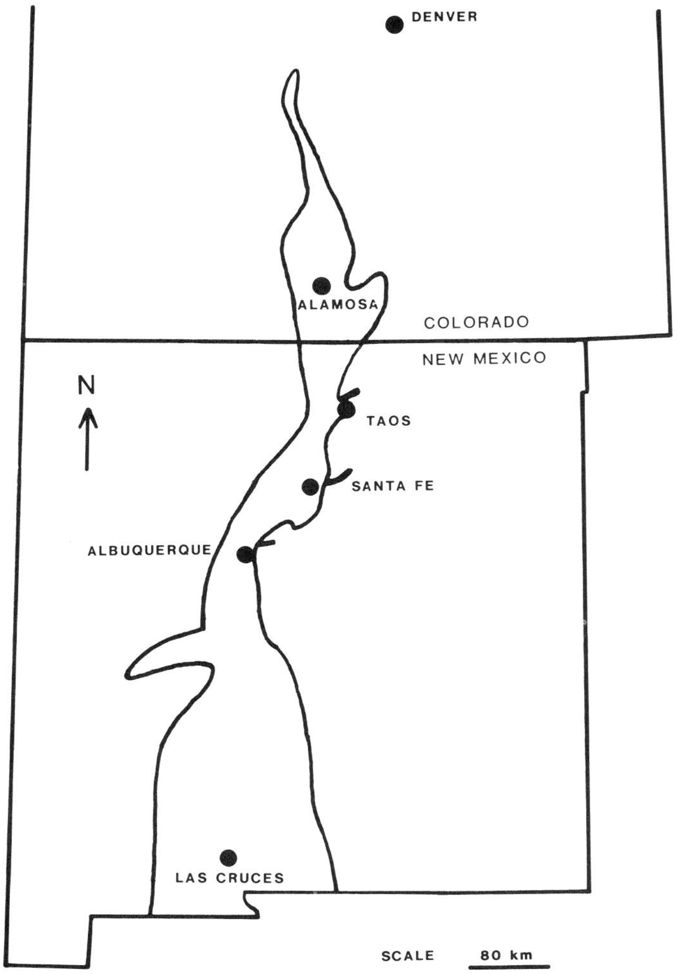

FIGURE 1. Location of fission-track-dating traverses in the Sandia Mountains near Albuquerque, in the Sangre de Cristo Mountains near Santa Fe (Santa Fe Range), and in the Sangre de Cristo Mountains near Taos (Taos Range). Rio Grande rift outlined by heavy, solid lines.

INTERPRETATION OF FISSION-TRACK DATES

Fission tracks in apatite anneal when exposed to relatively low temperatures for times on the order of millions of years (Naeser, 1979). When apatite cools from high temperatures during uplift and erosion, it passes through a temperature range where fission tracks are both annealed and retained (zone of partial stability). The temperature at which tracks effectively accumulate is termed the closure temperature (Haack, 1977). From experimental and field relations (Harrison and others, 1979), the effective closure temperature for apatite is estimated to be $100 + 25$°C for cooling acting over 1–10 m.y.

The approximate timing and rates of uplift and erosion are derived from measurement of the variation in apatite fission-track age as a function of elevation in mountain ranges. Since the rocks at higher elevations in a modern mountain range cooled before the rocks now at lower elevations, the apatite ages at the higher elevations are older. The ages at the higher elevations can be used as a minimum estimate of the time of initiation of uplift, and the difference between the ages of the rocks at the higher and lower elevations can be used to calculate an apparent uplift rate. Parrish (1983) defines apparent uplift rate as "the rate at which the critical isotherm (100°C) moves downward with respect to the rock column." The apparent uplift rate equals the uplift–erosion rate only when (1) the critical isotherms remain horizontal, (2) the geothermal gradient remains constant, and (3) uplift equals erosion (Parrish, 1983).

The recent (less than 15 m.y. old) apparent uplift rate can be estimated in the following manner. The approximate present elevation of the 100°C isotherm (and hence zero age) can be found from heat-flow data using the thermal-resistance method of Bullard (1939), given a reasonable estimate of surface temperature and assuming that the thermal conductivity decreases 10% in the 0–100°C range (Birch and Clark, 1940). The apparent uplift rate is calculated from the difference between the zero age recorded at this elevation and the fission-track age at the lowest exposed elevation in the range.

INTERPRETATION OF TRACK-LENGTH MEASUREMENTS

A problem often encountered in interpreting fission-track ages is deciding whether a fission-track age represents the time that a rock slowly cooled through the partial stability zone (Fig. 2a), or whether the age has been partially reset during a recent thermal event (Fig. 2b). Examination of the fission-track-length distribution (Lal and others, 1968) can sometimes be used to distinguish between the thermal histories depicted in Figure 2.

The expected track-length distributions for the two thermal histories are shown in Figure 2. The slow-cooling case is characterized by a unimodal histogram skewed toward the longer tracks (Fig. 2c), while the thermal-event case is characterized by a bimodal histogram (Fig. 2d).

RESULTS

Sandia Mountains

The Sandia Mountains are an east-tilted fault-block composed of Precambrian granitic and metasedimentary rocks overlain by Mississippian to Cretaceous sediments. Samples for dating were taken on the west side of the range from the Sandia granite (1.44 + 0.4 b.y.; Brookins and Majumdar, 1982) and from aplitic dikes that cut the granite.

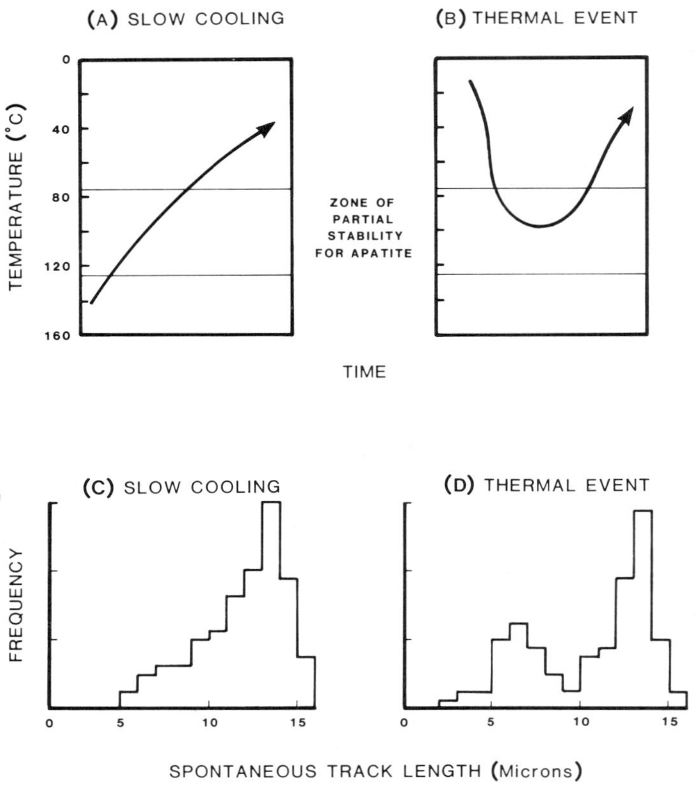

FIGURE 2. Time-temperature curves for **(A)** slow cooling and **(B)** thermal event. Expected track-length distributions for **(C)** slow cooling and **(D)** thermal event (after Gleadow and Duddy, 1982).

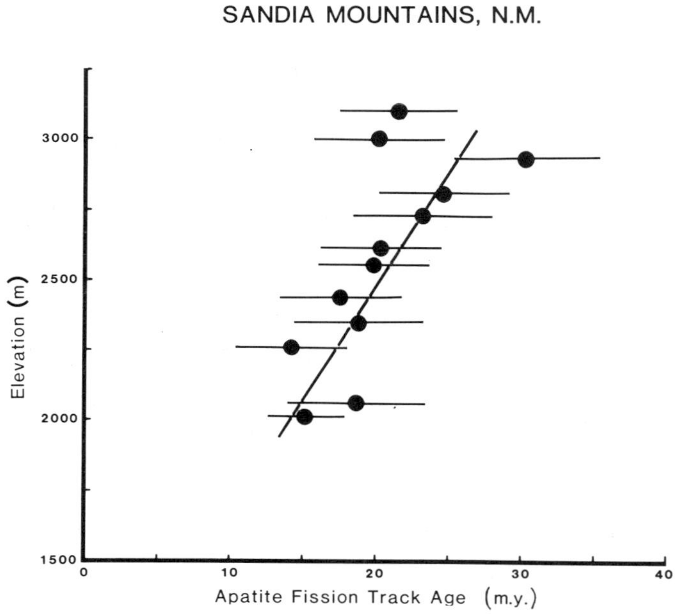

FIGURE 3. Apatite fission-track age vs. elevation for Sandia Mountains. Error bars represent two standard deviations. Solid line represents a least-squares fit to the data, yielding an apparent uplift rate of 81 m/Ma.

The apatite ages for this area (Fig. 3) show a clear relation with elevation, with an apparent uplift rate of 81 m/m.y. The track-length distributions (Fig. 4) suggest that the two highest points on the profile have been partially reset by thermal effects related to mineralization at the La Luz mine sometime in the late Tertiary.

The absence of latest Cretaceous and Paleocene rocks to the east and northeast of the range (Kelley and Northrop, 1975; Gorham and Ingersoll, 1979) implies that the Sandia region probably attained positive relief during the Laramide event. During the Eocene, the area was eroded to low relief; the Sandia Mountains apparently did not act as a source area for the Eocene Galisteo Formation, which crops out north and east of the range (Kelley and Northrop, 1975; Gorham and Ingersoll, 1979). The apatite fission-track age of 30 m.y. indicates that the Sandia granite was cooling during uplift and erosion in the Oligocene, at about the time that volcanism was active in the Cerrillos–Ortiz field 19–38 m.y. ago (Baldridge and others, 1980). Stearns (1953) pointed out that the Abiquiu(?) Formation is missing in the northern Sandia region, implying that the northern part of the range was uplifted in pre-Abiquiu(?) time (before 19–25 m.y. ago; Baldridge and others, 1980). This evidence is consistent with the fission-track-age data.

Two samples were collected to the south of the dating traverse at low elevations along the front of the range. The ages become progressively older to the south, suggesting that uplift and erosion during the early Tertiary were more active in the northern part of the range than they were to the south.

Santa Fe Range (Sangre de Cristo Mountains)

The Santa Fe Range is composed primarily of the Precambrian Embudo Granite (coarse-grained biotite granite, granitic gneiss, and pegmatitic leucogranite; Miller and others, 1963), which forms a large, composite batholith that intrudes older schists, gneisses, and amphibolites. Rb–Sr and K–Ar ages ranging between 1,235 and 1,673 m.y. have been determined for the Embudo Granite in the Picuris Range and near Truchas Peaks, north of the fission-track-dating traverse (Register and Brookins, 1979).

In contrast to the post-Eocene fission-track ages observed in the other areas examined during this study, the apatite ages for this profile (Fig. 5) range from Late Cretaceous to early Eocene, reflecting Laramide uplift. The apparent uplift rate for the region during the Laramide event is 125 m/m.y.

Uplift of the Santa Fe Range during the Laramide event coincides with the deposition of the Paleocene Nacimiento and Eocene San Jose (Cub Mesa Member) Formations in the San Juan Basin (Baltz, 1967), and the Eocene Galisteo Formation in the Galisteo Basin (Gorham and Ingersoll, 1979). During the Eocene, the Santa Fe Range was eroded to low relief (Baltz, 1978). No sediments were derived from the Santa Fe Range until renewed uplift occurred in the late Miocene, when the area acted as the source for the sediments in the Santa Fe Group (Galusha and Blick, 1971).

Taos Range (Sangre de Cristo Mountains)

Precambrian rocks that crop out in the Wheeler Peak–Gold Hill area in the Taos Range include tonalites, diorites, amphibolites, felsic volcanics and porphyries, and granites (Condie and McCrink, 1982). The U–Pb zircon dates for the tonalites and diorites are on the order of 1.75 b.y. (Condie and McCrink, 1982). The Precambrian rocks have been intruded by Tertiary granites related to the Latir volcanic field and the Questa caldera (Lipman, 1981). The 25.8–27.8-m.y.-old Rio Hondo pluton and the 22-m.y.-old Lucero stock (Rb–Sr dates from Hagstrum and Lipman, 1983) are exposed in the Wheeler Peak–Gold Hill area. Naeser (written comm. 1983) has determined apatite fission-track dates for both intrusions (Fig. 6, open symbols).

Fission-track samples for this study were collected from Precambrian outcrops in the vicinity of Wheeler Peak and from the Tertiary Rio Hondo pluton. The resulting ages (Fig. 6, closed symbols) correlate well with elevation, yielding an apparent uplift rate of 173 m/m.y.

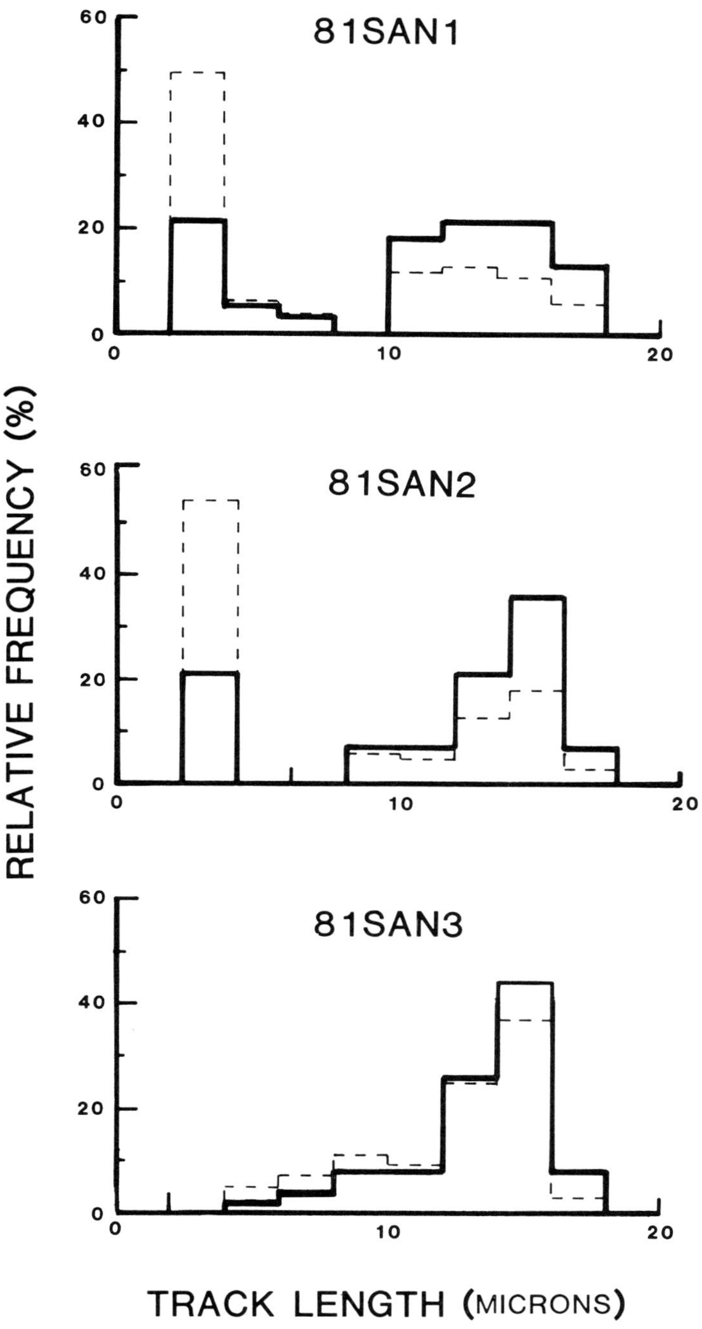

FIGURE 4. Track-length distributions for the three highest samples of the Sandia Mountain traverse (81SAN1, 3,095 m; 81SAN 3, 2,933 m). Solid line represents actual measurements, while dashed line represents the theoretical histogram corrected for measurement bias (Laslett and others, 1982).

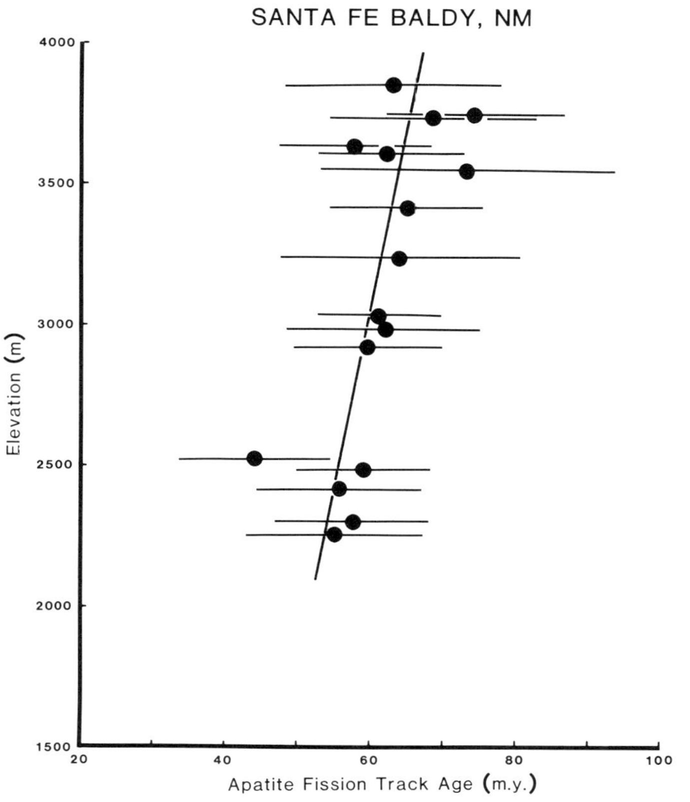

FIGURE 5. Apatite fission-track age vs. elevation for the Santa Fe Range. The least-squares fit gives an apparent uplift rate of 125 m/Ma.

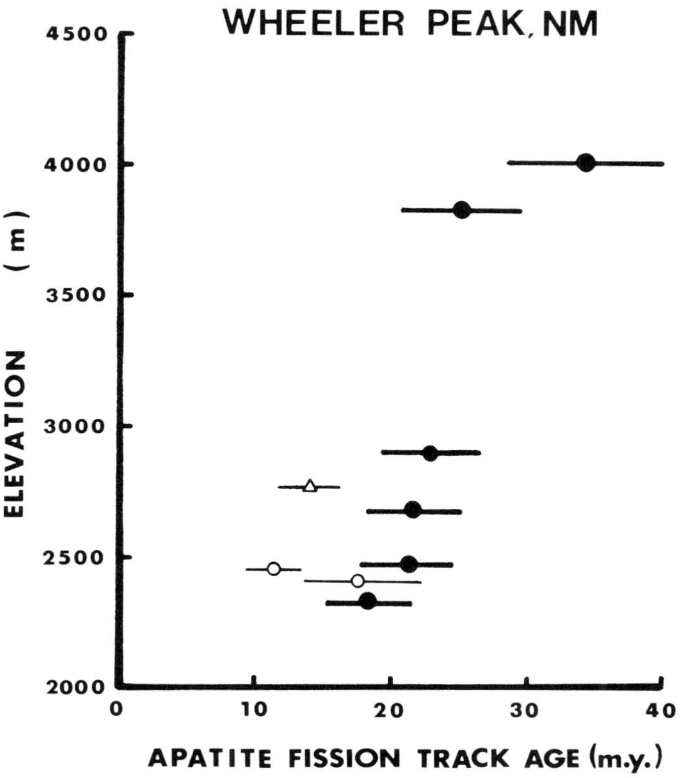

FIGURE 6. Apatite fission-track age vs. elevation for the Taos Range. Open symbols are data from Naeser (written comm. 1983); the open triangle is from the Lucero stock.

However, this correlation is deceptive because measurements of the track lengths in apatite from this traverse (Fig. 7) indicate that only the ages from the two samples at the highest elevations on Wheeler Peak (81WP1 and 81WP2) record uplift and erosion. The length distributions for the other four samples at lower elevations, which are from the Rio Hondo pluton and adjacent Precambrian rocks, are bimodal, suggesting thermal resetting. Apparently, the emplacement of the 22-m.y.-old Lucero stock has partially reset the apatite fission-track ages in the Rio Hondo pluton and the surrounding rocks. Hagstrum and Lipman (1983) and Johnson and others (1983) have also noted that the Rio Hondo pluton has been thermally altered by the Lucero intrusion. Consequently, the partially reset apatite fission-track ages cannot be used to study the uplift history of the area. The only usable dates for this region are the two ages from Wheeler Peak and the 14-m.y. age determined by Naeser for the Lucero stock (Fig. 6, open triangle). Since the Lucero stock age is probably an uplift age, it was used to estimate the apparent uplift rate of the area for the past 14 m.y. A detailed traverse through the Lucero stock is needed so that the more recent uplift history of the Taos Range can be constrained.

The Laramide uplift of the Taos Range portion of the Sangre de Cristo Mountains in the latest Cretaceous is recorded by the deposition of the Raton and Poison Canyon Formations in the Raton Basin (Baltz, 1965). Following the Laramide event, the area was eroded to low relief (McKinlay, 1956; Baltz, 1965). The fission-track ages from Wheeler Peak imply that uplift and erosion were in progress at least 35 m.y. ago; the earliest volcanism in the area occurred at about the same time (McKinlay, 1956).

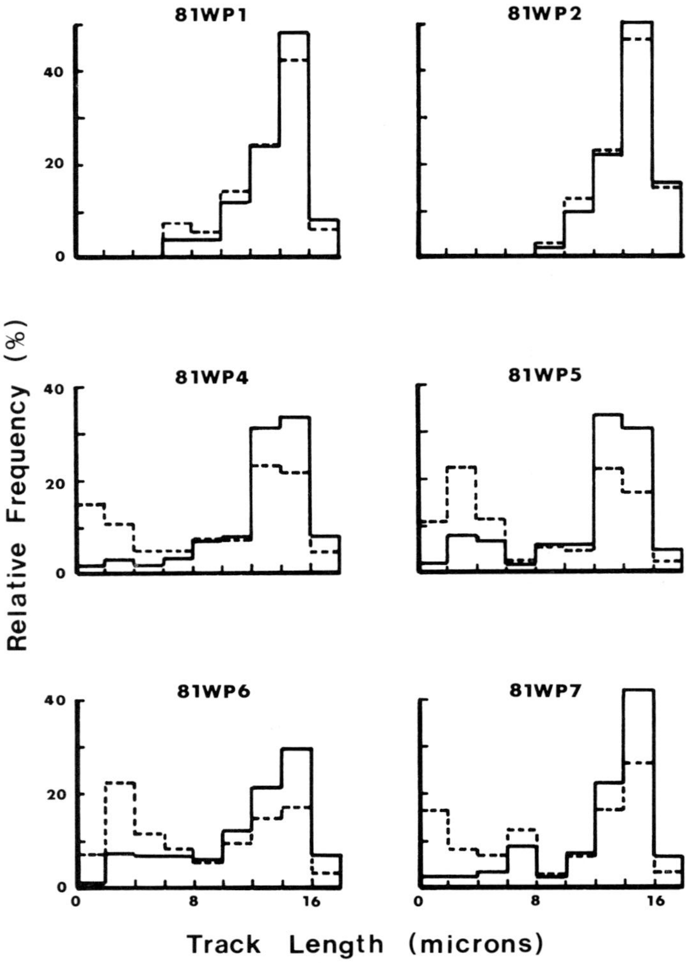

FIGURE 7. Track-length distributions from Taos Range, with 81WP1 at the highest elevation and 81WP7 at the lowest elevation. Symbols as in Figure 4.

DISCUSSION

Summary of Uplift Data for Northern New Mexico

Table 1 summarizes the uplift for the northern Rio Grande rift based on this and previously published studies. The apparent uplift rates derived from the fission-track data for each traverse for the past 14–15 m.y. were calculated using the heat-flow data of Reiter and others (1975) and Edwards and others (1978). The apparent uplift rate computed for the Santa Fe profile from the heat-flow and fission-track data is unusually low since it is averaged over 55 m.y.

Apatite fission-track ages presented by Lindsey and others (1983) for the eastern side of the Sangre de Cristo Mountains near Crestone, Colorado, range from 15.0 to 23.7 m.y. at elevations varying between 3,290 and 3,930 m. The apparent uplift rate for this area is about 70 m/m.y., which is comparable to the rate found in the Sandia Mountains over the same interval. An eastward tilt of the Sangre de Cristo Mountains is recorded by the ages from this traverse.

The uplift data of Axelrod and Bailey (1976) are based on studies of flora preserved at Bernalillo (near Albuquerque) and Skull Ridge (near Santa Fe). Axelrod and Bailey (1976) assumed that the decrease in mean annual temperature recorded by the flora was primarily due to rising elevation. Based on their data, they calculated that 1,200 m of epeirogenic uplift had occurred in these areas over the past 14 m.y. (rate = 85 m/m.y.). Unfortunately, the interpretation of Axelrod and Bailey is dubious because the possible effect of global climatic change on the mean annual temperature (Wolfe, 1978) was not properly considered (Baldridge, written comm. 1984).

Scott (1975) and Taylor (1975) found that 1,200 m of relative uplift had occurred in the northern Sangre de Cristo Mountains in Colorado since the deposition of the 7-m.y.-old Ash Hollow fauna. Their work suggests an uplift rate of 170 m/m.y. over a period of 7 m.y., which is the order of magnitude of the apparent uplift rates estimated from the heat-flow and fission-track data.

Paleogeography of Northern New Mexico

A schematic paleogeographic reconstruction of northern New Mexico (Fig. 8) was developed on the basis of published sedimentological studies and fission-track-age results of this study.

During the Laramide (Fig. 8a), the area currently occupied by the Sangre de Cristo Mountains, the Sandia Mountains, the Española Basin, the San Luis Basin, and the Brazos uplift was elevated to moderate relief (Baltz, 1978). This event is recorded by the apatite fission-track ages in the Santa Fe Range and by the sediments shed from this highland into the San Juan and Raton Basins.

TABLE 1. Summary of uplift data for Rio Grande rift in northern New Mexico and southern Colorado.

Location	Time (m.y.)	Rate (m/m.y.)	Source
Sandia Mtns. NM	15-30 0-15	81 230	this study
Santa Fe Range, NM	55-70 0-55	125 55	this study
Taos Range, NM	0-14	210	this study
Sangre de Cristo Mtns., CO	15-24	70	Lindsey and others, 1983
Española and Albuquerque Basins, NM	0-14	85	Axelrod and Bailey, 1976
Sangre de Cristo Mtns., CO	0-7	170	Scott, 1975 Taylor, 1975

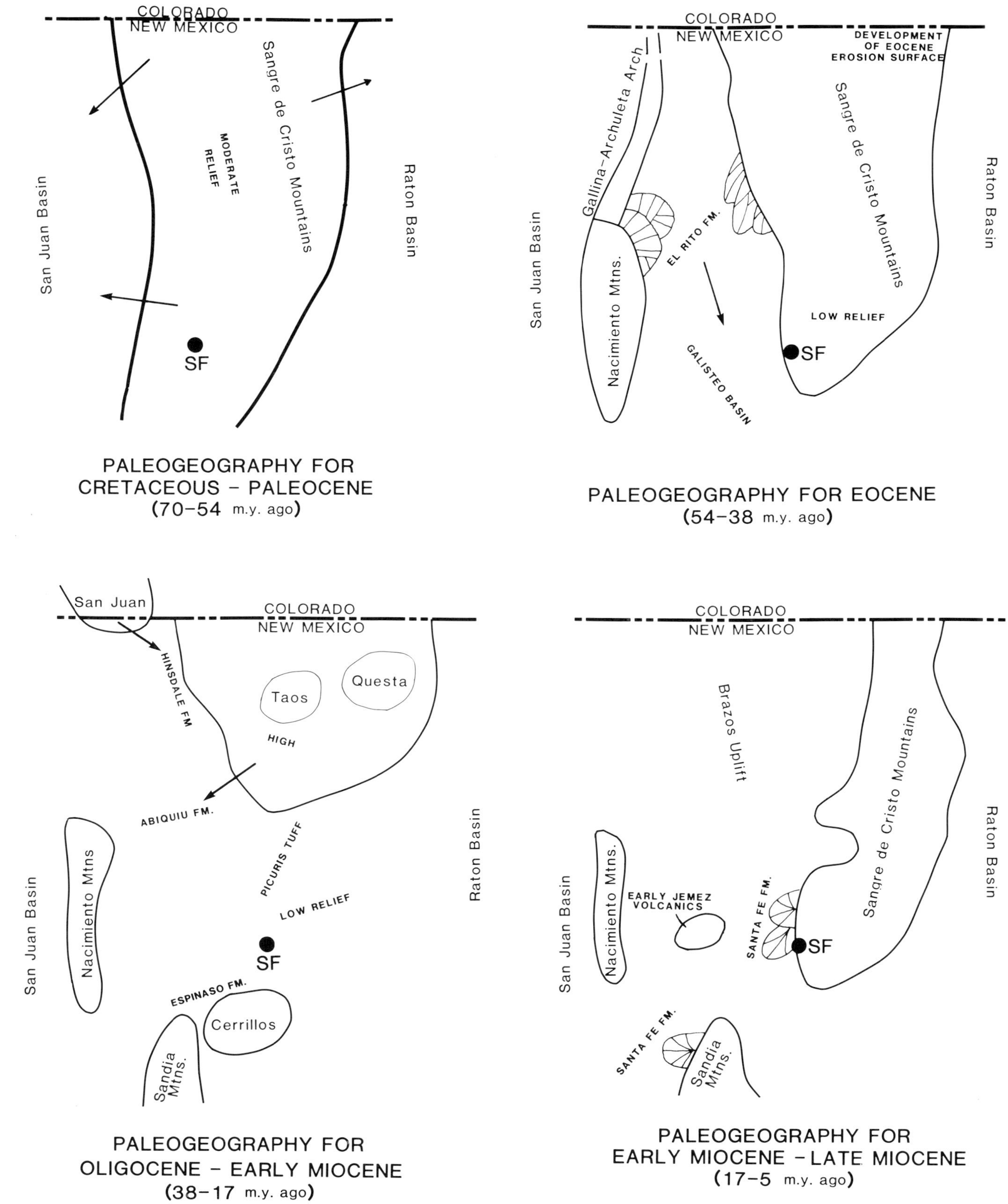

FIGURE 8. Paleogeography of northern New Mexico from fission-track-dating results and published geological studies (see text). Position of Santa Fe depicted by the large dot. After Baltz (1978), Vazzana and Ingersoll (1981), and Logsdon (1981).

The Galisteo Basin formed early in the Eocene (Fig. 8b). The Sangre de Cristo Mountains and the later Nacimiento Mountains were the source areas for the Galisteo Formation (Gorham and Ingersoll, 1979) and for the El Rito Formation (Logsdon, 1981). By late Eocene, the entire area was reduced to low relief (Baltz, 1978). The region near Santa Fe must have remained at low elevations until the culmination of rifting during the Miocene. Otherwise, the record of Laramide uplift indicated by the fission-track ages in the Santa Fe Range would not have been preserved.

The Oligocene to early Miocene was marked by abundant igneous activity, which is exhibited by the development of the San Juan (Lipman and others, 1978), Latir (Lipman, 1981), and Cerrillos–Ortiz (Stearns, 1953) volcanic fields (Fig. 8c). Evidence for early rift volcanism is preserved in a horst block in the Taos Plateau (Lipman and Mehnert, 1979). Volcaniclastic sediments shed off the volcanic highlands during this interval comprise the Los Pinos (Manley, 1981), Abiquiu (Vazzana and Ingersoll, 1981), Picuris (Cabot, 1938), and Espinaso (Kautz and others, 1981) Formations. The Taos Range and the Sandia Mountains were experiencing uplift and erosion at this time, as indicated by the apatite fission-track ages at the higher elevations of these two ranges. The correlation between the time that uplift began and the time that volcanism was active in these areas implies that the two events may be related. The uplifts on the eastern margin of the early rift were discontinuous.

Finally, in the middle Miocene (Fig. 8d) uplift was renewed in the Santa Fe Range, while uplift and erosion continued in the Taos Range and the Sandia Mountains. The Santa Fe Formation is composed of sediments from these uplifts (Galusha and Blick, 1971). Minor volcanism in the Jemez area occurred during this time (Bailey and others, 1969).

The fission-track data provide little information concerning events more recent than late Miocene. The heat-flow and fission-track data indicate that uplift rates for the past 15 m.y. have increased in the Sandia region, and the rates in the Taos Range are comparable to those in the Sandia area.

CONCLUSIONS

(1) The fission-track dates show that the Sandia Mountains began to rise at least 30 m.y. ago, and that the $100 + 25°C$ isotherm moved through the rock at an apparent rate of 81 m/m.y. until about 15 m.y. ago. The rate of cooling due to uplift and erosion approximately tripled (230 m/m.y.) in the past 15 m.y. The Taos Range was uplifting and eroding by at least 35 m.y. ago, and the apparent cooling rate for the past 14 m.y. in the Taos Range is about 210 m/m.y. In contrast to the other traverses studied, the fission-track dates from the Santa Fe Range record Laramide uplift and erosion at an apparent rate of 125 m/m.y.

(2) The paleogeographic reconstructions suggest that the relatively continuous uplifts that currently bound the eastern margin of the depression in northern and central New Mexico were not present before middle Miocene. The early Oligocene to middle Miocene uplifts in the Rio Grande region were much more discontinuous and seem to have been closely related to volcanic activity.

ACKNOWLEDGMENTS

We would like to thank Dr. C. W. Naeser for helping us set up our fission-track dating laboratory and sharing his data from the Questa area with us. We thank D. D. Blackwell, W. S. Baldridge, and A. W. Laughlin for their constructive reviews. This work was supported by an AAPG grant-in-aid and DOE grant AC03-81ER10973.

REFERENCES

Axelrod, D. I., and Bailey, H. P., 1976, Tertiary vegetation, climate, and altitude of the Rio Grande depression, New Mexico–Colorado: Paleobiology, v. 2, pp. 235–254.

Bailey, R. A., Smith, R. L., and Ross, C. S., 1969, Stratigraphic nomenclature of volcanic rocks in the Jemez Mountains, New Mexico: U.S. Geological Survey, Bulletin 1274-P, pp. P1–P19.

Baldridge, W. S., Damon, P. E., Shafiqullah, M., and Bridwell, R. J., 1980, Evolution of the central Rio Grande rift, New Mexico: new potassium–argon ages: Earth and Planetary Science Letters, v. 51, pp. 309–321.

Baltz, E. H., 1965, Stratigraphy and history of Raton Basin and notes on the San Luis Basin, Colorado–New Mexico: American Association of Petroleum Geologists, Bulletin, v. 49, pp. 2041–2075.

———, 1967, Stratigraphy and regional tectonic implications of part of Upper Cretaceous and Tertiary rocks, east-central San Juan Basin, New Mexico: U.S. Geological Survey, Professional Paper 552, 101 pp.

———, 1978, Resume of Rio Grande depression in north-central New Mexico; *in* Hawley, J. W. (ed.), Guidebook to the Rio Grande rift in New Mexico and Colorado: New Mexico Bureau of Mines and Mineral Resources, Circular 163, pp. 210–226.

Birch, F., and Clark, H., 1940, Thermal conductivity of rocks and its dependence upon temperature: American Journal of Science, v. 238, pp. 529–635.

Brookins, D. G., and Majumdar, A., 1982, The Sandia granite, New Mexico: biotite metamorphic and whole rock Rb–Sr ages: Isochron/West, no. 33, pp. 19–21.

Bullard, E. C., 1939, Heat flow in South Africa: Royal Society of London, Proceedings, ser. A, v. 173, pp. 474–502.

Cabot, E. C., 1938, Fault border of the Sangre de Cristo Mountains north of Santa Fe, New Mexico: Journal of Geology, v. 46, pp. 88–105.

Condie, K. C., and McCrink, T. P., 1982, Geochemistry of Proterozoic volcanic and granitic rocks from the Gold Hill–Wheeler Peak area, northern New Mexico: Precambrian Research, v. 19, pp. 141–166.

Edwards, C. L., Reiter, M., Shearer, C., and Young, W., 1978, Terrestrial heat flow and crustal radioactivity in northeastern New Mexico and southern Colorado: Geological Society of America, Bulletin, v. 89, pp. 1341–1350.

Galusha, T., and Blick, J. C., 1971, Stratigraphy of the Santa Fe Group, New Mexico: American Museum Natural History, Bulletin, v. 144, 127 pp.

Gleadow, A. J. W., and Duddy, I. R., 1982, Fission track lengths in the apatite partial stability zone and the interpretation of mixed ages: Workshop on Fission-track Dating, 5th International Conference on Geochronology, Cosmochemistry, and Isotope Geology, pp. 19–21.

Gorham, T. W., and Ingersoll, R. V., 1979, Evolution of the Eocene Galisteo Basin, north-central New Mexico: New Mexico Geological Society, Guidebook 30, pp. 214–224.

Haack, U., 1977, The closure temperature for fission track retention in minerals: American Journal of Science, v. 277, pp. 459–464.

Hagstrum, J. T., and Lipman, P. W., 1983, Paleomagnetic results from the structurally deformed Latir volcanic field and Questa caldera, New Mexico (abstract): Geological Society of America, Abstracts with Programs, v. 15, p. 380.

Harrison, T. M., Armstrong, R. I., Naeser, C. W., and Harakal, J. E., 1979, Geochronology and thermal history of the Coast Plutonic Complex, near Prince Rupert, British Columbia: Canadian Journal of Earth Sciences, v. 16, pp. 400–410.

Johnson, C. M., Lipman, P. W., and O'Neil, J. R., 1983, Sr, O, and H isotopic variations in the Questa magmatic system, New Mexico (abstract): Geological Society of America, Abstracts with Programs, v. 15, p. 380.

Kautz, P. F., Ingersoll, R. V., Baldridge, W. S., Damon, P. E., and Shafiqullah, M., 1981, Geology of Espinaso Formation (Oligocene), north-central New Mexico: summary: Geological Society of America, Bulletin, pt. I, v. 92, pp. 980–983.

Kelley, V. C., and Northrop, S. A., 1975, Geology of the Sandia Mountains, New Mexico: New Mexico Bureau of Mines and Mineral Resources, Memoir 29, 136 pp.

Lal, D., Muralli, A. V., Rojan, R. S., Tamnane, A. S., Lorin, J. C., and Pellas, P., 1968, Techniques for proper revelation and viewing of etch-tracks in meteoric and terrestrial minerals: Earth and Planetary Science Letters, v. 5, pp. 111–119.

Laslett, G. M., Kendall, W. S., Gleadow, A. J. W., and Duddy, I. R., 1982, Bias in measurement of fission-track length distributions: Nuclear Tracks, v. 6, pp. 79–85.

Lindsey, D. A., Johnson, B. R., and Andriessen, P. A. M., 1983, Laramide and Neogene structure of the northern Sangre de Cristo Range, south-central Colorado: Rocky Mountain Association of Geologists, Symposium, pp. 219–228.

Lipman, P. W., 1981, Volcano-tectonic setting of Tertiary ore deposits: Arizona Geological Society Digest, v. 14, pp. 199–211.

———, and Mehnert, H. H., 1979, The Taos Plateau volcanic field, northern Rio Grande rift, New Mexico; *in* Riecker, R. E. (ed.), Rio Grande rift: tectonics and magmatism: American Geophysical Union, Washington, D.C., pp. 289–311.

———, Doe, B. R., Hedge, C. E., and Steven, T. A., 1978, Petrologic evolution of the San Juan volcanic field, southwestern Colorado: Pb and Sr isotopic evidence: Geological Society of America, Bulletin, v. 89, pp. 59–82.

Logsdon, M. J., 1981, A preliminary basin analysis of the El Rito Formation (Eocene), north-central New Mexico: Geological Society of America, Bulletin, pt. I, v. 92, pp. 968–975.

Manley, K., 1981, Redefinition and description of the Los Pinos formation of north-central New Mexico: Geological Society of America, Bulletin, pt. I, v. 92, pp. 984–989.

McKinlay, P. F., 1956, Geology of Costilla and Latir Peak quadrangles, Taos County, New Mexico: New Mexico Bureau of Mines and Mineral Resources, Bulletin 42, 32 pp.

Miller, J. P., Montgomery, A., and Southerland, P. K., 1963, Geology of part of the southern Sangre de Cristo Mountains, New Mexico: New Mexico Bureau of Mines and Mineral Resources, Memoir 11, 106 pp.

Naeser, C. W., 1976, Fission-track dating: U.S. Geological Survey, Open-file Report 76-190, 50 pp.

———, 1997, Fission-track dating and geologic annealing of fission tracks; *in* Jager, E., and Hunziker, J. C. (eds.), Lectures in isotope geology: Springer-Verlag, New York, pp. 154–169.

Parrish, R. R., 1983, Cenozoic thermal evolution and tectonics of the Coast Mountains of British Columbia, 1. Fission-track dating, apparent uplift rates, and patterns of uplift: Tectonics, v. 2, pp. 601–632.

Register, M. E., and Brookins, D. G., 1979, Geochronology and rare-earth study of the Embudo Granite and related rocks: New Mexico Geological Society, Guidebook 30, pp. 155–158.

Reiter, M., Edwards, C. L., Hartman, H., and Weidman, C., 1975, Terrestrial heat flow along the Rio Grande rift, New Mexico and southern Colorado: Geological Society of America, Bulletin, v. 86, pp. 811–818.

Scott, G. R., 1975, Cenozoic surfaces and deposits in the southern Rocky Mountains: Geological Society of America, Memoir 144, pp. 248–277.

Stearns, C. E., 1953, Early Tertiary volcanism in the Galisteo–Tonque area, north-central New Mexico: American Journal of Science, v. 251, pp. 415–452.

Taylor, R. B., 1975, Neogene tectonism in south-central Colorado: Geological Society of America, Memoir 144, pp. 211–226.

Vazzana, M. E., and Ingersoll, R. V., 1981, Stratigraphy, sedimentology, petrology, and basin evolution of the Abiquiu Formation (Oligo-Miocene), north-central New Mexico: a summary: Geological Society of America, Bulletin, pt. I, v. 92, pp. 990–992.

Wolfe, J. A., 1978, A paleobotanical interpretation of Tertiary climates in the Northern Hemisphere: American Scientist, v. 66, pp. 694–703.

Rodeo at Taos Pueblo, ca 1920. Photo by Burt Harwood.

New Mexico Geological Society Guidebook, 35th Field Conference, Rio Grande Rift: Northern New Mexico, 1984

POTASSIUM–ARGON DATES FROM THE JEMEZ VOLCANIC FIELD: IMPLICATIONS FOR TECTONIC ACTIVITY IN THE NORTH-CENTRAL RIO GRANDE RIFT

JAMIE N. GARDNER and FRASER GOFF
Earth and Space Sciences Division, Los Alamos National Laboratory, Los Alamos, New Mexico 87545

INTRODUCTION

The Jemez Mountains volcanic edifice straddles the western margin of the Rio Grande rift which is locally expressed as three basins (Fig. 1). Volcanism in the Jemez Mountains has been intimately related to mid-Miocene through Quaternary tectonic activity of the Rio Grande rift (i.e., tectonic activity following the mid-Miocene volcanic hiatus of Chapin and Seager, 1975). Our work in the volcanic field has shown that tectonic activity has had a major influence on the dominant magmatic processes that were operative during various episodes of the volcanic field's development (e.g., Gardner, 1983, 1984). Hence, information regarding tectonic activity in the north-central rift can be inferred from studies of structures and timing of volcanism, with additional credibility provided by the petrology and petrogenesis of the volcanic products themselves. A discussion of the geochemical evidence for the petrogenetic models we use is beyond the scope of this paper; however, data supporting the petrogenetic models include major- and trace-element, microprobe, and Sr and O isotopic analyses (Gardner, 1982, 1983, 1984; Loeffler, 1984). In this paper we outline the geology and petrogenesis of the volcanic field, present new and existing radiometric dates, note variations of volcanism with respect to time, present some important relations of certain fault zones in the area, and summarize these data with emphasis on implications for tectonic activity in the north-central Rio Grande rift.

OUTLINE OF GEOLOGY AND PETROGENESIS

The Keres Group (Fig. 2), which comprises nearly half of the volume of the entire volcanic field, consists of rocks derived from magmas of two distinct sources: the upper mantle and the lower crust. Mafic magmas derived from the upper mantle differentiated through a broad spectrum of compositions, but gave rise to about 1,000 km^3 of two-pyroxene andesite. Magmas derived from partial melts of lower crust produced relatively minor volumes of chemically distinctive high-silica rhyolite. The Keres Group volcanism was accompanied by intense rifting, as suggested by field relations of mafic–intermediate rocks interbedded with widespread deposits of lahars and immature basin-fill gravels of

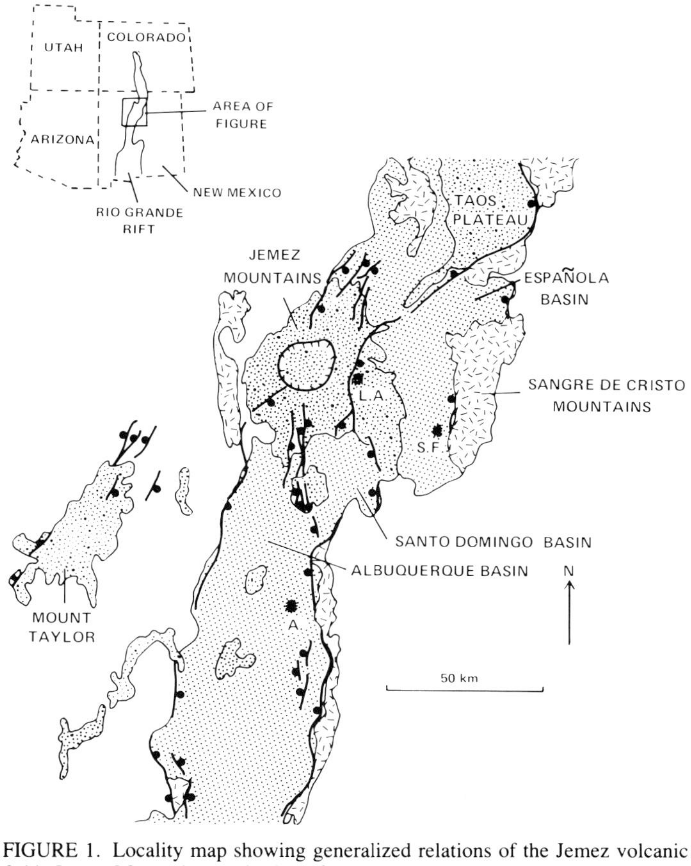

FIGURE 1. Locality map showing generalized relations of the Jemez volcanic field (Jemez Mountains) to basins of the north-central Rio Grande rift. Major fault zones are shown schematically. Random dash = Precambrian rocks; coarse, regular stipple = Tertiary–Quaternary rift-fill sediments; irregular stipple = Tertiary–Quaternary volcanic rocks. LA, SF, and A are Los Alamos, Santa Fe, and Albuquerque, respectively (modified from Baldridge and others, 1983).

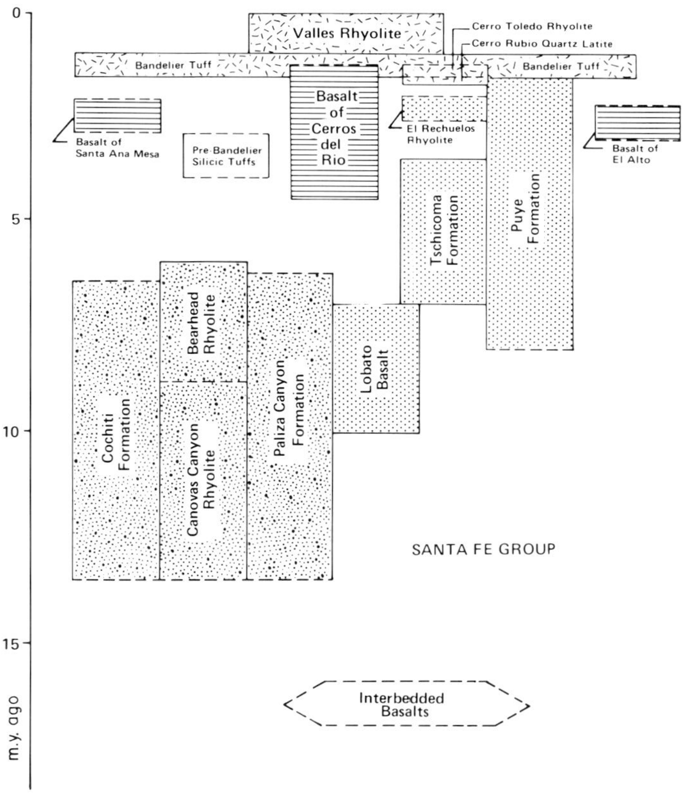

FIGURE 2. Generalized stratigraphic relations of major units in the vicinity of the Jemez volcanic field. Irregular stipple = Keres Group formations; coarse, regular stipple = Polvadera Group formations; random dash = Tewa Group formations; horizontally ruled pattern = young basalt fields, as indicated. Dashed lines indicate uncertainty. This figure is also a schematic south-to-north (left-to-right) section through the volcanic field (from Gardner, 1984; modified from Bailey and others, 1969).

the Cochiti Formation (Fig. 2), and fault control of the locations of high-silica rhyolite vents.

The best model for the genesis of the Tschicoma Formation (Fig. 2), whose latites and quartz latites dominate the 500 km^3 Polvadera Group (I. D. MacGregor, pers. comm. 1981), involves mixing of mafic and silicic magmas (Loeffler, 1984). End-member compositions for this mixing are approximated by basalt or basaltic andesite and high-silica rhyolite of the Keres Group. Smith and others (1970) indicate a paucity of faults in exposures of the Tschicoma Formation, and no apparent relations of vent locations to faults. While volcaniclastic debris derived from the Polvadera Group formed fan deposits of substantial volume (Puye Formation, Fig. 2) farther east in the Española Basin, field reconnaissance indicates virtually no significant basin-fill deposits intercalated with the Polvadera volcanic pile.

Mantle-derived magmas erupted at the basinward periphery of the main body of the volcanic field on the south, east, and north as the basalt fields of Santa Ana Mesa, Cerros del Rio, and El Alto, respectively (Figs. 2, 3) (Smith and others, 1970; Baldridge, 1979). The Tewa Group volcanism, dominated by the rhyolitic Bandelier Tuff, provided the culminating phase of the development of the Jemez volcanic field (Fig. 2). More than 500 km^3 of tuff erupted from a large, zoned magma chamber whose bulk composition was probably identical to the Tschicoma latitic compositions (Smith, 1979; Self and others, in prep.). Spatial and temporal overlap of the Bandelier magma and Tschicoma-like latite eruption supports this suggestion for the probable bulk composition of the Bandelier magma chamber, and strongly suggests parental relations of the Tschicoma Formation to the Tewa Group (Doell and others, 1968; Smith and others, 1970; Smith, 1979; Gardner, 1983, 1984; Self and others, in prep.).

POTASSIUM–ARGON DATES AND TIMING OF VOLCANISM

Twenty new potassium–argon dates (Table 1) have been obtained to constrain the timing of Keres Group volcanism. Sample F81-50, a basanite from a sequence of alkali basalts interbedded with lower Santa Fe Group arkosic sediments east of St. Peter's dome (Fig. 3), has yielded the oldest date in the area, of 16.5 ± 1.4 m.y. (see Table 1). While these alkali basalts are probably not petrogenetically related to volcanism in the Jemez volcanic field, they may be indicative of the onset of the thermal and tectonic events that have driven, or influenced, magmatism throughout development of the volcanic field (Gardner, 1984). Field relations together with the data of Table 1 (samples JG80-53 and F82-108) indicate the Keres Group volcanic activity began >13 m.y. ago. The Keres Group volcanism apparently waned by about 6 m.y. ago (samples JG80-49, JG8162, F82-77, and F82-117; Table 1), and overlapped with early Tschicoma Formation volcanism (Fig. 2).

The new dates (Table 1), a compilation of 95 available dates from the Jemez volcanic field and nearby areas (Table 2), and detailed geologic mapping, make possible the refinement of stratigraphic relations shown in Figure 2. Perhaps the most significant result of the refinement of stratigraphic relations is that the cyclic nature of volcanism implied by the formal stratigraphy is erroneous (see fig. 2 of Bailey and others, 1969; and Smith and others, 1970). Volcanic activity has been essentially continuous in the Jemez Mountains for the last 13 m.y. The Lobato Basalt volcanism occurred entirely within the time span of the Paliza Canyon activity, and, based on the dates and field relations, the boundary between the Canovas Canyon and Bearhead Rhyolites in Figure 2 is somewhat arbitrary.

Figure 4 is a graphic depiction of the compilation of dates plotted with respect to the predominant petrogenetic magma type outlined above. A number of features of Figure 4 are particularly noteworthy:

(1) While eruptions of the mantle-derived basaltic magmas spanned the volcanic field's entire >13 m.y. history, there was a gap in basaltic volcanism in the period 7–4 m.y. ago. Stratigraphic constraints on the dated basalts indicate that this gap is not the result of sampling bias. Those mantle-derived basalts that were erupted after the 7–4 m.y. lull in basaltic volcanism are the basalts of Santa Ana, Mesa, Cerros del Rio, and El Alto, which are distinctly peripheral to, and partially encircle, the main body of

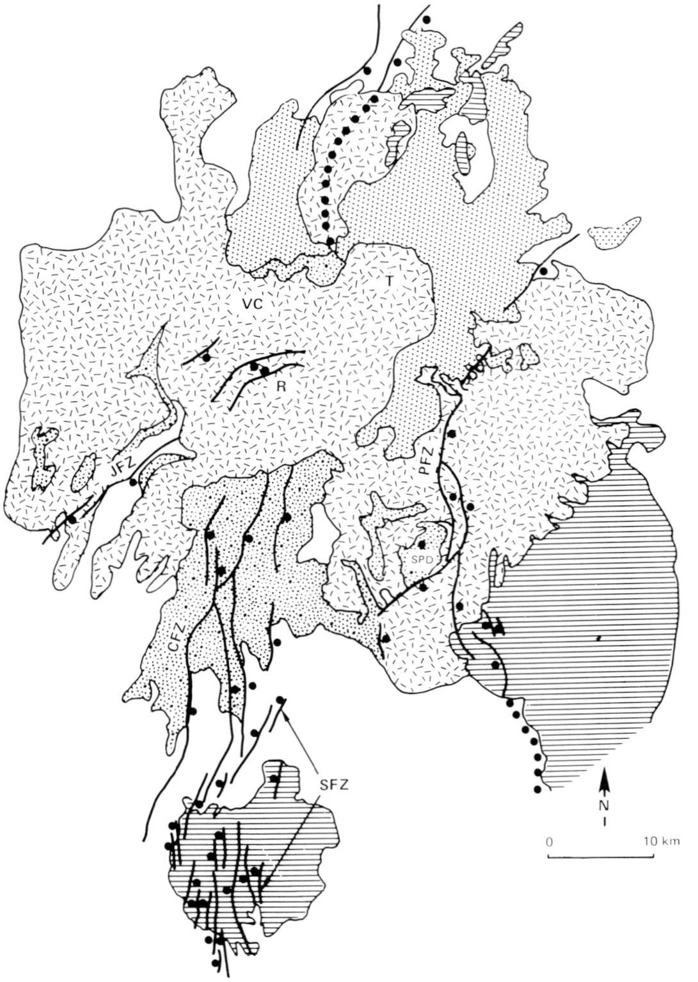

FIGURE 3. Generalized map showing distribution of major stratigraphic groups (patterns are the same as Figure 2) and the major fault zones in the Jemez Mountains. JFZ = Jemez fault zone; SFZ = Santa Ana Mesa fault zone; CFZ = Cañada de Cochiti fault zone; PFZ = Pajarito fault zone; VC = Valles caldera; R = resurgent dome of VC; T = Toledo embayment; and SPD = St. Peter's dome (from Gardner, 1984; modified from Smith and others, 1970).

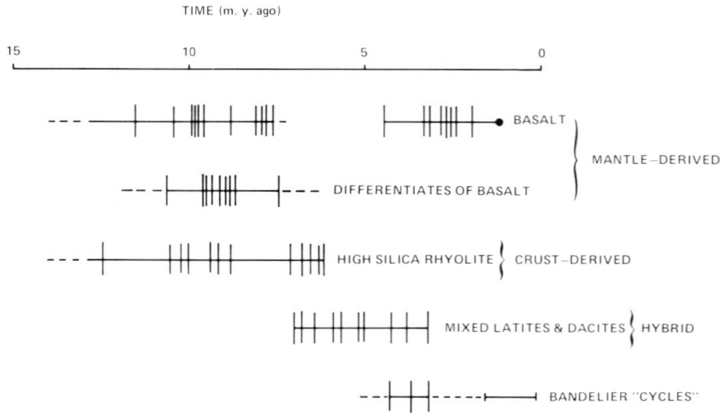

FIGURE 4. Dates of Jemez volcanic rocks (Table 2) plotted with respect to dominant magma type of Gardner (1982, 1983, 1984) and Loeffler (1984). Each vertical line represents a K–Ar date. Solid lines indicate continuous eruption of a given magma type, inferred from field relations, and dashed lines indicate uncertainty. Numerous dates from Tewa Goup—solid line for "Bandelier cycles"—not plotted for clarity.

TABLE 1. New K–Ar dates of rocks from the Jemez volcanic field and immediate vicinity.

Sample No.	Location	Rock Type	Material[2]	%K	Weight Sample, g	Moles Radiogenic ^{40}Ar	% Radiogenic ^{40}Ar	Age, Myr[1]	Comments	Lab[3]
JG80-44	35°49,8'N.Lat., 106°37.7' W. Long. from Cinder Cone @ 7735' elev., Ponderosa Quad.	Andesite	W/R	2.90	0.66078	3.539×10^{-11}	14	10.6 ± 1.4	Borrego Mesa	K
JG80-47c	SW 1/4, NE 1/4, Sec 24, T18N, R3E, 8880' elev., Redondo Peak Quad.	Rhyolite	W/R	4.07	0.46564	2.896×10^{-11}	24	8.79 ± 0.66	Dome near Cerro Pelado	K
JG80-48	SW 1/4, SW 1/4, Sec 27T18N, R3E; 1 km west of Cerro del Pino, Redondo Peak Quad.	Latite	Hbd	1.07	0.41310	7.763×10^{-12}	7	10.1 ± 2.5	Cerro del Pino	K
JG80-49	NW 1/4, SE 1/4, Sec 34, T18N, R3E; 1-3/4 km south of Cerro del Pino, Bear Springs Peak Quad.	Rhyolite	W/R	4.51	0.78161	3.785×10^{-11}	58	6.18 ± 0.09	Dome south of Cerro del Pino	K
JG80-53	Center Sec 2 T17N, R3E, 7720' elev., north wall of Paliza Cyn. Bear Springs Peak Quad.	Basalt	W/R W/R	1.27 1.14	0.48151 0.64379	1.223×10^{-11} 1.907×10^{-11}	24 10	11.5 ± 0.8 14.9 ± 2.7 (13.2 ± 1.24)	Lava flow overlies Conovas Canyon tuff	
JG80-55	NW 1/4, SE 1/4, NW 1/4, Sec 10; NW wall of Paliza Cyn. Bear Springs Peak Quad.	Andesite	W/R	3.07	0.69869	3.313×10^{-11}	6	8.88 ± 2.7	Plug in bottom of Paliza Canyon	K
JG81-50	35°42.9'N. Lat., 106°33' W. Long.; 1/4 km east of BM8009, Bear Springs Peak Quad.	Latite	Plag.	0.93	1.04681	1.252×10^{-11}	8	7.40 ± 1.6	Ruiz Peak area	K
JG81-62	35°46.1'N. Lat., 106°32.8' W. Long.; 1/2 km NW of Paliza Cyn., 1-3/4 km south of Cerro Pelado; Redondo Peak Quad.	Andesite	W/R	2.19	0.94874	2.297×10^{-11}	12	6.36 ± 0.95	Plug in Paliza Canyon	K
F80-79	Center Sec 7, T19N, R3E; Jemez Springs 15' Quad.	Latite	W/R*	2.44	0.71482	1.276×10^{-11}	7	4.21 ± 1.3	Tschicoma flow overlies Keres Group in west wall, Valles Caldera	K
F81-50	35°43.7' N. Lat., 106°22' W. Long. in Medio Canyon, Cochiti Dam Quad.	Basalt	W/R W/R W/R	0.964 1.020 1.220	0.4419 0.4608 0.4446	1.523×10^{-11} 1.381×10^{-11} 1.235×10^{-11}	11 16 10	16.5 ± 1.4	Basalt in lower Santa Fe Group beneath east side St. Peter's Dome	G G G
F82-33	35°46'N. Lat., 106°21' W. Long. 3/4 km southwest of Boundary Peak, Frijoles Quad.	Latite	W/R W/R	2.62 2.383	1.89414 0.57225	7.927×10^{-11} 23.175×10^{-12}	20 18.4	9.19 ± 0.74 9.77 ± 0.5 (9.48 ± 0.44)	Dome overlies Santa Fe Group, east side St. Peter's Dome	K G
F82-77	35°47.2' N. Lat. 106°22.3' W. Long., west side of conical dome.	Rhyolite	W/R	4.35	0.76275	3.587×10^{-11}	9	6.22 ± 1.2	Rabbit Hill Dome	K
F82-79	NW 1/4, SE 1/4, Sec 20, T18N, R5E, Bland Quad., east side Cochiti Canyon	Latite	W/R	2.70	0.5624	23.2×10^{-12}	23	8.8 ± 0.5	Dome in Upper Cochiti Canyon	G
F82-84	35° 44.5' N. Lat., 106° 24.8' W. Long., Cañada Quad., east side Cochiti Canyon	Rhyolite	W/R*	4.04	0.73427	4.808×10^{-11}	10	9.32 ± 1.6	Rhyolite intrudes andesite and capped by Peralta Tuff, Cochiti Canyon	K
F82-102	East edge, SE 1/4, Sec 5, T17N, R5E, Cañada Quad., east side Medio dia Canyon	Rhyolite	W/R	4.47	0.67200	4.828×10^{-11}	12	9.24 ± 1.5	Dome overlain by Peralta Tuff, Medio dia Canyon	K
F82-103	35°43.5' N. Lat., 106°22.7' W. Long., of Sanchez Canyon	Rhyolite	W/R	3.95	0.69057	4.964×10^{-11}	19	10.5 ± 0.9	Dome intrudes laharic deposits, west wall Sanchez Canyon	K
F82-108	35°43.5' N. Lat., 106°22.7' W. Long., bottom of Sanchez Canyon	Rhyolite	W/R W/R W/R	3.154 3.098 3.329	0.4614 0.4914 0.70743	3.359×10^{-11} 3.475×10^{-11} 4.734×10^{-11}	6 6 8	13.2 ± 3.3 11.6 ± 2.3 (12.4 ± 1.98)	Flow overlies tuff, Sanchez Canyon	G G K
GF82-117	35°44' N. Lat., 106°, 26.5' W. Long., bottom of Bland Canyon	Rhyolite	W/R	4.39	0.78136	3.888×10^{-11}	19	6.52 ± 0.59	Rhyolite intrudes Peralta Tuff, Bland Canyon	K
GF82-1	35°44' N. Lat., 106° 24.5' W. Long., west side Cerro Boletas	Rhyolite	W/R	4.08	0.77410	3.701×10^{-11}	23	6.74 ± 0.47	Rhyolite intrudes Peralta Tuff, west side Cerro Boltetas	K
F83-19	35°44.6' N. Lat., 106°22.8' W. Long., north side Cerro Picacho	Rhyolite	W/R	3.873	0.5322	25.51×10^{-12}	48.8	7.12 ± 0.4	Cerro Picacho Dome	G

EXPLANATION:
1. All dates calculated with decay and abundance constants of Steiger and Jager (1977).
2. Material analyzed: W/R = whole rock; W/R* = whole rock with magnetic fraction removed; Hbd = hornblende; plag = plagioclase.
3. Laboratory: K = D. Krummenacher, Dept. of Geology, San Diego State Univ., CA.
G = H. Krueger, Geochron Laboratories, MA.

the volcanic field (Fig. 3). Earlier basaltic volcanism, >13–7 m.y. old, comprises the basalts of the Paliza Canyon Formation and the Lobato Basalt of the Keres and Polvadera Groups, respectively (Fig. 2, Table 2).

(2) Magmas derived by differentiation of mantle-derived material were erupted early in the volcanic field's history, in the period >10–7 m.y.B.P. These magmas are represented predominantly by the andesites and less voluminous latites of the Paliza Canyon Formation of the Keres Group (Table 2).

(3) High-silica rhyolite magmas, derived from partial melts of lower crust, were continuously erupted in the period >13–6 m.y.B.P., early in the field's history. These crustal partial melts are the tuffs and domes of the Canovas Canyon and Bearhead Rhyolites of the Keres Group.

(4) Latites, quartz latites, and dacites formed by mixing of basalt or basaltic andesite and high-silica rhyolite magmas began to be erupted about 7 m.y. ago, at the same time as the beginning of the 7–4 m.y. lull in basaltic volcanism mentioned in (1) above. Furthermore, the onset of the eruption of these mixed latites and dacites was essentially contemporaneous with the cessation of eruption of differentiates of basalt (Paliza Canyon intermediate rocks) and the high-silica rhyolites discussed in (2) and (3) above, respectively.

(5) The onset of the Tewa Group volcanism ("Bandelier cycles") was essentially contemporaneous with the revival of basaltic volcanism at about 4 m.y.B.P., discussed in (1) above, and overlapped with the waning eruptions of mixed magmas discussed in (4).

FAULT ZONES

Figure 3 shows the four major fault zones in the immediate vicinity of the Jemez volcanic field as the Jemez, Santa Ana Mesa, Pajarito, and Cañada de Cochiti fault zones. The reader familiar with the general geology of the Jemez Mountains will note apparent discrepancies between the names we employ for fault zones and names used by some previous workers (e.g., Kelley and others, 1976; Kelley, 1978; Woodward and Ruetschilling, 1976; Woodward and others, 1977). It is important to realize that the names we use are applied to fault zones, not

TABLE 2. Compilation of dates of Jemez volcanic rocks only.

TEWA GROUP											
Valles Rhyolite		Tschirege Member (Bandelier Tuff)		Toledo Rhyolite		Otowi Member (Bandelier Tuff)		Pre-Bandelier Silicic Tuffs		El Alto, Santa Ana Mesa, and Cerros del Rio "Basalt Fields"	
Date	Ref.	Date	Ref.	Date	Ref.	Date	Ref.	Date	Ref.	Date	Ref.
0.13 ± 0.10	(10)										
0.434 ± 0.015	(4)	1.02 ± 0.04	(4)	1.22 ± 0.04	(4)	1.37 ± 0.04	(4)	3.15 ± 0.89	(9)	1.96 ± 0.06	(6)
0.494 ± 0.015	(4)	1.06 ± 0.03	(4)	1.25 ± 0103	(9)	1.44 ± 0.04	(4)	3.64 ±1.64	(9)	2.4 ± 0.3	(6)
0.502 ± 0.015	(4)	1.09 ± 0.03	(4)	1.28 ± 0.02	(9)	1.48 ± 0.09	(4)	4.2 ± 0.3(?)	(6)	2.5 ± 0.3	(1)
0.535 ± 0.015	(4)	1.1 ± 0.1	(2)	1.28 ± 0.07	(4)					2.6 ± 0.4	(1)
0.692 ± 0.015	(4)	1.19 ± 0.04	(4)							2.6 ± 0.4	(1)
0.707 ± 0.019	(4)	1.24 ± 0.05	(4)							2.6 ± 0.2	(6)
0.726 ± 0.015	(4)									2.78 ± 0.44	(6)
0.823 ± 0.074	(4)									2.76 ± 0.44	(7)
0.884 ± 0.028	(4)									2.8 ± 0.7	(8)
0.886 ± 0.019	(4)									2.8 ± 0.5	(8)
1.04 ± 0.05	(4)									2.8 ± 0.1	(6)
1.15 ± 0.03	(4)									2.8 ± 0.1	(6)
1.25 ± 0.11	(4)									3.1 ± 1.0	(8)
										3.2 ± 0.1	(2)
										4.4 ± 0.1	(2)

El Rechuelos Rhyolite		POLVADERA GROUP			
		Tschicoma Formation		Lobato Basalt	
Date	Ref.	Date	Ref.	Date	Ref.
2.07 ± 0.06*	(3)	3.15 ± 0.28	(6)	7.6 ± 0.4	(8)
		3.77 ± 0.12*	(3)	7.63 ± 0.16*	(3)
		4.2 ± 0.3	(6)	7.8 ± 0.5	(8)
		4.21 ± 1.3	(5)	7.8 ± 0.7	(8)
		5.0 ± 0.7	(6)	7.9 ± 0.5	(8)
		5.16 ± 0.13*	(3)	8.1 ± 0.2	(6)
		5.65 ± 0.82*	(3)		
		5.78 ± 0.18*	(3)	9.6 ± 0.3	(6)
		6.36 ± 0.95	(5)	9.6 ± 0.2	(2)
		6.82 ± 0.17*	(3)	9.7 ± 0.3	(2)
		6.87 ± 0.33*	(3)	9.8 ± 0.4	(1)
				9.9 ± 1.0	(8)

KERES GROUP									
		Paliza Canyon Formation							
Bearhead Rhyolite		(Latitic Rocks)		(Andesitic Rocks)		(Basaltic Rocks)		Canovas Canyon Rhyolite	
Date	Ref.	Date	Ref.	Date	Ref.	Date	Ref.	Date	Ref.
6.18 ± 0.09	(5)	7.4 ± 1.6	(5)			8.8 ± 0.3	(6)	9.24 ± 1.5	(5)
6.22 ± 1.2	(5)	8.8 ± 0.5	(5)	8.69 ± 0.38*	(3)	10.4 ± 0.5**	(6)	9.32 ± 1.6	(5)
6.52 ± 0.59	(5)	9.11 ± 0.19*	(3)	8.88 ± 2.7	(5)	13.2 ± 1.24	(5)	10.0 ± 0.3	(6)
6.74 ± 0.47	(5)	9.33 ± 0.19*	(3)	9.0 ± 0.3	(6)			10.2 ± 0.3	(6)
7.1 ± 0.2	(6)	9.48 ± 0.44	(5)	9.5 ± 2.5	(6)			10.5 ± 0.9	(5)
7.12 ± 0.4	(5)	10.1 ± 2.5	(5)	10.6 ± 1.4	(5)			12.4 ± 1.98	(5)
8.79 ± 0.66	(5)								

* = Date recalculated with decay and abundance constants of Steiger and Jager (1977) or by methods of Dalrymple (1979).

** = Date on basalt of Chamisa Mesa (Bailey et al., 1969).

References:
(1) Bachman and Mehnert (1978)
(2) Baldridge and others (1980)
(3) Dalrymple and others (1967)
(4) Doell and others (1968)
(5) This work
(6) Leudke and Smith (1978)
(7) Manley (1976b)
(8) Manley and Mehnert (1981)
(9) Self and others (in prep.)
(10) Marvin and Dobson (1979)

to individual faults. The Cañada de Cochiti and Santa Ana Mesa fault zones are commonly referred to as a single zone, also known as the San Felipe fault zone. However, we distinguish two fault zones based on age and sense of displacement: the Santa Ana Mesa fault zone is the active eastern boundary of the Albuquerque Basin south of the Jemez Mountains, and the Cañada de Cochiti zone is the inactive (see below) western boundary of the Española Basin east of, and containing, the Jemez Mountains. While the fault zones shown in Figure 3, except the Cañada de Cochiti zone, are currently seismically active (Cash and others, 1982), most important to this discussion are the Pajarito and Cañada de Cochiti zones.

The Cañada de Cochiti fault zone is a broad swath of north-trending normal faults that cuts the Keres Group rocks in the southern Jemez Mountains. The fault zone is obliterated by the Valles caldera, but is exposed in Mesozoic–Tertiary rocks north of a paleovalley eroded along the faults and filled with unfaulted Bandelier Tuff (dotted fault in Fig. 3) (Smith and others, 1970; Lawrence, 1979). Most faults in the Cañada de Cochiti fault have a down-to-the-east sense of displacement, and we estimate more than 500 m of cumulative offset across the fault zone. Because the volcanic and volcaniclastic deposits of the Keres Group are both cut by faults of this zone and thicken to the east across the fault zone, these faults were active during the Keres Group volcanism. Rhyolites of the Canovas Canyon and Bearhead Formations are intruded along these faults, but commonly the Bearhead domes (see Table 2) are not themselves faulted. Hence, faults of this zone originally represented the western boundary of the Española Basin of the Rio Grande rift and became inactive sometime prior to about 6 m.y. ago.

The Pajarito fault zone is a narrow band of north- and northeast-trending normal faults (Griggs, 1964; Smith and others, 1970; Golombek, 1981) which, we suggest, defines the present western boundary of the Española Basin of the Rio Grande rift. A very deep part of the Española Basin is adjacent to this fault zone, as suggested by more than 1,500 m of basin-fill sediments immediately east of the fault zone (Budding, 1978; Cordell, 1978; Goff and Grigsby, 1982). Faults of this zone show predominantly down-to-the-east sense of displacement and evidence of episodic or multiple faulting events indicated by progressively larger amounts of offset in older rock units (Griggs, 1964). The Pajarito fault zone offsets the 1.1-m.y.-old Bandelier Tuff by more than 100 m, and, as mentioned above, is currently seismically active (Griggs, 1964; Smith and others, 1970; Cash and others, 1982). Based in part on the work of Manley (1976a, 1979) in the northern Española Basin, Golombek (1981, 1983) and Golombek and others (1983) have suggested that the Pajarito fault zone became active since about 5 m.y. ago. Thus, fault activity bounding the western margin of the Española Basin apparently shifted from the Cañada de Cochiti fault zone, discussed above, to the Pajarito fault zone since roughly 6–5 m.y. ago.

The northeast-trending segment of the Pajarito fault zone southeast of St. Peter's dome appears to be older than portions of the fault zone further north, and is probably not directly related to the Española Basin (Figs. 1, 3). Movements on this segment of the fault have caused tilting of large blocks northwest of the fault trace with increasing amounts of tilt in progressively older rock units (Smith and others, 1970). Contrary to the conclusions of Golombek (1981; 1983), our detailed mapping corroborates these relations shown on the regional map of Smith and others (1970). Eocene Galisteo Formation rocks dip 25–45° NW, Santa Fe Group deposits dip 5–15° NW, and lower Keres Group rocks dip ≤10° NW (Gardner, 1984; Goff, in prep.), indicating this segment of the Pajarito fault zone is at least older than lower Santa Fe Group deposits at this locality (>16.5 m.y.B.P., see sample F81-50 in Table 1).

The northeast trend and substantially older age than portions of the fault zone further north suggest that this segment of the Pajarito fault zone may be associated with the northeast-trending Santo Domingo Basin. Alternatively, this older, northeast-trending segment of the fault zone may reflect the structural grain imparted to the area by the Jemez lineament, which is locally manifested as the Jemez fault zone, resurgent and subsurface structures in Valles caldera, and the Toledo embayment (Fig. 3) (Smith and others, 1970; Goff and Kron, 1980; Goff and Gardner, 1980; Goff, 1983; Gardner, 1983, 1984; Self and others, in prep.).

SUMMARY AND DISCUSSION

Following the mid-Miocene volcanic hiatus of Chapin and Seager (1975), revived rifting in the vicinity of the Jemez Mountains began about 16.5 m.y. ago as indicated by the age of the basanite within the lower Santa Fe Group (sample F81-50, Table 1) and the age of the northeast-trending segment of the Pajarito fault zone near St. Peter's dome. From 13 to 10 m.y.B.P. tectonic activity was probably intense. Basalts and high-silica rhyolites were vented during this period along faults of the Cañada de Cochiti zone; they are interbedded with immature basin-fill gravels and lahars of the Cochiti Formation. The Cochiti Formation deposits thicken dramatically from west to east into the rift across the Cañada de Cochiti fault zone. In the period 10–7 m.y.B.P., basalt, high-silica rhyolite, and differentiates of basalt were rapidly vented (Fig. 5) because open conduits were provided by on-going extension across the Cañada de Cochiti fault zone. Golombek and others (1983) suggest that intense tectonic activity in the area occurred around 10 m.y. ago, which for the most part agrees with our inferences from field relations and the fact that more than half the volume of the entire volcanic field was erupted during the 3 m.y. interval 10–7 m.y. ago (Fig. 5). Golombek (1981, 1983), however, suggests that tectonic activity around 10 m.y.B.P. was concentrated within a now-buried graben in the central Española Basin, east of the Pajarito fault zone, in order to account for the apparent discrepancy between estimates of maximum offset on the Pajarito fault zone and the tremendous thickness of basin-fill material immediately adjacent to the fault, as indicated by various geophysical studies. Golombek (1981, 1983) cites the gravity interpretation of Budding (1978), and Golombek and others (1983) cite the gravity model of Cordell (1979) as evidence of this hidden graben.

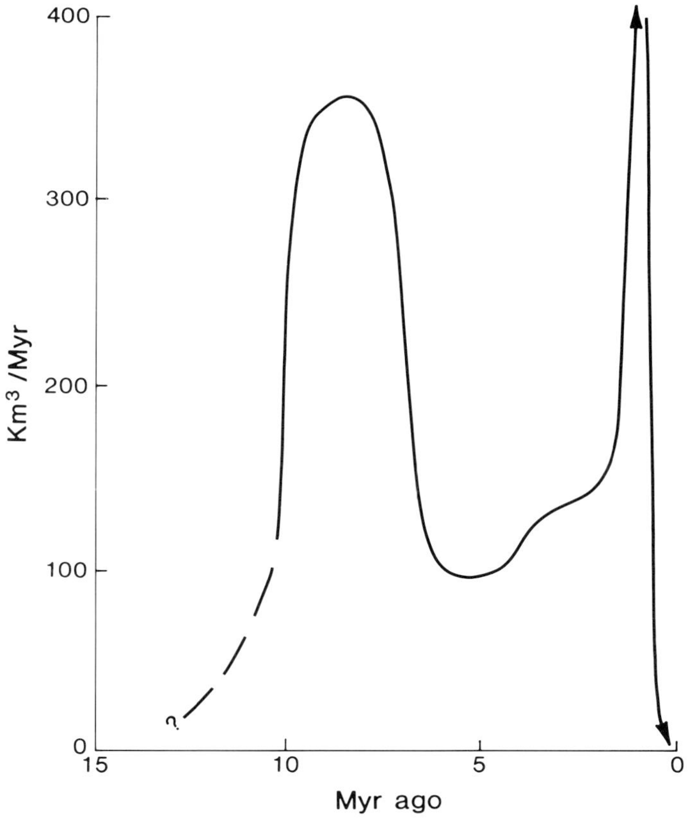

FIGURE 5. "Rate" of volcanism (volume per unit time) vs. time (m.y.B.P.) for Jemez volcanic activity. Plot goes off-scale with major Bandelier ash-flow eruptions (from Gardner, 1984).

Other geophysical studies in the same area (detailed gravity by Williams, 1979; seismic reflection by E. F. Homuth, pers. comm. 1979; see also summary in Goff and Grigsby, 1982, and Cordell, 1979, profile C–C′) do reveal intrabasin faults, but do not corroborate the existence of a large, buried graben. Furthermore, scrutiny of Budding's (1978) interpretive cross sections reveals that the position of the Pajarito fault zone is plotted incorrectly: the gravity low ("buried graben") is not some distance east of the Pajarito fault zone, it is *bounded* by the fault zone on the west. Our work indicates that displacement across the Cañada de Cochiti fault zone together with a post-4 m.y.B.P. displacement on the Pajarito fault zone can accommodate most of the 1,500 m of basin-fill sediment immediately east of the Pajarito fault zone. Hence, tectonic activity prior to 10 m.y.B.P. was probably not restricted to central parts of the Española Basin, but rather was concentrated within the Cañada de Cochiti fault zone, defining the western boundary of the basin.

At about 7–6 m.y.B.P., differentiates of basalt and high-silica rhyolite stopped being erupted, mixed magmas began to be eruped, and there was a sharp reduction in the "rate" of volcanic activity (Fig. 5). At about the same time, the beginning of the 7–4 m.y.B.P. lull in basaltic volcanism occurred, and the Cañada de Cochiti fault zone became inactive. Furthermore, the 7–4 m.y. lull (or a sharp reduction in the apparent volume of basaltic volcanism during this period) has been noted elsewhere in or near the rift (Baldridge and Perry, 1983; A. W. Laughlin, pers. comm. 1983; Seager and others, 1984) and may be indicative of events of regional significance. Because of the intimate association of basaltic volcanism and intense tectonism earlier in the volcanic field's history, we suggest that the gap in basaltic volcanism is indicative of a lull in tectonic activity. The reduction in the rate of volcanic activity and the petrogenetic transition from differentiates of basalt to hybrid latites at about 7 m.y. ago reflect a change in the dominant magmatic processes in response to the tectonic lull. Instead of being rapidly vented, as in earlier, more tectonically active stages of the field's development, pockets of basalt or basaltic andesite began to coalesce with pockets of high-silica rhyolite and continued to grow into large magma chambers from which small portions were erupted as hybrid latitic lavas. It is probable that these large, hybrid-magma chambers developed chemical and mineralogical zonation, and ultimately gave rise to the Bandelier Tuff (Smith, 1979; Self and others, in prep.). Spatial and temporal overlap of Bandelier-like and Tschicoma-like hybrid eruptions further suggests parental relations of Tschicoma magma chambers to the Tewa Group (Doell and others, 1968; Smith and others, 1970; Smith, 1979; Gardner, 1983, 1984; Self and others, in prep.).

Manley (1976a, 1979) suggests intense faulting during the period 5–3 m.y.B.P. in the Velarde graben in the northern Española Basin, and Golombek (1981, 1983) and Golombek and others (1983) estimate that the Pajarito fault zone became active around 5 m.y. ago. As a refinement of this estimate, we suggest that basin-bounding activity shifted from the Cañada de Cochiti zone to the north-trending part of the Pajarito fault zone at about 4 m.y.B.P. Golombek (1981, 1983) suggests that the position of the Pajarito fault zone is controlled by the location of an abrupt facies change from volcanic to volcaniclastic rocks. It is improbable that these deposits, which constitute a thin veneer relative to the sequence of units cut by the fault zone and the minimum depth of penetration of the fault (e.g., see Goff and Grigsby, 1982), could control the position of the fault. Instead, we suggest that during the 7–4 m.y. tectonic lull the development of large, hybrid-magma chambers within the crust beneath the center of the volcanic field caused a shift in basin-bounding faulting from the Cañada de Cochiti zone to the Pajarito zone, with revived tectonism at about 4 m.y.B.P. renewed basaltic volcanism at about 4 m.y. ago, caused by revived tectonic activity, gave rise to the basalt fields that are peripheral to, and partially encircle, the main body of the volcanic field. The peripheral positions of these young basalt fields may also indicate the existence of large magma chambers beneath the center of the volcanic field. Tectonic activity has continued from about 4 m.y. ago to the present, accompanied by the voluminous eruptions of Bandelier Tuff from the large, hybrid-magma chambers.

Golombek and others (1983) present rates of extension for the Española Basin which, in light of the new K–Ar dates presented herein and our evidence for episodic rifting for >16.5 m.y., are probably unrealistic. Golombek and others (1983) also provide an estimate of about 10% total extension across the basin. It is not clear from their discussion what data were used to generate this estimate, but if they did not include the movements on the Cañada de Cochiti fault zone, as may be inferred from the above discussion of the "buried graben," their estimate is much less than minimal.

The interval that we suggest as a tectonic lull (7–4 m.y. ago) is precisely the interval that is widely cited as one of rapid uplift throughout the Rio Grande rift (e.g., Chapin, 1979). The works of Axelrod and Bailey (1976), Taylor (1975), and Scott (1975) are cited in support of rapid uplift occurring 7–4 m.y. ago; however, these authors only state that uplift in the Rio Grande region must have occurred sometime *since* 7 m.y. ago. Manley (1979) presents evidence that relatively intense faulting occurred in the Velarde graben in the northern Española Basin sometime during 5–3 m.y.B.P. Therefore, our suggestion of the 7–4 m.y. tectonic lull is consistent with, and/or supported by, the works of Axelrod and Bailey (1976), Scott (1975), Taylor (1975), and Manley (1979).

ACKNOWLEDGMENTS

This work has been supported by the U.S. Department of Energy Office of Basic Energy Science and Division of Geothermal Energy, and the University of California Institute of Geophysics and Planetary Physics. We thank Dan Krummenacher and Donna Martin of San Diego State University and Geochron Laboratories for providing the potassium–argon dates. Our colleageus Scott Baldridge, Kim Manley, Bill Laughlin, Frank Perry, Mat Golombek, Bruce Loeffler, Bob Riecker, and Roy Bailey have provided encouragement, discussions, and/or data. Bjørn Larsen, Dave Vaniman, and an anonymous reviewer provided many helpful comments on this paper. Barbara Hahn patiently and cheerfully reworked the manuscript, and Cathy Beth Helmick helped photograph the figures that were prepared by John Tubb and crew.

REFERENCES

Axelrod, D. I., and Bailey, H. P., 1976, Tertiary vegetation, climate, and altitude of the Rio Grande depression, New Mexico–Colorado: Paleobiology, v. 2, pp. 235–254.

Bachman, G. O., and Mehnert, H. H., 1978, New K–Ar dates and the late Pliocene to Holocene geomorphic history of the central Rio Grande region, New Mexico: Geological Society of America, Bulletin, v. 89, pp. 283–292.

Bailey, R. A., Smith, R. L., and Ross, C. S., 1969, Stratigraphic nomenclature of volcanic rocks in the Jemez Mountains, New Mexico: U.S. Geological Survey, Bulletin 1274-P, 19 pp.

Baldridge, W. S., 1979, Petrology and petrogenesis of basaltic rocks from the central Rio Grande rift, New Mexico, and their relation to rift structure; *in* Riecker, R. E. (ed.), Rio Grande rift: tectonics and magmatism: American Geophysical Union, Washington, D.C., pp. 323–353.

———, and Perry, F. V., 1983, Spatial and temporal patterns of late Cenozoic basaltic volcanism of the central Rio Grande rift area: American Geophysical Union, Transactions, v. 64, no. 45, p. 753.

———, Bartov, Y., and Kron, A., 1983, Geologic map of the Rio Grande rift and southeastern Colorado Plateau, New Mexico and Arizona; *in* Riecker, R. E. (ed.), Rio Grande rift: tectonics and magmatism: American Geophysical Union, Washington, D.C., scale 1:500,000.

———, Damon, P. E., Shafiqullah, M., and others, 1980, Evolution of the central Rio Grande rift: new potassium–argon ages: Earth and Planetary Sciences Letters, v. 51, pp. 309–321.

Budding, A. J., 1978, Gravity survey of the Pajarito Plateau, Los Alamos and Santa Fe Counties, New Mexico: Los Alamos Scientific Laboratory, Report LA-7419, 15 pp.

Cash, D. J., Olsen, K. H., McFarland, N. J., and others, 1982, Earthquake catalog for northern New Mexico: Los Alamos National Laboratory, Report LA-9467-PR, 20 pp.

Chapin, C. E., 1979, Evolution of the Rio Grande rift: a summary; *in* Riecker, R. E. (ed.), Rio Grande rift: tectonics and magmatism: American Geophysical Union, Washington, D.C., pp. 1–6.

———, and Seager, W. R., 1975, Evolution of the Rio Grande rift in the Socorro and Las Cruces areas: New Mexico Geological Society, Guidebook 26, pp. 297–322.
Cordell, L., 1978, Regional geophysical setting of the Rio Grande rift: Geological Society of America, Bulletin, v. 89, pp. 1073–1090.
———, 1979, Gravimetric expression of graben faulting in Santa Fe County and the Española Basin, New Mexico: New Mexico Geological Society, Guidebook 30, pp. 59–64.
Dalrymple, G. B., 1979, Critical tables for conversion of K–Ar dates from old to new constants: Geology, v. 7, pp. 558–560.
———, Cox, A., Doell, R. R., and others, 1967, Pliocene geomagnetic polarity epochs: Earth and Planetary Science Letters, v. 2, pp. 163–173.
Doell, R. R., Dalrymple, G. B., Smith, R. L., and others, 1968, Paleomagnetism, potassium–argon ages, and geology of rhyolites and associated rocks of the Valles caldera, New Mexico: Geological Society of America, Memoir 116, pp. 211–248.
Gardner, J. N., 1982, Petrogenesis of the Jemez Mountains volcanic field: American Geophysical Union, Transactions, v. 63, p. 1131.
———, 1983, Tectonic controls of petrogenesis, Jemez volcanic field, New Mexico (abstract): Geological Society of America, Abstracts with Programs, v. 15, no. 5, pp. 380–381.
———, 1984, Tectonic and petrologic evolution of the Keres Group: implications for the development of the Jemez volcanic field, New Mexico: Ph.D. dissertation (in prep.), University of California at Davis.
Goff, F., 1983, Subsurface structure of Valles caldera: a resurgent cauldron in northern New Mexico (abstract): Geological Society of America, Abstracts with Programs, v. 15, no. 5, p. 381.
———, and Gardner, J. N., 1980, Geologic map of the Sulphur Springs area, Valles caldera geothermal system, New Mexico: Los Alamos Scientific Laboratory, Report LA-8634-MAP, scale 1:5,000.
———, and Grigsby, C. O., 1982, Valles caldera geothermal systems, New Mexico, U.S.A.: Journal of Hydrology, v. 56, pp. 119–136.
———, and Kron, A., 1980, In-progress geologic map of Cañon de San Diego, Jemez Springs, New Mexico, and lithologic log of Jemez Springs geothermal well: Los Alamos Scientific Laboratory, Report LA-8276-MAP, scale 1:12,000.
Golombek, M. P., 1981, Structural analysis of the Pajarito fault zone in the Española Basin of the Rio Grande rift, New Mexico: Ph.D. dissertation, University of Massachusetts, Amherst, 129 pp.
———, 1983, Geology, structure, and tectonics of the Pajarito fault zone in the Española Basin of the Rio Grande rift, New Mexico: Geological Society of America, Bulletin, v. 94, pp. 192–205.
———, McGill, G. E., and Brown, L., 1983, Tectonic and geologic evolution of the Española Basin, Rio Grande rift: structure, rate of extension, and relation to the state of stress in the western U.S.: Tectonophysics, v. 94, pp. 483–507.
Griggs, R. L., 1964, Geology and groundwater resources of the Los Alamos area, New Mexico: U.S. Geological Survey, Water-supply Paper 1753, 107 pp.
Kelley, V. C., 1978, Geology of the Española Basin, New Mexico: New Mexico Bureau of Mines and Mineral Resources, Geologic Map 48, scale 1:125,000.
———, Woodward, L. A., Kudo, A. M., and others, 1976, Guidebook to Albuquerque Basin of the Rio Grande rift, New Mexico: New Mexico Bureau of Mines and Mineral Resources, Circular 153, 31 pp.
Lawrence, J. R., 1979, Geology of the Cerro del Grant area, Rio Arriba County, north-central New Mexico: M.S. thesis, University of New Mexico, 131 pp.
Leudke, R. G., and Smith, R. L., 1978, Map showing distribution, composition, and age of late Cenozoic volcanic centers in Arizona and New Mexico: U.S. Geological Survey, Miscellaneous Investigations Series Map I-1091-A.
Loeffler, B. M., 1984, Major- and trace-element and strontium- and oxygen-isotope geochemistry of the Polvadera Group, Jemez volcanic field, New Mexico: implications for the petrogenesis of the Polvadera Group: Ph.D. dissertation, University of Colorado, Boulder, 259 pp.
Manley, K., 1976a, The late Cenozoic history of the Española Basin, New Mexico: Ph.D. dissertation, University of Colorado, 171 pp.
———, 1976b, K–Ar age determinations of Pliocene basalts from the Española Basin, New Mexico: Isochron/West, no. 16, pp. 29–30.
———, 1979, Stratigraphy and structure of the Española Basin, Rio Grande rift, New Mexico; *in* Riecker, R. E. (ed.), Rio Grande rift: tectonics and magmatism: American Geophysical Union, Washington, D.C., pp. 71–86.
———, and Mehnert, H. H., 1981, New K–Ar ages of Miocene and Pliocene volcanic rocks in the northwestern Española Basin, and their relationships to the history of the Rio Grande rift: Isochron/West, no. 30, pp. 5—8.
Marvin, R. F., and Dobson, S. W., 1979, Radiometric ages: compilation B, U.S. Geological Survey: Isochron/West, no. 26, pp. 3–32.
Scott, G. R., 1975, Cenozoic surfaces and deposits in the southern Rocky Mountains: Geological Society of America, Memoir 144, pp. 227–248.
Seager, W. R., Shafiqullah, M., Hawley, J. W., and others, 1984, New K–Ar dates from basalts and the evolution of the southern Rio Grande rift: Geological Society of America, Bulletin, v. 95, pp. 87–99.
Smith, R. L., 1979, Ash-flow magmatism: Geological Society of America, Special Paper 180, pp. 5–28.
———, Bailey, R. A., and Ross, C. S., 1970, Geologic map of the Jemez Mountains, New Mexico: U.S. Geological Survey, Miscellaneous Geological Investigations Map I-571, scale 1:125,000.
Stieger, R. H., and Jager, E., 1977, Subcommission on geochronology: convention on the use of decay constants in geo- and cosmochronology: Earth and Planetary Science Letters, v. 36, pp. 359–362.
Taylor, R. B., 1975, Neogene tectonism in south-central Colorado: Geological Society of America, Memoir 144, pp. 211–226.
Williams, L. M., 1979, Gravity study of the Los Alamos area, New Mexico: Los Alamos Scientific Laboratory, Report LA-8154-MS, 18 pp.
Woodward, L. A., and Ruetschilling, R. L., 1976, Geology of San Ysidro quadrangle, New Mexico: New Mexico Bureau of Mines and Mineral Resources, Geologic Map 37, scale 1:24,000.
———, DuChene, H. R., and Martinez, R., 1977, Geology of Gilman quadrangle, New Mexico: New Mexico Bureau of Mines and Mineral Resources, Geologic Map 45, scale 1:24,000.

The entire main line route of the Southern San Luis Valley Railroad is visible in this picture, running from the Colorado Aggregate Company yard behind the camera to Blanca in the distance.

QUATERNARY AND PLIOCENE FAULTING IN THE TAOS PLATEAU REGION, NORTHERN NEW MEXICO

STEPHEN F. PERSONIUS and MICHAEL N. MACHETTE
U.S. Geological Survey, Federal Center, Denver, Colorado 80225

INTRODUCTION

The Taos Plateau region of northern New Mexico contains numerous faults that offset sedimentary and volcanic rocks of Pliocene and Quaternary age. This report discusses these known or inferred faults and summarizes the work of Machette and Personius (unpublished mapping 1984). Although previous small-scale maps of the Rio Grande rift (Woodward and others, 1975; Baldridge and others, 1983) showed many of the features described in this report, our work focuses on faults and folds less than 5 m.y. old.

This report discusses some of the prominent faults in the Taos Plateau region and includes data on the amounts of vertical surface offset (fault throw), ages of faulted deposits, recency of fault movement, and morphology of fault scarps. Amounts of offset were determined in two general ways: (1) from direct field measurements of the net vertical offset of surfaces across the fault, and (2) from indirect measurements from 1:24,000-scale topographic maps, aerial photographs, and published data. The direct measurements probably are accurate to within 10% of the actual surface offsets, whereas the indirect measurements usually reflect the range of values that are common along a fault (for example, 5–20 m of surface offset).

Only faults that displace, or are suspected to displace, Pliocene or younger materials are shown in Figure 1. Many more faults may be present in the area. For example, Pliocene or younger faulting probably occurred, but cannot be demonstrated in areas where only Miocene or older rocks are exposed; most mapped faults in these areas are not shown. In other areas, deposition or erosion may have removed evidence of Pliocene and younger faulting.

Data used to compile Figure 1 are from published and unpublished geologic maps and literature, inspection of aerial photographs, and limited field investigations of Quaternary fault scarps. The geologic-map data are quite variable, ranging from reconnaissance maps at scales of 1:62,500 or smaller to detailed maps at scales of 1:24,000 or larger. Where several sources of data were available, we used the most recent data that focused on late Cenozoic structural and stratigraphic studies of an area.

Our approach of showing only faults for which direct evidence of Pliocene or younger movement exists yields a minimum number of faults. Readers using these data for hazards assessment, regional structural analysis, or other similar studies should be aware that many other Pliocene or younger faults may be, and probably are, present in the Taos Plateau.

GEOLOGIC SETTING

Because the main subject of this report is Pliocene and Quaternary faulting, we deal with the pre-Pliocene geology of the region by briefly discussing the basic structural setting of the Taos Plateau portion of the northern Rio Grande rift. Recent reports on the region include discussions of the general evolution of the Rio Grande rift by Chapin (1979); the guidebook of the Rio Grande rift compiled by Hawley (1978); a study of upper Cenozoic sedimentary and volcanic rocks of the Taos area by Lambert (1966); the geology of the Taos Plateau by Lipman and Mehnert (1979); and the geology of the Española Basin by Manley (1979). The following discussion is a broad overview of the late Cenozoic history of the Rio Grande rift in northern New Mexico.

The area covered in this report includes the northernmost part of the Española Basin, the southeastern edge of the Chama Basin, and the Taos Plateau (the southern part of the San Luis Basin). These basins are separated by two major structural elements: (1) the Brazos uplift, a southeast-trending ridge of Precambrian basement rock (Cordell, 1978) that terminates southward near Cerro Azul, and (2) the northeast-trending Embudo fault that parallels the Rio Grande in the southern part of the map area (Muehlberger, 1979). The Embudo fault is coincident with the Jemez lineament (Lambert, 1966; Cordell, 1978) where it traverses the Rio Grande rift. Outside of the rift, the Jemez lineament is marked by a northeast-trending alignment of late Miocene to Holocene volcanic centers (including the Jemez caldera) that extend from southwestern New Mexico to southeastern Colorado (Lambert, 1966).

The general structural pattern of the Taos Plateau region is that of a half graben with major displacement on the eastern margin of the rift, along the western front of the southern Sangre de Cristo Mountains. Gravity data suggest that this part of the rift is underlain by 5 km or more of basin-filling rocks (L. E. Cordell, unpubl. data cited *in* Lipman and Mehnert, 1979). Taking the topographic relief between the Taos Plateau and the Sangre de Cristo Mountains into account, Precambrian to lower Tertiary rocks may be offset by as much as 7–8 km along the east margin of the rift (Lipman and Mehnert, 1979). Conversely, the western margin of the rift is indistinctly marked by flows of the Servilleta Basalt that lap onto eastward-tilted pre-rift to early-rift volcanic and sedimentary rocks. A north–northwest-trending zone of small displacement, down-to-the-west antithetic normal faults that cut Pliocene Servilleta Basalt, forms the western boundary of the Rio Grande rift in the Taos Plateau region.

The central part of the region is dominated by the Taos Plateau volcanic field (Upson, 1939; Lambert, 1966; Lipman and Mehnert, 1979; Dungan and others, this volume). Tholeiitic lavas of the Pliocene Servilleta Basalt erupted from shield volcanoes, and are the most voluminous and widespread rock type on the plateau. About 150 m of flood basalt is exposed in the Rio Grande gorge, which bisects the southern part of the Taos Plateau. Here as many as 15 basalt flows form three sequences that are interbedded with sediment of the upper part of the Santa Fe Group (Lipman and Mehnert, 1979). Most K–Ar age determinations from flows of the Servilleta Basalt are concentrated in a narrow range from 3.6 to 4.5 m.y. (Ozima and others, 1967; Lipman and Mehnert, 1979). However, dates of 2.8 m.y. (Manley, 1979) and 4.6 m.y. (Baldridge and others, 1980) have been obtained on upper flows of the Servilleta Basalt in the Black Mesa and Rio Grande gorge areas, respectively. Despite these conflicting dates, Dungan and others (this volume) have chosen to use the chronology of Ozima and others (1967) and Lipman and Mehnert (1979) in their detailed examination of the Taos Plateau volcanic field.

The thick sequence of Santa Fe Group sediments within the northern Rio Grande rift provides a stratigraphic record of faulting during Miocene through Pleistocene time. The Tesuque and Chamita Formations of the lower part of the Santa Fe Group (Galusha and Blick, 1971; Manley, 1979) record the depositional response to Miocene uplift and extension; these and coeval units in the southern Rio Grande rift were deposited in a series of closed, fault-bounded basins.

Renewed extension, coupled with regional uplift and a subsequent increase in runoff in the early Pliocene (Chapin, 1979), resulted in establishment of the Rio Grande as a major integrated system that flowed southward through southern Colorado and New Mexico into northern Mexico. During the Pliocene and much of the Pleistocene, the ancestral Rio Grande and its tributary streams deposited a thick sequence of fluvial, bolson, and deltaic sediments in a series of elongate depressions

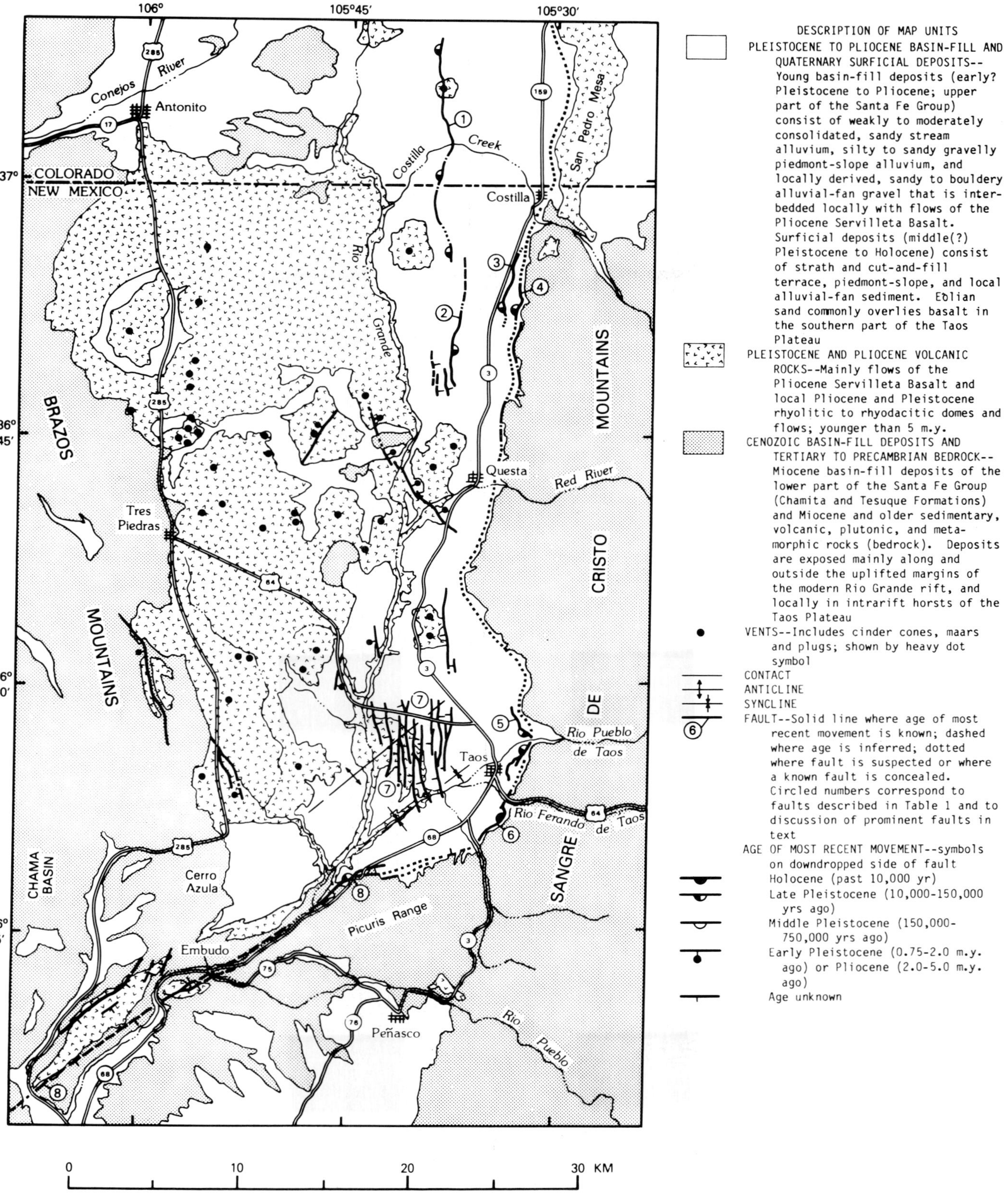

FIGURE 1. Map showing Pliocene and Quaternary faults in the Taos Plateau region, northern New Mexico.

along the length of the Rio Grande rift. These sediments are known collectively as the upper Santa Fe Group. In northern New Mexico the upper part of the Santa Fe Group is represented by the Pliocene Ancha and Puye Formations, and in the remainder of New Mexico it is represented by the Pliocene and Pleistocene Sierra Ladrones and Camp Rice Formations. Tilting of lower Santa Fe Group rocks during or before the early Pliocene created a strong unconformity that commonly separates the two parts of the Santa Fe Group. In the Española Basin, the upper part is composed of fanglomerates derived from opposite sides of the rift; the Puye Formation sediments were shed from volcanic highlands in the Jemez Mountains along the west side of the rift, and the Ancha Formation sediments were shed from the Precambrian-cored Sangre de Cristo Mountains east of the rift (Manley, 1979). Alluvium deposited by the Rio Grande in Pliocene time also is included in the upper part of the Santa Fe Group.

Along the upper reach of the ancestral Rio Grande (those parts in New Mexico and Colorado), deposition in the rift basins contained from 4–5 m.y. to perhaps 0.5 m.y. ago, at which time the upper and lower reaches of the Rio Grande were connected at El Paso, Texas. Base level along the upper Rio Grande dropped drastically as a result of this drainage integration. The Rio Grande and its tributaries subsequently cut deeply into basin-fill sediments in the rift and, in some places, into the underlying bedrock.

METHODS FOR DETERMINING AGES OF FAULTING

Two methods are used in this study to determine the ages of fault movement. The first method is a stratigraphic approach by which the ages of faulted and unfaulted deposits bracket the age of the most recent fault movement. Recurring episodes of movement often are recognized by comparing the cumulative amounts of displacement in deposits of different ages. A second method is to infer ages of undated fault scarps through a quantitative comparison of their morphology with morphologies of dated scarps. This latter method is discussed in the section entitled Scarp Morphology Methods.

Stratigraphic Methods

The distributions of young basin-fill deposits and volcanic rocks, surficial deposits, and some of the faults shown in Figure 1 are from maps by Lambert (1966), Kelley (1977), Manley and others (1978), and Lipman and Mehnert (1979), and from examination of aerial photographs. Pliocene and younger deposits were divided into four categories to provide age controls of fault movement. The oldest category, of Pliocene to early Pleistocene age (5.0–0.75 m.y.), mainly includes flows of the Servilleta Basalt, as well as local rhyolitic to rhyodacitic rocks and interbedded sediment. This age category also includes late Pliocene to early(?) Pleistocene aluvial-fan and piedmont-slope basin-fill deposits that immediately overlie basalt flows of the upper part of the Servilleta Basalt. The middle and late Pleistocene categories (750,000–150,000 and 150,000–10,000 yrs ago, respectively) include terrace, piedmont-slope, and alluvial-fan deposits graded to drainages that were incised into volcanic rocks of the Taos Plateau and into Miocene and older rocks in the Chama Basin west of the Taos Plateau. The youngest category includes Holocene (<10,000 yrs) alluvium of the lowest terraces, floodplains, and alluvial fans, and also a thin blanket of eolian sand on the southern Taos Plateau (Kelley, 1977).

Scarp-morphology Methods

Studies of slope-degradation processes and the morphology of young fault scarps by Wallace (1977) and recent demonstration of an empirical relation between the height of a fault scarp (H) and its maximum slope angle (θ) by Bucknam and Anderson (1979) have established one basis for determining the relative age of a fault scarp. We use this empirical relation to estimate the recency of fault movement for some Quaternary faults in the Taos Plateau region (see Machette, 1982, or Machette and McGimsey, 1983, for more detailed information about this dating method).

Our scarp nomenclature (Fig. 2) is modified from that of Bucknam and Anderson (1979). Measurements of scarp morphology are from detailed topographic surveys made along traverses perpendicular to the surface traces of faults. Each traverse yields a scarp profile (Fig. 2), from which measurements of H and θ (a data pair) were determined. Buchnam and Anderson (1979) showed that a significant correlation exists between the height (H) and the maximum slope angle (θ) of scarps formed by a single rupture event; a plot of data from a hypothetical scarp profile is illustrated in Figure 3. Data pairs are plotted as open circles and, where there are at least four data pairs, the equation for the line of best fit and the coefficient of determination (r^2) are included.

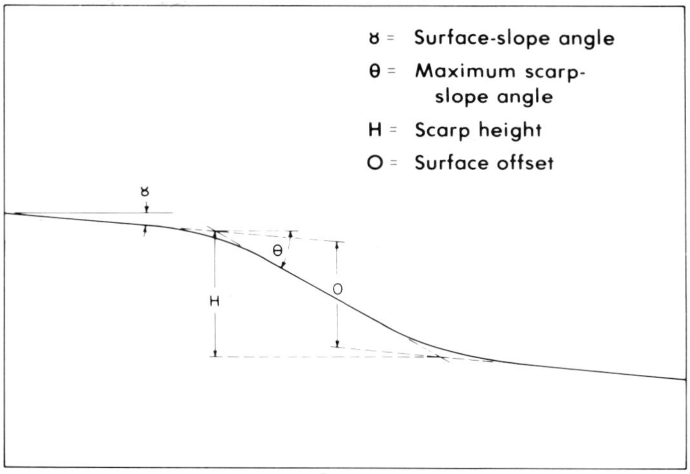

FIGURE 2. Diagrammatic profile of a hypothetical fault scarp (modified from Bucknam and Anderson, 1979, fig. 1).

We used the methods and results of Bucknam and Anderson (1979) and Machette (1982) to assign relative ages to fault scarps. The lines of best fit for two sets of scarp data were used for comparison in Figure 3 and in similar plots of scarp-morphology data (Figs. 6–10). The uppermost dashed line (labeled 5K in Fig. 3) represents combined data from two fault scarps in New Mexico: a fault near the Cox Ranch, just west of White Sands (most recent movement about 4,000 yrs ago; Gile and Machette *in* Seager, 1981; Machette, unpubl. data 1982) and segment C of the La Jencia fault near Magdalena (most recent movement about 5,000 yrs ago; Machette, 1982). The lowermost dashed line (labeled 15K in Fig. 3) is based on R. C. Bucknam's data (written

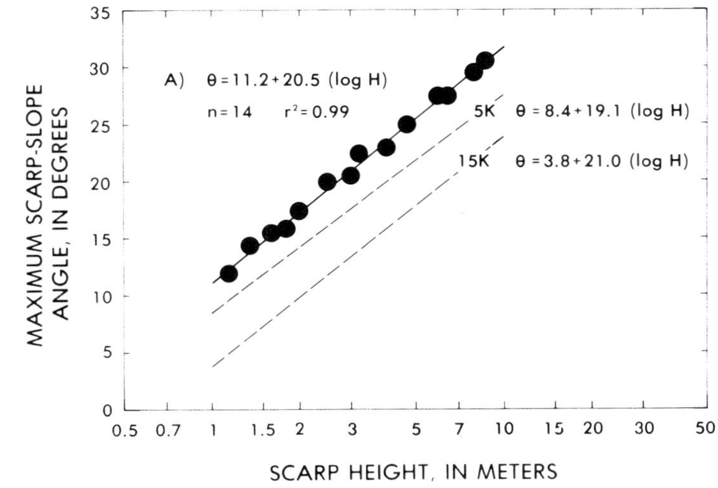

FIGURE 3. Plot of maximum scarp-slope angle (θ) against scarp height (H) for a hypothetical fault scarp. Also shown are the number of data pairs (n), the coefficient of determination (r^2), two reference lines (see text for sources) for age comparisons—5K (4,000–5,000 yrs ago) and 15K (15,000 yrs ago)—and equations for their lines of best fit.

comm. 1980) from the highest wave-cut shoreline of Lake Bonneville, in Utah, which is about 15,000 yrs old based on stratigraphic studies by Scott and others (1984). Although the Bonneville shoreline escarpment is an erosional feature, Bucknam and Anderson (1979) believe it is eroded in the same manner as a fault scarp. Factors (in addition to age) that affect the morphology of a scarp include the texture of the faulted deposits, scarp orientation, and the climate, biological activity, and topographic relief at the scarp site. Although we tried to choose profile sites that minimized these differences, we expected some variation in morphology between scarps of similar age. Therefore, we use the following criteria to assign relative ages to fault scarps: (1) scarps having data sets above the upper dashed line (5K) are considered to be Holocene in age (<10,000 yrs), (2) data sets that occur between the two dashed lines indicate scarps of Holocene or latest Pleistocene age, and (3) data sets that occur below the lower dashed line (15K) indicate scarps of late Pleistocene age (10,000–150,000 yrs old) or, less likely, of middle Pleistocene age (150,000–750,000 yrs old).

Some fault scarps in the region were formed by multiple episodes of surface rupture, as evidenced either by decreasing scarp heights in progressively younger deposits or by scarps that have discrete facets (that is, compound-slope angles) as recognized by Wallace (1977). These compound fault scarps are degraded fault scarps that have been refaulted one or more times (the older scarp slope is noted by the angle θ' in Fig. 4). We interpreted recency of faulting from many compound scarps by assuming that the steepest portion of the scarp (Hs in Fig. 4) is formed by the most recent faulting episode, and that Hs is a close approximation of the most recent offset. Therefore, plots of Hs versus θ for a compound scarp and plots of H versus θ for an adjacent single-event scarp should yield similar results. Compound fault scarps usually have Hs versus θ values (line A in Fig. 5) that plot to the left of Hm versus θ values (line B in Fig. 5). It is important to note that a fault scarp having a line of best fit with a low-slope value may indicate multiple episodes of faulting even though its resultant scarp does not exhibit compound slope angles.

PROMINENT QUATERNARY FAULTS

Most of the following information is based on field investigations conducted during 1980 and 1981, although some data were available from published results of others. The faults are keyed by numbers to Figure 1, and pertinent data for the faults are summarized in Table 1.

Mesita Fault (1)

The Mesita fault, first mapped by Colton (1976), extends from about 5 km north of Mesita cone in southern Colorado to the east flank of Ute Mountain in northern New Mexico. Only the southern 7 km of its 22-km trace are in New Mexico. The Mesita fault (fault 108 *in* Kirkham and Rodgers, 1981) forms a conspicuous west-facing scarp along the east side of the Sky Valley Ranch 7½′ quadrangle (Colorado), where it offsets the west flank of the Quaternary Mesita basalt cone (Tweto, 1979) about 13 m (Kirkham and Rodgers, 1981, p. 25). According to Burroughs (1978), flows of the Pliocene Servilleta Basalt are offset only 15–30 m by the Mesita fault. If so, then much of the fault's post-Servilleta offset is recorded in the 13-m-high scarps on Mesita cone. A strong calcic soil has formed in a 2- to 3-m-thick blanket of eolian sand and reworked basaltic cinders that overlies the basalt. Elsewhere on the Taos Plateau such soils are common in early(?) to middle Pleistocene alluvium.

South of Mesita cone, the fault is marked by 5- to 7-m-high south-trending scarps in early(?) to middle Pleistocene alluvium, and in New Mexico by 2- to 5-m-high fault scarps in middle to late(?) Pleistocene alluvium. The later scarps are discontinuous, moderately eroded, and dissected by streams; their gross morphology suggests they are late Pleistocene, but not latest Pleistocene (25,000–10,000 yrs) in age.

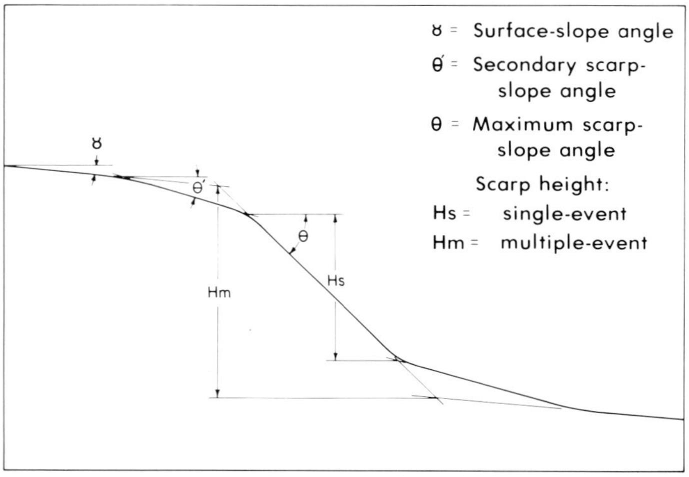

FIGURE 4. Diagrammatic profile of a hypothetical compound fault scarp.

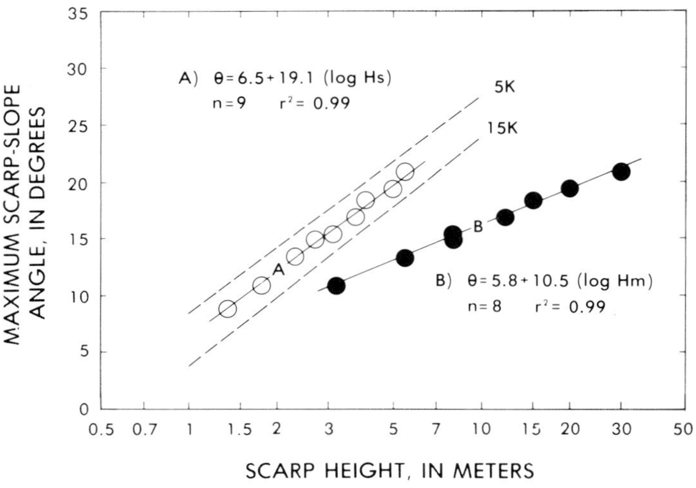

FIGURE 5. Plot of maximum scarp-slope angles (θ) against scarp heights (Hs and Hm) for a hypothetical compound fault scarp. Also shown are the number of data pairs (n), the coefficient of determination (r^2), and two reference lines (see text for sources) for age comparisons—5K (4,000–5,000 yrs ago) and 15K (15,000 yrs ago).

Sunshine Valley Faults (2)

Colton (1976) mapped a series of north-trending faults, here collectively named the Sunshine Valley faults, that extends from Sunshine Valley north between Guadaloupe and Ute Mountains. The longest of these suspected faults extends northward 22 km, from a south fork of Latir Creek to a point 6 km due east of Ute Mountain. Although Colton regarded these faults as the southern extension of the Mesita fault (1), we consider them to be separate because they are not continuous with the Mesita fault and have the opposite sense of movement, down to the east.

The Sunshine Valley faults cross an area underlain by late Pleistocene alluvium; however, 1–2 km west (downslope) of the main fault, elongate ridges of middle(?) Pleistocene alluvium occur 5–10 m above stream level. These ridges may be the result of offset along the Sunshine Valley faults and subsequent stream erosion through the uplifted, western block. The longest and easternmost of the Sunshine Valley faults is the only one that coincides with an obvious alignment of vegetation. Although we did not observe a topographic escarpment along this lineament, fine-grained sediment ponded on the east side of the lineament suggests that it coincides with a small fault scarp that now is largely buried. If the Sunshine Valley faults actually cut alluvium, their offsets must be small (less than 5 m). The three suspected faults that Colton (1976) mapped south of Sunshine and west of the main fault were not located in the field.

TABLE 1. Location, recency of movement, and size of some major faults in the Taos Plateau region. Ages: **h**, Holocene; **lp**, late Pleistocene; **mp**, middle Pleistocene; **ep**, early Pleistocene to Pliocene. Symbol **e** indicates scarp height estimated from 7½′ topographic map or aerial photograph. Types of fault movement are determined from scarp morphology, age determination of faulted deposits, and/or map pattern; all faults are normal except as noted.

Map No.	Fault name(s)	Location	Age of most recent movement	Measured scarp heights (m)	Age of displaced deposits	Fault length (km)	Type of fault movement	References
1	Mesita	E of Ute Mountain	lp	2-5e	ep-lp	22	Recurrent	Colton, 1976; Burroughs, 1978; Kirkham and Rodgers, 1981
2	Sunshine Valley (new name)	E of Sunshine, 14 km SSW of Costilla	lp?	<5e	mp?-lp	2-13	Unknown	Colton, 1976
3	Cedros Canyon (new name)	5-12 km S of Costilla E of NM Hwy 85	lp	2-14	mp-lp	7	Recurrent	Upson, 1939
4	Urraca Ranch (new name)	12 km S of Costilla at Urraca Ranch	lp	17	mp-lp?	2	Recurrent	Upson, 1939
5	Taos Pueblo (new name)							
	North part	2 km NE of Taos Pueblo	lp	2-18	ep-lp	5	Recurrent, segmented	None
	South part	2 km E of Taos, N to Rio Pueblo de Taos	lp	2-7	mp?-lp	5		None
6	Canon (new name)	E of Taos, from Canon S to Talpa	lp-h?	2-5	ep?-lp	6	Recurrent	Lambert, 1966, p. 47
7	Los Cordovas (new name)	N of Los Cordovas, 10 km W of Taos	ep-mp?	<15-30e	ep-mp?	<8-12	Unknown, many faults	Lambert, 1966, p. 47
8	Embudo (Picuris frontal fault)	Hondo Canyon, SW along Rio Grande, coincident with Jemez lineament	lp?	7-15	ep-mp	≥32 to 75	Recurrent, reverse to normal	Lambert, 1966, p. 47; Muehlberger, 1979, p. 77-82

Sangre de Cristo Fault Zone (Faults 3–6)

The Sangre de Cristo fault zone is one of the major Pliocene and Pleistocene structures of the Taos Plateau. The fault zone bounds the west side of the Sangre de Cristo Mountains along most of their 160-km length in southern Colorado and northern New Mexico. Scott (1970) first mapped many of the faults in this zone, and more recently Kirkham and Rodgers (1981) and McCalpin (1983) demonstrated that Holocene or late Pleistocene movement occurred along most of the Sangre de Cristo fault zone in Colorado. Information about the fault zone in New Mexico is limited mainly to our sparse scarp-morphology data from faults in the Costilla (Cedro Canyon fault, 3; Urraca Ranch fault, 4) and Taos areas (Taos Pueblo fault, 5; Cañon fault, 6). Our data suggest that the New Mexico portion of the Sangre de Cristo fault zone was less active in the late Quaternary than the portion of the fault zone in Colorado.

The Sangre de Cristo fault zone extends northward from New Mexico into southernmost Colorado along the west side of the San Pedro Mesa. Kirkham and Rodgers (1981) referred to this segment as the West San Pedro Mesa fault. San Pedro Mesa is capped by flows of 3- to 4-m.y.-old Servilleta Basalt that overlie poorly consolidated sediment of the Santa Fe Group; as a result, the West San Pedro fault is covered by landslide debris along most of its length. Burroughs (1978) estimated that basalts west of the fault at the base of San Pedro Mesa have been displaced at least 600 m; this offset corresponds to an average rate of uplift of 150–200 m/m.y. since the basalts were deposited. Immediately southeast of Costilla, the southernmost remnant of the basalt of San Pedro Mesa is uplifted 240–360 m above the adjacent plain of the Taos Plateau. Here the minimum average rate of uplift is 60–120 m/m.y.

Cedro Canyon Fault (3)

The surface trace of the Cedro Canyon fault extends from a point 5 km south of Costilla, New Mexico, south and south–southwest about 7 km. This fault is a westward splay of the main range-bounding fault, which is poorly expressed in this area. The fault is named for Cedro Canyon, the source area of a series of alluvial-fan deposits that are offset by the fault. Upson (1939, p. 731) first mapped the fault and showed its scarp in a photograph taken from a point 6.7 km south of Costilla. Here the fault forms a conspicuous 4- to 14-m-high scarp just east of New Mexico State Highway 3.

The Cedro Canyon fault offsets locally derived alluvial-fan debris of middle to late(?) Pleistocene and possibly early Pleistocene age. However, the latest Pleistocene to Holocene alluvial-fan sediments of Cedro Canyon bury the fault scarp at its north end. Although Upson (1939) considered the fault to be Holocene, stratigraphic evidence suggests that the most recent fault movement predated the Holocene; however, the scarp data suggest it need not be older than latest Pleistocene.

The Cedro Canyon fault scarps record multiple episodes of surface offset. The larger scarps are about 8–9 m high in old alluvium and have compound scarp-slope angles indicating young movement superposed on older, degraded scarps. South of the road to Urraca Ranch, single-event scarps less than 4 m high are in younger late(?) Pleistocene alluvium. Figure 6 shows the relation between both single (Hs)- and multiple-event (Hm) scarp heights and maximum scarp-slope angles (θ)

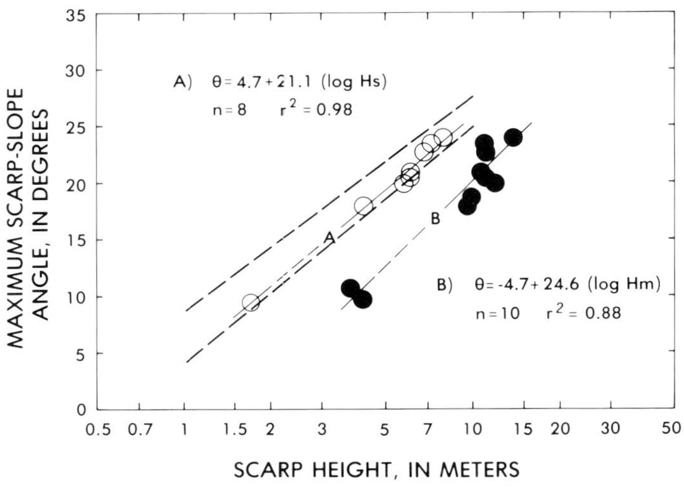

FIGURE 6. Plot of maximum scarp-slope angles (θ) against scarp heights (Hs and Hm) for the Cedro Canyon fault (3). Open circles (line A) are for most recent single-event part of scarp; filled circles (line B) are for entire multiple-event scarp.

for the Cedro Canyon fault. These data also suggest that the latest Pleistocene (10,000–25,00 yrs ago) movement produced scarps 2–8 m high (Hs values).

Urraca Ranch Fault (4)

Upson (1939, p. 731) mentioned a second fault in the Costilla area, here named the Urraca Ranch fault. This fault occurs at the mountain front and forms a prominent scarp across an alluvial fan at the mouth of Jaroso Canyon (La Jara Creek of Upson). The scarp is preserved a short distance adjacent to Jaroso Canyon; it is eroded or poorly developed farther to the north and south along the mountain front.

The soil in the fan alluvium is either early late Pleistocene or late middle Pleistocene in age, judging from the presence of a 0.5- to 1-m-thick, reddish-brown, clay-enriched (argillic) B horizon in the upper part of the alluvium. Upson (1939) reported a 70-ft (21-m)-high scarp, but the scarp we measured at Urraca Ranch is 16.5 m high (Hm) and has a maximum slope angle (θ) of 16–17°. These data indicate that this scarp is more degraded than similar scarps along the Cedro Canyon fault (Fig. 6). Although the scarp lacks a compound profile, its large height suggests several episodes of movement since the middle Pleistocene; the most recent movement may have occurred during late Pleistocene time, because the scarp still has a relatively steep slope.

Taos Pueblo Fault (5)

The Taos Pueblo fault forms a sinuous escarpment that trends roughly northwest and southwest, north and south of Taos Pueblo, respectively. The fault is separated into two segments by the Rio Pueblo de Taos where it emerges from the mountains. The heights of scarps vary from <2 to 18 m along the northern segment and from <2 to 7 m along the southern segment. The fault forms single, double, and compound scarps along a curving, concave-westward trace about 10 km long and 0.5–1.5 km basinward of the mountain front.

The northern segment of the fault typically has an 8- to 15-m-high scarp that locally splits into two closely spaced scarps, each with heights of 5–12 m. Due north of Taos Pueblo the fault trends northwest and cuts early(?) to middle Pleistocene coarse-grained fan alluvium. Farther north the fault parallels the mountain front and displaces late(?) Pleistocene alluvium less than 2 m. Scarp data from the northern segment (Fig. 7) show clear evidence of recurrent fault movement, the most recent of which produced 3- to 7-m-high scarps during the late Pleistocene.

The southern segment of the fault forms two nearly parallel southwest-trending scarps in middle to late Pleistocene piedmont-slope alluvium. These scarps are less than 2–3 m high, but join about 2 km south of Taos Pueblo to form a single scarp 5–6 m high (Hm, Fig. 8). The most recent rupture on the combined fault produced scarps 3–4 m high (Hs, Fig. 8) during the late Pleistocene, as indicated by the compound nature of the scarp. However, this most recent episode of movement may not be as young as that along the northern segment of the Taos Pueblo fault.

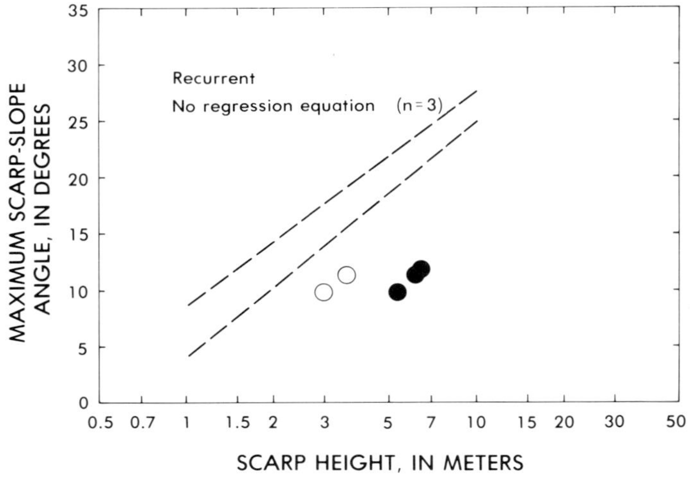

FIGURE 8. Plot of maximum scarp-slope angles (θ) against scarp heights (Hs and Hm) for the southern segment of the Taos Pueblo fault (5). Open circles are for the most recent, single-event part of scarp; filled circles are for the entire multiple-event scarp.

Cañon Fault (6)

The Cañon fault forms small scarps east and south of Cañon, a small town at the base of the Sangre de Cristo Mountains 1.6 km east–southeast of Taos. We traced these fault scarps from a point due east of Cañon, where the fault places Pleistocene alluvium in contact with downdropped Pennsylvanian rocks (Lambert, 1966), south and southwest to 1.5 km north of Talpa, a distance of about 6 km. South of the Rio Fernando de Taos the fault cuts late and middle to early(?) Pleistocene piedmont-slope alluvium. The scarp-morphology data (Fig. 9) shows multiple-event scarp heights (Hm) of 3.5 to 5.4 m and single-event scarp heights (Hs) of 1.8–3.4 m. The most recent movement on the Cañon fault was either early Holocene or latest Pleistocene in age.

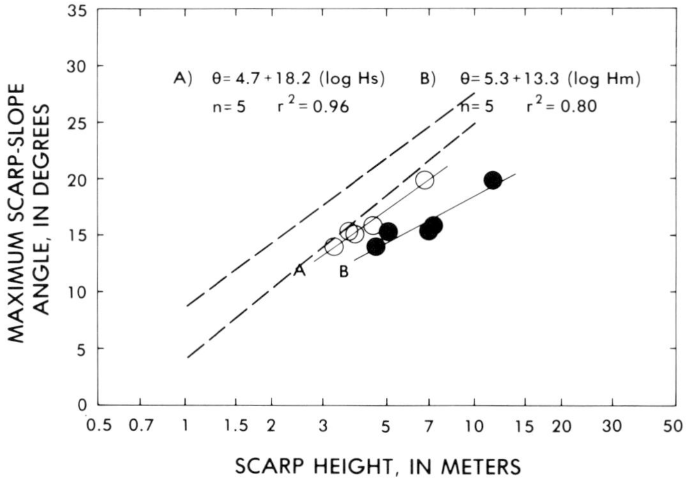

FIGURE 7. Plot of maximum scarp-slope angles (θ) against scarp heights (Hs and Hm) for the northern segment of the Taos Pueblo fault (5). Open circles (line A) are for most recent, single-event part of scarp; filled circles (line B) are for entire multiple-event scarp.

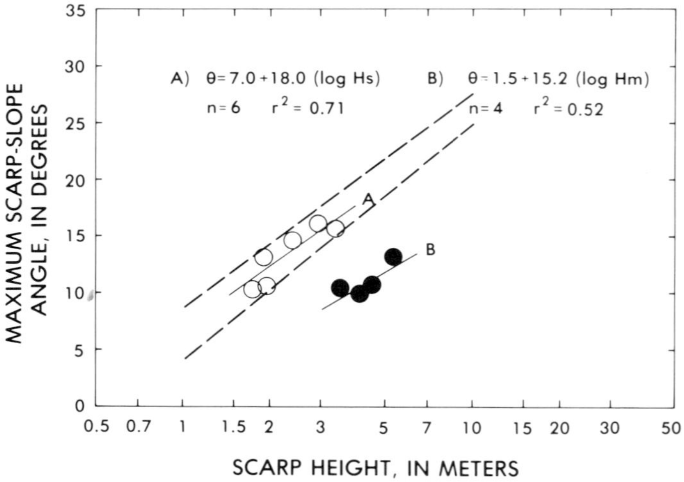

FIGURE 9. Plot of maximum scarp-slope angles (θ) against scarp (Hs and Hm) for the Cañon fault (6). Open circles (line A) are for most recent, single-event part of scarp; filled circles (line B) are for entire multiple-event scarp.

Los Cordovas Faults (7)

Los Cordovas faults are here named for a 5- to 8-km-wide zone of north-trending normal faults that extend about 12 km northward from the Rio Taos near Los Cordovas, New Mexico, to just south of U.S. Highway 64, which crosses the Rio Grande gorge west of Taos (Fig. 1). In areas of greatest offset, the faults place gravelly piedmont-slope alluvium of the uppermost Servilleta Formation (terminology of Lambert, 1966) down-to-the-west against stratigraphically lower 3- to 4-m.y.-old basalt (Lambert, 1966, p. 47). The alluvium probably is no older than late Pliocene or early Pleistocene, but clearly is older than the middle to late Pleistocene terrace alluvium found along major streams in the Taos area. Thus, movement on Los Cordovas faults may be as old as early Pleistocene, or as young as middle(?) Pleistocene.

Erosion of the piedmont gravel has produced west-facing scarps 15–30 m high, although the net offset is probably more. Most of the faults are poorly exposed because they are covered by alluvium that fills structurally controlled south-trending valleys.

Lambert (1966, fig. 1) showed only the more prominent Los Cordovas faults. Our photo reconnaissance confirms that most of the drainage in the area between Taos, the Rio Grande gorge, and the Rio Taos is structurally controlled by a combination of these north-trending faults and the southeast-dipping flank of the gorge arch (anticline) as mapped by Muehlberger (1979, fig. 2). Because Los Cordovas faults do not extend very far north of the axis of the gorge arch, the drainage in this area is directed to the west, down the piedmont slope toward the Rio Grande, rather than to the north. Near Los Cordovas, the Rio Taos follows a nearly straight course parallel to the axis of the gorge arch, suggesting that the river may overlie the axis of a previously unmapped syncline. Muehlberger (1979) suggested that the gorge arch (and by association our inferred syncline) resulted from northwestward-directed compression and thrusting of the Picuris Mountains block over the southern Taos Plateau along the Embudo fault (8) during the late Cenozoic.

Embudo Fault (8)

The Embudo fault extends 75 km along a general N 60° E trend and connects the northern part of the Española Basin with the southern part of the San Luis Basin. The few exposures we have seen along the northeastern part of the Embudo fault show high- to low-angle dip and reverse movement, whereas the movement along its southeastern part is inferred to be normal. The fault changes from normal to reverse sense of movement (the hinge point) near Embudo, where both northward and southward down-dip movement is recorded in the Servilleta Basalt. The southern part of the fault is considered to be the north-bounding structure of the Española Basin.

We feel that Muehlberger's detailed maps (1979) of road cuts near Hondo Canyon convincingly demonstrate that the Embudo fault has reverse movement. In the northernmost roadcut (Muehlberger, 1979, roadcut 1 of fig. 4) very coarse Pliocene alluvium of the Servilleta Formation (his terminology) is thrust over middle to early(?) Pleistocene post-Servilleta alluvium and fault-scarp colluvium. The Pleistocene age of the tectonically buried alluvium is inferred from the presence of a well-developed calcic soil having a discontinuous stage-III to stage-IV K horizon.

We measured two scarp profiles on the piedmont slope adjacent to and above Muehlberger's roadcut 1 (1979, fig. 3). Northeast of the road cut, the scarp has a height of about 15 m, whereas to the southwest it is only 7 m high (Hm, Fig. 10). However, the crest and backslope of the smaller scarp has been eroded and its height is therefore a minimum value. The morphology of the fault scarp is similar to that of scarps formed by late Pleistocene normal faults, although we cannot be certain that such comparisons are valid.

Northeast of Hondo Canyon, the Embudo fault splits into several parallel strands that form low, discontinous scarps in early to middle(?) Pleistocene sediment. This part of the fault is inferred to continue eastward near the base of the Picuris Mountains and to connect with the Cañon fault southeast of Ranchos de Taos.

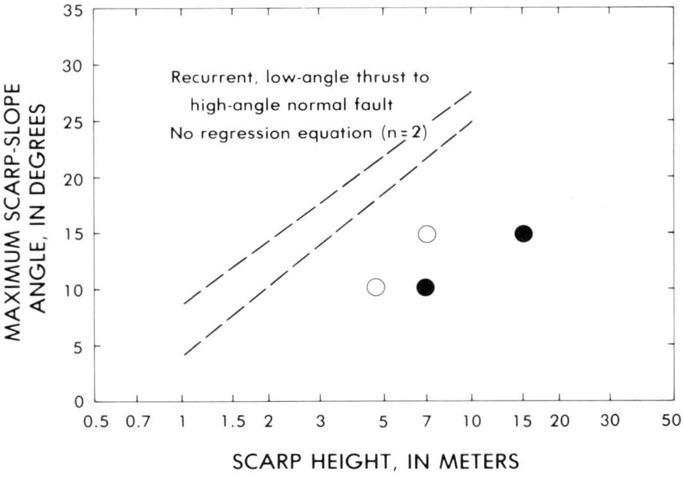

FIGURE 10. Plot of maximum scarp-slope angles (θ) against scarp heights (Hs and Hm) for the Embudo fault (8). Open circles are for the most recent, single-event part of scarp; filled circles are for entire multiple-event scarp.

CONCLUSIONS

Pliocene and Quaternary faults are abundant in the Taos Plateau region of the Rio Grande rift in northern New Mexico. These faults cut early(?) to latest Pleistocene sediments, Pliocene basalts, and Pliocene to early(?) Pleistocene basin-fill deposits. Morphologic, pedologic, and geologic data collected along some of the major fault scarps indicate that many were formed during the late Pleistocene (150,000–10,000 yrs ago), and some as recently as latest Pleistocene (25,000–10,000 yrs ago).

Our morphometric data suggest that significant surface rupturing of early Holocene or latest Pleistocene age may have occurred on the Cañon fault (6), which is here considered to be the southernmost segment of the Sangre de Cristo fault zone. Faults with evidence of latest Pleistocene surface rupturing include the Cedro Canyon fault (3) and the northern segment of the Taos Pueblo fault (5). In addition, faults along which the most recent surface ruptures date from the late Pleistocene include the Mesita (1) and Sunshine Valley (2) faults, the Urraca Ranch fault (4), the south segment of the Taos Pueblo fault (5), and possibly the Embudo fault (8). The Los Cordovas faults (7) are known only to be younger than the Pliocene to early(?) Pleistocene deposits which they displace; however, we suspect they may have experienced middle Pleistocene movement.

Fault scarps formed in middle to late Pleistocene surficial deposits in the region are as much as 18 m high; faults with recurrent movement generally have scarps higher than 7 m. The largest scarps are in Pliocene to early Pleistocene deposits along the Los Cordovas (7) and the Embudo (8) faults and early to middle Pleistocene alluvium along the Cedro Canyon fault (3); these scarps clearly are the product of recurrent movement, as evidenced by decreasing amounts of displacement in progressively younger deposits.

The Sangre de Cristo fault zone and the Embudo fault form the eastern and southeastern boundaries, respectively, of the Pliocene- to Pleistocene-age Rio Grande rift on the southern Taos Plateau. Farther south in the Española Basin, the Embudo and Pajarito faults (not discussed here) form the north and west margins of the rift, respectively. The Embudo fault lies coincident with the Jemez lineament, across which the rift abruptly changes geometry from the thick, east-dipping strata of the southern San Luis Basin to the thinner, west-dipping strata of the Española Basin.

Many faults on the Taos Plateau have displaced Pliocene and Quaternary deposits. In view of the distribution, size, and number of faults that could have histories of recurrent movement in this area, earthquakes and surface ruptures may have been associated with many tens of faults or fault segments during the last 5 m.y. Flows of Servilleta Basalt along the Sangre de Cristo fault zone near the Colorado–New Mexico border have been displaced a minimum of 240–360 m to perhaps as much as 600 m. These data indicate minimum average rates of uplift of 60–120 m/m.y. for the past 3–4 m.y.; these rates are comparable to those estimated for other major rift-bounding fault zones in central and southern New Mexico.

REFERENCES

Baldridge, W. S., Bartov, Y., and Kron, A., 1983, Geologic map of the Rio Grande rift and southeastern Colorado Plateau, New Mexico and Arizona; *in* Riecker, R. E. (ed.), Rio Grande rift—tectonics and magmatism: Washington D.C., American Geophysical Union, 2nd printing (1982), scale 1:500,000.

———, Damon, P. E., Shafiqullah, M., and Bridwell, R. J., 1980, Evolution of the central Rio Grande rift, New Mexico—new potassium–argon ages: Earth and Planetary Science Letters, v. 51, pp. 309–321.

Bucknam, R. C., and Anderson, R. E., 1979, Estimation of fault-scarp ages from a scarp-height–slope-angle relationship: Geology, v. 7, pp. 11–14.

Burroughs, R. L., 1978, Northern rift guide 2—Alamosa to Antonito, Colorado; *in* Hawley, J. W. (compiler), Guidebook to the Rio Grande rift in New Mexico and Colorado: New Mexico Bureau of Mines and Mineral Resources, Circular 163, pp. 33–36.

Chapin, C. E., 1979, Evolution of the Rio Grande rift—a summary; *in* Riecker, R. E. (ed.), Rio Grande rift—tectonics and magmatism: Washington, D.C., American Geophysical Union, pp. 1–5.

Colton, R. B., 1976, Map showing landslide deposits and late Tertiary and Quaternary faulting in the Fort Garland–San Luis area, Colorado–New Mexico: U.S. Geological Survey, Open-file Report 76-185, scale 1:250,000.

Cordell, L., 1978, Regional geophysical setting of the Rio Grande rift: Geological Society of America, Bulletin, v. 89, pp. 1073–1090.

Dungan, M. A., Muehlberger, W. R., Leininger, L., and others, 1984, Volcanic and sedimentary stratigraphy of the Rio Grande gorge, and the late Cenozoic geologic evolution of the southern San Luis Valley: New Mexico Geological Society, Guidebook 35 (this volume).

Galusha, T., and Blick, J. C., 1971, Stratigraphy of the Santa Fe Group, New Mexico: American Museum of Natural History, Bulletin 144, 127 pp.

Hawley, J. W. (compiler), 1978, Guidebook to the Rio Grande rift in New Mexico and Colorado: New Mexico Bureau of Mines and Mineral Resources, Circular 163, 241 pp.

Kelley, V. C., 1977, Geology of the Española Basin, New Mexico: New Mexico Bureau of Mines and Mineral Resources, Geologic Map 48, scale 1:125,000.

Kirkham, R. M., and Rodgers, W. P., 1981, Earthquake potential in Colorado—a preliminary evaluation: Colorado Geological Survey, Bulletin 43, 171 pp.

Lambert, P. W., 1966, Notes on the late Cenozoic geology of the Taos–Questa area, New Mexico: New Mexico Geological Society, Guidebook 17, pp. 43–50.

Lipman, P. W., and Mehnert, H. H., 1979, The Taos Plateau volcanic field, northern Rio Grande rift, New Mexico; *in* Riecker, R. E. (ed.), Rio Grande rift—tectonics and magmatism: Washington, D.C., American Geophysical Union, pp. 289–312.

Machette, M. N., 1982, Quaternary and Pliocene faults in the La Jencia and southern part of the Albuquerque–Belen Basins, New Mexico—evidence of fault history from fault-scarp morphology and Quaternary geology: New Mexico Geological Society, Guidebook 33, pp. 161–169.

———, and McGimsey, R. G., 1983, Map of Quaternary and Pliocene faults in the Socorro and western part of the Fort Sumner 1° × 2° quadrangles, central New Mexico: U.S. Geological Survey, Miscellaneous Field-studies Map MF-1465-A, scale 1:250,000, 12 pp.

Manley, K., 1979, Structure and stratigraphy of the Española basin, Rio Grande rift, New Mexico; *in* Riecker, R. E. (ed.), Rio Grande rift—tectonics and magmatism: Washington, D.C., American Geophysical Union, pp. 71–86.

———, Scott, G. R., and Wobus, R. (compilers), 1978, Preliminary geologic map of the Aztec 1° × 2° quadrangle, northwestern New Mexico and southwestern Colorado: U.S. Geological Survey, Open-file Report 78-466, scale 1:250,000.

McCalpin, J., 1983, Quaternary geology and neotectonics of the west flank of the northern Sangre de Cristo Mountains, south-central Colorado: Colorado School of Mines Quarterly, v. 77, no. 3, 97 pp.

Muehlberger, W. R., 1979, The Embudo fault between Pilar and Arroyo Hondo, New Mexico—an active intracontinental transform fault: New Mexico Geological Society, Guidebook 30, pp. 77–82.

Scott, G. R., 1970, Quaternary faulting and potential earthquakes in east-central Colorado: U.S. Geological Survey, Professional Paper 700-C, pp. C11–C18.

Scott, W. R., McCoy, W. D., Shroba, R. R., and Meyer, R., 1984, Reinterpretation of the exposed record of the last two cycles of Lake Bonneville, western United States: Quaternary Research, v. 20, no. 3, pp. 261–285.

Seager, W. R., 1981, Geology of the Organ Mountains and southern San Andres Mountains, New Mexico: New Mexico Bureau of Mines and Mineral Resources, Memoir 36, 97 pp.

Tweto, O., 1979, Geologic map of Colorado: U.S. Geological Survey, Special Map, scale 1:500,000.

Upson, J. E., 1939, Physiographic subdivisions of the San Luis Valley, southern Colorado: Journal of Geology, v. 47, pp. 721–736.

Wallace, R. E., 1977, Profiles and ages of young fault scarps, north-central Nevada: Geological Society of America, Bulletin, v. 88, pp. 1267–1281.

Woodward, L. A., Callender, J. F., and Zilinski, R. E., 1975, Tectonic map of the Rio Grande rift, New Mexico: Geological Society of America, Map and Chart Series MC-11, scale 1:1,000,000.

DIAGENESIS OF THE MISSISSIPPIAN ARROYO PEÑASCO GROUP, NORTH-CENTRAL NEW MEXICO

DANA S. ULMER and ROBERT L. LAURY
Department of Geological Sciences, Southern Methodist University, Dallas, Texas 75275

INTRODUCTION

The Arroyo Peñasco Group of the southern Sangre de Cristo Mountains comprises the oldest Paleozoic sediments found in north-central New Mexico. The purpose of this paper is to discuss the nature and significance of the varied and spectacular diagenetic fabrics preserved within this thin Mississippian section. Eight outcrops between Taos, Las Vegas, and Santa Fe (Fig. 1) were measured, described, and sampled. Samples from five of the outcrops, Cuchilla del Ojo, Ponce de Leon, Rio Grande del Rancho, Trout Springs, and Tererro, were studied in detail utilizing standard petrographic techniques and cathodoluminescence. The carbonate members of the Arroyo Peñasco Group have undergone dolomitization, dedolomitization, silicification, calcitization of evaporites, and brecciation. The complex diagenetic history of the Arroyo Peñasco Group carbonates resulted from periodic episodes of exposure prior to Pennsylvanian sedimentation.

PREVIOUS WORK

Much of the previous work on the Arroyo Peñasco Group in the Sangre de Cristo Mountains has concentrated on its biostratigraphic and facies relationships. A major problem surrounding the Arroyo Peñasco Group has been the stratigraphic nomenclature applied to the units. Figure 2 summarizes the evolution of the stratigraphic nomenclature for these Mississippian units. Armstrong (1955) first separated the rocks from the Sandia Formation and assigned a Mississippian Meramecian age on the basis of *Endothyra* sp. and *Pectogyra* sp. fauna. Baltz and Bachman (1956) and Baltz and Read (1960) disputed the Mississippian age of the units in the Sangre de Cristo and Sandia Mountains. They assigned a Devonian(?) age to the Espiritu Santo Formation on the basis of its fauna.

Baltz and Read (1960) did the first petrographic studies of the Arroyo Peñasco Group, relating the rock types to several periods of transgressions and regressions across the Mississippian platform. They proposed that the Macho Member was a result of karstification during one of the regressions.

Armstrong and Holcomb (1967), in their study of the Jemez, San Pedro, and Sangre de Cristo Mountains, found a faunal assemblage that suggested an Osagean to Meramecian age. Armstrong (1967) defined three depositional cycles based on the petrography. Cycle 1 was a transgressive sequence consisting of shallow-marine terrigenous clastics. The cycle-2 carbonates were a regressive sequence of intertidal to supratidal mudstones and evaporites. Cycle-3 sediments were a transgressive sequence reflecting a more open marine environment. Deposits of cycle 3 were arenaceous, echinodermal, oolitic, pelletal packstones, and wackestones. Armstrong considered the brecciated Macho Member a product of dissolution and collapse of evaporite units prior to Pennsylvanian sedimentation.

Recently completed M.S. theses by Williamson (1978) and Vaughn (1978) have provided information on the diagenesis of parts of the Arroyo Peñasco Group, namely the Del Padre Sandstone and the Macho Member of the Tererro Formation, respectively. The current research was initiated because a comprehensive study of the diagenetic textures and their relationship to Mississippian sedimentation within the entire Arroyo Peñasco Group had not been done previously.

ANALYTICAL METHODS

Standard petrographic thin sections were stained with a combined Alizarin Red–S and potassium-ferricyanide solution for dolomite and

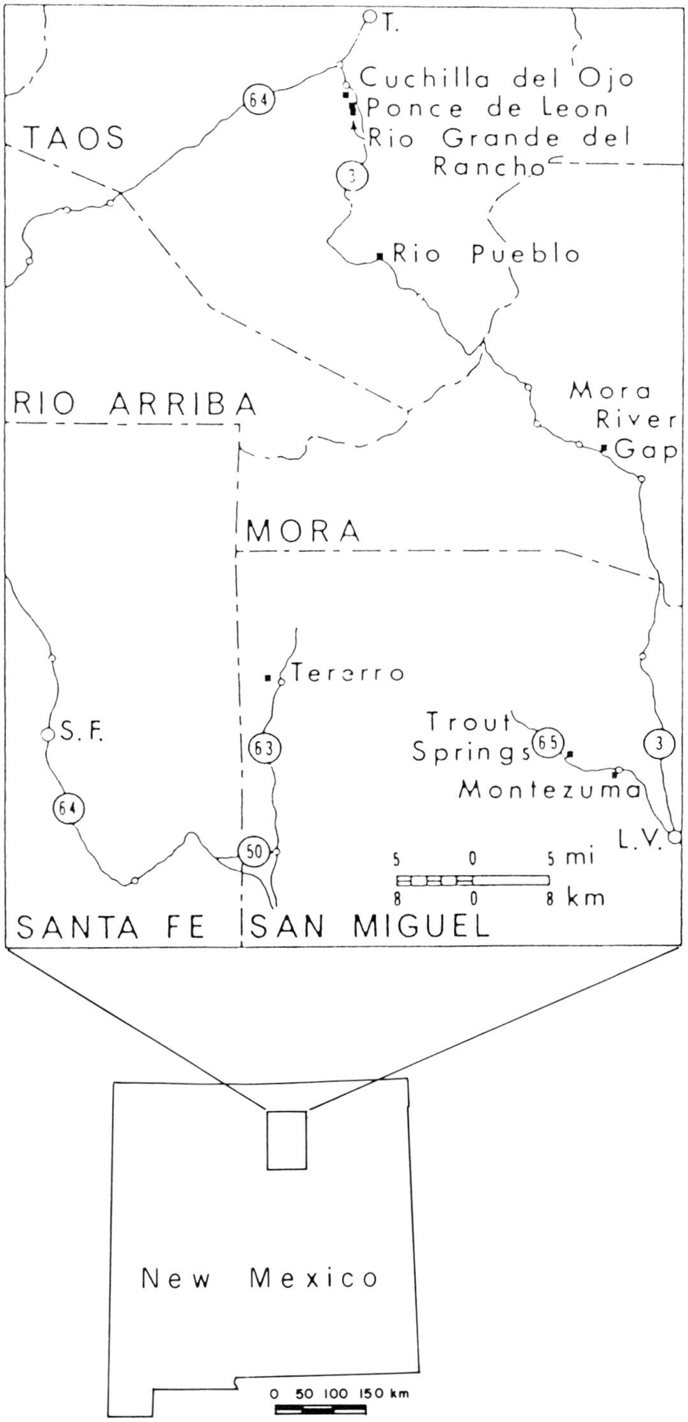

FIGURE 1. Location of the study area and measured sections of the Mississippian Arroyo Peñasco Group. T. – Taos, L.V. – Las Vegas, S.F. – Santa Fe.

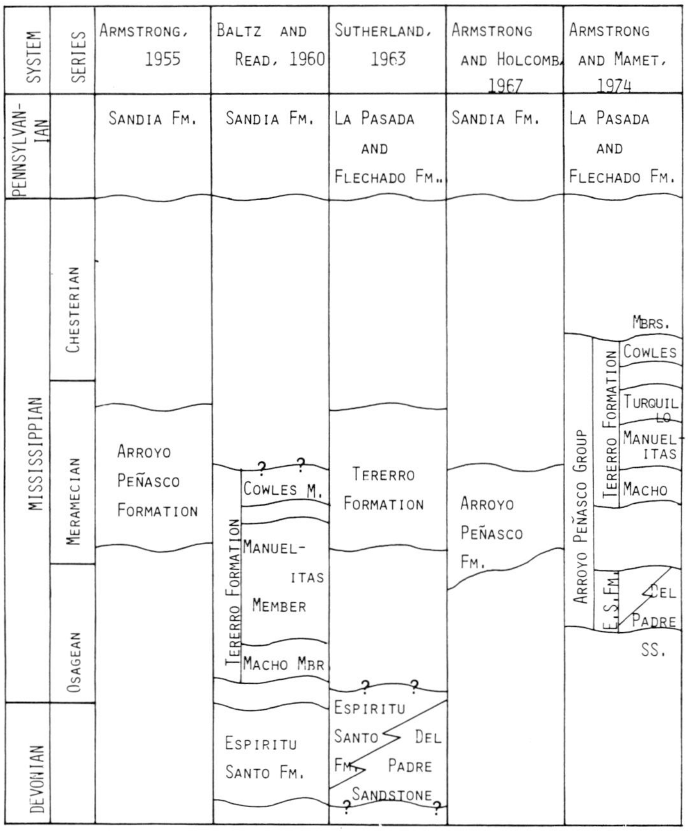

FIGURE 2. Evolution of the stratigraphic nomenclature of the Arroyo Peñasco Group.

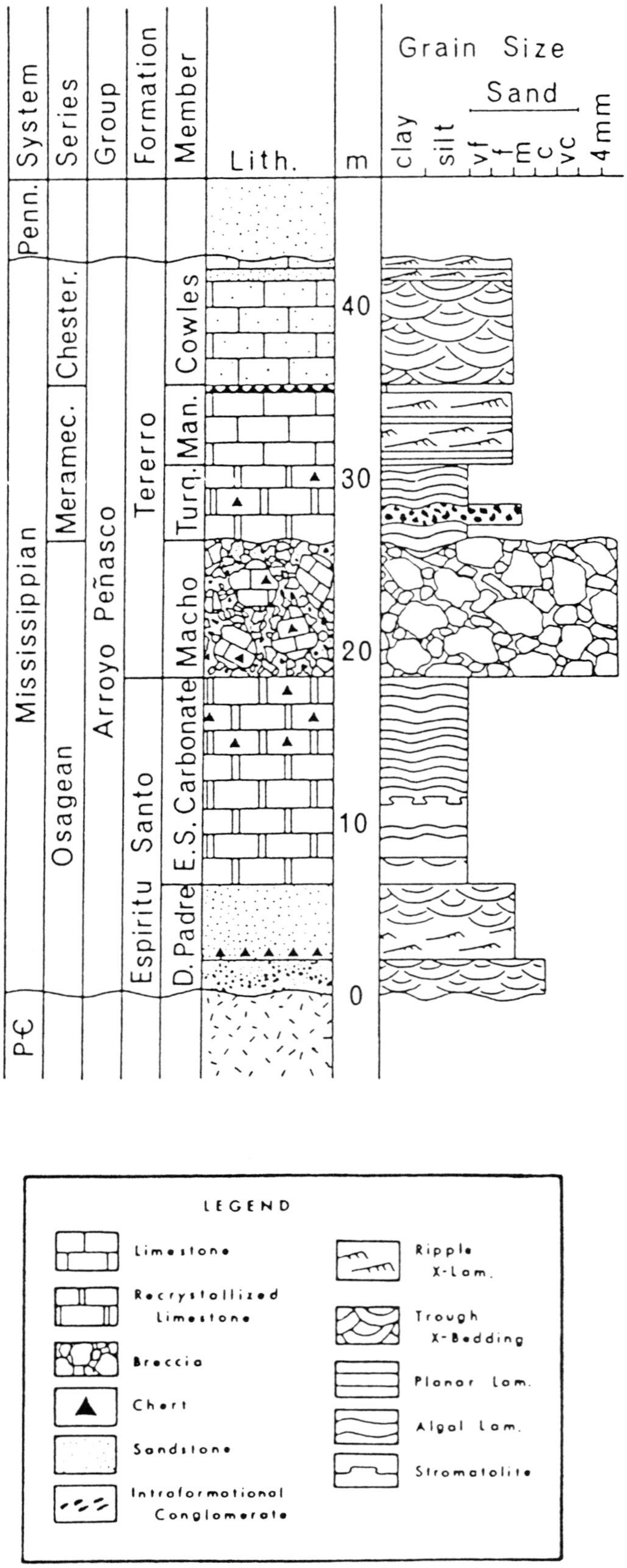

FIGURE 3. Idealized stratigraphic section of the Arroyo Peñasco Group. Meramec. – Meramecian, Chester. – Chesterian, D. – Del Padre, E.S. – Espiritu Santo, Turq. – Turquillo, Man. – Manuelitas.

iron-rich calcite (Evamy, 1963). Slabs were etched in a 10% hydrochloric acid solution for 40 sec to aid identification of the primary sedimentary textures.

Cathodoluminescent petrography was done with a Nuclide Corp. Luminoscope model ELM–2E on a Zeiss petrographic microscope. Cathodoluminescence is the emission of light produced by stimulating certain atoms in a specimen with a wide electron beam. The use of luminescence is a qualitative tool when not used in association with the electron microprobe. Iron and manganese are the most common trace elements responsible for luminescence, but Pb^{2+}, Ni^{2+}, Cs^{1+}, and rare earths also effect it (Machel, 1983). The intensity of the bands or zones produced by the excitation of the sample is a function of the Mn^{2+}/Fe^{2+} ratio, where manganese (also Pb^{2+} and Cs^{1+}) produces and iron (also Ni^{2+}) quenches luminescence, especially in dolomites.

STRATIGRAPHY

Overlying the highly weathered, low-relief Precambrian peneplain, the Arroyo Peñasco Group consists of the Espiritu Santo and Tererro Formations (Fig. 3). The Espiritu Santo Formation is subdivided into the lower Del Padre Sandstone and the upper carbonate member (Sutherland, 1963). Derived from Precambrian granites and schists, the sandstones were deposited during the late Osagean transgression across the platform (Armstrong and Holcomb, 1967). The Del Padre Sandstone represents a sequence of high-energy braided streams, beaches, and tidal and sabkha flats (Williamson, 1978). The Del Padre Sandstone filled the relict topography developed on the Precambrian peneplain and conformably grades upward into the crystalline limestones of the carbonate member (Fig. 4). Limestones of the carbonate member consist of algally laminated mudstones, wackestones, packstones, and gypsum horizons and nodules that were altered extensively during diagenesis. These sediments were deposited as part of a widespread supratidal to

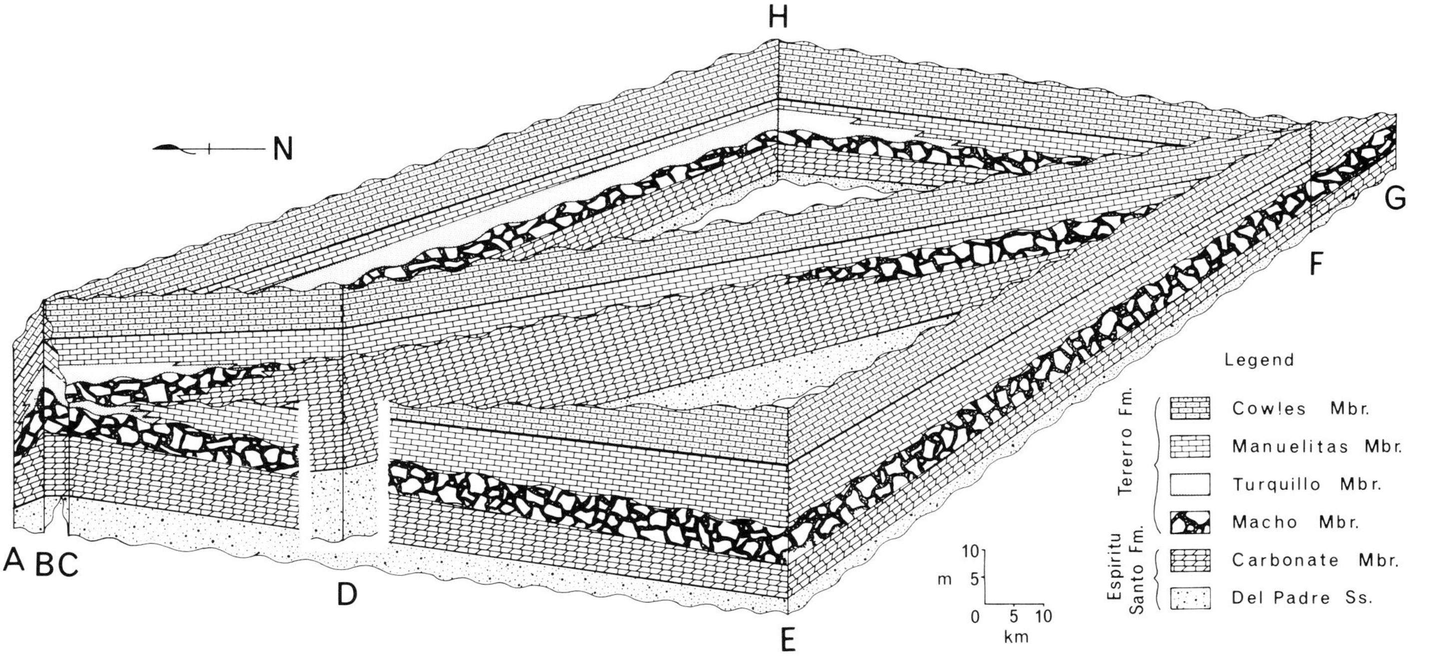

FIGURE 4. Fence diagram of the Arroyo Peñasco Group within the study area. Measured sections: A – Cuchilla del Ojo, B – Ponce de Leon, C – Rio Grande del Rancho, D – Rio Pueblo, E – Tererro, F – Trout Springs, G – Montezuma, H – Mora River Gap.

upper-intertidal complex with sporadic flooding producing laminae of bioclasts and intraclasts.

The Macho Member of the Tererro Formation is a cemented rubble packbreccia (Morrow, 1982). The breccia blocks are composed of pelletal mudstones, pelletal and oolitic packstones, and crystalline limestones. The crystalline limestones consist of calcitized evaporites and dolomites which will be discussed in the diagenesis section. The original sediments of the breccia were deposited in supratidal to intertidal environments. Dissolution of interbedded gypsum horizons during exposure resulted in collapse and brecciation. The Macho Member is present everywhere in the field area, except in the vicinity of the Rio Pueblo section (Fig. 4).

The Turquillo, Manuelitas, and Cowles Members of the Tererro Formation were deposited following brecciation. The Turquillo and Manuelitas Members fill the topography developed on the Macho Member. The Turquillo Member, found at the Ponce de Leon, Rio Grande del Rancho, and Mora River Gap sections, consists of algally laminated mudstones and wackestones, intraformational conglomerates, and crystalline limestones. The Turquillo Member was deposited in an intertidal environment. On the basis of the foraminifers present in this unit, Armstrong and Mamet (1979) concluded that Turquillo waters were of normal marine salinities.

The Manuelitas Member of the Tererro Formation represents a sequence of normal marine, subtidal, and shoaling to intertidal deposits that is laterally equivalent to the Turquillo Member. The Manuelitas Member is composed of pelloidal, bioclastic, and oolitic packstones and grainstones. Bioclasts commonly found within the unit are pelmatozoans, bryozoans, blue-green algae, sponge spicules, brachiopods, foraminifers, ostracods, and calcispheres.

The Cowles Member, composed of sandy, pelloidal, and bioclastic packstones and calcareous sandstones, represents the final transgression and regression of the Mississippian seas in north-central New Mexico. Prior to Pennsylvanian sedimentation, erosion removed parts of the Cowles Member (Fig. 3). At the Rio Grande del Rancho and Montezuma sections, both the Cowles and the Manuelitas Members have been removed by the extensive downcutting associated with the development of the Taos trough (Chapin, 1981).

DIAGENESIS

The Arroyo Peñasco Group has been altered extensively by diagenetic processes. Carbonates of the Espiritu Santo Formation have undergone pervasive alteration which includes dolomitization, dedolomitization, calcitization of gypsum, neomorphism, silicification of gypsum, and chertification resulting in the partial to complete recrystallization and replacement of these carbonates (Fig. 5A, B). The other members of the Arroyo Peñasco Group exhibit a less severe degree of recrystallization and replacement.

Dolomite

Varying according to locality, the amount of dolomite preserved within the carbonates of the Arroyo Peñasco Group ranges from less than one to over 75% (Table 1). Dolomite crystals range from 0.004 to 0.600 mm. The very finely crystalline, anhedral to subhedral dolomites replacing the algally laminated sediments are interpreted as formed penecontemporaneously under supratidal conditions by evaporative pumping analogous to that occurring today at Abu Dhabi, U.A.E. (McKenzie and others, 1980). Under the same hypersaline conditions, gypsum precipitation resulted in nodular and bedded gypsum, as well

TABLE 1. Percent dolomite preserved within the carbonate members of the Arroyo Peñasco Group. CDO – Cuchilla del Ojo, PDL – Ponce de Leon, RGR – Rio Grande del Rancho, TSP – Trout Springs, TRO – Tererro.

	CDO	PDL	RGR	TSP	TRO
Cowles Mbr.	1%	17%	-	1%	1%
Manuelitas Mbr.	25%	1%	-	1%	1%
Turquillo Mbr.	-	50%	75%	-	-
Macho Mbr.	4%	75%	56%	4%	1%
E.S. Carbonate Mbr.	10%	30%	50%	1%	30%

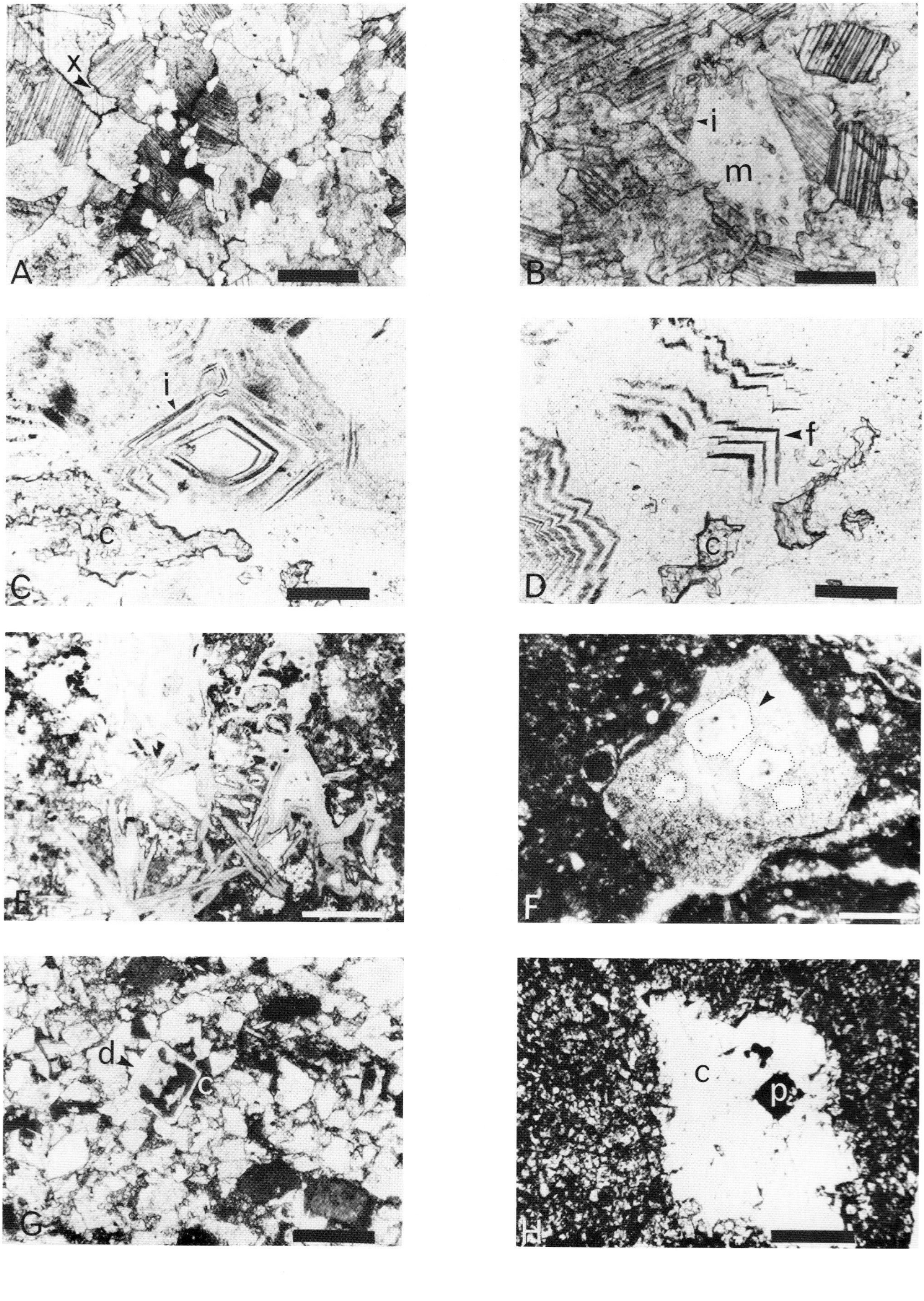
A
x
B
i
m
C
i
c
D
f
c
E
F
G
d
c
H
c
p

as scattered, euhedral gypsum (selenite) crystals within the dolomitized sediments. With increasing burial or sea-level rise, the sediments passed through the zone of hypersaline fluids to the mixing zone.

Dolomite that is limpid and euhedral is interpreted as forming in the meteoric phreatic zone. Porespace between the dolomite crystals is occluded by coarsely crystalline, poikolitic calcite spar. Euhedral dolomite crystals larger than 0.10 mm are interpreted as formed in the mixing zone, a result of co-mingling of meteoric and marine fluids (Choquette and Steinen, 1980).

The large diversity in crystal sizes and morphologies indicates dolomitization is the result of widely fluctuating pore-water salinities. Schizohaline conditions, as described by Folk and Siedlecka (1974), would produce the types of dolomites observed within the Arroyo Peñasco Group.

Cathodoluminescent petrography of dolomite at the Ponce de Leon and Tererro sections reveals a characteristic dull-red core with a brighter red or orange second-generation dolomite. This is interpreted as reflecting a change from oxidizing meteoric fluids (dull red) found under hypersaline conditions to marine and meteoric fluids (bright) found in the mixing zone. The orangish appearance of some of the dolomites may be produced by a higher calcium concentration in the crystal lattice.

Authigenic Silica

Discrete horizons of nodular chert were found at all localities examined. Chert nodules are the most abundant within the Espiritu Santo Carbonate and Manuelitas Members. The cherts are usually banded gray to black, with the darker bands containing more clays and organics and the lighter bands more carbonates. The contact between the chert and limestone is usually sharp, but the cherts may contain remnants of carbonate such as bioclasts, dedolomites, and calcitized evaporites. The chert nodules are interpreted as formed during initial dolomitization of the carbonates because of their occurrence on bedding planes.

The Coorong Lagoon model, developed by Peterson and von der Borch (1965) for inorganic precipitation of chert, is a plausible model to apply to the cherts within the Arroyo Peñasco Group. Periodic marine flooding and the associated bloom of organic material in ponded areas on the algal flats would raise the pH to values greater than 9.5. This would lead to dissolution of detrital quartz and the sinking of the dense, siliceous brines into the pond-floor sediments. Decay of organic material just beneath the sediment/water interface would cause the pH of the interstitial water to drop below a value of 7, triggering silica precipitation. The Arroyo Peñasco Group includes quartz and feldspar sands which provide an abundant source of silica. Silica dissolution is manifested by the presence of embayed quartz and partially replaced feldspar grains. Corroded quartz and feldspar are found in association with the horizons that have been silicified.

Other varieties of authigenic silica within the Arroyo Peñasco Group include euhedral megaquartz, anhedral megaquartz, chalcedonite, lutecite, and quartzine. Quartzine and lutecite are forms of length-slow chalcedony which have been reported as indicative of vanished evaporites (Folk and Pitman, 1971). Euhedral megaquartz is also interpreted as formed from evaporite precursors (Friedman and Shukla, 1980). Within the carbonates of the Espiritu Santo Formation and the Macho Member, silica replacing evaporites started out initially as euhedral megaquartz, and is sometimes followed by quartzine and/or lutecite (Fig. 5B). Carbonate inclusions are locally present within the megaquartz and chalcedony. Silicification of the evaporites occurred prior to the brecciation of the Macho Member, as indicated by broken clasts of megaquartz and length-slow chalcedony found within the matrix.

Friedman and Shukla (1980) described in the Silurian Lockport Formation of New York a similar occurrence of inclusion-free megaquartz, sometimes followed by length-slow chalcedony. They envisioned a situation where precipitation of sulfates at shallow depths increased the pH and dissolved detrital quartz grains. With a return to more normal marine to meteoric conditions, dissolution of the gypsum would decrease the pH and lead to precipitation of quartz. Oxygen-isotope data by Milliken (1979) for Mississippian silicified evaporites from Kentucky and Tennessee supported a meteoric origin for megaquartz and a mixing-zone origin for length-slow chalcedony. Within the Arroyo Peñasco Group, late dolomitization and silicification of evaporites could proceed under the same conditions prior to brecciation and recrystallization.

The upper members of the Tererro Formation, the Turquillo, Manuelitas, and Cowles, also contain quartz after evaporites. Within these Mississippian sediments the development of fibrous silica forms pseudomorphing evaporites, is more extensive, and occurs more commonly than does euhedral megaquartz. Much of the silica is associated with nonweathering-profile paleosilcretes. Summerfield (1983) recognized this type of silcrete in the Kalahari Basin in Africa, where it occurs as laterally extensive, partially to completely silicified beds a few meters thick. A similar silcrete formed at the top of the Manuelitas Member can be traced throughout the field area. The silcrete ranges from a massive chert at the Cuchilla del Ojo, Ponce de Leon, Tererro, and Mora River Gap sections to a horizon of very large chert nodules at the Rio Pueblo section, and to a horizon of silicified evaporite nodules at the Trout Springs section. Other more areally restricted horizons of silcrete occur within the Manuelitas Member. These horizons mark episodes of rapidly fluctuating sea level during Manuelitas deposition.

Length-slow chalcedony (both quartzine and lutecite) within the silcrete shows excellent zoning in both plane-polarized and reflected light. This is due to the presence of solid and fluid inclusions in the chalcedony. Hematite, calcite, and saline(?) fluid inclusions are the most common (Fig. 5C). In reflected light the fluid inclusions are milky white and the hematite inclusions are red. Gypsum pseudomorphs surrounded by pseudocubic or fortification zoning (Arbey, 1980), and dolomite pseudomorphs of silica and calcite are characteristic of these units (Fig. 5D). Length-fast chalcedony (chalcedonite), instead of length-slow chalcedony, is found replacing some of the evaporites at the Trout Springs section. Prior calcitization of evaporites, with subsequent replacement by silica, is a probable explanation for these anomalous silica forms. The presence of relict carbonate laminae within the chalcedony supports this hypothesis.

Length-slow chalcedony is found replacing pelmatozoan, brachiopod, and trilobite bioclasts within the Arroyo Peñasco Group (Fig. 4F). Milliken (1979) attributed the precipitation of length-slow chalcedony to sulfate-rich siliceous pore fluids. Because conditions would favor this theory within the Mississippian units, this is the probable mode of formation.

Dedolomitization and Calcitization of Gypsum

A major period of sea-level regression and climatic change occurred after the deposition of the interbedded carbonate and evaporite units of the Macho Member. This period of exposure was responsible for the dedolomitization and calcitization of evaporites within the Espiritu Santo Carbonates and, to a lesser degree, the Macho Member. Dedolomitization is the process by which dolomite is replaced by calcite (Fig.

← FIGURE 5. **A** – Recrystallized limestone containing quartz and feldspar grains. Stylolitic contacts (x) between calcite crystals (scale bar = 0.606 mm, plane-polarized light, Trout Springs–11, Espiritu Santo Carbonate). **B** – Recrystallized limestone and silicified gypsum (m). Note carbonate inclusions (i) within the megaquartz (scale bar = 0.138 mm, plane-polarized light, Cuchilla del Ojo–12, Espiritu Santo Carbonate). **C** – Length-slow chalcedony and megaquartz pseudomorphs after gypsum. Zoning produced by iron-oxide inclusions (i) (scale bar = 0.126 mm, plane-polarized light, c–calcite, Trout Springs–28, Manuelitas Member). **D** – Pseudocubic fortification zoning (f) (scale bar = 0.126 mm, plane-polarized light, c–calcite, Trout Springs–23, Manuelitas Member). **E** – Length-fast chalcedony and megaquartz pseudomorphs after gypsum (scale bar = 1.0 mm, plane-polarized light, Trout Springs–39, Cowles Member). **F** – Crinoid columnal partially replaced by length-slow chalcedony (scale bar = 0.486 mm, plane-polarized light, Trout Spring–19, Macho Member). **G** – Partial dolomite replacement. Calcite (c) appears dark due to staining (scale bar = 0.158 mm, plane-polarized light, Rio Grande del Rancho–27, Turquillo Member). **H** – Gypsum pseudomorph replaced by calcite (c) and pyrite (p) (scale bar = 0.253 mm, plane-polarized light, Ponce de Leon–5, Espiritu Santo Carbonate).

5G). During calcitization of evaporites, calcite replaced the gypsum (or selenite) within the carbonates of the Arroyo Peñasco Group (Figs. 5H, 6A). In the carbonates of the Espiritu Santo Formation and Macho Member, dedolomitization and calcitization of gypsum occurred penecontemporaneously. The amount of dedolomite within the sections is summarized in Table 2.

Low-relief supratidal and intertidal flats developed at this time resulted in the exposure of vast areas during minor fluctuations in sea level. Increased rainfall is thought to have accompanied an extended sea-level low, resulting in intrastratal solution and removal of $CaSO_4$ by undersaturated meteoric fluids. This led to formation collapse and brecciation which characterizes the Macho Member.

Back and others (1983), studying the Mississippian Pahasapa Limestone in the Black Hills of South Dakota and Wyoming, found dedolomitization and calcitization of evaporites to be related to a major exposure event. In their model, the presence of gypsum in the system is the major driving force of the dedolomitization process. Solution of the gypsum saturates or supersaturates the ground-water system with respect to calcium. This produces a decrease in the Mg^{2+}/Ca^{2+} ratio, which in turn causes calcite precipitation. Calcite precipitation depletes the aqueous system in CO_2 and lowers the pH. As a result, dolomite dissolves to accommodate the loss of CO_2 and to increase the Mg^{2+}/Ca^{2+} ratio. At low temperatures gypsum is the necessary driving component of the system, but at higher temperatures the reaction could occur without its presence.

Within the Arroyo Peñasco Group, the fluids derived from the overlying Macho Member were enriched, but undersaturated with respect to dissolved $CaSO_4$. These fluids migrated downsection until they intersected the water table. Once in the phreatic zone, the solutions continued to dissolve gypsum, which decreased the Mg^{2+}/Ca^{2+} ratio and supersaturated the solution with calcium. Thereafter, the system followed the model of Back and others (1983) until most of the accessible gypsum was consumed.

The top of the water table is delineated by the top of the Espiritu Santo Carbonate Member. All rocks above the water table were brecciated, whereas all units below the water table were recrystallized. From the nearly uniform thickness of the Espiritu Santo Carbonates we can infer that the water table was at a very uniform depth beneath the sabkha surface and that there was very little topographic relief prior to brecciation. Fluctuations in the water table are recorded at the base of some of the breccias by the coarsely recrystallized nature of the clasts and matrix. Recrystallization of parts of the Macho Member, especially at the Trout Springs section, indicates that recrystallization continued during the transgression that deposited the Turquillo and Manuelitas Members. Vaughn (1978) also recognized that the location of the water table was the main constraint over the formation of recrystallized limestones and breccia of the Espiritu Santo Formation and Macho Member.

The samples from the Espiritu Santo Carbonates and the Macho Member were examined under the luminoscope, as their primary diagenetic textures are no longer visible in transmitted light due to recrystallization. Luminescence revealed the presence of dedolomite and calcitized evaporites. Zoning within the pseudomorphs appears to be unrelated to surrounding calcites (Fig. 6B). If zoning was a result of calcitization, then the zoning should reflect the same diagenetic conditions found within the calcites. Because this is not the case, the zones may reflect the original trace-element chemistry of the precursor dolomite and gypsum.

Oglesby (1976) suggested that dolomite can incorporate approximately twice as much Mn^{2+} and Fe^{2+} within its crystal lattice as can calcite. With simultaneous grain-boundary dissolution of dolomite and reprecipitation of calcite, it may be possible to preserve the original trace-element composition of the dolomite within the calcite lattice (Fig. 7). The higher concentrations of Mn^{2+} and Fe^{+} inherited during dedolomitization would produce anomalous luminescence within the dedolomities (Fig. 6C). The presence of corroded zones (Fig. 6D) within the dedolomites also argues for the preservation of part of the original trace-element distribution of the dolomites.

TABLE 2. Percent dedolomite found within the carbonates of the Arroyo Peñasco Group. CDO – Cuchilla del Ojo, PDL – Ponce de Leon, RGR – Rio Grande del Rancho, TSP – Trout Springs, TRO – Tererro.

	CDO	PDL	RGR	TSP	TRO
Cowles Mbr.	1%	14%	-	1%	1%
Manuelitas Mbr.	1%	15%	-	1%	1%
Turquillo Mbr.	-	1%	1%	-	-
Macho Mbr.	21%	1%	2%	30%	1%
E.S. Carbonate Mbr.	53%	27%	27%	64%	6%

The calcitized evaporites generally are less luminescent than the surrounding dedolomites and calcites (Fig. 6E, F). This may reflect the lack of Mn^{2+} within the evaporite lattice. Zoning present in some the hexagonal to bladed crystals might actually reflect the chemistry of the pore fluids at the time of calcitization.

Another type of calcitized dolomite observed within the Espiritu Santo Carbonate Member is euhedral, limpid-dolomite rims surrounded by poikilitic calcite. Within some of the dolomite rims, remnants of the original, inclusion-rich dolomite core can be found forming a geopetal mound of rounded crystals. Folk and Siedlecka (1974) observed similar textures in the dolomites from Bear Island, Svalbard. Their interpretation is that the limpid dolomite formed during meteoric phreatic diagenesis and nucleated on cores of dolomite formed under hypersaline conditions. The hypersaline dolomites subsequently dissolved due to their greater solubility in meteoric phreatic environment (Fig. 8).

Cementation

Cementation of the packstones and grainstones of the upper members of the Tererro Formation involved early syntaxial overgrowths on pelmatozoans and blocky, isopachous calcite spar. Blocky calcite spar that coarsens toward the center of pore space follows the earlier cements. These types of cements are found throughout the Manuelitas and Cowles Members and are characteristic of meteoric and marine phreatic diagenesis (Longman, 1980).

Cathodoluminescence was employed to assess the distribution of trace elements (Mn^{2+} and Fe^{2+}) within these cements. The recrystallized limestones of the Espiritu Santo Formation and the Macho Member were also studied to see if their original trace-element distributions were preserved and, if so, how they compared to the upper units of the Arroyo Peñasco Group.

The seven cement zones identified within the Arroyo Peñasco Group are summarized in Figure 9. Nonluminescent zone-1 cements are interpreted as formed from oxidizing pore fluids of the meteoric phreatic lens having very low concentrations of Fe^{2+} and Mn^{2+} (Oglesby, 1976).

FIGURE 6. **A** – Zoned gypsum pseudomorphs (now calcitized) (scale bar = 0.606 mm, plane-polarized light, Ponce de Leon–float, probable Turquillo Member). → **B** – Dedolomite (dd) and cement zonation under cathodoluminescence (scale bar = 0.606 mm, Tererro–13b, Espiritu Santo Carbonate). **C** – Zoned dedolomite crystals under cathodoluminescence (scale bar = 0.606 mm, Trout Springs–10a, Espiritu Santo Carbonate). **D** – Cathodoluminescent view of corroded calcite pseudomorphs after dolomite (scale bar = 0.315 mm, Trout Springs–13, Espiritu Santo Carbonate). **E** – Calcitized gypsum under cathodoluminescent light (scale bar = 0.253 mm, Trout Springs–13, Espiritu Santo Carbonate). **F** – Zoned dedolomite and euhedral gypsum pseudomorphs under cathodoluminescence (scale bar = 0.315 mm, o–porosity, Ponce de Leon, Espiritu Santo Carbonate). **G** – Cement stratigraphy developed in ooid, pelmatozoan grainstone. Zones 1 through 4 (scale bar = 0.606 mm, Trout Springs–39, Cowles Member). **H** – Zones 1 through 6 cement stratigraphy (scale bar = 0.203 mm, Trout Springs–28, Manuelitas Member).

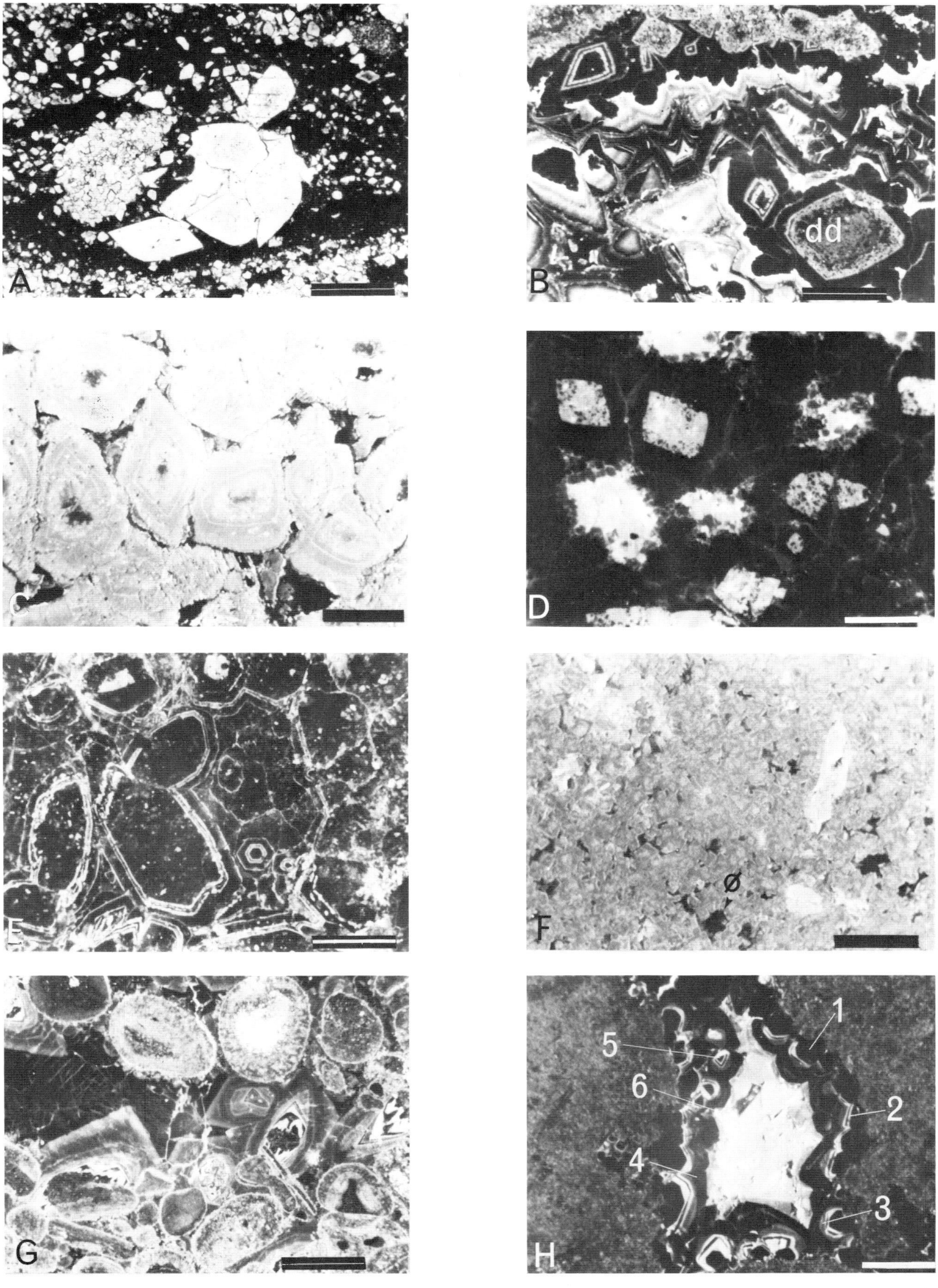
A
B
dd
C
D
E
F
ø
G
H
1
2
3
4
5
6

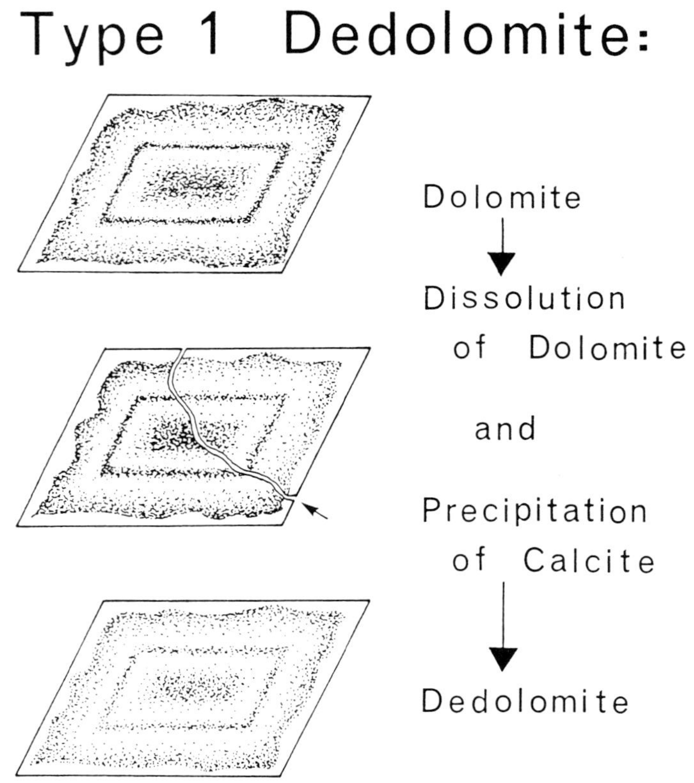

FIGURE 7. Type-1 dedolomite. Cathodoluminescent view of dolomite and dedolomite showing the coincident zone (arrow) of dolomite dissolution and calcite precipitation.

The bright and very dull to nonluminescent bands of zone-2 cements are interpreted as reflecting abrupt changes in the ground-water system. The paleoaquifer fluctuated between oxidizing meteoric (nonluminescent) and stagnant, reducing, marine pore waters (bright).

Dull to zoned-dull cements of zone 3 are interpreted as precipitated from reducing, marine pore fluids. The paleoquifer was not stagnant during zone-3 precipitation, as it was during precipitation of the bright cements (Grover and Read, 1983). Uplift and/or fall in sea level occurred between zone-3 and 4 cements. Nonluminescent zone-4 cements indicate a return to meteoric phreatic diagenesis within the paleoaquifer (Fig. 6G). Zone-4 and 5 cements reflect conditions similar to those that produced zone-1 and 2 cements.

The brightly luminescent zone-6 cements are interpreted as precipitated from reducing, marine pore waters in a stagnant paleoquifer (Fig. 6H). With continued burial, the fluids became increasingly reducing as a result of restricted fluid circulation. Under these conditions, the zoned-dull cements of zone 7 were precipitated.

A complete cement stratigraphy is not seen in every slide. This is due to differential rates of calcite cementation and the timing of porosity development. Fractures within the Arroyo Peñasco Group are filled by bright luminescing cements of zone 6 and zoned-dull cements of zone 7, reflecting their late origin.

The carbonates of the Espiritu Santo Formation and Macho Member are striking because both exhibit a cement stratigraphy that mimics the overlying packstones and grainstones of the Manuelitas and Cowles Members. These cements are found in porosity that developed prior to and during recrystallization and brecciation.

Pyrite and Other Mineralization

Most of the pyritization of the Arroyo Peñasco Group occurred after fracturing, as indicated by apparent co-precipitation of both calcite and pyrite in the fractures. The reducing conditions proposed for fracture-filling cements of zones 6 and 7 were also conducive to pyrite precipitation. Minor pyrite is associated with dedolomitization and calcitization of gypsum. Most of the pyrite has been oxidized to hematite and limonite during recent weathering, pseudomorphing the original euhedral crystal form.

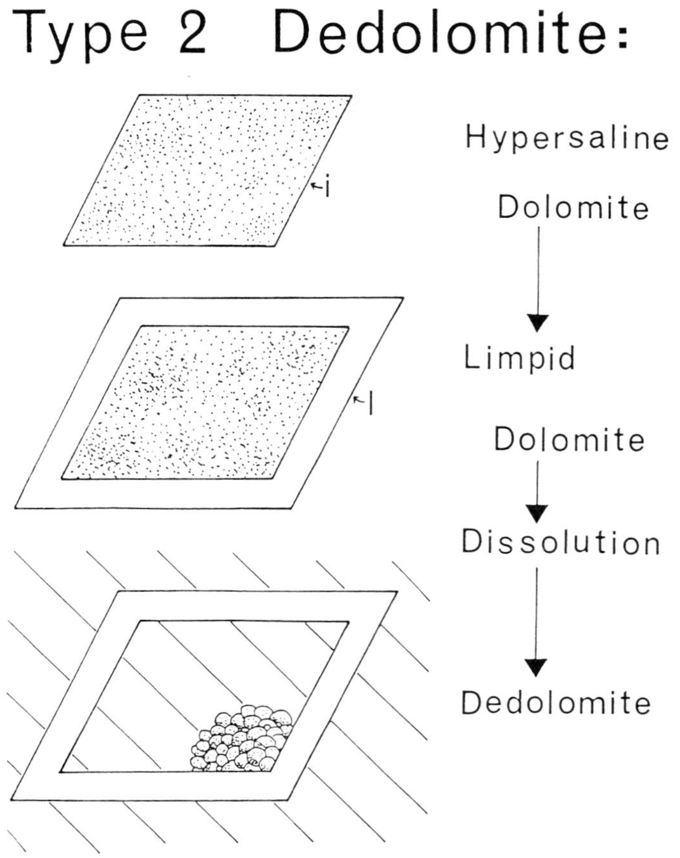

FIGURE 8. Type-2 dedolomite. Precipitation of two generations of dolomite (inclusion-rich = i, and limpid = l) followed by a period of meteoric attack and dissolution. Limpid and remnant inclusion-rich dolomite enclosed in poikilitic calcite after leaching.

Minor copper and lead mineralization also is found in the Arroyo Peñasco Group. This mineralization is interpreted as having resulted from hydrothermal activity during Laramide or more recent times. The friable sands of the Del Padre Sandstone at Cuchilla Del Ojo are cemented by a copper carbonate, giving the unit a pale-blue appearance. The Tererro and Trout Springs sections contain minor amounts of galena within the carbonates.

A second stage of Cenozoic karstification has produced caves within the Macho Member at the Trout Springs and Tererro sections. Due to its high porosity, the Macho Member acted as a conduit for migrating pore fluids, which makes it highly susceptible to dissolution.

CONCLUSIONS

The history of Mississippian sedimentation in northern New Mexico includes a minimum of four episodes of sea-level fluctuation. These events are accurately recorded by the dedolomites and calcitized gypsum of the Espiritu Santo carbonates, dissolution and brecciation of the Macho Member, and the paleosilcretes of the Manuelitas Member. In contrast, periods of submergence mark a return to normal sedimentation characterized by cementation, dolomitization, and chertification of the sediments (Fig. 10). The widespread correlability of certain chronologically constrained diagenetic fabrics reinforces the presence of an extensive, low-relief depositional surface for Mississippian strata.

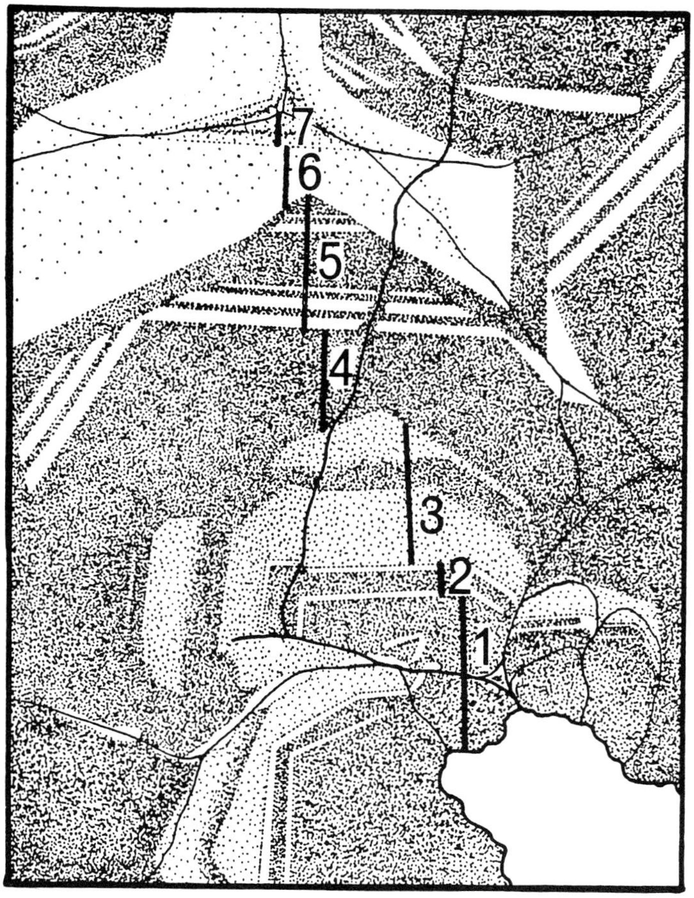

FIGURE 9. Schematic illustration of the cement stratigraphy found within the calcite spar of the Arroyo Peñasco Group. Numbers represent cement zones.

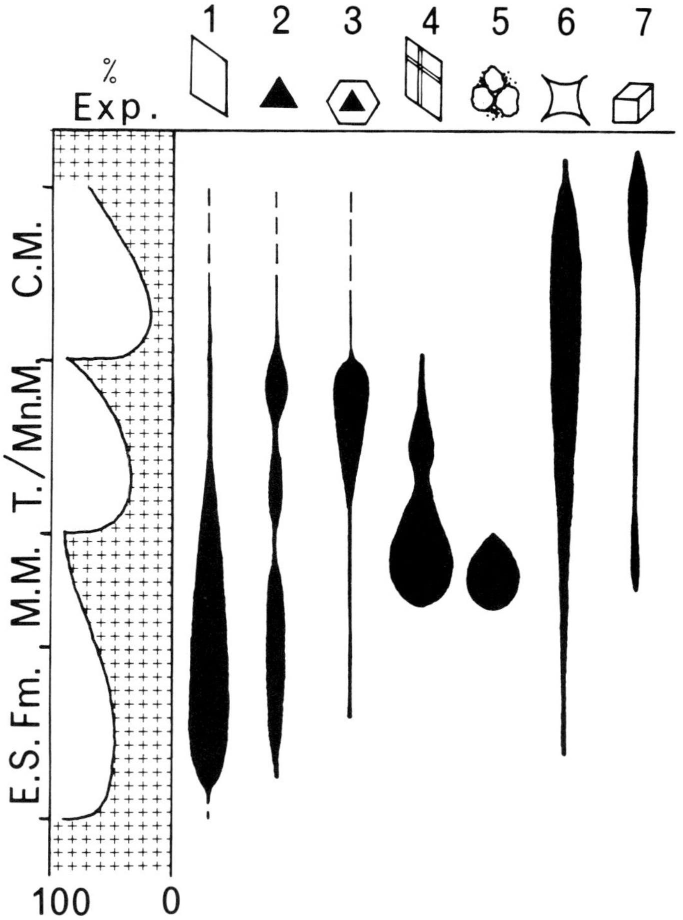

FIGURE 10. Summary diagram of the timing of diagenetic events. % Exp. – percent exposure, 1 – dolomitization, 2 – chertification, 3 – silicification of evaporites, 4 – dedolomitization, 5 – brecciation, 6 – cementation, 7 – pyritization, E.S.Fm. – Espiritu Santo Formation, M.M. – Macho Member, T./Mn. M. – Turquillo and Manuelitas Members, C.M. – Cowles Member.

ACKNOWLEDGMENTS

This study was supported by grants from Amoco Production Co., Houston, and the Southwestern Section of the American Association of Petroleum Geologists. The authors would also like to thank Mobil Research Lab for the use of their luminoscope and petrographic microscopes, R. Koepnick, D. Eby, and S. Dixon for discussions and instructions, J. Viglino for her helpful criticism of earlier drafts, and R. Harmon for pointing us in the right direction.

REFERENCES

Arbey, F., 1980, Les formes de la silica et l'identification des évaporites dans les formations silicifiès; *in* Les évaporites, méchanismes, diagenèse et applications: Centre de la Recherche, Exploration et Production Elf-Aquitaine, Bulletin, v. 4, pp. 309–365.

Armstrong, A. K., 1955, Preliminary observations on the Mississippian System of northern New Mexico: New Mexico Bureau of Mines and Mineral Resources, Circular 39, 42 pp.

———, 1967, Biostratigraphy and carbonate facies of the Mississippian Arroyo Peñasco Formation , north-central New Mexico: New Mexico Bureau of Mines and Mineral Resources, Memoir 20, 80 pp.

———, and Holcomb, L., 1967, An interim report on the Mississippian Arroyo Peñasco Formation of north-central New Mexico: American Association of Petroleum Geologists, Bulletin, v. 51, pp. 417–424.

———, and Mamet, B. L., 1974, Biostratigraphy of the Arroyo Peñasco Group, Lower Carboniferous (Mississippian), north-central New Mexico: New Mexico Geological Society, Guidebook 25, pp. 145–158.

———, and ———, 1979, The Mississippian System of north-central New Mexico: New Mexico Geological Society, Guidebook 30, pp. 201–210.

Back, W., and others, 1983, Process and rate of dedolomitization; mass transfer and ^{14}C dating in a regional aquifer: Geological Society of America, Bulletin, v. 94, pp. 1415–1429.

Baltz, E. H., and Bachman, G. O., 1956, Notes on the geology of the southeastern Sangre de Cristo Mountains, New Mexico: New Mexico Geological Society, Guidebook 7, pp. 96–108.

———, and Read, C. B., 1960, Rocks of Mississippian and probable Devonian age in Sangre de Cristo Mountains, New Mexico: American Association of Petroleum Geologists, Bulletin, v. 44, pp. 1749–1774.

Chapin, T. S., 1981, Geology of Fort Burgwin Ridge, Taos County, New Mexico: M.S. thesis, University of Texas at Austin, 150 pp.

Choquette, P. W., and Steinen, R. P., 1980, Mississippian non-supratidal dolomite, Ste. Genevieve Limestone, Illinois Basin; evidence for mixed-water dolomitization: Society of Economic Paleontologists and Mineralogists, Special Publication 28, pp. 163–169.

Evamy, B. D., 1963, The application of a chemical staining technique to study of dedolomitization: Sedimentology, v. 2, pp. 164–170.

Folk, R. L., and Pittman, J. S., 1971, Length-slow chalcedony; a new testament for vanished evaporites: Journal of Sedimentary Petrology, v. 41, pp. 1045–1058.

———, and Siedlecka, A., 1974, The schizohaline environment; its sedimentary and diagenetic fabrics as exemplified by late Paleozoic rocks of Bear Island, Svalbard: Sedimentary Geology, v. 11, pp. 1–15.

Friedman, G. M., and Shukla, V., 1980, Significance of authigenic quartz euhedra after sulfates; example from the Lockport Formation (Middle Silurian) of New York: Journal of Sedimentary Geology, v. 50, pp. 1299–1304.

Grover, G., Jr., and Read, J. F., 1983, Paleoaquifer and deep burial related cements defined by regional cathodoluminescent patterns, Middle Ordovician carbonates, Virginia: American Association of Petroleum Geologists, Bulletin, v. 67, pp. 1275–1303.

Longman, M. W., 1980, Carbonate diagenetic textures from near-surface di-

agenetic environments: American Association of Petroleum Geologists, Bulletin, v. 64, pp. 461–487.
Machel, H. G., 1983, Cathodoluminescence in carbonate petrography; some aspects of geochemical interpretation (abstract): American Association of Petroleum Geologists, Abstracts, pp. 119–120.
McKenzie, J. A., Hsu, K. J., and Schneider, J. F., 1980, Movement of subsurface waters under the sabka, Abu Dhabi, UAE, and its relation to evaporative dolomite genesis: Society of Economic Paleontologists and Mineralogists, Special Publication 28, pp. 11–30.
Milliken, K. L., 1979, The silicified evaporite syndrome—two aspects of silicification history of former evaporite nodules from southern Kentucky and northern Tennessee: Journal of Sedimentary Petrology, v. 49, pp. 245–256.
Morrow, D. W., 1982, Descriptive field classification of sedimentary and diagenetic breccia fabrics in carbonate rocks: Canadian Society of Petroleum Geologists, Bulletin, v. 30, pp. 227–229.
Oglesby, T. W., 1976, A model for the distribution of manganese, iron, and magnesium in authigenic calcite and dolomite cements in the Upper Smackover Formation in eastern Mississippi: M.S. thesis, University of Missouri, Columbia, 122 pp.
Peterson, N. M. A., and von der Borch, C. C., 1965, Chert; modern inorganic deposition in a carbonate precipitating locality: Science, v. 149, pp. 1501–1503.
Summerfield, M. A., 1983, Petrography and diagenesis of silcrete from the Kalahari Basin and Cape coastal zone, southern Africa: Journal of Sedimentary Petrology, v. 53, pp. 895–909.
Sutherland, P. K., 1963, Paleozoic rocks: New Mexico Bureau of Mines and Mineral Resources, Memoir 11, pp. 22–46.
Vaughan, F. R., 1978, The origin and diagenesis of the Arroyo Peñasco collapse breccia: M.S. thesis, State University of New York at Stony Brook, 70 pp.
Williamson, T. F., 1978, Petrology of the Lower Arroyo Peñasco (Mississippian), Taos County, New Mexico: M.S. thesis, University of Texas at Austin, 275 pp.

STRUCTURAL CONTROLS ON BASIN-MARGIN SEDIMENTATION: PENNSYLVANIAN TAOS TROUGH, NEW MEXICO, AND CONTEMPORARY BELIZE, CENTRAL AMERICA

PATRICIA WOOD DICKERSON
Conoco, Inc., P.O. Box 1959, Midland, Texas 79702

INTRODUCTION

Regional structural influences on mixed carbonate–siliciclastic basin-margin sedimentation in the Pennsylvanian Taos trough were analogous in several respects to influences on sedimentation in contemporary Belize. Common elements include episodically rising granitic highlands at the basin margins, supplying siliciclastic sediments, as well as faults and fault-bounded ridges active during carbonate-platform sedimentation and localizing reef or mound development.

Beyond the basin margins, however, the differences are significant. There was no wave-resistant organic barrier to snorkel along at the Pennsylvanian platform edge, as there is in Belize. Subsurface data are sparse, but apparently the carbonate platform of the southern Taos trough sloped gradually basinward. Additionally, whereas the Taos trough was relatively narrow and almost surrounded by highlands, the Gulf of Honduras east of Belize opens into the abyssal depths of the Cayman trench and the Caribbean Sea. There, siliciclastic sediments commonly bypass the shelf and are carried out through channels in the reef to be deposited at the foot of the steep slope or beyond.

The observations on basin-margin structure and sedimentation presented here are based upon several years' work in the Sangre de Cristo, as well as on a recent aerial and surface reconnaissance and underwater sampling in Belize.

TAOS TROUGH

Structural and Geographic Setting

Taos trough was part of the suite of Pennsylvanian to Early Permian basins and uplifts in New Mexico and Colorado generally referred to as the Ancestral Rockies. In places the older uplifts coincide with Cenozoic Rocky Mountain uplifts; in New Mexico, however, the Paleozoic features are found considerably east and west of the younger ones. A narrow extension of the Central Colorado trough, the Taos trough was a maximum of 140 km long and 165 km wide (Fig. 1; Baltz, 1978). Its depth varied in response to periodic movements of the surrounding highlands, the influences of which will be discussed below.

In contrast with the situation in Belize, northern New Mexico was not at the convergence of major crustal plates during the Pennsylvanian, although some workers suggest that the tectonism there may represent a response to late Paleozoic plate convergence farther to the southeast (Woodward and Ingersoll, 1979). Both the uplifts and the basins have foundations 1.5–1.7 b.y. old, primarily granitic and metamorphic continental crust; there is no evidence for late Paleozoic crust, either continental or oceanic, in the region.

Siliciclastic sedimentation predominated in the northern trough, where roughly 5,000 m of sediment accumulated. Casey's (1980) exemplary lithologic and paleoenvironmental study of the siliciclastic sediments should be consulted for details. Shallow-platform carbonate deposition characterized the southern part of the trough, and those strata attain thicknesses of about 1,000 m (Sutherland, 1963). Predictably, an area of mixed carbonate–siliciclastic sedimentation intervened. Kues (this guidebook) presents an excellent discussion of the evolution of Pennsylvanian stratigraphic nomenclature applied to this complex sedimentary suite.

In Pennsylvanian time, the area was at tropical latitude and the climate was one of high air and water temperatures and heavy rainfall. Those conditions resulted in rapid mechanical and chemical weathering of the granitic and metamorphic rocks that were sources of siliciclastic sediments. To the south, away from major siliciclastic influxes, the warm and well-oxygenated waters of moderately elevated pH favored extensive carbonate-platform development, and biologic productivity was generally high in the basin (Kues, this guidebook; Toomey, 1980; Sutherland and Harlow, 1973).

Uncompahgre Uplift

The principal structural element of the region during Pennsylvanian time was the Uncompahgre uplift, the eastern margin for part of which was the Pecos–Picuris fault zone (Fig. 2). That zone, intermittently active throughout the Pennsylvanian, had its origins in the Precambrian and continued its activity at least through Tertiary times (Miller and others, 1963; Robertson and Moench, 1979).

With each pulse of uplift along the west side of the trough, braided streams, alluvial fans, and fan-delta complexes carried coarse detritus progressively farther eastward into the basin (Casey, 1980; Figs. 3, 4). Headwaters of the streams were rising more or less continually and distances from headwaters to trough were not great; thus, braided streams developed, stream channels were relatively straight and confined, and the sediment load was coarse.

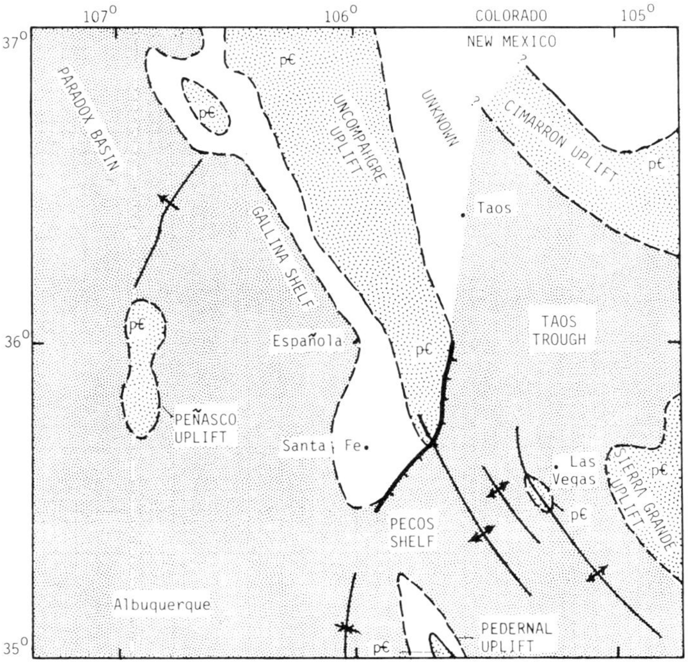

FIGURE 1. Pennsylvanian structural elements of north-central New Mexico. Portion of Pecos–Picuris fault zone coincident with eastern margin of Uncompahgre uplift shown with ticks on downthrown side (after Baltz, 1978; Woodward and Ingersoll, 1979).

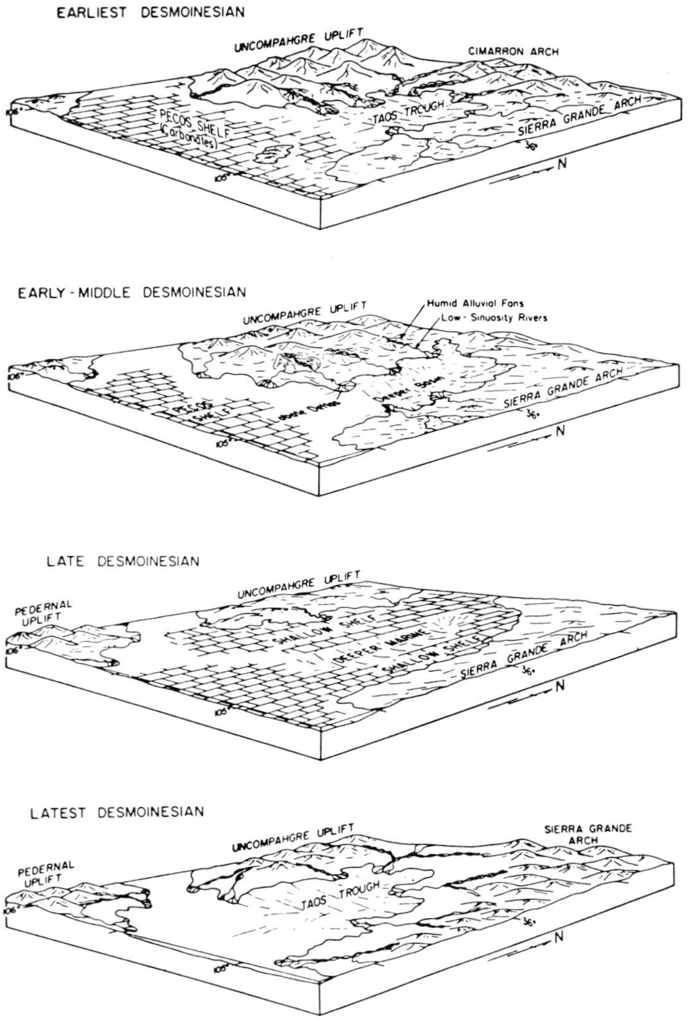

FIGURE 2. Desmoinesian paleogeography, north-central New Mexico (Casey, 1980).

FIGURE 3. Braided-stream-dominated humid alluvial-fan deposits (mid-Desmoinesian) about 4 km northwest of Tres Ritos along Rio Pueblo on N.M. 3. (For detailed description see Casey, 1980.)

Pedernal Uplift

From the Pedernal uplift, a broad platform extended north- and northeastward to the Uncompahgre and Sierra Grande uplifts (Fig. 2); about 1,000 m of primarily carbonate sediments accumulated on that platform (Sutherland, 1963). The Precambrian-granite-cored Pedernal uplift did not become a significant source of siliciclastic sediment until mid- to late Desmoinesian time and never rivaled the Uncompahgre in either elevation or volume of sediment supplied.

Although the Pedernal was quiescent during Early Pennsylvanian deposition, immediately east and subparallel to the Pecos–Picuris fault zone a series of fault-bounded ridges and basins formed concurrently with movement on the Pecos–Picuris zone (Booth, 1976, cited *in* Lisenbee, Woodward, and Connolly, 1979; Baltz and Bachman, 1956). Either erosion from, or nondeposition over, the ridges resulted in incomplete Pennsylvanian sections on those positive features. For example, one limestone unit (late Atokan–early Desmoinesian) that is approximately 235 m thick in the basin becomes only 45 m thick on one of the ridges north of Lamy and west of Apache Canyon fault (Booth, 1976, cited *in* Lisenbee, Woodward, and Connolly, 1979).

Where less prominent positive areas were present within the platform sequence, deposition may have been continuous, but higher-energy carbonate facies developed. Phylloid algal mounds flourished on shoals, scarps, and other slightly elevated areas of Pennsylvanian sea floors. The large mid-Desmoinesian algal mound at Montezuma (Toomey, 1980; Fig. 5) appears to have grown on an earlier north-trending Pennsylvanian ridge in the position now occupied by the Creston anticline. A sequence of stacked mounds exposed on Johnson Mesa is of similar trend; the possible influence of relict structure on mound development there and at other localities in the southern Sangre de Cristo is now being examined. Analogous colonization on elevated fault blocks within a carbonate platform takes place today in Belize and will be discussed below.

Northward along the western portion of the platform, progressively more clay, silt, and fine sand became mixed with the carbonate sediments. The sequence on Johnson Mesa comprises at least three cyclic sets of mixed siliciclastic/carbonate deposits; each set grades upward from phylloid algal mound to mound–debris shoal, marl, siltstone, and finally coal. Between cycles there was a brief depositional hiatus before submergence resulted in re-establishment of an algal community. In the region of mixed sedimentation, algae commonly colonized levees and bars of abandoned fan-delta lobes (Casey, 1980); Choi and Ginsburg (1982) describe similar colonization by corals occurring presently in the southern shelf lagoon of Belize.

Sierra Grande Uplift

The Sierra Grande was another Precambrian-granite-cored uplift, which formed the eastern margin of the Taos trough during the Pennsylvanian. The uplift was relatively stable, sloping gradually westward

FIGURE 4. Slope and basin calcareous siltstones and shales (mid-Desmoinesian) near crest of Holman Hill on N.M. 3. The thin, more competent sandstone beds are interpreted as prodelta turbidite deposits. (For detailed description see Casey, 1980.)

FIGURE 5. Phylloid algal mound (early Desmoinesian) near Montezuma, just northwest of Las Vegas, New Mexico. Length 183 m, height 18 m.

into the trough, until late Desmoinesian time when both the Sierra Grande and Pedernal uplifts were rejuvenated. Sands that were shed from the Sierra Grande mass are arkosic (microcline is particularly abundant) and much finer grained than those from the Uncompahgre and Cimarron source areas (Casey, 1980). The sand-body geometries on the east side of the basin are characteristic of a lower-relief coastal plain: terrigenous shelf, shallow marine, and fluvial–deltaic deposits. Carbonate buildups formed in places on the shelf. Following the late Desmoinesian pulse of uplift, an apron of coarser arkose was deposited on the flanks of the structure.

Cimarron Arch

By earliest Desmoinesian time the Cimarron arch had emerged as a major positive feature along the northern margin of the trough. From those highlands, as from the Uncompahgre to the west, coarse siliciclastic sediments were carried into the basin and deposited in braided-stream channels and fan deltas. This high block also supplied sediment in Late Pennsylvanian to Early Permian time, as south-flowing streams gradually filled the trough (Casey, 1980).

Thus, the Pennsylvanian sedimentary rocks of the Taos trough record the following depositional history: (1) Morrowan—fine-grained siliciclastic and platform carbonate sediments were deposited in the northern and southern parts, respectively, of a shallow epicontinental sea. (2) Atokan—coarser sands and gravels accumulated in a deepening depressions along the west side of the trough, while platform carbonate deposition took place along the south and east sides. (3) Desmoinesian (Fig. 2)—still coarser sands and gravels were shed from the west and north, while finer sands began to be supplied to the shelf and trough from the east; the trough axis migrated eastward and deepened further. The carbonate shelf persisted along the southern margin; then in the late Desmoinesian, a major marine transgression resulted in platform carbonate deposition throughout the basin. (4) Latest Desmoinesian through the remainder of the Pennsylvanian—the seas retreated southward and the trough was gradually filled with coarse-grained detritus by southward-flowing braided streams. Deltaic sedimentation continued into Early Permian time along the southern margin.

BELIZE

Structural and Geographic Setting

Belize, Central America, occupies the southeastern corner of the Yucatan Peninsula and is near the convergence of three crustal plates: the North American, the Caribbean, and the Cocos (Fig. 6). A major left-lateral strike–slip fault system (the Motagua–Polochic–Cayman trench system) separates the Yucatan continental block of the North American plate, on which Belize is located, from the Honduras continental block of the Caribbean plate (Molnar and Sykes, 1969).

The flat Yucatan platform, consisting of a sequence of Triassic to Early Jurassic red beds overlain by Cretaceous carbonates and evaporites, constitutes the northern half of Belize; the late Paleozoic Maya Mountains, composed of granite and minor low-grade metamorphic rocks, form the southern half (Fig. 7).

The continental margin comprises the shelf lagoon, with a maximum depth of 50 m, that extends some 260 km along the coast, as well as several elongate, fault-bounded ridges which generally support coral reefs. For example, the reef buildup on the ridge beneath Turneffe Island, the central of the three large ridges (Fig. 7), is as much as 1,219 m thick where penetrated in wells. Troughs between ridges are filled to varying extents with sediment (Silver and others, 1980; Dillon and Vedder, 1973)—both carbonate sediment from the shelf and siliciclastics carried from the coast down the slope and into the troughs.

Belize lies within the tropics at 15°45′–18°45′N latitude and is an area of high temperatures (24–27°C most of the year) and abundant rainfall (125–450 cm/yr). Structural controls on sedimentation can be

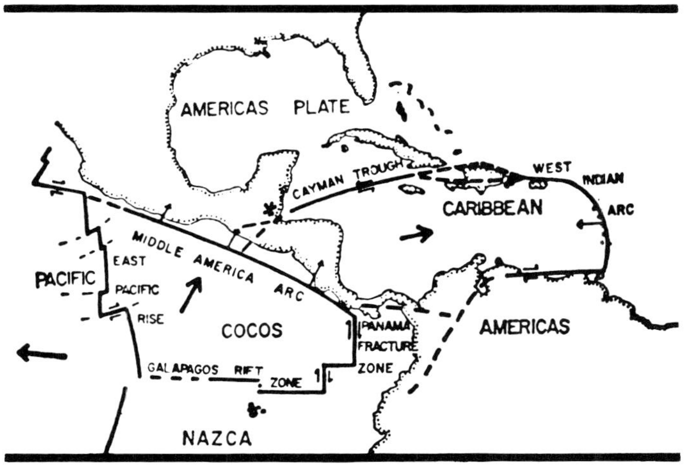

FIGURE 6. Location and plate-tectonic setting of Belize, Central America; asterisk on southeast side of Yucatan Peninsula marks position of Belize (Molnar and Sykes, 1969).

FIGURE 7. Block diagram illustrating physiographic features on- and offshore Belize (Dillon and Vedder, 1973). View is to the west. Maya Mountains are highlands in left center. The large ridge closest to the viewer is that of Glovers Reef–Lighthouse Reef. The middle ridge is that of Turneffe Island and the southern barrier. The distant ridge is that of Ambergris Cay and the northern shoreline.

readily observed, and carbonate and siliciclastic sedimentation proceed side by side. Indeed, to one schooled in the distinctness of carbonate and siliciclastic depositional environments it was enlightening to stand on an arkose beach composed of angular grains 2 or 3 mm across and to see the clear, aquamarine waters where carbonate deposition was taking place a scant hundred meters away. That is the depositional setting observed at Stann Creek, midway down the mainland coast near the mouths of two streams that drain the granitic highlands of the Maya Mountains.

Maya Mountains

As was the case for the Uncompahgre uplift along the western margin of the Taos trough, the Maya Mountains (elevation 1,200 m) consist of granite with some minor low-grade metamorphic rocks; mechanical and chemical weathering are rapid; and the streams draining the uplift are straight, have steep gradients, and carry coarse sediment. The mainland coastal beach mentioned in the preceding paragraph is about 20 km from the mountains. Much of the sediment is deposited in lobate deltas in the shelf lagoon, then reworked by strong longshore currents. A band of silt- and clay-laden water extends about 100 m into the lagoon from the mainland coast, beyond which carbonate deposition takes place. Lagoonal carbonates situated near the mouths of streams commonly contain quartz silt or clays formed during the breakdown of feldspars.

The shelf southeast of the Maya Mountains is an area of mixed carbonate–siliciclastic deposition. The southern lagoon is shallow and structurally unstable; therefore, any slight vertical tectonic adjustment can change the local depositional from a carbonate to a siliciclastic regime, or vice versa. Choi and Ginsburg (1982) document such a situation, in which patch reefs have developed on levees and channel-mouth bars of an abandoned delta lobe. An alternate interpretation is emerging, though, based upon subsequent drilling by those authors. The area they described in 1982 is part of an abandoned delta lobe; however, instead of colonizing on siliciclastic substrates, the patch reefs are growing on older carbonate mud banks (P. M. Harris, written comm. 1984).

Motagua–Polochic–Cayman Trench Faults

Along the southern shelf margin, faults of that system bound major ridges and troughs (Fig. 8; Dillon and Vedder, 1973). The sedimentary section varies greatly from ridge to trough and from one trough to the next, as with those of the southern margin of the Taos Basin. In contrast

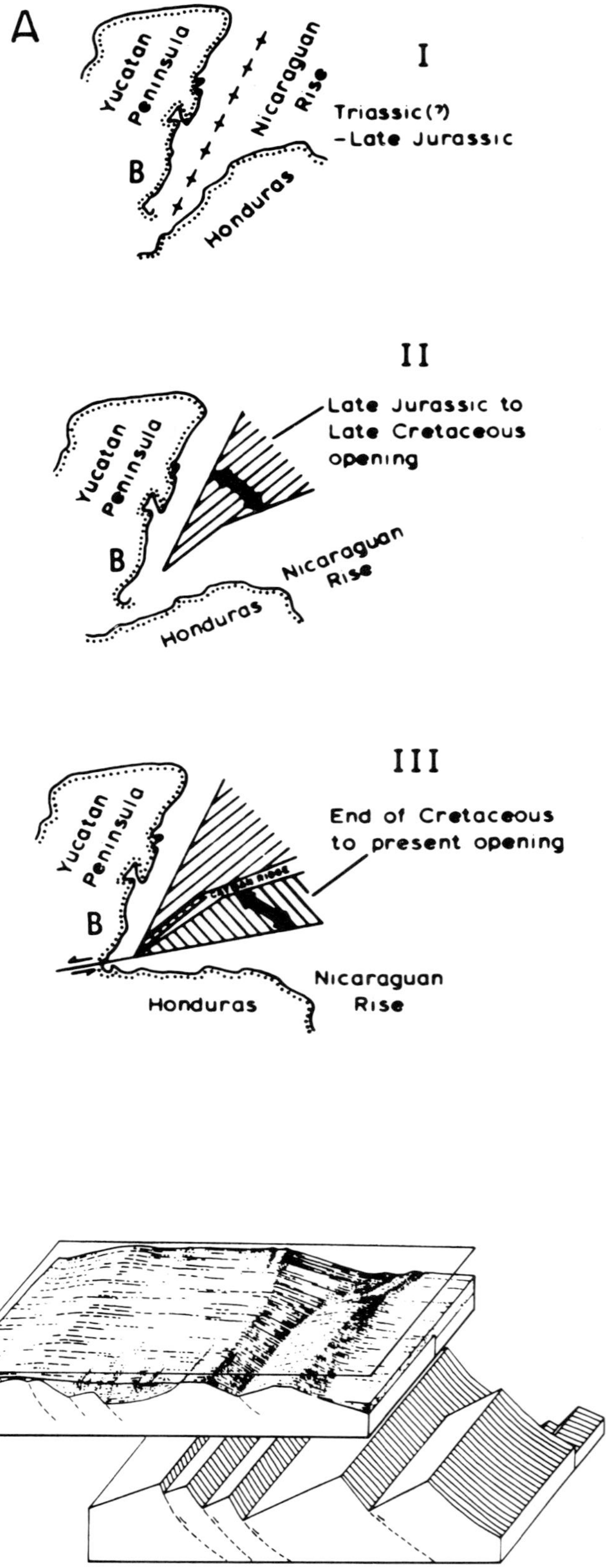

FIGURE 8. **A,** Widening of the bight of the Gulf of Honduras from Mesozoic to present. B = location of Belize (Dillon and Vedder, 1973). **B,** Formation of ridges during extension in the Gulf of Honduras (Dillon and Vedder, 1973). Ridge crests are favored sites for reef development, as waters are well oxygenated and within the photic zone; upwelling currents provide a continuous supply of nutrients.

with the Pennsylvanian ridges, however, those of Belize rise as much as 3,000 m above flanking trough floors.

More importantly, the fault-controlled ridges determine the positions of the barrier reef and large atolls (Fig. 7). The westernmost ridge forms the foundation for the northern portion of the barrier reef, Ambergris Cay, and the shoreline north of Ambergris. On the middle one are Turneffe Island and the southern barrier reef. The easternmost ridge extends beneath Glovers Reef and Lighthouse Reef.

Within the shelf, particularly in the southern lagoon, coral patch reefs develop on topographically higher areas, just as algae colonized higher areas of the Pennsylvanian shelf of the southern Taos trough. Where faults of the Motagua–Polochic system cut the shelf, rhombic blocks form between fault strands during strike–slip deformation. Coral communities thrive along the edges of elevated blocks where the waters are better oxygenated and increased current assures a continuing supply of nutrients. The array of straight-sided, generally rhombic patch reefs and atolls is particularly striking from the air.

CONCLUSIONS

Basin-margin sedimentation in the Pennsylvanian Taos trough took place under tropical climatic conditions; there, as in contemporary Belize, carbonate and siliciclastic deposition proceeded side by side, and large volumes of both types of sediment accumulated.

Episodic uplift of granitic highlands adjoining the Taos Basin, and rapid mechanical and chemical weathering under tropical conditions continually supplied siliciclastic sediments. Headwaters of streams draining the highlands were rising and distances to the basin were not great; thus, straight braided streams with steep gradients carried a load of coarse sediment into the trough. The same conditions prevail today, as straight streams with headwaters in the granitic Maya Mountains carry coarse arkosic sediments into the shelf lagoon of Belize. However, the Taos trough was confined among highlands, whereas the shelf lagoon of the Gulf of Honduras opens into the abyssal depths of the Caribbean Sea; thus, particles are commonly carried past the shelf and settle at or beyond the foot of the slope in a large ocean basin.

Formation of fault-bounded ridges and troughs during deposition on the southern shelf of the Pennsylvanian Taos Basin resulted in highly variable platform carbonate sequences from ridge to trough and from one trough top the next. Early Pennsylvanian ridges and fault scarps provided favored colonization sites for later Pennsylvanian algal buildups. Similarly, faults of the Motagua–Polochic–Cayman trench system bound ridges and troughs in the southern shelf of Belize. Elevated blocks, formed between fault strands during strike–slip deformation, provide colonization sites for thriving coral patch reefs. In contrast with those of the Taos trough platform, the ridges of the Belize margin rise as much as 3,000 m above the flanking trough floors; they control the positions of the barrier reef and large atolls along the entire shelf margin. The Taos trough platform apparently sloped gradually basinward and lacked a wave-resistant organic barrier, so the fauna was not treated to the sound of surf crashing on the reef.

ACKNOWLEDGMENTS

I am particularly thankful to Elmer Baltz for the grand tour he provided through the Pennsylvanian stratigraphy and structure of the Sangre de Cristo early in my work there; his willingness to discuss his knowledge of the area transcended even the fact that I resided in the state neighboring New Mexico on the southeast. Thanks, too, go to Burr Silver for a similarly thorough immersion (literally) in the geology of Belize. Ongoing dialogs with Bill Muehlberger on Sangre de Cristo structure and with Stan Frost on Belize sedimentation continue to enhance my understanding of those areas and the world at large. This paper benefited greatly from reviews by C. A. Ross and P. M. Harris, whose efforts I gratefully acknowledge.

REFERENCES

Baltz, E. H., 1978, Resume of Rio Grande depression in north-central New Mexico: New Mexico Bureau of Mines and Mineral Resources, Circular 163, pp. 210–227.

———, and Bachman, G. O., 1956, Notes on the geology of the southeastern Sangre de Cristo Mountains, New Mexico: New Mexico Geological Society, Guidebook 7, pp. 96–108.

Casey, J. M., 1980, Depositional systems and paleogeographical evolution of the late Paleozoic Taos trough, northern New Mexico; *in* Fouch, T. D., and Magathan, E. R. (eds.), Paleozoic paleogeography of the west-central United States: Rocky Mountain Section, Society of Economic Paleontologists and Mineralogists, Rocky Mountain Paleogeography Symposium I, pp. 181–196.

Choi, D. R., and Ginsburg, R. N., 1982, Siliciclastic foundations of Quaternary reefs in the southernmost Belize lagoon, British Honduras: Geological Society of America, Bulletin, v. 93, pp. 116–126.

Dillon, W. P., and Vedder, J. G., 1973, Structure and development of the continental margin of British Honduras: Geological Society of America, Bulletin, v. 84, pp. 2713–2732.

Kues, B. S., 1984, Pennsylvanian stratigraphy and paleontology of the Taos area, north-central New Mexico: This guidebook.

Lisenbee, A. L., Woodward, L. A., and Connolly, J. R., 1979, Tijeras–Cañoncito fault system—a major zone of recurrent movement in north-central New Mexico: New Mexico Geological Society, Guidebook 30, pp. 89–100.

Miller, J. P., Montgomery, A., and Sutherland, P. K., 1963, Geology of part of the southern Sangre de Cristo Mounains, New Mexico: New Mexico Bureau of Mines and Mineral Resources, Memoir 11, 106 pp.

Molnar, P., and Sykes, L. R., 1969, Tectonics of the Caribbean and Middle America regions from focal mechanisms and seismicity: Geological Society of America, Bulletin, v. 80, pp. 1639–1684.

Robertson, J. M., and Moench, R. H., 1979, The Pecos Greenstone Belt: a Proterozoic volcano-sedimentary sequence in the southern Sangre de Cristo Mountains, New Mexico: New Mexico Geological Society, Guidebook 30, pp. 165–174.

Silver, B. A., Neese, D., Koch, J., Goldhammer, R., and Stewart, G., 1980, The Belize carbonate complex: shelf margin model for hydrocarbon exploration: Institute for Energy Development, Tulsa, 212 pp.

Sutherland, P. K., 1963, Paleozoic rocks; *in* Geology of the southern Sangre de Cristo Mountains, New Mexico: New Mexico Bureau of Mines and Mineral Resources, Memoir 11, pp. 22–46.

———, and Harlow, F. H., 1973, Pennsylvanian brachiopods and biostratigraphy in southern Sangre de Cristo Mountains, New Mexico: New Mexico Bureau of Mines and Mineral Resources, Memoir 27, 173 pp.

Toomey, D. F., 1980, History of a Late Carboniferous phylloid algal bank complex in northeastern New Mexico: Lethaia, v. 13, pp. 249–267.

Woodward, L. A., and Ingersoll, R. V., 1979, Phanerozoic tectonic setting of Santa Fe country: New Mexico Geological Society, Guidebook 30, pp. 51–58.

A longer than usual mixed freight climbs slowly up the 4% grade of Barranca Hill from the Rio Grande valley (photo by Bob Richardson).

PENNSYLVANIAN STRATIGRAPHY AND PALEONTOLOGY OF THE TAOS AREA, NORTH-CENTRAL NEW MEXICO

BARRY S. KUES
Department of Geology, University of New Mexico, Albuquerque, New Mexico 87131

INTRODUCTION

The most extensively exposed sequence of Pennsylvanian strata in New Mexico occurs in the Sangre de Cristo Mountains, where it reaches a thickness of more than 2,000 m and includes sediments deposited from earliest (Morrowan) to latest (Virgilian) Pennsylvanian time. Outcrops of Pennsylvanian age extend continuously from near Pecos and Bernal in the south to near Eagle Nest and Elizabethtown in the north, a distance of about 135 km (Fig. 1). A large variety of depositional environments is present in this sequence, including basinal terrigenous mudstones, offshore marine shelf carbonates, deltaic marine and fluvial shales, sandstones and conglomerates, and arkosic continental clastic sediments, among others. Some of the marine units are highly fossiliferous, but with the notable exception of the brachiopods, these faunas have not been comprehensively studied. This paper summarizes the Pennsylvanian stratigraphy and paleontology of the Taos area, presents preliminary observations on a remarkably diverse Middle Pennsylvanian mollusc-dominated marine fauna currently under study by the author, and includes illustrations of some of the more conspicuous molluscan species present in this fauna. All specimens illustrated in this paper are in the University of New Mexico Department of Geology paleontological collections and bear UNM catalogue numbers.

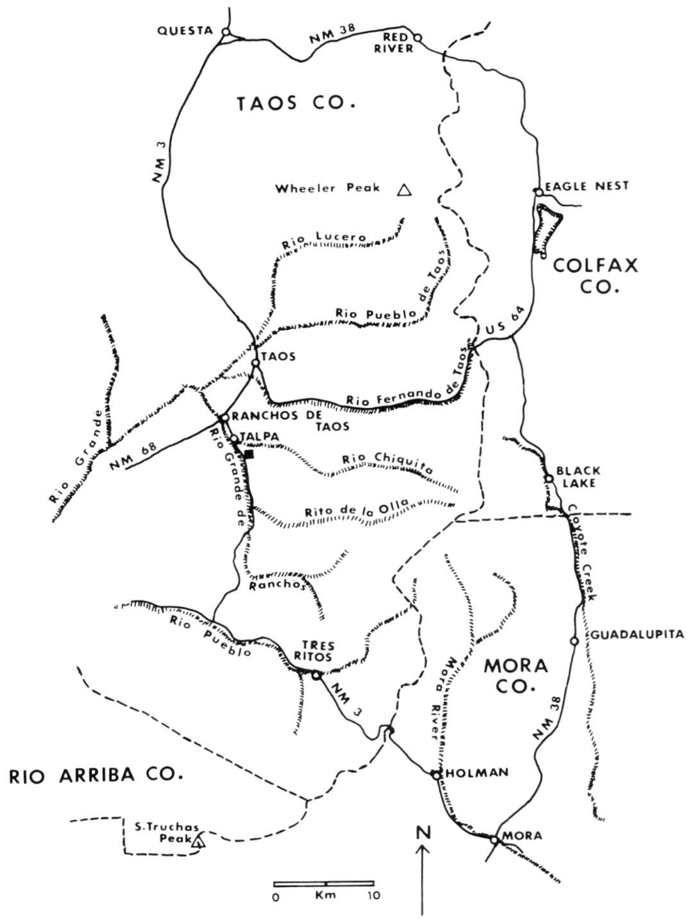

FIGURE 1. Reference map of Taos area. Black square indicates Flechado Formation locality discussed in text.

STRATIGRAPHY

The Carboniferous age of strata at the southern end of the Sangre de Cristo Mountains near Pecos was established by Jules Marcou, the first geologist to examine significant collections of fossils from the area (Marcou, 1856, 1858). Marcou recorded these strata as the "Mountain Limestone" (Lower Carboniferous), but Hall (1856) and Newberry (1861) soon recognized their actual "Coal Measures" or Pennsylvanian age. Although several other early workers also contributed information concerning the Pennsylvanian of the Santa Fe–Pecos area (see Sutherland and Harlow, 1973, for a summary), the first relatively detailed examination of Pennsylvanian stratigraphy and general geology of the Taos area was made by Stevenson (1881), as part of a regional study of the Sangre de Cristos in New Mexico and southern Colorado. Stevenson reported more than 1,000 m of Pennsylvanian strata in the Coyote Creek–Mora area, made a few observations on the Pennsylvanian near Taos, and recognized striking facies changes from Taos to Pecos, including an increase in marine limestones to the south.

Subsequently, no detailed stratigraphic analysis or formational subdivision of the Pennsylvanian in the Sangre de Cristos were attempted for more than half a century. Gruner (1920) reported general features of the stratigraphy in an area southwest of Wheeler Peak, and additional brief comments on the marine Pennsylvanian (termed the Magdalena Group or Formation) near Taos were presented by Darton (1928), Ray and Smith (1941), and Wanless (1947). Stratigraphic subdivision and correlation of the Pennsylvanian in the Taos and neighboring areas has proven difficult because of marked lateral facies changes and the isolation of Pennsylvanian exposures in this area from those in other regions (Sutherland, 1963). Beginning in the 1940's, it became customary to divide the Pennsylvanian into a basal, relatively thin, clastic-dominated Sandia Formation, overlain in ascending order by the "lower gray limestone" and "upper arkosic limestone" members of the Madera Formation, and finally by the arkosic continental clastics of the Pennsylvanian–Permian Sangre de Cristo Formation (e.g., Read and Wood, 1947; Brill, 1952; Baltz and Bachman, 1956; Baltz, 1965). The Sandia and Madera Formations had originally been defined in the Sandia Mountains, some 100 km to the southwest. Young (1946b) proposed the name Cortado Formation for the thick, predominantly shaly and arkosic cyclothemic sequence of Desmoinesian age in the Taos area, and Cieneguilla Formation for similar beds to the east near Black Lake, but these terms did not find general acceptance.

Concurrently, a general picture of the regional tectonic influences on Pennsylvanian deposition in north-central New Mexico began to be developed (Read and Wood, 1947; Brill, 1952; Baltz and Bachman, 1956; Wanek and Read, 1956; Sutherland, 1963; Baltz, 1965). Briefly, the area around Taos was found to be part of a north–south trending trough (variously called the Taos trough, Rowe–Mora Basin, or part of the Colorado–northern New Mexico geosyncline) between the adjacent Uncompahgre highland to the west and the Sierra Grande uplift somewhat farther away to the east. Thick sequences of marine, deltaic, and continental sediments, primarily eroded from the Uncompahgre highland, accumulated in this trough, while to the south (Santa Fe–Pecos–Las Vegas area) a much thinner sequence of mainly marine carbonate sediments was deposited on a shallow shelf (Pecos shelf of Sutherland, 1963). Marine sedimentation ceased in late Desmoinesian time in the Taos area, as the trough was filled with arkosic clastic sediments of the Sangre de Cristo Formation, but continued until nearly the end of the Pennsylvanian on the shelf before southward prograding Sangre de Cristo sediments covered it. The Pennsylvanian tectonic setting of the Taos trough was considered in detail by Sutherland (1963) and Casey (1980).

The Pennsylvanian stratigraphic nomenclature used widely today in the Sangre de Cristo Mountains (Fig. 2) was established by Sutherland (1963), following a detailed study of the area from Taos to Pecos. Sutherland rejected the use of Sandia and Madera Formations in this area and defined three new formations, the La Pasada, Flechado, and Alamitos. The La Pasada Formation, about 300 m thick, is primarily a cyclic carbonate unit of Morrowan to Desmoinesian age. It was deposited on the Pecos shelf and consists of shallow, nearshore shale, siltstone, and limestone units in its lower half and mainly neritic offshore carbonates in its upper part. The La Pasada was traced by Sutherland (1963) as far north as Jicarilla ridge, near the Truchas Peaks, about 40 km south of Taos. The Flechado Formation is the northern equivalent of the La Pasada, intertonguing with that unit near Jicarilla ridge, but thickening to about 750 m northward at its type locality along the Rio Pueblo in southern Taos County. The Flechado unconformably overlies Mississippian limestones of the Tererro Formation and consists primarily of olive, brown and dark-gray shales and siltstones, and low-feldspar sandstones and conglomerates; limestones represent only 5–10% of the formation (Sutherland, 1963).

In a study involving 85 measured sections in an area extending from the Rio Fernando (east of Taos) southward to the Rio Pueblo, Casey (1980) interpreted the Morrowan part of the Flechado as consisting of a basal fluvial–estuarine sandstone sequence overlain by a thick sequence of lignitic shales and siltstones representing shallow marine and prograding strand-plain environments. The Atokan to earliest Desmoinesian part includes distal alluvial-fan and braided-stream conglomerates, and coarsening-upward, stratigraphically complex fan-delta systems that prograded eastward and southeastward into the Taos trough as the Uncompahgre highland was uplifted. Sedimentation occurring in early to middle Desmoinesian time consisted of alluvial-fan sandstones and conglomerates dominated by braided-stream processes, low-sinuosity fluvial deposits, lobate, river-dominated delta systems, and various marine prodelta, shelf, and basinal mudstones and carbonates.

Overlying both the La Pasada and Flechado Formation is the Alamitos Formation (Sutherland, 1963). Approximately equivalent to the "upper arkosic limestone member" of the Madera, the Alamitos is composed chiefly of arkosic sandstones, with lesser amounts of shales, siltstones, and limestones. In general, the lower half of the formation has a greater abundance of clastics, whereas the upper part has a relatively high amount of limestone. In the Pecos area the Alamitos is about 400 m thick and spans Desmoinesian to Virgilian time, but to the north it thickens to 1,200 m and represents only the late Desmoinesian. The northernmost exposures of the Alamitos studied by Sutherland (1963) are along the north side of the Rio Pueblo.

Throughout the southern Sangre de Cristos, the Alamitos is overlain by the nonmarine Sangre de Cristo Formation, which consists of poorly sorted arkosic sandstones and conglomerates interbedded with red shales and siltstones. The Sangre de Cristo Formation increases in thickness northward, from about 150 m near Pecos to 725 m east of Mora, to

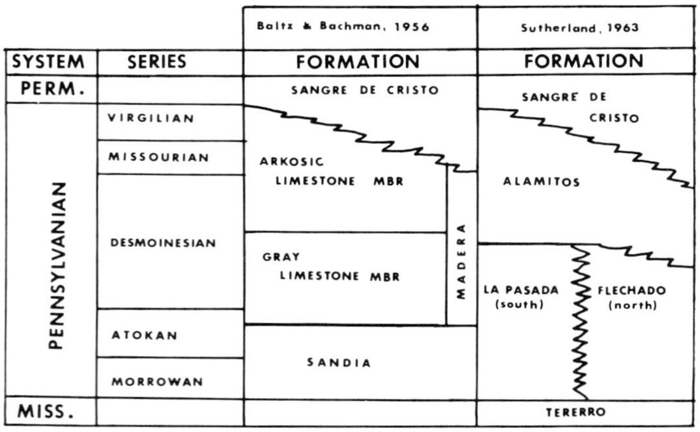

FIGURE 2. Stratigraphic nomenclature of Taos to Las Vegas area (after Casey, 1980).

2,875 m just north of the New Mexico–Colorado boundary (Brill, 1952). About 1,515 m were measured in the Moreno Valley east of Taos by Wanek and Read (1956). In the Pecos area the basal Sangre de Cristo interfingers with beds of Virgilian age and is probably mainly of Early Permian age, but to the north it overlies marine units of late Desmoinesian age (Sutherland, 1963). The presence of *Callipteris conferta* and other fossil plants indicates an Early Permian age for at least part of the formation in the Moreno Valley (Wanek and Read, 1956). Lungfish burrows and other Early Permian vertebrate remains were reported from near the top of the Sangre de Cristo Formation near the village of Pueblo, 38 km south of Pecos (Vaughn, 1964).

North of Taos, in the Eagle Nest area, Pennsylvanian strata of Atokan and Desmoinesian age reach an estimated thickness of 1,500 m, but formation names used in the south have not been applied; the entire Pennsylvanian is assigned to the undivided Magdalena Group (Clark, 1966; Clark and Read, 1972). The sequence is composed of alternating coarse-grained gray to brown sandstones, thick beds of brown, gray, purple, and pale-red shales, and minor gray, locally oolitic or nodular limestones. The Sangre de Cristo Formation overlies the Magdalena Group in the Eagle Nest area and reaches a thickness of about 1,600 m.

PREVIOUS PALEONTOLOGICAL STUDIES

Geologists on several early exploring expeditions in the 1850's and 1860's observed and collected fossils from the Pennsylvanian of the southern Sangre de Cristos (Sutherland and Harlow, 1973), but their efforts concentrated in the Santa Fe and Pecos areas, near the Santa Fe Trail. The first collections of Pennsylvanian invertebrate fossils from near Taos were made, ironically, by the vertebrate paleontologist Edward D. Cope, while a member of the Wheeler Survey of 1874. Cope (1875, p. 66) followed the valley of Rancho Creek (now the Rio Grande de Ranchos) for 13 km, noting thick sandstone deposits containing plant fossils and reporting "limestone of the Carboniferous period, which is at the top thin-bedded and alternating with dark-colored shales." Cope found in these beds "great numbers of characteristic fossils, weathered out and beautifully preserved, including Echinoderms, Crinoids, small pieces of *Orthoceras, Goniatites,* Spirifers and other brachiopods, with Gastropods resembling *Trochus, Turritella,* and *Nerita.*" These fossils were almost certainly collected from the Flechado Formation, at or near the highly fossiliferous shales south of Talpa.

A few years later, during the summers of 1878 and 1879, J. J. Stevenson surveyed the geology of the Sangre de Cristos and made collections of Pennsylvanian fossils from several localities in the Taos area, including Taos Peak, along the Rio Fernando de Taos, at the head of Mora Creek, and in the area of Coyote Creek and Black Lake (Stevenson, 1881). These collections, and those made by Cope, were turned over to C. A. White, who listed nearly 100 taxa and described nine new species, mainly from Cope's locality (White, 1881). One of the most distinctive gastropods was named *Murchisonia copei* after its discoverer; later Girty made it the type species of a new genus *Taosia,* now considered to be a subgenus of *Stegocoelia* (Fig. 3.33).

White (1881, p. 19) remarked that this fauna showed "the contemporaneous prevalence of a greater profusion of specific forms than has hitherto been discovered in any one Carboniferous locality so far westward and southward," and perceptively suggested that the high number of species observed was probably just a small part of the total species that actually existed in the Taos area, "as indicated by the fact that at the single locality of Taos alone, Professor Cope obtained 50 species . . . and he was not then making a special investigation in that direction." Recent work by me at or near the locality south of Talpa where Cope made his collection (discussed below) has supported White's suspicions: more than 170 invertebrate species, mainly molluscs, have so far been identified.

Subsequent paleontological information on the Taos area, with the exceptions mentioned below, consists of lists of fossils identified from various Pennsylvanian units (Table 1). The first modern study of Taos area paleontology was presented by Young (1945a), who in an unpublished doctoral dissertation reported or described nearly 150 species of

TABLE 1. Papers reporting lists of fossils from Pennsylvanian marine units in the Taos area, after 1900.

Gruner, 1920	Five brachiopods and one coral from Wheeler Peak area.
Ray & Smith, 1941	Late Pennsylvanian plant assemblage, 3.2 km south of Agua Fria Creek, Moreno Valley.
Brill, 1952	Corals, brachiopods, molluscs, and a trilobite from south end of Moreno Valley, about 4.8 km north of Black Lake.
Wanek & Read, 1956	Fusulinids from Mora-Holman area.
Tschanz et al., 1958	Corals, brachiopods, bryozoans, and molluscs of Middle Pennsylvanian age from upper part of "Magdalena Group," and several marine invertebrate taxa and plants of Late Pennsylvanian age from lower Sangre de Cristo Formation, Coyote area.
Sutherland, 1963	Brachiopods from several localities south of Taos.
Clark & Read, 1972	About 70 taxa of invertebrates from 60 localities in Eagle Nest area.
Sutherland & Harlow, 1973	Fusulinid and conodont zones (along with descriptions of brachiopods), Sangre de Cristo Mountains.

Pennsylvanian invertebrates of which 36 were considered new. Several new nautiloids and a new genus and species of giant scaphopod, *Prodentalium raymondi* (Fig. 4.31), were described in publications (Young, 1942, 1945b), and the stratigraphy and brachiopod fauna were briefly discussed in abstracts (Young, 1941, 1946a, b), but the majority of the taxa described as new species in the dissertation were never published. About 35 species of gastropods alone were reported from near Talpa. Later, Sutherland and Harlow (1973) comprehensively described the Pennsylvanian brachiopods of the Sangre de Cristos, a total of 94 species. Though most of their collections were made from the Santa Fe–Nambe–Pecos area, several localities in Taos County were also reported.

Most recently, Casey (1980) summarized the Atokan and Desmoinesian faunal communities that existed in the Taos area. Casey recognized a productid community in prodeltaic facies, a productid–*Composita* community farther offshore, a molluscan–*Mesolobus* community in nearshore deltaic embayments, and transitional brachiopod–crinoid and phylloid algae communities in shallow, abandoned fan-delta platform environments where carbonates were accumulating. Basinal environments within the Taos trough were characterized by few invertebrate body fossils but numerous trace fossils of nonselective deposit feeders. Though Casey's communities were not quantitatively defined and probably represent informal associations of species rather than true paleocommunities, his work represents the first attempt to interpret the distribution and associations of the diverse Pennsylvanian marine faunas of the Taos area, and to tie them in with specific depositional environments.

PALEONTOLOGY OF FLECHADO FORMATION NEAR TALPA

An excellent and characteristic example of a highly fossiliferous, nearshore-marine, dominantly dark-gray shale–siltstone facies of the Flechado Formation occurs along the east side of New Mexico State Highway 3, about 2.5 km south of the village of Talpa. The steeply dipping section exposed here is more than 150 m thick and consists of several relatively thin conglomeratic to medium-grained, locally trough crossbedded sandstone units separated by thicker intervals of dark-gray shales and siltstones. Following the interpretation given by Casey and Scott (1979) for sequences in the immediate vicinity, the sandstone beds represent delta-front and fluvial-channel sediments associated with fan-delta lobes that prograded intermittently across shallow, muddy, marine environments.

The slope-forming shales and siltstones that form the greater part of the thickness of this sequence are abundantly fossiliferous. The fossils erode readily from these relatively soft units and their preservation is generally very good. Based on the brachiopod species present and their ranges in the Pennsylvanian of the Sangre de Cristos as determined by Sutherland and Harlow (1973), the age of this section is early Desmoinesian. My intention here is to briefly characterize this unusual fauna, illustrate some of the most common or conspicuous gastropod and bivalve species (Figs. 3, 4) (see Sutherland and Harlow, 1973, for descriptions and illustrations of most of the brachiopods), and summarize some aspects of ongoing investigations.

About 170 species of marine invertebrates have thus far been identified from this locality, in collections totaling more than 15,000 specimens. The list of species below provides a general measure of the relative diversity of major groups.

COELENTERATA
- *Paraconularia crustula*
- *Syringopora*? sp.
- solitary rugose corals, 2 spp.

BRYOZOA
- *Fistulipora* sp.
- *Prismopora triangulata*

BRACHIOPODA
- *Lingula carbonaria*
- *Orbiculoidea capuliformis*
- *O. missouriensis*
- *O.*? sp. (very large)
- *Crania* sp.
- *Rhipidomella carbonaria*
- *Derbyia* cf. *crassa*
- *Plicochonetes*? sp.
- *Mesolobus striatus*
- *M.* cf. *profundus*
- *Chonetinella jeffordsi*
- *Leptalosia*? sp.
- *Kozlowskia haydenensis*
- *Desmoinesia "missouriensis"*
- *D.* cf. *muricatina*
- *Sandia*? sp.
- *Echinaria* cf. *knighti*
- *Buxtonia* sp.
- *Antiquatonia hermosana*
- *Linoproductus* sp.
- *Wellerella* sp.
- *Hustedia "mormoni"*
- *Composita subtilita*
- *Cleiothyridina pecosii*
- *Crurithyris* cf. *planoconvexa*
- *Neospirifer cameratus*
- *N. tewaensis*?
- *Anthracospirifer curvilateralis chavezae*
- *A. "opimus"*?
- *Punctospirifer* sp.
- *Phricodothyris perplexa*
- *Beecheria* cf. *bovidens*

MOLLUSCA—BIVALVIA
- *Nuculopsis anodontoides*
- *N.* aff. *anodontoides*
- *N. croneisi*
- *N. girtyi*
- *N.*? sp.
- *Paleyoldia glabra*
- *Phestia arata*
- *P.* cf. *bellistriata*
- *Solemya (Janeia) radiata*
- *Parallelodon carbonarius*?
- *P. hexacostata*
- *P.* aff. *tenuistriatus*
- cf. *Parallelodon*
- *Promytilus* sp.
- *Myalina (M.)* cf. *arbala*
- *Leptodesma (L.)* aff. *longa*
- *Acanthopecten meeki*
- *Clavicosta echinata*
- *Limipecten* cf. *morsei*
- Aviculopectinidae indet.
- *Streblochondria tenuilineata*
- *Chaenocardia ovata*
- *Posidonia*? n.sp.
- *Paleolima* sp.
- *Schizodus amplus*
- *S. curtus*
- *S.* aff. *ulrichi*
- *Astartella compacta*
- *A. concentrica*

A. newberryi
A. varica
Cypricardinia carbonaria
Edmondia anodontoides
E. ovata?
Wilkingia terminale
W. n.sp.?
Chaenomya sp.
Sanguinolites costata
S. aff. *costata*
cf. *S.*? *elongatus* of Ciriacks, 1963

MOLLUSCA—ROSTROCONCHIA
Pseudoconocardium cf. *obliquum*
P. n.sp.?

MOLLUSCA—GASTROPODA
Bellerophontacea
Bellerophon (*B.*) *crassus*
B. (*B.*) aff. *planodorsatus*
B. (*B.*) *wewokanus*
B. (*Pharkidonotus*) *percarinatus*
Knightites (*Cymatospira*) *montfortianus*
K. (*Retispira*) *"tenuilineata"*
Euphemites "carbonarius"
E. cf. *enodis*
E. nodocarinatus (=*Bellerophon inspeciosus* of White, 1881)
Euomphalacea
Straparollus (*Amphiscapha*) cf. *catilloides*
S. (*A.*) cf. *reedsi*
S. (*Euomphalus*) *plummeri*
S. (*E.*) *pernodosus*
Pleurotomariacea
Trepospira (*T.*) *illinoisensis*
Euconospira aff. *turbiniformis*
E. n.sp.
Spiroscala n.sp.
cf. *Schwedagonia*
Glabrocingulum (*G.*) *grayvillense*
G. (*G.*) spp.
G. (*Ananias*) sp.
Worthenia tabulata
Gosseletina spironema
G. cf. *spironema*
Platyzona n.sp.
Shansiella carbonaria
S. beckwithana
Porcellia sp.
Glyptotomaria (*Dictyotomaria*) *quadrilineata*
G. (*D.*) aff. *scitula*
G. (*D.*) n.sp.?
Paragoniozona multilirata
P. cf. *nodolirata*
Phymatopleura nodosus
Trochonematacea
Amaurotoma aff. *humerosa*
A.? n.sp.
Platyceratacea
Platyceras (*Orthonychia*) *parvum*
Strophostylus girtyi
Microdomatacea
Microdoma? n.sp.
Eucochlis perminuta
Anomphalacea
Anomphalus verruculiferus
A. sp.
Neritacea
Naticopsis (*N.*) aff. *judithae*
N. (*Jedria*) *meeki*
N. (*J.*) *ventrica*
Trachydomia whitei
T. sp.
Oncochilus vaningeni
Murchisoniacea
Murchisonia (*M.*?) sp.
M. (*Donaldospira*?) sp.
M. (new subgen.?) n.sp.
Stegocoelia (*S.*) aff. *wortheni*
S. (*Taosia*) *copei*
S. (*T.*) cf. *percostata*
S. (*T.*) spp.
Goniasma lasallensis
G.? sp.
Loxonematacea
Plocezyga (*P.*) cf. *corona*
P. (*P.*) aff. *tenuilirata*
P. (*P.*) spp.
P.? (*Gamizyga*) *illineata*
Pseudozygopleura (*P.*) spp.
Stephanozyga subnodosa
S. n.sp.?
Cerithiacea
Orthonema inornatum
O. cf. *salteri*
O. n.spp.
Callispira quinquecostata
C.? sp.
Subulitacea
Leptoptygma subtilistriatum
L. virgatum
L. sp.
Ceraunocochlis fulminata
Strobeus regularis
S. brevis
S. paludinaeformis
S. primigenius?
S. n.sp.
Meekospira peracuta
Girtyspira sp.
Pyramidellacea
Donaldina stevensana
D. cf. *robusta*
Streptacis cerithiformis
S. aff. *scalpta*

MOLLUSCA—SCAPHOPODA
"Dentalium" sublaeve
Prodentalium raymondi

MOLLUSCA—CEPHALOPODA
Brachycycloceras sp.
Metacoceras sp.
Mooreoceras sp.
Pseudorthoceras knoxense

TRILOBITA
Ditymopyge sp.

ECHINODERMATA
crinoid stem fragments and plates
Archaeocidaris triplex
echinoid spines and plates

VERTEBRATA
Petalodus? sp.

FIGURE 3. Gastropods from the Flechado Formation, Taos area. All specimens are from the locality south of Talpa discussed in text. **1–2,** *Bellerophon* (*B.*) *crassus*, top and side views, UNM 7223, x1; **3–5,** *Bellerophon* (*B.*) *wewokanus*, top, side, and apertural views, UNM 7224, x1; **6–7,** *Bellerophon* (*Pharkidonotus*) *percarinatus*, top and side views, UNM 7225, x1; **8–10,** *Euphemites nodocarinatus*, apertural, top, and side views, UNM 7226, x1; **11,** *Knightites* (*Cymatospira*) *montfortianus*, top view, UNM 7227, x1.5; **12–13,** *Knightites* (*Retispira*) *"tenuilineata,"* side and top views, UNM 7228, x2; **14–15,** *Euphemites "carbonarius,"* top and side views, UNM 7229, x2; **16,** *Glabrocingulum* (*G.*) *grayvillense*, apertural view of an eroded specimen, UNM 7230, x1.5; **17,** *Spiroscala* n.sp., adapertural view, UNM 7015, x2; **18,** *Paragoniozona multilirata*, slightly oblique adapertural view, UNM 7119, x3; **19–20, 50,** *Straparollus* (*Amphiscapha*) cf. *catilloides*, top, basal, and side views, UNM 7231, x1.5; **21,** *Phymatopleura nodosus*, adapertural view, UNM 7099, x3; **22,** *Shansiella carbonaria*, adapertural view, UNM 7058, x2; **23,** *Trepospira* (*T.*) *illinoisensis*, top view, UNM 7193, x1; **24,** *Platyzona* n.sp., apertural view, UNM 7069, x2; **25,** *Platyceras* (*Orthonychia*) *parvum*, side view, UNM 7232, x1; **26–27,** *Worthenia tabulata*, adapertural and oblique top views, UNM 7233, x1.5; **28–29,** *Straparollus* (*Euomphalus*) *plummeri*, top and basal views, UNM 7234, x1.5; **30,** *Gosseletina* cf. *spironema*, oblique top view of a laterally crushed specimen, UNM 7027, x1; **31–32,** *Anomphalus verruculiferus*, top view, UNM 7235, and basal view, UNM 7236, both x2; **33,** *Stegocoelia* (*Taosia*) *copei*, apertural view, UNM 7237, x1; **34,** *Microdoma*? n.sp., adapertural view, UNM 7238, x2; **35,** *Plocezyga*? (*Gamizyga*) *illineata*, apertural view, UNM 7239, x1.5; **36–37,** *Strophostylus girtyi*, top and apertural views of a laterally compressed specimen, UNM 7240, x1; **38,** *Glyptotomaria* (*Dictyotomaria*) *quadrilineata*, slightly oblique adapertural view, UNM 7079, x2; **39–40,** *Trachydomia whitei*, adapertural view, UNM 7241, and apertural view, UNM 7242, both x1.5; **41,** *Strobeus primigenius*?, apertural view, UNM 7243, x1; **42,** *Strobeus regularis*, adapertural view, UNM 7244, x1; **43,** *Strobeus brevis*, apertural view, UNM 7245, x1; **44,** *Strobeus*? n.sp., apertural view, UNM 7246, x1; **45,** *Strobeus paludinaeformis*, apertural view, UNM 7247, x1; **46,** *Stephanozyga subnodosa*, adapertural view, UNM 7248, x1; **47,** *Pseudozygopleura* (*P.*) sp., apertural view, UNM 7249, x1; **48,** *Meekospira peracuta*, apertural view, UNM 7250, x1.5; **49,** *Callispira quinquecostata*, adapertural view, UNM 7251, x1.

1
2
3
4
5
6
7
8
9
10
11
12
13
14
15
16
17
18
19
20
21
22
23
24
25
26
27
28
29
30
31
32
33
34
35
36
37
38
39
40
41
42
43
44
45
46
47
48
49
50

This list is preliminary and is based almost entirely on specimens collected from the surface of outcrops. Some bulk sediment samples have been screen-washed, and the concentrates contain large numbers of minute fossils, particularly fusulinids and gastropods, that have yet to be studied. In addition, some gastropod taxa identified here only to the generic or subgeneric level include numerous species; for example, there are probably 10 or 12 as yet unidentified species of the small, high-spired gastropod *Pseudozygopleura* (*Pseudozygopleura*) (Fig. 3.47). Thus, the total number of species at this locality is almost certainly well over 200.

Gastropods are especially common and diverse in this fauna, to an extent found in few other Pennsylvanian localities in North America. Although many species are uncommon to rare, some, such as *Worthenia tabulata* (Fig. 3.26–27), *Glabrocingulum* (*Glabrocingulum*) *grayvillense* (Fig. 3.16), *Trachydomia whitei* (Fig. 3.39–40), *Anomphalus verruculiferus* (Fig. 3.31–32), *Microdoma*? n.sp. (Fig. 3.34), *Stegocoelia* (*Taosia*) *copei* (Fig. 3.33), and *Bellerophon* (*Pharkidonotus*) *percarinatus* (Fig. 3.6–7) are abundant. One or two of these common species may dominate numerically a particular interval of the shale sequence. Further, preliminary examination of sieved sediment samples shows that in some units the number of minute gastropods (those with shells less than 2 mm high, representing small species and juveniles of larger species) per unit volume of shale exceeds the number of larger gastropods by at least a factor of five. The majority of gastropod species in this fauna are present also in contemporaneous units in the midwestern United States, but some, such as *Anomphalus verruculiferus* (Fig. 3.31–32), *Glyptotomaria* (*Dictyotomaria*) *quadrilineata* (Fig. 3.38), *Paragoniozona multilirata* (Fig. 3.18), and *Callispira quinquecostata* (Fig. 3.49) have been reported only from New Mexico, Colorado, or west Texas. Several undescribed species are also present—not surprising in view of the fact that very few studies of New Mexico Pennsylvanian gastropods have been done.

Brachiopods are far less diverse and less abundant than gastropods in most parts of this sequence. However, some shale units contain high numbers of the small chonetid *Mesolobus striatus*, which may represent as much as 35% of all megainvertebrates present. Bivalves are generally less abundant than gastropods or brachiopods, and most bivalve species are uncommon to rare. Nuculoids, especially *Nuculopsis girtyi* (Fig. 4.1), and the several species of *Astartella* (Fig. 4.24–26) are locally very common, however, and comprise about 57% and 35%, respectively, of the bivalve specimens from this locality. Other molluscan groups are uncommon, although moderate numbers of scaphopod and orthoconic nautiloid shell fragments are present in some beds. Rostroconchs are represented by a total of 10 specimens; the single polyplacophoran valve from this locality was described by Kues (1983). Solitary rugose corals are present through most of the sections and appear to be most numerous in beds dominated by *Mesolobus*. Crinoids, echinoids, bryozoans, and trilobites are very minor elements of this mollusc-dominated fauna. Systematic descriptions and more comprehensive documentation of the gastropod, bivalve, rostroconch, and scaphopod taxa are in preparation.

Although the above observations are generally true for the locality as a whole, examination of collections made from each shale interval between a pair of sandstone units and from different levels within each interval suggests that considerable variation in the relative abundances of taxa exists through the sequence. Several assemblages, each characterized by different dominant species as well as by different relative abundances of subsidiary species, appear to have occupied nearshore-marine environments during deposition of the shale at this locality. Quantitative analysis of the constituent species in each assemblage must await completion of taxonomic studies, but the assemblages observed qualitatively very likely represent subsets of some of the broad, generalized "communities" recognized by Casey (1980) through the Flechado of the Taos area. Only slight differences in color, grain size, and other lithologic features of the shales are discernible at outcrop; more detailed examination of the physical characteristics of the shale intervals is planned. At present, it appears that relatively subtle differences in the environment, related to proximity to the shifting fan-delta lobes, affected the abundance of many mollusc species, resulting in significant variations in the proportions of taxa living in various prodelta- and interdelta-embayment environments.

While detailed consideration of the paleobiogeographic relationships of the Flechado mollusc fauna is premature, some general comments may be made. During the Pennsylvanian, a long, north–south trending line of connected positive areas (Front Range, Apishapa, and Sierra Grande uplifts) existed to the east of the present Sangre de Cristo Mountains in Colorado and northeastern New Mexico (Mallory, 1972; Chronic, 1979; Armstrong et al., 1979). This would have prevented any direct connection of Flechado faunas with contemporaneous marine faunas present to the northeast, for example in Kansas. Sutherland and Harlow (1973) noted a rather poor correlation of Desmoinesian brachiopod species in the Sangre de Cristos with Midcontinent species. A continuous band of marine Pennsylvanian carbonate and clastic facies is present in the subsurface south of the Sierra Grande uplift, from near Las Vegas through the Tucumcari Basin and into the Texas panhandle (Roberts et al., 1976). Midcontinent species present in north-central New Mexico are thus probably derived from the extensive marine environments that prevailed in north-central Texas during the Pennsylvanian. The Middle Pennsylvanian gastropod and bivalve faunas of north-central Texas are diverse; however, they have not been extensively studied, and further comparisons thus cannot be made. In view of the restriction of Pennsylvanian marine environments in the Taos area between uplifts to the east and west, and the possible partial separation of the Taos trough (or Rowe–Mora Basin) from the central Colorado seaway by a westward extension of the Sierra Grande uplift (Cimarron arch of Baltz, 1965), the presence of some endemic mollusc species restricted to north-central New Mexico is not unlikely. Several apparently endemic brachiopods were reported by Sutherland and Harlow (1973).

ACKNOWLEDGMENTS

I am grateful to Mr. LeGrand Smith for showing me the Talpa locality and later conveying to me his large collections of fossil invertebrates from the Taos area for study. His continued interest, along with that of fellow fossil enthusiasts Lud and Lois Gritzo, who loaned specimens from their collection, represent the ideal collaboration between amateur and professional paleontologists. I also thank Elmo Brown and Pat Dickerson for reviewing the manuscript and offering helpful suggestions on its improvement.

FIGURE 4. Bivalves, scaphopods, and rostroconchs from the Flechado Formation, Taos area. All specimens are from the locality south of Talpa discussed in text, → unless otherwise indicated. **1,** *Nuculopsis girtyi,* left valve, UNM 8787, x1.5; **2,** *Nuculopsis anodontoides,* left valve, UNM 8788, x1.5; **3,** *Paleyodia glabra,* left valve, UNM 8789, x1.5; **4,** *Paleyodia glabra,* anterior part of left valve, UNM 8790, x2; **5,** *Phestia bellistriata,* anterior part of right valve, UNM 8791, x1; **6,** *Phestia arata,* left valve, UNM 8792, x1.5; **7,** *Promytilus* sp., left valve, UNM 8793, x1.5; **8,** *Posidonia*? n.sp., left valve, UNM 8794, x2; **9,** *Myalina* (*M.*) cf. *arbala,* part of left valve, UNM 8795, x0.5; **10,** *Leptodesma* (*L.*) cf. *longa,* left valve, UNM 4511, x1; **11,** *Parallelodon hexacostata,* right valve, UNM 8797, x1.5; **12,** *Parallelodon carbonarius*?, anterior part of right valve, UNM 8798, x1; **13,** *Acanthopecten meeki,* left valve, UNM 8799, x1.5; **14,** *Clavicosta echinata,* left valve, UNM 8800, x1.5; **15,** *Streblochondria tenuilineata,* right valve, UNM 8801, x1.5; **16,** *Chaenocardia ovata,* left valve, UNM 8802, x1.5; **17,** *Edmondia ovata*?, slightly oblique view of right valve having ventral periphery partially covered by matrix, UNM 8803, x1; **18,** *Edmondia anodontoides,* left valve, UNM 8804, x1.5; **19,** *Schizodus amplus,* left valve, UNM 3878, from Taos Canyon, x0.5; **20,** *Schizodus* aff. *ulrichi,* right valve, UNM 8805, x1.5; **21, 22,** *Schizodus curtus,* right valve, x3, and top view, x2, UNM 8806; **23,** *Sanguinolites costata,* anterior part of left valve, UNM 8807, x1; **24,** *Astartella newberryi,* left valve, UNM 8808, x2; **25,** *Astartella concentrica,* left valve, UNM 8809, x2; **26,** *Astartella varica,* left valve, UNM 8810, x1; **27,** *Wilkingia terminale,* left valve, UNM 8811, x1; **28,** *Chaenomya* sp., left valve, UNM 8812, x1; **29,** cf. *Sanguinolites*? *elongatus,* left valve, UNM 8813, x1; **30,** *"Dentalium" sublaeve,* fragment of shell, UNM 8815, x1; **32–34,** *Pseudoconocardium* cf. *obliquum,* posterior, top, and left side views, UNM 8816, x1.5.

1
2
3
4
5
6
7
8
9
10
11
12
13
14
15
16
17
18
19
20
21
22
23
24
25
26
27
28
29
30
31
32
33
34

REFERENCES

Armstrong, A. K., Kottlowski, F. E., Stewart, W. J., and others, 1979, The Mississippian and Pennsylvanian (Carboniferous) Systems in the United States—New Mexico: U.S. Geological Survey, Professional Paper 1110-W, pp. W1–W27.

Baltz, E. H., Jr., 1965, Stratigraphy and history of the Raton Basin and notes on the San Luis Basin, Colorado–New Mexico: American Association of Petroleum Geologists, Bulletin, v. 49, pp. 2041–2075.

———, and Bachman, G. O., 1956, Notes on the geology of the southeastern Sangre de Cristo Mountains, New Mexico: New Mexico Geological Society, Guidebook 7, pp. 96–108.

Brill, K. G., Jr., 1952, Stratigraphy in the Permo-Pennsylvanian zeugogeosyncline of Colorado and northern New Mexico: Geological Society of America, Bulletin, v. 63, pp. 809–880.

Casey, J. M., 1980, Depositional systems and basin evolution of the late Paleozoic Taos trough, northern New Mexico: Texas Petroleum Research Committee, Report no. UT 80-1, 236 pp.

———, and Scott, A. J., 1979, Pennsylvanian coarse-grained fan deltas associated with the Uncompahgre uplift, Talpa, New Mexico: New Mexico Geological Society, Guidebook 30, pp. 211–218.

Chronic, J., 1979, The Mississippian and Pennsylvanian (Carboniferous) Systems in the United States—Colorado: U.S. Geological Survey, Professional Paper 1110-V, pp. V1–V25.

Ciriacks, K. W., 1963, Permian and Eotriassic bivalves of the middle Rockies: American Museum of Natural History, Bulletin, v. 125, art. 1, 100 pp.

Clark, K. F., 1966, Geology of the Sangre de Cristo Mountains and adjacent areas between Taos and Raton, New Mexico: New Mexico Geological Society, Guidebook 17, pp. 56–65.

———, and Read, C. B., 1972, Geology and ore deposits of Eagle Nest area, New Mexico: New Mexico Bureau of Mines and Mineral Resources, Bulletin 94, 152 pp.

Cope, E. D., 1875, Report on the geology of that part of northwestern New Mexico examined during the field-season of 1874: Annual report upon the geographical explorations and surveys west of the one hundredth meridian, in California, Nevada, Nebraska, Utah, Arizona, Colorado, New Mexico, Wyoming, and Montana [Wheeler Survey], Appendix LL of the annual report of the chief of engineers for 1875, pp. 61–97.

Darton, N. H., 1928, "Red Beds" and associated formations in New Mexico, with an outline of the geology of the state: U.S. Geological Survey, Bulletin 794, 356 pp.

Girty, G. H., 1939, Certain pleurotomariid gastropods from the Carboniferous of New Mexico and Texas: Washington Academy of Science, Journal, v. 29, pp. 21–36.

Gruner, J. W., 1920, Geologic reconnaissance of the southern part of the Taos Range, New Mexico: Journal of Geology, v. 28, pp. 731–742.

Hall, J., 1856, Description and notices of the fossils collected upon the route; *in* Whipple, A. W., Report of explorations for a railway route near the 35th parallel: U.S. 33rd Congress, 2nd Session, Senate Executive Document 78 and House Executive Document 91, v. 3, pt. 4, pp. 99–105.

Kues, B. S., 1983, A chiton from the Flechado Formation (Middle Pennsylvanian) of northern New Mexico: New Mexico Journal of Science, v. 23, pp. 19–22.

Mallory, W. M., 1972, Regional synthesis of the Pennsylvanian System: Geologic atlas of the Rocky Mountain region, Rocky Mountain Association of Geologists, Denver, Colorado, pp. 111–127.

Marcou, J., 1856, Resume of a geological reconnaissance extending from Napoleon, at the junction of the Arkansas with the Mississippi, to the Pueblo de Los Angeles, in California; *in* Whipple, A. W., Report of explorations for a railway route near the 35th parallel: U.S. 33rd Congress, 2nd Session, Senate Executive Document 78 and House Executive Document 91, v. 3, pt. 4, pp. 165–175.

———, 1858, Geology of North America, with two reports on the prairies of Arkansas and Texas, the Rocky Mountains of New Mexico, and the Sierra Nevada of California: Zurich, 144 pp.

Newberry, J. S., 1861, Geological report; *in* Ives, J. C., Report upon the Colorado River of the West: U.S. 36th Congress, 1st Session, Senate Executive Document 77, pt. 3, 154 pp.

Ray, L. L., and Smith, J. F., Jr., 1941, Geology of the Moreno Valley, New Mexico: Geological Society of America, Bulletin, v. 52, pp. 177–210.

Read, C. B., and Wood, G. H., Jr., 1947, Distribution and correlation of Pennsylvanian rocks in the late Paleozoic sedimentary basins of northern New Mexico: Journal of Geology, v. 55, pp. 220–236.

Roberts, J. W., Barnes, J. J., and Wacker, H. J., 1976, Subsurface Paleozoic stratigraphy of the northeastern New Mexico basin and arch complex: New Mexico Geological Society, Guidebook 27, pp. 141–152.

Stevenson, J. J., 1881, Report upon geological examinations in southern Colorado and northern New Mexico during the years 1878 and 1879: Report upon United States geographical surveys west of the one hundredth meridian [Wheeler Survey], v. 3, supplement, 420 pp.

Sutherland, P. K., 1963, Paleozoic rocks: New Mexico Bureau of Mines and Mineral Resources, Memoir 11, pp. 22–53.

———, and Harlow, F. H., 1973, Pennsylvanian brachiopods and biostratigraphy in southern Sangre de Cristo Mountains, New Mexico: New Mexico Bureau of Mines and Mineral Resources, Memoir 27, 173 pp.

Tschanz, C. M., Laub, D. C., and Fuller, G. W., 1958, Copper and uranium deposits of the Coyote district, Mora County, New Mexico: U.S. Geological Survey, Bulletin 1030-L, pp. 343–398.

Vaughn, P. P., 1964, Evidence of aestivating lungfish from the Sangre de Cristo Formation, Lower Permian of northern New Mexico: Los Angeles County Museum of Natural History, Contributions in Science, no. 80, 8 pp.

Wanek, A. A., and Read, C. B., 1956, Third day, Taos to Eagle Nest and Elizabethtown: New Mexico Geological Society, Guidebook 7, pp. 82–95.

Wanless, H. R., 1947, Regional variations in Pennsylvanian lithology: Journal of Geology, v. 55, pp. 237–253.

White, C. A., 1881, Report on the Carboniferous invertebrate fossils of New Mexico: Report upon United States geographical surveys west of the one hundredth meridian [Wheeler Survey], v. 3, supplement, appendix, 38 pp.

Young, J. A., Jr., 1941, Pennsylvanian Brachiopods from New Mexico: Geological Society of America, Bulletin, v. 52, p. 1978.

———, 1942, Pennsylvanian Scaphopoda and Cephalopods from New Mexico: Journal of Paleontology, v. 16, pp. 120–125.

———, 1945a, Paleontology and stratigraphy of the Pennsylvanian strata near Taos, New Mexico: Unpublished Ph.D. dissertation, Harvard University, 314 pp.

———, 1945b, *Huanghoceras bryani,* new name: Journal of Paleontology, v. 19, p. 319.

———, 1946a, Cyclothems, the key to Lower Pennsylvanian diastrophism in northern New Mexico: Geological Society of America, Bulletin, v. 57, p. 1246.

———, 1946b, Pennsylvanian strata near Taos, New Mexico: preliminary report: Geological Society of America, Bulletin, v. 57, pp. 1246–1247.

Note added in proof

A recent paper that deals comprehensively with the late Paleozoic stratigraphy of an area from Las Vegas to Mora, along the southeastern side of the Sangre de Cristo Mountains, deserves mention, although it was received too late to be incorporated into the main discussion. It is Baltz, E. H., and Myers, D. A., 1984, *Porvenir Formation (new name)—and other revisions of nomenclature of Mississippian, Pennsylvanian, and Lower Permian rocks, southeastern Sangre de Cristo Mountains, New Mexico:* U.S. Geological Survey, Bulletin 1537-B, 39 pp. In the area covered, the term Sandia Formation is retained for rocks of mainly Atokan age, and the Madera Formation is raised to a Group including the Desmoinesian Porvenir Formation, overlain by the Alamitos Formation, of Desmoinesian to (locally) Wolfcampian age.

DEPOSITIONAL ENVIRONMENTS OF THE PERMO-PENNSYLVANIAN SANGRE DE CRISTO FORMATION, COYOTE CREEK DISTRICT, MORA COUNTY, NEW MEXICO

C. ELMO BROWN
Placid Oil Company, 410 17th Street, Suite 2000, Denver, Colorado 80202

INTRODUCTION

The Sangre de Cristo Formation is a Pennsylvanian–Permian clastic unit which was deposited in the Paleozoic Rowe–Mora Basin of northern New Mexico and south-central Colorado. Bachman and Read (1952), Zeller and Baltz (1954), Tschanz and others (1958), and May and others (1977) studied the Sangre de Cristo Formation in the Coyote Creek district, an inactive copper-mining area located approximately 8 mi northeast of Mora and 2 mi southeast of Guadalupita in Mora County, north-central New Mexico (Fig. 1). These earlier studies focused on the uranium-rich zones within the formation. A detailed description of the entire Sangre de Cristo Formation within the district, however, was lacking.

The purpose of this study is to describe and classify the various depositional units of the Sangre de Cristo Formation found within the Coyote Creek district. Classification is based on lateral and vertical variations in geometry, lithology, texture, and sedimentary structure. Variations are documented by measured sections, sketches, and sediment-transport direction measurements from better outcrop exposures. From this information, a depositional history can be formulated.

Present Structural Setting

The Coyote Creek district lies within a Laramide imbricate-thrust zone which marks the eastern edge of the Sangre de Cristo Mountains (May and others, 1977). A major fault, with displacement of several thousand feet, separates Precambrian metasedimentary rocks from Pennsylvanian sedimentary rocks (Fig. 2). The structural strike of the sedimentary rocks and minor thrust faults found within them is generally to the north. Dip of beds in the vicinity of the thrust zone varies from nearly vertical to slightly overturned (Zeller and Baltz, 1954). Near the center of Coyote Creek valley, dips abruptly flatten to 7–14° east. Zeller and Baltz (1954) mapped a thrust fault separating the steep dips on the west from the flatter dips to the east; however, no evidence for this fault was found on measured section SDC-1. It appears that only folding of the rocks occurred in this part of the valley.

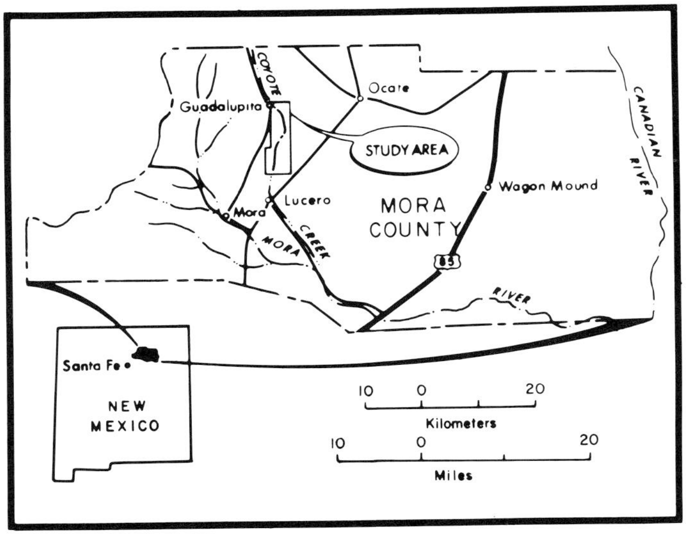

FIGURE 1. Location of study area (modified from Tschanz and others, 1958).

DEPOSITIONAL SYSTEMS OF THE SANGRE DE CRISTO FORMATION

Tschanz and others (1958) divided the Sangre de Cristo Formation in the Coyote Creek district into six lithologic units (Fig. 3). Each unit is characterized by many abrupt facies changes and by interfingering unit boundaries which are difficult to delineate. It is uncertain how persistent these units are due to the lack of detailed mapping.

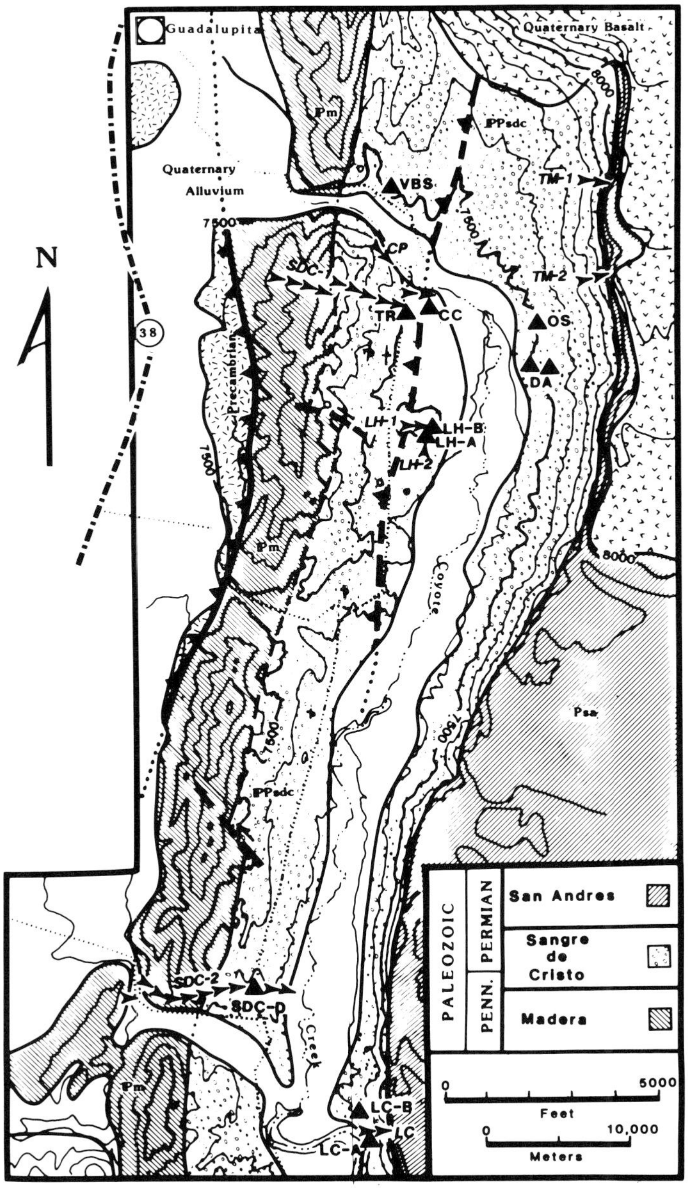

FIGURE 2. Topographic and geologic map of Coyote Creek district showing measured section and sketch locations (modified from Zeller and Baltz, 1954).

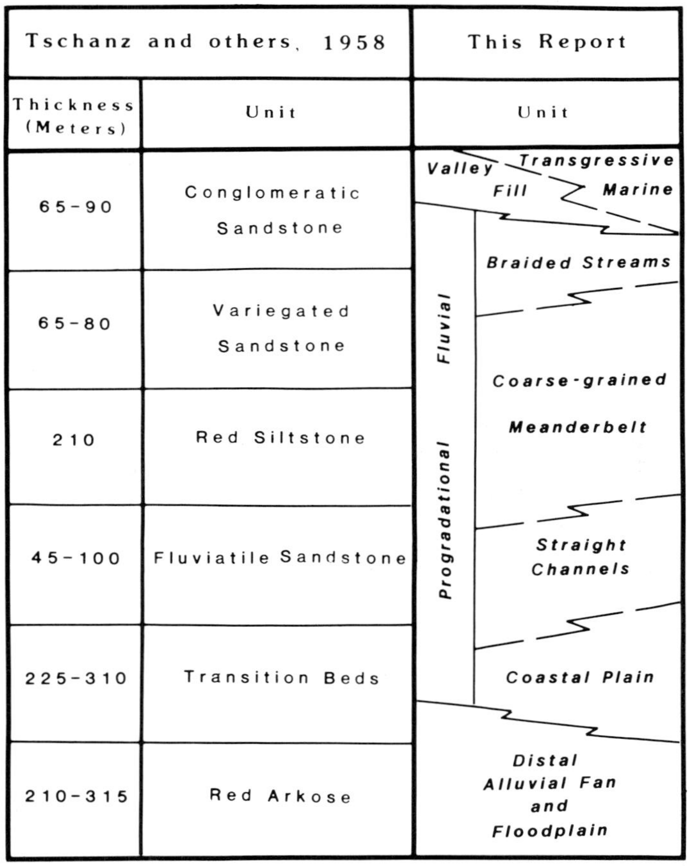

FIGURE 3. Classification of Sangre de Cristo units within Coyote Creek district.

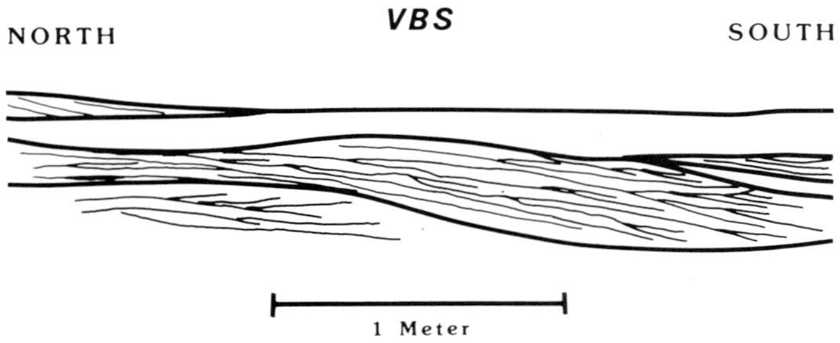

FIGURE 4. Sketch of distal alluvial-fan, sheetflood sandstone. In this and all following figures, blacked-in boxes denote clay clasts and open circles denote gravel.

For the purpose of this study, the Sangre de Cristo Formation is divided into three genetically different depositional systems. In ascending order they are: (1) distal alluvial fan with associated floodplain, (2) progradational fluvial system, and (3) shallow-marine bar and valley fill.

DISTAL ALLUVIAL-FAN AND ASSOCIATED FLOODPLAIN SYSTEM

Deposits of the distal alluvial-fan and associated floodplain depositional system occupy the lower 400–650 m of the formation where outcrop attitude is nearly vertical to slightly overturned. Approximately 30% of the system consists of red to light-brown, arkosic-sandstone beds which form distinct parallel ridges. Separating these ridges are wide, covered areas which represent finer-grained facies.

Distal Alluvial-fan Facies

Ridge-forming sandstones in the lower part of the Sangre de Cristo Formation are coarse- to very coarse-grained, submature, subrounded to angular arkoses. Some of the beds have pebbles of granite and feldspar, either scattered throughout or concentrated along scour surfaces. Angular, unweathered potassium-feldspar grains indicate short transport distances.

The most common sandstone type is represented by sheetlike arkoses which are dominated by small-scale, low-angle, trough crossbeds with very little basal scour (Fig. 4). Pebbles are often scattered throughout with occasional heavy minerals or pebbles lining trough scours. Large grain size and lack of textural variability make internal stratification difficult to see. Paleocurrent direction is interpreted to be to the south, parallel to trough axes.

The small scale and nonerosional nature of the crossbedding suggest that these numerous sandstone packages were deposited in fairly shallow water with very low turbulence and high sedimentation rates, indicating sheetflood deposition (Collinson, 1978). These sheetflood sediments are interpreted to represent deposition in a southward-prograding distal alluvial-fan environment similar to the terminal-fan model defined by Friend (1978). The terminal-fan model is characterized by a distributary channel pattern (Allen, 1965) which eventually disappears down-fan. This loss of channelization is produced by a decrease in discharge caused by several factors including water loss into permeable alluvium and a high evapotranspiration rate. A high evapotranspiration rate would also inhibit bank stabilizing plant growth in the distal fan while increasing the rate of runoff and sediment transport in the source area.

It is believed that when sediment-loaded floodwaters emerged from the channelized system to the unconfined distal alluvial fan, water depth and velocity were reduced, thus decreasing the capacity of the water to carry sediment. This resulted in rapid sheetflood sedimentation.

Floodplain Facies

Approximately 70% of the distal alluvial-fan and associated floodplain system is fine-grained sediment. Even though exposures are limited, lithologies appear to be primarily orange to red, micaceous, silt and sandy silt with some thin beds of limestone, maroon and brown shale, and red, coarse-grained sandstone. The maroon and brown shales occur commonly near the contact with the overlying fluvial system and are often lenticular. Limestone beds are light gray to reddish gray, often silty, unfossiliferous, very hard, nodular micrites. These beds increase in number and thickness to the south and are interpreted as having formed in "K"-horizon caliche zones (Gile and others, 1965) or in highly alkaline playa lakes similar to those described by Goudie (1973). In either case, micrite precipitation requires an environment with infrequent clastic sedimentation.

Friend (1978) stated that the bulk of distal terminal-fan deposits is usually siltstone and calcareous marlstone. Therefore, fine-grained sediments within this lower interval of the Sangre de Cristo Formation are interpreted as floodplain and playa sediments associated with the distal alluvial-fan sheetflood sands.

PROGRADATIONAL FLUVIAL SYSTEM

The middle 500–700 m of the Sangre de Cristo Formation in the study area is interpreted as having been deposited in a bedload-dominated fluvial system. This system is divided into two facies assemblages. The lowermost is a coastal-plain sequence where straight to slightly meandering fluvial-channel sandstones constitute only a minor percentage of the primarily fine-grained package. Gradationally overlying this facies assemblage is a channel-dominated interval which prograded over the coastal-plain sequence. Only a minor percentage of fine-grained sediment is preserved in the channel-dominated system. It is believed that this upward decrease in fine-grained sediment is a result of an increase in channel-migration rate and a decrease in aggradation rate.

Coastal-plain-facies Assemblage

This 130-m-thick facies assemblage consists of approximately 70% silt and shale, with lesser amounts of limestone, fine- to medium-grained sandstone, and very coarse-grained sandstone. Outcrops are limited to a few small pits dug by Tschanz and his co-workers during their studies of copper mineralization in this part of the Sangre de Cristo Formation (1958).

From these outcrops, three interfingering facies are evident: (1) distal alluvial-fan sheetflood sands, (2) fluvial facies with straight to slightly meandering channels and associated fine-grained overbank sediments, and (3) lacustrine and marsh sediments. Sheetflood sands represent a gradational contact with the underlying distal alluvial-fan system. The upper 60 m of the coastal-plain assemblage contains no sheetflood deposits.

Fluvial Deposits

Over 60% of the coastal-plain system is interpreted as having been deposited by fluvial processes with fine-grained overbank sediments constituting the bulk of the deposits. A very few sandstones exhibit the characteristics needed to classify them as channel-deposited sandstones (Fig. 5). These sandstone beds contain coarse- to very coarse-grained arkosic sand, some clay clasts (3 cm maximum width), and abundant pebbles (10 cm maximum width) which line scours and bedding planes. Large, low-angle, trough crossbeds and large tabular crossbeds are the dominant bedforms. Overlying some of these coarse sand units is a fining-upward sequence consisting of poorly laminated to structureless, occasionally root-mottled, sand and shaly sand, and poorly laminated silt.

Channel sandstones are interpreted as straight to slightly meandering bedload deposits. The low sand/shale ratio can be attributed to the small size of channels (less than 7 m at bankfull depth), the small amount of channel migration, and, most importantly, the apparent high rate of vertical floodplain accretion (Leeder, 1978).

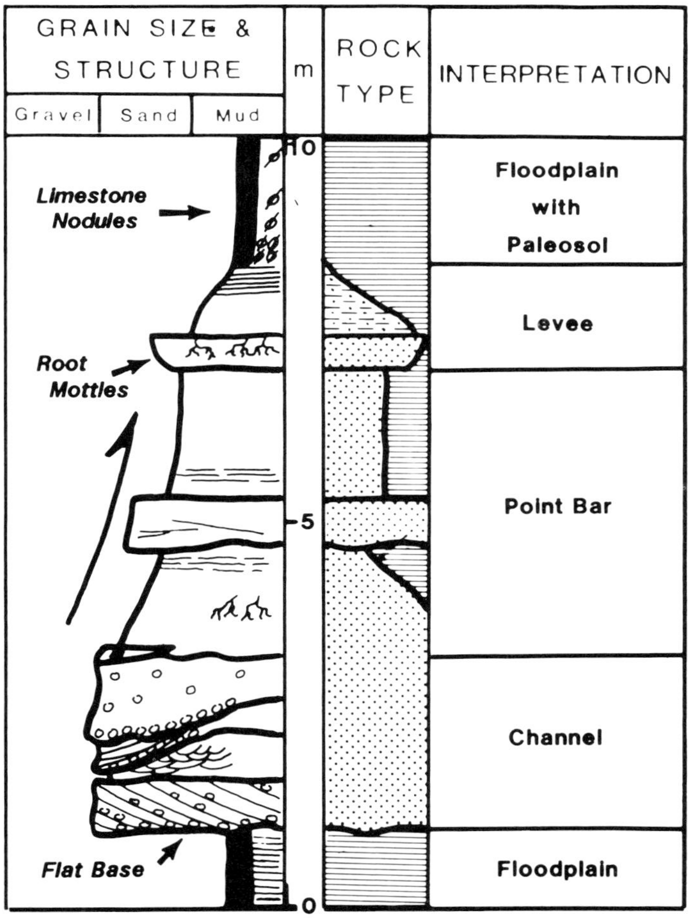

FIGURE 5. Measured section at outcrop TR showing a rare bedload channel sandstone and associated point bar in the coastal-plain-facies assemblage.

Overbank sediments are varied, ranging from laminated to blocky floodplain silts and shales to the crossbedded silts and sands of natural levees and crevasse splays (Fig. 6). Crevasse-splay deposits are the most common sand bodies within the fluvial facies (at least in number, if not in total thickness). These deposits exhibit much the same characteristics as splays described by Reineck and Singh (1975). Even though not one single event can be laterally examined for any distance, lateral variations can be postulated from scattered outcrops.

Lacustrine and Marsh Sediments

Approximately 15% of the coastal-plain-facies assemblage is interpreted as lake or marsh deposits. The lacustrine–marsh facies within the lower 80 m of the assemblage is characterized by copper-rich lenses up to 100 m across, consisting of black to gray-green shales with occasional thin, nodular micrites (Fig. 7). The highest copper percentage is found within the black shales where copper minerals have replaced plant fragments representing *Lebachia, Calamites,* and *Lepidodendron*. As the copper ore seems to be related to syngenetic production of hydrogen sulfide in a highly organic, reducing environment, it is thought that the shales were deposited in small, stagnant ponds. The close areal relationship of limestones with the shales indicates that micrite production may also be lacustrine-related.

The nonfluvial section in the upper 50 m of the coastal-plain-facies assemblage is characterized by cycles of brown to greenish-gray to grayish-green shales grading upward into reddish-brown, silty limestones or limy siltstones. These sediments eventually grade into a very hard, conchoidally fractured, light-gray, slightly silty to pure micritic limestone. Lateral continuity of these units is unknown. The thickness of the limestone beds in the area of measured section SDC-1 averages 1 m. One limestone in measured section SDC-2, however, is approximately 15 m thick, suggesting a possible thickening and coalescing of the limestones to the south.

These unfossiliferous micrites were interpreted by May and others (1977) as having been deposited in a lacustrine environment. Indeed, the limestones seem to fit the lacustrine criteria set forth by Pettijohn (1957) in that they are dense, nodular micrites which are broken by what are interpreted as calcite-filled shrinkage, or syneresis, cracks. Krumbein and Sloss (1951) indicated that lacustrine deposits may also be powdery marls; a similar example was located in measured section CP.

It is thought that the small, reducing basins represented by the shale lenses in the lower 80 m of the coastal-plain-facies assemblage eventually coalesced into larger lakes which probably existed for long periods. The lack of copper ore in the clastic lacustrine cycles probably reflects more thorough aeration of the water column by increased wind and wave action as the fetch of the lakes increased.

Channel-dominated-facies Assemblage

Finer-grained sediments that dominate the coastal-plain-facies assemblage grade upward into the sandstone deposits of a channel-dominated-facies assemblage. Three channel types are identified as having progressed across the area during deposition of this coarser-grained interval, in ascending order: (1) small, slightly meandering channels, (2) point-bar and chute-bar deposits of coarse-grained meandering streams, and (3) cross-cutting channel deposits of braided streams. Total thickness of the assemblage is estimated at 180 m, approximately half of the thickness postulated by Tschanz and others (1958) for equivalent units.

Small, Slightly Meandering Channels

Medium- to coarse-grained sandstones comprise approximately 50% of the lower 65 m of the channel-dominated-facies assemblage. Sandstones usually occur as single channels less than 2 m thick. Medium-scale, low-angle, trough crossbedding is the major sedimentary structure with lesser amounts of tabular and epsilon crossbedding (Fig. 8). The base of the channels is highly erosional and commonly lined with clay or limestone clasts. Some channels show fining-upwards textural trends grading into root-mottled, silty or shaly sands. Interbedded fine-grained

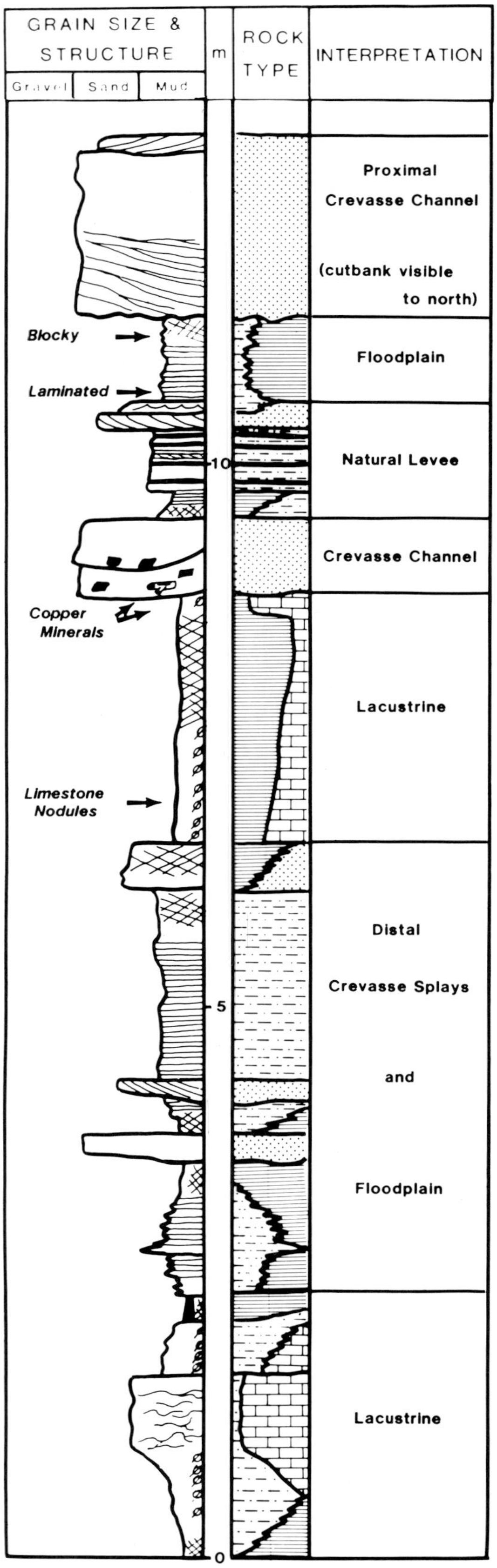

FIGURE 6. Measured section CP showing fine-grained facies of coastal-plain assemblage.

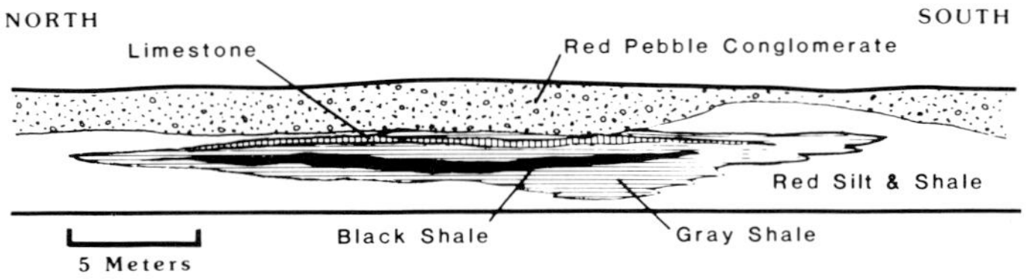

FIGURE 7. Copper-rich shale and limestone lense from the coastal-plain assemblage (from Tschanz and others, 1958).

sediments are primarily laminated to blocky, light-green to dark-brown, silty shales.

The origin of these sediments is considered to be similar to the low-sinuosity fluvial deposits in the underlying coastal-plain-facies assemblage. The major differences are the increase of channel sandstones and the decrease of crevasse-splay and lacustrine deposits in the younger, dominately coarse-grained unit.

Point- and Chute-bar Deposits

Up-section from the straight to meandering channel facies are at least 60 m of fluvial deposits of which approximately 60% is channel sandstone. Some of these sandstones contain characteristics similar to the point-bar and chute-bar deposit described by McGowen and Garner (1970) from the coarse-grained meanderbelts of the Colorado River in Texas.

Two types of point-bar deposits were recognized in the Sangre de Cristo Formation (Fig. 9). The first type grades upward from medium-grained sandstone with small trough and contorted crossbedding into finer-grained, epsilon-crossbedded sandstone. The top is structureless and is capped by what is interpreted as a crevasse-splay deposit.

The second type of point bar consists entirely of trough cross-stratification. Bedset size decreases upward, as does grain size. A crevasse-splay deposit also caps this particular section.

It is possible that these two types of point bars are actually end points in a depositional spectrum. The trough crossbeds in both types represent lower point-bar deposition associated with more turbulence near the main channel. In the case of outcrop DA (Fig. 9) it appears that sand waves have migrated out of the channel and onto the bar surface. Decrease in bedset thickness can be related to decreasing water depths associated with bar growth and migration. Silt drapes were deposited during low water levels.

Epsilon crossbeds represent interval accretion on an upper point-bar surface. The upper point bar was sufficiently above low-water stand that clay-drape deposition was not possible. Even though the entire bar was crossbedded at one time, subsequent plant growth on the uppermost surface destroyed all sedimentary structures.

A diagnostic feature of Recent coarse-grained meanderbelt systems is the formation of chute-bar complexes. At least one chute bar was recognized in this section (Fig. 10). The chute-channel deposit is coarse- to very coarse-grained sandstone with lenses of gravel and clay clasts up to 10 cm across. The chute bar itself consists of steeply dipping avalanche faces lined with gravel and clay clasts. These large tubular crossbeds are capped by planar topset beds.

Braided-stream Channels

The upper 100 m or more of the prograding fluvial facies is dominated by cross-cutting sandstone channels. Less than 10% of this interval is fine-grained overbank material. The individual sandstone bodies are coarse- to very coarse-grained (Fig. 11). Most outcrops show no vertical textural trends; however, a few of the sandstone units fine upward.

Most braided-stream sandy facies of Miall (1977, 1978) were recognized in the section with large trough, large planar, and horizontal cross-stratification predominating. Erosional scours and scour fill are also common. Major scour surfaces are often lined with gravel and clay clasts up to 10 cm across. Silt- and clay-draped reactivation surfaces are also present. Siliceous concretions are abundant in the fine-grained

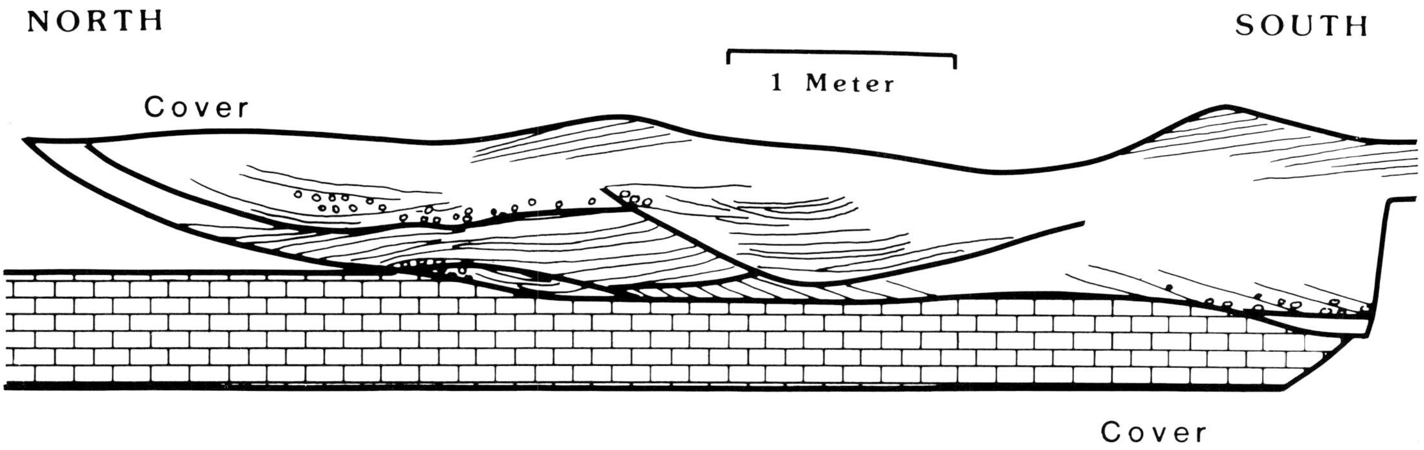

FIGURE 8. Sketch of a singular, medium-grained channel sandstone cutting into a lacustrine limestone at outcrop CC.

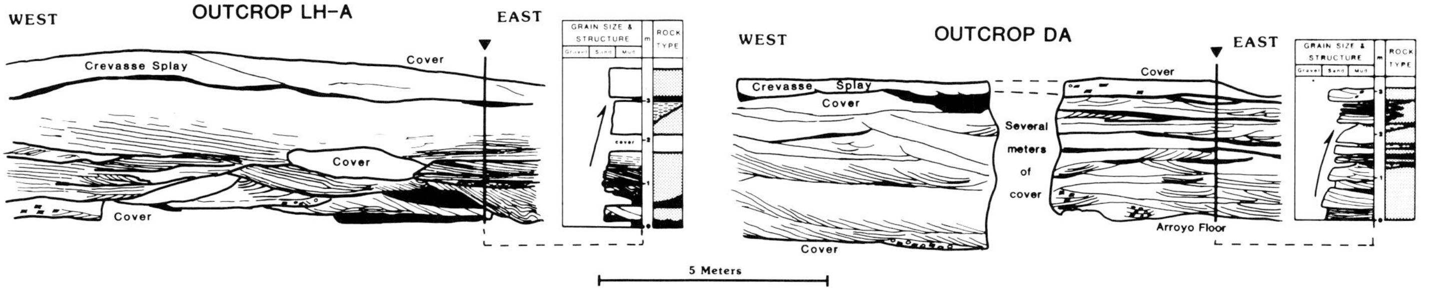

FIGURE 9. Comparison of two point-bar deposits in the progradational fluvial-facies assemblage.

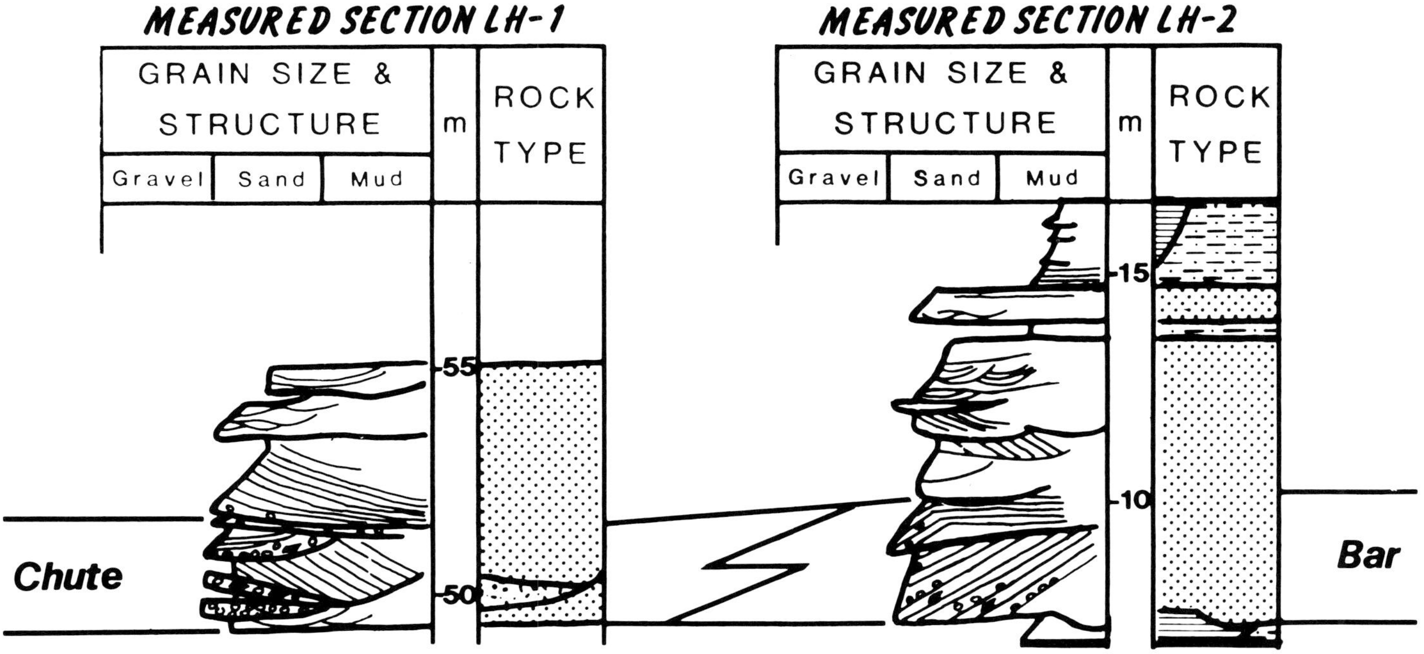

FIGURE 10. Laterally equivalent measured sections interpreted as representing chute-bar deposits.

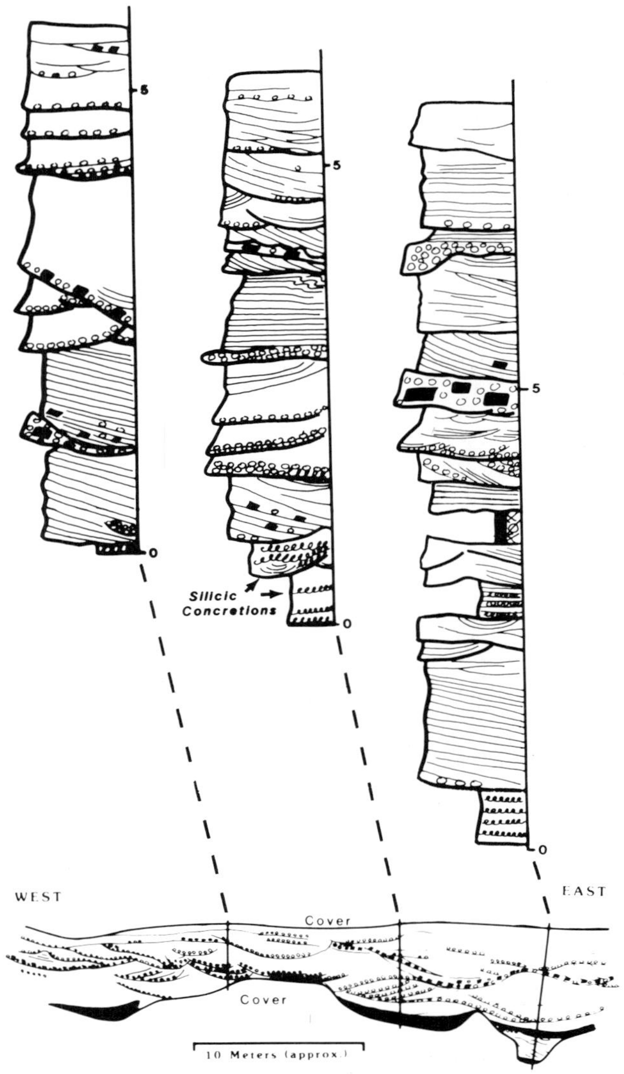

FIGURE 11. Braided-stream deposit at outcrop LC-B.

overbank material and are locally present along some scour surfaces. These concretions are interpreted as having been precipitated during post-burial alteration.

The lack of preserved overbank material in this interval is related to a decrease in rate of subsidence. This allowed braided fluvial channels to migrate across the area and erode channel-stabilizing fine-grained sediments.

VALLEY-FILL AND TRANSGRESSIVE MARINE SEQUENCE

A 12-m-thick conglomerate is present at the top of the Sangre de Cristo Formation in measured section LC (Fig. 12). According to Miall's classification, the primary facies is massive gravel with minor occurrences of trough and planar crossbedding. Since equivalent gravels were not recognized in measured sections TM-1 and TM-2 to the north, it is apparent that this accumulation is lenticular. This gravel lens is interpreted as a valley-fill sequence related to a rapid rise in sea level. Sea-level rise can be illustrated by the large offshore-bar complex of the Glorieta Member of the San Andres Formation which directly overlies the valley-fill sequence. The gravel, with individual clasts attaining diameters of 20 cm, was deposited when the capacity of the water to carry sediment was diminished by the rise in base level.

A thin conglomerate, possibly representing another valley-fill deposit, occurs approximately 18 m below the thick gravel lens. The

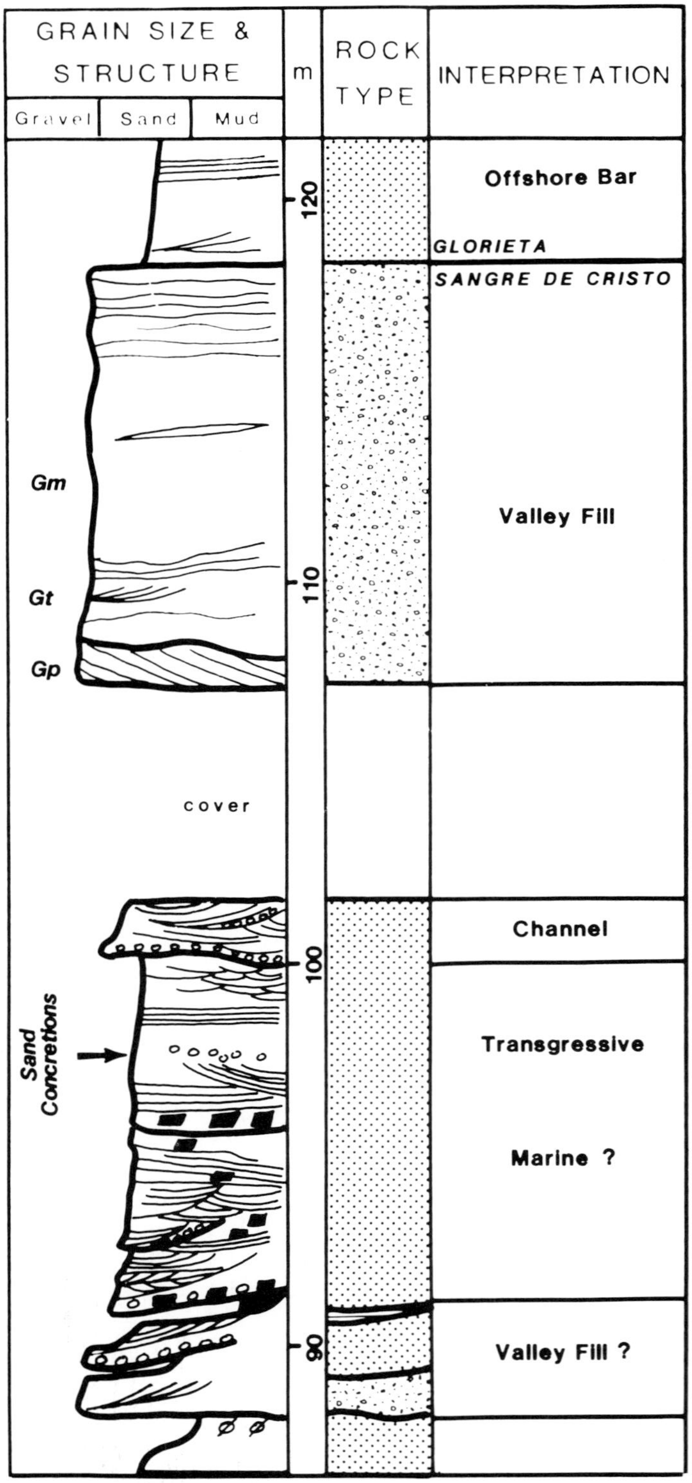

FIGURE 12. Measured section LC showing valley fill and possible transgressive marine units at the top of the Sangre de Cristo Formation. Recognized facies are massive gravel (Gm), trough-crossbedded gravel (Gr), and planar-crossbedded gravel (Gp).

medium-grained, low-angle trough- and planar-bedded sediments overlying the thin gravel deposit may also represent a transgressive marine sand.

DEPOSITIONAL HISTORY

The Rowe–Mora Basin was part of a Paleozoic trough containing both the Maroon and Central Colorado Basins, which extended north-

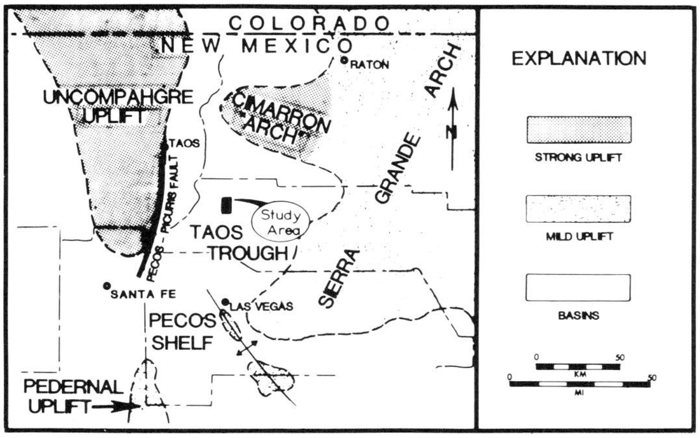

FIGURE 13. Pennsylvanian paleogeography of the Taos trough area (after Casey, 1980b).

ward approximately 600 km into Moffat County, Colorado (Brill, 1952). Baltz (1965) postulated that the Rowe–Mora Basin was separated from the Central Colorado Basin by the northwest-trending Cimarron arch. Sutherland (1963) divided the Rowe–Mora Basin into two distinct regions based on dominant sediment type. To the north was the rapidly subsiding, clastics-filled Taos trough, and to the south the stable, carbonate Pecos shelf (Fig. 13). The Coyote Creek district lies within the Taos trough, about 20 km east of the basinal axis (Bachman, 1953).

Sutherland (1963) believed that the Taos trough was separated from the Uncompahgre uplift to the west by the Pecos–Picuris fault, a reactivated Precambrian fault zone. Large displacement on this fault caused the Uncompahgre uplift to become a major sediment contributor during much of the Pennsylvanian (Foster and others, 1972; Mallory, 1972).

To the east, the Taos trough sloped gently onto the ancestral Sierra Grande uplift. The importance of this positive area, in terms of sediment production, is masked by a lack of subsurface data and the near-coincident position of the Tertiary Sierra Grande arch (Casey, 1980a). Roberts and others (1976) believed that the Sierra Grande uplift was the major source area for the Taos trough during the Late Pennsylvanian. However, this uplift was probably only a mild, northeast-trending feature supplying minor amounts of sediments until it was buried in the Early Permian.

The Sangre de Cristo Formation is a Late Pennsylvanian–Early Permian deposit which asymmetrically filled the Taos trough. Deposits thin southward from a maximum thickness of 1,700 m near the Cimarron arch, 15 km north of the study area (May and others, 1977). The formation represents a north–south progradational unit which for the most part intertongues with, or conformably overlies, the Madera Formation. The basal boundary is considered time-transgressive, ranging from upper Desmoinesian in the north to lower Wolfcampian in the south. In the Coyote Creek district the basal contact is upper Desmoinesian or possibly lower Missourian (Casey, 1980a; Tschanz and others, 1958).

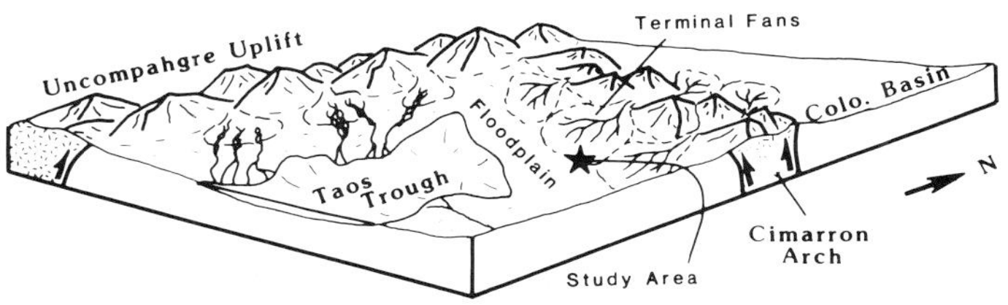

FIGURE 14. Paleogeography of the Coyote Creek region during alluvial-fan deposition of the Sangre de Cristo Formation.

By classifying the separate environments of the Sangre de Cristo Formation and identifying the sedimentary processes controlling those environments, a depositional history of the area can be formulated. The resultant depositional history suggests that Sangre de Cristo sedimentation in the Coyote Creek area was strongly influenced by the rapid uplift, erosion, and eventual burial of the Cimarron arch.

Distal Alluvial-fan Deposition

Early Sangre de Cristo sediments suggest that final uplift of the Cimarron arch began during the upper Desmoinesian, resulting in an initiation of arkosic alluvial-fan deposition (Fig. 14). Grain size decreased within the fans from proximal boulder conglomerates near the arch (May and others, 1977) to distal siltstones and marlstones with thin pebbly sandstone in the study area. Progradation of the fans displaced the fluvial, deltaic, and marine deposits of the time-equivalent Madera Formation to the south. Lack of major variation within the alluvial-fan-facies interval suggests that sedimentation rates were nearly equal to the subsidence.

Coastal-plain Deposition

By the time the coastal-plain-facies assemblage was being deposited, uplift on the Cimmaron arch had essentially halted, thus reducing source potential for the alluvial fans (Fig. 15). Eventually, the arch was eroded to a point where fan deposition ceased in the Coyote Creek area.

The fine-grained coastal-plain sediments, representing minor channels, marsh, and lake deposits, were probably sediments on a large delta plain. It is believed that sediment was fed from the Colorado Basin into the Taos trough through a breach of the still partially positive Cimarron arch. By the end of the coastal-plain depositional sequence, sedimentation was being outpaced by subsidence. At this time, the shoreline of the Taos Sea was probably only a few kilometers to the south.

Progradational Fluvial Sequence

Finally, the Cimarron arch adjacent to the Coyote Creek area was breached by the fluvial systems of the Colorado Basin (Fig. 16). As subsidence rates decreased, bedload fluvial systems prograded across the area into the Taos trough. Fluvial facies consisted of distal, straight to slightly meandering distributary channels which coalesced both updip and up-section into larger, coarse-grained meanderbelt channels.

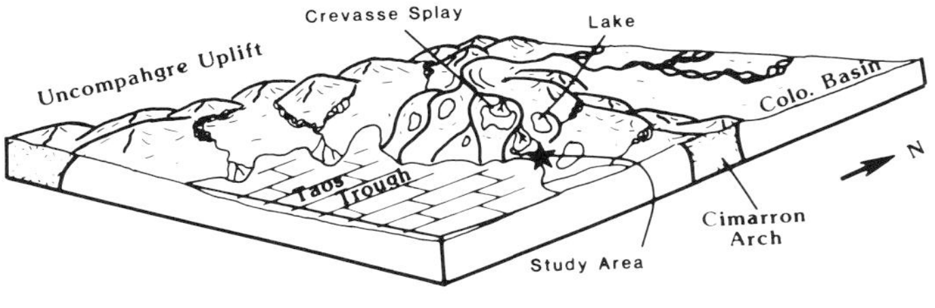

FIGURE 15. Paleogeography of the Coyote Creek region during coastal-plain-facies deposition of the Sangre de Cristo Formation.

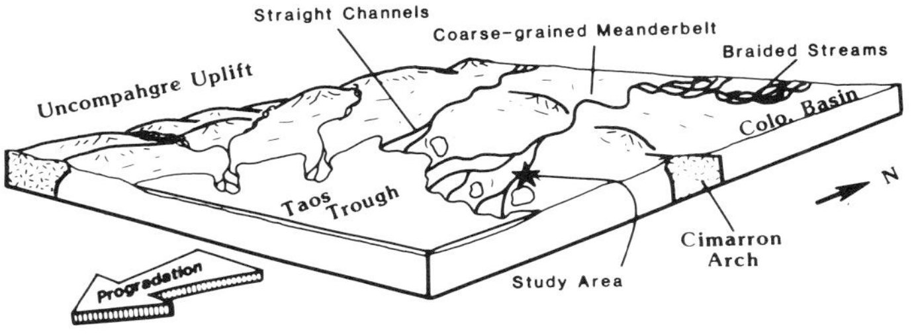

FIGURE 16. Paleogeography of the Coyote Creek region during fluvial progradation of the Sangre de Cristo Formation.

Sinuosity and migration of the channels seem to be directly related to subsidence and aggradation rates. Ultimately, these rates decreased, resulting in the deposition of rapidly migrating braided-stream channels which eroded any bank-stabilizing material. By the end of Sangre de Cristo deposition, the Rowe–Mora Basin had been filled with sediment. Glorieta seas transgressed across the area beveling the top of the Sangre de Cristo Formation to a flat surface, thus closing the sedimentary history of the Paleozoic Rowe–Mora Basin.

ACKNOWLEDGMENTS

Appreciation is extended to Placid Oil Company for allowing me the opportunity to publish this paper. Technical services and advice were provided by Leslee Howard and her drafting staff at Placid Oil. Many thanks go to my wife, Kathy Brown, without whose typing skills and moral support this paper would not have been completed. And, finally, a big *thank you* to the people of the Coyote Creek area for allowing me not only to study their rocks, but also to share a little bit of their lives. Siempre Coyote!

REFERENCES

Allen, J. R. L., 1965, A review of the origin and characteristics of Recent alluvial sediments: Sedimentology, v. 5 (special issue), pp. 91–191.

Bachman, G. O., 1953, Geology of a part of northwestern Mora County, New Mexico: U.S. Geological Survey, Oil and Gas Investigations Map OM-137.

———, and Read, C. B., 1952, Uranium bearing copper deposits near Guadalupita, Mora County, New Mexico: U.S. Geological Survey, Open-file Report TEM-435, 9 pp.

Baltz, E. H., 1965, Stratigraphy and history of Raton Basin and notes on San Luis Basin, Colorado–New Mexico: American Association of Petroleum Geologists, Bulletin, v. 49, no. 11, pp. 2041–2075.

Brill, K. G., Jr., 1952, Stratigraphy in the Permo-Pennsylvanian zeugogeosyncline of Colorado and northern New Mexico: Geological Society of America, Bulletin, v. 65, pp. 809–880.

Casey, J. M., 1980a, Depositional systems and basin evolution of the Late Paleozoic Taos trough, northern New Mexico: Ph.D. dissertation, University of Texas at Austin.

———, 1980b, Depositional systems and paleogeographic evolution of the Late Paleozoic Taos trough, northern New Mexico; *in* Paleozoic paleogeography of the west-central United States: Rocky Mountain Association of Geologists, Paleogeography Symposium no. 1, pp. 181–196.

Collinson, J. D., 1978, Vertical sequence and sand body shape in alluvial sequences; *in* Fluvial sedimentology: Canadian Society of Petroleum Geologists, Memoir 5, pp. 577–586.

Foster, R. W., Frentress, R. M., and Riese, W. C., 1972, Subsurface geology of east-central New Mexico: New Mexico Geological Society, Special Publication 4, 22 pp.

Friend, P. F., 1978, Distinctive features of some ancient river systems; *in* Fluvial sedimentology: Canadian Society of Petroleum Geologists, Memoir 5, pp. 531–542.

Gile, L. H., Peterson, F. F., and Grossman, R. B., 1965, The K horizon; a master soil horizon of carbonate accumulation: Soil Science, v. 99, pp. 74–82.

Goudie, A., 1973, Duricrusts in tropical and subtropical landscapes: Clarendon Press, Oxford, 174 pp.

Krumbein, W. C., and Sloss, L. L., 1951, Stratigraphy and sedimentation (1st edition): W. H. Freeman and Company, San Francisco, 487 pp.

Leeder, M. R., 1978, A quantitative stratigrahic model for alluvium, with special reference to channel deposit density and interconnectedness; *in* Fluvial sedimentology: Canadian Society of Petroleum Geologists, Memoir 5, pp. 587–596.

McGowen, J. H., and Garner, L. E., 1970, Physiographic features and stratification types of coarse-grained point bars: modern and ancient examples: Sedimentology, v. 14, pp. 77–111.

Mallory, W. W., 1972, Pennsylvanian arkose and the ancestral Rocky Mountains; *in* Geologic atlas of the Rocky Mountain region: Rocky Mountain Association of Geologists, pp. 131–132.

May, R. T., Strand, J. R., Reid, B. E., and Phillips, W. R., 1977, Preliminary study of favorability for uranium of the Sangre de Cristo Formation in the Las Vegas Basin, northeastern New Mexico: U.S. Department of Energy, Grand Junction, Colorado, 46 pp.

Miall, A. D., 1977, A review of the braided river depositional environment: Earth Science Review, v. 13, pp. 1–62.

———, 1978, Lithofacies types and vertical profile models in braided river deposits; a summary *in* Fluvial sedimentology: Canadian Society of Petroleum Geologists, Memoir 5, pp. 597–604.

Pettijohn, F. J., 1957, Sedimentary rocks (2nd edition): Harper and Row, New York, 718 pp.

Reineck, H. E., and Singh, I. B., 1975, Depositional sedimentary environments: Springer-Verlag, 439 pp.

Roberts, J. W., Barnes, J. J., and Wacker, H. J., 1976, Subsurface Paleozoic stratigraphy of the northeastern New Mexico basin and arch complex: New Mexico Geological Society, Guidebook 27, pp. 141–151.

Sutherland, P. K., 1963, Paleozoic rocks; *in* Geology of part of the Sangre de Cristo Mountains, New Mexico: New Mexico Bureau of Mines and Mineral Resources, Memoir 11, pp. 22–46.

Tschanz, C. M., Laub, D. C., and Fuller, G. W., 1958, Copper and uranium deposits of the Coyote district, Mora County, New Mexico: U.S. Geological Survey, Bulletin 1030-L, pp. 343–398.

Zeller, H. D., and Baltz, E. H., 1954, Uranium-bearing copper deposits in Coyote Creek district, Mora County, New Mexico: U.S. Geological Survey, Circular 334, 11 pp.

CORRELATION OF EOCENE ROCKS OF THE NORTHERN RIO GRANDE RIFT AND ADJACENT AREAS: IMPLICATIONS FOR LARAMIDE TECTONICS

SPENCER G. LUCAS
Department of Geology, University of New Mexico, Albuquerque, New Mexico 87131

INTRODUCTION

The oldest Tertiary rocks in the northern Rio Grande rift are of Eocene age. These rocks—the Galisteo, El Rito, and Blanco Basin Formations—represent continental sediments that were shed from complex uplifts and deposited by rivers in adjoining basins as the result of an essentially amagmatic basement deformation, the Laramide orogeny (Chapin and Cather, 1981; Lucas and Ingersoll, 1981). For many years, the age of the basal Tertiary rocks in the northern Rio Grande rift was imprecisely known. However, recent stratigraphic, vertebrate paleontological, and sedimentological studies now permit a more refined correlation of these units, a correlation that suggests that there was a middle Eocene pulse of Laramide deformation in northern New Mexico. This paper reviews these recent studies, presents a revised correlation of the Eocene rocks of the northern Rio Grande rift, and briefly discusses the implications of this new correlation for the Laramide tectonic history of northern New Mexico.

GALISTEO FORMATION

The Galisteo Formation (Hayden, 1869) consists of as much as 1,300 m of fluvial sandstone, mudstone, and conglomerate exposed in north-central New Mexico between Albuquerque and Santa Fe (Fig. 1) (Stearns, 1943; Lucas, 1982). During the Wasatchian (early Eocene), most of the lower part of the Galisteo Formation was deposited in a rapidly subsiding basin centered near Cerrillos by primarily high-sinuosity meandering streams that flowed south and southwestward from the Brazos–Sangre de Cristo uplift (Gorham, 1979; Gorham and Ingersoll, 1979). However, by Duchesnean (late Eocene) time, sediments of the Galisteo Formation were being deposited in a broader basin by streams of generally lower sinuosity that flowed south and southwestward from the Brazos–Sangre de Cristo uplift and eastward from the Nacimiento uplift (Gorham, 1979; Gorham and Ingersoll, 1979). The Wasatchian interval of the Galisteo Formation contains fossils of palms and ferns (Lee and Knowlton, 1917), and of gars, soft-shelled turtles, and mammals that were amphibious browsers (*Coryphodon*), rooters (*Ectoganus*), and small herbivores or omnivores (*Hyopsodus, Hyracotherium, Microsyops*) (Lucas, 1982). These fossils suggest the presence of a warm, wet, heavily vegetated environment in north-central New Mexico during the Wasatchian (Lucas and Schoch, 1981). However, by Duchesnean time, a subtropical, seasonally arid climate, and the presence of parkland in north-central New Mexico is suggested by palynomorphs (Leopold and MacGintie, 1972) and fossils of ruminant artiodactyls, hyracodontid rhinoceroses, hyaenodontid creodonts, and tortoises from the upper part of the Galisteo Formation (Lucas, 1982; Lucas and Schoch, 1981).

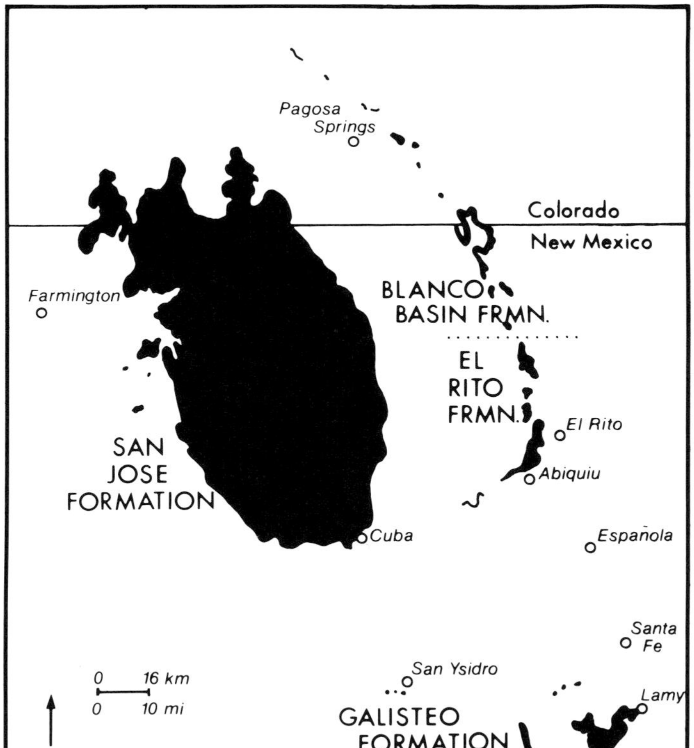

FIGURE 1. Distribution of the Galisteo, El Rito, San Jose, and Blanco Basin Formations in northern New Mexico and southern Colorado. From Dane and Bachman (1965) and Tweto (1979).

The fossil mammals from the Galisteo Formation provide a firm basis for age assignment. The Cerrillos local fauna (Lucas and Kues, 1979), from an interval 369–424 m above the base of the Galisteo Formation near Cerrillos, is of middle Wasatchian (Lysitean) age (Lucas, 1982; Lucas and others, 1981). The Tonque local fauna, from the upper part of the Galisteo Formation in the Cerrillos area, Hagan Basin, and Rio Puerco fault zone, is of Duchesnean age (Lucas 1982; Lucas and others, 1981). Thus, the fossil mammals of the Cerrillos local fauna support correlation of most of the lower part of the Galisteo Formation with the Almagre local fauna of the Regina Member of the San Jose Formation in the San Juan Basin (Fig. 2) (Lucas, 1982; Lucas and others, 1981). The Tonque local fauna is equivalent to most (or all) of the mammal-bearing interval of the Baca Formation in west-central New Mexico (Fig. 2) (Lucas, 1982, 1983). Assignment of a Duchesnean age to the Tonque local fauna is also consistent with radiometric dates from the predominantly volcaniclastic Espinaso Formation, which overlies the Galisteo Formation in the Cerrillos–Hagan Basin area. The oldest K–Ar dates reported from clasts within the Espinaso are 34.3 ± 0.8 m.y. (10 m above the base: Kautz and others, 1981, table 1), 34.6 ± 0.7 m.y. (200 m above the base: Kautz and others, 1981, table 1) and 37.8 ± 1.2 m.y. (lower part of the type section: Weber, 1971). These dates indicate that the contact between the Galisteo and Espinaso Formations in north-central New Mexico is at, or very close to, the Eocene–Oligocene boundary (Fig. 2).

Despite the paleontological and radiometric data that bear on the age of the Galisteo Formation, the precise age of Galisteo strata below the Wasatchian mammal-bearing interval and between the Wasatchian and Duchesnean mammal-bearing intervals is somewhat unclear. The 369 m of Galisteo Formation below the Wasatchian mammal-bearing interval in the Cerrillos area are mostly medium- to coarse-grained conglomeratic sandstone (Lucas, 1982) that unconformably overlies the Upper Cretaceous Mesaverde Group. Elsewhere, the basal Galisteo Formation unconformably overlies either the Mesaverde Group or the Upper Cretaceous Mancos Shale (Beaumont, 1979; Lucas, 1982). On

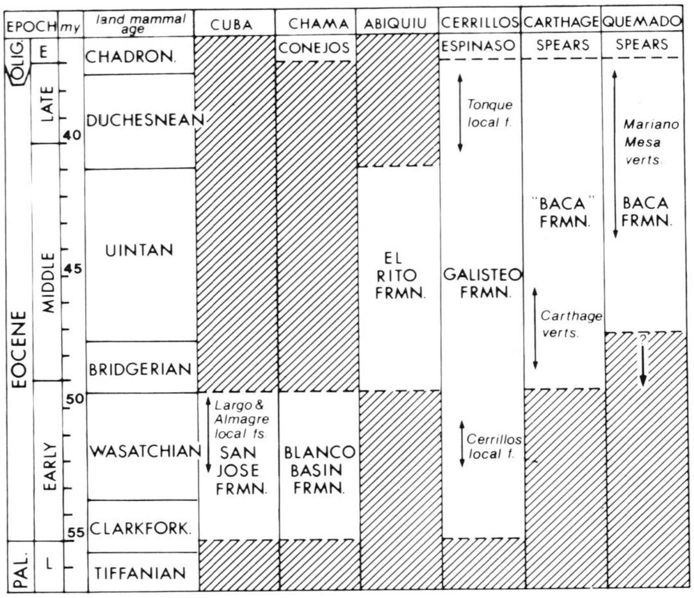

FIGURE 2. Correlation of the Eocene formations of northern New Mexico with the Baca Formation in the Baca Basin and the "Baca Formation" in the Carthage–La Joya Basin. Baca correlations based on Lucas (1983, 1984) and Lucas and others (1982, 1983). Eocene time scale from Berggren and others (1978), Hardenbol and Berggren (1978), Lowrie and Alvarez (1981), and Prothero and others (1982).

the basis of lithologic similarity, and because Wasatchian mammals from Galisteo mudstones above the basal sandstones are temporally equivalent to Wasatchian mammals from the Regina Member of the San Jose Formation, a tentative correlation of the basal sandstones of the Cuba Mesa Member of the San Jose and the Galisteo Formation sandstones below the Wasatchian mammal-bearing interval seems reasonable (Fig. 2).

Numerous unconformities are present within the Galisteo Formation between the Wasatchian mammal-bearing interval and the Duchesnean mammal-bearing interval (Gorham, 1979; Gorham and Ingersoll, 1979). Clearly, this interval spans much of the Eocene (medial Wasatchian to Duchesnean, about 10 m.y.), but it seems unlikely that Galisteo strata within this interval represent a continuous record of this period (Lucas, 1982; Lucas and Ingersoll, 1981).

BLANCO BASIN FORMATION

The Blanco Basin Formation (Cross and Larsen, 1935) consists of as much as 175 m of arkosic sandstone, conglomerate, and silty mudstone exposed as isolated erosional remnants south of the San Juan Mountains in Colorado and west of the Tusas Mountains in New Mexico (Fig. 1) (Larsen and Cross, 1956; Muehlberger and others, 1960; Dunn, 1964a; Muehlberger, 1967, 1968). The work of Dunn (1964a) indicates that the Blanco Basin Formation represents a "complex of alluvial fan, mudflow and pediment gravel deposits" (p. 38) derived from a complex metamorphic terrane intruded by silicic and alkalic rocks to the north and east of the present San Juan and Tusas Mountains. Red mudstones of the Blanco Basin Formation appear to be multicyclic red beds derived from Mesozoic and Paleozoic red beds of the source terrane (Dunn, 1964a).

No fossils are known from the Blanco Basin Formation, but on the basis of lithology, stratigraphic position, and inferred depositional history most workers (e.g., Van Houten, 1957; Baltz, 1965, 1967, 1978; Muehlberger, 1967) have considered it to be a correlative of the San Jose Formation of the San Juan Basin and, therefore, of early Eocene age. Dunn (1964a, 1964b, 1965) provided the most explicit documentation of this correlation based on the following observations and conclusions:

(1) The mineralogy of basal sandstones (Cuba Mesa Member) of the San Jose Formation is virtually identical to that of the Blanco Basin Formation (also see Muehlberger, 1967).

(2) In Colorado, the Blanco Basin unconformably overlies the Paleocene upper member of the Animas Formation. [In New Mexico, there is an angular unconformity between the Blanco Basin and rocks of Triassic (Chinle) to Late Cretaceous (Lewis) age: Muehlberger and others, 1960.] In the San Juan Basin of northwestern New Mexico, the San Jose Formation unconformably overlies the Paleocene Nacimiento Formation (Baltz, 1967). The Nacimiento Formation grades into, and intertongues with, the upper member of the Animas Formation in the northern San Juan Basin, and it is clear that the upper Animas is a piedmont facies proximal to the basinal facies of the Nacimiento Formation (Baltz, 1967). Therefore, the Blanco Basin and San Jose Formations occupy the same stratigraphic position above rock units of equivalent age.

(3) Crossbedding in the Blanco Basin Formation indicates transport from northeast to the southwest, towards the San Juan Basin. The work of Baltz (1967), who built upon earlier work by Dane (1946), and preliminary paleocurrent data gathered by Larry Smith of the University of New Mexico (oral comm. 1984) indicate a north to northeasterly source area for at least the basal Cuba Mesa Member of the San Jose Formation. The Blanco Basin and San Jose Formations are texturally similar, although the Blanco Basin is generally coarser-grained than the San Jose. Therefore, the Blanco Basin is readily interpreted as an upslope, piedmont facies that gave way basinward to the basinal San Jose.

(4) Cobbles of the Blanco Basin Formation present on top of Archuleta Mesa suggest earlier continuity of the Blanco Basin across the Gallinas–Archuleta arch, the structural feature that presently divides the San Juan Basin (and San Jose Formation) to the west from the northern Chama Basin (and Blanco Basin Formation) to the east.

Therefore, it seems most reasonable to consider the Blanco Basin Formation to be of early Eocene age, a correlative of the San Jose Formation. However, more work will be needed to determine whether or not the Blanco Basin Formation is equivalent to the entire San Jose Formation, or only part of the San Jose.

The Conejos Formation, of Oligocene age (Lipman and others, 1970; Butler, 1971), disconformably overlies the Blanco Basin Formation (Dunn, 1964a; Muehlberger, 1967). Cross and Larsen (1935) and Larsen and Cross (1956) described this contact as an "apparent conformity" and, because they assigned a Miocene age to the Conejos, tentatively assigned an Oligocene age to the Blanco Basin Formation. However, Epis and Chapin (1975) attributed the near concordance between the Blanco Basin and Conejos to the development of a prevolcanic, late Eocene erosional surface of low relief that developed on the upper surface of the Blanco Basin Formation prior to Conejos deposition.

EL RITO FORMATION

The El Rito Formation (Smith, 1938) consists of as much as 120 m of predominantly brick-red quartzite conglomerate, quartzose sandstone, and breccia (Fig. 3A) exposed as isolated erosional remnants in northern New Mexico from Cerro Pedernal eastward to the western flank of the Tusas Mountains (Smith, 1938; Budding and others, 1960; Smith and others, 1961; Bingler, 1968a, 1968b, 1968c; Logsdon, 1981). Analysis by Logsdon (1981) indicates that the El Rito is an alluvial-fan deposit derived from debris flows and braided streams which had their headwaters on the Precambrian crystalline terrane of the Brazos–Sangre de Cristo uplift and on the Gallina–Archuleta arch and Nacimiento uplift. Logsdon's (1981) work thus supports earlier suggestions by Baltz (1978), Gorham (1979), and Gorham and Ingersoll (1979) that the El Rito and Galisteo depositional basins were at some time continuous, and thus that the El Rito is an upslope, piedmont facies of at least part of the Galisteo Formation. Logsdon (1981) also noted extensive diagenetic effects in the El Rito that are typical of Cenozoic desert alluvium, including pore fillings of calcite, zeolite, and hematite, replacement of grains, feldspar alterations, and the removal of heavy minerals. These features suggest that diagenesis of the El Rito Formation may have taken place under arid to semiarid conditions (Logsdon, 1981).

FIGURE 3. Basal conglomerates of the El Rito and Galisteo Formations. **A,** Brick-red conglomerate and sandstone of the El Rito Formation (TE) unconformably overlying buff sandstone of the Upper Jurassic Morrison Formation (JM) near Abiquiu, New Mexico. **B,** Brown conglomerate of the Galisteo Formation (TG) unconformably overlying buff sandstone of the Upper Cretaceous Menefee Formation (KM) near Windmill Hill in the Rio Puerco fault zone, New Mexico.

As with the Blanco Basin Formation, no fossils are known from the El Rito Formation. Nevertheless, based on its lithology, stratigraphic position, and inferred depositional history, the consensus (e.g., Baltz, 1978; Logsdon 1981; Lucas and Ingersoll, 1981) is that it is a correlative of the Galisteo Formation. This correlation, however, is not very precise because it indicates a Wasatchian–Duchesnean time span of more than 15 m.y. for the El Rito Formation. The El Rito Formation unconformably overlies rocks that range in age from Precambrian to Late Cretaceous (Mancos) (Budding and others, 1960; Smith and others, 1961; Bingler, 1968c), and is unconformably overlain by rocks as old as Oligocene (Conejos) (Manley, 1981). Therefore, precise age limits are not placed on the El Rito Formation by its stratigraphic position. Muehlberger (1968) asserted that near Brazos Peak in the northern Tusas Mountains the El Rito and Blanco Basin Formations "interfinger," which might indicate that they are time equivalent (Lucas and Ingersoll, 1982). However, Muehlberger's assertion actually reflected his uncertainty as to whether or not an outcrop north of Canones Box, interpreted as Blanco Basin Formation, might actually pertain to the El Rito Formation; no actual interfingering of the two units was ever documented (Muehlberger, 1967).

Despite its lack of fossils, the El Rito Formation may be correlated more precisely by the use of "event stratigraphy" (Ager, 1973, p. 63). Event stratigraphy is the correlation of depositional (or tectonic) events when neither biostratigraphic nor strictly lithological correlations are possible. An event-stratigraphic correlation of the El Rito Formation necessitates the identification of sediments in the Galisteo Formation that were produced by the tectonic event responsible for El Rito deposition. I identify the basal Galisteo conglomerate in the Rio Puerco fault zone (near San Ysidro, Fig. 1) and medial Galisteo conglomerates in the Hagan Basin–Cerrillos area as the Galisteo sediments that correspond to the El Rito depositional event. This identification is based on the following observations:

(1) The basal unit of the Galisteo Formation in the Rio Puerco fault zone is a polymodal, mostly clast-supported conglomerate with a generally unordered fabric (Fig. 3B). Cobbles of this conglomerate are predominantly quartzite. This conglomerate is overlain by predominantly red mudstone of the Galisteo Formation that contains Duchesnean mammals (Lucas, 1982). Imbrication of pebbles in locally ordered fabric of this conglomerate and crossbedding in Galisteo sandstones above the red, mammal-bearing mudstones indicate a northwestern source for these sediments, the Nacimiento uplift (Gorham, 1979; Gorham and Ingersoll, 1979). It should be noted that Slack (1973, 1975) and Slack and Campbell (1976) erroneously identified these Galisteo strata in the Rio Puerco fault zone as San Jose Formation, and therefore concluded that the Nacimiento uplift acted as a source area for these sediments during the latest Paleocene–early Eocene (Lucas, 1980, 1982).

(2) Medial conglomerates and conglomeratic sandstones of the Galisteo Formation in the Hagan Basin (lower white-conglomeratic sandstone member, lower buff-conglomerate and sandstone member, and middle white-conglomeratic sandstone member of Gorham, 1979, and Gorham and Ingersoll, 1979) and Cerrillos area (units 50–52, 62, and 66–67 of Lucas, 1982, fig. 4) have conglomeratic portions that are generally polymodal, mostly clast-supported conglomerates that usually have unordered fabric and significant amounts of quartzite cobbles. These conglomerates are stratigraphically below Duchesnean mammal-bearing mudstones and sandstones. Crossbedding and pebble imbrications indicate that the conglomerates were derived from a source to the northwest, the Nacimiento uplift (Gorham, 1979; Gorham and Ingersoll, 1979). Each of these conglomerates has an erosional base, and correlation within the Hagan Basin indicates that these bases are intraformational unconformities with locally significant stratigraphic relief (Gorham, 1979; Gorham and Ingersoll, 1979).

(3) Paleocurrents from the lower part of the Galisteo Formation below these conglomerates indicate a northern source area (Brazos–Sangre de Cristo geanticline), not a source area to the northwest (Nacimiento uplift) during this primarily (or entirely?) Wasatchian phase

of Galisteo deposition (Gorham, 1979; Gorham and Ingersoll, 1979).

(4) As noted above, Dunn's (1964a) work suggests continuity of the Blanco Basin and San Jose Formations across the Gallina–Archuleta arch during at least part of the Wasatchian. Thus, this uplift could not have been a source for El Rito sediments until after the Wasatchian, or during the late Wasatchian (Logsdon, 1981).

(5) Preliminary paleocurrent analysis of the Cuba Mesa Member of the San Jose Formation (L. Smith, oral comm. 1984) does not indicate a southeastern or eastern source area (Nacimiento uplift) for these early Eocene sediments; a northern source area is indicated.

(6) "The red-bed phase of the Galisteo Formation that crops out along the escarpments southwest of Lamy, New Mexico, and along U.S. Highway 85 at La Bajada Hill suggests a close similarity between the El Rito Formation and the upper part of the Galisteo Formation" (Galusha and Blick, 1971, p. 34).

(7) Baltz (1967) noted that the medial sandstone member of the San Jose Formation, the Llaves Member, thins and fines westward, and contains cobbles and pebbles of bluish and gray metaquartzite similar to the metaquartzites of the Brazos uplift. He thus suggested derivation of the Llaves Member from the Brazos–Sangre de Cristo uplift. This would mean that there were probably no highlands (Nacimiento uplift, Gallina–Archuleta arch) between the San Juan Basin and the Brazos–Sangre de Cristo uplift during the Wasatchian.

(8) The paleontological evidence cited above suggests that the Duchesnean upper part of the Galisteo Formation was deposited under seasonally arid climatic conditions. If Logsdon's (1981) suggestion that El Rito diagenesis took place under an arid or semiarid climate is correct, this diagenesis may have taken place during the Duchesnean, after the termination of El Rito deposition.

The above observations support the idea that during at least most of the Wasatchian, the Nacimiento uplift and Gallinas–Archuleta arch did not supply sediment to the San Juan or Galisteo Basins. It is during the post-Wasatchian interval of Galisteo deposition that sandstones and conglomerates derived from source areas to the west and northwest appear. The El Rito Formation, which received some sediment from the west and northwest, thus is probably equivalent to these post-Wasatchian Galisteo sandstones and conglomerates. The appearance of seasonally dry climates during the Duchesnean in the Galisteo Basin suggests that El Rito deposition may have been essentially over by this time. Thus, a Bridgerian–Uintan age assignment for the El Rito Formation seems most reasonable (Fig. 2).

IMPLICATIONS FOR LARAMIDE TECTONIC HISTORY

The idea that Laramide deformation in northern New Mexico was the result of primarily northeastward movement of the Colorado Plateau so that it impinged upon the basement buttresses of the Rocky Mountain foreland, finally resulting in significant wrench tectonism, is widely accepted (e.g., Kelley, 1955; Sales, 1968; Baltz, 1967, 1978; Woodward and Callender, 1977; Chapin and Cather, 1981; Lucas and Ingersoll, 1981). Although some earlier workers suggested an Eocene culmination of Laramide deformation in northern New Mexico (e.g., Baltz, 1967, 1978), it was Chapin and Cather (1981) who more specifically stated that the sharpest pulse of Laramide deformation began about 55 m.y. ago in latest Paleocene–earliest Eocene time. According to Chapin and Cather (1981), it was at this time that the majority of the Laramide Echo Park basins (including the Galisteo–El Rito and Carthage–La Joya Basins in New Mexico) formed as en-echelon, asymmetric downwarps and tilted and subsided blocks between strands of the wrench-fault system. However, the correlations presented here and the recently published paleontological evidence on the age of the "Baca Formation" in the Carthage–La Joya Basin (Lucas and others, 1982, 1983) suggest that the Laramide Echo Park basins of north- and south-central New Mexico did not form until the middle Eocene, at, or just after, 50 m.y. They thus suggest that there was a major pulse of Laramide deformation of the Rocky Mountain foreland during the middle Eocene.

By Paleocene time, crowding of the Colorado Plateau against the foreland produced the basement-cored Brazos–Sangre de Cristo uplift

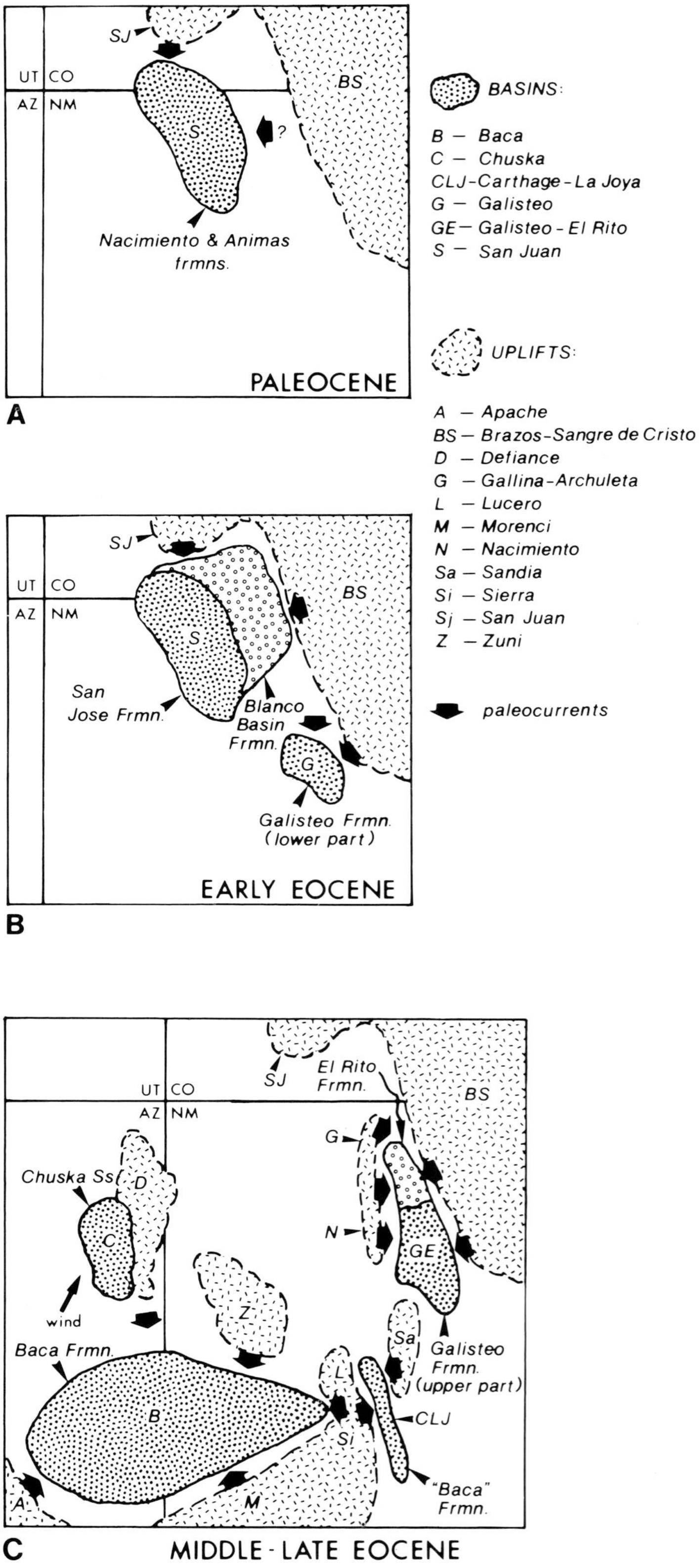

FIGURE 4. Three stages of Laramide deformation and sedimentation of the Rocky Mountain foreland of northern New Mexico. **A**, Paleocene; **B**, early Eocene; **C**, middle–late Eocene. Presence of Sandia uplift during the middle–late Eocene is conjectural. Based on the arguments in the text and on Baltz (1978), Chapin and Cather (1981), Lucas and Ingersoll (1981), and Cather (1983a, 1983b).

and, on the Colorado Plateau, the monoclinal San Juan uplift (Kelley, 1955; Tweto, 1975; Chapin and Cather, 1981). Paleocene sediments shed from these uplifts were deposited in the San Juan Basin (Fig. 4A): Ojo Alamo, Nacimiento, and Animas Formations (Baltz, 1967; Powell, 1973; Sikkink, 1983). Movement of the Nacimiento uplift began just before the Eocene (Baltz, 1967), but no sediments were shed from the Nacimiento uplift at this time. The beginning of the Eocene saw major upthrusting of the San Juan and Brazos–Sangre de Cristo uplifts and the consequent enlargement of source terranes. These larger terranes were the sources of thick, generally coarse-grained sediments deposited in the rapidly subsiding San Juan (Blanco Basin and San Jose Formations) and Galisteo (lower part of Galisteo Formation) Basins (Fig. 4B). Red beds of Mesozoic and late Paleozoic age contributed significant amounts of sediment to these basinal deposits and account for the multicyclic red beds that characterize their finer-grained intervals (Dunn, 1964a; Baltz, 1967). However, it was not until middle Eocene time that new, en-echelon, Echo Park basins formed in the foreland (Fig. 4C). The Galisteo–El Rito Basin was superimposed in part on the older Galisteo Basin, and the Carthage–La Joya Basin formed between the Sierra and Sandia uplifts. At, or slightly after, this time, deposition began in the Baca Basin along the southern margin of the Colorado Plateau, and the San Juan Basin was elevated to form an area of non-deposition (Fig. 4C). The eolian Chuska Sandstone, definitely older than middle Oligocene (Naeser, 1971; Hackman and Olson, 1977) and possibly of late Eocene age, was deposited just west of the San Juan Basin by winds from the south (Fig. 4C) (Wright, 1956; Trevena and Nash, 1978).

Identification of a middle Eocene pulse of Laramide deformation in the Rocky Mountain foreland of northern New Mexico is at slight variance with the conclusions of Chapin and Cather (1981). Nevertheless, it should be pointed out that basinal sedimentation may have lagged behind the initiation of compression. If this lag had a duration of as much as 5 m.y., then Chapin and Cather's (1981) identification of a sharp pulse of Laramide deformation beginning about 55 m.y. is upheld. However, the timing of this pulse of Laramide deformation is only as precise as the correlation of the synorogenic sediments it produced. The new correlation of Eocene rocks of the northern Rio Grande rift and adjacent areas presented here needs additional, especially paleontological, corroboration. If it is accepted, it does suggest that a major pulse of Laramide deformation that produced the wrench-faulted, Echo Park-type basins of the Rocky Mountain foreland of New Mexico occurred in the middle Eocene rather than at the beginning of the Eocene.

ACKNOWLEDGMENTS

I am especially grateful to the following people for information related to this paper: S. M. Cather, T. W. Gorham, R. V. Ingersoll, L. D. McFadden, and L. N. Smith. I also thank M. B. Lucas and A. Lucas, without whom this paper could not have been completed. Finally, comments on an earlier version of this paper by C. E. Chapin and R. V. Ingersoll improved its content and clarity.

REFERENCES

Ager, D. V., 1973, The nature of the stratigraphical record: John Wiley and Sons, New York, 114 pp.

Baltz, E. H., Jr., 1965, Stratigraphy and history of Raton Basin and notes on San Luis Basin, Colorado–New Mexico: American Association of Petroleum Geologists, Bulletin, v. 49, pp. 2041–2075.

———, 1967, Stratigraphy and regional tectonic implications of part of Upper Cretaceous and Tertiary rocks, east-central San Juan Basin, New Mexico: U.S. Geological Survey, Professional Paper 552, 101 pp.

———, 1978, Resume of Rio Grande depression in north-central New Mexico: New Mexico Bureau of Mines and Mineral Resources, Circular 163, pp. 210–228.

Beaumont, E. C., 1979, Geology of the Cerrillos coal field, Santa Fe County, New Mexico: New Mexico Geological Society, Guidebook 30, pp. 269–274.

Berggren, W. A., McKenna, M. C., Hardenbol, J., and Obradovich, J. D., 1978, Revised Paleogene polarity time scale: Journal of Geology, v. 86, pp. 67–81.

Bingler, E. C., 1968a, Geologic map of El Rito quadrangle, Rio Arriba County, New Mexico: New Mexico Bureau of Mines and Mineral Resources, Geologic Map 20, scale 1:24,000.

———, 1968b, Geologic map of the Valle Grande Peak quadrangle, Rio Arriba County, New Mexico: New Mexico Bureau of Mines and Mineral Resources, Geologic Map 21, scale 1:24,000.

———, 1968c, Geology and mineral resources of Rio Arriba County, New Mexico: New Mexico Bureau of Mines and Mineral Resources, Bulletin 91, 158 pp.

Budding, A. J., Pitrat, C. W., and Smith, C. T., 1960, Geology of the southeastern part of the Chama Basin: New Mexico Geological Society, Guidebook 11, pp. 78–92.

Butler, A. P., Jr., 1971, Tertiary volcanic stratigraphy of the eastern Tusas Mountains, southwest of the San Luis Valley, Colorado–New Mexico: New Mexico Geological Society, Guidebook 22, pp. 289–300.

Cather, S. M., 1983a, Laramide Sierra uplift: Evidence for major pre-rift uplift in central southern New Mexico: New Mexico Geological Society, Guidebook 34, pp. 99–101.

———, 1983b, Lacustrine deposits of the Eocene Baca Formation, western Socorro County, New Mexico: New Mexico Geological Society, Guidebook 34, pp. 179–185.

Chapin, C. E., and Cather, S. M., 1981, Eocene tectonics and sedimentation in the Colorado Plateau–Rocky Mountain area: Arizona Geological Society Digest, v. 14, pp. 173–198.

Cross, W., and Larsen, E. S., 1935, A brief review of the geology of the San Juan region of southwestern Colorado: U.S. Geological Survey, Bulletin 843, 138 pp.

Dane, C. H., 1946, Stratigraphic relations of Eocene, Paleocene, and latest Cretaceous formations of eastern side of San Juan Basin, New Mexico: U.S. Geological Survey, Oil and Gas Investigations Preliminary Chart 24.

———, and Bachman, G. O., 1965, Geologic map of New Mexico: U.S. Geological Survey, scale 1:500,000.

Dunn, D. E., 1964a, Evolution of the Chama Basin and Archuleta anticlinorium, eastern Archuleta County, Colorado: Ph.D. thesis, University of Texas at Austin, 114 pp.

———, 1964b, Tectonic significance of the Blanco Basin Formation: Geological Society of America, Special Paper 76, p. 271.

———, 1965, Evolution of the Chama Basin and Archuleta anticlinorium, eastern Archuleta County, Colorado (abstract): Dissertation Abstracts, v. 25, pp. 5211–5212.

Epis, R. C., and Chapin, C. E., 1975, Geomorphic and tectonic implications of the post-Laramide, late Eocene erosion surface in the southern Rocky Mountains: Geological Society of America, Memoir 144, pp. 45–74.

Galusha, T., and Blick, J. C., 1971, Stratigraphy of the Santa Fe Group, New Mexico: American Museum of Natural History, Bulletin, v. 144, 127 pp.

Gorham, T. W., 1979, Geology of the Galisteo Formation, Hagan Basin, New Mexico: M.S. thesis, University of New Mexico, Albuquerque, 136 pp.

———, and Ingersoll, R. V., 1979, Evolution of the Eocene Galisteo Basin, north-central New Mexico: New Mexico Geological Society, Guidebook 30, pp. 219–224.

Hackman, R. J., and Olson, A. B., 1977, Geology, structure, and uranium deposits of the Gallup 1 × 2′ quadrangle, New Mexico and Arizona: U.S. Geological Survey, Miscellaneous Investigations Map I-981, scale 1:250,000.

Hardenbol, J., and Berggren, W. A., 1978, A new Paleogene numerical time scale; *in* Cohee, G. V., Glaessner, M. F., and Hedberg, H. D. (eds.), Contributions to the geologic time scale: American Association of Petroleum Geologists, pp. 213–234.

Hayden, F. V., 1869, Preliminary field report of the U.S. Geological Survey of Colorado and New Mexico: U.S. Geological Survey, 3rd Annual Report, 155 pp.

Kautz, P. F., Ingersoll, R. V., Baldridge, W. S., Damon, P. E., and Shafiqullah, M., 1981, Geology of the Espinaso Formation (Oligocene), north-central New Mexico: Geological Society of America, Bulletin, pt. II, v. 92, pp. 2318–2400.

Kelley, V. C., 1955, Regional tectonics of the Colorado Plateau and relationships to the origin and distribution of uranium: University of New Mexico, Publications in Geology, no. 5, 120 pp.

Larsen, E. S., and Cross, W., 1956, Geology and petrology of the San Juan region, southwestern Colorado: U.S. Geological Survey, Professional Paper 258, 303 pp.

Lee, W. T., and Knowlton, F. H., 1917, Geology and paleontology of the Raton Mesa and other regions in Colorado and New Mexico: U.S. Geological Survey, Professional Paper 101, 450 pp.

Leopold, E. B., and MacGintie, H.D., 1972, Development and affinities of Tertiary floras in the Rocky Mountains; *in* Graham, A. (ed.), Floristics and paleofloristics of Asia and eastern North America: Elsevier, New York, pp. 147–200.

Lipman, P. W., Steven, T. A., and Mehnert, H. H., 1970, Volcanic history of the San Juan Mountains, Colorado, as indicated by potassium–argon dating: Geological Society of America, Bulletin, v. 81, pp. 2329–2352.

Logsdon, M. J., 1981, A preliminary basin analysis of the El Rito Formation (Eocene), north-central New Mexico: Geological Society of America, Bulletin, pt. I, v. 92, pp. 968–975.

Lowrie, W., and Alvarez, W., 1981, One hundred million years of geomagnetic polarity history: Geology, v. 9, pp. 392–397.

Lucas, S. G., 1980, Stratigraphy and age of the Galisteo Formation and the structural history of the Rio Puerco fault zone (abstract): Geological Society of America, Abstracts with Programs, v. 12, p. 279.

———, 1982, Vertebrate paleontology, stratigraphy, and biostratigraphy of Eocene Galisteo Formation, north-central New Mexico: New Mexico Bureau of Mines and Mineral Resources, Circular 186, 34 pp.

———, 1983, The Baca Formation and the Eocene–Oligocene boundary in New Mexico: New Mexico Geological Society, Guidebook 34, pp. 187–192.

———, 1984, *Protitanotherium* (Mammalia, Perissodactyla) from the Eocene Baca Formation, west-central New Mexico: New Mexico Journal of Science, in press.

———, and Ingersoll, R. V., 1981, Cenozoic continental deposits of New Mexico: an overview: Geological Society of America, Bulletin, pt. I, v. 92, pp. 917–932.

———, and Kues, B. S., 1979, Vertebrate biostratigraphy of the Eocene Galisteo Formation, north-central New Mexico: New Mexico Geological Society, Guidebook 30, pp. 225–229.

———, and Schoch, R. M., 1981, Paleoenvironments in north-central New Mexico during the early and late Eocene (abstract): Geological Society of America, Abstracts with Programs, v. 13, p. 500.

———, Sullivan, R. M., and Logan, T. R., 1983, Glyptosaurine lizard from Eocene Baca Formation, south-central New Mexico: New Mexico Geology, v. 5, pp. 77–78.

———, Schoch, R. M., Manning, E., and Tsentas, C. P., 1981, The Eocene biostratigraphy of New Mexico: Geological Society of America, Bulletin, pt. I, v. 92, pp. 951–967.

———, Wolberg, D. L., Hunt, A., and Schoch, R. M., 1982, A middle Eocene titanothere from the Baca Formation, south-central New Mexico: Journal of Paleontology, v. 56, pp. 542–545.

Manley, K., 1981, Redefinition and description of the Los Pinos Formation of north-central New Mexico: Geological Society of America, Bulletin, pt. I, v. 92, pp. 984–989.

Muehlberger, W. R., 1967, Geology of Chama quadrangle, New Mexico: New Mexico Bureau of Mines and Mineral Resources, Bulletin 89, 114 pp.

———, 1968, Geology of Brazos Peak quadrangle, New Mexico: New Mexico Bureau of Mines and Mineral Resources, Geologic Map 22, scale 1:24,000.

———, Adams, G. E., Longgood, T. E., and St. John, B. E., 1960, Stratigraphy of the Chama quadrangle, northern Rio Arriba County, New Mexico: New Mexico Geological Society, Guidebook 11, pp. 93–102.

Naeser, C. W., 1971, Geochronology of the Navajo–Hopi diatremes, Four Corners area: Journal of Geophysical Research, v. 76, pp. 4978–4985.

Powell, J. S., 1973, Paleontology and sedimentation models of the Kimbeto Member of the Ojo Alamo Sandstone; *in* Fassett, J. E. (ed.), Cretaceous and Tertiary rocks of the southern Colorado Plateau: Four Corners Geological Society, Durango, pp. 111–122.

Prothero, D. R., Denham, C. R., and Farmer, H. G., 1982, Oligocene calibration of the magnetic polarity time scale: Geology, v. 10, pp. 650–653.

Sales, J. K., 1968, Crustal mechanics of Cordilleran foreland deformation: a regional and scale model approach: American Association of Petroleum Geologists, Bulletin, v. 52, pp. 2016–2044.

Sikkink, P. G. L., 1983, Depositional environment of the Tertiary Ojo Alamo Sandstone, San Juan Basin, New Mexico (abstract): Geological Society of America, Abstracts with Programs, v. 15, p. 308.

Slack, P. B., 1973, Structural geology of the northeast part of the Rio Puerco fault zone, Sandoval County, New Mexico: M.S. thesis, University of New Mexico, Albuquerque, 74 pp.

———, 1975, Tectonic development of the northeast part of the Rio Puerco fault zone, New Mexico: Geology, v. 3, pp. 665–668.

Slack, P. B., and Campbell, J. A., 1976, Structural geology of the Rio Puerco fault zone and its relationship to central New Mexico tectonics: New Mexico Geological Society, Special Publication 6, pp. 46–52.

Smith, C. T., Budding, A. J., and Pitrat, C. W., 1961, Geology of the southeastern part of the Chama Basin: New Mexico Bureau of Mines and Mineral Resources, Bulletin 75, 57 pp.

Smith, H. T. U., 1938, Tertiary geology of the Abiquiu quadrangle, New Mexico: Journal of Geology, v. 46, pp. 933–965.

Stearns, C. E., 1943, The Galisteo Formation of north-central New Mexico: Journal of Geology, v. 5, pp. 301–319.

Trevena, A. S., and Nash, W. P., 1978, Provenance and correlation of some mid-Tertiary (?) eolian sandstone units, New Mexico and Arizona (abstract): Geological Society of America, Abstracts with Programs, v. 10, p. 240.

Tweto, O., 1975, Laramide (Late Cretaceous–early Tertiary) orogeny in the southern Rocky Mountains: Geological Society of America, Memoir 144, 44 pp.

———, 1979, Geologic map of Colorado: U.S. Geological Survey, scale 1:500,000.

Van Houten, F. B., 1957, Appraisal of Ridgway and Gunnison "tillites," southwestern Colorado: Geological Society of America, Bulletin, v. 68, pp. 383–388.

Weber, R. H., 1971, K–Ar ages of Tertiary igneous rocks in central and western New Mexico: Isochron/West, no. 1, pp. 33–45.

Woodward, L. A., and Callender, J. F., 1977, Tectonic framework of the San Juan Basin: New Mexico Geological Society, Guidebook 28, pp. 209–212.

Wright, H. E., Jr., 1956, Origin of the Chuska Sandstone, Arizona–New Mexico: a structural and petrographic study of a Tertiary eolian sediment: Geological Society of America, Bulletin, v. 67, pp. 413–434.

MIOCENE STRATIGRAPHIC RELATIONS AND PROBLEMS BETWEEN THE ABIQUIU, LOS PINOS, AND TESUQUE FORMATIONS NEAR OJO CALIENTE, NORTHERN ESPAÑOLA BASIN

S. JUDSON MAY
Arco Exploration Company, Box 5540, Denver, Colorado 80217

INTRODUCTION

The complex stratigraphic relations of the Abiquiu Formation, the mid-Miocene part of the Los Pinos Formation, and the Tesuque Formation are locally well exposed in the northern Española Basin near Ojo Caliente (Fig. 1). Exposures in this area show that much of the Los Pinos in the Ojo Caliente–Petaca area is equivalent in stratigraphic level and age to the Chama–El Rito Member of the Tesuque Formation (Fig. 2) and most or all of the Abiquiu Formation in the Española Basin. In the Ojo Caliente area the Abiquiu and Los Pinos represent a large southwest-sloping fan complex of volcaniclastic conglomerate and sandstone derived from rhyolitic and andesitic terranes to the north and northeast (Manley, 1981). The Tesuque Formation records a large arkosic accumulation of sand carried mainly west and south from Precambrian rocks of the Sangre de Cristo uplift. In the Ojo Caliente area these two alluvial systems merged and intertongued in a complex manner, producing the Chama–El Rito Member of the Tesuque Formation which contains interbedded volcaniclastic gravels and arkosic sandstones of the two systems.

ABIQUIU FORMATION

The Abiquiu Formation (Abiquiu Tuff of Smith, 1938) consists of light-gray, tuffaceous sandstone and minor volcaniclastic conglomerate exposed near the town of Abiquiu and distributed widely throughout the Abiquiu 30-min quadrangle. Only the uppermost part of the Abiquiu is exposed at Ojo Caliente and is composed of andesite- and rhyolite-pebble conglomerate containing sparse pebbles of Precambrian quartzite, metarhyolite, gneiss, and schist. The Abiquiu of this report is restricted to the lower 0–80 m of the roughly 420 m mapped as Abiquiu by Smith (1938) in the Ojo Caliente area. The remaining upper 360 m of volcaniclastic conglomerate and tuffaceous sandstone is considered to be the Los Pinos Formation (Cordito Member?) based on its stratigraphic position and approximate lithologic similarity with the Cordito to the north.

Lithology

Near Ojo Caliente, the Abiquiu contains mainly white, very light-gray, light-brown, and minor red and green layers of tuffaceous sandstone. In the upper 20 m are a few layers of andesite- and rhyolite-pebble conglomerate, sparse fluvial crossbeds, laminar beds and channel-fill structures. Sparse, discontinuous seams of mudstone occur locally. In the lower 40 m, several dark lenses of conglomerate and breccia contain subangular cobbles and fragments of locally derived Precambrian metarhyolite. These lenses are lithologically similar to the Ritito Conglomerate north of La Madera. The best local exposure is 3 km west–southwest of Ojo Caliente in a 70 m section along the north wall of Arroyo El Rito (Fig. 1).

The sandstone is variable in composition, consisting of medium to coarse-grained, subangular to subrounded quartz, K-feldspar and plagioclase, with a few granular fragments of Tertiary sedimentary and volcanic rocks and Precambrian metamorphic rock. Traces of opaque minerals and altered biotite flakes also occur. Larger angular fragments and granules of metarhyolite and pink feldspar occur in minor amounts throughout the sandstone, but generally decrease upward in size and number. The fine-grained interstitial debris in most layers has a tuffaceous appearance.

The Abiquiu appears to thicken greatly from a maximum of 90 m near Ojo Caliente to over 300 m to the west and southwest in the Abiquiu area. The basal contact with the pre-Miocene rocks may drop as much as 200 m because of pre-Abiquiu topography, allowing for the greater thickness to the west.

Contacts

Throughout most of the western Española Basin the Abiquiu unconformably overlies pre-Miocene rocks. In the hills west of Ojo Caliente the Abiquiu overlies a mature, early Miocene topography eroded on Precambrian crystalline rock (see third-day road log, this guidebook). This surface locally exhibits up to 300 m of relief. The Abiquiu probably represents the initial major influx of clastic debris that filled the Española Basin.

Smith (1938) and Smith and others (1961) recognized a gradational upper contact of the Abiquiu with the overlying Santa Fe Formation. However, in the Ojo Caliente area, the contact they recognized is considered here to be the upper contact of the Los Pinos Formation. Under the more restricted definition of the Abiquiu used here, the Abiquiu grades upward into the Los Pinos in the Ojo Caliente area. To the south and southwest it probably grades upward into the Chama–El Rito Member of the Tesuque Formation of the Santa Fe Group (Fig. 2), although this relationship can be seen only west of El Rito Creek. Galusha and Blick (1971) also reported a slight angular unconformity between the Abiquiu and Chama–El Rito near the "Ojo Caliente intrusive" southwest of the town, but the beds under the unconformity at this location are the tuff-ring facies of the much younger Ojo Caliente tuff ring (May, 1980) and are located in the middle part of the Chama–El Rito (Figs. 1, 2).

In the Ojo Caliente area the only exposure of the upper contact is 3 km west of town. Here the formation grades up into the Los Pinos through a 40-m section. The very light-gray tuffaceous sandstones gradually turn darker gray and the rhyolite- and andesite-pebble conglomerates gradually increase to the point where they make up a major part of the section. The contact is chosen at approximately 15 m above the lowest volcanic-pebble conglomerate or where the conglomerates make up more than 25% of the section. Although a few of the conglomerates are locally erosional at their bases, the contact exhibits neither significant erosion nor angularity between the underlying and overlying beds.

Age

The Abiquiu ranges from 22 to 17–18 m.y. in age (early Miocene) near Ojo Caliente. Basalt flows of the Hinsdale Formation lie between the Abiquiu and the underlying Precambrian rocks along the north wall of Arroyo El Rito and have been K–Ar dated by Baldridge and others (1980) at 22.1 ± 0.6 m.y. These flows intertongue with thin discontinuous seams of light-gray tuffaceous sandstone, indicating Abiquiu deposition during eruption of the flows. However, the lower part of the Abiquiu in its type area is probably lower stratigraphically and older than the Abiquiu in the Ojo Caliente hills, implying that it is also partly upper Oligocene. Vazzana (1980) inferred a 27–26 m.y. age for the base of the formation to the west.

Age of the upper contact is well constrained locally. Four km west of Ojo Caliente, a small olivine-nephelinite flow on the south slope of Cerro Negro yielded a K–Ar date of 18.9 ± 0.7 m.y. (Baldridge and

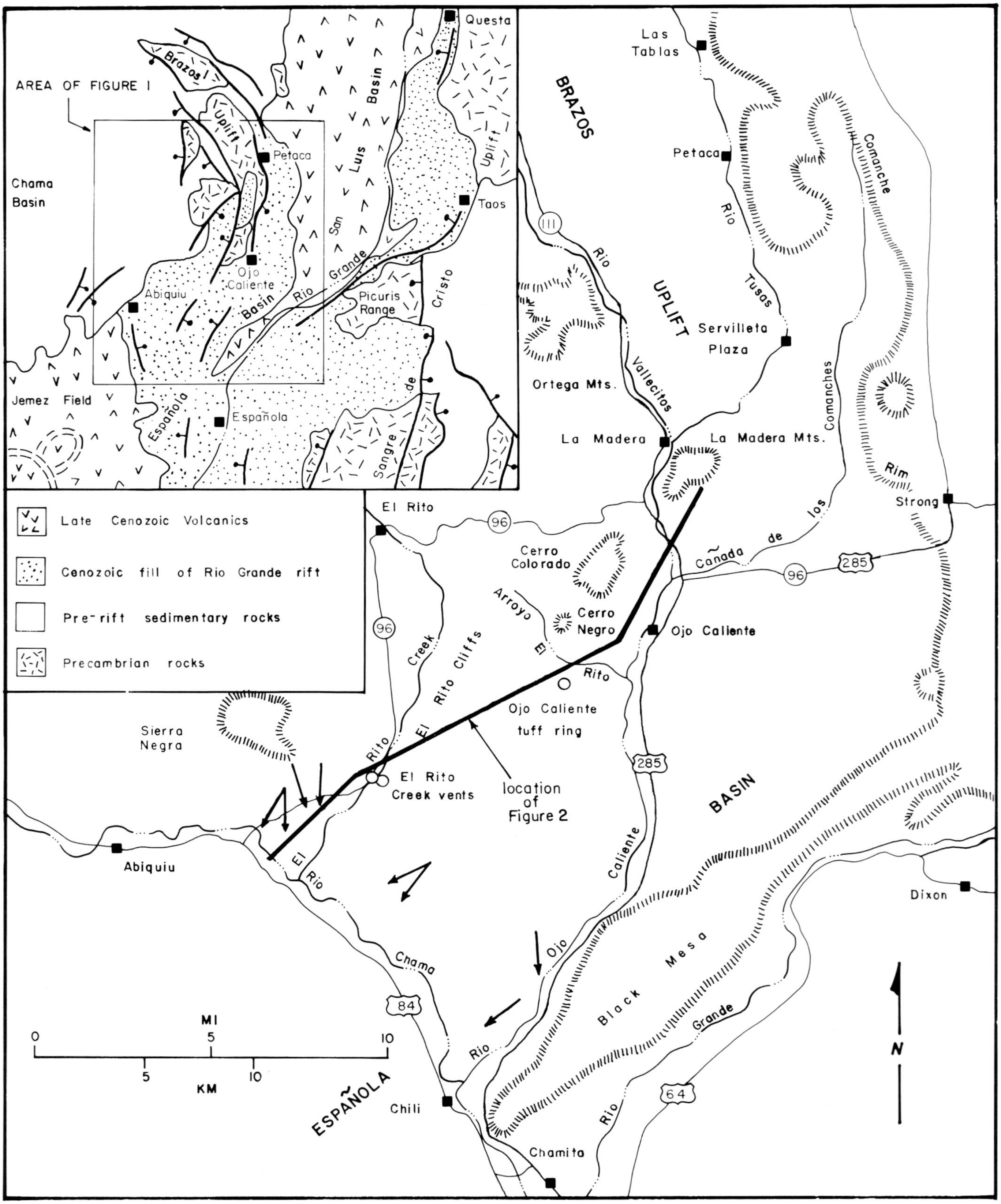

FIGURE 1. Index map of northern Española Basin and southern Brazos uplift showing areas discussed in text. Arrows in southern part of map show paleochannel directions in the Chama–El Rito Member of the Tesuque Formation.

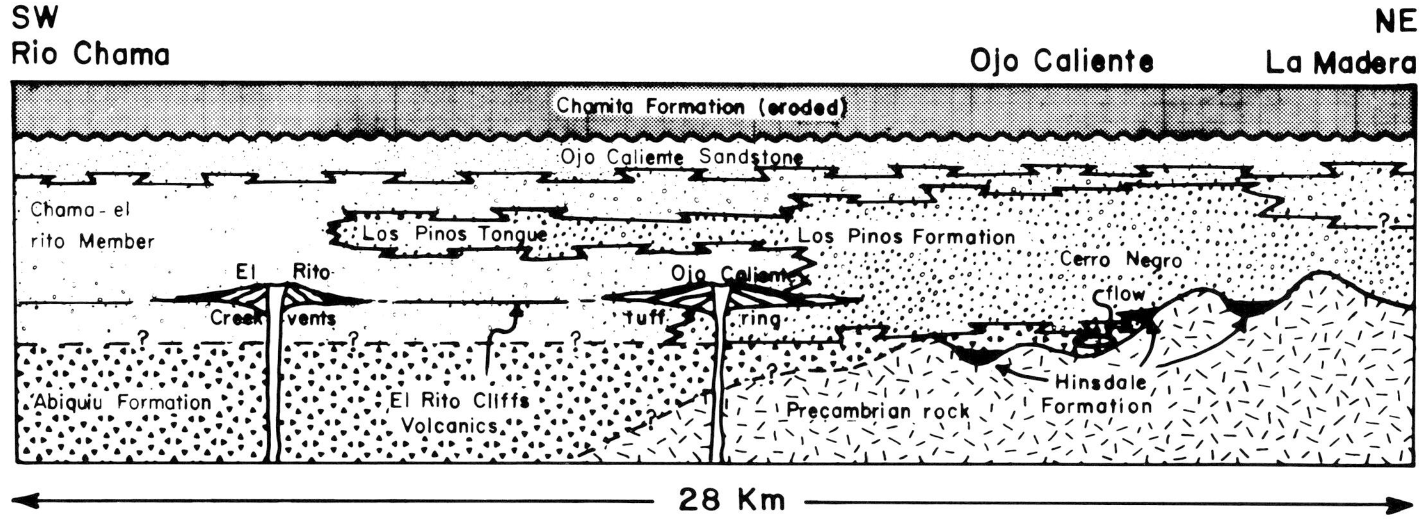

FIGURE 2. Generalized northeast–southwest stratigraphic section of Miocene rocks from La Madera to Rio Chama. Thicknesses not to scale.

others, 1980); this flow occurs approximately 40 m below the top of the Abiquiu, implying an approximate 18 m.y. age for the top. Manley and Mehnert (1981) reported a 17.3 ± 0.8 m.y. K–Ar age on a dacitic clast approximately 10 m below the top of the Abiquiu Formation 7 km northeast of Abiquiu. They also reported a 15.9 ± 0.9 m.y. K–Ar age on a dike that cuts the Abiquiu 6 km north of Sierra Negra, thereby bracketing the age of the upper contact between 17.3 and 15.9 m.y. Since the dike is probably of the same age as the mid-Chama–El Rito volcanic field of May (this guidebook), the 17.3 m.y. date may approximate the age of the upper contact most closely in this area.

LOS PINOS FORMATION

The Los Pinos Formation, first named Los Pinos Gravel by Atwood and Mather (1932), is composed of conglomerate, sandstone and ash-flow tuff. Interbedded basalt flows are part of the Hinsdale Formation (Lipman and Mehnert, 1975). On the east side of the Tusas Mountains, the formation forms an eastward-thickening wedge of volcaniclastic debris carried south by Miocene and Pliocene streams flowing out of the late Tertiary San Juan volcanic field (Lipman and Mehnert, 1975).

The Los Pinos (Cordito Member?) at Ojo Caliente was first mapped and correlated with the Conejos Formation by Atwood and Mather (1932). These deposits were also assigned to the Abiquiu Tuff by Smith (1938). Three to five km south of Ojo Caliente, the Los Pinos grades laterally into the Chama–El Rito Member of the Tesuque Formation, which overlies the Abiquiu Formation in the Abiquiu area.

Bingler (1968) first mapped the Los Pinos at Ojo Caliente, but included the upper 80 m of Abiquiu in his basal Los Pinos. In this report the Los Pinos is restricted to those volcaniclastic conglomerates and sandstones (approximately 360 m thick) that overlie the light-gray tuffaceous sandstones of the Abiquiu Formation west of Ojo Caliente and underlie a thin tongue of light-brown to pale-orange siltstones, sandstones, and gray volcanic conglomerates of the Chama–El Rito 1 km east of town. The Los Pinos is also recognized as a northward-thickening tongue, 130–200 m thick, in the upper half of the Chama–El Rito along the north-trending cliffs east of El Rito Creek (Figs. 1, 2).

Lithology

In the Ojo Caliente area, the Los Pinos Formation is primarily purplish-gray volcanic-pebble conglomerate with numerous interbeds of slightly tuffaceous pink, light-brown, and pale-orange siltstone and sandstone that are similar to the Chama–El Rito. The basal 30 m of the formation contain numerous locally derived angular fragments of metarhyolite, granite, amphibolite gneiss, and muscovite schist where it overlies hills of Precambrian rock.

The conglomerates consist of granules, pebbles, and cobbles of maroon and bluish-gray rhyolite and rhyolitic welded tuff, containing eutaxitic structure and phenocrysts of quartz. Clasts of light-gray quartz latite(?) and andesite porphyry are also abundant; both contain phenocrysts of hornblende, light-gray feldspar, and sparse biotite and quartz. Minor granules and pebbles of soft, light-pink to light-gray andesite and rhyolitic tuff breccia and a few zones with rounded boulders of porphyritic andesite(?) and rhyolite (0.3–1.0 m in diameter) imply one or more nearby volcanic sources for some of the debris. Minor angular fragments of Precambrian metarhyolite, amphibolite, schist, and pink feldspar throughout the Los Pinos are slightly larger in average size than the more rounded volcanic clasts and are most common in the lower third of the section.

Conglomerates are 0.1–3.0 m thick and generally have erosional bases exhibiting 0.1–0.3 m of relief. Fluvial crossbeds, slightly imbricated cobbles, and granular to pebbly volcaniclastic and arkosic sandstone fill numerous paleochannels. Some conglomerates grade laterally into granular sandstone, and in many areas the sandstones and conglomerates are finely interbedded.

A tongue of Los Pinos conglomerate in the Chama–El Rito Member along the cliffs east of El Rito Creek (Fig. 1) is similar to the Los Pinos conglomerate at Ojo Caliente but contains more clasts of Precambrian rock. In the southern part of the tongue, the conglomerate contains about 75% rounded pebbles and cobbles of andesite, rhyolite, quartz latite, and dacite(?), and 25% subangular pebbles of quartzite, metarhyolite, and gneiss. As the tongue is traced northward along the cliffs, the pebbles of Precambrian rock gradually become larger (locally 0.3–1.0 m in diameter) and eventually comprise up to 90% of the debris in some layers near the latitude of Ojo Caliente. No northward increase in grain size or angularity of the volcanic clasts is noticeable.

Light-brown, pink, and pale-orange siltstones and sandstones that are interbedded with the Los Pinos conglomerates generally consist of quartz, feldspar, and minor, small lithic grains in a slightly tuffaceous matrix. The layers are loosely cemented, evenly bedded, and are approximately similar in appearance to those of the Chama–El Rito although the amount of tuffaceous debris decreases slightly to the south.

Contacts

The Los Pinos unconformably overlies the Precambrian rocks near Ojo Caliente along a surface of high relief, with stratigraphically higher levels of the Los Pinos overlying what were the higher elevations on the hills before Neogene east-tilting of the region. In areas where the elevation on the surface of Precambrian rocks was low enough, the Abiquiu Formation underlies the Los Pinos along a gradational contact.

The upper contact of the Los Pinos grades up into a thin tongue of upper Chama–El Rito 1 km east of Ojo Caliente and is located at the uppermost major gravel layer in the pale-orange siltstones and sandstones of the Chama–El Rito. Because most gravel layers are not continuous for more than 1 km, the uppermost major gravel occurs at slightly different levels in the section from one place to another, and the contact rises and falls as it is traced laterally.

North of Ojo Caliente, the upper contact with the Chama–El Rito is more complex. At Ojo Caliente, the Los Pinos conglomerates (360 m thick) appear to be at approximately the same stratigraphic level as the middle to upper(?) Cordito Member north of La Madera, but are separated from it, at least in their upper part, by a laterally intervening zone of the Chama–El Rito exposed on the north side of Cañada de los Comanches (Figs. 1, 2). Northward of Ojo Caliente, the volcaniclastic conglomerates in the upper Los Pinos diminish abruptly and the upper contact drops in the section near the confluence of Cañada de los Comanches and Rio Ojo Caliente. Exposures along N.M. 96 for the first 3 km north of Cañada de los Comanches contain more volcaniclastic conglomerates than the overlying Chama–El Rito to the east, suggesting that this gradational contact continues north at a lower stratigraphic level (parallel to the highway) for at least a few kilometers. However, poor exposures between this area and La Madera make it difficult to demonstrate that the Los Pinos at Ojo Caliente is physically continuous with the Cordito to the north. The relationship is better exposed 8–10 km west–northwest of Ojo Caliente in the hills east of El Rito, where a complex intertonguing relationship occurs between the Los Pinos, Chama–El Rito and Abiquiu.

The Los Pinos exposed at Ojo Caliente grades south and upward into the Chama–El Rito Member of the Tesuque Formation. Three to five km south of Ojo Caliente the number of conglomerate layers diminishes and the more arkosic pale-orange siltstones and sandstones of the Chama–El Rito dominate the section from here south into the Española Basin. A few volcaniclastic conglomerates extend many kilometers to the south, and tongues of pink and brown sandstone, up to 25 m thick (typical of the Chama–El Rito), extend north beyond La Madera, thereby making this laterally gradational contact difficult to locate precisely. Of all contacts in the Tertiary section, the contact between the Los Pinos and Chama–El Rito is the most subject to reinterpretation or relocation by other workers, depending on where the most significant lithologic change is thought to occur.

The tongue of Los Pinos in the cliffs east of El Rito Creek grades south into the upper half of the Chama–El Rito and is thought to be a west–southwest extension of the thick gravels at Ojo Caliente. The lower contact of the tongue, although gradational, occurs as a relatively sharp upward increase in the abundance of conglomerate near the base of the cliffs. The contact appears to drop in the section northward near the latitude of Ojo Caliente.

The upper contact of the tongue is exposed very poorly a short distance east of the cliffs, but with difficulty can be traced to the north along a large easterly dip-slope in the upper Chama–El Rito. At the latitude of Ojo Caliente this contact rises stratigraphically at the expense of the overlying Chama–El Rito and is only about 35 m below the base of the Ojo Caliente Sandstone, similarly to the situation east of Ojo Caliente.

Age

Age of the Cordito(?) at Ojo Caliente ranges from 18 to approximately 12 m.y. (mid-Miocene). The base of the member is approximately 40 m stratigraphically above an olivine-nephelinite flow dated at 18.9 ± 0.7 m.y. by Baldridge and others (1980), and is probably 17–18 m.y. old.

The approximate 12 m.y. age for the upper contact is not well constrained near Ojo Caliente. Southeast of town the upper contact is about 110 m higher stratigraphically than the 15.3 m.y. old Ojo Caliente tuff ring (May, this guidebook) and very approximately 250 m lower than basalt flows in the lower Chamita Formation near Chili (24 km south) dated at 9.6 ± 0.2 m.y. by Baldridge and others (1980). The middle of the Los Pinos tongue exposed along the cliffs east of El Rito Creek is about 170 m above a 15.3 m.y. old volcanic vent (date by Ekas and others, this guidebook) exposed along the creek, and is probably 13–14 m.y. old.

In southern Colorado, the Los Pinos underlies, and is interbedded with, basalts of the Hinsdale Formation, from which Lipman and Mehnert (1975) reported radiometric ages of 25–5 m.y. Bingler (1968) also reported a 25.9 m.y. date from a welded rhyolite tuff in the lower part of the Cordito Member west of Petaca. The Los Pinos is therefore upper Oligocene to uppermost Miocene in the eastern Tusas Mountains and is age-equivalent to the Abiquiu, Tesuque, and Chamita Formations to the south.

HINSDALE FORMATION

Two exposures of basalt flows occur between the Precambrian and Tertiary rocks in the hills west of Ojo Caliente; one at Arroyo El Rito, the second near Cerro Colorado. The basalts (formerly Jarita Basalt of Butler, 1946; Barker, 1958; May, 1980; Baldridge and others, 1980) are now included in the Hinsdale Formation in accordance with Lipman and Mehnert (1975), who assigned all interlayered basalts of the Los Pinos Formation to the Hinsdale. The flows are at slightly different levels in the Tertiary section and appear to be small, isolated flows that were extruded into Miocene valleys from several local vents now buried by the Los Pinos and Abiquiu Formations.

The Arroyo El Rito exposure, the lower of the two, is 4 km S70W of Ojo Caliente. There a series of four(?) superimposed flows overlies the Precambrian rocks and underlies the Abiquiu Formation (Figs. 1, 2). The greatest thickness of the flows is 30 m where they have filled a valley eroded on the Precambrian rocks. Elsewhere, the thickness is generally 15–20 m. Lateral extent of the flows is uncertain, but probably is greater than that seen in the exposures. To the north, stratigraphically higher parts of the Abiquiu directly overlie the Precambrian. Some flows are separated from each other by thin, discontinuous zones of light-gray, tuffaceous sandstone (0–1.2 m thick) similar to the Abiquiu Formation.

Chemical analyses and norm calculations for the flows indicate a quartz-tholeiite composition with 53% SiO_2 and 3.8% normative quartz (Baldridge and others, 1980). Most major-element oxides are within 1% of those reported by Barker (1958) for his Jarita Basalt at Tusas Creek in the Las Tablas quadrangle. Similar SiO_2 values at 51 and 52% were reported by Lipman and Mehnert (1975) and Lipman and others (1978) for basalts in the Hinsdale Formation.

Several flows of the basalt also occur on the southeast flank of Cerro Colorado between the Precambrian rock and the overlying Los Pinos Formation. The basalt is 0–35 m thick and is exposed intermittently for a distance of 3 km (Fig. 2). The dark-gray flows are deeply weathered olive green and purple in most outcrops and locally exhibit spheroidal weathering. Petrographically, the flows are approximately similar to the flows along Arroyo El Rito, but are approximately 65 m higher in the section. They could be slightly younger than the flows at Arroyo El Rito or could have erupted at the same time, but at an elevation 65 m higher on the Precambrian rocks, thereby not being buried until the basal Los Pinos was deposited.

TESUQUE FORMATION (CHAMA–EL RITO MEMBER)

The Chama–El Rito Member of the Tesuque Formation was named by Galusha and Blick (1971) for buff to pink sandstone, siltstone, and interbedded purplish-gray, lenticular layers of volcaniclastic conglomerate exposed in the badlands north of Rio Chama, 5 km east of its confluence with El Rito Creek. These beds are confined to the western and northern parts of the Española Basin and grade north into the Cordito(?) Member of the Los Pinos near Ojo Caliente.

Lithology

The Chama–El Rito is primarily a sequence of arkosic siltstone, sandstone, and lesser volcaniclastic conglomerate. Minor beds of white and pale yellowish-green airfall tuff and tuffaceous sandstone, thin beds of reddish-brown silty clay, and very sparse layers of white limestone occur mainly in the upper 200 m in the badlands north of Rio Chama, implying lower paleoslopes to the southwest.

South of Ojo Caliente the Chama–El Rito is 460–550(?) m thick and typically occurs in widely spaced, fault-controlled exposures in the structurally higher areas. It also forms a tongue (35–60 m thick) between the Los Pinos Formation and Ojo Caliente Sandstone 1 km east of Ojo Caliente. Two km north of Ojo Caliente the member thickens abruptly to more than 150 m along a laterally gradational contact with the underlying Los Pinos conglomerates and is well exposed in the badlands on the north side of Cañada de los Comanches.

The bulk of the Chama–El Rito is a sequence of silty to medium-grained, moderately to well-sorted, arkosic to lithic sandstone beds colored light-orange brown, orange-pink, and salmon. Layers are medium-bedded to massive, with even, parallel bedding. Mineral composition is variable, but typical amounts are 50–70% quartz, 20–30% plagioclase and K-feldspar, 5–15% volcanic and lesser plutonic rock fragments, 1–3% opaque minerals, and 1–2% other miscellaneous grains including secondary(?) clay. In some beds, the sand-sized fraction contains glassy and tuffaceous grains (20–35%). The sandstones are typically friable to crumbly, slightly tuffaceous, and weakly cemented with minor clay, hematite, and calcite. Interbeds of loosely cemented, reddish-brown, sandy siltstone (0.1–3.0 m thick) occur in the south near Rio Chama along with very minor beds of dune-like sandstone.

Volcaniclastic conglomerates occur throughout the Chama–El Rito section, but a few systematic variations are noticeable. Conglomerates comprise up to 60% of the Los Pinos at Ojo Caliente, but become thinner, fewer, and slightly finer-grained as they are traced south and southwest into the Chama–El Rito. They make up less than 10% of the section in the south near Rio Chama. In most areas, the conglomerate are sparse or absent in the upper 25–35 m. Five to six km southwest of Ojo Caliente, an ill-defined gravelly zone (90–120 m thick) in the Chama–El Rito is about 70 m stratigraphically above the Ojo Caliente tuff ring of May (this guidebook) and is probably equivalent to the tongue of Los Pinos along the El Rito cliffs.

The conglomerates contain subangular to rounded granules, pebbles, and small cobbles of rhyolite ash-flow tuff, light-gray porphyritic andesite, dacite, quartz latite(?), and Precambrian blue and gray quartzite, brown metarhyolite, granite, quartz, and green to black gneiss and amphibolite. Rock types are given above in very approximate order of decreasing abundance, although relative amounts are variable throughout the region and no systematic study of the variations was attempted. Clasts of the Tertiary volcanics are by far the most abundant except near hills of Precambrian rock and are generally more rounded than those of the Precambrian. The gravels are poorly sorted and generally have a light-gray or light-brown sandy matrix rich in quartz and lesser feldspar, rock fragments, and minor tuffaceous debris. A detailed petrographic study of the conglomerates, as also suggested by Galusha and Blick (1971), will provide needed additional insight on the source of these deposits (Ekas and others, and Dethier and Martin, this guidebook).

Geometry of the conglomerates is variable. Most layers grade laterally and upward into beds of nonvolcanic sandstone. Many have sharp erosional bases and, in some areas, a subtle upward-fining grain size. Some form continuous sheet-like layers (3–6 m thick) that are traceable for up to 1 km (especially in the south near Rio Chama) and that have only 0.1–1.0 m of erosional relief at their bases. Within these larger tabular layers are numerous small stringers (0.1–1.0 m thick) of slightly imbricated pebbles and cobbles that extend laterally 5–30 m, but do not scour the underlying beds.

Numerous paleochannels (generally 2–8 m wide) exhibit 0.1–0.6 m of erosional relief at their bases, although a few cut down as much as 1.0–1.2 m. These channels generally form at the base of the larger tubular layers. A few also occur as small, lenticular channels (3–15 m across) that are isolated in the sandstone. The bottom of a few of the more deeply entrenched channels is lined with lag zones of imbricated cobbles. In some channels, trough-shaped crossbeds in the sandstone dip 5–20° into the channel axes, approximately conforming to the floor of the channels. Foreset beds are sparse. Few, if any, slump structures occur on the channel walls, and flute casts and ripple marks in the finer-grained layers are very sparse.

Paleotrend measurements on eight channels show a high directional variability and an average S20W direction (Fig. 1). Directions are determined by:

(1) pebble imbrication in paleochannels,
(2) dip direction of foreset crossbeds, and
(3) correlation of channels between outcrops.

Channels west of El Rito Creek trend almost due south, whereas those to the east near lower Rio Ojo Caliente trend more to the southwest. This convergence may indicate that the axis of the basin trended north–south during the mid-Miocene and was near the longitude of lower El Rito Creek. Possible minor east-tilting of the west side of the basin may have caused this southerly diversion of the southwest-flowing streams near Ojo Caliente.

Several thin, widely spaced layers of white, very light-gray and pale yellowish-green airfall tuff and tuffaceous sandstone occur in the upper 90 m of the Chama–El Rito and lower 60 m of the overlying Ojo Caliente Sandstone. The tuffs first appear about 6–8 km south of Ojo Caliente and increase slightly in abundance southward toward Rio Chama and beyond.

In the badlands west of lowermost Rio Ojo Caliente, several white and very light-gray limestones (0–1.5 m thick) occur 45–125 m below the top of the Chama–El Rito, implying low paleoslopes and proximity to the basin center. Most layers are traceable for 90–1000 m, and, in some areas, several discontinuous beds occur at various levels in the section. Although no fossils were identified, the beds appear to be intensely bioturbated by worm(?) tubes and contain mudcracks(?). In thin section, the tubes are filled with mosaic, sparry calcite and fibrous chert. The matrix is calcareous mud (micrite). Scattered throughout the mud are sparse rounded quartz grains (eolian?) and possible fragments of fossil algae and $CaCO_3$. The sparse limestones may be deposits of small playas or shallow depressions of standing saline water on the interfluvial plains and are absent to the north near Ojo Caliente.

Contacts

The Chama–El Rito grades down into the Abiquiu Formation throughout most of the area; grades north into the Cordito(?) Member of the Los Pinos Formation; and grades up into the overlying Ojo Caliente Sandstone as a series of tongues of orange-brown siltstone of the Chama–El Rito and light-brown eolian sand of the Ojo Caliente Sandstone.

Age

In the Ojo Caliente–Rio Chama area, the Chama–El Rito is laterally equivalent to the Cordito(?) Member of the Los Pinos. Here the Cordito(?) has an age range of 18–12 m.y. and establishes an approximately similar age range for the Chama–El Rito in this area. A 15.3 m.y. old volcanic horizon occurs in lower middle part of the Chama–El Rito between Ojo Caliente and lower El Rito Creek (Ekas and others, and May, this guidebook).

CORRELATION OF UNITS

Correlation of the Miocene units is complicated by contacts that intertongue, grade laterally and are time-transgressive as a result of evolving uplifts, and by basin geometry and source areas. Figure 2 summarizes the stratigraphic relationships between La Madera and the Rio Chama.

The lower 0–80 m of section west of Ojo Caliente consist of white tuffaceous sandstone (here considered to be upper Abiquiu) which is probably similar in stratigraphic level and age to part of the upper Abiquiu in its type area. The lower few meters of metarhyolite conglomerate and breccia in the section at Ojo Caliente may be physically continuous with, but younger than, the 90 m thick basal granitic gravel of the Abiquiu in the type area. This unit is also tentatively correlated with the Ritito Conglomerate to the north, which Vazzana (1980) considered to be basal Abiquiu. The Abiquiu in the type area probably thinned to the north, as it was deposited on the hilly topography of the former Laramide Brazos uplift in the Ojo Caliente area during the early Miocene.

The Abiquiu as defined here is age-equivalent with, and appears to grade northeast into, the lower Cordito in the El Rito–Ortega Mountain area, based on limited reconnaissance work (May, 1980). However, careful study of Abiquiu – Los Pinos – Chama–El Rito relationships is still needed in this critical area of intertonguing. The Upper Member of the Abiquiu Formation (of Vazzana, 1980; Vazzana and Ingersoll, 1981) contains clasts of ash-flow tuff similar to those of the Cordito, implying that the conglomerates of the upper Abiquiu are southwestern tongues of the Cordito (Manley, 1981). However, 3 km west of Ojo Caliente the Abiquiu clearly grades upward into the Los Pinos along a contact that is about 18 m.y old. South of Ojo Caliente the Abiquiu underlies, and in most places grades up into, the Chama–El Rito.

Reconnaissance work north of Ojo Caliente suggests that the 360 m of volcaniclastic conglomerate at Ojo Caliente [mapped as upper Abiquiu Tuff by Smith (1938), but considered here to be Cordito(?)] is approximately equivalent in stratigraphic level and probably in age to the middle part of the Cordito Member as defined by Barker (1958) in the Las Tablas quadrangle and Butler (1946, 1971) in the eastern Tusas Mountains. These conglomerates are thought to be coarser-grained, stratigraphically higher and younger than the Abiquiu in its type area near Abiquiu, and grade laterally to the south into the Chama–El Rito Member of the Tesuque Formation.

SOURCE AREAS AND DEPOSITIONAL ENVIRONMENTS

During the Miocene, the Ojo Caliente area was near the southern end of a large alluvial apron derived from volcanic terranes in the San Juan Mountains and the Questa–Taos area (Lipman and Mehnert, 1975; Manley, 1981). The Brazos uplift, which was still a low northwest-trending highland of Precambrian rock, was gradually being buried by Abiquiu and Los Pinos alluvium at this time. In broadest terms, this volcanic system merged and intertongued with a dominantly arkosic alluvial system (Tesuque Formation) brought west and southwest from the Sangre de Cristo Mountains.

The Abiquiu Formation formed a broad alluvial deposit derived from streams originating in a rhyolitic to andesitic volcanic region to the north and northeast (Vazzana and Ingersoll, 1981). The tuffaceous alluvium buried a widespread erosion surface formed on hills of Precambrian rock in the western Brazos uplift and on lower Tertiary and Mesozoic sedimentary rocks to the southwest where relief was more subdued. The debris entered a mildly sagging basin (early Española Basin) whose boundaries were wider than its present fault boundaries, as indicated by the presence of Abiquiu on the eastern flanks of the Nacimiento uplift and at Cerro Pedernal, both of which are west of the fault margin of the basin. Paleochannel measurements of the lower Abiquiu indicate southwesterly stream flow, and measurements in the upper Abiquiu indicate southeasterly flow away from the San Juan volcanic field along the western side of the Brazos highland, perhaps as a result of early east-tilting of the western Española Basin (Vazzana, 1980; Vazzana and Ingersoll, 1981). Clasts of ash-flow tuff are similar to those of the Los Pinos, possibly implying a common source area and that the Abiquiu may have simply been a more distal, finer-grained part of the early Los Pinos depositional complex (Manley, 1981).

The Los Pinos formed a widespread alluvial deposit derived from areas consisting of rhyolitic and andesitic ash-flow tuffs and flows. Several distant and local volcanic sources are indicated. In the southeastern San Juan Mountains, the Los Pinos forms a debris apron around late-stage volcanic cones and domes along the northern rim of the Platoro caldera (Lipman and Steven, 1970) and "accumulated as great coalescing alluvial fans along the east front of the San Juan Mountains; accumulation was a response to eastward tilting of the Oligocene volcanic sequence concurrent with structural development of the San Luis Valley graben" (Lipman, 1975). According to Manley (1981), P. W. Lipman identified clasts of rhyolitic ash-flow tuff in the Los Pinos that are similar to volcanic units in the Questa area to the northeast. In the eastern Tusas Mountains, Manley also noted clasts derived from local outcrops of ash-flow tuff that erupted from the Questa area. A few large cobbles and boulders of light-gray, rhyolitic ash-flow tuff and porphyritic andesite in the Ojo Caliente area also indicate local erosional sources for a small amount of the debris.

Manley (1981) noted a greater amount of porphyritic, intermediate-composition volcanic clasts in the Chama–El Rito than in the Cordito, which contains abundant clasts of rhyolitic ash-flow tuff, and suggested different sources for the volcanic debris in the two units. Although the Los Pinos conglomerates at Ojo Caliente are similar in appearance to the Cordito and contain clasts of rhyolitic ash-flow tuff, more petrographic study of the clasts is needed to positively identify them as the Cordito Member.

Near Ojo Caliente, the Los Pinos streams maintained southwesterly courses as they flowed around the southern end of the former Laramide Brazos uplift. This is indicated by several lines of evidence:

(1) thickness, abundance, and grain size of conglomerate layers in the Los Pinos (and laterally equivalent Chama–El Rito) decrease to the south and southwest;
(2) imbricated pebbles in the Los Pinos fluvial channels have a slight tendency to dip northeast in many exposures (no systematic study was made, however);
(3) on the southeastern flank of Cerro Colorado, erosional debris of Hinsdale basalt flows in the basal Los Pinos occurs southwest of Hinsdale exposures;
(4) abundance, grain size, and angularity of quartzite, metarhyolite, and gneissic rubble in the Los Pinos tongue at El Rito cliffs increase northward, culminating southwest of Cerro Colorado and La Madera Mountain, which formed major topographic highs of Precambrian rock during Los Pinos deposition.

By 18 m.y. ago, shortly after eruption of the olivine-nephelinite flow at Cerro Negro, the Abiquiu alluvial system was buried by the Los Pinos (Cordito Member?) in the Ojo Caliente area. Shortly thereafter, light-brown and pale-orange beds of arkosic sandstone occurred in Los Pinos gravels of the Ojo Caliente area and were abundant in the Chama–El Rito to the south. This change at about 18 m.y. might be a slightly delayed effect of uplift of the Sangre de Cristo Range 1–2 m.y. earlier, which according to Manley (1981) was already the site of a volcanic edifice in the Questa area supplying debris to the lower Los Pinos Formation. Local advance of the Los Pinos gravels over the finer-grained Abiquiu system at Ojo Caliente occurred simultaneously with introduction of the arkosic Chama–El Rito above the tuffaceous Abiquiu farther south at this time. This implies uplift of an already present volcanic terrane in the Sangre de Cristos to the northeast near Questa and simultaneous uplift of a granitic and sedimentary terrane farther south along the range.

The Chama–El Rito south of Ojo Caliente accumulated in the more central part of the Española Basin as it was simultaneously receiving arkosic sand and andesitic to rhyolitic gravel from the two source areas. The south- and southwest-flowing streams were probably shallow and wide, forming a braided distributary(?) system, as indicated by the lateral extent of many gravel layers and the scarcity of cut-and-fill structures more than 0.6 m deep. Sparse lenticular sandstone layers at various levels in the conglomerates may represent small sandbars between the migrating channels. Ripple marks and fluvial dunes are sparse and upper flow-regime bedforms absent. A slight southwesterly decrease in average grain size and in the number of cut-and-fill channels also indicate a slight southwesterly decrease in stream turbulence and competence.

The numerous fine-grained sandstones of the Chama–El Rito probably were deposited on broad, flat areas between the distributary channels (interfluvial plains), but depositional mechanisms are uncertain. The sandstone layers may have been deposited by short-lived surges of sand-laden water overflowing the channels, forming sheet floods across the interfluvial plains. In support of this interpretation, the sandstones are finer-grained and more evenly bedded than the conglomerate layers and have a laterally gradational relationship with some of the conglomerates. An eolian origin for some of the fine-grained sandstones and siltstones is supported by good sorting, a lack of upward-fining texture, and a massive appearance similar to loess. Therefore, paleoslopes south of Ojo Caliente gradually flattened into the Española Basin during the mid-Miocene, as indicated by lacustrine limestones in the south, convergent paleostream trends, and finer grain size.

ACKNOWLEDGMENTS

The above text is modified from part of my Ph.D. dissertation done at the University of New Mexico under direction of L. A. Woodward and supported by USGS grants 14-08-0001-G-255 and 14-08-0001-G-348, New Mexico Bureau of Mines and Mineral Resources grant 900-60-84, and New Mexico Energy Research and Development grant 117-69-B. The paper was critically reviewed and improved by K. Manley and R. V. Ingersoll.

REFERENCES

Atwood, W. W., and Mater, K. F., 1932, Physiography and Quaternary geology of the San Juan Mountains, Colorado: U.S. Geological Survey, Professional Paper 166, 176 pp.

Baldridge, W. S., Damon, P. E. Shafiqullah, M., and Bridwell, R. J., 1980, Evolution of the Rio Grande rift, New Mexico; new potassium-argon ages: Earth and Planetary Science Letters, v. 51, pp. 309–321.

Barker, F., 1958, Precambrian and Tertiary geology of Las Tablas quadrangle, New Mexico: New Mexico Bureau of Mines and Mineral Resources, Bulletin 45, 104 pp.

Bingler, E. C., 1968, Geology and mineral resources of Rio Arriba County, New Mexico: New Mexico Bureau of Mines and Mineral Resources, Bulletin 91, 158 pp.

Butler, A. P., 1946, Tertiary and Quaternary geology of the Tusas–Tres Piedras area, New Mexico (abstract): Geological Society of America, Bulletin, v. 57, no. 12, p. 1183.

———, 1971, Tertiary volcanic stratigraphy of the eastern Tusas Mountains, southwest of the San Luis Valley, Colorado–New Mexico: New Mexico Geological Society, Guidebook 30, pp. 289–300.

Galusha, T., and Blick, J. C., 1971, Stratigraphy of the Santa Fe Group, New Mexico: American Museum of Natural History, Bulletin, v. 144, art. 1, 127 pp.

Lipman, P. W., 1975, Evolution of the Platoro caldera complex and related volcanic rocks, southeastern San Juan Mountains, Colorado: U.S. Geological Survey, Professional Paper 852, 128 pp.

———, and Mehnert, H. H., 1975, Late Cenozoic basaltic volcanism and development of the Rio Grande depression in the southern Rocky Mountains: Geological Society of America, Memoir 144, pp. 119–154.

———, and Steven, T. A., 1970, Reconnaissance geology and economic significance of the Platoro caldera, southeastern San Juan Mountains, Colorado: U.S. Geological Survey, Professional Paper 700-C, pp. C19–C29.

———, Doe, B. R., Hedge, C. E., and Steven, T. A., 1978, Petrologic evolution of the San Juan volcanic field, southwestern Colorado: Pb and Sr isotope evidence: Geological Society of America, Bulletin, v. 89, pp. 59–82.

Manley, K., 1981, Redefinition and description of the Los Pinos Formation of north-central New Mexico: Geological Society of America, Bulletin, v. 92, pp.984–989.

May, S. J., 1980, Neogene geology of the Ojo Caliente–Rio Chama area, Española Basin, New Mexico: Ph.D. thesis, University of New Mexico, Albuquerque, 204 pp.

Muehlberger, W. R., 1979, The Embudo fault between Pilar and Arroyo Hondo, New Mexico: an active intracontinental transform fault: New Mexico Geological Society, Guidebook 30, pp. 77–82.

Smith, C. T., Budding, A. J., and Pitrat, C. W., 1961, Geology of the southeastern part of the Chama Basin: New Mexico Bureau of Mines and Mineral Resources, Bulletin 75, 57 pp.

Smith, H. T. U., 1938, Tertiary geology of the Abiquiu quadrangle, Rio Arriba County, New Mexico: Journal of Geology, v. 46, pp. 933–965.

Steinpress, M. G., 1981, Neogene stratigraphy and structure of the Dixon area, Española Basin, north-central New Mexico: summary: Geological Society of America, Bulletin, pt. 1, v. 92, pp. 1023–1026.

Vazzana, M. E., 1980, Stratigraphy, sedimentary petrology and basin evolution of the Abiquiu Formation, north-central New Mexico: M.S. thesis, University of New Mexico, Albuquerque, 115 pp.

———, and Ingersoll, R. V., 1981, Stratigraphy, sedimentology, petrology and basin evolution of the Abiquiu Formation (Oligo-Miocene), north-central New Mexico: summary: Geological Society of America, Bulletin, pt. 1, v. 92, pp. 990–993.

Barranca was the station at the top of Barranca Hill. Stages for Taos and Ojo Caliente connected with the trains. Now only a stock pen and the abandoned grade (visible in the foreground) remain.

THE CHAMA–EL RITO MEMBER OF THE TESUQUE FORMATION, ESPAÑOLA BASIN, NEW MEXICO

LESLIE M. EKAS[1], RAYMOND V. INGERSOLL[1], W. SCOTT BALDRIDGE[2], and MUHAMMAD SHAFIQULLAH[3]
[1]Department of Earth and Space Sciences, University of California, Los Angeles, California 90024; [2]Earth and Space Sciences Division, MS J978, Los Alamos National Laboratory, Los Alamos, New Mexico 87545; [3]Laboratory of Isotope Geochemistry, Department of Geosciences, University of Arizona, Tucson, Arizona 85721

INTRODUCTION

The Chama–El Rito is the lowest member of the Tesuque Formation on the north and west sides of the Española Basin (Galusha and Blick, 1971; Manley, 1979; May, 1979; Tedford, 1981). The Tesuque Formation and the underlying Abiquiu Formation mark the earliest deposition of sediments in the Española Basin of the Rio Grande rift (Galusha and Blick, 1971). The Chama–El Rito Member is a thick sequence (440–460 m; May, 1980) of arkosic and volcaniclastic sandstones and mudstones with interbedded volcanic-rich conglomerates. It includes sparse airfall tuffs and limestones to the southeast and volcanic vents with associated debris-flow deposits (May, 1980). The sediments were deposited in alluvial channels on a broad plain at the distal edge of an alluvial fan. They are relatively undisturbed, showing faulting with minor offset and minor eastward tilting. In general, they are poorly lithified and eroded to a badlands topography. Since the sediments are relatively flat-lying, the best exposures are in areas of rapid erosion.

Hayden (1869) first applied the name "Santa Fe Marls" (now called the Santa Fe Group). Spiegel and Baldwin (1963) defined the Tesuque Formation in the southern part of the type area of the Santa Fe Group. Galusha and Blick (1971) further subdivided the Santa Fe Group into two formations (Tesuque and Chamita), and subdivided these formations into members. The Chama–El Rito and Ojo Caliente Members in the western part of the basin are laterally equivalent to the Nambé, Skull Ridge, and Pojoaque Members in the Tesuque Formation to the east. Kelley's (1978) map of the Española Basin delineates the extent of the Chama–El Rito Member. May (1980) mapped the Chama–El Rito and surrounding units in detail and described the lithology.

This paper summarizes stratigraphic, sedimentologic, and petrographic data from a recent study of the Chama–El Rito (Ekas, 1984), and presents new radiometric and chemical data on volcanic clasts of intermediate composition and interbedded basaltic-vent deposits. This information provides new constraints on the paleotectonic and paleoenvironmental development of the Española Basin in particular, and the northern Rio Grande rift in general.

REGIONAL SETTING

The southwest part of the Española Basin is covered by the Jemez volcanic complex; the basin is bordered by the Brazos uplift (Tusas Mountains) to the north, the Taos Plateau to the northeast, and the Sangre de Cristo Mountains to the east and southeast (Fig. 1). The Española Basin is one of a series of approximately northeast-trending structural and topographic depressions that lie along the Rio Grande valley from central Colorado to southern New Mexico (Chapin and Seager, 1975). This depression resulted from crustal extension that produced the Rio Grande rift (Chapin, 1979; Chapin and Seager, 1975; Lucas and Ingersoll, 1981; Riecker, 1979).

The Española Basin probably became a depression during Abiquiu deposition (Vazzana and Ingersoll, 1981). The Abiquiu and related units [Picuris Formation and Bishops Lodge Member of the Tesuque Formation (Manley, 1979)] occur stratigraphically below the Chama–El Rito Member in exposures on both sides of the Española Basin. The Chama–El Rito Member extends to the western edge of the Española Basin, but is not exposed along the east side except in the Dixon area (e.g., Steinpress, 1981). Deposition of the east-derived Tesuque members (Nambé, Skull Ridge, and Pojoaque) concurrently with the north-derived Chama–El Rito and their interfingering in the center of the Española Basin suggest that the Española Basin had become a down-dropped area by the beginning of Tesuque deposition.

AGE

Based on land-mammal fossils, Tedford (1981) assigned the Chama–El Rito Member middle Hemingfordian to early Clarendonian ages (corresponding approximately to 18–10 m.y. ago). This age is older than Galusha and Blick (1971) suggested, but more in agreement with Manley's (1979) suggestion of 20 to 10 m.y.

Our date on a basalt bomb (basanite, Tables 1–3) from a vent (Fig. 2) in the lower part of the middle Chama–El Rito (May, 1980; this guidebook) is 15.3 ± 0.4 m.y., in agreement with Tedford's (1981) age estimates. Manley and Mehnert's (1981) date on a basalt from the same location is 13.5 ± 2.8 m.y.

STRATIGRAPHY AND SEDIMENTOLOGY

The Chama–El Rito Member is composed primarily of tan arkosic sandstones and mudstones with interbedded gray, volcanic-rich gravel lenses (Fig. 3). The sandstones are thinly to thickly bedded, and the red mudstones form very thin beds. In most exposures the bedding features are difficult, if not impossible, to detect because the lithology and grain size are homogeneous and the rocks are poorly lithified. Where bedding features are evident, the sandstones and mudstones are laterally extensive and parallel-bedded, but individual layers are difficult to trace due to nondistinctive character of individual layers. The gray volcani-

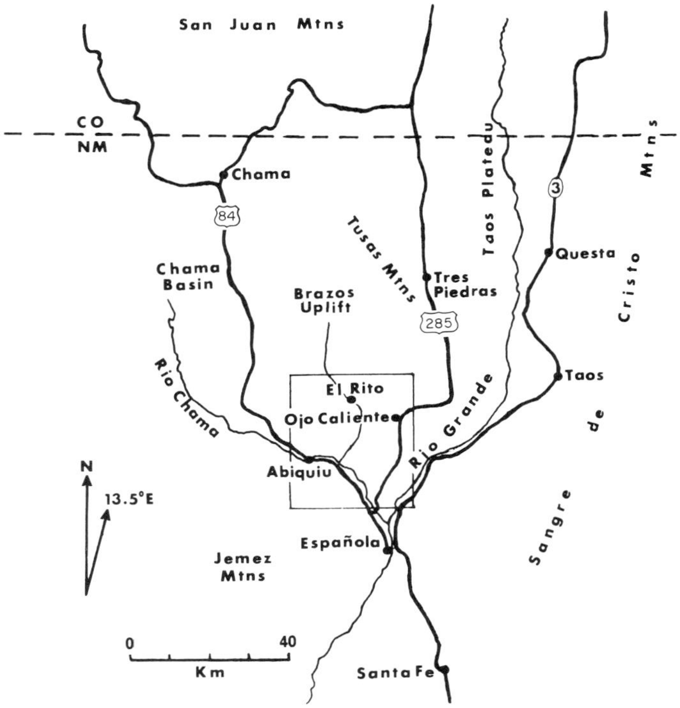

FIGURE 1. Location map of study area. Area of Figure 2 is indicated by rectangle.

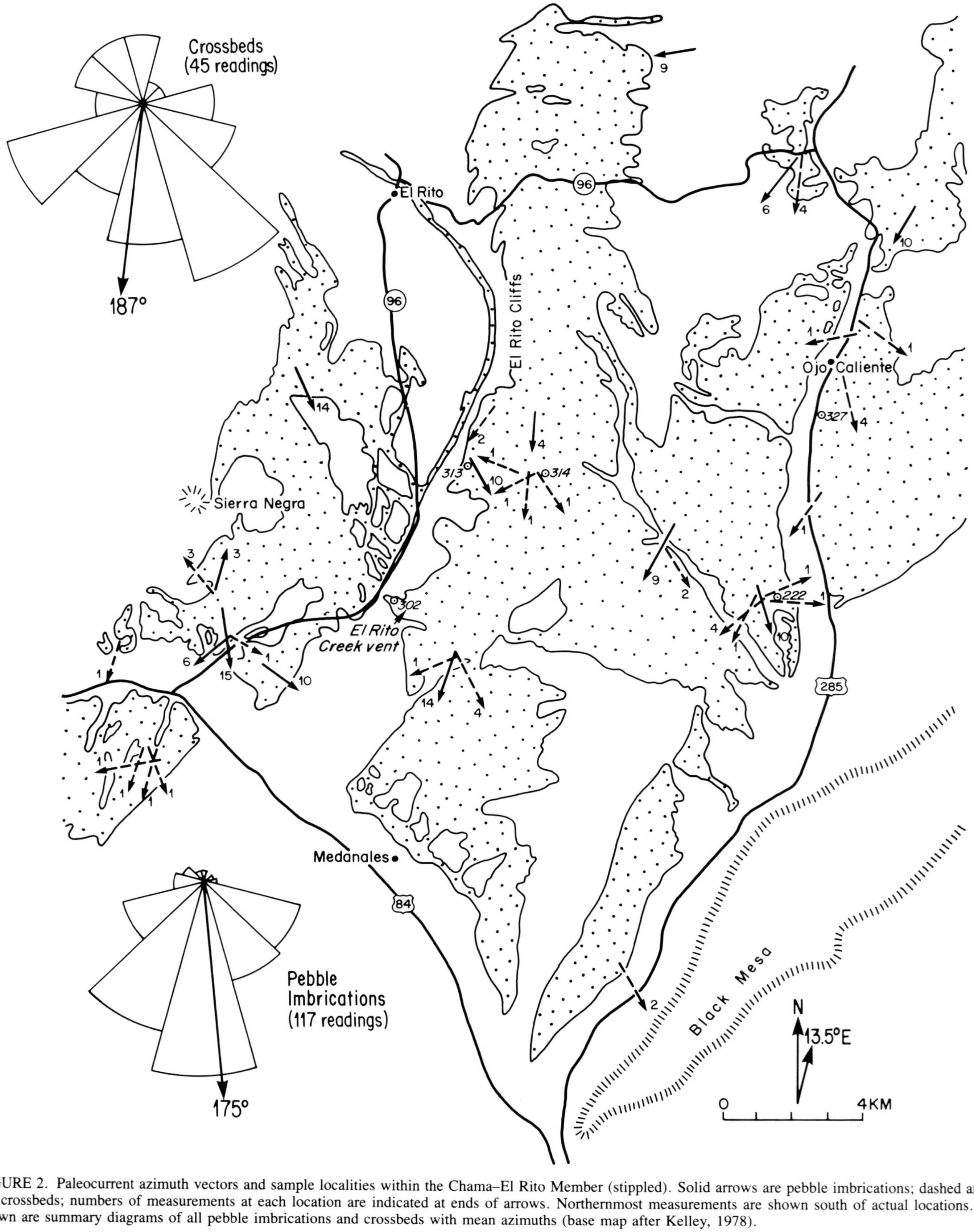

FIGURE 2. Paleocurrent azimuth vectors and sample localities within the Chama–El Rito Member (stippled). Solid arrows are pebble imbrications; dashed arrows are crossbeds; numbers of measurements at each location are indicated at ends of arrows. Northernmost measurements are shown south of actual locations. Also shown are summary diagrams of all pebble imbrications and crossbeds with mean azimuths (base map after Kelley, 1978).

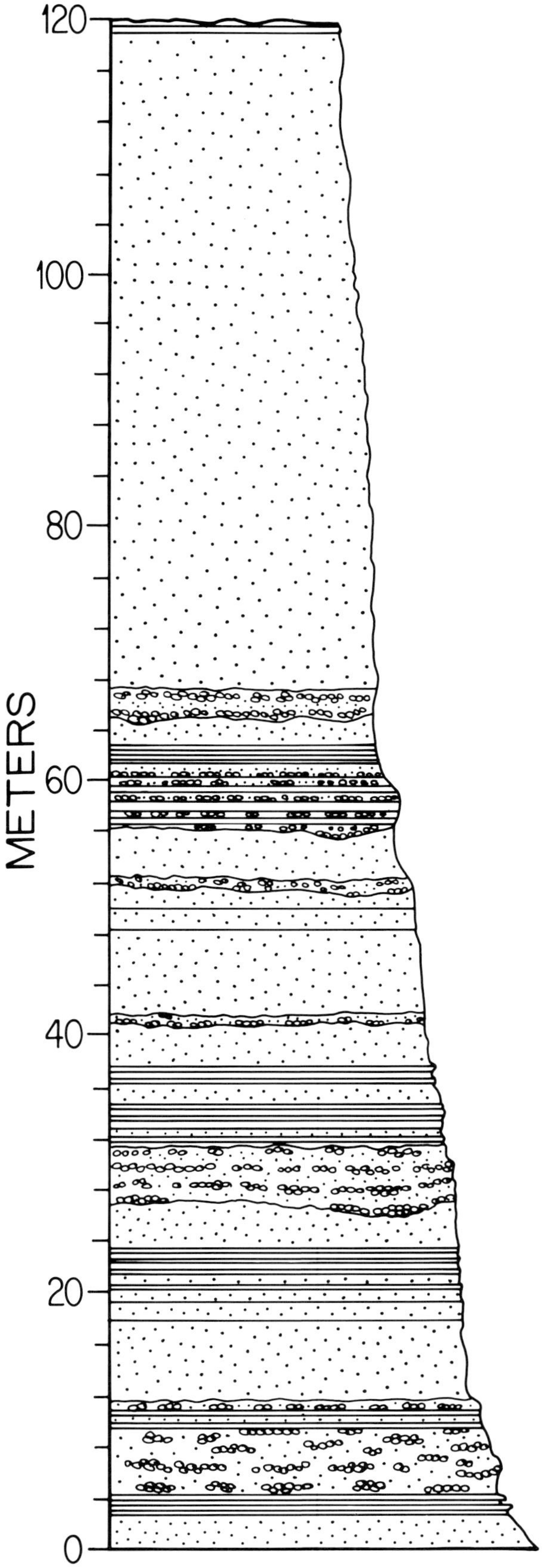

FIGURE 3. Typical section of Chama–El Rito Member: lines = mudstones, dots = sandstones, circles = conglomerates.

TABLE 1. Weight-percent compositions and Barth–Niggli cation norms (cation %) of clasts and basalt bomb from the Chama–El Rito. See Table 3 for sample descriptions and ages.

	222	302	313	314	327
SiO_2	60.15	43.45	58.89	66.46	64.30
TiO_2	0.69	1.92	0.79	0.70	0.69
Al_2O_3	15.22	13.31	18.66	14.37	14.66
Fe_2O_3	4.04	7.47	4.83	2.56	3.57
FeO	0.32	5.06	0.26	1.12	0.23
MnO	0.05	0.23	0.06	0.05	0.05
MgO	1.24	10.34	1.89	1.78	1.42
CaO	6.66	11.95	4.23	4.14	5.73
Na_2O	3.52	2.56	5.07	4.03	3.90
K_2O	2.98	0.72	4.46	3.25	3.38
P_2O_5	0.28	1.19	0.52	0.25	0.21
TOTAL	95.15	98.20	99.66	98.61	98.14
Mg-value †					
Q	15.74	-	2.28	20.06	17.11
or	18.67	4.34	26.17	19.56	20.47
ab	33.52	12.34	45.22	36.86	35.90
an	17.96	23.17	14.89	11.74	12.83
ne	-	6.67	-	-	-
wo	2.02	-	-	-	1.40
di	7.26	23.52	0.87	6.02	8.04
hy	-	-	4.75	2.00	-
ol	-	23.42	-	-	-
mt	-	1.27	-	1.28	-
ilm	0.61	2.73	0.49	0.99	0.45
hm	2.99	-	3.34	0.96	2.55
sp	0.62	-	0.90	-	0.81
ap	0.62	2.54	1.08	0.53	0.45

† Calculated based on Fe^{3+}/Fe^{2+} = 0.1

Analyses by x-ray fluorescence. FeO by spectrophotometric determination of ICl reaction (Bower, 1984).

TABLE 2. Trace-element compositions (PPM) of clasts and basalt bomb from the Chama–El Rito. See Table 3 for sample descriptions and ages.

	222	302	313	314	327	1σ error (rel. %)
V	43	182	61	34	33	13
Cr	19	188	18	25	22	4
Co	14	49	19	29	25	6
Ni	<75	257	<75	<75	<75	5
Zn	54	108	48	46	38	6
Rb	36	22	81	70	65	3
Sr	627	1166	901	592	584	5
Y	14	27	16	6	20	6
Zr	162	200	201	156	156	7
Ba	1149	1270	1360	977	987	10
La	50	63	41	36	27	10
Ce	76	170	135	63	83	13

TABLE 3. Descriptions, locations, and ages of clasts and basalt bomb from the Chama–El Rito. Constants used: $\lambda_\beta = 4.963 \times 10^{-10}\ yr^{-1}$, $\lambda_e = 0.581 \times 10^{-10}\ yr^{-1}$, $\lambda = 5.544 \times 10^{-10}\ yr^{-1}$, $^{40}K/K = 1.167 \times 10^{-4}$ atom/atom.

SAMPLE NUMBER	SAMPLE DESCRIPTION & LOCATION	K-ANALYSES: INDIVIDUAL % K	K-ANALYSES: MEAN	RADIOGENIC $^{40}Ar \times 10^{-12}$m/g: INDIVIDUAL ANALYSES	RADIOGENIC $^{40}Ar \times 10^{-12}$m/g: MEAN	% ATMOSPHERIC ^{40}AR: INDIVIDUAL ANALYSES	% ATMOSPHERIC ^{40}AR: MEAN	AGE IN MILLION YEARS
UAKA 83-185 (222)	Biotite in hornblende dacite. Phenocrysts of plagioclase, hornblende, quartz, and biotite. Microphenocrysts of magnetite and sphene. Much secondary carbonate. Clast from channel-lag deposit in Chama-El Rito Member, Tesuque Formation, Lyden Quad., Rio Arriba County, New Mexico Lat. 36°3.50'N, Long. 106°3.39'W	7.211 7.144 7.117 7.192 7.149 7.196	7.162	272.3 272.4 272.6 272.4 272.5	272.4	33.4 33.0 33.4 33.3 33.2	33.3	21.8±0.5
UAKA 83-186 (302)	Groundmass of basanite. Phenocrysts of olivine. Bomb from vent within Chama-El Rito Member, Tesuque Formation, Medanales Quad., Rio Arriba County, New Mexico Lat. 36°14.20'N, Long. 106°11.50'W	0.4663 0.4660 0.4571 0.4567 0.4596	0.4611	12.20 12.40 12.37 12.06 12.25	12.26	42.8 41.9 42.1 43.3 42.3	42.5	15.27±0.39
UAKA 83-184 (313)	Biotite in andesite. Phenocrysts of plagioclase, hornblende (replacing augite), and biotite. Magnetite microphenocrysts. Clast within alluvial deposit of Chama-El Rito Member, Tesuque Formation, El Rito Quad., Rio Arriba County, New Mexico Lat. 36°16.37'N, Long. 106°9.40'W	7.370 7.341 7.350 7.378 7.365	7.361	277.5 277.0 278.2 280.1 279.4 278.4 280.3 279.0 279.0 277.9 278.4 279.0	278.7	40.2 40.1 40.0 40.0 40.1 40.1 40.8 41.0 41.4 41.6 41.5 41.4 41.4	40.7	21.7±0.5
UAKA 83-187 (327)	Biotite in hornblende dacite. Phenocrysts of plagioclase, hornblende, quartz, and biotite. Microphenocrysts of magnetite and sphene. Clast from tongue of Los Pinos Formation gravels in the Chama-El Rito Member, Tesuque Formation, El Rito Quad., Rio Arriba County, New Mexico Lat. 36°16.34'N, Long. 106°8.48'W	6.860 6.852 6.816 6.888 6.866	6.856	269.5 268.4 268.3 268.3 266.8 268.3	268.3	37.3 37.4 38.8 39.2 39.8 38.7	38.5	22.6±0.5
UAKA 83-187 (327)	Biotite in hornblende dacite. Phenocrysts of plagioclase, hornblende, biotite, and quartz. Microphenocrysts of magnetite and sphene. Much secondary carbonate. Clast in channel-lag deposit of Los Pinos Formation that is interbedded with Chama-El Rito Member, Tesuque Formation, on the east Ojo Caliente Quad., Rio Arriba County, New Mexico Lat. 36°17.30'N, Long. 106°2.58'W	7.253 7.295 7.316 7.308 7.342 7.356 7.383 7.330	7.323	290.0 287.8 288.6	289.2	40.2 40.5 40.5	40.4	22.6±0.5

clastic units contain broad lenses of pebbly sand to conglomerate (1 cm to 7 m thick). These lenses represent alluvial channels because they usually have erosional bases, channel-lag deposits, crossbeds, overall upward-fining grain size, graded bedding within beds, and minor lateral extent (e.g., Bull, 1972). The sandstones and mudstones show few sedimentary structures; they represent the waning stages of flooding and overbank deposition in distributary systems. Some sands between channels may be eolian. At the top of the Chama–El Rito in the southeast and in the middle of the area north of Sierra Negra, some of the sand is clearly eolian. These sands are well sorted and compositionally homogeneous, and have large, wedge-shaped sets of crossbeds similar to the overlying Ojo Caliente sandstones.

In general, the amount of coarse-grained material in the Chama–El Rito decreases to the south. In addition to eolian sands, a few limestone beds occur in the southeastern region. These characteristics corroborate a northern source for the alluvial complex, with more distal parts in the south.

At the El Rito Cliffs and west of Ojo Caliente (Fig. 2), May (1980) mapped tongues of the Cordito Member of the Los Pinos Formation (Manley, 1981) interfingering with the Chama–El Rito. The Los Pinos Formation contains mainly gravel, whereas the majority of the Chama–El Rito is sandstone. In addition, the Los Pinos gravels in this region contain more Precambrian-derived clasts than does the Chama–El Rito (May, 1980). However, the Chama–El Rito also shows a slight enrichment in Precambrian-derived debris, such as quartzites and metasedimentary and metavolcanic clasts where it interfingers with the Los Pinos. In general, however, the petrology and provenance of the Cordito Member of the Los Pinos and the Chama–El Rito are indistinguishable (see below). The Cordito Member is interpreted to represent the more proximal part of the Chama–El Rito fan complex.

Around the vent near El Rito Creek (Fig. 2) the rocks are dominantly maroon, tuffaceous material mixed with abundant basaltic spatter and bombs interbedded with more typical Chama–El Rito sandstone (May, 1980). To the southwest of the vent, on the southeast side of Sierra Negra, is a fault-bounded outcrop of similar lithologic facies. Interbedded with the arkosic sandstones are red to maroon, matrix-supported breccias. The debris is dominantly basaltic and tends to be very poorly sorted. These characteristics suggest that these units are debris-flow deposits (e.g., Johnson, 1970) that may have been derived from the vent or another buried or eroded source. Basaltic cobbles south and southwest of the vent indicate that regional paleocurrent directions are towards the south.

PALEOCURRENTS

Dip azimuths of crossbeds and pebble imbrications were measured in the Chama–El Rito Member to resolve paleocurrent directions. One hundred sixty-two directions were measured throughout the section over the entire area. The regional dip was removed using a stereonet and the original paleocurrent azimuths were resolved. The mean paleocurrent azimuth of each area and the number of readings are plotted in Figure 2.

The vector mean for all pebble-imbrication azimuths (Fig. 2) is almost directly south (175°; standard deviation of 52). Two problems in measuring pebble imbrications were encountered. First, there were few cobble layers where measurement could be made. About 80–90% of the Chama–El Rito is sand-sized or finer. Second, most of the clasts are volcanic and tend to be ellipsoidal, thereby decreasing the tendency for imbrication relative to flat clasts.

The summary diagram for the crossbed azimuths (Fig. 2) shows more variation than that for the pebble imbrications. The mean azimuth is 187° (standard deviation of 84). Crossbed directions are more variable probably because they represent overbank deposits as well as main channel directions.

All the paleocurrent data indicate a southerly direction of flow. This is consistent with the general decrease in grain size to the south, the south and southwestward transportation of basaltic debris from the vent near El Rito Creek, and the presence of limestones in the southeast.

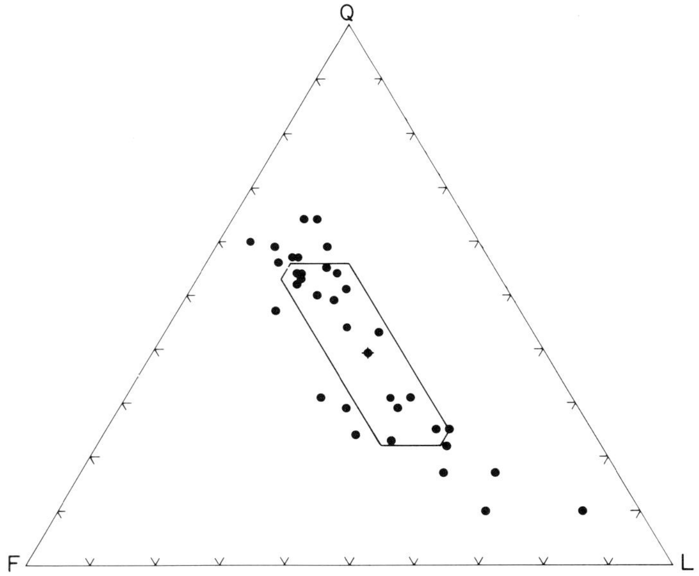

FIGURE 4. Quartz (Q)–feldspar (F)–lithics (L) triangle for sand petrology of the Chama–El Rito Member. Circle with cross and hexagon indicate mean and standard deviation, respectively, for all samples (see Ingersoll, 1978).

PETROLOGY AND PROVENANCE

The provenance of the Chama–El Rito Member was determined by combining the following information: paleocurrent data; sand-petrology data; pebble–cobble compositions; and geochemistry and age dates of volcanic clasts. The methods outlined by Dickinson (1970), Ingersoll (1978), and Ingersoll and others (1984) were used to study sand petrology. Since the sandstones tend to be poorly lithified or completely unlithified, the samples were impregnated; then they were stained for potassium feldspar and plagioclase. A statistical point count was made of each section using a petrographic microscope, and the results were recalculated as illustrated in Figures 4–7. Ekas (1984) discusses the methods and results in detail.

The sand composition can be divided into two groups, although a continuum of compositions probably exists. The tan, medium- to fine-grained, moderately well-sorted sands are arkosic, with quartz com-

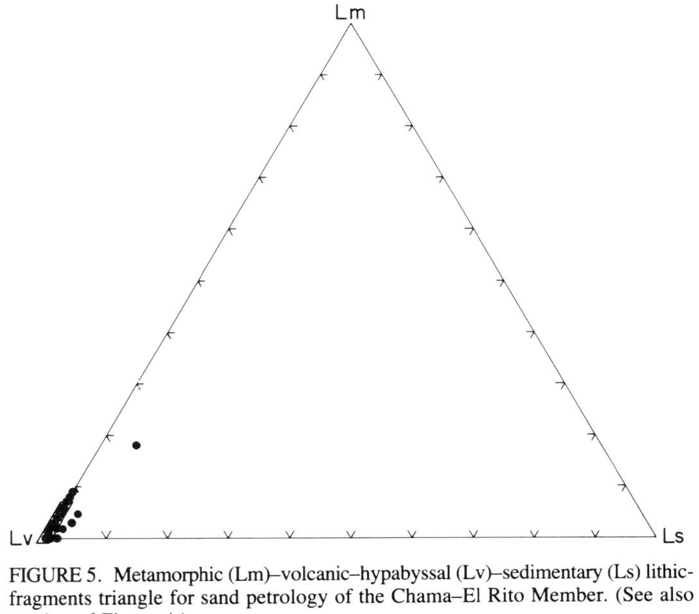

FIGURE 5. Metamorphic (Lm)–volcanic–hypabyssal (Lv)–sedimentary (Ls) lithic-fragments triangle for sand petrology of the Chama–El Rito Member. (See also caption of Figure 4.)

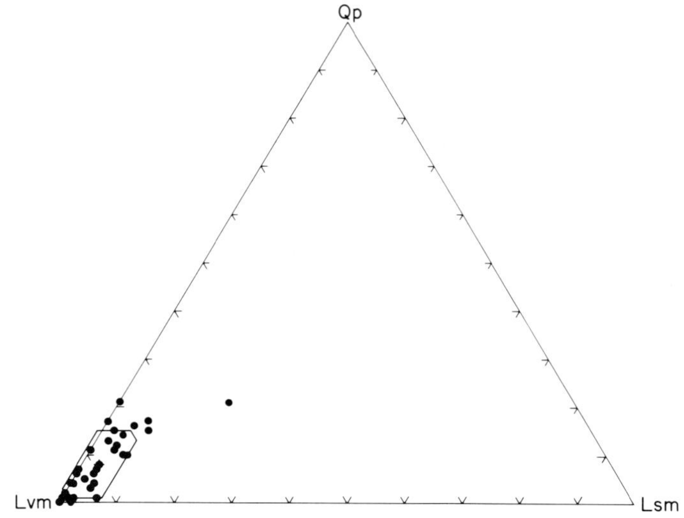

FIGURE 6. Polycrystalline-quartz (Qp)–volcanic–metavolcanic (Lvm)–sedimentary–metasedimentary (Lsm) lithic-fragments triangle for sand petrology of the Chama–El Rito Member. (See also caption to Figure 4.)

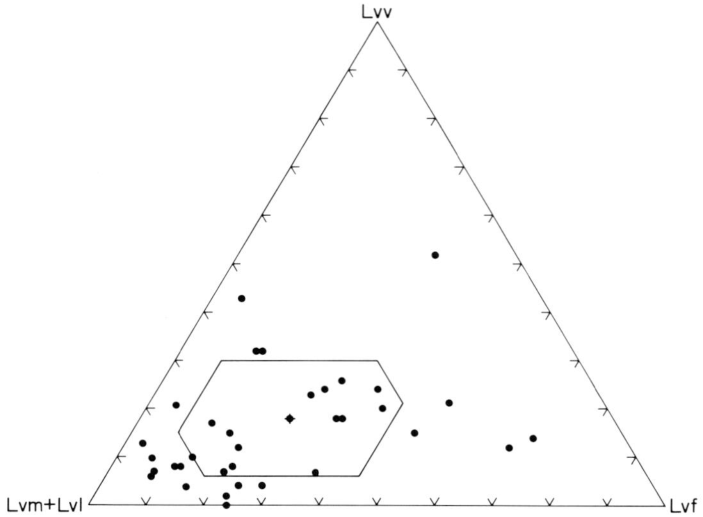

FIGURE 7. Vitric (Lvv)–microlitic-lathwork (Lvm + Lvl)–felsitic (Lvf) volcanic-lithics triangle for sand petrology of the Chama–El Rito Member. (See also caption of Figure 4.)

prising 47–66%, feldspars 23–37%, and lithic fragments 9–24% of the QFL populations. The gray, medium- to coarse-grained, poorly sorted sands are volcaniclastic, with 30–81% lithic clasts, 10–31% quartz, and 9–39% feldspars (Fig. 4). Volcaniclastic detritus is dominant in the lithic populations (Figs. 5, 6). The volcanic lithic fragments have been subdivided based on their textural characteristics (e.g., Dickinson, 1970) (Fig. 7). There are generally two groups of volcanic lithic types: one group is rich in microlitic and lathwork textures (generally intermediate compositions), corresponding to the gray volcaniclastic sands, whereas the other group is a combination of microlitic, lathwork, and felsitic textures, representing a wide range of compositions and corresponding to the arkosic sands. Petrology of samples in the Los Piños Formation north of Ojo Caliente shows the same compositions.

Compositions of the conglomerates were determined by making point counts on a 100-point grid at 31 stations in the area and throughout the section. Volcanics are the most common lithic types (Figs. 8, 9). There is an enrichment of basaltic clasts south of the vent. In general though, dacites and andesites are dominant, and rhyolitic tuffs and other silicic volcanics of varying compositions form significant components. Quartzites (Q) form a significant minority of clasts, but are less than 23% of each sample.

The petrology and lithology of the Chama–El Rito suggest that a primary sediment sources was a volcanic terrane. Paleocurrent information supports northern sources. Possible sources of volcanic rocks to the north are the San Juan field (Colorado), the Latir volcanic field (Questa area), and intermediate rocks of the Taos Plateau. In order to determine the source terrane, we analyzed and dated four representative coarse clasts from various locations (Fig. 2). (Note that two of the clasts, 314 and 327, come from interbedded "Los Pinos" gravels, thus confirming a common source for the Los Pinos and the Chama–El Rito.) The ages are very similar, ranging from 21.7 to 22.6 ± 0.5 m.y. (Table 3). Three of the samples are hornblende dacites and the fourth is an andesite. The most probable source for these volcanic rocks is beneath the Taos Plateau. Lipman and Mehnert (1979, table 1) describe "early rift volcanics" of the Brushy Mountain area as "compositionally diverse intermediate to silicic" volcanic rocks that were erupted between 25

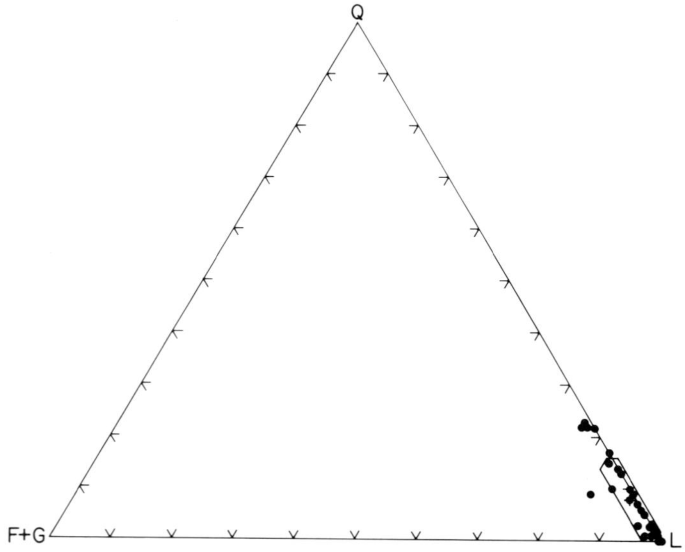

FIGURE 8. Quartz–quartzite (Q)–felspar–granitic (F + G)–other lithics (L) triangle for conglomerate counts of the Chama–El Rito Member. (See also caption of Figure 4.)

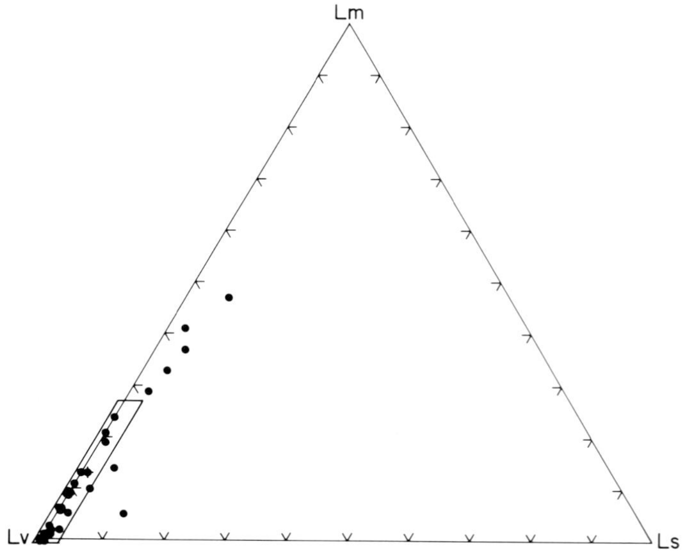

FIGURE 9. Metamorphic (Lm)–volcanic-hypabyssal (Lv)–sedimentary (Ls) lithics triangle for conglomerate counts of the Chama–El Rito Member. (See also caption of Figure 4.)

and 22 m.y. ago. These dates and compositions coincide well with those of the four analyzed clasts [67L-69A,74L-214 (Lipman and Mehnert, 1979, table 3) match well the geochemistry of the four clasts (Table 1)]. Due to the amount of alteration and secondary mineralization in the clasts from the Chama–El Rito, a closer comparison is not possible. Volcanic rocks of similar composition in the San Juan volcanic field are too old to have been the source of the clasts (e.g., Lipman and others, 1970). The Latir volcanic field near Questa may also be too old. Lipman (1983) reported volcanic ages of 35–26 m.y., although intrusive rocks are younger (P. W. Lipman, oral comm. 1984); thus, volcanic rocks young enough to have been sources for the clasts could have been present in the Latir field and subsequently been eroded. In this case, clasts in the Chama–El Rito may be the only known record of this field (which may have been continuous with the early-rift volcanics of the Taos Plateau area).

The main source of the arkosic component of the petrology is probably the Sangre de Cristo Mountains. The quartz and feldspar in the more arkosic samples are fine- to medium-grained and are better sorted than the more volcaniclastic samples, thus signifying transport from a more distal source, possibly by a combination of alluvial and eolian processes. Furthermore, the arkosic samples have higher proportions of K feldspar (lower P/F ratio). In contrast, plagioclase is the dominant feldspar in the volcaniclastic sands. This enrichment in K feldspar (predominantly microcline) and quartz along with their scarcity in the volcanic petrology suggest ultimate granitic, metavolcanic, and quartzitic sources (e.g., Precambrian rocks of the Sangre de Cristo and Tusas Mountains) for the arkosic sands.

CONCLUSIONS

The Chama–El Rito Member of the Tesuque Formation is a thick sequence of arkosic sandstones and mudstones with interbedded volcaniclastic pebbly sandstones and conglomerates. These sediments were deposited on the distal edges of an alluvial fan whose source was to the north. A vent in the middle of the section has an age of 15.3 m.y, in agreement with age estimates of 18–10 m.y. (Tedford, 1981) and 18–12 m.y. (May, this guidebook).

The sand and gravel petrology indicates two source terranes: a volcanic source, and one rich in quartz and feldspar. The volcanic clasts have dominantly intermediate compositions (dacite to andesite). The arkosic component of the sandstones was derived from granitic, metavolcanic, and quartzitic sources (possibly Sangre de Cristo and Tusas Mountains), although eolian transport from distant sources cannot be discounted. Four andesitic to dacitic clasts have K–Ar ages of approximately 22 m.y. The combined paleocurrent, petrologic, and chronologic data from the Chama–El Rito imply derivation from early rift volcanics of the Taos Plateau, and possibly the Questa area, and secondary source areas of Precambrian rocks in the Sangre de Cristo and Tusas Mountains.

ACKNOWLEDGMENTS

The Institute of Geophysics and Planetary Physics at Los Alamos National Laboratory provided financial support for the research of Ekas and Ingersoll. Support for Baldridge was provided by the Office of Basic Energy Sciences of the U.S. Department of Energy. We thank N. Bower for aid with chemical analysis. The geochronologic work of Shafiqullah was supported by NSF Grant EAR-8206432. K. Manley and S. J. May critically reviewed the manuscript.

REFERENCES

Bower, N. W., 1984, Simple spectrophotometric determination of ferrous iron in twelve French geochemical reference standards: Geostandards Newsletter, v. 8, pp. 61–62.

Bull, W. B., 1972, Recognition of alluvial-fan deposits in the stratigraphic record: Society of Economic Paleontologists and Mineralogists, Special Publication 16, pp. 63–83.

Chapin, C. E., 1979, Evolution of the Rio Grande rift—a summary; *in* Riecker, R. E. (ed.), Rio Grande rift: tectonics and magmatism: American Geophysical Union, Washington, D.C., pp. 1–5.

———, and Seager, W. R., 1975, Evolution of the Rio Grande rift in the Socorro and Las Cruces areas: New Mexico Geological Society, Guidebook 26, pp. 297–321.

Dickinson, W. R., 1970, Interpreting detrital modes of graywacke and arkose: Journal of Sedimentary Petrology, v. 40, pp. 695–707.

Ekas, L. M., 1984, The Chama–El Rito Member of the Tesuque Formation, Española Basin, north-central New Mexico, M.S. thesis, University of California at Los Angeles, in prep.

Galusha, T., and Blick, J. C., 1971, Stratigraphy of the Santa Fe Group, New Mexico: American Museum of Natural History, Bulletin, v. 144, art. 1, pp. 1–127.

Hayden, F. V., 1869, Preliminary field report of the U.S. Geological Survey of Colorado and New Mexico: U.S. Geological Survey, Third Annual Report, 155 pp.

Ingersoll, R. V., 1978, Petrofacies and petrologic evolution of the Late Cretaceous fore-arc basin, northern and central California: Journal of Geology, v. 86, pp. 335–352.

———, Bullard, T. F., Ford, R. L., and others, 1984, The effect of grain size on detrital modes: a test of the Gazzi–Dickinson point-counting method: Journal of Sedimentary Petrology, v. 54, pp. 103–116.

Johnson, A. M., 1970, Physical processes in geology: Freeman Cooper and Company, San Francisco, 577 pp.

Kelley, V. C., 1978, Geology of Española Basin, New Mexico: New Mexico Bureau of Mines and Mineral Resources, Geologic Map 48.

Lipman, P. W., 1983, The Miocene Questa caldera, northern New Mexico: relation to batholith emplacement and associated molybdenum mineralization; *in* The genesis of Rocky Mountain ore deposits: changes with time and tectonics: Denver Region Exploration Geologists Symposium, Proceedings, pp. 133–148.

———, and Mehnert, H. H., 1979, The Taos Plateau volcanic field, northern Rio Grande rift, New Mexico; *in* Riecker, R. E. (ed.), Rio Grande rift: tectonics and magmatism: American Geophysical Union, Washington, D.C., pp. 289–311.

———, Steven, T. A., and Mehnert, H. H., 1970, Volcanic history of the San Juan Mountains, Colorado, as indicated by potassium–argon dating: Geological Society of America, Bulletin, v. 81, pp. 2329–2352.

Lucas, S. G., and Ingersoll, R. V., 1981, Cenozoic continental deposits of New Mexico: an overview: Geological Society of America, Bulletin, v. 92, pt. I, pp. 917–932, pt. II, pp. 1807–1981.

Manley, K., 1979, Stratigraphy and structure of the Española Basin, Rio Grande rift, New Mexico; *in* Riecker, R. E. (ed.), Rio Grande rift: tectonics and magmatism: American Geophysical Union, Washington, D.C., pp. 71–86.

———, 1981, Redefinition and description of the Los Pinos Formation of north-central New Mexico: Geological Society of America, Bulletin, v. 92, pt. I, pp. 984–989.

———, and Mehnert, H. H., 1981, New K–Ar ages for Miocene and Pliocene volcanic rocks in the northwestern Española Basin and their relationships to the history of the Rio Grande rift: Isochron/West 30, pp. 5–8.

May, S. J., 1979, Neogene stratigraphy and structure of the Ojo Caliente–Rio Chama area, Española Basin, New Mexico: New Mexico Geological Society, Guidebook 30, pp. 83–88.

———, 1980, Neogene geology of the Ojo Caliente–Rio Chama area, Española Basin, New Mexico: Ph.D. thesis, University of New Mexico, Albuquerque, 204 pp.

Riecker, R. E. (ed.), 1979, Rio Grande rift: tectonics and magmatism: American Geophysical Union, Washington, D.C., 438 pp.

Spiegel, Z., and Baldwin, B., 1963, Geology and water resources of the Santa Fe area, New Mexico: U.S. Geological Survey, Water-supply Paper 1525, 258 pp.

Steinpress, M. G., 1981, Neogene stratigraphy and structure of the Dixon area, Española Basin, north-central New Mexico: Geological Society of America, Bulletin, v. 92, pt. I, pp. 1023–1026, pt. II, pp. 2553–2671.

Tedford, R. H., 1981, Mammalian biochronology of the late Cenozoic basins of New Mexico: Geological Society of America, Bulletin, v. 92, pt. I, pp. 1008–1022.

Vazzana, M. E., and Ingersoll, R. V., 1981, Stratigraphy, sedimentology, petrology and basin evolution of the Abiquiu Formation (Oligo-Miocene), north-central New Mexico: Geological Society of America, Bulletin, v. 92, pt. I, pp. 990–992, pt. II, pp. 2401–2483.

Taos Pueblo, ca 1920. Photo by Burt Harwood.

GEOLOGY AND STRUCTURE ALONG PART OF THE NORTHEAST JEMEZ MOUNTAINS, NEW MEXICO

DAVID P. DETHIER and BLAKE A. MARTIN
Department of Geology, Williams College, Williamstown, Massachusetts 01267

INTRODUCTION

The Miocene sedimentary fill and interlayered volcanic deposits of the central Rio Grande rift record tectonic and volcanic events in surrounding source areas with considerable clarity. Deformation of the fill and changes in sedimentary textures, lithologies, and paleocurrent directions give the most complete record of late Cenozoic tectonic evolution within the rift.

The central Rio Grande rift of Chapin (1979) initially evolved as a structural sag along Laramide structures in middle Oligocene time, after most arc volcanism in nearby areas had ceased (Baldridge and others, 1980). There is no evidence that the early rift was bounded by major faults (Baltz, 1978), but adjacent mountain areas provided sediment to the internally drained basin. Until at least mid-Miocene time, the western margin of the Española Basin was defined by faults along a part of the Nacimiento uplift presumably now buried by Jemez volcanic rocks (Golombek and others, 1983). Minor faulting marked the eastern margin along the flank of the Sangre de Cristo Range.

Basin-filling sediment of late Oligocene to mid-Miocene age is mainly volcaniclastic debris and arkosic sandstone deposited on alluvial fans and exposed on the margins of the Española Basin as the Abiquiu, Los Pinos, and Picuris Formations (Manley, 1979). By about 11 m.y. ago, the western margin of the Española Basin had shifted eastward, volcanic activity was underway in the Jemez Mountains, and faulting disrupted the Oligocene and Miocene units. The Española Basin tilted westward in mid- to late Miocene time along faults presumably now covered by Jemez volcanic rocks (Golombek and others, 1983), and in latest Miocene or early Pliocene the Velarde graben formed in the central part of the basin (Manley, 1979). The graben is thought to be the most active part of the Española Basin at present and is bounded by the Pajarito fault on the west, an unnamed NE-trending fault on the north, and an ill-defined set of north-trending faults located east of the Rio Grande (Manley, 1979; Golombek and others, 1983).

Inferences about the nature and timing of the mid-Miocene tectonic event and subsequent development of the Velarde graben are based primarily on the stratigraphy and structure of the Santa Fe Group. Decades of paleontologic and stratigraphic studies on the Santa Fe sediment were summarized in a monumental work by Galusha and Blick (1971), who divided the sediment into the Tesuque and Chamita Formations. The Tesuque Formation was subdivided into the following members (ascending): Nambé, Skull Ridge, Pojoaque, Chama–El Rito, and Ojo Caliente Sandstone. With the exception of the Ojo Caliente, an eolian unit, all other members of the Tesuque are considered to be alluvial-fan deposits. Studies by May (1980, this guidebook) on the Chama–El Rito, by Steinpress (1981) on the Santa Fe near Dixon, and by Manley (1979; 1981) on other parts of the Española Basin have contributed greatly to our understanding of mid-Miocene stratigraphy. However, detailed stratigraphic information and paleocurrent measurements are sparse in many key areas of the basin, particularly in critical areas near the Jemez Mountains. Smith and others (1970), for instance, did not subdivide the Santa Fe Group, and Kelley (1978) mapped the area east of the NE Jemez Mountains incorrectly as Ojo Caliente Sandstone. We report results of a detailed study on the Santa Fe Group exposed in a narrow horst at the east edge of the northeast Jemez Mountains. The upthrown block is exposed a few kilometers north of the northwest corner of the Velarde graben, near the intersection of the Jemez lineament (Chapin, 1979) and the Parajito fault. We tentatively correlate the Santa Fe exposed in the horst and nearby areas with the upper part of the Chama–El Rito Member of the Tesuque Formation.

SETTING

Location and Topography

We studied an area some 20 km northwest of Española, New Mexico (Fig. 1), where the Santa Fe Group disappears beneath upper Miocene volcanic rocks. Exposures in the field area are relatively good because the Rio del Oso, Arroyo del Palacio, and smaller tributary arroyos that drain the NE Jemez Mountains have dissected the poorly consolidated sediment into a series of gravel-capped mesas. Local relief is as much as 600 m and topography is rugged.

Geology

Santa Fe Group

Correlation and dating of the Santa Fe Group has been hampered by the general similarity of most lithologic units, lack of marker beds, time-transgressive contacts, and disagreements about nomenclature. We follow the usage of Manley (1979, 1981). Manley (1979) and May (1980) interpreted Santa Fe sequences (Fig. 2) in the northern Española Basin as distal-fan deposits constructed generally to the south–southwest. May (1980) measured a composite section (up to 550 m) of the Chama–El Rito Member 20 to 30 km northeast of the study area. He described arkosic sandstone cut by channels filled with volcanic gravel and at least one maar and associated tuffaceous deposits. May (1980, this guidebook) suggested that much of the Chama–El Rito Member is the distal equivalent of a volcanic-fan complex (the Cordito Member of the Los Pinos Formation) derived from volcanic rocks in northern New Mexico and southern Colorado. Ekas and others (this guidebook) dated a basalt flow interlayered with the middle Chama–El Rito at 15.3 ± 0.4 m.y. Manley and Mehnert (1981) showed that the upper part of the unit near Chili must be older than the ~10-m.y.-old flows that overlie the Ojo Caliente Sandstone. Abiquiu Formation (Vazzana and Ingersoll, 1981) underlies the Chama–El Rito 30 km north of the study area and its upper surface is dated at about 17 m.y.B.P. The Chama–El Rito is thus bracketed between 17 and 10 m.y. ago near the study area.

The fluvial Chama–El Rito sandstone grades up into the eolian Ojo Caliente Sandstone in a zone 30 to 50 m thick (May, 1980). The Ojo Caliente is a massive to planar-crossbedded, medium-grained sandstone with sparse discontinuous interbeds of silt and altered tephra, deposited by winds blowing from the southwest (Dethier and Manley, in press). The dune deposits are at least 200 m thick in their principal outcrop area. The Ojo Caliente is younger than the Chama–El Rito and is overlain by basal Chamita Formation, which is probably about 10 m.y. old in the Chili area. The contact between the units appears gradational in some areas, and is clearly unconformable in others.

The Chamita Formation is the principal fill in the Velarde graben; it was deposited between 10 and about 5 m.y. ago in its type area (Manley, 1979). Medium- to fine-grained, massive sandstone, gravelly channel deposits, silty sand, and lenses of altered tephra are typical of this unit, which was deposited on alluvial fans from northerly and northeasterly sources (Manley, 1979; Dethier and Manley, in press). In the vicinity of the Chili quadrangle the Chamita is slightly coarser-grained than the Chama–El Rito, contains more clasts from eastern source areas and more altered tephra; it is difficult to distinguish sandstones from the two units. Chamita stratigraphy is critical to establishing the amount of throw on the Velarde fault, which appears to be an extension of an unnamed NE-trending fault in the Chili quadrangle.

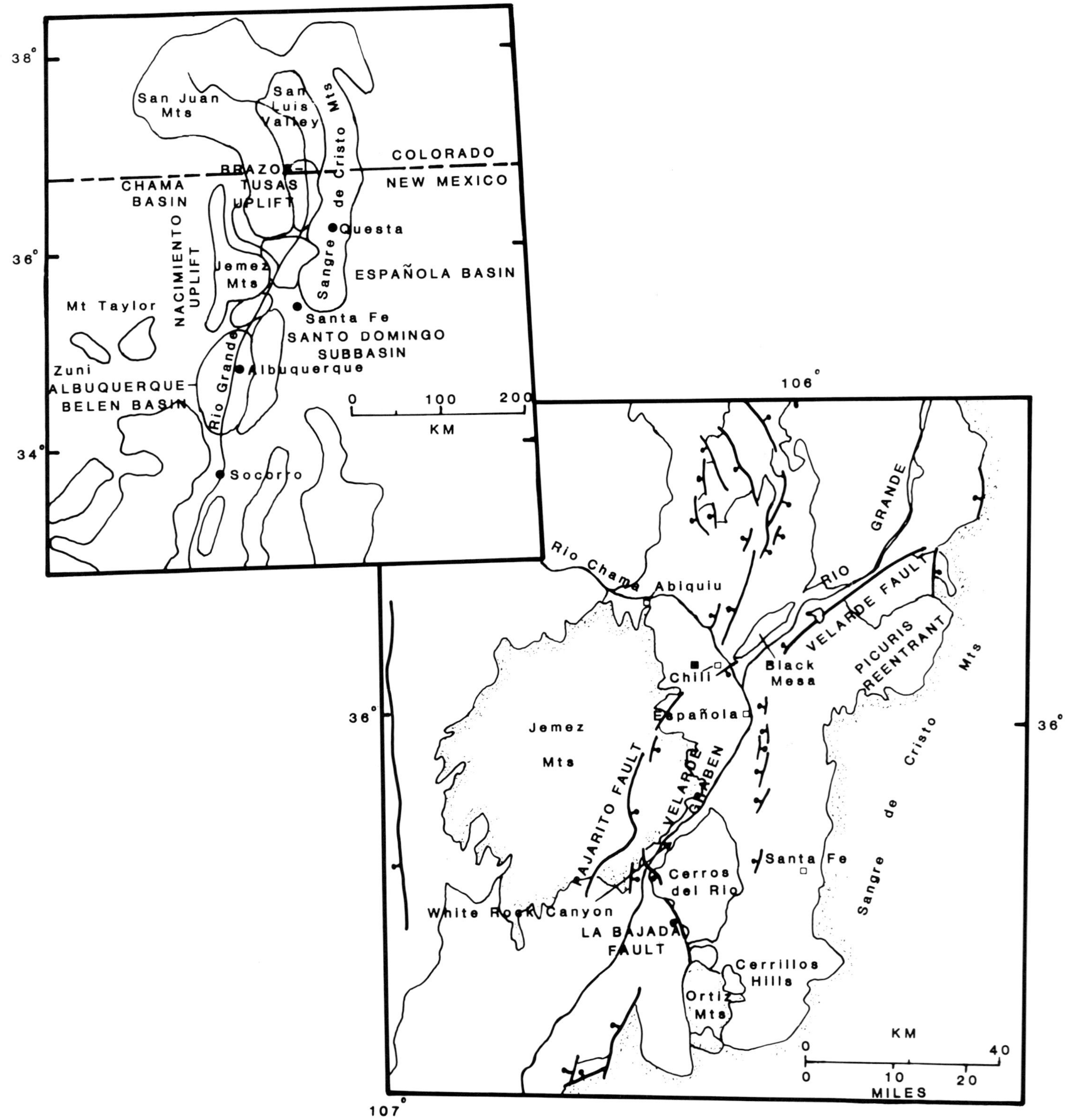

FIGURE 1. Map showing location of the Rio Grande rift and relationship of the study area to geographic features of the Española Basin (after Manley, 1979). Study area is indicated by a square near Chili.

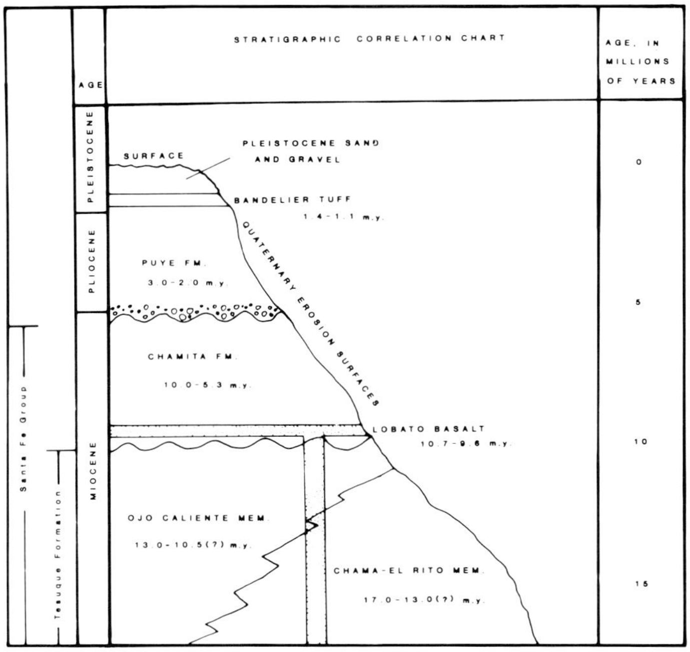

FIGURE 2. Stratigraphic names and relations for the Santa Fe Group and younger rocks, northern Española Basin, New Mexico (modified from Manley, 1979, 1981, and Baldridge and others, 1980).

Lobato Basalt

Olivine-basalt flows, dikes, sills, and maar deposits were mapped in contact with the Santa Fe Group in the Chili quadrangle (Dethier and Manley, in press). Most of the flows in this area probably originated at Clara Peak or Lobato Mesa, and proximity of the dikes to Lobato Mesa suggests that all these rocks should be included with the Lobato Basalt. Four published K–Ar dates (Baldridge and others, 1980; Manley and Mehnert, 1981) and three new dates reported below suggest that a significant episode of basaltic volcanism occurred in the NE Jemez Mountains between about 10.7 and 9.5 m.y. ago. Flows are interlayered in the basal 50 m of the Chamita Formation. With one exception, flows and dikes of Lobato age are not exposed south of the NE-trending fault that marks the northern edge of the Velarde graben.

Pliocene and Quaternary Deposits

The upper surface of the Chamita Formation is unconformably overlain by a resistant cobble gravel that marks the base of the Puye Formation, which is dated at about 3 m.y. near the Rio Grande; the upper part of this Jemez-derived volcanic fanglomerate is about 2 m.y. old (Manley, 1979). The lower and upper Bandelier Tuff, dated at 1.4 and 1.1 m.y., respectively, overlie the Puye Formation along the east flank of the Jemez Mountains, but only a few isolated remnants of lower Bandelier airfall and tuff are preserved near Lobato Mesa. Basinwide erosion commenced sometime after 1.1 m.y.B.P. and base level lowered to within 100 m of present levels by 0.2 m.y. ago (Dethier and Demsey, this guidebook).

METHODS

The 20 km^2 study area straddles the boundary between the Chili and Vallecitos quadrangles and contains fault-bounded sections of undivided Santa Fe Group (Dethier and Manley, in press). We measured eight representative sections, the thickest being 100 m, and made paleocurrent measurements on crossbedded gravel, imbricated pebbles, and channels at 17 locations. At eight stations we counted 100 pebbles in gravel-rich channels, and we made visual estimates of pebble composition at numerous outcrops. We collected oriented samples of poorly consolidated sandstone, most of them from darker-colored, lithic-rich units, and studied thin sections cut from 21 epoxy-impregnated samples as well as sections of vitric tephra interbedded with silty sandstone.

Petrographic techniques followed methods suggested by Dickinson (1970). Four hundred points were counted on each section (Van der Plas and Tobi, 1965) and grains categorized as quartz, feldspar, lithic, or miscellaneous. Each class was subdivided using standard techniques. Lithic volcanic clasts were subdivided as glass-rich, rhyolite/dacite, latite/andesite, and basalt. This allowed easy comparison to the pebble counts, as most of the pebbles at each site were volcanic.

DATA

Figure 3 shows the geology of the study area and paleocurrent measurements from near the Rio del Oso and from Chama–El Rito exposures 6 km to the northwest, near the mouth of the Arroyo del Palacio. Near La Sotella and La Utah, the Chama–El Rito Member is exposed only in the narrow, upthrown block.

Lithologic Description

The Chama–El Rito Member is at least 140 m thick in the study area and grades up into the Ojo Caliente Sandstone over a thickness of 10 to 40 m (Fig. 4). The base of the Chama–El Rito is not exposed in the study area. Arkosic, massive to planar-crossbedded layers of sand, silty fine-grained sand, clayey-silt beds less than 1 m thick, and sparse channels of pebbly volcanic sand dominate the unit. The channels cut locally as deep as 2 m into underlying beds, but most contacts suggest an aggrading environment. Thin beds of clayey silt, and calcareous and chert-rich sediment containing abundant stem casts and burrows indicate playa-lake deposition. Discontinuous beds and lenses of fresh to altered vitric and lithic tephra up to 1.5 m thick are associated with fine-grained sediment. Isolated sections of dune sand less than 15 m thick are common in the upper Chama–El Rito.

The Chama–El Rito Member typically contains layers of sandstone (60–70%), siltstone (15–20%), and pebbly conglomerate (5–10%). Sandstones range from coarse-grained and granular to silt-rich sand. The sandstones are moderately well sorted, but most contain at least 10% silt. Sandstone colors are typical of the Santa Fe Group: grayish-orange to moderate red. Gravel beds are gray–green to grayish-orange and silt to silty-clay beds are gray–green to reddish-brown.

Petrology

Most sediment in the Chama–El Rito Member is arkosic, but almost all gravel and much of the sand in gravel channels and lenses are volcanic. Figures 5a and 5b summarize pebble counts and sandstone petrology. The gravel is more than 90% volcanic. The largest clasts are andesites, but most of the pebbles are glass or silicic volcanic rock. Few clasts originated from Precambrian or Paleozoic source rock. The most distinctive lithology is a reddish to purple, densely welded tuff that contains collapsed, devitrified pumice and abundant phenocrysts of blue sanidine, quartz, and plagioclase. Steinpress (1981) reported similar clasts in the Chama–El Rito Member in the Dixon area. Descriptions of volcanic clasts in the Los Pinos Formation to the north (May, 1980) are similar to those in the study area, but May reported as much as 25% clasts of Precambrian quartzite, metavolcanics, and gneiss, lithologies nearly absent in the study area.

The sandstones studied in thin section are volcanic arenites that average 50% quartz, 10% feldspar, and 40% lithic and miscellaneous fragments calculated on a matrix-free basis. Most of the lithic fragments are silicic tuffs or volcanic glass. Plagioclase/total feldspar ratios average 60%. Polycrystalline quartz/quartz ratios average 16.4%. Grains in the volcanic sandstones are angular to subrounded. Quartz grains show little secondary overgrowth, and alteration of the feldspars is variable.

K–Ar dates from local basaltic units indicate a minimum age of 10.7 m.y. for the top of the Chama–El Rito in this area. A basalt dike intrudes a palagonitic breccia in basal Chamita Formation near the Rio del Oso, 1 km east of the map area. The dike gave an age of 10.7±0.3 m.y., a Lobato Basalt flow interbedded with basal Chamita in the southern map area gave a K–Ar date of 10.3±0.3 m.y., and an olivine-basalt flow in the Chamita Formation 4 km southeast of the SE corner of the field area gave a K–Ar age of 9.6±0.2 m.y. (M. J. Aldrich, pers. comm. 1984).

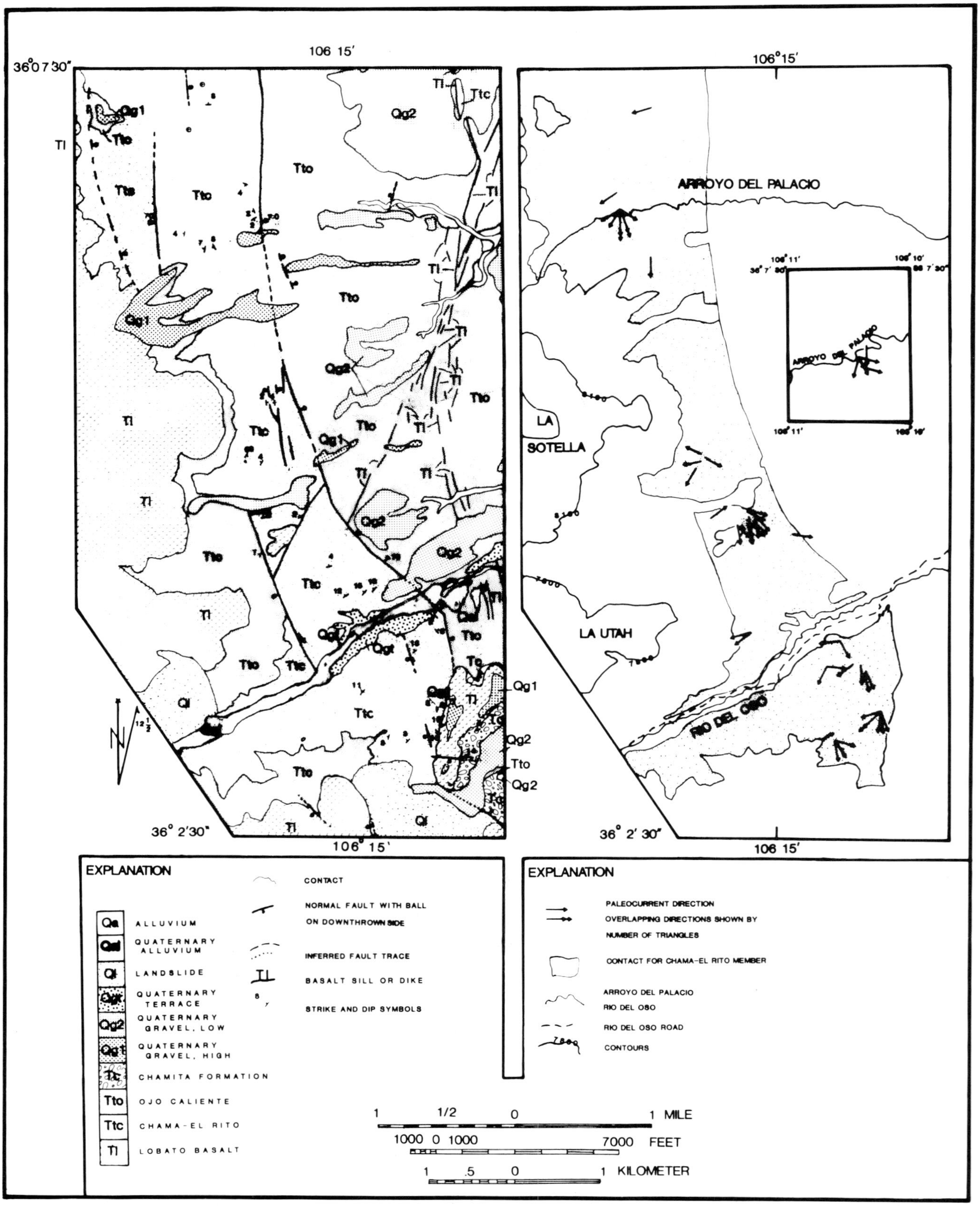

FIGURE 3. Geologic map and paleocurrent map for the Chama–El Rito Member of the Tesuque Formation for part of the Chili and Vallecitos quadrangles, New Mexico.

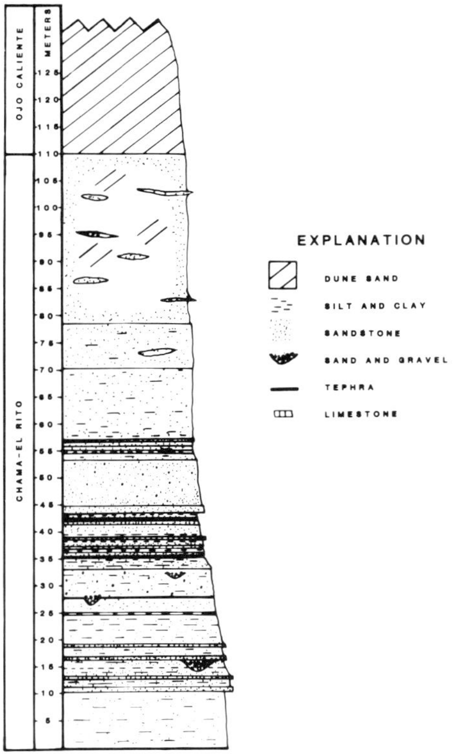

FIGURE 4. Composite stratigraphic column for the Chama–El Rito Member of Tesuque Formation east of Lobato Mesa, New Mexico.

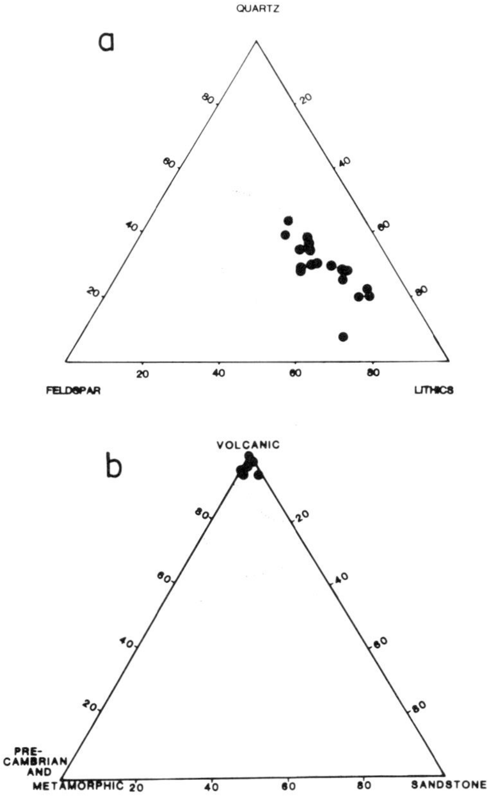

FIGURE 5. Triangular diagrams showing lithologies of the Chama–El Rito in the study area. The dotted fields outline values reported for the Chama–El Rito by Steinpress (1981). **a,** Petrography of 20 sandstone samples. **b,** Pebble lithologies from eight sites.

Paleocurrent Directions

Paleocurrent directions range from ENE to W, but the most common direction is SSE (Fig. 6). Direction does not correlate with stratigraphic position. Dips on units that enclose crossbedded gravels are gentle, so no corrections were applied to the measurements. However, rotation of fault blocks may account for some of the variation in values. One SE transport direction was measured on a 10-m-thick sequence of subangular, matrix-supported, gray debris flows and poorly sorted gravel composed entirely of volcanic detritus. The volcaniclastic tongue fills channels cut into sandstone about 70 m below the Chama–El Rito/Ojo Caliente contact. We did not observe similar monolithologic deposits elsewhere in the study area, but they suggest a nearby volcanic source.

Faults

Faults with displacement of more than 5 m are shown in Figure 3. The major eastern fault zone drops the Ojo Caliente a minimum of 350 m, whereas motion on the west side of the narrow horst is at least 70 m. The eastern fault apparently truncates the lower Bandelier Tuff on a mesa south of Arroyo del Palacio, but does not break latest Pleistocene surfaces. The fault continues to the south and disappears beneath upper Quaternary gravel south of Rio del Oso. A NW-striking fault exposed 1 km south of this area drops the Lobato at least 70 m to the northeast. Planes on the eastern fault dip from 70° E to vertical, and slickensides preserved at two locations plunge to the north at 25 to 40°. Thick flows of Lobato Basalt overlie the Ojo Caliente in the western portion of the map area, and are interlayered with the basal Chamita Formation to the south. A single sill more than 20 m thick intrudes the Chama–El Rito Member at the western margin of the Rio del Oso dike swarm. The swarm of vertical dikes curves from N 20° E to N 20° W from the north to south end of the field area. Dikes at the western edge of this swarm end against major faults.

DISCUSSION

Stratigraphic relations, textures, petrology, and radiometric dates indicate that the undivided Santa Fe of Dethier and Manley (in press) is the upper part of the Chama–El Rito Member of the Tesuque Formation as mapped by May (1980). Conglomerate and sandstone petrology and paleocurrent indicators imply that the area received sediment carried by low-gradient streams from sources to the northeast and northwest. Silicic airfall tephra likely originated from sources located to the south and west, a direction suggested by the eolian crossbeds in the Chama–El Rito. The Chama–El Rito is extensively faulted and intruded by basaltic dikes trending NNE to NNW, but most extension in the area

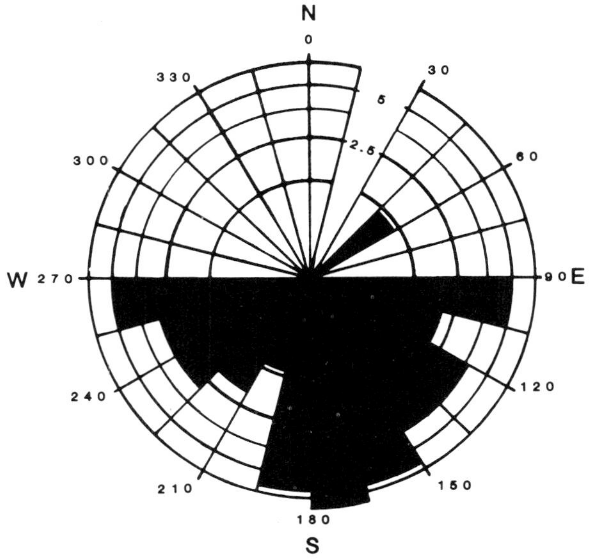

FIGURE 6. Rose diagram showing 50 paleocurrent directions measured in the Chama–El Rito Member of Tesuque Formation in the area outlined in Figure 3.

apparently occurred after deposition of the Ojo Caliente and before about 9.5 m.y. ago.

The Chama–El Rito Member was deposited by braided streamflow in broad, braided channels, by sheetfloods, and by streamflow on the outer reaches of broad alluvial fans (May, 1980). Interbedded playa deposits and dunes, and vertebrate fossils reported by Galusha and Blick (1971) indicate a semiarid environment. Sandy fluvial sediment displays laminar bedding, ripple lamination, angular crossbeds, upward-fining cycles and pebble imbrication typical of distal-fan deposits. Abundant beds of silt and silty sandstone, minimal erosion, lacustrine deposits, and lack of debris-flow deposits indicate a low-energy environment. Deposits are lithologically similar to, but finer-grained than, those mapped by May (1980) at similar stratigraphic levels 30 km to the northwest and by Steinpress (1981) 40 km to the east. Sandstone and conglomerate petrography show that little Precambrian detritus reached the Rio del Oso area in mid-Miocene time despite the broad zones of those rocks now exposed to the northwest and northeast. These sources were apparently buried, or their contribution overwhelmed, by volcanic detritus from the north and northeast. Paleocurrent measurements imply that the axis of the Española Basin lay to the south or slightly east of the study area during upper Chama–El Rito deposition. Our measurements indicate more southerly and southeasterly transport than those reported by May (1980).

Vazzana and Ingersoll (1981) proposed that the upper Abiquiu Formation was deposited by southeast-flowing streams that transported intermediate and silicic volcanic debris. They suggested a source in the San Juan volcanic field, but Manley (1981) believed that the Abiquiu is equivalent in age and source area to the Cordito Member of the Los Pinos Formation. The dominant felsic lithologies of the Chama–El Rito in the study area are typical of the Cordito Member. Southwest transport directions in other areas and sparse andesitic cobbles may indicate reworking of some portion of the Los Pinos, but lithologies of gravels transported to the SE and SW are not consistently different. The Jemez volcanic field was active by about 13 m.y. ago, producing felsic airfall tuffs that accumulated in the study area and in the central Española Basin (Manley and Naeser, 1977). However, relief must have been low in the northern Jemez region, because there is only local indication of eastward transport and no major deposits of Jemez volcanic rocks. Rocks of the Chamita Formation mapped near Clara Peak contain some locally derived clasts, but through at least Miocene time transport in the northeast Jemez Mountains was to the south and sediment was derived from the north and northeast.

Major normal faults drop the Ojo Caliente adjacent to the Chama–El Rito Member in the area of interest. Some fault motion postdates eruption of the Lobato Basalt. North- and northeast-trending structures that cut Lobato Mesa (Smith and others, 1970) imply that north-trending horsts and grabens formed to the west as well. Faults west of Abiquiu strike southwest and exhibit about 670 m of throw after 7.6 m.y.B.P., but no indication of earlier activity (Manley and Mehnert, 1981). However, western dikes of the Rio del Oso swarm end against the horst-bounding structures, indicating at least some motion along the faults before 9.7 m.y. ago (Baldridge and others, 1980). The north end of the eastern fault offsets the Bandelier Tuff less than 15 m. Thick Quaternary cover and poor exposures prevent tracing the fault to a post-Lobato fault at the south end of the map area. However, some connection is likely. Deposits we mapped show no evidence for growth faults.

At least some post-Lobato motion is required for the longest fault in the study area, but, unlike the Cañones fault system to the west (Manley and Mehnert, 1981), faults in the study area show evidence for activity between about 12 and 9.6 m.y. ago. Most other faults in the Rio del Oso area are parallel to those that define the horst, trend between N 20° E and N 20° W, but offsets are generally less than 20 m. The Rio del Oso dike swarm and a group near Abiquiu intruded weak zones during late Miocene extension (Bachman and Mehnert, 1978), but there is no evidence for offset across the three best-exposed dikes in the study area. However, their aggregate width exceeds 35 m.

Major extension occurred in the Rio del Oso area between about 11 and 9.5 m.y. ago, and at least some extension continued until late Quaternary time. We have no evidence that large amounts of offset occurred on local faults after 9 m.y. ago. However, local and regional evidence (Golombek and others, 1983) indicates that the Pajarito fault and possibly the unnamed NE-trending fault were active and formed the northwestern boundary for the Velarde graben by about 7 m.y. ago. Manley and Mehnert's (1981) data suggest major motion on NE-striking, east-side-down faults after 7.6 m.y. ago northwest of the Velarde graben. This fault system may have formed the western boundary of the northwest Española Basin until late Pliocene time, but sometime after 9.5 and before 4 m.y. ago the axis of deposition shifted to the Velarde graben. Additional field studies and dating in the northeast Jemez Mountains will improve our understanding of the fault kinematics that produced this shift.

ACKNOWLEDGMENTS

Portions of this work were supported by the Los Alamos National Laboratory and by the Bronfman Science Center at Williams College. We gratefully acknowledge the field studies, support, and encouragement of Kim Manley, Jim Aldrich, and Scott Baldridge, and careful reviews of an earlier version of this paper by Judson May and Ray Ingersoll.

REFERENCES

Bachman, G. O., and Mehnert, H. H., 1978, New K/Ar dates and the late Pliocene to Holocene geomorphic history of the central Rio Grand region, New Mexico: Geological Society of America, Bulletin, v. 89, pp. 283–292.

Baldridge, W. S., Damon, P. E., Shafiqullah, M., and Bridwell, R. J., 1980, Evolution of the central Rio Grande rift, New Mexico: new potassium–argon ages: Earth and Planetary Science Letters, v. 51, pp. 309–321.

Baltz, E. H., 1978, Resume of Rio Grande depression in north-central New Mexico: New Mexico Geological Society, Guidebook 29, pp. 210–228.

Chapin, C. E., 1979, Evolution of the Rio Grande rift—a summary; *in* Riecker, R. E. (ed.), Rio Grande rift: tectonics and magmatism: American Geophysical Union, Washington, D.C., pp. 1–5.

Dickinson, W. R., 1970, Interpreting detrital modes of graywacke and arkose: Journal of Sedimentary Petrology, v. 40, pp. 695–707.

Dethier, D. P., and Manley, K., 1984, Geologic map of the Chili quadrangle, Rio Arriba County, New Mexico: U.S. Geological Survey, Miscellaneous Field Studies Map, in press.

Galusha, T., and Blick, J. C., 1971, Stratigraphy of the Santa Fe Group, New Mexico: American Museum of National History, Bulletin, v. 144, no. 1, 128 pp.

Golombek, M. P., 1983, Geology, structure and tectonics of the Pajarito fault zone in the Española Basin of the Rio Grande rift, New Mexico: Geological Society of America, Bulletin, v. 94, pp. 192–205.

———, McGill, G. E., and Brown, L., 1983, Tectonic and geologic evolution of the Española Basin, Rio Grande rift: structure, rate of extension, and relation to the state of stress in the western United States: Tectonophysics, v. 94, pp. 483–707.

Kelley, V. C., 1978, Geology of the Española Basin, New Mexico: New Mexico Bureau of Mines and Mineral Resources, Geologic Map 48.

Manley, K., 1979, Stratigraphy and structures of the Española Basin, Rio Grande rift, New Mexico; *in* Riecker, R. E. (ed.), Rio Grande rift: tectonics and magmatism: American Geophysical Union, Washington, D.C., pp. 71–86.

———, 1981, Redefinition and description of the Los Pinos Formation of north-central New Mexico: Geological Society of America, Bulletin, pt. I, v. 92, pp. 984–989.

———, and Mehnert, H. H., 1981, New K–Ar ages for Miocene and Pliocene volcanic rocks in northwestern Española Basin and their relationships to the history of the Rio Grande rift: Isochron/West, no. 30, pp. 5–8.

May, J., 1980, Neogene geology of the Ojo Caliente, Rio Chama Area, Española Basin, New Mexico: Ph.D. dissertation, University of New Mexico, Albuquerque, 205 pp.

Smith, R. L., Bailey, R. A., and Ross, C. S., 1970, Geologic map of the Jemez Mountains, New Mexico: U.S. Geological Survey, Miscellaneous Geologic Investigations Map I-571.

Steinpress, M. G., 1981, Neogene stratigraphy and structure of the Dixon area, Española Basin, north-central New Mexico: Geological Society of America, Bulletin, pt. II, v. 92, pp. 2553–2671.

Van der Plas, L., and Tobi, A. C., 1965, A chart for judging the reliability of point counting results: American Journal of Science, v. 263, pp. 87–90.

Vazzana, M. E., and Ingersoll, R. V., 1981, Stratigraphy, sedimentology, petrology, and basin evolution of the Abiquiu Formation (Oligo-Miocene), north-central New Mexico: summary: Geological Society of America, Bulletin, pt. I, v. 92, pp. 990–992.

A PALEONTOLOGICAL ANALYSIS OF THE ALAMOSA FORMATION (SOUTH-CENTRAL COLORADO: PLEISTOCENE: IRVINGTONIAN)

KAREL L. ROGERS
Science Division, Adams State College, Alamosa, Colorado 81102

INTRODUCTION

The floor of the San Luis Valley of south-central Colorado is made up of Alamosa Formation sediments. The Alamosa Formation overlies Miocene Santa Fe Formation (Siebenthal, 1910). In all places where the glaciers of the Sangre de Cristo Mountains extended far enough, they deposited their moraines on top of the Alamosa Formation. The type locality of the Alamosa Formation (Siebenthal, 1910) is an outcrop called Hansen Bluff (Fig. 1), a feature created by the Hansen Bluff fault, a northern segment of the La Sauses fault where it is offset by movement of the Manassa fault. The bluff has receded somewhat eastward because of lateral erosion by the Rio Grande, which has since moved its course westward from the bluff. All wells west of the Hansen Bluff fault are artesian (C. A. Repenning, written comm. 1984).

At the type locality, fossils of fresh-water snails, vertebrate fragments, pollen, and plant remains have been reported (Lochman-Balk and Bruning, 1971). Some of the vertebrate fragments were identified as *Camelops* sp. (R. C. Peterson, oral comm.), and Price (1971) published a short list of pollen types.

Extensive fossil collection at the Hansen Bluff locality was not carried out until the summers of 1979, 1980, and 1981, when the author led crews of 12 or more to recover fossils using techniques of Hibbard (1949) and refinements of those techniques described by Waters (1978). Matrix was labeled as it was taken from the bluff, transported to the washing site in burlap bags, dried in large frames (as the matrix dried, much of the dirt dropped through the screening in the frame), transferred to a small washing box with its label, gently washed, allowed to dry, transferred to a paper bag with the label, transported to the lab, and then sorted under a dissecting microscope into small plastic boxes labeled as to location and bag number. This technique allowed for control on fossil breakage, location, and on crew-member performance. All but the smallest of fossils, such as pollen and ostracodes, were recovered in this manner.

Fossils were recovered from a 2-km-long section of Hansen Bluff (Fig. 2). Five measured sections were dug and sampled for fossils. Only one unit could be correlated with certainty along the length of the bluff. In addition, concentrations of fossils were dug from seven localities, some of which were made up of more than one unit.

The fossil assemblage has been given the name Alamosa Local Fauna after the county and town of its origin. Fossils will reside at the University of Colorado Natural History Museum, Boulder.

Recent climate of the San Luis Valley is arid to semiarid, with prevalent sunshine, cool temperatures, and frequent southwest winds. Mean annual temperature is 41.6°F (5.3°C), with temperatures ranging from −50°F (−45.5°C) to 93°F (33.9°C). Mean annual precipitation is 17.35 cm. Temperatures at the valley floor are frequently cooler than those in the surrounding mountains because of temperature inversions. Cloudy days tend to moderate temperatures when they occur, although the modern cimate has only about 80 days/year classified as cloudy by the National Weather Service.

AGE OF SEDIMENTS

Sediments outcropping at Hansen Bluff are considered to be late Irvingtonian in age based upon the presence of Bishop Ash dated at 0.74 m.y. (Izett and Naeser, 1976); the Brunhes–Matuyama magnetic reversal (E. E. Larsen, unpubl.); and the presence of *Microtus paroperarius, Pitymys meadensis,* and *Equus scotti* (E. Anderson and C.

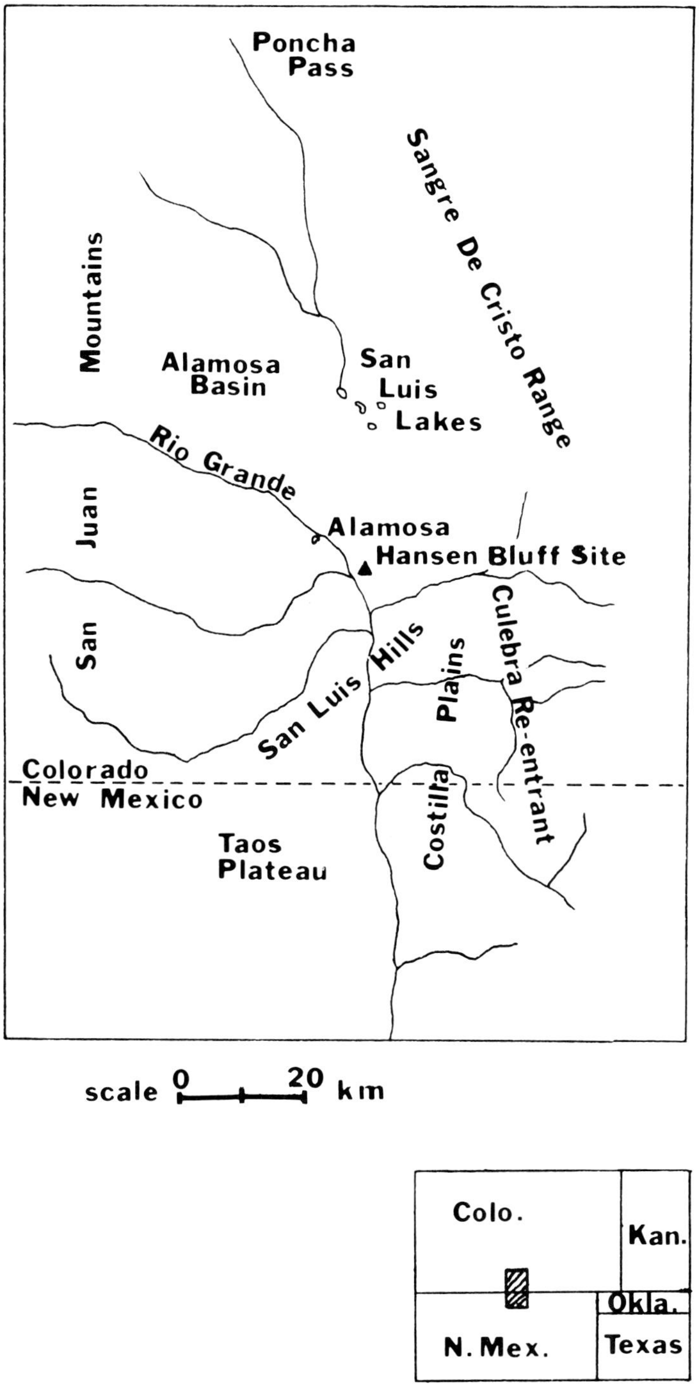

FIGURE 1. Location of the Hansen Bluff site from which the Alamosa Local Fauna has been recovered.

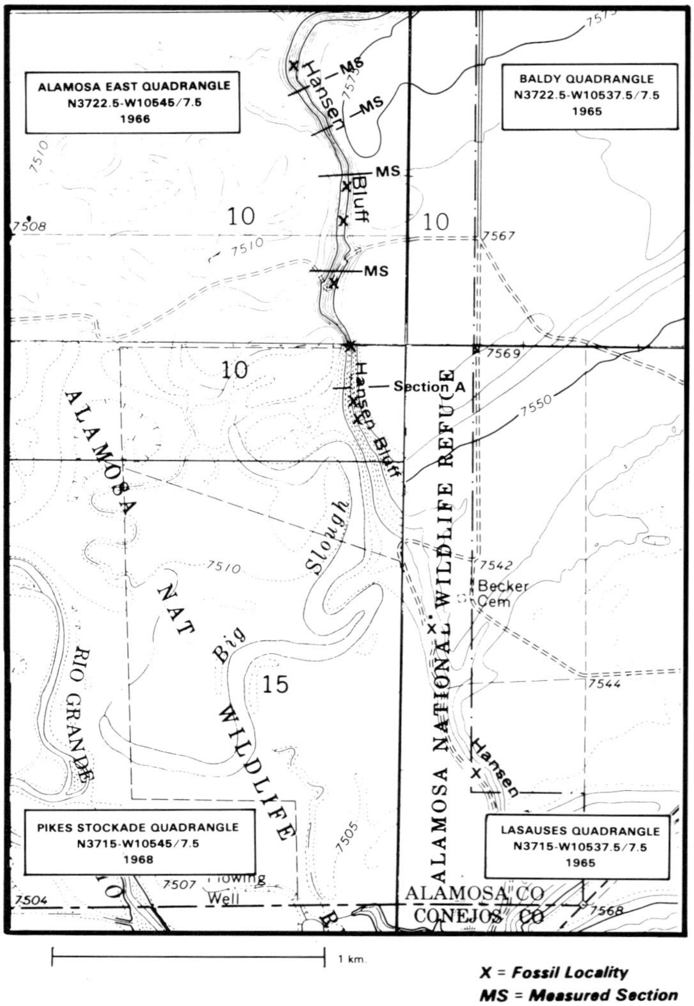

FIGURE 2. Locations of five measured sections and seven fossil-bearing localities along Hansen Bluff. The Alamosa Local Fauna has been recovered from a section of Hansen Bluff 2 km long and nearly 17 m high.

A. Repenning, unpubl.). This would correlate with the pre-Bull Lake deposits described by McCalpin (1982) in the northern part of the San Luis Valley. According to C. A. Repenning (written comm. 1984), deposits from the Sacagawea Ridge glaciation in Wyoming are cut by a terrace containing the type "O" tuff (Richmond, 1976) and are thus older than 600,000 years. This ash has not been found in the Alamosa region, but the age of sediments at the top of the section is considered to be about 0.6 m.y. based upon paleontological arguments.

Measured section A correlated with paleontological, paleomagnetic, and sedimentation data is shown in Figure 3. The Brunhes–Matuyama magnetic reversal occurs in unit 14c. All sediments above are magnetically normal and all sediments below are magnetically reversed (E. E. Larsen, unpubl.). At least two separate ashes in the sediments of the bluff have been located, but only the Bishop Ash has been identified as yet (G. A. Izett, oral comm.). Bishop Ash has not been located in section A, but its level in other sections 0.5–1.0 km away correlates with the oldest normally magnetized units above the magnetic reversal in unit 14c (Fig. 3).

A core extending 400+ cm below the bottom of sediments shown in Figure 3 was sampled for paleomagnetic information, but the Jaramillo and the youngest Bandelier Tuff were not located (E. E. Larsen, written comm. 1983). Sediments at the bottom of section A are considered to be less than 0.90 m.y. old based on paleontological arguments as well as on failure to find the normal Jaramillo subchronology (C. A. Repenning, written comm. 1984).

MAJOR SIGNIFICANCE OF FOSSILS

The following information is based upon unpublished information which will appear in detailed form in a group paper (Rogers, K., Anderson, E., Brown, T., Forester, R. M., Hall, S. A., Izett, G. A., Larsen, E. E., Repenning, C. A., Smith, G. R., and Taylor, D. W., in prep.). Authors of results are given with the fossil groups.

Pollen: Information is based upon data and conclusions of S. A. Hall (unpubl.). The Alamosa Local Fauna is noteworthy in that it is the first middle Pleistocene pollen sequence in the western United States with an absolute age. Spores, pollen, and fresh-water algae are abundant up to unit 14c, indicating that the site was first a marsh, then a lake. Units up to and including unit 9 in section A have pollen assemblages representing a sage–grassland valley-floor vegetation with pine–spruce forest in the adjacent mountains. Above units 9 through 14c pollen indicate a down-elevation movement of forest vegetation, possibly associated with the onset of glaciation, but not necessarily with lowered annual temperature. Possibly temperatures were moderated, and cloud cover and moisture were increased. Units above 14c are devoid of fossil pollen.

Bryozoans: Information is based upon the identifications of E. Kaufman (oral comm.). Fossils of reworked Permian bryozoans were found at localities correlating with units 7 and 8 of section A and in unit 10 of section A. They were also found in a number of other units of the other measured sections. These fossils indicate that the source area for many of the sediments is the Whiskey Creek Formation in the Sangre de Cristo Mountains.

Ostracodes: Information is based upon the data and conclusions of R. M. Forester (unpubl.). Units 12 and 14c of section A contain concentrations of ostracodes. The lower ostracode bed is reversely magnetized and contains species that occur today in the San Luis Valley, but not together in the same bodies of water, indicating some difference in paleoclimate. Water abundance was most prominent in late winter and spring, with cold winters and warm summers. These ostracodes occur in short-lived ponds.

The upper ostracode bed in unit 14c, which is just above the Brunhes–Matuyama magnetic reversal at 0.75 m.y., contains species characteristically limited in northern distribution by winter temperature, which today occur in shallow lakes. Possibly heavy cloud cover during this time acted to moderate temperatures and reduce evaporation from water bodies. There is evidence in the ostracode fauna to indicate a trend toward warmer, drier summers from the bottom to the top of the bed.

Molluscs: Information is based upon the data and conclusions of D. W. Taylor (unpubl.). Aquatic snails with a more northern or higher than modern distribution were present in sediments correlating with unit 12 in section A. Other snails and clams in the fauna were species more characteristic of the area today, living in ponds, sloughs, or backwaters that fluctuate seasonally, or even dry up entirely. The molluscan assemblage of the Alamosa Local Fauna is most similar to that of the Cudahy Fauna in Meade County, southwestern Kansas.

Fish: Information is based upon data and conclusions of G. R. Smith (oral comm.). Four genera of fish, *Catostomus, Gila, Dionba,* and *Salmo,* are present and abundant at Hansen Bluff. *Catostomus* sp. is present as two forms; an early form present in the bottom of the section was replaced by a late form below the level of the ash in unit 14c. The two forms do not have an ancestral–descendent relationship. *Gila* sp. is present in the lower part of the section up to and including unit 14b; it is an indicator of warm water. Fossils of *Gila* sp. represent fish over 40 cm in length, much larger than modern *Gila,* which today ranges north to northern New Mexico. *Dionba* sp., today restricted to Mexico, is present at fossil localities corresponding to units 7 and 8 in section A. This species is an indicator of warm water. The uppermost units of the bluff contain remains of *Salmo,* an indicator of cold, well-oxygenated water. No living relatives of these trout are known today.

Reptiles and Amphibians: These fossils have been identified and interpreted by the author. Sixteen species of reptiles and amphibians

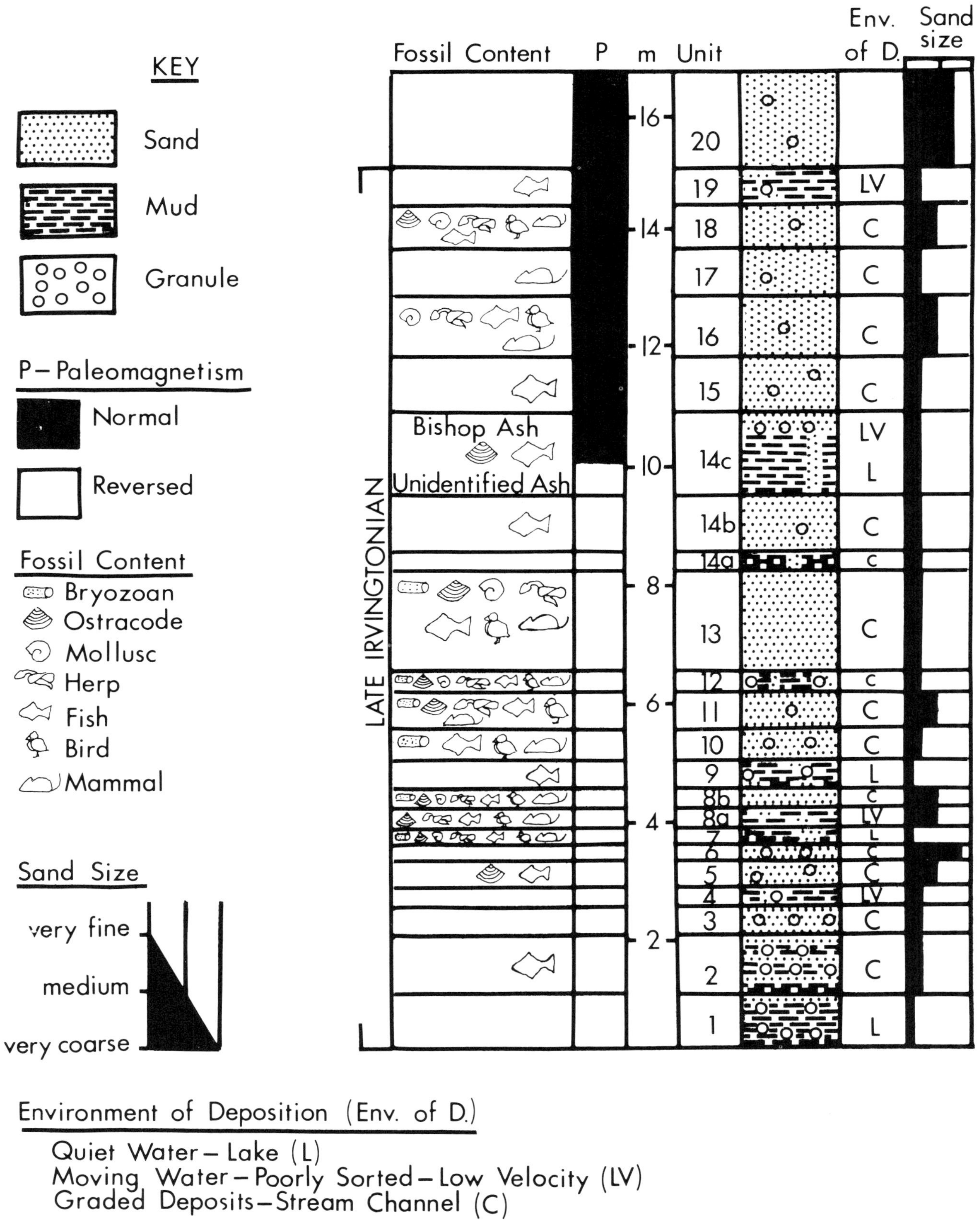

FIGURE 3. Measured section A at Hansen Bluff, Alamosa Formation, south-central Colorado. Greatest depth of the section is within Alamosa Formation.

have been identified, one of which is being described as a new species. Three of the species occur today southeast of the Sangre de Cristo Mountains, but not in the San Luis Valley. These southeastern species are present only at fossil localities corresponding to units 7 and 8 of section A.

Three faunal groupings can be recognized in the Alamosa Local Fauna: (1) the bottom of the section in units up to and including unit 9, (2) the middle of the section including units 11 through 13, and (3) the top of the section including units 15 through 19 in section A.

The oldest faunal grouping indicates relatively warm and dry times because of the presence of Front Range species that are today bounded by mean annual temperatures 5–8°F warmer than the San Luis Valley, and by the presence of neotenic tiger salamander fossils that are interpreted to indicate warm, poorly oxygenated water that did not support fish (Rogers, in press). Dry conditions are indicated by the presence of spadefoot toads.

The middle fossil grouping is consistent with what would be expected in cool, moist conditions in the San Luis Valley. Tiger salamanders were metamorphosing, as would be expected when conditions on land were more suitable for terrestrial amphibians (Rogers, in press), and the abundant *Bufo* fossils from low in the section are replaced by the more water-dependent genus *Rana*. Species with a modern range from south and east of the mountains are absent.

The upper fossil grouping seems to indicate a cold, dry climate because *Bufo* is again abundant, horned toads first appear in the deposits, very few *Rana* fossils are present, and spadefoot toads are again abundant. Salamanders from this level of the section were metamorphosing, apparently because of favorable terrestrial conditions (Rogers, in press).

Birds: Much of the information below is based upon the identifications and conclusions of J. A. Feduccia (written comm. 1980). Fossils of a duck the size of the lesser scaup and a coot, close to the living American coot, were found at localities corresponding to unit 18 in section A. Fossils similar in size and morphology to the whistling swan and to a larger than living version of the American avocet were identified from a fossil locality corresponding to units 7 and 8 in section A. Passeriform elements are abundant, but are not identifiable. All of these birds could be found in a fresh-water lake environment. In addition, one fossil of an Irvingtonian grebe is in the process of being identified to species by researchers at the University of Michigan.

Mammals: Information is based upon the data and conclusions of E. Anderson (unpubl.) and C. A. Repenning (oral comm.). Of the 22 species of mammals represented, at least nine species are extinct. There is a distinct faunal similarity to the Cudahy Fauna of Meade County, Kansas, with at least eight species present in both the faunas. The fauna consists primarily of small rodents, with very few carnivores represented. Pocket gophers of the genus *Geomys* are common throughout the section. Modern species of *Geomys* do not occur near Alamosa; the nearest modern taxon is found about 200 km northeast of Alamosa County, on the Front Range in Pueblo County. *Equus scotti* is represented by teeth and phalanges, and *Camelops* sp. is represented by metapodials. *Camelops* is known only from localities correlating with unit 18 in section A. *Microtus paroperarius* and *Pitymys meadensis* are most common in sediments corresponding to units 7 and 8 in section A, but are present throughout the entire section. *Microtus paroperarius, Pitymys meadensis, Ondatra annectens, Synaptomys meltoni, Neotoma ozarkensis,* and *Equus scotti* are indicative of Irvingtonian.

According to Repenning (written comm. 1984), the mammalian fauna does not vary greatly from top to bottom of the Alamosa Formation exposed on Hansen Bluff. One remarkable exception is apparent in the size of *Microtus paroperarius*. The first lower molars of *M. paroperarius* from the lower part of the exposures are larger than at any other published locality and average 3.5 mm in length; in the highest part of the sections these teeth average 2.9 mm in length and are comparable in size to the type material from the Cudahy Fauna of Kansas (600,000 years old). A parallel trend is known to occur in the complication of the last upper molar of most microtines. The last upper molars of *M. paroperarius* from the base of the section are more primitive than has been observed elsewhere, while those from the top of the section are complicated and comparable to the condition of this species from the Cudahy Fauna. Because of the presence of the Bishop Ash and the base of the Brunhes paleomagnetic anomaly in the section, these larger and more primitive populations of *M. paroperarius* from the lower part of the section are demonstrably older, possibly 300,000 years older, than the population in the Cudahy Fauna. It is suggested, because of the similarity of populations from the uppermost part of the Hansen Bluff sections to those of the Cudahy Fauna, that these younger faunas may approach 600,000 years in age. The cause of this progressive decrease in size of *M. paroperarius* is not known, although populations (presumably of an even younger age) with even smaller first lower molars than those in the Cudahy Fauna are known. The species becomes extinct at the end of the Irvingtonian mammalian age, about 450,000 years ago, and is replaced by immigrant *M. pennsylvanicus,* which is still extant. No post-Irvingtonian index species are recognized with the exception of species of *Microtus* that are believed to have been inadvertently included in samples due to slope wash of modern material.

Sedimentation: Information is based upon the data and conclusions of T. Brown (written comm. 1983). Sedimentary analysis was done by dry sieve and settling tube, and resulted in definition of three major environments of deposition: quiet water, such as a lake or pond; fluvial, low-velocity currents; and fluvial stream-channel deposits. Sediments of section A are primarily channel deposits with thin layers of low-energy sediments in units 19, 14c, 8, and 4, and lake deposits in units 14c, 9, 7, and 1 (Fig. 3).

CONCLUSIONS

The following conclusions are those of the author and do not necessarily correspond with those of other workers, although the author's conclusions have been modified extensively due to comments by C. A. Repenning (written comm. 1984).

Sediments cropping out at Hansen Bluff probably represent a span of time from about 0.6 to nearly 0.90 m.y. and include two interglacials and one glacial period. The glacial period was probably not colder, on the average, than is the modern climate as judged by the composition of the fauna. Increased cloud cover during that time may have resulted in moderated temperature extremes as well as greater humidity, as indicated by the pollen, ostracodes, reptiles, and amphibians. The glacial period, however, was probably colder, as well as wetter, than the preceding interglacial, judging by the loss of poikilotherms that occur today only on the Front Range where modern temperatures are warmer than in the San Luis Valley. The last interglacial is indicated by a drying trend, probably brought on by decreased cloud cover, but without much evidence of warming climate except for evidence of the ostracodes that suggest warmer, dryer summers in the higher parts of unit 14c and increasing seasonality. During the glaciation, species of molluscs occurring today at higher elevations or with a more northern modern distribution are represented in the fauna.

Overall, the entire fauna is very r-selected compared to low-altitude faunas, resulting in low diversity, high abundance of individual species, and in species being represented primarily by small, immature individuals. This is characteristic of unstable environments with frequent harsh conditions and periodic die-offs. Possible evidence of thermal stress is shown in some species by increased or decreased body size in homeotherms or poikilotherms, respectively. During the glacial period the fauna shows some increase in r-selection, with decreased faunal diversity in coldblooded forms.

Boellstorff (1978) indicates that the type Nebraskan till at Afton, Iowa, is reversely magnetized and older than the 0.7-m.y.-old Hartford Ash, but younger than the 1.2-m.y.-old Coleridge Ash. He also indicates that the type Aftonian paleosol contains both the Hartford Ash and the restricted Pearlette Ash (type "O" of 0.6 m.y. and, in Kansas, containing in its base the Cudahy Fauna). The presence of the 0.74-m.y.-old Bishop Ash in several of the sections at Hansen Bluff, and of the change from Brunhes normal polarity to Matuyama reversed polarity slightly down section from the position of this ash, clearly suggests that the central part of the type Alamosa Formation is correlative with the type Nebraskan till. This part of the section, including units 10 through the lower part of 14c shows evidence of cooler, moister summers (frogs, salamanders, and possibly molluscs) and milder, but wetter, winters. The overlying units of the type Alamosa Formation, probably beginning in the younger half of unit 14c, seem to be a correlative of the type Aftonian paleosol, even though the type "O" Pearlette Ash has not been found in the San Luis Valley, since *Microtus paroperarius* is reduced in size to that known from the Cudahy Fauna. The underlying "interglacial," unit 9 and older, has not been named in North America. Uplift of Hansen Bluff along the Hansen Bluff fault would appear to have taken place during the Aftonian and possibly just before 600,000 years ago. No later deposition took place on the bluff with the exception of wind-blown silt and a conspicuous caliche zone.

ACKNOWLEDGMENTS

This research was funded by the National Geographic Society grants 2115-79, 2165-80, and 2328-81, and I thank the Society for its generosity. I am most grateful to E. Anderson, T. Brown, J. A. Feduccia, R. M. Forester, S. A. Hall, G. A. Izett, E. E. Larsen, C. A. Repenning, G. R. Smith, and D. W. Taylor for agreeing to work with me on this project, identifying and interpreting fossils. Thanks are also due to the many students who spent long hours hauling, washing, and picking. Personnel of the Alamos and Monte Vista Wildlife Refuges were very helpful, especially Mel Nail. The illustrations were prepared by John Leak and Lynn Slocum.

REFERENCES

Boellstorff, J., 1978, North American Pleistocene stages reconsidered in light of probable Pliocene–Pleistocene continental glaciation: Science, v. 202, pp. 305–307.

Hibbard, C. W., 1949, Techniques of collecting microvertebrate fossils: University of Michigan, Contributions from the Museum of Paleontology, v. 8, no. 2, pp. 7–19.

Izett, G. A., and Naeser, C. W., 1976, Age of the Bishop Tuff of eastern California as determined by the fission-track method: Geology, v. 4, pp. 587–590.

Lochman-Balk, C., and Bruning, J. E., 1971, Lexicon of stratigraphic names used in south-central Colorado and northern New Mexico, San Luis Basin; *in* James, H. L. (ed.), Guidebook of the San Luis Basin, Colorado: New Mexico Geological Society, Guidebook 22, pp. 101–111.

McCalpin, J., 1982, Quaternary geology and neotectonics of the west flank of the northern Sangre de Cristo Mountains, south-central Colorado: Colorado School of Mines Quarterly, v. 77, no. 3, pp. 1–89.

Price, C. R., 1971, Preliminary paleopalynological analysis of Alamosa Formation sediments; *in* James, H. L. (ed.), Guidebook of the San Luis Basin, Colorado: New Mexico Geological Society, Guidebook 22, pp. 219–220.

Richmond, G. M., 1976, Pleistocene stratigraphy and chronology in the mountains of western Wyoming; *in* Mahaney, W. C. (ed.), Quaternary stratigraphy of North America: Dowden, Hutchinson, and Ross, Stroudsburg, Pennsylvania, pp. 353–379.

Rogers, K. L., 1984, Facultative metamorphosis in a series of high altitude fossil populations of *Ambystoma tigrinum* (Irvingtonian: Alamosa County, Colorado): Copeia, in press.

Siebenthal, C. E., 1910, Geology and water resources of the San Luis Valley, Colorado: U.S. Geological Survey, Water-supply Paper 210, 128 pp.

Waters, B., 1978, Modified bulk sampling of fossils: Curator, v. 21, no. 2, pp. 136–148.

San Luis Valley Southern's No. 105 heads south over the Rattlesnake Canyon trestle with a string of empty "refers" for vegetable shipments. This engine was torn down and abandoned in the early 1950's (photo by Bob Richardson).

The Rattlesnake Canyon trestle is in fairly good condition after 74 years, but is no longer in use.

VOLCANIC AND SEDIMENTARY STRATIGRAPHY OF THE RIO GRANDE GORGE AND THE LATE CENOZOIC GEOLOGIC EVOLUTION OF THE SOUTHERN SAN LUIS VALLEY

M. A. DUNGAN[1], W. R. MUEHLBERGER[2], L. LEININGER[2], C. PETERSON[2], N. J. McMILLAN[1], G. GUNN[1], M. LINDSTROM[3], and L. HASKIN[3]

[1]Department of Geological Sciences, Southern Methodist University, Dallas, Texas 75275; [2]Department of Geological Sciences, University of Texas at Austin, Austin, Texas 78713; [3]Department of Earth and Planetary Sciences, Washington University, St. Louis, Missouri 63130

INTRODUCTION

The Taos Plateau volcanic field (TPVF) is one of several dominantly basaltic volcanic fields erupted along the Jemez lineament over the last 10 m.y. These volcanic fields comprise a regional magmatic event that was coincident with regional uplift and tectonic rejuvenation of the Rio Grande rift postdating late Miocene time. TPVF volcanism was dominated by extensive floods of the highly fluid olivine tholeiites (1,500 km^2) of the Servilleta Basalt. Andesites and dacites erupted from shields scattered throughout the field are lesser in volume, but coeval with the basalts. The main phase of basaltic and intermediate volcanism (~4.5–3.5 m.y.) was succeeded by a period of low-volume alkalic basalts (>1.8 m.y.).

Post-volcanic erosion by the Rio Grande has cut a canyon up to 180 m deep along the southern edge and northward through the TPVF. Exposures the length of the canyon reveal the interplay between volcanism, sedimentation, and deformation during an approximately one-to-two-million-year period. Earlier history is exposed only along the southeastern margin of the TPVF and in the Tusas Mountains to the west of the study area. The widespread floods of Servilleta Basalts, which appear to have been erupted from vents located along the western margin of the field, record voluminous, short-lived, effusive episodes interspersed with longer intervals during which gravel fans prograded to the west from the Sangre de Cristo Mountains and north–northwest from the Picuris Mountains. Olivine andesites and two-pyroxene dacites erupted from cones both east and west of the Rio Grande and Red River gorges are also locally interlayered with basalts. The entire volcanic section was tilted to the southeast during and following volcanic activity. Deformation during and after volcanism is also recorded by reoccurring movements on intra-rift and rift-boundary faults, differential subsidence of the narrow Velarde and Taos grabens, and local warping of the volcanic units. Local zones of reverse faulting along the eastern segment of the Embudo fault and warping of the plateau surface (Taos graben) demonstrate a shift to local compression late in the evolution of the basin.

REGIONAL SETTING

Crustal extension and related volcanism associated with the inception of the Rio Grande rift are dated at ~28–25 m.y. in both the southern (Chapin and Seager, 1975; Seager and others, 1984) and northern (Lipman and Mehnert, 1975) segments of the rift. The volcanism characteristic of the early phase of rifting is in part basaltic, which distinguishes it from the calc-alkaline intermediate rocks of the Oligocene San Juan and Mogollon–Datil volcanic fields that preceded rifting. Large intermediate to silicic centers are an important component of the initial pulse of rift-related volcanism, but these too are distinct from the Oligocene calc-alkaline intermediate volcanism due to the presence of alkali-rich, high-SiO_2 rhyolites (e.g., Questa caldera, Lipman, 1983). Recent studies of basins formed in the early phase of rifting indicate a pronounced NW trend to the faults associated with them.

The establishment of N-trending, high-angle faults separating deep, narrow, asymmetric grabens during post-Miocene uplift and faulting contrasts with the NW trends typical of the earlier phase of extension, and clearly indicates a major reorientation of the regional stress field in conjunction with tectonic rejuvenation of the Rio Grande rift. Basaltic volcanism within the axis and on the flanks of the rift became voluminous and widespread *after* 10 m.y. and is, therefore, temporally associated with the observed tectonic rejuvenation. The oldest volcanics in all the relevant fields have not been dated, but several did not begin until 7–8 m.y. (Stormer, 1972; O'Niell and Mehnert, 1980) and others initiated activity still more recently.

The Jemez lineament is defined by Plio-Pleistocene volcanic fields extending along a line from central Arizona to northeastern New Mexico. It crosses the rift at a high angle along the Embudo fault zone, the boundary between the San Luis and Española Basins. Structural analysis of Proterozoic rocks in the Sangre de Cristo and Picuris Mountains indicates that the lineament is a zone of Precambrian structural weakness (Nielsen and others, 1983). The TPVF lies along the Rio Grande rift axis, but is slightly offset to the north of the Jemez trend. The other major fields of the Jemez lineament, from SW to NE the Springerville, Zuni, Mt. Taylor, Jemez, Ocate, and Raton–Clayton fields, are located partly or entirely on the rift flanks outside the axial depressions. The spacing of Jemez lineament volcanic fields outside the axial region of the rift (measured center-to-center) is 100–140 km. In contrast, the

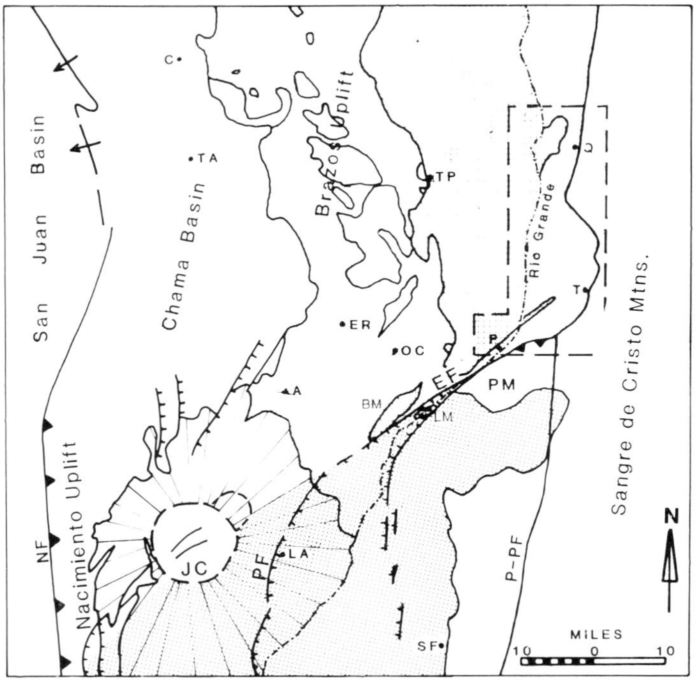

FIGURE 1. Index map of northern New Mexico showing location of study area (region outlined by dashed lines in Figure 2) relative to the San Luis (horizontal ruling) and Española (shaded) Basins of the Rio Grande rift and surrounding regional elements. Symbols: TPVF–Taos Plateau volcanic field (vertical ruling); JC–Jemez caldera and its volcanic field (radial patterns); BM–Black Mesa; LM–La Mesita; EF–Embudo fault zone; PM–Picuris Mountains; P-PF–Pecos–Picuris fault; SF–Santa Fe; LA–Los Alamos; A–Abiquiu; ER–El Rito; OC–Ojo Caliente; T–Taos; TP–Tres Piedras; Q–Questa; TA–Tierra Amarilla; C–Chama; NF–Nacimiento fault zone; P–Pilar; unpatterned areas expose pre-rift rocks.

TPVF lies midway between the Jemez and Ocate fields, which are approximately 100 km apart along the Jemez zone. Thus, the position and structural setting of the TPVF is unique among the Jemez zone volcanic fields. Furthermore, the TPVF lies in an actively subsiding basin which buried its older parts. In most other fields along the zone, the inception of volcanism is datable because erosion has accompanied uplift of the rift flanks, leaving the volcanic rocks perched on mesas.

The Jemez volcanic field, which began erupting at about 10–13 m.y., is a highly diverse assemblage which includes the entire spectrum from basalt to rhyolite (Ross and others, 1961; Gardner and others, 1984). It is unusual among the Jemez zone volcanic fields in that rhyolite comprises a significant proportion of the eruptives. The remaining fields are dominantly basaltic, with a subordinate component of intermediate rocks (andesites to latites). Pyroclastic deposits and rhyolites are rare, and are altogether absent in several of the volcanic fields. Recent studies, while demonstrating some contrasts among the various volcanic fields, have also served to emphasize similarities (Dungan and others, 1981b; Dungan and others, 1983a; Condit, 1983; Aubele and Crumpler, 1983; Baldridge and others, 1983).

Each volcanic field embraces some diversity in basaltic-magma types, but the flank eruptives are more alkaline and more diverse. Activity appears to have peaked in the interval 5–3 m.y., during which time tholeiitic to transitional basalts with diktytaxitic groundmass textures were the dominant type. Andesites and more silicic rocks of high-K, calc-alkaline affinity were erupted coevally with basalts of the period of peak activity.

The TPVF lavas were erupted in an actively subsiding basin which underwent internal deformation during and after volcanism. Although volcanism ceased within the San Luis Basin at approximately 1.8–1.5 m.y., eruption of basalts continued into the Quaternary on the rift flanks, and evidence for ongoing deformation on the TPVF and elsewhere indicates that the rift remains an active tectonic province. The apparent waning of volcanism beginning at 2–3 m.y., following a regional peak at 5–3 m.y., suggests that the late Cenozoic rejuvenation of activity peaked in the late Pliocene, and that a progressively less active system may be predicted for the future.

PREVIOUS WORK

The geomorphology and general geology of the Taos–Questa area were first described by Lambert (1966). The nomenclature and stratigraphic relationships of Cenozoic sedimentary units in the Española Basin and vicinity have been discussed by Galusha and Blick (1971), Manley (1976, 1979), and Steinpress (1980). The Cenozoic tectonic and rifting history has been extensively studied by Kelley (1952, 1954, 1977, 1979) and Manley (1976, 1979). Baltz (1978) presented a summary of the Phanerozoic history of the northern rift.

Analyses of TPVF lavas were published by Wells (1937), Larson and Cross (1956), and Aoki (1967), whose work complemented the study of Ozima and others (1967). The dating and paleomagnetic investigation of Ozima and others demonstrated that at least two magnetic reversals occurred during Servilleta time which ranged from 4.5–3.5 m.y. Lipman and Mehnert (1979) also published a map, chemical analyses of all rock types present in the TPVF, and provided additional dates which are generally consistent with those of Ozima and others. Manley (1979) and Baldridge and others (1980) have published single K–Ar dates of 2.8 m.y. and 4.6 m.y., respectively, on stratigraphically young Servilleta flow units.

Baldridge and others (1980) have criticized the Ozima and others (1967) chronology on two counts. The lack of a mass discrimination correction in the earlier work yields an older and narrower range on ages (5.0–4.5 m.y.) which bring these data into essential agreement with the more recent data. The second point made by Baldridge and others (1980) is that only three, rather than six, magnetic reversals are represented by the volcanic section. This apparent discrepancy could as easily be explained by a lack of eruptions during three of the magnetic intervals as a mismatch between the K–Ar chronology and the magnetic stratigraphy. We argue below that the periods occupied by Servilleta volcanism were short compared to the intervals between eruptive episodes. Clearly, additional dates on the complete stratigraphic section are needed to satisfactorily resolve these discrepancies. We have continued to use the Lipman and Mehnert/Ozima and other chronology here, because the combined studies span the entire range of activity. However, we acknowledge that these ages may be 20–30% too young.

Zimmerman and Kudo (1979) also studied andesites and dacites, and generated additional major- and trace-element analyses. A select suite of mafic samples was analyzed (including mineral chemistry and some Sr- and Nd-isotopic analyses) by Dungan and others (1981a). Recent major-element, trace-element, mineral-chemistry analyses plus isotopic analyses and petrologic modeling by Dungan and co-workers (Dungan and others, 1983a, b, c; McMillan and Dungan, 1982; McMillan and Willaims, 1983; Williams and Murthy, 1979; McMillan and others, 1983; Moorbath and others, 1983; Thompson and Dungan, 1982) have greatly expanded the analytical data base.

STRUCTURAL EVOLUTION OF THE SOUTHERN SAN LUIS VALLEY

The Taos Plateau volcanic rocks and associated sediments (see map in Lipman and Mehnert, 1979) lie at the southern end of the San Luis Basin (SLB), one of the major axial grabens of the Rio Grande rift. The SLB is terminated at its south end by the Embudo fault zone, which separates the east-tilted SLB from the west-tilted Española Basin and from the upthrown Picuris Mountain block. Although the Abiquiu embayment is topographically part of the Española Basin, it exhibits the same east tilt as the SLB. This is demonstrated by the succession of broad, east-sloping terraces of the El Rito River drainage, between Abiquiu and El Rito.

These similar tilts suggest that the Abiquiu embayment is a western extension of the SLB. However, the structural relationship is complicated by the outcrop of Precambrian quartzite at Cerro Azul (west of the map area, see map in Lipman and Mehnert, 1979). Gravity data in Cordell (1976) indicate that Cerro Azul is the surface expression of an E–W basement ridge north of the Embudo fault which at least in part separates the SLB from the Abiquiu embayment. Like the Abiquiu region, the Tusas Mountains to the north are east-tilted sediments and interbedded volcanics which were deposited within an earlier, broader rift basin. The uplifted Tusas Mountains are now the topographic margin of the southern SLB, but are not separated from it by a major fault. The seismically active boundary between the San Juan and Chama Basins (Cash and others, 1983) is probably the hinge of the broad, east-tilted region that is bounded on the east by the Taos fault.

North-trending, down-to-the-east faults and monoclines that parallel the Rio Grande define the western border of the main downthrown graben block (Taos graben) which lies between the Sangre de Cristo Mountains on the east and a major intra-rift horst on the west. Although the horst lies largely in the subsurface (Cordell, 1978), it is exposed in two places west of the Rio Grande on the Taos Plateau (Brushy Mountain and Timber Mountain) and extensively in southern Colorado as the San Luis Hills. Rocks exposed on the horst are early Miocene intrusions, lavas, and pyroclastic rocks associated with the early phase of rifting (Lipman and Mehnert, 1979; Thompson and Dungan, 1982). Much of the Sangre de Cristo Mountain front has no Quaternary fault scarp along it within the study area, yet it marks the position of the main border fault (Taos fault) of the SLB (and the Taos graben) on the eastern margin of the Taos Plateau. The northwest-trending frontal segment of the mountain front and the Red River fault zone (Fig. 2) may be a reactivation of a northwest-trending fault system present in the Questa caldera (Lipman, 1983).

The Embudo zone is seismically active (Cash and others, 1983), and there is a sag in a releveling line along U.S. Highway 84 where it crosses the fault southwest of Black Mesa. Post-Pliocene movements are demonstrated by the fault scarps along the base of the Picuris Mountains, in the paleodepression tilted by Servilleta Basalts at La Mesita and Black Mesa where they straddle the fault zone (and are displaced by it), and in the extra thickness of the Chamita Formation under Black Mesa.

The history of motion on the Embudo fault zone is complex. Left

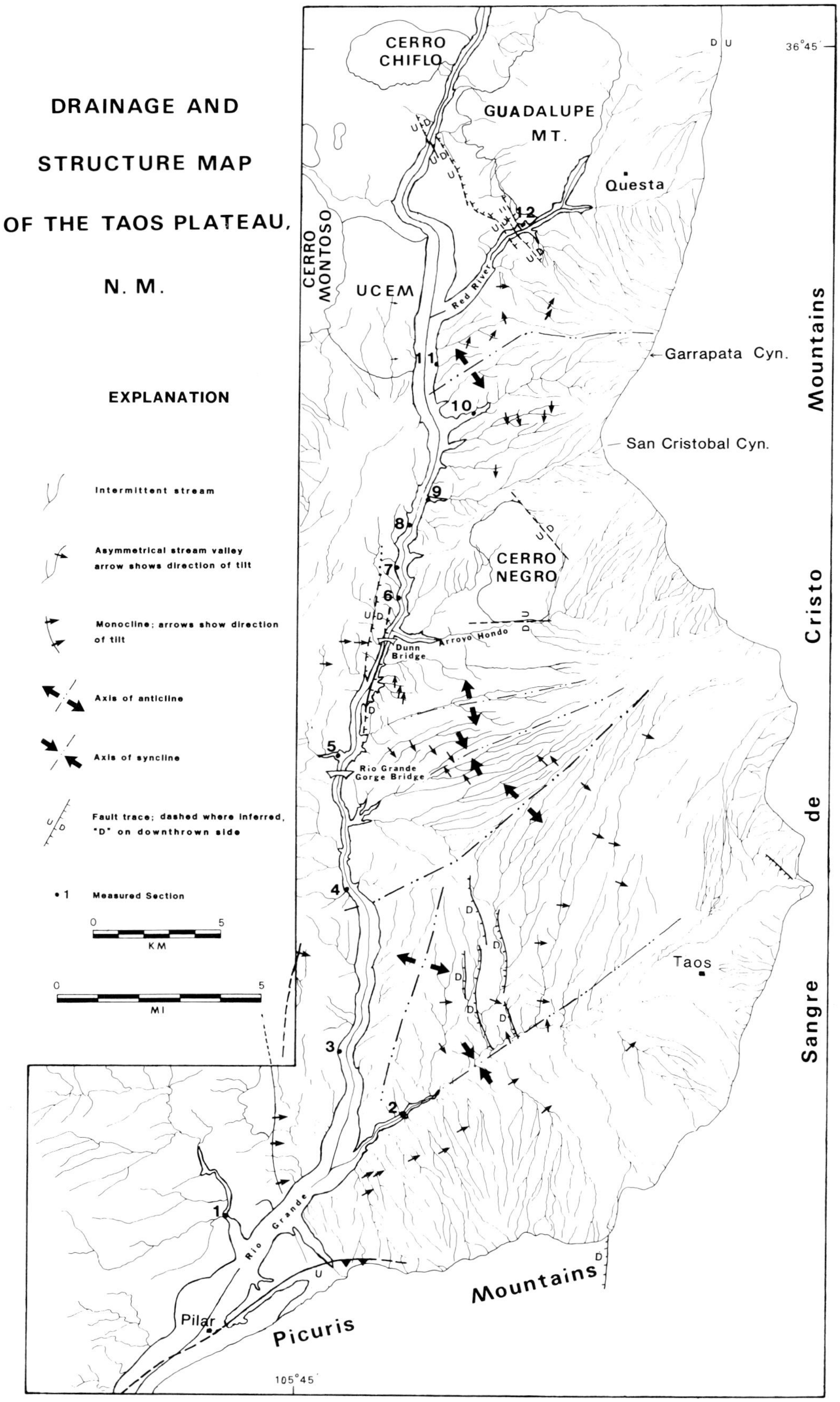

FIGURE 2. Drainage and structure map of the Taos Plateau. Location of measured sections of Figure 7 shown by numbers. Modified from Peterson (1981).

slip along the Embudo is required by the opening of the San Luis and Española Basins. Early Santa Fe sedimentary rocks of the Bradley conglomerate are broken by a conjugate fracture system which suggests left-lateral wrenching with transcurrent movement oriented N 60° E, parallel to the Embudo fault trend (Leininger, 1982). Left-lateral slip along the Embudo fault is a late development in the history of the rift. Early rifting was dominated by simple extension and dip–slip motion. The transcurrent motion began during the Pliocene and is synchronous with the major uplift of the Picuris Mountains in that epoch. The Embudo fault at present has a hinge point near Pilar. East of Pilar the south side is moving up and over the north (Taos Plateau) side. Analysis of the fractured Bradley conglomerate clasts plus the microfaulted and sheared Ojo Caliente dunes in the Pilar area support a late compressive (reverse faulting) and strike–slip stage (left-lateral offset) (Muehlberger, 1979; Leininger, 1982). West of Pilar the movement is slightly divergent, resulting in a narrow, downthrown fault block that opens southwestward into the Española Basin. The north boundary fault crosses the Rio Grande and appears to split into two branches. One apparently continues S 60° W through the crestal graben of the Jemez caldera (i.e., along trend of the Jemez lineament), and the other turns southward as the Pajarito fault (Golombek, 1983; Los Alamos fault zone of Kelley, 1954) that marks the west side of the Velarde graben (Manley, 1979). The south boundary fault of the narrow downthrown block turns southward through Española and marks the east side of the Velarde graben.

Active subsidence of the Velarde graben is asymmetrical with a sense opposite to that characterizing the SLB. Strata within the Velarde graben are tilted to the west and movement is dominantly along the Pajarito fault. Thus, the Velarde graben is the counterpart of the Taos graben in its structural position across the Embudo fault zone which can be regarded as an intracontinental transform (Muehlberger, 1979) connecting two narrow zones of active extension.

The Pajarito fault scarp exhibits a large magnitude of displacement (100 m in 1.1. m.y.) which is comparable to the rate determined by Lipman and Mehnert (1979) for San Pedro Mesa (500 m in 3 m.y.). Although movements along most of the Taos fault (and the Sangre de Cristo frontal fault in general) are difficult to assess quantitatively, we infer similar large displacements throughout its entire length. The highest scarp within the Taos Plateau proper is along the Dunn Bridge fault (35 m in 3.5 m.y.) which probably marks the structurally lowest point in the Taos graben. Its total length is but a small fraction of the Pajarito or Taos faults. All other fault scarps in the region are short, with generally low displacements (meters to a few tens of meters).

Volcanism in the Española Basin straddles the Velarde graben. The relatively small Cerros del Rio field (Baldridge, 1979) lies to the east of the Pajarito fault; the much larger Jemez volcanic field lies to the west of the Pajarito. Additional basaltic volcanic centers lie to the south along the offset to the Albuquerque Basin (San Felipe Mesa). Vents in the Taos Plateau volcanic field are scattered throughout the southern San Luis Basin. None of the post-Miocene vents lie to the east of the Taos fault, although the coeval Ocate volcanic field lies 40 km to the east on the boundary between the Sangre de Cristo Mountains and the High Plains. The crestal graben of the resurgent dome of the Valles caldera lies along the Jemez trend, suggesting some Jemez trend control. However, there is very little indication that the structures which define boundaries between major structural blocks on the Taos Plateau play a role in controlling the locations of vents.

MID- TO LATE CENOZOIC SEDIMENTATION HISTORY

Thick sections of pre- and early-rift volcanic rocks and early-rift alluvial deposits west of the Taos Plateau, and in the Questa volcanic center to the east, indicate that the early phase of rifting (>20 m.y.) was not accompanied by uplift of the region (Lipman and Mehnert, 1979). Although the early-rift volcaniclastic deposits are not exposed in the study area, the Picuris Formation, a late Oligocene (to early Miocene?) tuffaceous sand and gravel deposit, has been mapped to the east along the northern Picuris Mountain border (Montgomery, 1953), and the Abiquiu Formation, which thickens southward into the Abiquiu region (Smith, 1938), occurs to the west.

Gravity data indicate that the greatest subsidence (7–8 km) occurred in the deep Taos graben along the eastern edge of the southern SLB (Cordell, 1978; Lipman and Mehnert, 1979). Five kilometers of alluvial sediments shed off the Sangre de Cristo Mountains filled the basin as it dropped down along the Taos fault (Cordell, 1978). The Rio Grande flows along the western boundary of the Taos graben. Its position is in part controlled by apparently long-lived normal faults associated with this boundary, and in part by alluvial fans prograding to the west from the mountain front. Exposures in the gorge reveal depositional systems only as old as late Pliocene.

The early history of this region is exposed along the uplifted southern margin of the Taos Plateau within the Embudo fault zone where the Picuris Mountains have been overriding the plateau since deposition of the early Pliocene Cieneguilla Member of the Santa Fe Group. The oldest Tertiary unit in the study area is the early Miocene(?) Bradley conglomerate member (informal name) of the Tesuque Formation (Fig. 3). This unit is exposed in a small area at the north base of the Picuris Mountains southeast of Pilar. The unit is composed of debris flows in which the dominant clast lithologies are granitic and quartzitic. This deposition appears to record the presence of an alluvial-fan surface fed by a source area 7–15 km south of the study area. The Bradley conglomerate represents the initial emergence of the southern Picuris Mountains (Leininger, 1982).

Approximately 18 to 13 m.y. B.P. the depositional regime changed and detritus shed off the uplifting Sangre de Cristo Mountains covered

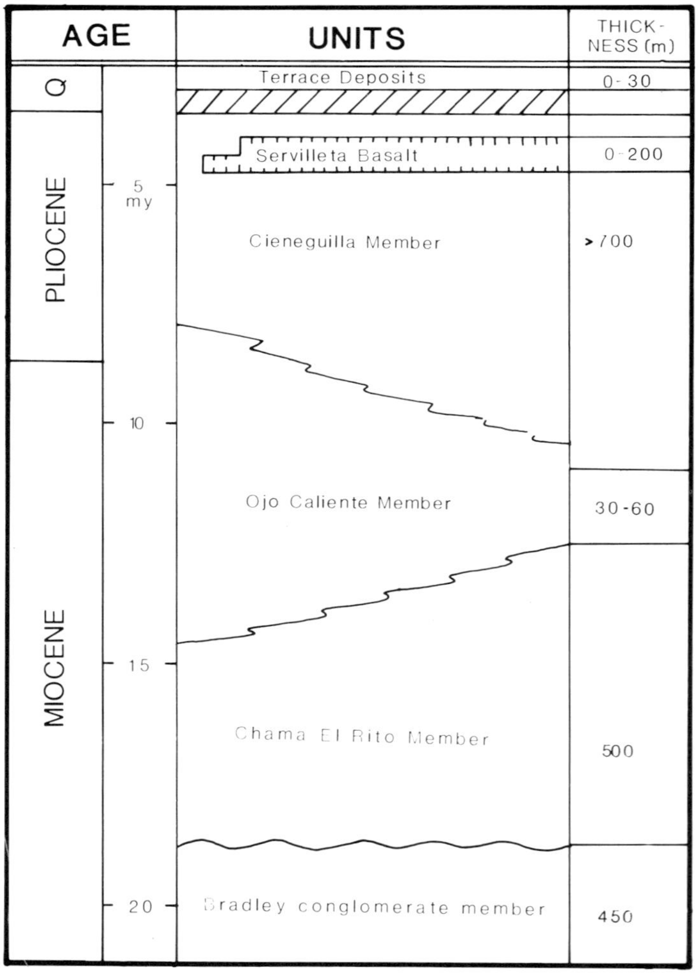

FIGURE 3. Cenozoic stratigraphy exposed in the study area. Except for the terrace deposits, all units are part of the Tesuque Formation, Santa Fe Group. Details of interfingering between the Cieneguilla Member, Servilleta Basalt (formerly Servilleta Formation), and other volcanic types are shown in Figures 5 and 7. Q–Quaternary.

the study area and extended southwestward well into the Española Basin. The southwesterly flowing braided streams of the Chama–El Rito alluvial fan buried the Bradley conglomerate. Streams scoured upper Bradley deposits, producing an unconformable surface of modest relief. The predominantly volcaniclastic Chama–El Rito unit has upward progressively more detritus of Precambrian quartzite derived from the emerging Picuris Hills south of the study area. Steinpress (1980) recognized a Dixon Member which overlies the Chama–El Rito near La Mesita (15 km SW of Pilar) that is characterized by predominantly Paleozoic detritus derived from the southeast, especially in the coarse fractions.

Deposition of the Ojo Caliente Member probably began around 15 m.y. ago. Prevailing westerly winds deflated the distal Chama–El Rito fan surface and other exposures along the western SLB margin, transporting the eolian sediment into the southern SLB. Early Ojo Caliente sands interfingered with Chama–El Rito fluvial sands and gravels (Fig. 3). Eolian sedimentation eventually dominated, perhaps reflecting a climatic change to a more arid environment. The upper Ojo Caliente Member interfingered with Cieneguilla Member sands and gravels that were shed north from the Picuris Hills rising nearby (Fig. 4).

Cieneguilla Member sedimentation was a result of uplift and erosion of the Picuris Range beginning at 8–10 m.y. Lower Cieneguilla deposits contain a high percentage of volcanic pebbles and thick sand lenses. This indicates that earlier Tesuque Formation deposits (Chama–El Rito Member and Ojo Caliente Member) were initially a major source for the lower Cieneguilla beds. Middle and upper Cieneguilla deposits were derived entirely from the Precambrian quartzite and schistose rocks of the Picuris Mountains and became progressively more coarse through the late Miocene and Pliocene, reflecting an increase in relief between the SLB and Picuris Mountains. In mid- to late Cieneguilla time a major uplift and tilting event juxtaposed coarse, proximal facies alluvial-fan sediments on top of early distal sandy-facies sediment, concurrently with Servilleta volcanism (Leininger, 1982).

LATE CENOZOIC VOLCANISM

Lipman and Mehnert (1979) classified and mapped the basin units of the TPVF and dated main phases of mafic and intermediate activity at between 4.5 and 2.5 m.y. (see also Ozima and others, 1967; Baldridge and others, 1980). Younger alkalic lavas, in part xenocrystic, postdate the eruption of the Servilleta Basalt. The Cerro Chiflo quartz latite is much older (10.2 m.y.) than the main phase of activity and is distinct from younger, intermediate-composition lavas. The No Agua rhyolite domes on the western edge of the Taos Plateau erupted during the main phase of activity, but do not seem to be petrologically related to the basalts, andesites, and dacites.

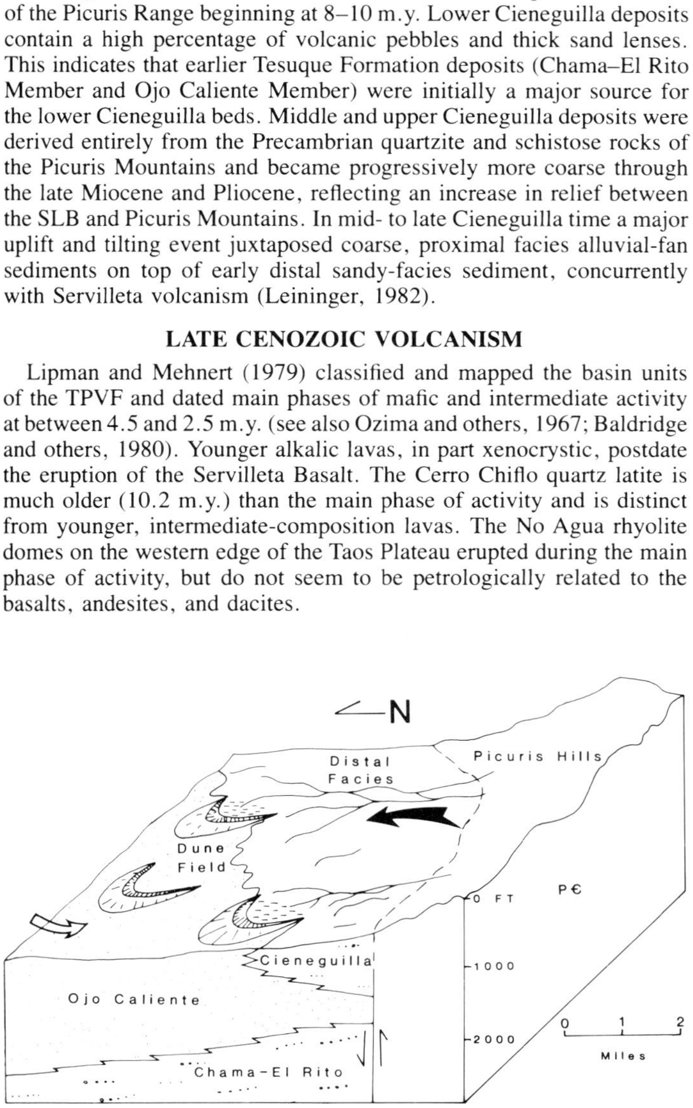

FIGURE 4. Block diagram of Pilar area during upper Ojo Caliente Member deposition. The dune sands interfinger with, and are reworked by, Cieneguilla Member braided streams from the rising Picuris Hills. Open arrow is prevailing wind direction; solid arrow is stream-sediment transport direction. From Leininger (1982).

Lipman and Mehnert (1979) have called attention to the zoned distribution of vents on the Taos Plateau. Basalt shields on the western margin of the TPVF are partially surrounded on the east by olivine-andesite shields. Dacite cones, built in part on earlier andesite shields (San Antonio, Ute, Guadalupe), form an outer semicircular ring circumscribing all other vents. Most of the younger, alkalic lavas erupted from cinder cones localized near the greatest concentration of older shields. This organized pattern suggests a unified magmatic system during much of the Taos Plateau's volcanic history. The present study is an attempt to provide a temporal perspective that complements the spatial relationships revealed in Lipman's map.

Below we summarize the compositional, petrographic, and flow features that characterize the different units exposed within the Rio Grande and Red River gorges. One of the units is a new informal designation (San Cristobal lavas) used in subdividing the Servilleta Basalt (Figs. 6, 7). Consequently, we describe these rocks in more detail than the intermediate (andesite and dacite) or alkalic rocks which are discussed in Lipman and Mehnert (1979). Readers are asked to refer to that article plus Dungan and others (1981a) for additional analyses and descriptive materials, and for a geologic map of the Taos Plateau.

Terminology

The volcanic-rock nomenclature used in this paper follows, with one exception, the terminology proposed by Lipman and Mehnert (1979). We have substituted dacite for rhyodacite and extended the compositional range (60–67.5% SiO_2) to which this name is applied on the basis of additional analyses. The distinctions between olivine andesite and dacite are discussed below. Lipman and Mehnert (1979) advocated substitution of the term Servilleta Basalt for Servilleta Formation and restriction of the former to the diktytaxitic olivine tholeiites which are the dominant unit of the TPVF. We concur with this usage and introduce the informal term San Cristobal lavas (SCL) to refer to a package of flows which is similar to, but distinct from, the Servilleta olivine tholeiites sensu stricto in terms of composition, petrography, and flow structure.

The major-element analyses of intermediate-composition lavas from the TPVF that were available to Lipman and Mehnert (1979) enabled them to distinguish two mineralogically distinct groups, an olivine-andesite rock type with 56–59% SiO_2, and more evolved "rhyodacites" which are characterized by pyroxene phenocrysts. The olivine-phyric rocks are dominantly in the range 56–58% SiO_2, but the entire range extends from 54 to 60% SiO_2. We retain the term olivine andesite for these rocks. The pyroxene-phyric rocks form a coherent group that ranges in SiO_2 from around 60 to 67.5%. The rocks with >65% SiO_2 are comparatively rich in plagioclase and a few contain minor hornblende, but these samples appear to form a compositional continuum even though the majority still fall within the 62–64%-SiO_2 limits originally defined for "rhyodacite." We have chosen to apply the term dacite to the entire spectrum of pyroxene-phyric rocks ranging from 60 to 67.5% SiO_2 and have subdivided this range into low-SiO_2 (<64) and high-SiO_2 (>64) dacites. The entire suite, from andesite to dacite, falls into the high-K series of Gill (1981) and Pecerrillo and Taylor (1976). Both classifications use 65% SiO_2 as the boundary between andesite and dacite. Application of the name rhyodacite to rocks with less than 65% SiO_2 would be contradictory to the current usage of this term, particularly in light of the relatively high Mg/(Mg + Fe) of the TPVF low-SiO_2 dacites.

Silicic Alkalic Basalt

Lipman and Mehnert (1975) proposed the term silicic alkalic basalt (SAB) for Cenozoic mafic lavas of the Rocky Mountains which are distinct from both typical olivine tholeiites and nepheline-normative alkali-olivine basalts. SAB lavas of the TPVF are characterized by comparable to higher SiO_2 than typical tholeiites (50–53% SiO_2), higher alkalies (especially K_2O), and much higher concentrations of incompatible trace elements such as LREE, Rb, Nb, Zr, Th, and U. The trace-element signatures of SABs generally resemble those of alkali-olivine basalts. Their higher SiO_2 and higher alkalies result in normative compositions characterized by minor *hy* or more rarely minor *ne*. Even

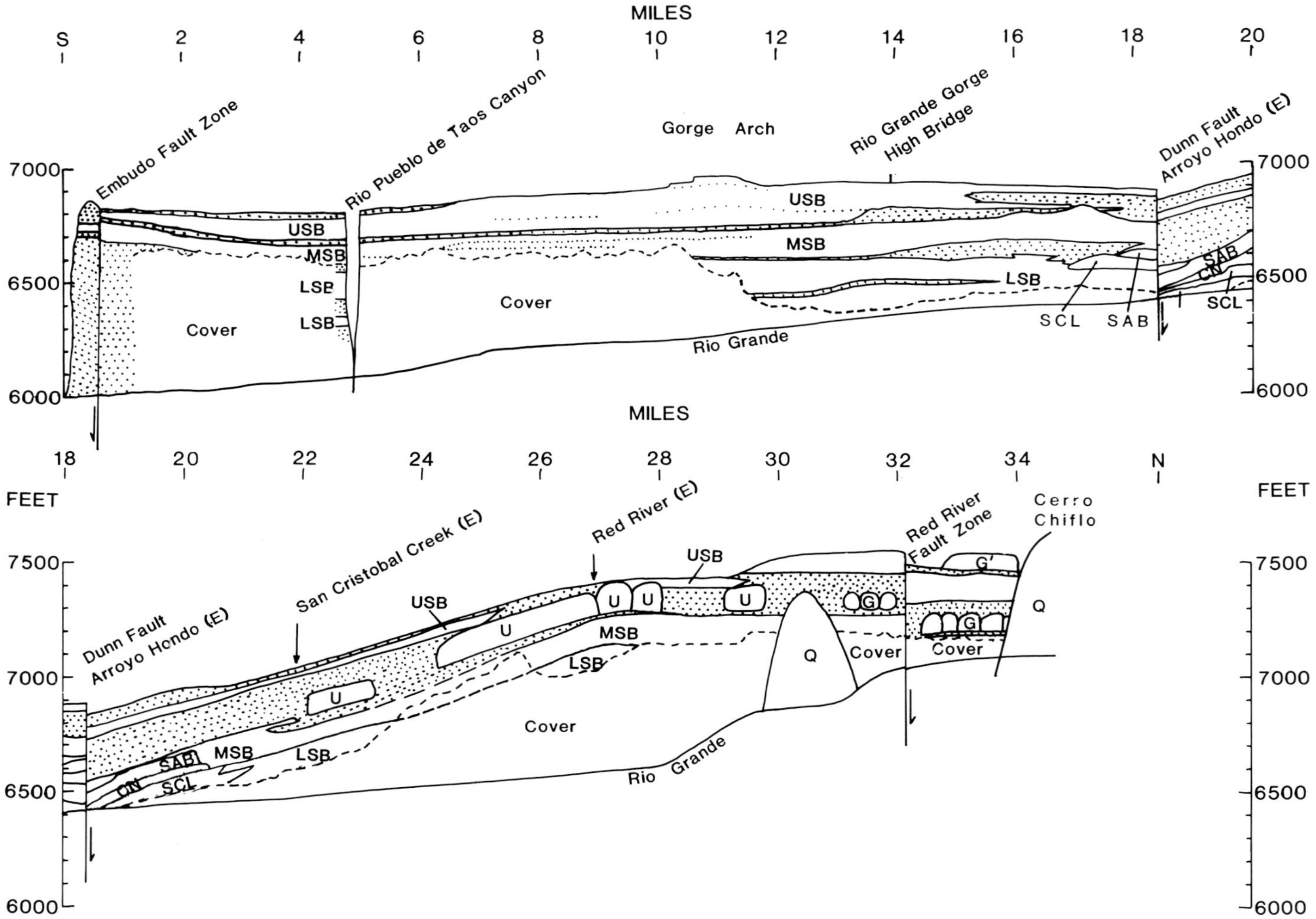

FIGURE 5. Generalized stratigraphy of Rio Grande gorge from Pilar (south end) to Cerro Chiflo (north end), showing sedimentary intervals (dot pattern) and volcanic intervals (labeled units). Symbols: Servilleta Basalt divided into lower (LSB), middle (MSB), and upper (USB) units; SAB–silicic alkalic basalt; CN–Cerro Negro dacite; SCL–San Cristobal basaltic lavas; U–dacite from unnamed cerrito east of Cerro Montoso (flows rim only west wall of canyon); G–older dacite from Guadalupe Mountain; G′–younger dacite from Guadalupe Mountain; Q–quartz latite from pyroclastic vent exposed in the Rio Grande gorge and flows from Cerro Chiflo; E–location of rivers entering the Rio Grande from the east. Modified from Leininger (1981) for strip from Pilar to High Bridge, and Peterson (1981) from High Bridge to Cerro Chiflo.

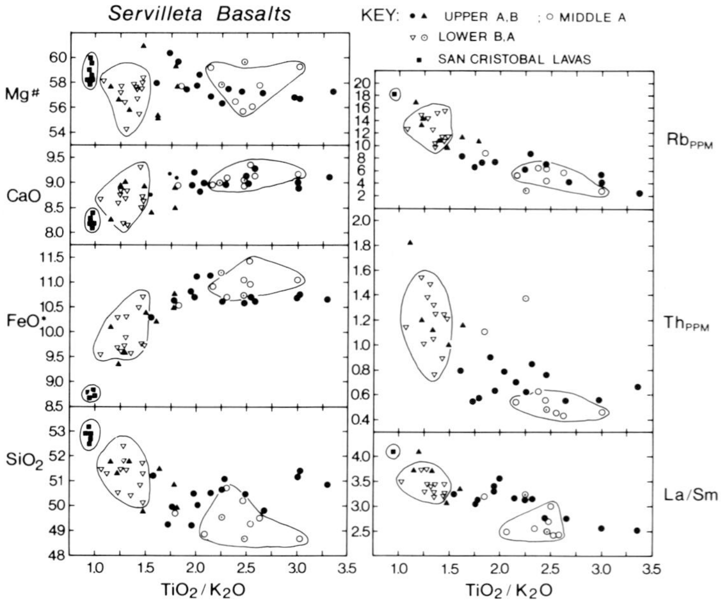

FIGURE 6. Servilleta Basalt and San Cristobal lava, chemical variations for select major and trace elements. ($Mg\# = 100\ Mg/(Mg + Fe^{2+})$). Circled fields are SCL (■), LSB type B (∇) and MSB type A (0). Closed symbols are types A (0) and B (Δ) from the USB. Two samples of type A within LSB are denoted by 0. Note the large variations in incompatible elementts such as Rb and Th in comparison to the relatively minor range of major elements, and the nearly constant Mg#. These relations eliminate simple fractional crystallization of a single parent as the primary source of major- and trace-element variations.

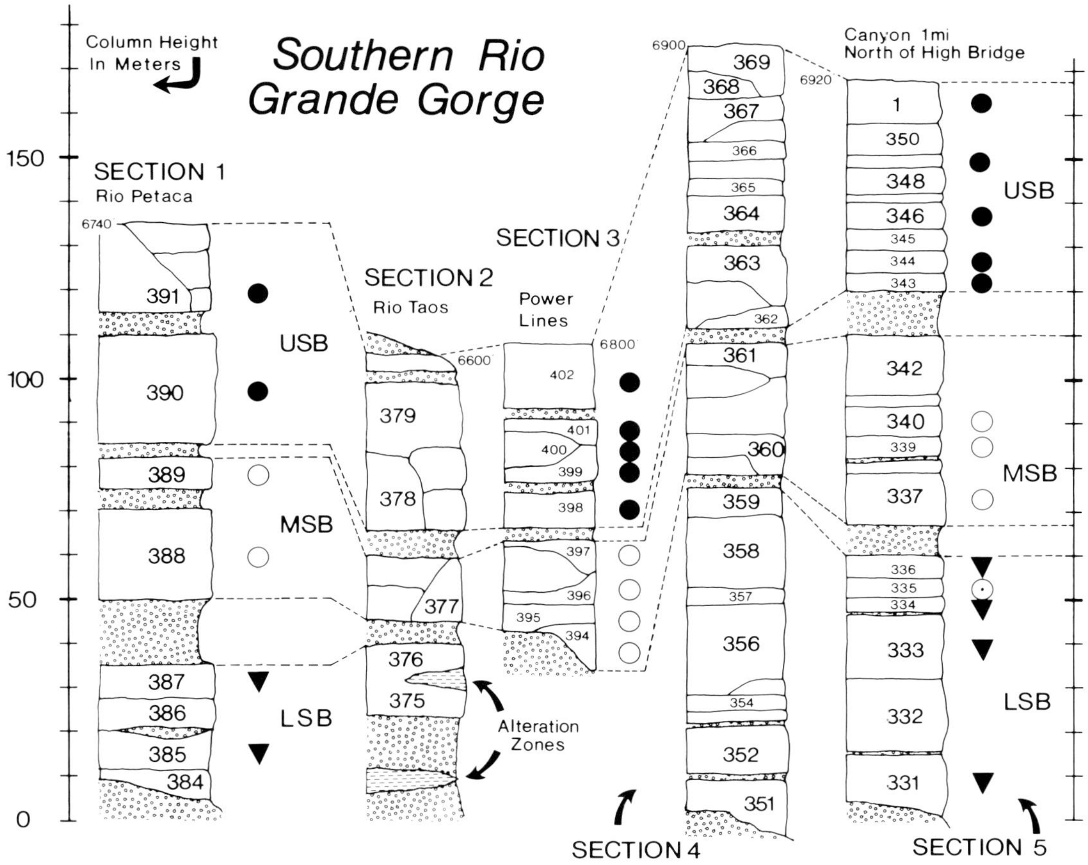

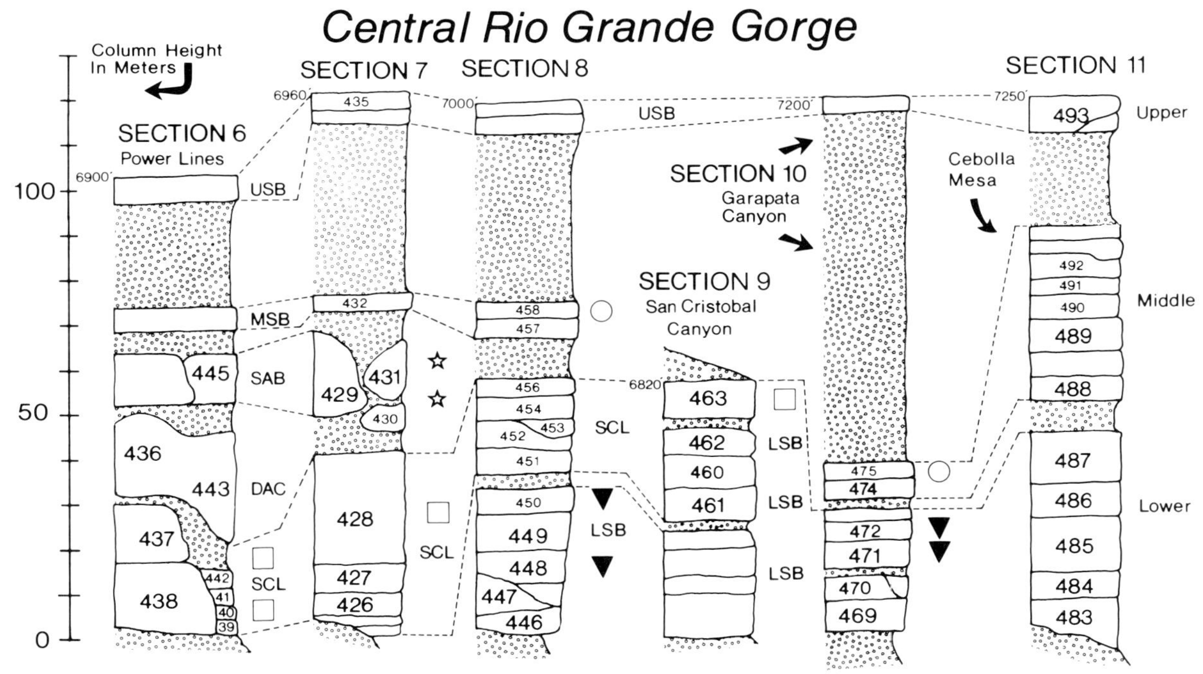

FIGURE 7. Measured stratigraphic columns for mixed volcanic/sedimentary sections from the southern and central segments of the Rio Grande gorge, Taos Plateau. Locations of sections shown in Figure 2. Sampled flows are numbered, symbols to the right of some sections indicate chemical data were used to establish the magma type. Symbols are the same as in Figure 6 except for LSB type B, where filled triangles are used instead of open triangles. Elevations of the rim of the gorge at each section are given in feet adjacent to each column.

more SiO_2-rich lavas, which may be related to SABs, are the so-called xenocrystic basaltic andesites (XBA) which have SiO_2 as high as 57% and abundant xenocrysts of quartz and sodic plagioclase. XBA flows of the TPVF are closely associated in space and in time with SAB lavas and have similar trace-element concentrations. Most of the SAB and XBA flows of the TPVF are late-stage, post-Servilleta eruptives associated with monogenetic strombolian cones. The majority of vents are located along the western margin of the TPVF in the vicinity of San Antonio Mountain—No Agua–La Segita Peaks (see map in Lipman and Mehnert, 1979). In contrast to the thin and extensive pahoehoe lavas of the Servilleta, the SAB and XBA flows are thicker aa lavas with substantial basal breccias.

The only recognized SAB lavas within the Rio Grande gorge are a group of closely related flows that occupy a narrow stratigraphic horizon between the lower and middle Servilleta. This group typically overlies the San Cristobal lavas (see below) and has a very similar, but more limited geographic extent. Thin sedimentary interbeds separate the SAB lavas from both the underlying and overlying volcanic units. The SAB "package" is more than one-flow thick at only one locality where we have observed several flow termini (section 7, Fig. 7), and consists of a single thick (10–15 m) flow through most its extent. The most extensive of these flows is exposed at road level at Dunn Bridge.

The SAB unit exposed within the gorge is chemically and petrographically distinct from all other SAB lavas. It has relatively high SiO_2 (52.5 wt%) K_2O and the LREE are lower (e.g., RG-15 in Table 1, #9; or TP-7 in Dungan and others, 1981a). Various flow units within the package have different ratios of 100 Mg/(Mg + Fe) ranging from 56 to 52, indicating variable degrees of olivine fractionation. All the flows contain olivine phenocrysts, but the less differentiated variants contain more abundant, coarse-grained olivine phenocrysts of a more magnesian composition (Fo_{83} vs. Fo_{75}) than the more evolved flow. Despite evidence for variable olivine fractionation, the flows are closely similar in all other compositional parameters. The groundmass texture is intergranular. Plagioclase phenocrysts in the SAB lavas exhibit a complex zoning pattern and abundant small crustal xenoliths (gabbros, norites, anorthosites) are present.

TABLE 1. Select major- and trace-element analyses of basaltic rocks.

	1	2	3	4	5	6	7	8	9
Sample	RG.335	RG.340	RG.25	RG.346	RG.21	RG.91	RG.336	RG.440	RG.15
Section #	5	5	B	5	CC	LS	5	6	DB
Unit	LSB	MS3	USB	USB	USB	USB	LSB	LSB	-
Magma Type	A	A	A	A	B	B	B	SCL	SAB
SiO_2	48.69	48.73	50.83	50.67	51.81	51.70	51.34	53.20	52.01
TiO_2	1.25	1.27	1.21	1.21	1.21	1.21	1.11	1.13	1.55
Al_2O_3	16.36	16.44	16.41	16.45	16.46	15.77	16.77	16.49	16.94
Fe_2O_3	1.93	2.79	1.13	3.40	2.13	2.92	2.36	9.85	3.27
FeO	8.94	8.50	9.64	7.53	7.72	7.52	7.43	-	5.87
MnO	0.17	0.18	0.17	0.17	0.16	0.16	0.16	0.14	0.15
MgO	7.98	8.02	7.20	6.88	6.14	6.94	6.80	6.12	5.64
CaO	9.07	9.21	9.16	8.99	9.00	8.26	8.72	8.23	7.74
Na_2O	3.13	3.04	3.03	2.96	3.16	3.13	3.22	3.21	4.08
K_2O	0.51	0.42	0.36	0.54	0.91	1.04	1.04	1.29	1.71
P_2O_5	0.17	0.18	0.16	0.19	0.26	0.21	0.18	0.23	0.35
H_2O	0.40	0.66	-	0.93	-	-	0.90	-	-
Total	98.60	99.44	99.30	99.90	98.96	98.86	100.05	99.89	99.33
FeO*	10.68	11.01	10.66	10.59	9.64	10.15	9.55	8.86	8.81
Mg#	59.7	59.1	57.2	56.3	55.8	57.5	58.5	57.8	55.9
TiO_2/K_2O	2.45	3.02	3.36	2.24	1.33	1.16	1.07	0.88	0.90
Ta	0.48	0.40	n.a.	0.52	n.a.	n.a.	0.51	0.70	n.a.
La	7.73	7.71	7.63	10.19	16.4	14.0	12.23	18.25	22.8
Ce	19.2	18.5	17.4	24.0	36.5	33.0	28.00	39.1	50.4
Sm	3.11	3.06	3.00	3.21	4.35	3.71	3.45	4.00	5.29
Eu	1.11	1.13	1.03	1.10	1.35	1.18	1.18	1.29	1.58
Tb	0.52	0.54	0.62	0.57	0.72	0.67	0.57	0.66	0.79
Yb	1.79	1.95	1.90	1.90	2.10	1.96	2.03	1.95	2.33
Lu	0.28	0.29	0.31	0.28	0.32	0.31	0.32	0.31	0.40
Hf	2.26	2.16	2.14	2.75	2.99	1.84	2.74	3.40	2.24
Th	0.48	0.46	0.66	0.63	1.12	2.95	1.15	1.60	4.41
Sc	23.4	24.0	24.0	23.0	24.0	22.5	23.2	22.4	23
Cr	227	195	185	164	148	199	172	183	32
Ni	151	146	136	140	104	175	113	95	-
Nb	7.4	6.8	5.4	7.3	10.1	8.8	8.5	13.5	20.3
Zr	89	84	78	91	120	105	112	132	172
Y	18.9	20.6	20.5	19.8	22	20.6	20.6	20.1	25.2
Sr	371	354	280	323	410	343	428	534	616
Rb	4.3	2.7	3.6	6.5	11.5	17.2	12.6	18.1	26.7
Ba	234	270	195	295	440	-	367	568	625
V	154	139	-	146	-	-	132	137	-
La/Sm	2.49	2.52	2.54	3.17	3.77	3.77	3.54	4.56	4.31
La/Yb	4.32	3.95	4.01	5.36	7.81	7.14	6.02	9.36	9.79
La/Nb	1.05	1.13	1.41	1.40	1.63	1.59	1.44	1.35	1.12
Zr/Rb	20.7	31	21.7	14.0	10.4	6.1	8.9	7.3	6.4
Nb/Rb	1.72	2.5	1.50	1.12	0.88	0.51	0.67	0.75	0.76
Zr/Nb	12.0	12.3	14.4	12.5	11.9	11.9	13.2	9.8	8.5

Section locations on Figure 2; CC = Gorge rim near Cerro Chiflo, LS = LaSegita Peaks, DB = Dunn Bridge

Intermediate-composition Lavas

Dacite lavas flowed westward from the two double cones—Cerro Negro and Guadalupe Mountain (Fig. 2)—and are exposed in the Rio Grande gorge several kilometers north of Dunn Bridge and in the vicinity of Cerro Chiflo, respectively. The upper Red River gorge is incised through a mixed section, comprising Servilleta basalt, olivine andesite, and low-SiO_2 dacite that form a composite shield beneath the south end of Guadalupe Mountain. The intermediate lavas interfinger with Servilleta Basalts to the west, and are absent in the lower Red River gorge. A vent located west of the Red River gorge and east of Cerro Montoso (UCEM–*u*nnamed *c*errito *e*ast of Cerro *M*ontoso) is the source of a single, large dacite flow which crops out at or near the top of the west wall of the gorge for a length of 12 km.

The Cerro Negro, Red River, and Guadalupe dacites exposed in the gorge are comparable in composition and phenocryst mineralogy to dacite lavas exposed on the Cerro Negro and Guadalupe cones, and to those that comprise other dacitic shields on the Taos Plateau. The dominant phenocrysts are orthopyroxene and augite, although plagioclase phenocrysts are present in some rocks in minor abundance, and plagioclase microphenocrysts are widespread. Some low-SiO_2 dacites also contain relict, resorbed olivines jacketed by euhedral orthopyroxene phenocrysts. The UCEM dacite is unique among the TPVF intermediate lavas in occurrence (single flow vs. shield) and composition—anomalously low concentrations of incompatible trace elements (except Rb) more akin to the Servilleta Basalts. The UCEM flow is a high-SiO_2 dacite which is aphyric and exceptionally glassy.

TPVF dacites typically exhibit pronounced internal foliations, commonly expressed as well-developed platy zones. The flow fronts of the TPVF dacites are typically splayed into radial, finger-like extensions. The Cerro Negro and Guadalupe dacite flows exposed in the gorge have flow directions at approximately right angles to the canyon. As a consequence, the exposures of the dacite flows in the gorge occur as a series of isolated-finger cross sections with prominent concentric platy foliations. These fingers are as much as 50–100 m in diameter in the Rio Grande gorge south of Cerro Chiflo, where Guadalupe Mountain flows are exposed beneath a cap of upper Servilleta (Fig. 5).

The olivine andesites that comprise the middle portion of the stratigraphic section in the Red River gorge (Fig. 9) are also atypical in several respects compared to andesites which comprise other shields on the TPVF. These lavas are characterized by exceptionally large skeletal-olivine phenocrysts (2–5 mm) of Fo_{80} composition. However, the outstanding feature of these lavas is the relatively wide spectrum of compositions present in a single section. Most olivine-andesite shields are remarkably homogeneous in the compositions of lavas exposed, whereas the Red River andesites range from 54–57% SiO_2. Moreover, there is a regular increase in SiO_2 from the base of the section to the

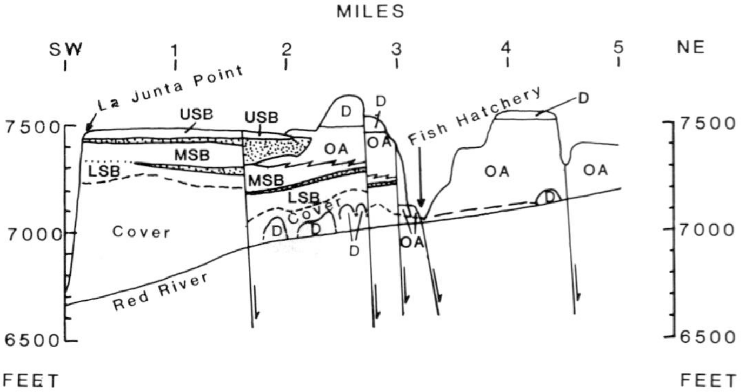

FIGURE 8. Generalized stratigraphy in Red River gorge for 5 mi upstream from its junction with the Rio Grande at La Junta point. Symbols as in Figure 6 except for: OA–olivine andesite; D–dacite. From McMillan (1984).

top which can be modeled as the result of pre-eruptive, two-component mixing between Servilleta Basalt and the low-SiO_2 dacite that caps the Red River section (McMillan and Dungan, 1982).

The Cerro Chiflo quartz latite and an apparently related pyroclastic cone exposed in the gorge to the south are the lone representatives of late Miocene volcanism in the TPVF. Cerro Chiflo is also compositionally and mineralogically distinct from later TPVF intermediate lavas in that it is more evolved ($SiO_2 = 67.5$, $MgO < 1.5$ wt%), and contains abundant hornblende and biotite phenocrysts and crustal xenoliths.

San Cristobal Lavas

The San Cristobal lavas (SCL) are an extreme chemical variant of the Servilleta Basalts (Table 1, #8); they are characterized by high SiO_2 (52.5–53.5), high K_2O (1.25 wt%), and high concentrations of incompatible trace elements (REE, Rb, Zr, Nb, Th, Hf, Sr, and Ba). The SCL comprise a package of flows that interfinger with the upper portion of the lower Servilleta between Manby Hot Springs and San Cristobal Creek. Near the middle of this range the entire section bounded by sedimentary interbeds consists of SCL. To the north and south of this area the proportion of typical lower Servilleta flows progressively increases until the SCL no longer are present. This geometry suggests that the SCL were erupted from a discrete vent contemporaneously with the younger package of lower Servilleta.

The SCL tend to form thick (15–25 m), massive flows, although several vertical sections are composed of multiple thin flows (<3 m). All SCL exhibit pahoehoe flow tops. However, the distinctive coarse ophitic–diktytaxitic groundmass texture and vesicular segretations typical of the Servilleta Basalt sensu stricto are absent. The SCL groundmass texture is intergranular. Plagioclase phenocrysts are rare, but the groundmass is very plagioclase-rich. Olivine phenocrysts are comparatively abundant (15%), coarse (2–4 mm), and most are skeletal to some degree. Chromite inclusions in olivine phenocrysts are more abundant than in typical Servilleta flows.

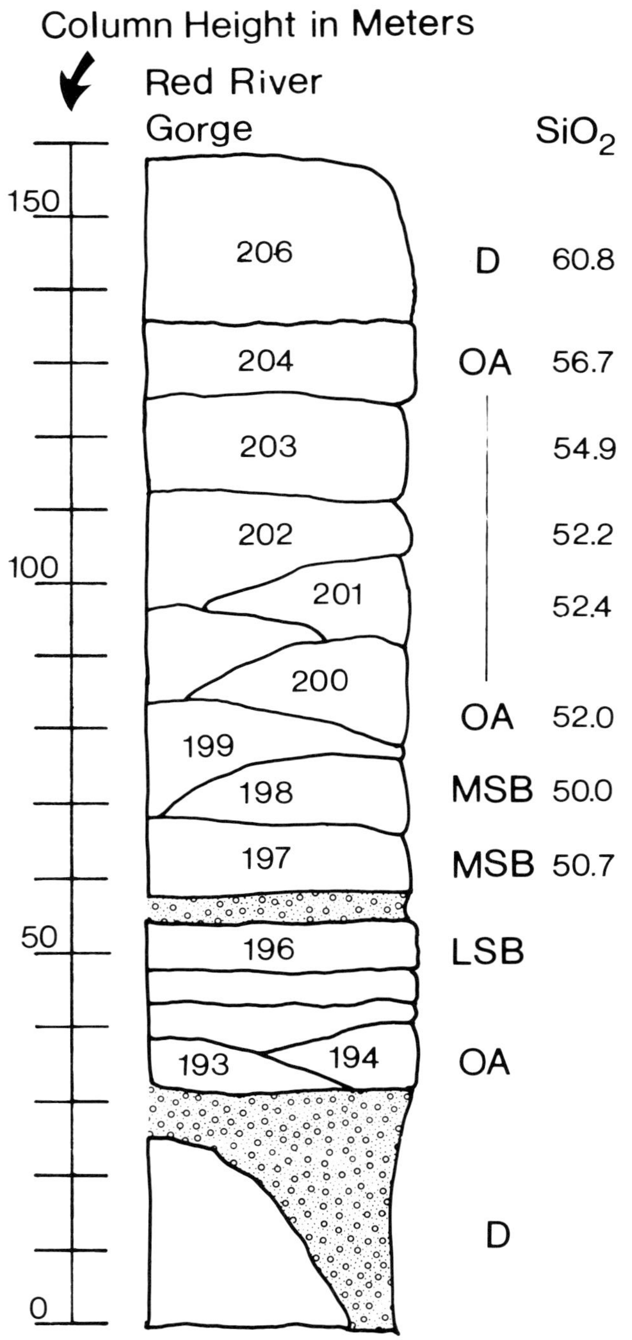

FIGURE 9. Measured stratigraphic column for a mixed volcanic/sedimentary section in the Red River gorge west of the Fish Hatchery fault. Rock types and SiO_2 contents are noted where chemical data are available.

Servilleta Basalts

Introduction

The Servilleta Basalts (SB) are by far the dominant volumetric component of the TPVF. Our stratigraphic studies indicate that most of the SB were erupted in three major pulses that were short in duration compared to the intervening periods represented by sedimentary interbeds (Peterson, 1981). The stratigraphic units that correspond to these three eruptive periods are referred to as lower (LSB), middle (MSB), and upper (USB) Servilleta. The youngest subunits of the USB are two petrographically distinctive variants which are restricted in areal extent to the recognized vent areas on the western side of the TPVF and do not appear at the gorge rim. Geochemical studies indicate that the SB are rather uniform in most major elements (except K_2O) and somewhat more variable in many trace elements (especially Rb, Nb, Sr, Ba, Zr, Th, and LREE). We have concluded that most of the variation observed among Servilleta and SCL samples is the product of mixing between a basaltic end member and andesite magma (Dungan and others, 1983c). Variable degrees of fractionation of the basaltic parent cannot produce the observed compositional spectrum. It also appears implausible that these variations are due to source-region heterogeneity or variable degrees of melting of the same source. In the sections below we briefly summarize some of the general characteristics of Servilleta lavas and then discuss the temporal variations as revealed in the gorge stratigraphy.

Petrography, Mineral Chemistry, and Segregation Features

Olivine phenocrysts (1–3 mm in the long dimension) are present in almost all SB in modest amounts (5–15%). A few flows within the upper Servilleta are aphyric, containing abundant olivine microphenocrysts (<<1 mm). Most of the olivine-phyric SB also contain small, iron-rich groundmass olivines in addition to the phenocrysts. The morphology of the phenocrysts varies from equant to moderately skeletal. We have observed a correspondence between olivine morphology (and size), groundmass texture, and stratigraphic unit. However, there is very little systematic variation in the compositions of olivine-phenocryst cores, as virtually all core compositions in phenocrysts >1 mm fall in the narrow range of Fo_{83}–Fo_{80} (most are Fo_{82-81}). Olivine-phenocryst rims and groundmass grains are zoned to compositions as iron-rich as Fo_{50}, and the microphenocrysts in the aphyric SB flows are also iron-richer (Fo_{75-65}) than the large phenocrysts.

Many olivine phenocrysts contain chromian-spinel inclusions, indicating that it was also an early crystallizing phase. Spinel inclusions within olivine-phenocryst cores are fairly uniform within a single sample and are similar among flows. Spinels partly or completely included in the groundmass are low-Cr, low-Al titanomagnetites. Ilmenite is also present in the groundmass, but is never present as inclusions except as partial enclosures in phenocryst rims. Plagioclase is virtually devoid of chromian-spinel inclusions.

Plagioclase is sparse as a phenocryst phase, but is ubiquitous as microphenocrysts and microlites in glassy-flow margins. The cores of the phenocrysts and microphenocrysts vary from An_{71} to An_{65}. There is a close correspondence between plagioclase composition and bulk rock composition, particularly in terms of bulk rock K_2O and Or-content in the plagioclase. Low K_2O rocks (<0.80) typically have Or <1.0 and higher maximum An-contents than rocks with higher K_2O. Augite is the other important mineral in the groundmass, but it does not occur as phenocrysts.

The groundmass textures in the interiors of SB flows range from intergranular (uncommon) to coarse ophitic–diktytaxitic. Diktytaxitic groundmass texture is a distinguishing feature of the Servilleta Basalts and is present throughout the USB and MSB. Diktytaxitic textures are also present in some LSB flows, but some units (especially the SCL) have groundmass in which ophitic augite is poorly developed or absent and, therefore, have groundmass textures approaching intergranular. The lavas containing non-diktytaxitic textures have poorly developed vesicular segregations (see below), and coarse olivine phenocrysts that commonly are skeletal and include greater than typical quantities of chromian spinel.

One consequence of the diktytaxitic groundmass textures characteristic of the Servilleta olivine tholeiites is the development of vesicular segregations (cf. Kuno, 1960). These occur as vertical pipes (5–10 cm in diameter) and horizontal sheets (up to 30–40 cm thick in thick flows). The vertical pipes are largely confined to the lower one-third to one-half of the flows and commonly terminate upward in a horizontal sheet. The thickest and most extensive sheet-like segregation usually occurs at the top of the zone where the pipes terminate. The sheets decrease in thickness, length, and spacing upward, toward the vesicular pahoehoe-flow tops. The mineralogy of the segregations is the same as in the surrounding flow interiors, but segregations are mesostasis-rich, highly enriched in incompatible trace elements, and depleted in Ni, Cr, Sr, and Sc compared to the matrix. Together with the highly vesicular nature of these features, compositional contrasts between the matrices and segregations indicate that the segregations represent the effects of post-emplacement, in-situ differentiation which occurred in response to movement of residual liquid through the crystalline framework. The pipes represent diapirs rising through the porous framework in response to a density contrast between the crystals and the gas-rich residual liquid. The sheets fill horizontal fractures produced by contraction during cooling.

The vesicular segregations are best developed in flows with coarse ophitic–diktytaxitic texture and least developed in flows with ophitic–intergranular textures, but all true SB contain them. The geochemical trends defined by the trace-element enrichments and depletions in the segregations are superimposed on the primary, pre-eruption trends, resulting in some scatter in the data. Major elements are slightly affected as well by the development of the segregations.

Chemical Stratigraphy

Samples from volcanic sections exposed in the Rio Grande and Red River gorges and in the canyon of Rio Petaca have been analyzed by XRF and INAA. Chemical data exist for some or all of the flows present in sections 1 (Rio Petaca), 12 (Red River), 3, 5, 7, 8, 9, and 10 (Rio Grande gorge, see Fig. 7). Most of the samples have relatively complete data sets produced by merging major-element analyses by XRF (Washington University, Johnson Space Center, and University of Massachusetts) with trace elements (Zr, Rb, Nb, Y, Sr, Ba, Ni, Cr) by XRF (University of Massachusetts) and by INAA (REE, Th, Sc, Hf, and Cr—Washington University). Some trace elements were not analyzed in some samples, and for section 3 we have only INAA data.

We have subdivided the Servilleta Basalts exposed in the gorge and on the Taos Plateau surface into two chemical types termed A and B. These distinctions are based on a combination of major- and trace-element criteria. A number of criteria can be used to distinguish types A and B in the LSB and MSB. One excellent discriminant is TiO_2/K_2O (Fig. 6). Type-B flows have distinctively higher K_2O contents (plus higher Rb, Ba, Sr, Nb, Zr, LREE, Th, and Hf). Type B is also slightly higher in SiO_2 but lower in TiO_2, FeO, and CaO. The sampled MSB lavas all have type-A compositions (e.g., Table 1, #2). The LSB stratigraphic unit is dominantly composed of type B and interfingering San Cristobal lavas (SCL) (Table 1, #7, #8). One notable exception is a single type-A lava present within the LSB in section 5 (Table 1, #1). Little or no overlap in composition for most elements occurs between types A and B lavas of the LSB and MSB. The smallest degree of overlap occurs for incompatible trace elements. The SCL are similar to type B, but are even more extreme in the above-mentioned enrichments and depletions in comparison to type A. The range of TiO_2/K_2O in the type-B Servilleta-SCL is 1.0–1.5, whereas the same ratio in the type-A lavas is 1.75–2.75.

The USB overlaps in composition with both the type-A and type-B Servilleta and extends the range of TiO_2/K_2O variation to 3.3 (Table 1, #3, 4, 5, 6). There is no clustering of compositions into groups in the USB; rather, a continuum is defined. USB lavas that fall within the 1.25–2.5 TiO_2/K_2O range are difficult to classify as either type A or B because of variations in elements that do not exhibit overlap between lower and middle type A and type B. Thus, the distinction between types A and B in the USB has less meaning than for the lower two units.

The USB has been analyzed more than the lower and middle units because of more widespread exposures. Thus, additional analyses might show more variability within the lower units. A suggestion of such variability is provided by a few of the relatively small number of samples analyzed from the northern segment of the gorge. Specifically, the apparent type-A sample from the MSB at Garapata Canyon, where the stratigraphic relations are straightforward, is transitional in composition between the well-defined groups, and a similar sample from apparent LSB at section 12 is also anomalous with respect to some elements. Nonetheless, in the southern TPVF, where we have better stratigraphic and chemical control, the correlations are excellent and the variability within stratigraphic units for many elements is small compared to the differences between adjacent units.

Many of the measured sections contain thin sedimentary interbeds within the units defined as LSB, MSB, and USB on the basis of physical and chemical stratigraphic criteria. The LSB in particular contains a laterally extensive sedimentary interbed that has been recognized in all measured sections, in which the LSB is exposed between San Cristobal Creek (section 9) and Rio Petaca (section 1). Because the LSB is absent between sections 4 and 1, and occurs below river level at Dunn Bridge, we cannot be absolutely sure that this thin interbed is the same unit in all sections, but all indications are that it is. Peterson (1981) separated the LSB into two units, but our chemical data demonstrate that the flows on both sides of this sedimentary unit are nearly identical type-B lavas. Thus, the LSB was erupted during at least two episodes separated by a significant time break, and the bulk of the LSB lavas (except SCL) was derived from a homogeneous magma reservoir which was available for successive eruptions.

The single, petrographically distinct flow of type-A composition within the LSB at section 5 (Fig. 7, RG-335) is of uncertain importance stratigraphically. We lack sufficient control north and south of section 5 to ascertain how extensive and volumetrically important this subunit may be. Its importance to our petrologic model of Servilleta evolution is large, because it demonstrates that the type-A magma, and especially the low-K_2O (also low incompatible trace element) variant of type A, is present in all three units. A low-K_2O type-A basalt (RG-25 or TP-1 in Dungan and others, 1981a) has also been sampled from the USB (Table 1, #3).

The MSB is the most homogeneous of all the sampled units. It also is usually the thinnest in measured sections and apparently lacks sedimentary interbeds. On the basis of our data it appears that the MSB is a batch of basalt that erupted in a single event with a minimum of mixing with intermediate magmas.

The USB is compositionally the most variable and includes the distinctive petrographic variants confined to the vent areas. Several thin sedimentary interbeds, which are less extensive laterally than the unit seen within the LSB, are present in the southern gorge. The USB is laterally, and in some cases vertically, heterogeneous. Although most of the samples from the southern plateau are intermediate in composition between the type-A and type-B lavas of the MSB and LSB, they are classified as type A. The type-B lavas of the USB are located primarily along the western margin of the plateau (extending southward to Black Mesa) and especially in close proximity to the vent areas. The three southern Servilleta vent areas are located east and northeast of Tres Piedras, New Mexico, and directly south of Cerro del Aire (see map in Lipman and Mehnert, 1979). These are covered by very coarse diktytaxitic lavas characterized by distinctive, large plagioclase laths (3–5 mm in length). The La Segita shield (Lipman and Mehnert, 1979) and the surrounding lavas derived from this vent area are distinctive because of the presence of xenocrysts. Some flows contain orthopyroxene, others contain quartz and sodic feldspar. The xenocrystic flows are dominantly type-B lavas. Crossey's (1979) detailed geochemical studies of one of the La Segita lavas exposed in the roadcut just west of northern No Agua peaks, and of the coarse diktytaxitic lavas exposed in roadcuts just east of Tres Piedras (Petaca Creek), demonstrate significant within-flow heterogeneity over and above what can be accounted for by in-situ differentiation that leads to development of the vesicular segregations. We regard these variations as primary heterogeneities resulting from incomplete homogenization during mixing. The variability that characterizes the entire USB is thought to reflect heterogeneities on a much larger scale in the batch of magma that was erupted during three to perhaps five separate events.

RED RIVER GORGE

An extensive NW-trending fault system is exposed in the Red River gorge and to the northwest along the Rio Grande gorge. The Red River volcanic section is comprised of a rare composite sequence in which dacite, andesite, and basalt are interbedded (Fig. 9). A portion of this sequence exhibits evidence for magma mixing between basalt and dacite to produce hybrid olivine andesites ranging from 54% to 57% SiO_2 (McMillan and Dungan, 1982).

The lowest unit exposed at Red River (west of New Mexico State Fish Hatchery on upthrown side of major fault) is a dacite which underlies basalt of the LSB. This is the only exposure of dacite older than the LSB. Elsewhere, stratigraphic relations suggest that andesite and dacite volcanism overlapped primarily with the USB and to a lesser extent with the MSB. However, this may be an apparent relationship because most of the andesite and dacite vents lie west of the gorge. Basalts of the LSB, plus interbedded andesites, overlie a sedimentary unit containing cobbles and pebbles of typical Precambrian lithologies, indicating a throughgoing, west-flowing drainage. The thin sedimentary interbed overlying the LSB lacks coarse material and Precambrian detritus in particular.

Basalts of the MSB, which apparently flowed eastward into the vicinity of the present Red River gorge, interfinger with a dominantly olivine-andesite section that thickens to the east and is increasingly dominated by andesite until at the Fish Hatchery only two basalt flows at the base of the package are present. The andesite flows apparently were erupted in rapid succession from a vent located just east of the Fish Hatchery. A thick dacite flow, erupted from another vent west of the Fish Hatchery that lies along the trend of several faults, caps the section. This dacite apparently predates the Guadalupe volcanoes which in part post-date and in part pre-date the USB. The andesites laterally equivalent to the MSB are the hybrids produced by mixing of basalt and the low-SiO_2 dacite that caps the section. No other dacite analyzed from the Taos Plateau serves as a suitable end member. These andesites are distinctive in containing large (2–5 mm) skeletal-olivine phenocrysts. The USB laps onto this andesite section only along the western margin.

These andesites apparently blocked the drainage of the ancestral Red River at the time they were erupted. Initially, gravels accumulated behind the volcanic pile and to the west of the Sangre de Cristo Mountains. Eventually, this basin was filled with fluvial and lacustrine sediments as indicated by remnant gravel caps on topographic highs adjacent to the Red River gorge. When the basin was topped off, the fan was diverted to the south relative to its previous position. Some of this diversion may have been guided by movement (west side up) on the Fish Hatchery fault.

South of the Red River gorge, the MSB is capped by a gravel unit that thickens tremendously over a distance of several miles. This relationship is obscured on the west side of the gorge by the UCEM flow which partially occupies this stratigraphic interval, but it is well displayed on the east side between the Red River gorge and San Cristobal Canyon. This sediment package is the diverted fan.

Subsequent to USB time, the sediment-transport route to the south was blocked, either by other coalescing fans prograding westward from the Sangre de Cristo Mountains or by continued movement on the Fish Hatchery fault. The ancestral Red River was localized between the postulated southern topographic high and the dacite flow capping the gorge at section 12. At some point, downcutting into the USB and headward erosion re-established the Red River gorge in its present position.

LA MESITA AND BLACK MESA

La Mesita and Black Mesa are Servilleta Basalt-capped mesas which occur as outliers in close proximity to the southern end of the Taos Plateau. Black Mesa is an elongated, cigar-shaped mesa which extends southwestward from the southern tip of the Taos Plateau. It lies to the north of the Velarde graben and the present Rio Grande. La Mesita is a much smaller mesa which lies to the south of the Rio Grande at the north end of Black Mesa. The Embudo fault has produced displacements of as much as 50 m (down to the north) in basalts capping the south end of La Mesita. Structural relations at the north end of La Mesita are more complex, suggesting a local scissors motion in the Embudo fault. A set of NE-trending, en-echelon normal faults on Black Mesa is dominantly associated with displacements down to the southwest (Kelley, 1978). These structures demonstrate tectonic activity subsequent to upper Servilleta time and define the northeastern end of the Velarde graben, the actively subsiding portion of the Española Basin.

Black Mesa and La Mesita are at the same elevation as the USB lavas that mark the southern tip of the Taos Plateau, and almost certainly were once contiguous with it. Headward erosion by the Rio Grande during the Pleistocene, controlled by continued faulting and subsidence of the Velarde graben, resulted in the isolation of these outliers. The basalt section at the southern end of the Taos Plateau consists of two to three thin olivine-phyric USB flows typical of the lavas that form the western margin of the TPVF south of Tres Piedras (e.g., Comanche Rim). The flows combine to form a section less than 10 m in thickness. The volcanic sections on Black Mesa and La Mesita are up to 40 m thick and are in part petrographically distinct from the olivine-phyric lavas at the southern tip of the Taos Plateau. Gunn (1980) has measured the thickness of volcanic sections on La Mesita and at the northern end of Black Mesa. Slumping on the north side of La Mesita and structural complications associated with the Embudo fault limit the accuracy of the data locally. However, the thicknesses shown on the isopach map in Figure 10 are minimum estimates. Figure 10 illustrates that the

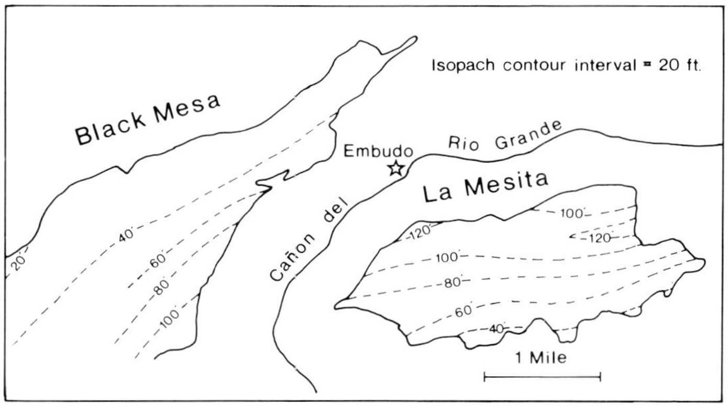

FIGURE 10. Basalt-thickness isopach map of La Mesita and Black Mesa from Gunn (1981). Minimum estimates of basalt thickness shown in feet.

thicknesses of the two volcanic outliers range from 7–15 m on their southern (La Mesita) and northern (Black Mesa) rims to approximately 30–35 m within the axis of the Velarde graben. The isopach contours approximately parallel the trends of normal faults associated with the Embudo–Velarde system. Even thicker volcanic sections were probably present to the north of La Mesita, but have been removed by the Rio Grande.

Basalts present on La Mesita and the southern rim of Black Mesa are distinctive aphyric lavas with fine-grained groundmass textures easily distinguished from the olivine-phyric, coarsely diktytaxitic lavas on the Taos Plateau and on the northern rim of Black Mesa. Olivine microphenocrysts in the aphyric lavas are more iron-rich (Fo_{76-60}) than the typical coarse olivine phenocrysts (Fo_{83-80}) observed elsewhere. In the areas of thickest section on La Mesita there are up to six aphyric flows which exhibit a progressive upward increase in thickness from less than 2 m to about 10 m.

The distinctive aphyric SB variety containing relatively Fe-rich olivine microphenocrysts that comprises La Mesita is present within the USB at sections 1, 2, 3, and 4. In addition to being absent at the southern tip of the Taos Plateau, aphyric lavas are not present in section 5. The presence of the aphyric lavas to the northeast of La Mesita, but not directly to the north (except on the southern rim of Black Mesa), suggests that the aphyric flows represent a single large eruption that initially moved to the southeast from vent areas near Tres Piedras and then turned southwestward along the depression between the SE-tilted plateau surface and the upthrown Picuris block. The anomalous thickness of USB at La Mesita and southern Black Mesa was produced by ponding in the actively subsiding depression associated with the Velarde graben. The monogenetic unit present on La Mesita indicates that the graben was not accessible to other SB units, probably because northward prograding fans served as a topographic barrier and southeast tilting of the plateau during SB time forced lavas eastward. Only when the fans were topped by earlier eruptive units were the aphyric lavas able to overcome the barrier and flow southwest. The coarse olivine-phyric lavas of the southern TPVF probably post-dated the aphyric unit and banked up against the lobe of aphyric flows filling the Embudo–Velarde depression.

SUMMARY AND CONCLUSIONS

Lateral variations in the thickness of SB units are controlled by the positions of alluvial fans, a shadowing effect produced by other volcanoes or the topographic high associated with the mid-graben horst of the San Luis Valley, distance from the vent, and local ponding in topographic lows. La Mesita is an example of ponding in an actively subsiding, fault-bounded basin. The thick section near the Rio Grande gorge bridge lies between major alluvial fans, south of the topographic obstruction of the horst, and in relatively close proximity to the vent areas. Thinning of the volcanic units to the north and south is caused by a combination of the factors listed above.

The LSB abruptly terminates against a thick gravel unit in the vicinity of the gorge arch, and the upper and middle units thin due to the topographic obstruction of this ancestral Rio Taos–Picuris alluvial fan. The MSB persists several miles farther south (e.g., section 3), but is largely absent southwest of Rio Petaca. The effect of prograding alluvial fans is well displayed in Rio Petaca which flows south at approximately a right angle to the fan axis. Two km north of the confluence of the Rio Petaca and the Rio Grande the section is thickened, from about 30 m at the rim to 125 m where the canyon narrows, by the addition and individual thickening of lower units. As it does in most areas of the southern plateau, the USB flowed over the fan with approximately constant thickness.

North of the Rio Grande gorge bridge, substantial thinning of the USB, MSB, and LSB resulted from the presence of the ancestral Arroyo Hondo fan and the shadowing effects of the horst, Cerro Montoso, and UCEM. LSB relations are complicated by the SCL shield, which probably vented comparatively near the gorge, and by the syn- to post-volcanic syncline which places all or much of the LSB below river level between Dunn Bridge and section 8. The SB is greatly thinned from section 9 to an area 1 km north of section 10, where all three units abruptly thicken. This attenuation is due to the ancestral Red River fan and the shadow effect of the horst.

In most areas where high-angle faults cut the volcanic section, the thickness of volcanic units is closely similar on both sides of the faults. Notable exceptions occur where the USB is decreased from five flows to one across the Embudo fault, the MSB is truncated against the fault, and in the Red River gorge where the LSB and MSB are slightly thickened on the downthrown side. Large changes in thickness of interbedded sedimentary units are noted wherever these faults are exposed. A particularly good example occurs at the Dunn Bridge fault, where the sedimentary package between MSB and USB increases 17 m across the fault.

Chemical data combined with our physical stratigraphy indicate that the eruption of the Servilleta Basalts occurred in three major episodes which probably were very shortlived compared to the time intervals between eruptive periods. Although the USB is apparently heterogeneous in comparison to the MSB and LSB, each unit is characterized by a distinctive chemical signature. Flows within the major units are commonly separated by thin sedimentary interbeds (0.5–3 m). In most cases the basalt compositions on either side of these breaks are very similar. This suggests that major eruptive episodes may have lasted up to several hundred years and that magma reservoirs of nearly constant composition persisted over this time.

The chemical variation with time is clearly not consistent with a progressive temporal evolution for the SB. Petrologic arguments based on the positive correlations between SiO_2, K_2O, and virtually all incompatible trace elements, combined with the presence of the low-K, low-SiO_2 variant of magma type A in all three units, suggest that source-region controls do not dominate chemical variability in the Servilleta. The nearly constant $Mg/(Mg+Fe)$ of the SB convincingly rules out fractional crystallization of a single parental basalt as the source of these variations. Mixing of the low-K basalt end member with andesitic magmas accounts very well for the range of compositions observed. The efficacy of mixing mafic and intermediate magmas in the TPVF is demonstrated at Red River where olivine andesites were produced by mixing of basalt and low-SiO_2 dacite magmas. Most TPVF andesites and dacites are the products of repeated episodes of mixing and fractionation combined with assimilation of a crustal anatectic melt.

Post-volcanic deformation of the TPVF is demonstrated by faulting along the Embudo fault zone and by other north or northwest-trending faults that occur in and around the Rio Grande gorge. The present location of the river is controlled by a combination of overall east-tilting of the plateau, westward prograding alluvial fans, and the local control exerted by the faults, which in turn are surficial manifestations of the deep Taos graben. Other drainages, especially those east of the Rio Grande gorge, have been greatly modified by contemporary warping of the plateau surface resulting from compressive deformation. Local compression within the axial graben of the rift is generated in this case by opposite rotations of rhomb-shaped crustal blocks, and is well displayed in the Embudo zone which is compressive at its northeastern end (NE of Pilar), but is clearly opening south of Embudo where the Velarde graben is undergoing active extension.

ACKNOWLEDGMENTS

This manuscript benefited from reviews by Scott Baldridge, Peter Lipman, and Phil Kyle. The research was supported by the National Aeronautics and Space Association grants to Dungan and Muehlberger, the National Science Foundation grants EAR-8026451 and EAR-8218734 to Dungan, a New Mexico Geological Society grant to McMillan, and by the Institute for the Study of Earth and Man (SMU).

We thank Kay Schutza for assistance in preparation of the manuscript.

REFERENCES

Aoki, K., 1967, Petrography and petrochemistry of latest Pliocene olivine-tholeiites of the Taos area, northern New Mexico, U.S.A.: Contributions to Mineralogy and Petrology, v. 14, pp. 191–203.

Aubele, J. C., and Crumpler, L. S., 1983, Geology of the central and eastern parts of the Springerville–Show Low volcanic field, Arizona (abstract): Geological Society of America, Abstracts with Programs, v. 15, no. 5, p. 303.

Baldridge, W. S., 1979, Petrology and petrogenesis of Plio-Pleistocene basaltic rocks from the central Rio Grande rift, New Mexico, and their relation to rift structure; *in* Riecker, R. E. (ed.), Rio Grande rift: tectonics and magmatism: American Geophysical Union, pp. 323–353.

———, Damon, P. E., Shafiqullah, M., and Bridwell, R. J., 1980, Evolution of the central Rio Grande rift, New Mexico: new potassium-argon ages: Earth and Planetary Science Letters, v. 51, pp. 309–321.

———, Perry, F. V., Gladney, E. S., and Shafiqullah, M., 1983, Late Cenozoic basaltic volcanism of the central Rio Grande rift area, New Mexico (abstract): Geological Society of America, Abstracts with Programs, v. 15, no. 5, p. 381.

Baltz, E. H., 1978, Resume of Rio Grande depression in north-central New Mexico; *in* Hawley, J. W. (ed.), Guidebook to Rio Grande rift: New Mexico Bureau of Mines and Mineral Resources, Circular 163, pp. 210–228.

Cash, D. J., Olsen, K. H., Stewart, J. N., and Wolff, J. J., 1983, Earthquake catalogue for northern New Mexico: Los Alamos National Laboratory, Progress Report LA-9782-PR, 16 pp. (Earthquakes for September 1973–December 1982, fig. 5, p. 15.)

Chapin, C. E., and Seager, W. R., 1975, Evolution of the Rio Grande rift in the Socorro and Las Cruces areas: New Mexico Geological Society, Guidebook 26, pp. 297–321.

Condit, C. D., 1983, Field, petrologic, and petrochemical data for the western Springville–Show Low volcanic field, Arizona (abstract): Geological Society of America, Abstracts with Programs, v. 15, no. 5, p. 303.

Cordell, L., 1976, Aeromagnetic and gravity studies of the Rio Grande graben in New Mexico between Belen and Pilar: New Mexico Geological Society, Special Publication 6, pp. 62–70.

———, 1978, Regional geophysical setting of the Rio Grande rift in southern New Mexico: Geological Society of America, Bulletin, v. 89, pp. 1073–1089.

Crossey, L. J., 1979, Intraflow chemical variations in Servilleta Basalts, Taos Plateau, New Mexico: M.S. thesis, Washington University, St. Louis, Missouri, 105 pp.

Dungan, M. A., and others, 1981a, Continental rift volcanism; *in* Basaltic volcanism on the terrestrial planets: Lunar and Planetary Institute, the Basaltic Volcanism Study Project, pp. 108–131.

———, Phelps, D., Stormer, J. C., Blanchard, D. P., Nielsen, R. L., McMillan, N. J., and Thompson, R. A., 1981b, Late Cenozoic volcanism of the northern New Mexico–southern Colorado portion of the Rio Grande rift: Lunar and Planetary Institute, Proceedings of Conferences on Processes of Planetary Rifting, pp. 137–140.

———, Nielsen, R. L., and Phelps, D., 1983a, Spatial and temporal variation patterns in late Cenozoic basaltic volcanism, NE Jemez lineament, northern New Mexico (abstract): Geological Society of America, Abstracts with Programs, v. 15, no. 5, p. 379.

———, Thompson, R. A., McMillan, N. J., Nielsen, R. L., and Williams, S., 1983b, Time-dependent basalt crust interactions in the RGR and the role of assimilation/fractional crystallization (AFC) in the origin of calc-alkaline magmas in a rift setting (abstract): Geological Society of America, Abstracts with Programs, v. 15, no. 5, p. 379.

———, Haskin, L., Lindstrom, M. M., and Cadiff, N. J., 1983c, Petrology and geochemistry of the Servilleta Basalts, Taos Plateau volcanic field, northern New Mexico (abstract): Geological Society of America, Abstracts with Programs, v. 15, no. 5, p. 302.

Galusha, T., and Blick, J. C., 1971, Stratigraphy of the Santa Fe Group, New Mexico: American Museum of Natural History, Bulletin, v. 144, art. 1, 127 pp.

Gardner, J. N., Loeffler, B. M., and Goff, F., 1984, Tectonic and geologic framework of the Jemez volcanic field, New Mexico (abstract): Proceedings of the ISEM Conference on Open Magmatic Systems, ed. Dungan, M. A., Grove, T. L., and Hildreth, W., Institute for the Study of Earth and Man, Dallas, p. 52.

Gill, J., 1981, Orogenic andesites and plate tectonics: Springer-Verlag, 390 pp.

Golombek, M. P., 1983, Geology, structure, and tectonics of the Pajarito fault zone in the Española Basin of the Rio Grande rift, New Mexico: Geological Society of America, Bulletin, v. 94, pp. 192–205.

Gunn, G. R., 1980, Volcanic stratigraphy and structure of the La Mesita–Black Mesa area, southeast San Luis Valley, north central New Mexico: M.S. thesis, Southern Methodist University, Dallas, 76 pp.

Kelley, V. C., 1952, Tectonics of the Rio Grande depression of central New Mexico: New Mexico Geological Society, Guidebook 3, pp. 92–105.

———, 1954, Tectonic map of part of the upper Rio Grande rift, New Mexico: U.S. Geological Survey, Oil and Gas Inventory Map OM-157.

———, 1977, Geology of the Albuquerque Basin, New Mexico: New Mexico Bureau of Mines and Mineral Resources, Memoir 33, 60 pp.

———, 1978, Geology of the Española Basin, New Mexico: New Mexico Bureau of Mines and Mineral Resources, Geologic Map 48.

———, 1979, Tectonics, middle Rio Grande rift, New Mexico; *in* Riecker, R. E. (ed.), Rio Grande rift: magmatism and tectonics: American Geophysical Union, pp. 57–70.

Kuno, H., 1960, High-alumina basalt: Journal of Petrology, v. 1, pp. 121–145.

Lambert, W., 1966, Notes on the late Cenozoic geology of the Taos–Questa area, New Mexico: New Mexico Geological Society, Guidebook 17, pp. 43–50.

Larson, E. S., Jr., and Cross, W., 1956, Geology and petrology of the San Juan region, southwestern Colorado: U.S. Geological Survey, Professional Paper 258, 303 pp.

Leininger, R. L., 1982, Cenozoic evolution of the southernmost Taos Plateau, New Mexico: M.S. thesis, University of Texas at Austin, 110 pp.

Lipman, P. W., 1983, The Miocene Questa caldera, northern New Mexico: relation to batholith emplacement and associated molybdenum mineralization; *in* The genesis of Rocky Mountain ore deposits: changes with time and tectonics: Proceedings of the Exploration Geologists Society Symposium, Denver Region, pp. 133–148.

———, and Mehnert, H. H., 1975, Late Cenozoic basaltic volcanism and development of the Rio Grande depression in the southern Rocky Mountains: Geological Society of America, Memoir 144 pp. 119–154.

———, and ———, 1979, The Taos Plateau volcanic field, northern Rio Grande rift, New Mexico; *in* Riecker, R. E. (ed.), Rio Grande rift: tectonics and magmatism: American Geophysical Union, pp. 289–312.

Manley, K., 1976, The late Cenozoic history of the Española Basin, New Mexico: Ph.D. dissertation, University of Colorado, 171 pp.

———, 1979, Stratigraphy and structure of the Española Basin, Rio Grande rift, New Mexico; *in* Riecker, R. E. (ed.), Rio Grande rift: magmatism and tectonics: American Geophysical Union, pp. 289–312.

McMillan, N. J., and Dungan, M. A., 1982, Origin of olivine andesite by mixing of basaltic and dacitic magmas, Red River gorge, Taos Plateau volcanic field, Rio Grande rift, New Mexico (abstract): Geological Society of America, Abstracts with Programs, v. 14, no. 7, p. 563.

———, and Williams, S., 1983, Open-system magmatic evolution: origin of andesites and dacites, Taos Plateau volcanic field, N.M. (abstract): Geological Society of America, Abstracts with Programs, v. 15, no. 5, p. 432.

———, Moorbath, S., and Dungan, M. A., 1983, Isotopic constraints on open-system magmatic evolution, Taos Plateau volcanic field (TPVF), New Mexico (abstract): Geological Society of America, Abstracts with Programs, v. 15, no. 6, p. 641.

Montgomery, A., 1953, Pre-Cambrian geology of the Picuris Range, north-central New Mexico: New Mexico Bureau of Mines and Mineral Resources, Bulletin 30, 89 pp.

Moorbath, S., Dungan, M. A., Harmon, R. S., McMillan, N. J., and Hoefs, J., 1983, Pb-, O-, and Sr-isotopic evidence for mantle and lower crustal involvement in the genesis of intermediate lavas of the Taos Plateau volcanic field (TPVF): EOS, v. 64, no. 45, p. 7531.

Muehlberger, W. R., 1979, The Embudo fault between Pilar and Arroyo Hondo, New Mexico; an active intracontinental transform fault: New Mexico Geological Society, Guidebook 30, pp. 77–82.

Nielsen, K. C., Bell, D. A., and Norman, M. D., 1983, Precambrian basement fabrics near the northeast-trending Jemez lineament, north-central New Mexico: EOS, v. 64, p. 857.

O'Niell, M. J., and Mehnert, H. H., 1980, Late Cenozoic volcanism, uplift and erosion, southern Sangre de Cristo Mountains, New Mexico (abstract): Geological Society of America, Abstracts with Programs, v. 12, no. 6, p. 300.

Ozima, A., Kono, M., Kaneoka, I., Kinoshita, H., Nagata, T., Larson, E. E., and Strngway, D. W., 1967, Paleomagnetism and potassium-argon ages of some volcanic rocks from the Rio Grande gorge, New Mexico: Journal of Geophysical Research, v. 72, pp. 2615–2622.

Peccerillo, A., and Taylor, S. R., 1976, Geochemistry of Eocene calc-alkaline volcanic rocks from the Kastamonu area, northern Turkey: Contributions to Mineralogy and Petrology, v. 58, pp. 63–81.

Peterson, C. M., 1981, Late Cenozoic stratigraphy and structure of the Taos Plateau, northern New Mexico: M.S. thesis, University of Texas at Austin, 57 pp.

Ross, C. S., Smith, R. L., and Bailey, R. A., 1961, Outline of the geology of the Jemez Mountains, New Mexico: New Mexico Geological Society, Guidebook 12, pp. 139–143.

Seager, W. R., Shafiqullah, M., Hawley, J. W., and Marvin, R. F., 1984, New K/Ar dates from basalts and the evolution of the southern Rio Grande rift: Geological Society of America, Bulletin, v. 95, pp. 87–99.

Smith, H. T. U., 1938, Tertiary geology of the Abiquiu quadrangle, New Mexico: Journal of Geology, v. 46, pp. 933–965.

Steinpress, M. G., 1980, Neogene stratigraphy and structure of the Dixon area, Española Basin, north-central New Mexico: M.S. thesis, University of New Mexico, 127 pp.

Stormer, J. C., Jr., 1972, Ages and nature of volcanic activity on the southern High Plains, New Mexico and Colorado: Geological Society of America, Bulletin, v. 83, pp. 2443–2448.

Thompson, R., and Dungan, M. A., 1982, Calc-alkaline stratovolcanoes along the Rio Grande rift axis? Field and petrological evidence (abstract): Geological Society of America, Abstracts with Programs, v. 14, no. 2, p. 631.

Wells, R. C., 1937, Analyses of rocks and minerals from the laboratory of the United States Geological Survey 1914–36: U.S. Geological Survey, Bulletin 878, 134 pp.

Williams, S., and Murthy, V. R., 1979, Sources and genetic relationships of volcanic rocks from the northern Rio Grande rift: Rb-Sr and Sr-Nd evidence: EOS, v. 60, p. 407.

Zimmerman, C., and Kudo, A. M., 1979, Geochemistry of andesites and related rocks, Rio Grande rift, New Mexico; *in* Riecker, R. E. (ed.), Rio Grande rift: tectonics and magmatism: American Geophysical Union, pp. 355–381.

EXHUMED MID-MIOCENE VOLCANIC FIELD IN THE TESUQUE FORMATION, NORTHERN ESPAÑOLA BASIN, NEW MEXICO

S. JUDSON MAY
ARCO Exploration Company, Box 5540, Denver, Colorado 80217

INTRODUCTION

A small field of basaltic tuff rings, cones, minor flows, and dikes is locally exposed for a distance of 20 km between the town of Ojo Caliente and the lower El Rito Creek area in the northern Española Basin (Fig. 1). The field is one of the few interruptions in a mid-Miocene volcanic lull identified throughout much of the Rio Grande rift (Chapin, 1979). These tephra eruptions occurred during deposition of the middle Chama–El Rito Member of the Tesuque Formation and are closely associated in terms of mineral and textural composition, stratigraphic position, age, and possibly eruptive mechanism. North–northeast-trending dikes and minor growth faults contemporaneous with volcanism imply a local N70W-oriented axis of least principal stress during mid-Miocene time. A possible genetic relationship with the Hinsdale Basalt (of Lipman and Mehnert, 1975) is suggested on the basis of approximately similar mineralogy, stratigraphic position, and age.

OJO CALIENTE TUFF RING

The Ojo Caliente tuff ring (May, 1980) is the best exposed of the middle Chama–El Rito volcanic centers and is located 3 km southwest of Ojo Caliente in the area where the mid-Miocene part of the Los Pinos Formation grades south into the Chama–El Rito (Fig. 1). Other vents elsewhere in the Chama–El Rito occur within 40 m of the same stratigraphic level (May, Miocene Stratigraphic Relations etc., fig. 2, this guidebook). A circular vent, 1 km across, containing bombs, inward-dipping basaltic tephra, and concentric dikes, is surrounded by a tuff-ring facies of outward-dipping basaltic tuff (Fig. 2). The tuff ring tapers radially outward into the Los Pinos and Chama–El Rito beds and is bounded by angular unconformities at the top and base. The tuff ring is separated from the vent facies by a funnel-shaped angular unconformity that dips 24–50° into the vent (Fig. 3). Deeper stratigraphic levels of the tuff ring are exposed in the north and northwest because of post-volcanic tilting to the southeast and deep erosion of Arroyo El Rito. As a result, higher beds of the Chama–El Rito in the south still overlie the higher slopes on the tuff ring. A N20W-trending fault cutting the southern part of the vent is downdropped to the west, allowing preservation of Chama–El Rito beds that were stratigraphically high enough to overlie the vent. Here, the vent margin (Fig. 2) is inferred beneath the Chama–El Rito. Exposures of the vent facies 50–75 m north of the dotted line indicate proximity to the vent margin and contain steep northerly dips and a concentric dike similar to dikes near the vent margin elsewhere.

Vent Facies

The vent facies is a mottled maroon to gray, basaltic lapilli tuff and is confined completely to the vent. Ill-defined bedding results from numerous flattened basaltic bombs and indistinct layering in the tuffaceous matrix. Inward dips as steep as 36° near the vent margin flatten to less than 10° in the middle of the vent.

The tuff is a poorly sorted mixture of angular, scoriaceous lapilli in an ashy matrix. In some areas, the tuff grades into zones of extremely scoriaceous and amygdaloidal, cryptocrystalline basalt and basaltic glass. The basaltic lapilli are composed of fine-grained augite and traces of olivine(?) in dark-brown tachylite, which also contains traces of finely divided magnetite and hematite, and many calcite-filled amygdules. Plagioclase is scarce to absent. The more crystalline basaltic lapilli contain numerous flow-banded microlites of plagioclase with lesser augite grains in dark-brown tachylite. A few angular cognate blocks of fine-grained basalt are scattered throughout the tuff and may be fragments of basalt explosively torn from the walls of the underlying conduit. Sparse exotic debris includes angular fragments of polycrystalline quartz with feathery biotite grains (Ortega Quartzite?) brought up from the Precambrian basement, and pulverized fragments of Tertiary latite and rhyolite from the underlying Chama–El Rito beds. In some areas, light-colored tuffaceous debris similar to the tuff-ring facies forms inward-dipping layers, 0–20 m thick, immediately inside the vent unconformity, and may represent large masses of debris on the vent rim which slumped back into the vent during eruption.

Numerous oval, spindle-shaped, and ropey basaltic bombs are scattered throughout the tuffaceous matrix and have a maximum size of 1–2 m (Fig. 4). Many appear to have been tapered by flight through the air and have light-gray, vesicular rims.

Numerous dark-gray, inward-dipping basalt dikes, concentric to the vent margin, are most abundant within the outer 60 m of the vent. The concentric dikes, 1–10 m thick, dip inward at angles equal to, or greater than, the obscurely bedded vent facies. Grain size is slightly larger than that of the basalt bombs in the vent. Most dikes are only slightly vesicular and exhibit flow flotation and amygdules aligned parallel to both contacts. On a small scale, the contacts have irregular shapes and a few apophyses, but no scoriaceous zones. Some dikes are severely contorted, implying slumping or internal movement of the vent material during cooling. Although some of these "dikes" could have been shallow lava lakes ponded inside the vent, the evidence indicates that most of them are intrusive into the vent facies. One radial dike north of the vent may have fed the small flow at the base of the tuff-ring facies on the north side of Arroyo El Rito (Fig. 2). The flow, 0–1 m thick, wedges out eastward in the tuff-ring facies, into a horizon containing coarse lapilli erupted contemporaneously with the flow.

Petrographically, the flow, dikes, and vent bombs are identical in thin section and are probably chemically similar to the glassy and frothy basalt of the vent facies, but faster cooling of the vent facies seems to have caused less complete crystallization. The most crystalline samples contain scattered, glomerophyric aggregates of augite (0.25–1.0 mm) and smaller isolated grains of augite. The groundmass consists mainly of tiny, randomly oriented laths of plagioclase (0.2 mm) and sparse olivine (0.1 mm) suspended in dark-brown to black basaltic glass. Some plagioclase laths and augite grains are zoned. The nonflow-banded plagioclase may have crystallized after emplacement. Calcite- and natrolite(?)-filled vesicles occur in most samples.

Tuff-ring Facies

The surrounding tuff-ring facies outside the vent is a circular, fan-shaped pile of bedded basaltic tuff that tapers radially from a thickness of 75 m at the vent margin to about 25 m in its westernmost exposure and approximately 10 m at its eastern end. Original thicknesses may have been slightly greater, especially near the vent where erosion probably continued until the Chama–El Rito buried the higher flanks of the volcano. The height-to-width ratio of the structure was approximately 1:25 prior to erosion (lower than most cinder cones).

Although Galusha and Blick (1971) described these beds as the Abiquiu Tuff, a local and genetic relationship with the vent is clear. The relatively non-systematic pattern of dip directions in the vent and tuff-ring facies (Fig. 2) partly results from later, Pliocene warping and 9° of southeastward tilting into the Española Basin. When these dips are

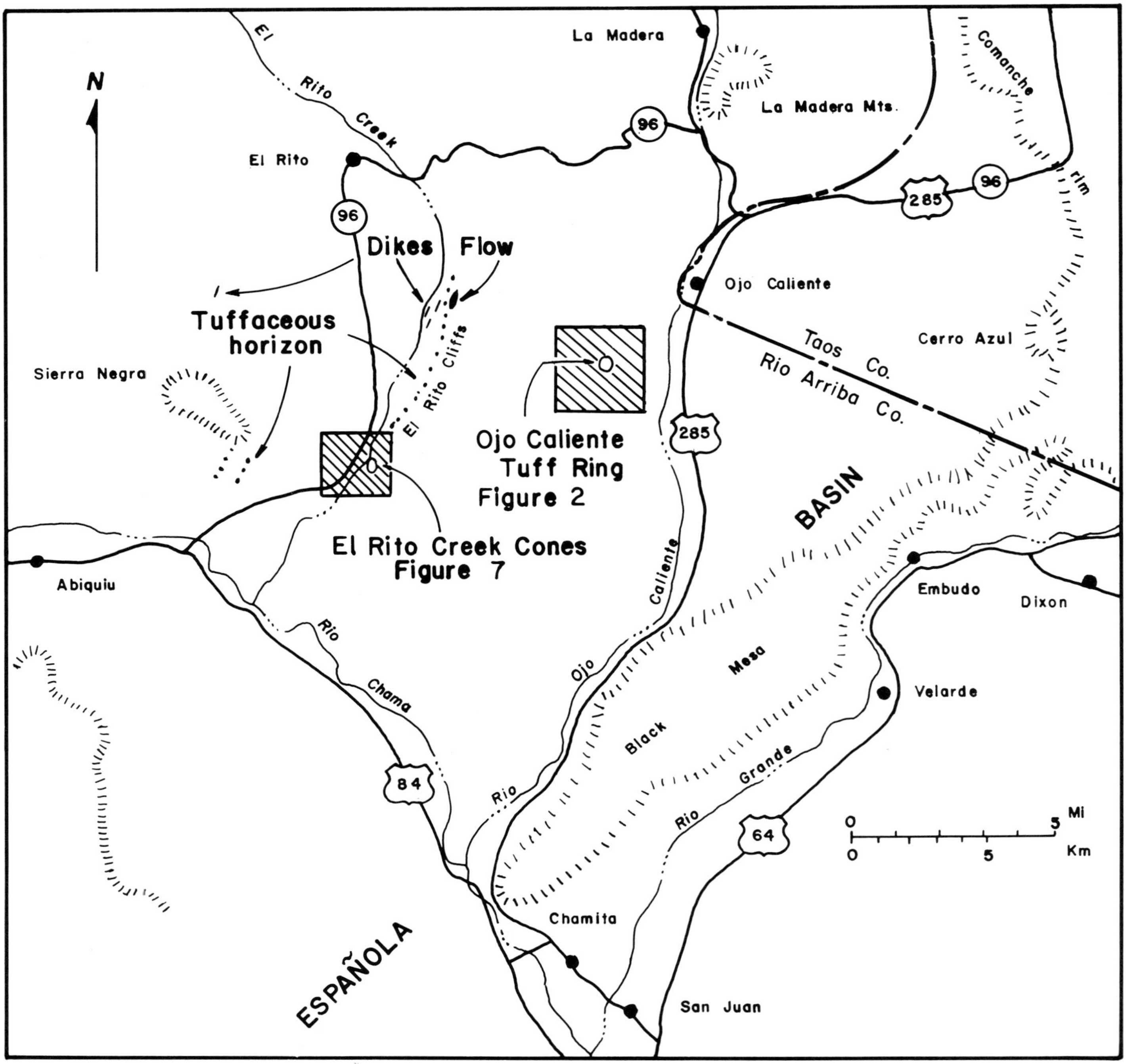

FIGURE 1. Map showing location of mid-Chama–El Rito volcanic centers in northern Española Basin.

rotated out using a stereonet, a more systematic pattern develops in which the tuff ring dips radially outward and the vent facies dips concentrically inward. This systematic pattern is most evident on the northwest side of the vent (Fig. 5). A general steepening of dips near the vent margin is seen in both the vent and tuff-ring facies, as is typical of tuff rings and maare (Heiken, 1971; Lorenz, 1973). Few dips exceeded 25° prior to regional tilting.

The unconformity between the top of the tuff ring and the overlying Chama–El Rito is more angular than the one at the base and gradually becomes more discordant (3–25°) as it is traced toward the vent because of the tuff ring's steeper dips in the vent area. The unconformity itself also slopes away from the vent, with higher beds of the Chama–El Rito lying at successively higher levels on the tuff ring, indicating that the volcano formed a low hill during deposition of the Tertiary beds. The unconformity at the base of the tuff ring is less angular, but where traced away from the vent, the tuff ring appears to lie on successively higher beds in the Los Pinos Formation.

The tuff-ring facies is mainly pale greenish-brown, basaltic air-fall tuff. Bedding is even and parallel. Subtle graded bedding occurs in some layers. Very sparse, discontinuous layers of foreset crossbeds dip away from the vent, possibly indicating rare base surges during the eruption or minor streams flowing down the flanks of the volcano. Sparse layers of fine-grained lapilli occur locally in the section. Average grain size decreases slightly away from the vent. Sparse bomb sags containing fragments of the Precambrian basement and Tertiary volcanic cobbles of the Chama–El Rito occur within 200 m of the vent margin

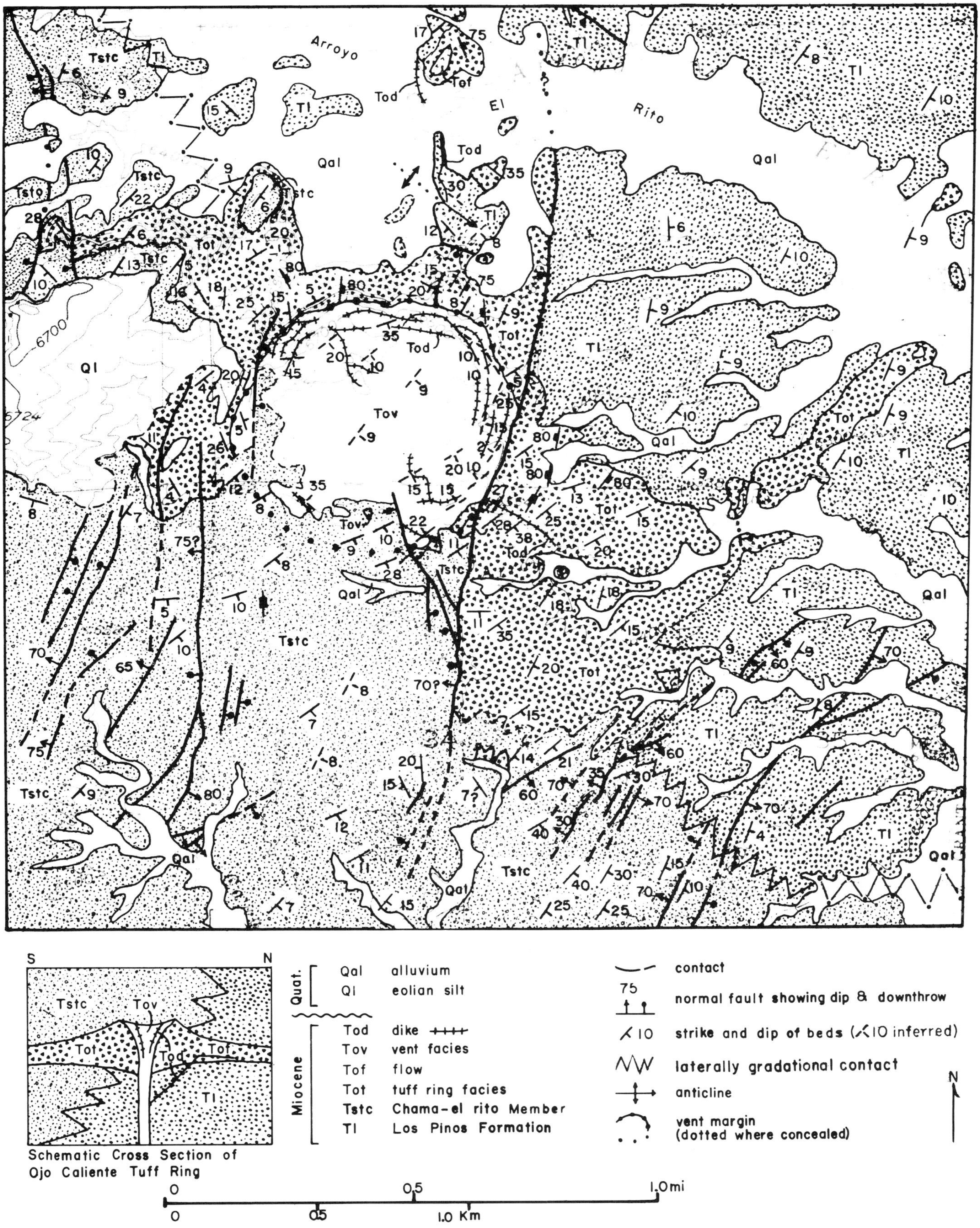

FIGURE 2. Geologic map of Ojo Caliente tuff ring. Modified from May (1980).

FIGURE 3. Angular unconformity along northwestern vent margin of Ojo Caliente tuff ring. Inward-dipping vent facies (Tov) truncates outward-dipping tuff-ring facies (Tot) (SW¼ sec. 27, T24N, R8E).

FIGURE 4. Ropey basalt bomb in tuffaceous matrix of Ojo Caliente tuff-ring vent (SE¼ sec. 27, T24N, R8E).

(Fig. 6). Ropey bombs and angular blocks of basalt similar to those in the vent facies are very sparse.

The finer-grained fraction of the tuff consists of a heterogeneous mixture of poorly sorted scoriaceous lapilli, devitrified glassy shards, fragmented crystals of clinopyroxene (augite?), sparse angular chips of Precambrian quartzite and metarhyolite, and rounded granules of Tertiary rhyolite, andesite, latite, and sandstone. Secondary epidote and minor chlorite are common alteration products of the basaltic debris and are probably responsible for the tuff's pale greenish-brown color.

Relationships to Tertiary Section

Near the top, base, and distal ends, the debris is mixed with fluvial volcaniclastic debris of the Los Pinos and Chama–El Rito. Poorly sorted, rounded pebbles of rhyolite and latite, and angular, vesicular clasts of basalt are scattered through a locally crossbedded sandy matrix of quartz, orthoclase, microcline, clinopyroxenes of mixed origin, and tuffaceous debris. The basal 15–20 m of overlying Chama–El Rito have a light-brown, tuffaceous appearance from reworking of the tephra. Layers of basaltic tephra are interbedded with typical Chama–El Rito beds overlying the tuff ring, especially in the distal areas. The tuff-ring facies is approximately 275 m higher in the section than basalt flows of the Hinsdale Formation (formerly Jarita Basalt) 3 km southwest of Ojo Caliente that were K–Ar dated at 22.1 ± 0.6 m.y. by Baldridge and others (1980).

Large veins of milky opalite and white calcite occupy northeast-trending joints in the tuff-ring facies 100–200 m east of the vent. The opalite may have formed from silica released by devitrification of the glassy debris as hydrous fluids migrated through the tuff.

Local Deformation

Minor structural deformation on a local scale accompanied the eruption. A small growth fault, 1 km east of the vent, is located at the base of the tuff ring and several minor ones in the tuff-ring facies are exposed beneath the vent unconformity on the northwestern side of the vent. An angular unconformity is exposed in the lower tuff ring along the west-trending cliffs 1 km northwest of the vent. More subtle unconformities elsewhere can be traced for short distances through the tuff-ring facies.

Eruptive Mechanisms

Eruptions of the tuff ring were probably explosive despite the relative scarcity of dune-like structures and the apparent lack of palagonite

FIGURE 5. View south toward northwestern part of the Ojo Caliente tuff ring, showing dark vent facies (Tov) at upper left and light-colored tuff-ring facies (Tot) which dips to the right (away from vent). At far right, flat-lying Chama–El Rito (Tstc) overlies the tuff ring in angular unconformity (SE¼ sec. 27, T24N, R8E).

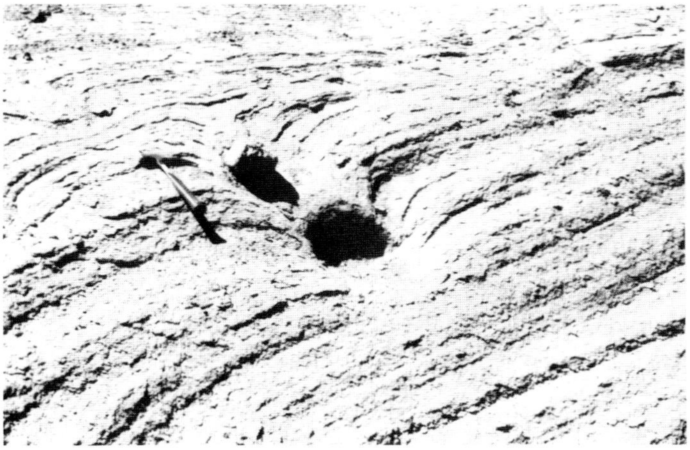

FIGURE 6. Bomb sags in tuff-ring facies of Ojo Caliente tuff ring, 180 m northeast of vent. Cavity was once occupied by a block, since weathered out. Another block of andesite (above) occupies sagged bedding. Two other bomb sags are at left (SE¼ sec. 27, T24N, R8E).

which otherwise might indicate the presence of water necessary for phreatomagmatic eruptions. Considerable fragmentation and horizontal dispersal of the ejecta distinguish this vent from cinder cones as indicated by:

(1) A fairly low topographic profile and low outward dips of the tuff-ring facies;
(2) very minor presence of possible base-surge crossbeds;
(3) sparse bomb sags;
(4) the abundance of glassy, juvenile debris;
(5) widely scattered fragments of the underlying Chama–El Rito and Precambrian basement.

The fact that the tuff ring probably erupted in the central part of a large Miocene basin in which numerous streams were depositing a thick pile of porous sand and gravel may imply the presence of abundant interstitial ground water necessary for a phreatomagmatic eruption. Aubele and others (1976) noted the presence of many similar volcanic structures along the Rio Grande rift.

EL RITO CREEK TUFFACEOUS SPATTER CONES

Two tuffaceous spatter cones of basaltic composition are exposed 12 km southwest of Ojo Caliente along El Rito Creek (Figs. 1, 7). These vents occur in the lower part of the middle Chama–El Rito and are probably 0–40 m stratigraphically lower than the Ojo Caliente vent. A tuffaceous spatter facies and a surrounding fluvial volcaniclastic facies are recognized. The two vents are partly covered by the Chama–El Rito beds and Quaternary alluvium of El Rito Creek. Mineral composition, texture, general appearance, and age of the tuffaceous spatter facies are nearly identical to those of the vent facies of the Ojo Caliente tuff ring, implying a similar magmatic source. The El Rito Creek vents are smaller than that of the Ojo Caliente tuff ring and the surrounding fan of air-fall is less well exposed at the current level of erosion.

Ekas and others (this guidebook) report a 15.3 ± 0.4 m.y. K–Ar date on a basaltic bomb derived from the vent on the east side of the creek. This age determination is significant because it probably dates the entire volcanic field, thereby providing age-control through the mid-Miocene part of the section for a distance of 20 km between Ojo Caliente and the lower El Rito Creek area.

Tuffaceous Spatter Facies

The western part of a tuffaceous spatter cone is exposed along the west side of N.M. 96 in a northeast-trending elliptical outcrop (Fig. 7). The eastern side of the exposure is in contact with downfaulted beds of the Chama–El Rito that are stratigraphically 30–60 m higher than the cone. The debris is an obscurely bedded accumulation of maroon, basaltic tuff mixed with abundant, dark-gray basaltic spatter and egg-shaped bombs slightly flattened parallel to layering in the tuff. The obscure layers appear to dip radially away from a point near the highway, and probably represent material aerially deposited on the outer flanks of a volcano whose vent is now beneath El Rito Creek alluvium. The debris occurs as ill-defined, layered units that probably resulted from either eruptive surges causing variations in the amount of basaltic spatter that accumulated on the upper flanks of the volcano, or possibly from minor lahars of volcanic debris that moved down the side of the volcano. The debris appears to dip to the west beneath the green, red, and light-brown tuff and tuffaceous sandstone in the volcaniclastic facies surrounding the vent, indicating burial of the cone by the tuff. The presence of very sparse bomb sags in the tuff and obscure layering that dips radially away from the cone imply a local source for the tuff in the vicinity of the vent.

The eastern tuffaceous spatter cone, east of the creek, appears to overlie the center of a bowl-shaped structural depression in the underlying volcaniclastic facies (Fig. 7). Dips in the surrounding volcaniclastic facies range from 9 to 45° inward. Only the northern halves of the cone and depression are exposed. Because of later southwestward tilting of the area, the Chama–El Rito still unconformably overlies the southwestern half of the volcano and depression. A small outlier of Chama–El Rito also unconformably overlies the western side of the vent margin. To the southeast, the lower Ojo Caliente Sandstone is downfaulted against the depression by a later fault.

The northern half of a circular vent, 240 m across, is identified in the cone by steep inward dips in the obscurely bedded tuffaceous spatter facies and by slightly greater amounts of ropey bombs and red, scoriaceous basalt. Obscure eastward dips a few meters to the east of the vent margin are thought to be tuffaceous spatter and flows erupted from the vent and deposited on the outer slopes of the cone. Because the outer flanks of the cone appear to stratigraphically overlie the volcaniclastic facies, this cone may be slightly higher in the section and younger than the western cone. The scoriaceous basalt of the vent has a dark, glassy, extremely vesicular groundmass containing red to opaque hematite and finely divided magnetite. Suspended in the groundmass are tiny laths of plagioclase and grains of augite. Amygdules are mainly filled with calcite and a few are lined with tridymite(?).

Volcaniclastic Facies

The spatter cones are surrounded by volcaniclastic facies of dark-green, red, and light-brown tuffaceous sandstone and lesser air-fall tuff derived from the volcanoes. The debris grades laterally outward and intertongues with beds of the Chama–El Rito. The relationship is seen best on the east side of Arroyo del Perro 2.5–3 km north of Highway 96, where a thick section of green tuffaceous sandstone grades north into light-brown, arkosic sandstone.

Most tuffaceous layers consist of variegated, poorly sorted fluvial sandstone with minor amounts of locally derived basaltic debris. Typical constituents are grains of quartz (45–65%), feldspar (25–30%), granules and pebbles of andesite and latite (5–10%), basaltic ash and lapilli (0–20%), fragments of Tertiary sandstone and Precambrian metamorphic rock (0–3%), opaque minerals (0–2%), and variable secondary calcite. Chips and blocks of locally derived basalt are most common near the vents. Fluvial crossbeds and channel fills are widespread.

Several other isolated outcrops of the volcaniclastic facies are in fault contact with the Chama–El Rito 6 km to the west (Fig. 1). The beds are mainly red and green tuffaceous sandstone with minor rhyolite pebbles and much secondary calcite, and are overlain by a slight angular unconformity. Correlation with the volcaniclastic facies of El Rito Creek is likely, but is difficult to demonstrate because of poor intervening exposures. Interbedded layers of andesite- and rhyolite-pebble conglomerate and arkosic sandstone of the Chama–El Rito occur at various levels in the volcaniclastic facies and indicate continued deposition of the Chama–El Rito during volcanism and erosion of the tuffaceous debris.

The abundance of arkosic sandstone and sparse andesite pebbles in the volcaniclastic facies indicate that the Chama–El Rito streams were

Quat.	Qal	alluvium
	Qt	terrace deposits
Miocene	Tsto	Ojo Caliente Sandstone
	Tet	volcaniclastic facies
	Tev	tuff-spatter facies
	Tstc	Chama-el rito beds

- contact
- 75 normal fault showing dip and downthrow
- 10 strike and dip of beds
- 10 inferred strike and dip of beds
- vent margin (dotted where concealed)

FIGURE 7. Geologic map of El Rito Creek tuffaceous spatter cones, 5 km northeast of Rio Chama on N.M. Highway 96. Modified from May (1980).

able to maintain their southwesterly course through the basaltic field in spite of the accumulating tuffaceous debris. Some layers of greenish tuff may have formed small fans that rerouted the streams around the lower flanks of the volcanoes, for a short time, but otherwise contributed only moderate amounts of debris to what was mainly Chama–El Rito alluvium. Minor north–northeast-trending growth faults 1 km N20W of the western El Rito Creek vent and a slight angular unconformity at the top of the volcaniclastic facies in some outcrops (Fig. 8) indicate local deformation and erosion during this volcanic episode.

EL RITO CLIFFS VOLCANIC ROCKS

Northeast of the El Rito Creek vent, a greenish-gray volcaniclastic horizon in the lower–middle Chama–El Rito can be traced northeastward for several kilometers between El Rito Creek and an area of high cliffs to the east (Fig. 1). Although the zone is laterally continuous, its thickness and abundance of volcanic debris varies greatly, implying the presence of several widely spaced volcanic sources buried locally in the Chama–El Rito. Overall appearance, composition, and stratigraphic level of the debris are similar to the Ojo Caliente tuff-ring facies and the volcaniclastic facies of the El Rito Creek vents. A small flow and several northeast-trending dikes are petrographically similar to the basaltic debris in the tuffaceous horizon and are probably of similar age and genesis.

The volcaniclastic facies, 10–30 m thick in most areas, is mainly light greenish-gray and greenish-brown fluvial sandstone containing variable amounts of reworked basaltic tuff and debris. Numerous sandy layers contain fluvial crossbeds and sparse channel-fill structures. Locally contorted layers indicate soft-sediment deformation by slumping or possible minor tectonic deformation associated with the eruptions. A persistent angular unconformity seems to characterize the upper contact of the volcaniclastic facies, but is much more subtle than the ones at the Ojo Caliente tuff ring and El Rito Creek vents.

Composition of the clasts is variable, but the sand-sized fraction is mainly poorly sorted, subrounded quartz and feldspar. Less abundant clasts include broken crystals of plagioclase, magnetite, sparse augite(?), devitrified glass shards and secondary epidote, sericite, and calcite. The coarser debris includes scattered angular clasts of slightly amygdaloidal basalt, Precambrian metarhyolite, and sparse pebbles of Tertiary rhyolite, dacite, and andesite.

A small, dark-gray amygdaloidal basalt flow, 1 m thick, overlies the volcaniclastic facies (Fig. 1) and appears to have been fed by a small dike nearby (SW1/4 sec. 24, T24N, R7E). Proximity to an explosive vent is indicated by numerous amygdaloidal basaltic bombs and angular blocks of Precambrian metarhyolite. A metarhyolite block, 0.3 m in diameter, occupies 0.5 m deep bomb sag in the underlying sandstone.

Several N20E-trending dikes 0.5 km southwest of the flow (Fig. 1) occur in the underlying Chama–El Rito and are petrographically similar to the flow. The basalt contains nonflow-banded, partly zoned, fine-grained laths of plagioclase (15%), and tiny magnetite specks (1%) in a slightly vesicular tachylite groundmass. The dikes and flow were probably contemporaneous.

FIGURE 8. Angular unconformity between volcaniclastic facies (Tet) of the El Rito Creek tuffaceous spatter cones and overlying tan Chama–El Rito (Tstc). Two meters high ledge above the unconformity is a well-cemented Chama–El Rito conglomerate layer (also in background at extreme left) along east wall of Arroyo del Perro; 1.5 km northwest of El Rito Creek vents.

MID-MIOCENE STRESS FIELD

Dike trends within the volcanic field indicate that the tensional stress field responsible for the pervasive north–northeast-trending latest Miocene–Pliocene faults in the northwestern part of the Española Basin may have been in operation by at least mid-Miocene time. The N20E-trending dikes associated with the El Rito cliffs volcaniclastic horizon are exposed in the lower Chama–El Rito beds, but were probably contemporaneous with mid-Chama–El Rito deposition. Petrographic similarity of the dikes with the small nearby flow in the Chama–El Rito implies a similar age. A north–northeast-trending basaltic dike north of Sierra Negra (Fig. 1) yielded a K–Ar age of 15.9 ± 0.9 m.y. (Manley and Mehnert, 1981) and is probably related to the volcanic field. Three km west of Ojo Caliente, several vertical dikes trend N30E away from the slightly older Cerro Negro plug (see geologic map of Stop 4, third-day road log, this guidebook) and are approximately 18 m.y. old (May, 1980). Several north–northeast-trending growth faults in an area 1 km northwest of the El Rito Creek vents also parallel the north–northeast-trending mid-Miocene dikes of the region. These faults and dikes are consistent with a nearly horizontal, approximately N70W-oriented axis of least principal stress in the mid-Miocene. However, the lack of major faults of this age in the area and throughout the basin indicates that the magnitude of extension was probably very slight at this time. In the latest Miocene and Pliocene, following deposition of most or all of the Tesuque Formation, this stress field must have intensified, producing the pervasive N25E-trending post-Tesuque faults of the region and tilting the rocks southeast into the basin. The three dikes that radiate outward from the Ojo Caliente tuff ring (Fig. 2) may indicate a locally dilational stress environment centered around that vent during eruption.

This local west–northwest direction of extension in the mid-Miocene is suggested for the northwestern part of the Española Basin and differs somewhat from more regional interpretations by Zoback and Zoback (1980) and Golombek and others (1983), who have proposed a regional west–southwest direction of extension during mid-Miocene time and a change to west–northwest extension about 10 m.y. ago. Their work is based on a broader range of information throughout the Basin and Range province and the Rio Grande rift.

This local west–northwest-extending domain within the larger west–southwest-extending Basin and Range province may have been mechanically influenced by north–northeast-trending Precambrian tectonic trends or Laramide structural grain parallel to the southern end of the Brazos uplift. Possibly the north–northeast-trending dikes and faults of the mid-Miocene volcanic field signify the presence of north–northeast-trending pre-Miocene fractures that were simply close enough to being perpendicular to a west–southwest axis of extension to allow access to the rising basaltic magma.

CONCLUSION

About 15 m.y. ago, a small field of basaltic tuff rings and tuffaceous spatter cones erupted in the Ojo Caliente–El Rito Creek area during deposition of the middle part of the Chama–El Rito and may be related to basalts of the Hinsdale Formation in the San Luis Basin and Tusas Mountains. The abundance of tephra in the volcanics probably resulted from explosive magma interaction with near-surface ground water in the Chama–El Rito. Subtle angular unconformities, dikes, and a few growth faults near the volcanic centers indicate local structural deformation and possible minor local west–northwest crustal extension accompanying the eruptions. This volcanic horizon dips gently southeast into the subsurface toward the axis of the Española Basin and its lateral extent is not known. A detailed magnetic survey in the northern part of the basin might further delineate the limits of this field.

ACKNOWLEDGMENTS

The above text is modified from a chapter of my Ph.D dissertation done at the University of New Mexico under direction of L. A. Woodward and supported by USGS grants 14-08-0001-G-255 and 14-08-0001-G-348, New Mexico Bureau of Mines and Mineral Resources grant 900-60-84, and New Mexico Energy Research and Development grant 117-69-B. This paper was critically reviewed and improved by W. S. Baldridge, M. P. Golombek, R. V. Ingersoll, and K. R. Geitzenauer.

REFERENCES

Aubele, J. C., Crumpler, L. S., Loeber, K. N., and Kudo, A. M., 1976, Maare and tuff rings of New Mexico: New Mexico Geological Society, Special Publication 6, pp. 109–114.

Baldridge, W. S., Damon, P. E., Shafiqullah, M., and Bridwell, R. J., 1980, Evolution of the central Rio Grande rift, New Mexico: new potassium–argon ages: Earth and Planetary Science Letters, v. 51, pp. 309–321.

Chapin, C. E., 1979, Evolution of the Rio Grande rift—a summary; *in* Riecker, R. E. (ed.), Rio Grande rift: tectonics and magmatism: American Geophysical Union, Washington, D.C., pp. 1–5.

Galusha, T., and Blick, J. C., 1971, Stratigraphy of the Santa Fe Group, New Mexico: American Museum of Natural History, Bulletin, v. 144, art. 1, 127 pp.

Golombek, M. P., McGill, G. E., Brown, L., 1983, Tectonic and geologic evolution of the Española Basin, Rio Grande rift: structure, rate of extension and relation to the state of stress in the western United States; *in* Morgan, P., and Baker, B. H. (eds.), Processes of continental rifting. Lunar and Planetary Institute topical conference on the processes of continental rifting, Dec. 3–5, 1981: Tectonophysics, v. 94, nos. 1–4, pp. 483–507.

Heiken, G. H., 1971, Tuff rings: examples from the Fort Rock–Christmas Lake Valley Basin, south-central Oregon: Journal of Geophysical Research, v. 76, no. 23, pp. 5615–5626.

Lipman, P. W., and Mehnert, H. H., 1975, Late Cenozoic basaltic volcanism and development of the Rio Grande depression in the southern Rocky Mountains: Geological Society of America, Memoir 144, pp. 119–154.

Lorenz, W., 1973, On the formation of maars: Bulletin of Volcanology, v. 37, pp. 183–204.

Manley, K., and Mehnert, H. H., 1981, New K–Ar ages for Miocene and Pliocene volcanic rocks in the northwestern Española Basin and their relationships to the history of the Rio Grande rift: Isochron/West, no. 30, pp. 5–8.

May, S. J., 1980, Neogene geology of the Ojo Caliente–Rio Chama area, Española Basin, New Mexico: Ph.D. thesis, University of New Mexico, Albuquerque, 204 pp.

Zoback, M. L., and Zoback, M., 1980, State of stress in the conterminous United States: Journal of Geophysical Research, v. 85, no. B11, pp. 6113–6156.

PROTEROZOIC ROCKS OF THE TAOS RANGE, SANGRE DE CRISTO MOUNTAINS, NEW MEXICO

JOHN C. REED, Jr.
U.S. Geological Survey, Denver, Colorado 80225

INTRODUCTION

Proterozoic basement rocks in the core of the Taos Range are invaded by Tertiary plutons and are partly mantled by Tertiary volcanic rocks related to the Miocene Questa caldera. Laramide thrust faults and high-angle faults related both to the Rio Grande rift and to the Questa caldera cut the basement rocks (Lipman, 1983). The Proterozoic rocks include supracrustal rocks of volcanic, volcaniclastic, and sedimentary origins and a variety of plutonic rocks, all of which have undergone amphibolite-grade regional metamorphism. Most of the rocks have conspicuous foliations; many have distinct mineral lineations produced by shearing and recrystallization during and after metamorphism.

Much of the Taos Range (Fig. 1) was mapped by McKinlay (1956, 1957) and Clark and Read (1972), but these studies did not focus on the Precambrian rocks. Condie (1980) made a reconnaissance of the Precambrian rocks in the southern part of the range, and McCrink (1982) and Condie and McCrink (1982) published a detailed description and geochemical study of the well-exposed sequence of supracrustal and intrusive rocks in the Gold Hill–Wheeler Peak area. Mapping of the Precambrian terrane by the U.S. Geological Survey began in 1980 in conjunction with a long-range study of the Questa caldera and as part of the evaluation of the mineral-resource potential of Wilderness and Wilderness Study Areas in the Taos Range. A preliminary map of the southern part of the range has been published (Reed and others, 1983) and a map of the entire area is in preparation. A few preliminary lead–uranium ages are available on zircon from the Proterozoic rocks. All the ages quoted in this report are from Bowring and others (1984) except as noted. The uncertainties in the ages quoted are generally less than 10 m.y. A paper fully reporting the analytical results on which the ages are based is currently in preparation.

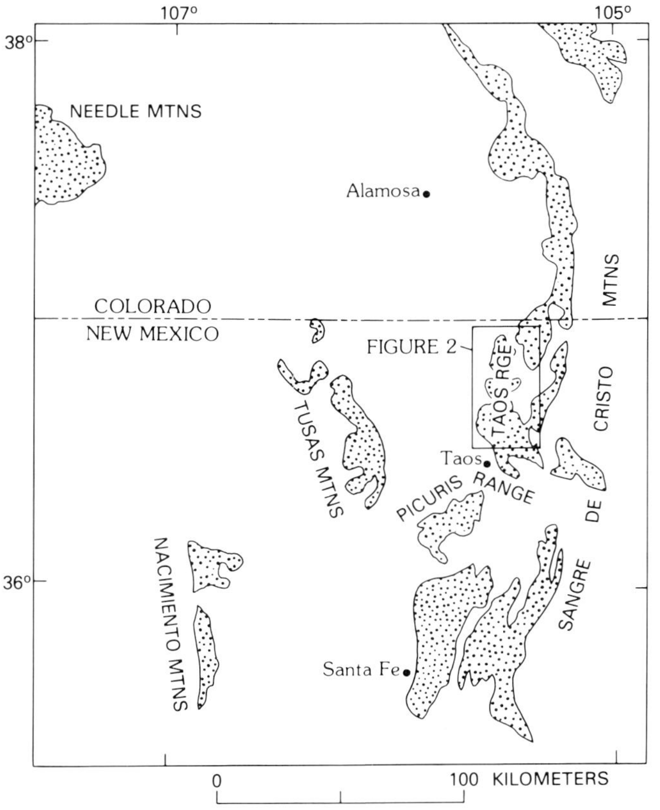

FIGURE 1. Index map showing location of the Taos Range and other areas of Precambrian outcrop in northern New Mexico and southern Colorado (stippled).

SUPRACRUSTAL ROCKS

Layered-gneiss Sequence

Interlayered feldspathic gneiss, biotite and hornblende gneiss, hornblende gneiss, and amphibolite comprise much of the Proterozoic terrane south of the Questa caldera, and similar rocks are widespread to the north and east (Fig. 2). The mafic gneisses are generally fine- to medium-grained and consist of various proportions of quartz, oligoclase–andesine, blue-green hornblende, smoky-brown biotite, epidote, and magnetite. Compositional layers range in thickness from a few millimeters to several meters and are generally parallel to foliation. In many places layers tend to be lens-shaped and discontinuous, suggesting that much of the layering represents transposed fold limbs. Locally, outcrop-scale isoclinal folds (Fig. 3) with limbs sheared off parallel to layering confirm this inference.

Occasional lenses a few centimeters to several meters thick of thinly laminated ferruginous quartzite, magnetite ironstone, and quartz–epidote–calcite marble are interleaved with the mafic gneisses. South of Gold Hill a lens of chert and chert breccia 20 m thick and 200 m long occurs in mafic gneiss of the layered gneiss sequence. In a few places lenses of muscovite or muscovite–biotite schist a few meters thick are interleaved with the biotite and hornblende gneisses south of the caldera. Some of these layers contain kyanite, sillimanite, and staurolite, but the paragenetic relations among these minerals have not been established.

Between Rio Hondo and the south margin of the caldera much of the supracrustal sequence is fine-grained, light-gray, greenish-gray or pink felsic gneiss which contains scattered 2–5 mm ovoid grains of bluish-gray quartz and 1–5 mm laths of feldspar. The rocks consist of a microcrystalline groundmass of granoblastic quartz, plagioclase (An_{5-15}), potassic feldspar, epidote, and scattered flakes of brown biotite partly altered to chlorite. The large quartz grains consist of single, optically continuous crystals or subgrains with nearly parallel orientations. The rounded and embayed shapes of the grains suggest that they are partly resorbed phenocrysts (Gresens and Stensrud, 1974). The feldspar laths are either well-twinned cloudy plagioclase (An_{10-15}) or grid-twinned microcline with irregular blotches of albite. In some rocks, biotite flakes and sphene aggregates in 1–2 mm irregular clots give the rock a spotted appearance.

The felsic gneiss contains layers of amphibole–biotite gneiss, amphibole gneiss, and calc-silicate gneiss. In a few places, thinly layered felsic gneiss displays tangential crossbeds with individual laminations marked by heavy mineral concentrations. Irregular blocks of amphibolite, some with relict basaltic textures and chilled margins, are locally common. These may represent volcanic bombs, boulders, or fragments of mafic dikes or sills dismembered during deformation. Near their contacts, felsic and mafic gneisses are interlayered on all scales and in

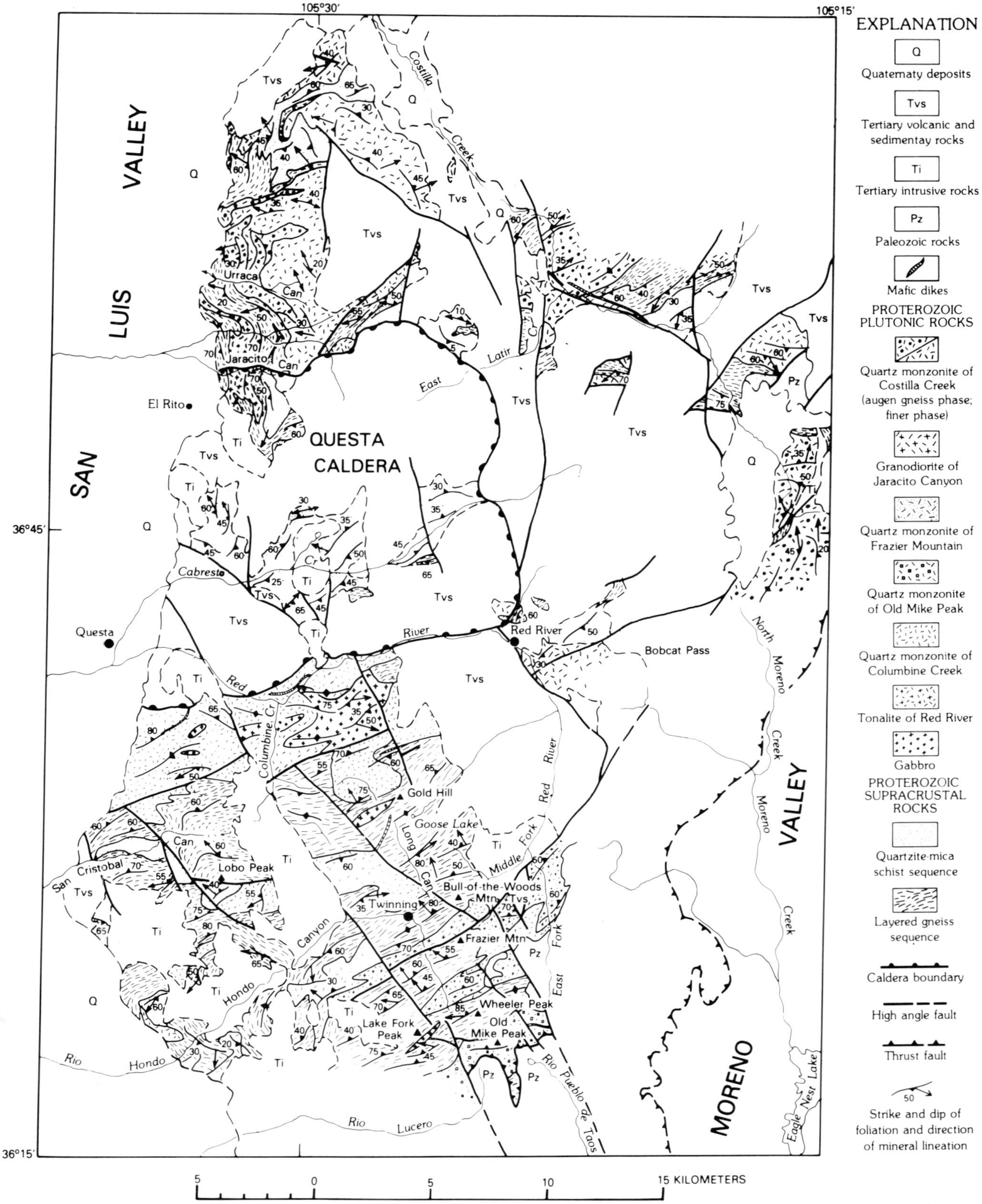

FIGURE 2. Simplified geologic map of the Precambrian rocks of the Taos Range.

FIGURE 3. Breccia along Tertiary fault zone in layered mafic gneiss on ridge 2.5 km southwest of Twinning. Note rootless isoclinal-fold noses in gneiss indicated by arrows. Scale is 18 cm long.

all proportions, and thin layers and lenses of porphyritic felsic gneiss occur locally throughout the supracrustal succession.

Between Gold Hill and Bull-of-the-Woods Mountain mafic and felsic gneisses are interlayered with lustrous, fine-grained phyllitic schist composed chiefly of quartz, muscovite, and chlorite. The schist commonly contains abundant irregular sieve-textured aggregates of rusty-weathering calcite or dolomite surrounded by aggregates of biotite. A few irregular grains of albite can be identified in some of the schist. The schist contains interbeds a few centimeters to several meters thick of fine-grained, very thinly laminated specularite ironstone, greenish-gray to white calcareous quartzite, and white siliceous marble. A few layers in the schist contain small angular fragments of fine-grained felsic rocks some of which resemble flattened lapilli. The schist also contains lenses of massive, fine-grained calcareous greenstone as much as 200 m thick. Most of the greenstone consists of a felted mosaic of actinolite, chlorite, epidote, and calcite with scattered aggregates of quartz and albite.

The layered gneisses are believed to represent a sequence of predominantly volcaniclastic rocks interlayered with flows, tuffs, and sedimentary rocks. Compositions of many of the amphibolite layers are basaltic, and some have well-preserved basaltic textures. Locally, amygdaloidal volcanic breccia is preserved (Fig. 4). The porphyritic felsic gneiss has a composition close to that of rhyolite. Some of it may represent flows or hypabyssal intrusives, but, in many places, the porphyritic felsic gneiss has features suggesting that it is a reworked tuff or tuff-breccia (Fig. 5). Many of the biotite–hornblende gneisses have compositions similar to that of andesite, but no volcanic textures have been identified. Some of these rocks have indistinct graded bedding, which indicates that they may have been derived from graywackes.

FIGURE 4. Mafic volcanic breccia in layered-gneiss sequence on ridge about 2 km south of Gold Hill. Dark fragments are amygdaloidal greenstone; matrix is epidote-rich greenstone. Amygdules are filled with quartz and albite. Note fine-grained nonamygdaloidal borders on many fragments. Pencil is 15 cm long.

Reliable indicators of facing directions are rare in the layered-gneiss sequence, and nowhere has the stratigraphic order been clearly established. There is a suggestion of a coherent stratigraphy in the area between Gold Hill and Bull-of-the-Woods Mountain where detailed mapping by McCrink (1982) shows no major repetition of stratigraphic units. He concluded that the sequence faces north, on the basis of scour-and-fill structures in one outcrop and the occurrence of fragments of felsic rocks in adjacent mafic layers. On the other hand, crossbedding and graded bedding in the felsic volcaniclastic rocks north of Gold Hill suggest that the sequence faces south. Relict cumulus layers along the northwest contact of the metagabbro sill west of Gold Hill also suggest a south-facing sequence. Study of oriented thin sections of the graded laminations in the chert lens and the ironstone layers may yield further evidence of facing direction.

Zircon from a felsic volcaniclastic rock near Gold Hill has an age of 1,750 Ma. This is probably a good estimate of the depositional age of the layered-gneiss sequence.

Quartzite–Mica-schist Sequence

Quartzite interlayered with mica schist and mica gneiss comprise most of the exposed floor of the Questa caldera, and similar rocks are extensive to the north and northeast. Quartzite with little or no associated micaceous rocks crops out in a narrow belt in the layered gneisses between the mouth of San Cristobal Canyon and Lobo Peak, and in several small areas to the south. Similar quartzite is mapped and described by Clark and Read (1972) and Condie (1980) in the upper reaches of the Rio Pueblo de Taos and between Bobcat Pass and North Moreno Creek, but neither of these areas was accessible during the present study.

FIGURE 5. Layered felsic gneiss about 2 km northeast of Gold Hill. Coarser layers contain abundant laths of white feldspar and ovoid grains of blue quartz. A few of the finer layers are crossbedded. Scale is 18 cm long.

The quartzite is white to blue-gray, sugary to vitreous, and ranges from massive to well layered on scales of a few centimeters to several meters. Layering is marked by differences in color, grain size, or mica content and is parallel to bedding as indicated by heavy mineral streaks. Near the mouth of San Cristobal Canyon the quartzite contains a 2-m-thick layer of quartz-pebble conglomerate. Pebbles are angular to subrounded and are as much as 10 cm in diameter; the pebbles consist entirely of gray, brown, or white quartz or quartzite. Pebble layers have not been found elsewhere.

Rocks interleaved with the quartzite are coarse-grained mica schist, medium-grained mica gneiss, and, very locally, gray calcite marble. The mica schist and gneiss are strongly foliated and rudely to well layered. Typically, they consist predominately of quartz, biotite, and muscovite. Most lack plagioclase, but some contain as much as 20% oligoclase–andesine (An_{15-25}). Some of the schist in the caldera south of El Rito contains abundant disseminated graphite. Mica schists and gneisses in and north of the caldera contain sillimanite, andalusite, and cordierite in apparent equilibrium. Similar rocks south of the caldera contain various combinations of kyanite, staurolite, sillimanite, and cordierite, but the paragenetic relations between these minerals are not clear.

Layering in the micaceous rocks is defined by differences in color, grain size, and proportions of quartz and micaceous minerals. In several areas the schist contains conspicuous, randomly oriented porphyroblasts of light-green muscovite as much as 5 cm in diameter. Muscovite in the porphyroblasts is only slightly warped and has sieve-textured margins containing abundant poikilitic inclusions of matrix minerals.

Contacts between the layered gneisses and the quartzite and mica-schist sequence are generally conformable. Although intrusive rocks rarely cut the quartzite sequence, perhaps due to differences in mechanical properties, there is no conclusive evidence for an intrusive event predating deposition of the quartzite sequence. Neither is there any apparent difference in structural or metamorphic history between the layered-gneiss sequence and the quartzite–mica-schist sequence. These relations suggest that no major unconformity separates the two sequences. Because indicators of facing direction are sparse in both sequences, their relative age cannot be definitely established. Where dips of layering are gentle and structure is relatively simple, the layered-gneiss sequence generally underlies the quartzite sequence. Basement rocks in the downdropped block within the Questa caldera consist almost entirely of quartzite and mica schist, which also suggests that this sequence once overlay the layered gneisses in an extensive area. These relations, although tenuous, suggest that the quartzite–schist sequence is the younger.

The only place where facing indicators are relatively common is in the narrow belt of quartzite between the mouth of San Cristobal Canyon and Lobo Peak where crossbedding faces consistently south. The structure of the area has not been worked out in detail. The quartzite may occupy the trough of a complex syncline with the south limb structurally removed. A simpler interpretation is that the quartzite represents a large interbed within the layered-gneiss sequence. If so, the facing direction is consistent with that inferred for the layered gneisses at Gold Hill along strike to the east.

PLUTONIC ROCKS

The supracrustal rocks are invaded by a diverse array of plutonic rocks ranging in composition from pyroxenite to granite. All the plutonic rocks are of calc-alkaline affinity; the felsic rocks are all weakly peraluminous, and the intermediate and mafic rocks are generally metaluminous. In the following discussion the plutonic rocks are described using igneous-rock names, although all the rocks have been deformed and regionally metamorphosed along with the supracrustal rocks and most display conspicuous foliation.

Gabbro and Ultramafic Rocks

Gabbro, now metamorphosed to coarse-grained amphibolite and amphibole gneiss, forms several irregular bodies enclosed by younger quartz monzonite south of the Red River and numerous dikes and sills cutting the layered gneisses. The two largest sills are the sill west of Gold Hill and the sill south of Lake Fork Peak, the former 500 m thick and the later about 200 m thick. Ultramafic rocks occur as layers and lenses in gabbro, as isolated pods and lenses in the supracrustal rocks, and as small dikes and sills cutting the supracrustal rocks. McCrink (1982) reported that ultramafic rocks cut dikes and sills of gabbro near Gold Hill, but that deformed bodies of ultramafic rocks are cut by gabbro near Wheeler Peak.

The gabbro is generally a medium- to coarse-grained, dark-green to greenish-gray rock composed of rudely equidimensional clots of blue-green hornblende set in a mosaic of plagiocase (An_{30-50}), epidote, and quartz with a few flakes of chestnut-brown biotite and minor amounts of calcite. In smaller bodies the rock is medium- to fine-grained and displays chilled margins against enclosing supracrustal rocks. Foliation is best developed in the finer-grained gabbro; coarse-grained gabbro is commonly massive or only very rudely foliated. Original ophitic or intergranular textures are faintly preserved in some rocks, and a few have relict cumulus layering (Fig. 6). A few ultramafic rocks contain relict grains of olivine. Zircon from gabbro in the sill west of Gold Hill has an age of about 1,741 Ma.

Tonalite of Red River

Medium- to coarse-grained, strongly foliated biotite–hornblende tonalite forms several elongated northeast–southwest trending bodies in the headwaters of the East Fork of Red River north and east of Frazier (Frazer) Mountain and Wheeler Peak. The tonalite typically consists of a mosaic of recrystallized plagioclase, quartz, and biotite, studded with 0.5–1 cm ovoid porphyroclasts of well-twinned and faintly zoned plagioclase (An_{30-35}). Hornblende forms large, irregular, sieve-textured grains and scattered smaller grains in the mosaic. In a few rocks traces of the original igneous texture are preserved. Locally the tonalite grades imperceptibly into gabbro which closely resembles the main masses of gabbro previously described. Contacts of the tonalite with the layered mafic gneiss are generally sharp. Slabs and blocks of layered gneiss are locally abundant in the tonalite, and a few dikelets of tonalite cut the enclosing gneiss.

Zircon from tonalite northeast of Frazier Mountain has an age of about 1,750 Ma. The gradation from tonalite to gabbro suggests that the rocks may be essentially of the same age, and uncertainties in the age determinations (S. A. Bowring, personal comm. 1983) do not preclude this possibility.

Quartz Monzonite of Columbine Creek

Gray to pink, gneissic quartz monzonite crops out in a large pluton south of the Red River between the town of Red River and the mountain front to the west. Similar rock makes up a number of smaller plutons which cut the supracrustal rocks as far south as Rio Hondo. The rock

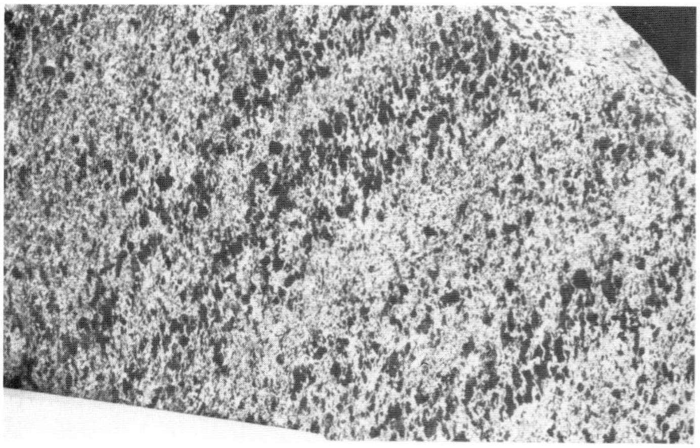

FIGURE 6. Cumulus layering in gabbro (now metamorphosed to amphibolite). Foliation transects layering at about 60°. Specimen from unmapped body about 5 km east of confluence of East Latir Creek and Costilla Creek. Height of specimen shown is about 18 cm.

is weakly to moderately well foliated and in outcrop appears medium- to coarse-grained. It consists of a fine-grained inequigranular mosaic of sutured quartz, plagioclase (An_{5-15}), and grid-twinned microcline, with scattered irregular dark clots about 5 mm in diameter. The clots consist of crudely aligned, undeformed flakes of chloritized biotite and skeleton grains of epidote. The distribution of biotite–epidote clots and quartz segregation lenses in the generally fine-grained mosaic gives the impression that the rock is medium- or coarse-grained. Near Tertiary plutons recrystallization has locally destroyed the foliation, resulting in a vaguely mottled, light-gray to pink, massive sugary rock.

The quartz monzonite contains abundant inclusions of layered mafic gneiss, amphibolite, and gabbro, and a few inclusions of felsic volcanic rocks. Dikes of quartz monzonite cut the gabbro body east of the mouth of Columbine Creek. A 30-m-thick concordant body of felsic gneiss in the quartzite sequence just south of the summit of Lobo Peak is interpreted as a sheared sill of quartz monzonite.

Zircon from the quartz monzonite along the mountain front south of Questa has an age of 1,730 Ma.

Quartz Monzonite of Old Mike Peak

Pink, medium- to coarse-grained, weakly to moderately well-foliated quartz monzonite forms an elongate pluton that extends southwest from the headwaters of the East Fork of Red River into the valley of Rio Lucero. The map by Condie (1980) shows that the pluton is extensively exposed in the lower reaches of the valley of Rio Lucero, but this area is currently inaccessible as it lies within the lands of the Taos Pueblo. The rock consists of 0.5–1 cm crystals of untwinned, faintly perthitic potassic feldspar and altered zoned plagioclase (cores AN_{15}; rims near An_5) in a mosaic of recrystallized feldspar and quartz. The rock is slightly less silicic and more potassic than the quartz monzonite of Columbine Creek and differs from it chiefly in its weaker foliation, in the character of the potassic feldspar, and in the conspicuous zoning of the plagioclase.

The contact of the quartz monzonite north and west of Old Mike Peak is generally sharp and parallel to layering in the adjacent layered gneiss. A few discordant dikes of quartz monzonite cut the wallrocks, and the intrusive contains scattered blocky inclusions of layered gneiss that contain internal folds. An attempt to date the quartz monzonite was only partly successful due to a very small yield of zircon. The results show that the age of the rock is about 1,700 Ma, but do not establish the place of the quartz monzonite in the intrusive sequence.

Quartz Monzonite of Frazier Mountain

Fine-grained, strongly foliated, pink to gray felsic intrusive rocks form several semiconcordant plutons in a belt extending southwestward from the lower part of the East Fork of Red River to the mountain front near the mouth of Hondo Canyon. Similar rocks form dikes and sills ranging in thickness from a few centimeters to several meters in both the surrounding mafic gneiss and in the tonalite of Red River. The rocks in these bodies are chiefly quartz monzonite, but range from granite to granodiorite. The rocks typically consist of porphyroclasts of untwinned perthitic potassic feldspar and strongly zoned plagioclase (An_{15-17} cores; An_{5-10} rims) in a mosaic of recrystallized quartz, albite, epidote, and postkinematic biotite.

The rocks in these bodies resemble the quartz monzonite of Old Mike Peak in mineralogy and major-element chemistry. The character of the potassic feldspar and the zoning of the plagioclase are also similar. They differ in grain size, texture, degree of development of foliation, and in the common occurrence of allanite and muscovite as accessories in the Frazier Mountain rocks. The two rocks probably were comagmatic and their minor compositional and textural differences have been emphasized by cataclasis and recrystallization. A dike of the quartz monzonite cutting tonalite along the East Fork of Red River has an age of about 1,699 Ma.

Granodiorite of Jaracito Canyon

Fine- to coarse-grained, strongly foliated gray granodiorite forms a semiconcordant pluton along the mountain front east and north of El Rito. The granodiorite commonly contains abundant inclusions of the enclosing gneisses, some of which display internal foliation discordant to foliation in the granodiorite. The granodiorite consists of equidimensional grains of zoned plagioclase (An_{30} cores; An_{22} rims) in a mosaic of smaller recrystallized plagioclase, potassic feldspar, and quartz. Ragged undeformed flakes of dark smoky-brown biotite and irregular poikilitic grains of olive to grass-green hornblende are aligned parallel to the foliation. The granodiorite has an age of about 1,678 Ma, appreciably younger than the tonalite of Red River. Similar granodiorite from one of a number of small unmapped bodies near the mouth of Hondo Canyon has an age of 1,689 Ma. Analytical uncertainties (S. A. Bowring, personal comm. 1983) are large enough that the rocks could be contemporaneous.

Quartz Monzonite of Costilla Creek

Gneissic quartz monzonite comprises a considerable part of the exposed basement north and northeast of the Questa caldera. Because the basement rocks are broken by numerous Tertiary faults and are extensively mantled by Tertiary rocks, it is not clear whether the various isolated outcrop areas of quartz monzonite are parts of a single large pluton or represent several independent plutons. The quartz monzonite has two textural phases. The most extensive phase is fine- to medium-grained, light gray to pink, and weakly to moderately foliated; the other phase is coarse-grained, moderately to strongly foliated, and is studded with large potassic-feldspar augen. Both phases contain small concordant lenses of pegmatite composed chiefly of quartz and gray potassic feldspar and containing 1–5 cm aggregates of magnetite. Contacts between the finer-grained quartz monzonite and the augen gneiss are gradational over distances of a few meters to several tens of meters; the two phases are interleaved on scales of a few centimeters, and the augen gneiss becomes finer-grained and contains progressively fewer augen until the two phases become indistinguishable.

Contacts of the quartz monzonite with the enclosing rocks are concordant with layering in the wallrocks and foliation in the plutonic rock. Contacts of the augen-gneiss phase with wallrocks are generally sharp, but those of the finer phase are less well defined because the enclosing rocks commonly consist of amphibolite and amphibole gneiss interleaved with fine-grained, pink, felsic gneiss, much of which probably represents sills of quartz monzonite, but some of which may represent felsic volcanic rocks or arkosic sediments. The finer-grained quartz monzonite consists of an inequigranular mosaic of strained quartz, grid-twinned microcline, zoned plagioclase (An_{16-25} cores; An_{5-10} rims), and crudely aligned flakes of deep-smoky-brown biotite. The augen gneiss is very similar except for the presence of large porphyroclasts of microcline (many of which are now aggregates of grain fragments), and for the concentration of slightly coarser biotite flakes into anastomosing folia that enclose lenses of recrystallized quartz and feldspar; these lenses give the rock a coarse-grained appearance in hand specimen.

The augen-gneiss phase has an age of about 1,644 Ma; several samples of the finer phase have been collected but have not yet been analyzed.

MAFIC DIKES

Mafic dikes cut the foliated Proterozoic rocks throughout the range, but are truncated by the unconformity at the base of Mississippian strata near Wheeler Peak. Many of the dikes have approximately east–west trends and some can be traced for several kilometers. Most of the dikes are medium- to fine-grained diabase, but a 60-m-thick dike north of Costilla Creek is hornblende–pyroxene diorite. This dike has margins and apophyses of diabase similar to the diabase in the other mafic dikes. Most of the dikes have chilled margins and display diabasic or ophitic textures. Three whole-rock samples of the diorite dike along Costilla Creek fit closely on a rubidium–strontium isochron indicating an age of about 500 Ma, but a fourth sample of somewhat altered rock from the dike margin falls significantly off the isochron (Z. E. Peterman, personal comm. 1984).

METAMORPHIC AND STRUCTURAL HISTORY

All the basement rocks were regionally metamorphosed and deformed during the Early or Middle Proterozoic. Effects of shearing during the

Laramide orogeny and contact metamorphism and brecciation during the mid-Tertiary are superimposed on the Proterozoic mineral assemblages and structures.

Mineral assemblages in the pelitic metasedimentary rocks and in the mafic gneisses and intrusive rocks indicate Proterozoic metamorphism in the lower-grade part of the amphibolite facies. Particularly significant are the occurrence of kyanite, staurolite, sillimanite, and cordierite in pelitic rocks south of the caldera, the occurrence of the possible equilibrium assemblage sillimanite–andalusite–cordierite in the pelitic rocks in and north of the caldera, and the equilibrium assemblage of blue-green hornblende–calcic plagioclase–biotite in the mafic rocks throughout most of the range. The phyllitic schist and associated mafic rocks between Gold Hill and Bull-of-the-Woods Mountain contain chlorite–actinolite–albite–epidote assemblages, indicating that in this area metamorphism was only to greenschist facies.

The earliest period of deformation of the supracrustal rocks produced the isoclinal folds now preserved as small intrafolial fold noses and sheared out limbs in the supracrustal rocks. The principal foliation is parallel to layering except very locally in the noses of these isoclines. The foliation and mineral lineation in both supracrustal and plutonic rocks are marked by flattened, elongate lenses of recrystallized quartz and feldspar and crudely aligned but undeformed grains and aggregates of micaceous minerals and hornblende. The mineral lineation lies on the foliation planes at various angles to the axes of the early isoclines. The foliation and lineation are inferred to be the results of cataclasis and recrystallization. The lack of deformation of the metamorphic minerals (Fig. 7) indicates that the climax of regional metamorphism post-dated significant movement along foliation planes. Because the youngest dated plutonic rock, the quartz monzonite of Costilla Creek, displays the same foliation and lineation, these structures, most of the isoclinal folds, and the regional metamorphism must be younger than 1,644 Ma. However, folds contained in blocky inclusions in the quartz monzonite of Old Mike Peak and rotated foliated inclusions in the granodiorite of Jaracito Canyon show that some folds and some of the foliation formed earlier. Wobus and Hedge (1982) have dated a post-tectonic granite porphyry stock at Tusas Mountain, about 50 km west of the Taos Range, at between 1,500 and 1,430 Ma. This age provides a younger limit for metamorphism and formation of the principal foliation.

More open folds, probably of several generations (Norman and Nielsen, 1982), are superimposed on early isoclines, but superimposed folds are rarely seen in single outcrops. It is these later folds that are responsible for much of the complexity of the pattern of foliation and lineation trends (Fig. 2). Some of the open folds have poorly developed crenulation cleavage with recrystallized mica parallel to the axial planes, which suggests that they formed during the waning stages of regional metamorphism.

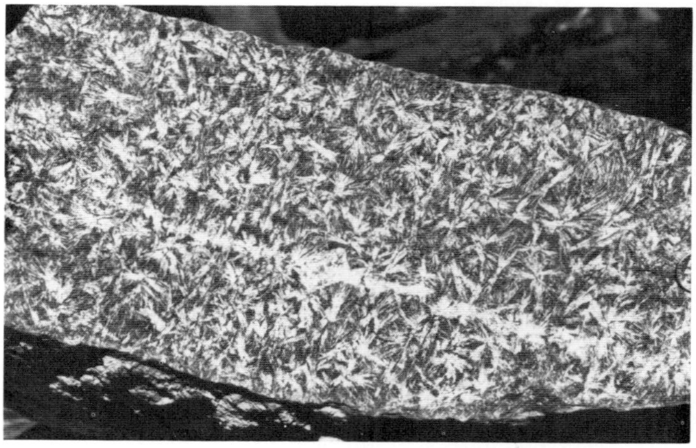

FIGURE 7. Sillimanite rosettes on foliation plane in micaceous quartzite. Rosettes are about 3 cm in diameter. White streak is a scratch. Specimen from about 4 km east of confluence of East Latir Creek and Costilla Creek.

It has been suggested (Reed and others, 1983) that the broad belt of steeply dipping, northeast-trending foliation south of the Questa caldera is an expression of the Jemez zone, a northeast-trending zone of crustal weakness of Precambrian ancestry that has been an important control of late Cenozoic volcanism and tectonism (Mayo, 1958; Lipman and Mehnert, 1979). Most of the foliation in the belt is not related to a younger crosscutting shear zone, but local zones of phyllonite and silicified blastomylonite a few meters to a few tens of meters thick lie parallel to the steep foliation and apparently mark ductile shear zones. Thus, the broad belt of steep foliation may be due to rotation of the earlier regional foliation during movement along narrower and more local shear zones related to the Jemez zone.

Tertiary faults in the basement rocks are marked by zones of shattered rocks, breccia (Fig. 3), and microbreccia. The breccias consist of angular fragments of wallrock in a dark-green to gray mortar of quartz, feldspar, chlorite, sericite, and limonite. Fragments in the microbreccia include individual grains of plagioclase—some with warped, broken, and offset twin lamellae.

CONCLUSIONS

The principal conclusions of the present investigation of the Precambrian rocks of the Toas Range are:

(1) The supracrustal rocks include a sequence of layered mafic and felsic gneisses derived from volcanic, volcaniclastic, and associated sedimentary rocks, and a sequence of quartzite and pelitic schist derived from nonvolcanic shelf sediments. In general, the quartzite–schist sequence overlies the volcanic sequence, but, locally, quartzite and schist are interlayered in the volcanic sequence.

(2) The protoliths of the layered gneisses are of Early Proterozoic age (about 1,750 Ma) and included volcanic and volcaniclastic rocks ranging in composition from rhyolite to basalt. Volcanic rocks of intermediate composition have not been identified, but many of the mafic gneisses have bulk compositions close to andesite and may have been derived from graywacke with an appreciable volcanic component.

(3) Calc-alkaline plutonic rocks, the oldest of which were essentially contemporaneous with the volcanic rocks, were emplaced in the supracrustal rocks over a span of only about 120 m.y., a span comparable to that required for emplacement of the various plutons of the central Sierra Nevada batholith (Bateman, 1983). The plutonic rocks include large volumes of quartz monzonite, smaller volumes of granodiorite and tonalite, and still smaller volumes of granite, gabbro, and ultramafic rocks. There seems to be no simple relation between age of emplacement and composition, although there is a general progression from ultramafic and mafic rocks through intermediate compositions to potassic quartz monzonite. There is a general decrease in age of emplacement from south to north, counter to the regional trend noted by Condie (1982), perhaps due to local structural complexity or to the small size of the area investigated. No major body of post-tectonic Proterozoic granite exists in the Taos Range as suggested by Condie and McCrink (1982), although the possibility of dikes or very small, poorly exposed plutons cannot be excluded.

(4) Isoclinal folding of the supracrustal rocks and cataclasis and recrystallization of both plutonic and supracrustal rocks to form the principal regional foliation took place during the early stages of regional metamorphism between 1,644 and 1,430 Ma. Regional metamorphism of most of the rocks took place at temperatures near the aluminum-silicate triple point. The occurrence of kyanite in the rocks south of Questa caldera suggest that P–T gradients were lower there than in the rocks now exposed in and north of the caldera, where andalusite occurs (Grambling, 1981).

(5) Notation of the regional foliation into near parallelism with narrow northeast-trending zones of ductile shear post-dated regional metamorphism. These shear zones may be related to the Jemez zone.

REFERENCES

Bateman, P. C., 1983, A summary of critical relations in the central part of the Sierra Nevada batholith, California, U.S.A.; *in* Roddick, J. A. (ed.), Circum-Pacific plutonic terranes: Geological Society of America, Memoir 159, pp. 241–254.

Bowring, S. A., Reed, J. C., Jr., and Condie, K. C., 1984, U–Pb geochronology of Proterozoic volcanic and plutonic rocks, Sangre de Cristo Mountains, New Mexico (abstract): Geological Society of America, Abstracts with Programs, v. 16, no. 4, p. 216.

Clark, K. F., and Read, C. B., 1972, Geology and ore deposits of the Eagle Nest area, New Mexico: New Mexico Bureau of Mines and Mineral Resources, Bulletin 94, 152 pp.

Condie, K. C., 1980, Precambrian rocks of the Red River–Wheeler Peak area, New Mexico: New Mexico Bureau of Mines and Mineral Resources, Geologic Map 50, scale 1:48,000.

———, 1982, Plate-tectonic model for Proterozoic continental accretion in the southwestern United States: Geology, v. 10, pp. 37–42.

———, and McCrink, T. P., 1982, Geochemistry of Proterozoic volcanic and granitic rocks from the Gold Hill–Wheeler Peak area, northern New Mexico: Precambrian Research, v. 19, pp. 141–166.

Grambling, J. A., 1981, Kyanite, andalusite, sillimanite and related mineral assemblages in the Truchas Peaks region, New Mexico: American Mineralogist, v. 66, pp. 702–722.

Gresens, R. L., and Stensrud, H. L., 1974, Recognition of more metarhyolite occurrences in northern New Mexico and a possible Precambrian stratigraphy: Mountain Geologist, v. 11, no. 3, pp. 109–124.

Lipman, P. W., 1983, The Miocene Questa caldera, northern New Mexico; relation to batholith emplacement and molybdenum mineralization; *in* The genesis of Rocky Mountain ore deposits—changes with time and tectonics: Society of Exploration Geologists, Denver Region, Symposium Proceedings, pp. 133–148.

———, and Mehnert, H. H., 1979, The Taos Plateau volcanic field, northern Rio Grande rift, New Mexico: *in* Riecker, R. E. (ed.), Rio Grande rift: tectonics and magmatism: American Geophysical Union, pp. 289–311.

Mayo, E. B., 1958, Lineament tectonics and some ore districts in the Southwest: Mining Engineering, v. 10, no. 11, pp. 1169–1175.

McCrink, T. P., 1982, Precambrian geology of the Taos Range, Taos County, New Mexico: Unpublished M.S. thesis, New Mexico Institute of Mining and Technology, 123 pp.

McKinlay, P. K., 1956, Geology of Costilla and Latir Peak quadrangles, Taos County, New Mexico: New Mexico Bureau of Mines and Mineral Resources, Bulletin 42, 32 pp.

———, 1957, Geology of the Questa quadrangle, Taos County, New Mexico: New Mexico Bureau of Mines and Mineral Resources, Bulletin 53, 23 pp.

Norman, M. D., and Nielsen, K. C., 1982, Polyphase deformation of a Proterozoic sequence in the Taos Range, northern New Mexico (abstract): Geological Society of America, Abstracts with Programs, v. 14, no. 3, p. 134.

Reed, J. C., Jr., Lipman, P. W., and Robertson, J. R. [M], 1983, Geologic map of the Latir Peak and Wheeler Peak Wildernesses and Columbine–Hondo Wilderness Study Area, Taos County, New Mexico: U.S. Geological Survey, Miscellaneous Field Studies Map MF-1570-B, scale 1:50,000.

Wobus, R. A., and Hedge, C. E., 1982, Redefinition of the Precambrian Tusas Mountain and Tres Piedras granites, north-central New Mexico: Mountain Geologist, v. 19, no. 4, pp. 105–114.

Taos Plaza, northeast corner, at the July Fiesta, ca 1938. Photographer unknown.

PRECAMBRIAN GEOLOGY AND METALS POTENTIAL OF THE TWINING–GOLD HILL AREA, TAOS RANGE, NEW MEXICO

M. DEWITT DAGGETT III
Noranda Exploration, Inc., P.O. Box 50326, Tucson, Arizona 85703

INTRODUCTION

Early Proterozoic supracrustal and intrusive rocks crop out in the northern Sangre de Cristo Mountains from east of the town of Taos to north of the New Mexico/Colorado state line. Recent studies of this terrane in the Taos Range segment of the Sangre de Cristo Mountains refine previously published information on the area. This paper presents an overview of the Precambrian supracrustal geology of the southern Taos Range, describes in detail the geology of a small area north of Twining, and discusses the region's potential for deposits of volcanogenic metals.

Reed (this guidebook) outlines the Precambrian geology of the entire range, based on recent U.S. Geological Survey mapping (Reed and others, 1983). McCrink (1982) mapped portions of the Gold Hill and Wheeler Peak areas and conducted a geochemical study of the rocks (Condie and McCrink, 1982).

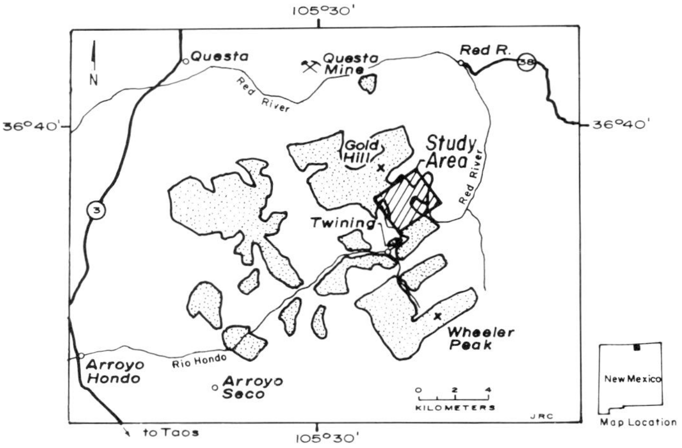

FIGURE 1. Index map showing Precambrian supracrustal rocks (stippled) in the southern Taos Range and the area discussed in this paper (outlined).

GEOLOGY

Overview

The Proterozoic supracrustal section of northern New Mexico comprises two major lithologic units: (1) a suite of mafic and felsic metavolcanic rocks with intercalated volcaniclastic and epiclastic metasedimentary rocks, and (2) a metasedimentary sequence dominated by quartzite, but also containing metamorphosed conglomerate, shale, arkose, and rare felsic tuff. In the areas of Picuris and Truchas Peaks to the south, these sequences are the Vadito Group and the Ortega Group, respectively (Montgomery, 1963; Long, 1976; Robertson and Moench, 1979). The Vadito Group is the older of the two, although locally it structurally overlies Ortega Group quartzite (Grambling and Codding, 1982; Holcombe and Callender, 1982). Felsic volcanic rocks from the southern Taos Range yield U–Pb zircon ages of 1,750 m.y. (Bowring and others, 1984).

In the Taos Range, metavolcanic rocks dominate the supracrustal assemblage exposed between the Taos Pueblo and Red River, while a quartzite and mica-schist sequence constitutes most of the supracrustal outcrop north of Red River (Reed, this guidebook). Proterozoic gabbroid to granitic plutons, and Tertiary granitic stocks and volcanic rocks related to the Questa caldera, intrude and cover older rocks throughout the range (Reed and others, 1983). Numerous high-angle faults further define the outcrop distribution of supracrustal sequences.

The most extensive exposure of metavolcanic and associated metasedimentary rocks in the area lies in the southern part of the Taos Range, between Gold Hill and Wheeler Peak (Fig. 1). A 2.5 by 3.5 km portion of this sequence, situated east and north of Twining between Gold Hill and Bull-of-the-Woods Mountain, is the focus of this study. The area was reconnaissance-mapped at scales of 1:24,000 and 1:12,000, and locally mapped in detail at a scale of 1:6,000. Geologic control is good to fair in places, but is lacking in others owing to thick soil and forest cover.

Textures and primary structures are surprisingly well preserved in metavolcanic and metasedimentary rocks, in spite of amphibolite-facies metamorphism and intense deformation (Reed, this guidebook, and McCrink, 1982). These textures allow the determination of protoliths in most cases, and rocks are named accordingly in the descriptions and discussion which follow. The prefix "meta" is omitted. Reference to megascopic metamorphic appearance is made in descriptive sections of this report. Primary structures also allow the determination of an apparent stratigraphy; the following section describes rocks of the immediate study area on that basis, from oldest to youngest.

Rock Types

Basaltic Flows and Tuff

A mafic volcanic sequence composed of basaltic flows, tuff, and tuff breccia crops out from Gold Hill Ridge east to the headwaters basin of the Red River (Fig. 2). The sequence is 1,000 m thick on Gold Hill Ridge. Basalt flows occur on the margins of the mafic sequence, and thin flows are interbedded with tuff in the middle portions. Individual flows are 5–30 m thick where flow margins can be discerned. Flow rock is fine-grained, mostly massive, with poorly developed schistosity, and commonly is vesicular or amygdaloidal. Flow breccia is preserved at one locality and in a few places along the east wall of Long Canyon there are faint suggestions of small pillows. Deformation makes this interpretation tenuous.

Basaltic tuff and tuff breccia make up the middle two-thirds of the mafic sequence. Lithic fragments and broken plagioclase phenocrysts in a foliated, fine-grained, dark-green groundmass characterize the tuff; fragments and phenocrysts constitute 0.5–15% of the rock (fig. 4 in Reed, this guidebook). Thin sections reveal a groundmass of green to blue-green hornblende needles mixed with calcite and minor epidote and biotite. Plagioclase phenocrysts are altered to albite and calcite. The tuff contains rare subhedral to euhedral hornblende phenocrysts partially altered to chlorite. Groundmass hornblende is largely unaltered. Basaltic-tuff breccia is best exposed at the 11,800-ft-contour level along Gold Hill Ridge. At this locality a fine-grained, probably ash-rich basaltic groundmass contains up to 40% lapilli- and block-sized clasts of amygdaloidal basalt. The breccia is monolithologic overall, but contains accidental fragments of fine-grained felsite and coarse-grained granitic rock. Basaltic clasts are flattened in the plane of layering and clast shapes, and edge textures indicate that many were viscous

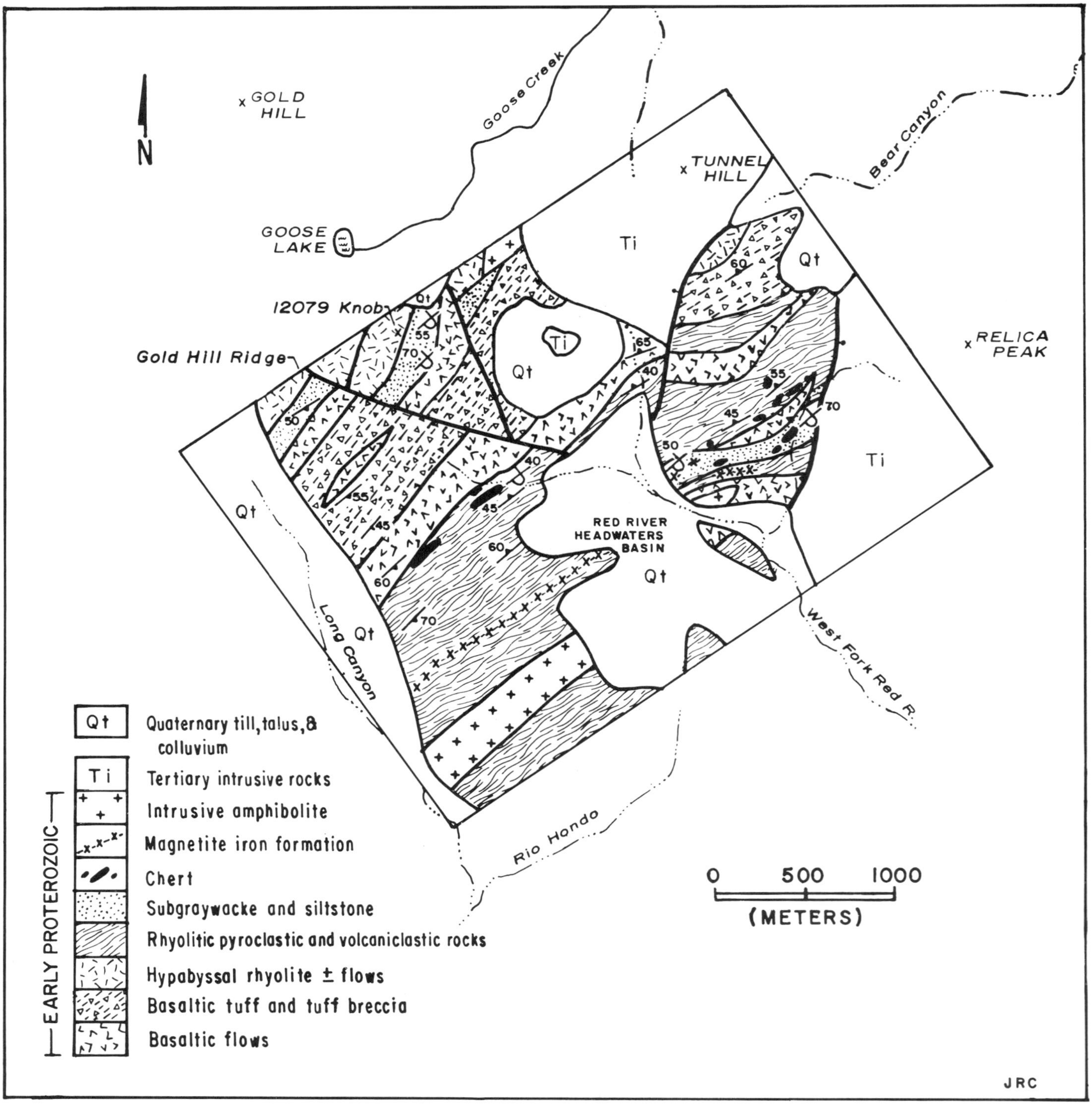

FIGURE 2. Simplified geologic map of a portion of the Gold Hill to Bull-of-the-Woods section.

when incorporated into the groundmass. The tuff breccia is more than 100 m thick. Outcrops display large-scale grading from clast-rich to clast-poor layers. These textures and features suggest that this rock originated as an agglomerate.

Rhyolitic Tuff

Fine-grained felsic tuff and volcaniclastic sedimentary rocks crop out south of the basaltic sequence. The felsic section is more than 1,000 m thick and contains crystal-lapilli tuff, lithic-lapilli tuff, very fine-grained, ash-rich tuff, and reworked tuff. Poorly to well foliated quartz–sericite phyllite and schist are the dominant lithologies in the section. Individual layers contain up to 10% potassium feldspar and/or quartz phenocrysts, 15% lithic lapilli, and 25% lapilli-sized fragments of unknown protolith. These latter fragments are mineralogically identical to the groundmass and may be pumice. Some individual tuff layers are graded over a few tens of meters from lithic-rich bottoms to ash-rich tops. Where observed, the contacts between fine-grained tops and lithic- or lapilli-bearing bottoms of overlying tuff units are sharp. Mafic and felsic dikes and sills intrude the tuff section.

Massive Rhyodacite

Massive rhyodacite crops out along Gold Hill Ridge west of Goose Lake, around the head of Long Canyon, and locally within the rhyolitic tuff section. This rock is gray to pink, mostly massive, and weakly foliated to nonfoliated. The rock contains 1–20% potassium feldspar and plagioclase phenocrysts in an aphanitic sericite–quartz groundmass. Phenocrysts are euhedral to subhedral and 1–4 mm in size. Subrounded quartz phenocrysts comprise up to 20% of the phenocryst population, but are rare to absent in many of these rocks. This rock type probably originated as hypabyssal sills and dikes, as evidenced by its massive character and mostly unbroken phenocrysts. Some of this rock possibly formed as flows.

Sedimentary Rocks

A 100-m-thick subgraywacke section occurs in the headwaters basin of the West Fork of the Red River. These sedimentary rocks are metamorphosed to biotite–quartz schist, chlorite–quartz schist, minor chlorite schist, and minor sandstone. Muscovite is present in most rock types. Garnet and calcite occur in some, and the rocks contain a trace to 10% magnetite and pyrite. Mappable rock units are 1–4 m thick; layering within these units is a few centimeters thick.

One hundred meters of laminated to thin-bedded siltstone and very fine-grained quartz sandstone and graywacke underlie the basaltic section along and east of Gold Hill Ridge. The rocks contain three distinct grain-size fractions, most in separate laminae. Grading is evident in a few layers; small-scale trough crossbedding is present at a single locality along Gold Hill Ridge. Centimeters-thick layers of magnetite, magnetite and quartz, epidote, and epidote and calcite are interbedded with the clastic rocks.

Chert and Iron Formation

Chert and iron formation occurs at a number of localities in the Gold Hill area (Fig. 2). Rhyolitic tuff and subgraywacke host chert along Gold Hill Ridge and in the headwaters basin of the Red River. The chert is massive and cryptocrystalline, and forms pods and lenses 3–10 m thick and up to 100 m long. The chert contains a trace to 8% disseminated and fracture-bound pyrite and most chert outcrops are iron-stained. A large chert body crops out at the 11,600-ft level on, and east of, Gold Hill Ridge. This chert occurs as discontinuous lenses for over 600 m of strike length; individual lenses are as much as 70 m thick. The chert is white to light gray and has a fragmental texture, with angular to subrounded fragments ranging in size from pebbles to cobbles. The fragmental texture occurs along the south side of the lenses, the presumed stratigraphic top of the unit; the northern, basal parts of the lenses are massive. I interpret this fragmental rock as a syngenetic breccia. Fine-grained, disseminated tourmaline is an accessory mineral in the chert. Waxy, pale-green sericite, identified by x-ray occurs in clots and veins.

Oxide-facies iron formation, composed of massive magnetite or magnetite and quartz, occurs in both the rhyolitic tuff and sedimentary sections. Most iron-formation layers are 1–8 m thick and continuous for a few tens of meters along strike. Bedding in the iron formation varies from thin and continuous, with sharp bedding contacts, to massive and structureless. The largest iron formation extends in outcrop and float for 1,500 m, from the bottom of Long Canyon across Gold Hill Ridge and to the steep slope above the Red River. This unit is more than 30 m thick and produces a strong high on the regional aeromagnetic map (USGS Open-file Report 82-952).

Sulfide-facies iron formation, stratabound stringers of fine-grained pyrite in a cryptocrystalline siliceous groundmass, is present at two localities within rhyolitic tuff. Pyrite constitutes up to 25% of the rock in selected hand samples. Overall, the 3–5 m thick sulfide-bearing layers contain 5–15% pyrite. Chlorite, garnet, magnetite, and tourmaline are accessory minerals. Locally, this rock contains small lithic fragments with the same mineralogical makeup as the groundmass.

Structure and Pseudostratigraphy

Primary Structures

Primary sedimentary structures are preserved in a few places in the area and all indicate that stratigraphic tops are to the south. Observed features include small-scale crossbedding in sandy layers north of the basalt on Gold Hill Ridge, graded bedding in sedimentary rocks south of Goose Lake and in the headwaters basin of the Red River, outcrop-scale grading in lithic-lapilli rhyolite tuff in the eastern part of the area, and the position of a stockwork feeder zone on the north side of chert lenses west of Relica Peak (discussed below). These data agree with facing directions reported by Reed (this guidebook) from other localities in the range, but disagree with data reported by McCrink (1982). Overall, the data indicate stratigraphically younger rocks to the south, with felsic tuff overlying the mafic sequence. However, transposition of asymmetric folds can produce the same apparent results (see Callender, 1983, p. 143).

Pseudostratigraphy

In terranes where transposition is recognized, the regional map distribution of rock units defines a pseudostratigraphy (Callender, 1983). Transposition is clearly evident on a small scale in the Gold Hill area (this report) and to the south near Wheeler Peak (Callender, 1983, fig. 3). Map-scale folds were not identified in this or previous studies in the area (McCrink, 1982), though intense polyphase deformation is characteristic of the range (Norman and Nielsen, 1982).

In recognition of the structural complexities, map-scale lithologic layering may no longer reflect the primary stratigraphic succession. On the other hand, a major lithologic unit, such as the thick basalt sequence, probably behaved coherently during strain. While transposition has altered the details of bedding and stratigraphic relationships within some rock types, the pseudostratigraphic order of the major rock units may not differ significantly from the original, truly stratigraphic sequence.

Folding and Transposition

The Proterozoic rocks in the Taos Range are intensely and multiply deformed, a characteristic common to Proterozoic rocks throughout northern New Mexico (Callender, 1983). Layering in the rocks south of Gold Hill is defined by bedding and parallel to subparallel foliation. Layering strikes northeast and dips are moderate to steep. Layering east of Gold Hill Ridge is overturned to the north, based on the primary structures cited above (Fig. 2). The map patterns of major rock units parallel the northeast trend of outcrop-scale layering.

Evidence of isoclinal folding and bedding transposition is well preserved in sedimentary sequences in the headwaters of the Red River and near the 12,079-ft knob south of Goose Lake. Rocks at these localities display isoclinal-fold noses from a few centimeters to 0.5 m across. The noses are often isolated from one or both fold limbs. Near the 12,079-ft knob, centimeters-thick epidote + calcite layers are spectacularly boudinaged. Mafic and felsic volcanic rocks do not show the small-scale effects of the intense deformation as clearly as the sedimentary rocks do, although fold noses, cleavage, and attenuated bedding are locally evident.

Alteration

Mapping identified several modes of moderate to intense sericitic and silicic alteration associated with chert lenses and iron formation. Stockworks, stringers, and veins of quartz, accompanied by pervasive silicification, underlie chert lenses in the eastern part of the area. Silicification is present for 150 m on the north or down-section side of the lenses and textures, and field relationships argue convincingly that the zone of silicification is a feeder structure for exhalative chert.

Sericitic alteration is more widespread. White mica replacing chlorite in felsic tuff and mafic rocks, muscovite and sericite along fractures, massive muscovite schist in shear zones, and waxy-green sericite in chert lenses are manifestations of this alteration type. An envelope of sericitic alteration surrounds the silicified zone described above. Restricted zones of intense sericitic alteration occur stratigraphically beneath chert and iron formation east of Gold Hill Ridge. Sericite schist at these latter localities contains up to 30% disseminated tourmaline.

Widespread chloritic alteration is not evident in the area. Where developed, it is spotty and weak. Mafic rocks locally exhibit carbonate alteration.

DISCUSSION

Geologic Synthesis

The supracrustal succession from Gold Hill to Bull-of-the-Woods is a sequence of submarine, felsic, dominantly pyroclastic rocks with interlayered basaltic flows and tuff and minor epiclastic sedimentary rocks. Because the range exposes only remnants of the supracrustal succession, the mafic/felsic ratio of the regional terrane cannot be ascertained. The apparent base of the sequence south of Gold Hill is massive rhyodacite, which originated in part as hypabyssal intrusions and possibly in part as flows. Siltstone and fine-grained graywacke with interbedded magnetite iron formation overlie the rhyodacite and reflect a period of volcanic quiescence. This period of sedimentation was followed by explosive submarine mafic volcanism manifested by basaltic flows, tuff, and tuff breccia or agglomerate. Clast size in the agglomerate implies relative proximity to a vent.

A thick section of rhyolitic tuff and reworked tuff caps the basaltic rocks. Much of this pyroclastic sequence was probably deposited as debris and slump flows off the flanks of underwater volcanic edifices. Data from well-studied submarine volcanic complexes indicate that this is a common, if not universal, feature in the subaqueous volcanic environment (Fiske, 1963; Fiske and Matsuda, 1964). Outcrop-scale grading of lithic fragments and ash, and interbedded iron formation and sedimentary rocks are further evidence of a subaqueous environment. Along the eastern edge of the area, sedimentary rocks followed by additional rhyolite and basalt cap the felsic-tuff section.

Regional Potential for Deposits of Volcanogenic Metals

The submarine mafic and felsic volcanic rocks in the southern Taos Range constitute a permissive geologic environment for deposits of volcanogenic metals. Several lines of evidence indicate that the seafloor hydrothermal-processes requisite for the formation of volcanogenic deposits operated in the Precambrian section between Gold Hill and Bull-of-the-Woods. These include the pyrite-bearing chert lenses and adjacent feeder zone in the eastern part of the area, chert and syngenetic chert breccia on Gold Hill Ridge, exhalative-type iron formation at multiple locations in the section, two occurrences of exhalative stratabound pyrite, and associated stratabound tourmaline—a common mineral constituent of volcanogenic systems (Slack, 1982). The hydrothermal systems which gave rise to these specific features were iron-rich but sulfur and base-metal poor (oxide-facies iron formation), or were relatively cool and lacking in all metals (chert). The

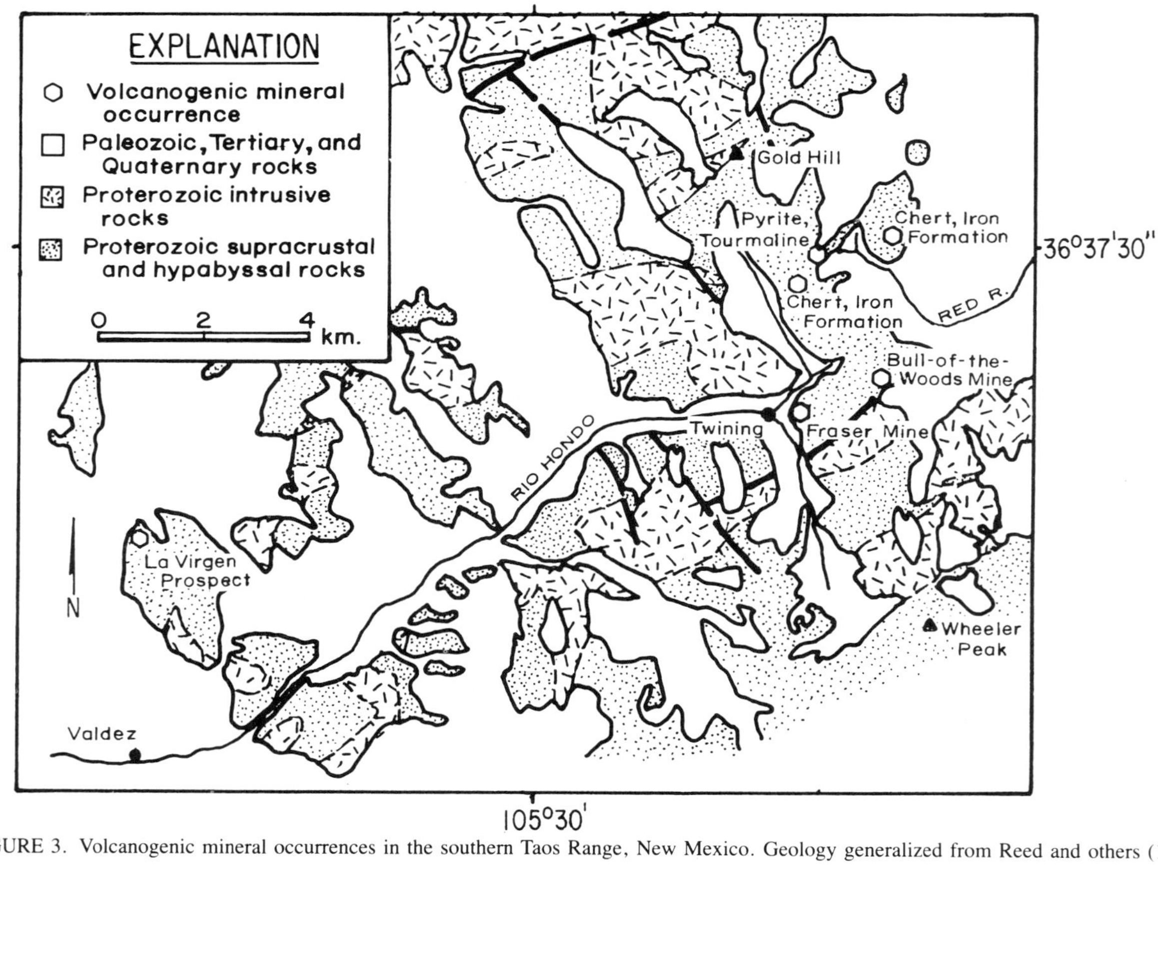

FIGURE 3. Volcanogenic mineral occurrences in the southern Taos Range, New Mexico. Geology generalized from Reed and others (1983).

chert, iron formations, local sedimentary component of the section, and lack of extensive felsic flows and flow/dome complexes reflect a distal volcanogenic setting for the Gold Hill area (Hutchinson and others, 1971).

At least two other volcanogenic-type mineral occurrences are known in the southern Taos Range (Fig. 3). Magnetite iron formation and pyritic chert crop out in the vicinity of the Fraser mine east of Twining (Noranda reconnaissance, 1982). These are associated with sheared rhyolitic rocks and chlorite, sericite, and talc schists which are part of a larger sheared and mineralized zone extending from Twining to Bull-of-the-Woods Mountain (Fulp and Woodward, 1981). Mines and prospects in this zone pursue copper sulfides and secondary copper minerals. The iron formation and chert are exhalative in origin; a volcanogenic origin for the copper is equivocal (Ludington and others, 1983, p. 8). North of the mouth of Hondo Canyon, the La Virgen Prospect exposes a small body of massive sphalerite with subordinate pyrite, chalcopyrite, and galena (J. Loghry, oral comm. 1983). Biotite schist and chlorite–sericite schist host the sulfides. Boulders of pegmatitic amphibolite are present in dump material. The Precambrian host rocks are invaded and largely destroyed by the Tertiary Rio Hondo quartz monzonite, but the morphology of the sulfide body and the mineralogy of the immediate host rocks imply a volcanogenic origin for the occurrence.

These volcanogenic-type mineral shows attest to good geologic potential for massive sulfide deposits in the Precambrian terrane of the Taos Range. The Taos Range supracrustal rocks are in many respects similar to other Precambrian terranes in southern Colorado and northern New Mexico which host volcanogenic deposits and prospects (Fulp and Woodward, 1981; Riesmeyer and Robertson, 1979; Sheridan and Raymond, 1977). Known occurrences in these areas contain both base and precious metals. However, the lack of kilometer-scale continuity of exposure of potential host rocks and the area's present and projected land status decrease the exploration appeal of the Taos area Precambrian terrane.

REFERENCES

Bowring, S. A., Reed, J. C., Jr., and Condie, K. C., 1984, U–Pb geochronology of Proterozoic volcanic and plutonic rocks, Sangre de Cristo Mountains, New Mexico (abstract): Geological Society of America, Abstracts with Programs, v. 16, no. 4, p. 216.

Callender, J., 1983, Transposition structures in Precambrian rocks of New Mexico: New Mexico Geological Society, Guidebook 34, pp. 143–146.

Condie, K. C., and McCrink, T. P., 1982, Geochemistry of Proterozoic volcanic and granitic rocks from the Gold Hill–Wheeler Peak area, northern New Mexico: Precambrian Research, v. 19, pp. 141–166.

Fiske, R. S., 1963, Subaqueous pyroclastic flows in the Ohanapecos Formation, Washington: Geological Society of America, Bulletin, v. 74, pp. 391–406.

———, and Matsuda, T., 1964, Submarine equivalents of ash flows in the Tokiwa Formation, Japan: American Journal of Science, v. 262, pp. 76–106.

Fulp, M. S., and Woodward, L. A., 1981, Precambrian metallic mineralization in New Mexico: New Mexico Geology, v. 3, no. 3, pp. 33–42.

Grambling, J. A., and Codding, D. B., 1982, Stratigraphic and structural relationships of multiply deformed Precambrian metamorphic rocks in the Rio Mora area, New Mexico: Geological Society of America, Bulletin, v. 93, pp. 127–137.

Holcombe, R. J., and Callender, J. F., 1982, Structural analysis and stratigraphic problems of Precambrian rocks of the Picuris Range, New Mexico: Geological Society of America, Bulletin, v. 93, pp. 138–149.

Hutchinson, R. W., Ridler, R. H., and Suffel, G. G., 1971, Metallogenic relationships in the Abitibi Belt, Canada: Canadian Institute of Mining and Metallurgy, Transactions, v. 74, pp. 106–115.

Long, P. E., 1976, Precambrian granitic rocks of the Dixon–Peñasco area, northern New Mexico: Ph.D. thesis, Stanford University, 533 pp.

Ludington, S., Briggs, J. P., and Robertson, J. M., 1983, Mineral resource potential of the Columbine–Hondo Wilderness Study Area, Taos County, New Mexico: U.S. Geological Survey, Miscellaneous Field Studies Map MF-1570-A, scale 1:50,000.

McCrink, T. P., 1982, Precambrian geology of the Taos Range, Taos County, New Mexico: M.S. thesis, New Mexico Institute of Mining and Technology, 123 pp.

Montgomery, A., 1963, Precambrian rocks; *in* Miller, J. P., Montgomery, A., and Sutherland, P. K. (eds.), Geology of part of the southern Sangre de Cristo Mountains, New Mexico: New Mexico Bureau of Mines and Mineral Resources, Memoir 11, pp. 7–21.

Norman, M. D., and Nielsen, K. C., 1982, Polyphase deformation of a Proterozoic sequence in the Taos Range, northern New Mexico (abstract): Geological Society of America, Abstracts with Programs, v. 14, no. 3, p. 134.

Reed, J. C., Jr., Lipman, P. W., and Robertson, J. M., 1983, Geologic map of the Latir Peak and Wheeler Peak Wildernesses and Columbine–Hondo Wilderness Study Area, Taos County, New Mexico: U.S. Geological Survey, Miscellaneous Field Studies Map MF-1570-B, scale 1:50,000.

Riesmeyer, W. D., and Robertson, J. M., 1979, Precambrian geology and ore deposits of the Pecos mine, San Miguel County, New Mexico: New Mexico Geological Society, Guidebook 30, pp. 175–179.

Robertson, J. M., and Moench, R. H., 1979, The Pecos Greenstone Belt: a Proterozoic volcano-sedimentary sequence in the southern Sangre de Cristo Mountains, New Mexico: New Mexico Geological Society, Guidebook 30, pp. 165–173.

Sheridan, D. M., and Raymond, W. H., 1977, Preliminary data on some Precambrian deposits of zinc–copper–lead sulfides and zinc spinel (gahnite) in Colorado: U.S. Geological Survey, Open-file Report 77-607, 27 pp.

Slack, J. F., 1982, Tourmaline in Appalachian–Caledonian massive sulfide deposits and its exploration significance: Applied Earth Science, Institution of Mining and Metallurgy, Transactions (B), v. 91, pp. B81–B89.

The San Luis Valley Southern No. 106 heads south at San Acacio in 1955, probably to pick up a load of scoria at Mesita. This locomotive was retired soon afterward, and now reposes in the Colorado Railroad Museum (photo by Bob Richardson).

AN OVERVIEW OF THE PRECAMBRIAN GEOLOGY OF THE TUSAS RANGE, NORTH-CENTRAL NEW MEXICO

REINHARD A. WOBUS
Department of Geology, Williams College, Williamstown, Massachusetts 01267

INTRODUCTION

The Tusas Range is a northwest-trending portion of the Brazos uplift of north-central New Mexico (Fig. 1). It is underlain predominantly by Precambrian crystalline and supracrustal rocks that are flanked and locally mantled by Tertiary and Quaternary volcanic and clastic sedimentary strata. Its subdued and partly forested ridges include some of the highest elevations between the Rio Grande depression to the east and the Chama Basin to the west, including Jawbone Mountain (3,256 m), Burned Mountain (3,106 m), Tusas Mountain (3,092 m), and Kiowa Mountain (2,968 m). The range loses its identity to the southeast at La Jarita Mesa, and to the northwest the Precambrian rocks become more thoroughly buried by volcanic rocks of the San Juan volcanic field in the Brazos Peak 15-min quadrangle.

The Precambrian (Proterozoic) rocks that form the core of the Tusas Range are of three general categories: bimodal metavolcanic and associated volcaniclastic metasedimentary rocks, a thick section of vitreous quartzite and interlayered pebble conglomerate, and several small stocks ranging in composition from quartz diorite to granite (Fig. 2). Except for the youngest of these stocks (the Tusas Mountain Granite), all of the Precambrian rocks have undergone regional metamorphism ranging from upper greenschist to middle amphibolite facies and are weakly to moderately foliated. Folding about northwest-plunging axes was probably concurrent with regional metamorphism, but much of the high-angle faulting in the area is Tertiary and younger.

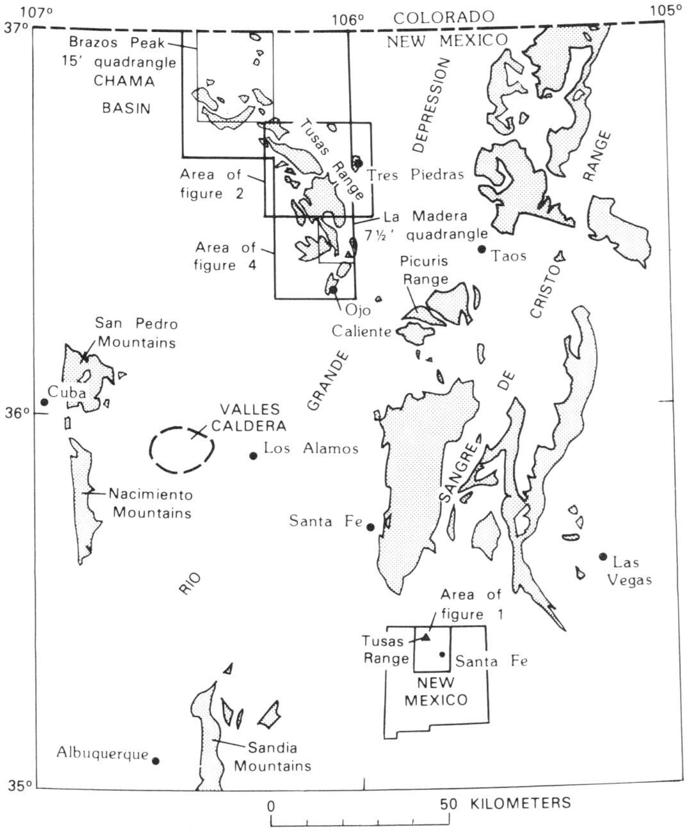

FIGURE 1. Distribution of Precambrian rocks (stippled pattern) in north-central New Mexico, showing location of Tusas Range and boundaries of Brazos Peak 15-min quadrangle and La Madera 7½-min quadrangle.

PREVIOUS WORK

Many geologists have studied the Precambrian rocks of the Tusas Range and have reached a variety of conclusions concerning the stratigraphy of the layered metamorphic rocks and the history of igneous intrusion. In the first reconnaissance study of the area, Just (1937) interpreted the mafic metavolcanic rocks to be the oldest unit, overlain by a single quartzite and conglomerate sequence and interlayered with the felsic metavolcanics. Jahns (1946) restated Just's conclusions in his report on the Petaca pegmatite district. Barker (1958) mapped a large part of the Tusas Range in his study of the Las Tablas 15-min quadrangle (now subdivided into the Burned Mountain, Mule Canyon, Canon Plaza, and Las Tablas 7½-min quadrangles). He concluded that there were two quartzite sequences with the metavolcanic rocks in between. Gresens and Stensrud (1974) returned to Just's earlier stratigraphic interpretation of a single quartzite unit overlying the metavolcanics. Burns and Wobus (1983) and Manley and Wobus (1982a) agreed with the single-quartzite stratigraphy of Gresens and Stensrud, Jahns, and Just in a re-study of the Tusas Range as part of the U.S. Geological Survey's mapping of the Aztec $1 \times 2°$ quadrangle (Manley and others, 1978; Wobus and Manley, 1982; Manley and Wobus, 1982a, 1982b).

The plutonic rocks and the Precambrian geochronology of the Tusas Range have been discussed by Barker and Friedman (1974), Gresens (1975), Long (1972), Maxson (1976), Wobus and Dobson (1981), and Wobus and Hedge (1982). The economic geology of the Precambrian rocks was included in a report on Rio Arriba County by Bingler (1968).

ROCK UNITS

Mafic Metavolcanic Rocks and Associated Metasedimentary Rocks

Originally named the Hopewell Series by Just (1937), these rocks were renamed the Moppin Metavolcanic Series by Barker (1958). They are best exposed in the creek valley southwest of Hopewell Lake and along Hopewell Ridge, where they form slab-like outcrops with a well-defined cleavage that is usually parallel to schistosity. The most common rocks in the unit are greenschists (chlorite schist and phyllite) and silver-gray sericite schists, interlayered with sericitic metagraywacke and porphyritic metabasalt with highly altered plagioclase phenocrysts. Rare interbeds of sericitic metaconglomerate with flattened, rod-like cobbles and of purplish, magnetite-rich schist also occur. The metabasalt locally contains well-preserved pillow structures in the Brazos Peak 15-min quadrangle northwest of the Tusas Range.

Also included within this rock unit are exposures of altered amphibolite (metabasalt) south of Kiowa Mountain, mapped by Barker (1958) as a part of the Kiawa Mountain Formation. Chemical analyses of all the metabasalts, provided by Barker and Friedman (1974), show that they are of tholeiitic affinity.

Felsic Metavolcanic Rocks and Associated Metasedimentary Rocks

Just (1937) named the felsic metavolcanic rocks the Vallecitos Rhyolites, interpreting them as rhyolite and trachyte flows that are inter-

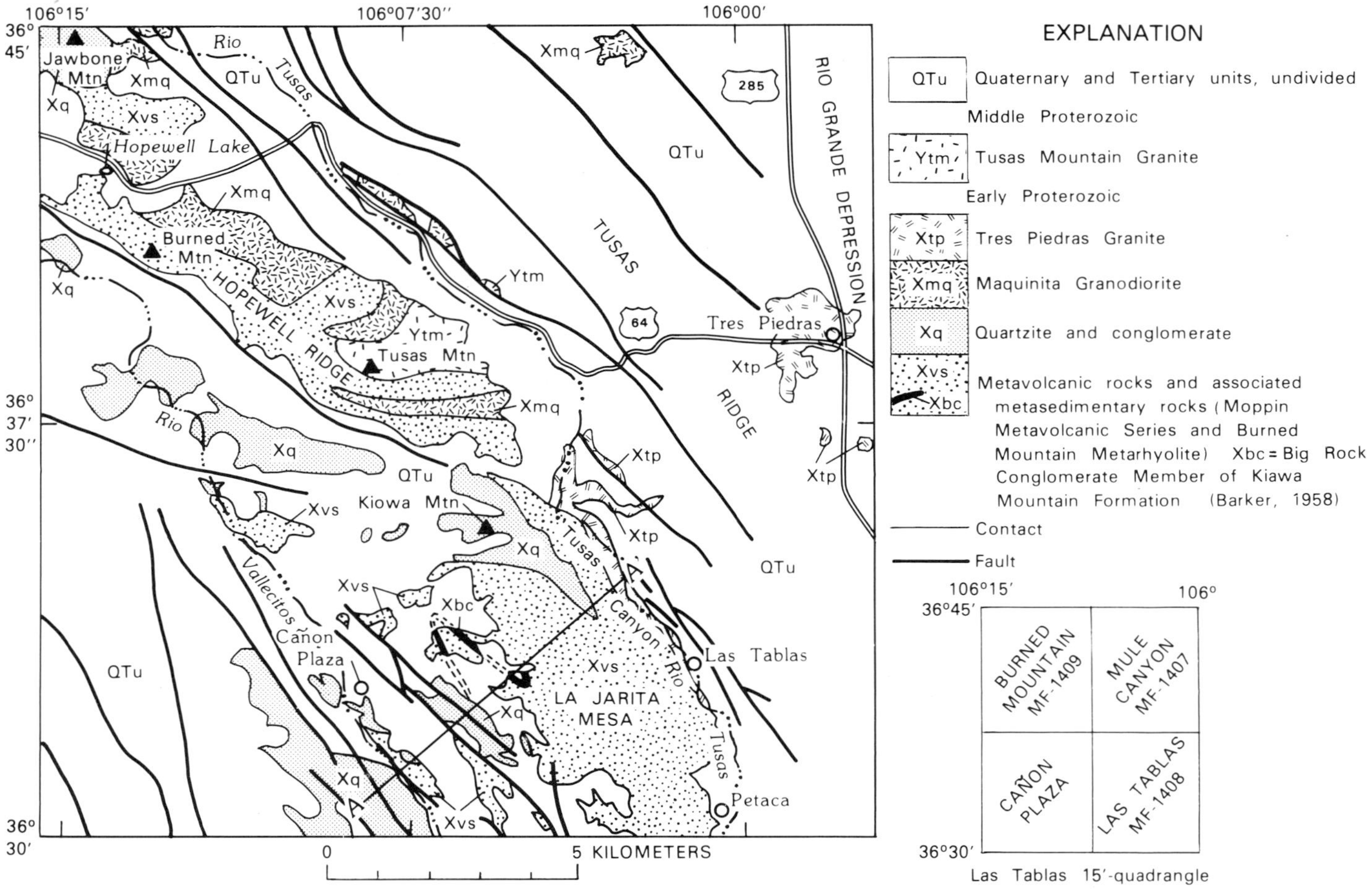

FIGURE 2. Generalized geologic map of the Tusas Range emphasizing Precambrian units. After Barker (1958), Manley, and Wobus (1982a, 1982b), and Wobus and Manley (1982).

layered with mafic rocks of the Hopewell Series and with some of the quartzites. Barker (1958) renamed these rocks the Burned Mountain Metarhyolite and noted that some of the metarhyolite formed sills and dikes that cut the mafic metavolcanic units (Moppin), whereas the rest represented flows and ash-flows interlayered with the mafic rocks near the top of the Moppin.

The typical metarhyolite, best exposed at Burned Mountain, on the south side of Jawbone Mountain, and on the western edge of La Jarita Mesa, is pink to gray to dark red. Texturally it is very fine-grained porphyritic with conspicuous rounded phenocrysts of red or gray quartz 1–2 mm in diameter. Many outcrops also show relic eutaxitic fabric, with devitrified flattened pumice fragments indicative of ash-flow origin. The rock has been dated as 1,775–1,725 m.y. by L. T. Silver of the California Institute of Technology, using the U–Pb method on zircon (Barker and Friedman, 1974).

Gresens and Stensrud (1974) recognized that much of the rock on La Jarita Mesa that Barker had mapped as quartzite was actually altered metarhyolite. Some of this rock retains remnants of quartz and feldspar phenocrysts, whereas the rest, though chemically identical to the porphyritic variety, is fine-grained, massive to weakly foliated, and probably represents metamorphosed silicic tuff. The most severely altered metarhyolite occurs on the southeastern side of La Jarita Mesa where the rock has been hydrothermally converted, with varying degrees of metasomatism, to quartz–muscovite schist (Petaca Schist of Just, 1937, and Barker, 1958). This area surrounds the Petaca pegmatite district (Jahns, 1946), where radiometric ages for altered metarhyolite and pegmatitic muscovite define a $1{,}425 \pm 25$ m.y. Rb–Sr isochron (Long, 1972).

Discontinuous beds of micaceous conglomerate interlayered with the felsic metavolcanics on La Jarita Mesa probably were deposited in a braided alluvial environment during the time of explosive volcanic activity (K. A. Eriksson, Virginia Polytechnic Institute, written comm. 1982). These conglomerate beds were named the Big Rock Conglomerate Member of the Kiawa Mountain Formation by Barker (1958); they contain elongated cobbles and pebbles of milky quartz, red and black quartz, gray–green phyllite, and white quartz–muscovite schist that is probably altered metarhyolite. Although the conglomerate beds are only a few meters to tens of meters thick, they form important marker horizons in the otherwise monotonous terrane of altered metarhyolite on La Jarita Mesa (Manley and Wobus, 1982a).

Quartzite and Conglomerate

Quartzite and interlayered pebble conglomerate in the Tusas Range were originally named Ortega Quartzite by Just (1937), who believed the unit to be younger than the metavolcanic rocks. The name was derived from the Ortega Mountains, southwest of the Tusas Range, which are underlain almost entirely by quartzite and minor interbeds of conglomerate. Dark laminations in the white bluish-gray quartzites contain hematite, kyanite, and viridine (manganian andalusite) and serve to accentuate bedding planes and crossbedding.

Barker (1958) accepted the name Ortega for quartzites he thought were older than the metavolcanic rocks, and he defined the Kiawa Mountain Formation to include supposedly younger quartzites that form the conspicuous ridge that trends northwest from Kiowa Mountain. Barker described the Kiawa Mountain Formation quartzite as ". . . dense, vitreous, light bluish-gray quartzite that contains irregularly dis-

TABLE 1. Comparison of stratigraphic interpretations of layered Proterozoic metamorphic rocks, Tusas Range, north-central New Mexico.

	Just (1937)	Barker (1958)	Wobus and Hedge (1982)
Massive porphyritic leucogranite at Tusas Mountain	TUSAS GRANITE	TRES PIEDRAS GRANITE	TUSAS MOUNTAIN GRANITE
Foliated biotite quartz monzonite to granite at Tres Piedras and Tusas Canyon	TUSAS GRANITE	TRES PIEDRAS GRANITE	TRES PIEDRAS GRANITE
Foliated grano-diorite to quartz diorite	TUSAS GRANITE	MAQUINITA GRANODIORITE	MAQUINITA GRANODIORITE

tributed pebbly layers . . . ," noting that it was ". . . similar to the Ortega quartzite exposed west of the Rio Vallecitos . . ." and that ". . . indeed, these two quartzites are essentially indistinguishable in outcrop" (Barker, 1958, p. 31).

Gresens and Stensrud (1974) and Burns and Wobus (1983) advocated a return to the single-quartzite stratigraphy of Just, arguing on the basis of locally well-preserved crossbedding that all of the quartzites are younger than the metavolcanic rocks and that the units are, in fact, structurally connected by large-scale folding. A comparison of the stratigraphic interpretations for the supracrustal rocks of the Tusas Range is found in Table 1.

Intrusive Rocks

Three kinds of intrusive rocks form small stocks that cut the layered metamorphic rocks of the Tusas Range. Foliated granodiorite to quartz diorite occurs primarily along Hopewell Ridge, foliated quartz monzonite to granite outcrops in Tusas Canyon and in and near the village of Tres Piedras, and a stock of porphyritic leucocratic granite that is nearly structureless underlies Tusas Mountain. As shown in Table 2, Just (1937) called all of these rocks Tusas Granite, but Barker (1958) abandoned that term; he re-named the latter two occurrences Tres Piedras Granite and named the granodiorite to quartz diorite the Maquinita Granodiorite. Wobus and Hedge (1982) provided field and isotope evidence that the granite porphyry at Tusas Mountain is considerably younger than the granite at Tres Piedras and in Tusas Canyon, proposing the name Tusas Mountain Granite for that occurrence.

For details on the petrography and chemistry of the intrusive rock units, see Barker (1958) and Wobus and Hedge (1982).

TABLE 2. Proterozoic rocks of the Tusas Range, north-central New Mexico.

Just (1937)	Barker (1958)	Wobus (this report)
Vallecitos Rhyolites	Burned Mountain Metarhyolite	
Petaca Schist phase of Ortega Quartzite		
Ortega Quartzite	Kiawa Mountain Formation*: upper quartzite Jawbone Conglomerate Member* amphibolite lower quartzite Big Rock Conglomerate Member*	Vitreous quartzite and interlayered pebble conglomerate (formerly assigned to Kiawa Mountain Formation and Ortega Quartzite but probably all Ortega Quartzite)
Hopewell Series	Moppin Metavolcanic Series	Bimodal metavolcanics and associated volcaniclastic metasedimentary rocks (Burned Mountain Metarhyolite and Moppin Metavolcanic Series; may include amphibolite of Kiawa Mountain Formation)
	Ortega Quartzite Petaca Schist (muscovitized parts of Ortega Quartzite and Kiawa Mountain Formation)	

* Denotes names formally adopted in Barker, 1970

Maquinita Granodiorite

The Maquinita is dark-gray or greenish-gray (tan where weathered), medium-grained, generally well-foliated biotite granodiorite to quartz diorite. It is locally highly sheared and altered, particularly near Hopewell Lake. It intrudes the Moppin Metavolcanic Series at many locations along Hopewell Ridge and contains rounded to elongate inclusions of Moppin near these contacts. Its age has been determined radiometrically to be 1,725–1,675 m.y. by L. T. Silver (Barker and Friedman, 1974).

Tres Piedras Granite

The Tres Piedras Granite is well exposed in inselberg-like outcrops in and near the village of Tres Piedras and in the steep walls of Tusas Canyon. It is pink to buff, fine- to medium-grained quartz monzonite, generally well foliated and often weathering to coarse grus. Its contacts with the layered metamorphic rocks in Tusas Canyon are concordant, appearing as a series of thin sills within the altered metarhyolite it intrudes. No other contacts with Precambrian rock units are exposed.

The Tres Piedras Granite was probably emplaced during the same intrusive event as the Maquinita Granodiorite. Both foliated rock types bear the structural imprint of the single period of regional metamorphism that affected their wall rocks, and such syntectonic plutons in the southern Rocky Mountains characteristically give radiometric ages of about 1,700 m.y. (Hedge and others, 1977). The only published dates for the Tres Piedras Granite are those by Maxon (1976), but they are problematic because his sampling included rocks at Tusas Mountain that are

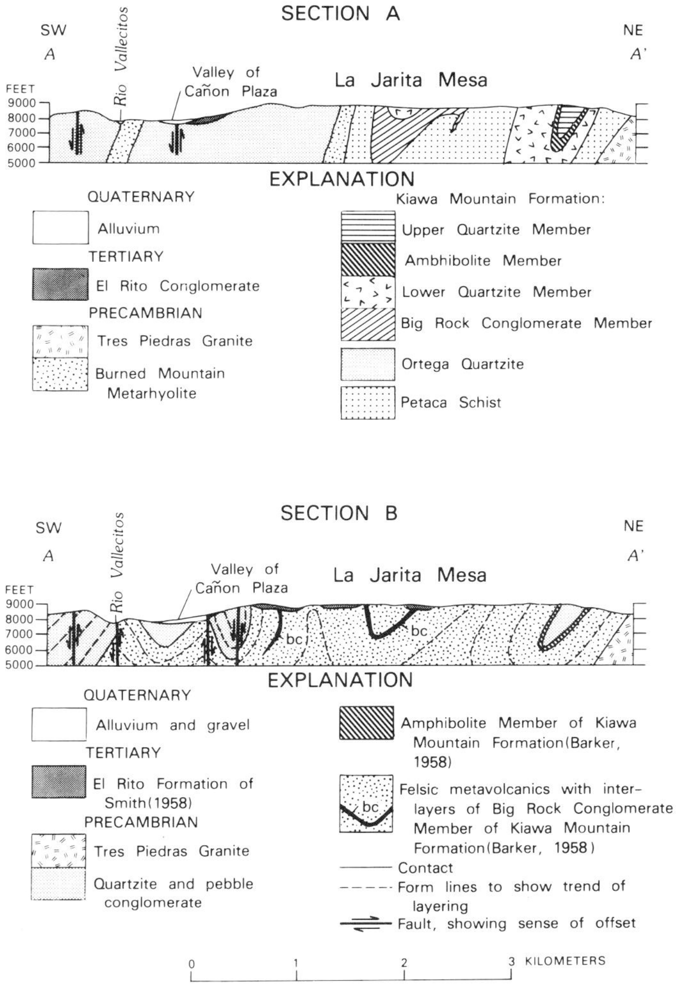

FIGURE 3. Comparison of geologic cross sections along line AA′ in Figure 2, showing differences in interpretation of Precambrian structure and stratigraphy in southwestern part of Tusas Range. Section A from Barker (1958) and Section B after Manley and Wobus (1982a).

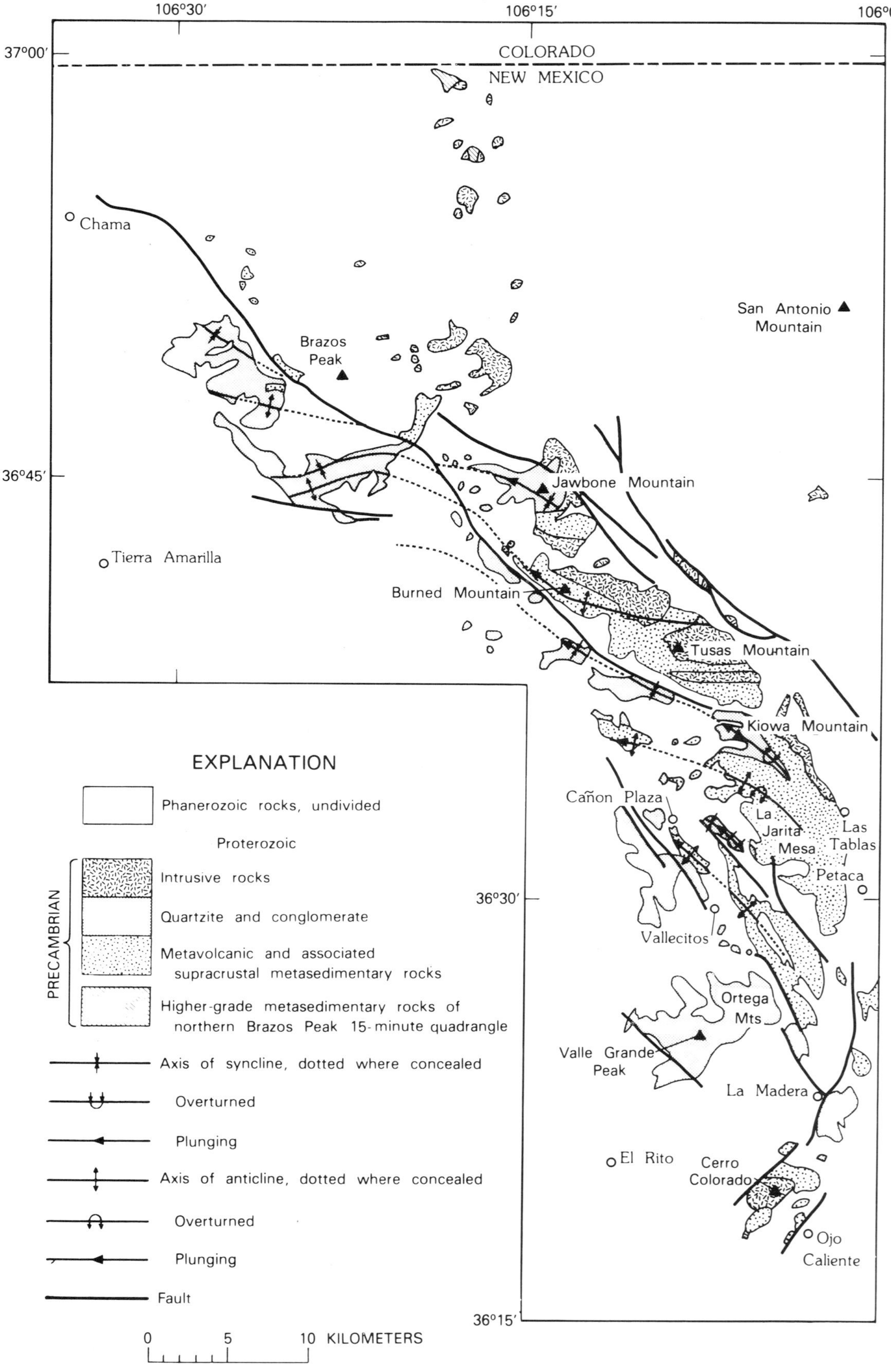

FIGURE 4. Small-scale geologic map of Precambrian rocks in eastern part of Aztec 1 × 2° quadrangle, New Mexico, showing trends of major fold axes. Revised from Manley and others (1978).

clearly of another intrusive generation (Wobus and Hedge, 1982). His U–Pb ages on zircon define a chord that intersects the concordia curve at 1,654 m.y. and 98 m.y., with a concordant data point at 1,654 m.y. that could serve as a minimum value for the age of the Tres Piedras Granite.

Tusas Mountain Granite

This unit was named by Wobus and Hedge (1982) for the single stock of porphyritic granite that underlies Tusas Mountain and vicinity. The rock is characteristically light gray to pink and fine- to medium-grained porphyritic; it is generally structureless except for a weak primary-flow foliation preserved locally near, and parallel to, the margins of the stock. Phenocrysts of rounded, dark-gray quartz and subhedral pink microcline are 1–5 mm in diameter, and fluorite and muscovite are common accessory minerals. The rock is chemically similar to the average alkali granite and is enriched by a suite of trace elements (fluorine, beryllium, lithium, and tin) that are indicative of late vapor-phase activity during its cooling history. Rb–Sr whole-rock analyses and U–Pb dating of zircon from the Tusas Mountain Granite indicate an age between 1,500 and 1,430 m.y. (Wobus and Hedge, 1982).

STRUCTURAL INTERPRETATION

Structural complexity and metamorphic grade increase from north to south in the Tusas Range. Through most of the area, metamorphic foliation or schistosity appears to be parallel to original sedimentary or volcanic layering, and crossbedding is locally well preserved in the quartzite and conglomerate. Near the southern edge of the map, however (particularly west of the Rio Vallecitos), and further south in the Ortega Mountains, more intense deformation of former crossbedding in the quartzite has rendered these primary structures useless for stratigraphic and structural purposes, as recognized by Bingler (1965) in the La Madera 7½-min quadrangle.

On La Jarita Mesa northwest of Petaca, intense metasomatic alteration of former felsic metavolcanic rocks and associated metasedimentary units has nearly obliterated their original layering and converted them to a quartz–sericite schist (Petaca Schist of Just, 1937, and Barker, 1958) with a strong northwest-trending schistosity that may or may not parallel original layers. Only the preservation of quartz phenocrysts attests to the original nature of the rocks in that area.

The most obvious structures of the Tusas Range are a series of northwest-plunging anticlines and synclines, some overturned, that are defined by repetition of strata and by the preservation of crossbedding in the quartzites. Some of these folds were recognized by Barker (1958), but reinterpretation of his data and new mapping west of La Jarita Mesa require a re-evaluation of his structural and stratigraphic conclusions for the region as a whole (Manley and Wobus, 1982a). The differences are best shown in Figure 3, which is a comparison of NE–SW cross sections along line AA′ in Figure 2. This remapping provides structural evidence for a single quarzite unit overlying the metavolcanic rocks in the Tusas Range. Using additional mapping by Muehlberger (1968) in the Brazos Peak 15-min quadrangle to the northwest and by Bingler (1968) in the La Madera 7½-min quadrangle to the south, it is possible to portray the folded nature of quartzites and metavolcanic rocks through most of the exposed Precambrian terrane west of the Rio Grande rift in northern New Mexico (Fig. 4). Folds of about the same wavelength probably continue through the Ortega Mountains, but transposition of bedding due to shearing and the lack of continuous marker horizons in the quartzites there have precluded the delineation of fold axes.

Faults in the Tusas Range are generally high-angle and northwest-trending. They cut Precambrian and Tertiary rocks alike, offsetting units as young as the Pliocene Servilleta Basalt (Manley and Wobus, 1982b); many were likely developed during the evolution of the Rio Grande rift.

SUMMARY AND CONCLUSIONS

The Precambrian supracrustal rocks of the Tusas Range consist of bimodal metavolcanic rocks (Moppin Metavolcanic Series and Burned Mountain Metarhyolite) overlain by a thick sequence of quartzite and conglomerate. These units are repeated by folding several times through the region, with the older metavolcanic rocks being exposed only in the cores of large eroded anticlines. A radiometric age of 1,775–1,725 m.y. for the Burned Mountain Metarhyolite indicates that the supracrustal rocks are of late Early Proterozoic age; rocks of significantly greater age are not known anywhere in the southern Rocky Mountains.

The layered metamorphic rocks are roughly contemporaneous with other supracrustal assemblages (quartzites overlying metavolcanic rocks) in northern New Mexico and southern Colorado. These include the Picuris Range (Holcombe and Callender, 1982) and the Rio Mora portion of the Sangre de Cristo Mountains (Grambling and Codding, 1982) in New Mexico, and the Needle Mountains in southwestern Colorado (Burns and Wobus, 1983). Condie (1982) recognized this same general trend in rocks of similar age throughout the Southwestern United States, suggesting that this suite of rocks may have been generated by the opening and closing of successive back-arc basins underlain by continental crust.

The intrusive rocks of the Tusas Range are likewise similar in age and structural style to those recognized elsewhere in the southern Rocky Mountains. Both the Maquinita Granodiorite and the Tres Piedras Granite are typical of syntectonic plutons emplaced prior to or during the single major regional metamorphic event about 1,700 m.y. ago.

The Tusas Mountain Granite, on the other hand, was intruded during a younger, anorogenic plutonic episode about 1,500–1,400 m.y. ago. This event was even more widespread than the 1,700 m.y. plutonism and is recognized from Labrador to California (Silver and others, 1977). The Tusas Mountain Granite is more alkalic, however, than granites typical of this event (Anderson, 1983). Significantly, the Tusas Mountain stock represents the only dated Precambrian pluton west of the Rio Grande in northern New Mexico that was emplaced during the major period of pegmatite formation and pervasive hydrothermal alteration about 1,400 m.y. ago.

ACKNOWLEDGMENTS

Fred Barker and Dick Jahns introduced me to the Tusas Range during a Penrose Conference field trip in 1971. Glenn Scott and Kim Manley of the U.S. Geological Survey were stimulating and helpful colleagues during the mapping of the Aztec 1 × 2° quadrangle, and I am grateful to numerous students from Williams College who helped as field assistants during the four summers devoted to that mapping. I also benefited from discussions in the field with Jack Reed, Carl Hedge, and Mike Williams.

REFERENCES

Anderson, J. L., 1983, Proterozoic anorogenic granite plutonism of North America: Geological Society of America, Memoir 161, pp. 133–154.

Barker, F., 1958, Precambrian and Tertiary geology of Las Tablas quadrangle, New Mexico: New Mexico Bureau of Mines and Mineral Resources, Bulletin 45, 104 pp.

———, 1970, Ortega quartzite and the Big Rock and Jawbone Conglomerate Members of the Kiawa Mountain Formation, Tusas Mountains, New Mexico: U.S. Geological Survey, Bulletin 1294-A, pp. A21–A22.

———, and Friedman, I., 1974, Precambrian metavolcanic rocks of the Tusas Mountains, New Mexico: New Mexico Geological Society, Guidebook 25, pp. 115–117.

Bingler, E. C., 1965, Precambrian geology of La Madera quadrangle, Rio Arriba County, New Mexico: New Mexico Bureau of Mines and Mineral Resources, Bulletin 80, 132 pp.

———, 1968, Geology and mineral resources of Rio Arriba County, New Mexico: New Mexico Bureau of Mines and Mineral Resources, Bulletin 91, 158 pp.

Burns, L. K., and Wobus, R. A., 1983, Correlations and revisions of Precambrian stratigraphy, Needle Mountains, southwest Colorado, and Tusas Mountains, north-central New Mexico (abstract): Geological Society of America, Abstracts with Programs, v. 15, no. 5, p. 424.

Condie, K. C., 1982, Plate-tectonics model for continental accretion in the Southwestern United States: Geology, v. 10, pp. 37–42.

Grambling, J. A., and Codding, D. B., 1982, Stratigraphy and structural relationships of multiply deformed Precambrian metamorphic rocks in the Rio Mora area, New Mexico: Geological Society of America, Bulletin, v. 93, pp. 127–137.

Gresens, R. L., 1975, Geochronology of Precambrian metamorphic rocks, north-

central New Mexico: Geological Society of America, Bulletin, v. 86, pp. 1444–1448.
———, and Stensrud, H. L., 1974, Recognition of more metarhyolite occurrences in northern New Mexico and a possible Precambrian stratigraphy: Mountain Geologist, v. 11, pp. 109–124.
Hedge, C. E., Houston, R. L., Tweto, O. L., Reid, R. R., Harrison, J. E., and Peterman, Z., 1977, The Precambrian of the Rocky Mountain region (abstract): Geological Society of America, Abstracts with Programs, v. 9, no. 7, p. 1010.
Holcombe, R. J., and Callender, J. F., 1982, Structural analysis and stratigraphic problems of Precambrian rocks of the Picuris Range, New Mexico: Geological Society of America, Bulletin, v. 93, pp. 138–149.
Jahns, R. H., 1946, Mica deposits of the Petaca district, Rio Arriba County, New Mexico: New Mexico Bureau of Mines and Mineral Resources, Bulletin 25, 294 pp.
Just, E., 1937, Geology and economic features of the pegmatites of Taos and Rio Arriba Counties, New Mexico: New Mexico School of Mines, Bulletin 13, 73 pp.
Long, L. E., 1972, Rb–Sr chronology of Precambrian schist and pegmatite, La Madera quadrangle, northern New Mexico: Geological Society of America, Bulletin, v. 83, pp. 3425–3432.
Manley, K., and Wobus, R. A., 1982a, Reconnaissance geologic map of the Las Tablas quadrangle, Rio Arriba County, New Mexico: U.S. Geological Survey, Map MF-1408.
———, and ———, 1982b, Reconnaissance geologic map of the Mule Canyon quadrangle, Rio Arriba County, New Mexico: U.S. Geological Survey, Map MF-1407.
———, Scott, G. R., and Wobus, R. A., 1978, Preliminary geologic map of the Aztec 1° × 2° quadrangle, northwestern New Mexico: U.S. Geologic Survey, Open-file Report 78-466, 7 pp.
Maxon, J. R., 1976, Age and implications of the Tres Piedras Granite, north-central New Mexico (abstract): Geological Society of America, Abstracts with Programs, v. 8, no. 5, p. 608.
Muehlberger, W. R., 1968, Geology of Brazos Peak quadrangle, New Mexico: New Mexico Bureau of Mines and Mineral Resources, Geologic Map 22.
Silver, L. T., Bickford, M. E., Van Schmus, W. R., Anderson, J. L., Anderson, T. H., and Medaris, L. G., Jr., 1977, The 1.4–1.5 b.y. transcontinental anorogenic plutonic perforation of North America (abstract): Geological Society of America, Abstracts with Programs, v. 9, no. 7, pp. 1176–1177.
Smith, H. T. U., 1938, Tertiary geology of the Abiquiu quadrangle, New Mexico: Journal of Geology, v. 46, no. 7, pp. 933–965.
Wobus, R. A., and Dobson, P. F., 1981, Proterozoic intrusive history of the Tusas Mountains, north-central New Mexico (abstract): Geological Society of America, Abstracts with Programs, v. 13, no. 4, p. 230.
———, and Hedge, C. E., 1982, Redefinition of the Precambrian Tusas Mountain and Tres Piedras Granites, north-central New Mexico: Mountain Geologist, v. 19, pp. 105–114.
———, and Manley, K., 1982, Reconnaissance geologic map of the Burned Mountain quadrangle, Rio Arriba County, New Mexico: U.S. Geological Survey, Map MF-1409.

STRATIGRAPHIC SUMMARY AND STRUCTURAL PROBLEMS OF PRECAMBRIAN ROCKS, PICURIS RANGE, NEW MEXICO

PAUL W. BAUER
Department of Geology, University of New Mexico, Albuquerque, New Mexico 87131

INTRODUCTION

The Picuris Range, located in north-central New Mexico 20 km southwest of Taos (Fig. 1), is a Precambrian-cored projection pointing westward from the southern Sangre de Cristo Mountains. All of the Precambrian rocks in the range are of Proterozoic age. They are bounded on the east by Cenozoic and Paleozoic rocks, and on the other three sides by sediments of the Rio Grande rift and volcanic rocks of Tertiary age.

Mineral assemblages in the Proterozoic rocks indicate medium-grade (amphibolite facies) peak metamorphic conditions near 530°C, 3.7 kb (Holdaway, 1978), at or near the Al_2SiO_5 triple point.

Several major stratigraphic and structural problems remain to be solved in the highly deformed Proterozoic rocks of the Picuris Range. This report discusses several of these problems.

PREVIOUS WORK

Two major groups of Proterozoic rocks are present in the Picuris Range and adjoining uplifts: a metamorphosed clastic sequence without significant volcanic rocks (Ortega Group), and a metamorphosed volcanic sequence (Vadito Group). There are two hypotheses concerning the relative age of these sequences in the Picuris Range. One holds that the Ortega Group is younger than the Vadito Group (Just, 1937; Gresens and Stensrud, 1974; Holcombe and Callender, 1982); the other maintains that the Vadito Group is younger than the Ortega Group (Montgomery, 1953; Nielsen, 1972; Long, 1976; Scott, 1980).

Although there has been considerable controversy over the stratigraphic sequence in the Precambrian rocks of northern New Mexico, the most recent work (Holcombe and Callender, 1982; Grambling and Codding, 1982; McCarty, 1983) has suggested that the Ortega rocks overlie the Vadito. Radiometric dates from rocks of the Truchas Range are consistent with this interpretation (Grambling, pers. comm. 1984).

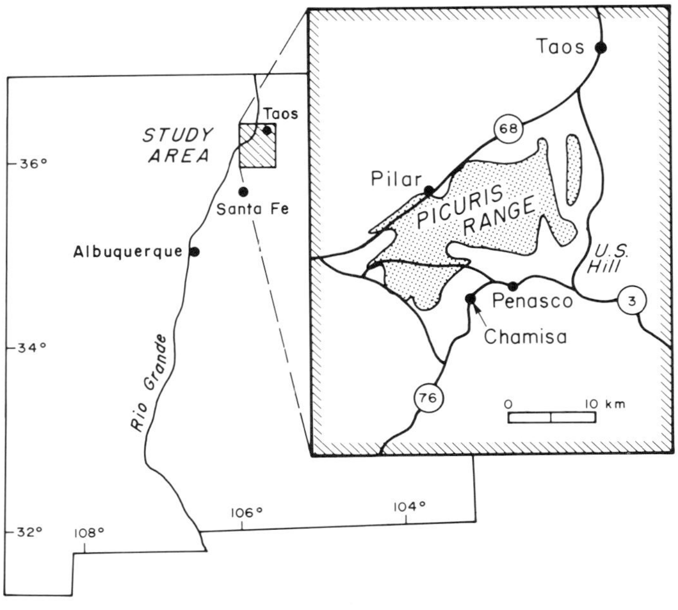

FIGURE 1. Location map showing the Picuris Range, New Mexico.

STRATIGRAPHY

The Proterozoic sequence in the Picuris Range consists of two thick, lithologically dissimilar sequences of rocks separated by a major fault (Fig. 2). The base of neither sequence is exposed. In the Picuris Range, the two groups may be separated by a fault-bounded slice of rock (the Piedra Lumbre Formation) that is not totally part of either group.

In the southern Picuris Range, the Vadito Group consists of a lower unit of interbedded amphibolites, metamorphosed volcaniclastic rocks, quartz–muscovite–biotite schists, and phyllites which are overlain by quartzites and conglomerates of the Marquenas Quartzite. Farther north lie fine-grained, banded, garnet–biotite–muscovite phyllites, metasiltstones, schists, and calc-silicates of the Piedra Lumbre Formation.

The Ortega Group consists of the massive, clean, crystalline Ortega Quartzite overlain by interbedded pelitic schists and quartzites of the Rinconada Formation, and by fine-grained, carbonaceous black slates and phyllites of the Pilar Phyllite. In the northern Picuris Range, above the Pilar, is a unit that resembles the Piedra Lumbre and is here informally called the schist of Hondo Canyon. The youngest Proterozoic rocks are coarse-grained quartz monzonite and granodiorite of the Embudo Granite, and coarse-grained granitic pegmatites.

Miller and others (1963) and Grambling and Codding (1982) have suggested correlations between Truchas and Rio Mora rock units and lithologically similar units in the Picuris. In the Truchas and Rio Mora areas, stratigraphy and geochronology suggest that Ortega is younger than Vadito. By inference this should also be true in the Picuris area. U–Pb zircon dates of about 1,700 m.y. for the felsic schist below the Ortega Quartzite (L. T. Silver, unpubl. data 1984) yield a maximum age for deposition of the Ortega Quartzite.

STRATIGRAPHIC PROBLEMS

Although Nielsen (1972) correlated the Piedra Lumbre Formation of the southern Picuris Range with the schist of Hondo Canyon in the northern Picuris Range, they differ in several aspects. Whereas the Piedra Lumbre Formation is mostly a fairly uniform fine-grained muscovite–biotite phyllite, the schist of Hondo Canyon additionally contains interlayered black-slate and black-phyllite horizons which range in thickness from a few to tens of meters, plus a significant thickness of light-colored, thinly bedded, micaceous quartzite near its base. The contact between Piral Phyllite and Piedra Lumbre in the southern Picuris Range is fairly abrupt, and is characterized by a strongly sheared phyllonitic zone. In contrast, the same contact in the northern Picuris Range is gradational, with various calc-silicate layers in the transitional zone rather than phyllites. These differences together with tentative structural differences (described below) indicate that either the Piedra Lumbre and Hondo Schist units are not correlative, or that the major fault described in the southern Picuris above the Pilar Formation does not exist above the Pilar in the northern Picuris. It seems appropriate to avoid applying the term Piedra Lumbre to these northern rocks until their relationship to the Piedra Lumbre at its type locality is better understood.

A second stratigraphic problem concerns the micaceous, quartz-rich rock that Montgomery (1953) termed the Rio Pueblo Schist. It is structurally below Ortega Quartzite in the Pilar area, and a similar rock occupies the same structural position near U.S. Hill. Gresens and Stensrud (1974) have argued that these rocks are metarhyolite sequences upon which the Ortega Quartzite was deposited. If so, then that contact may separate a major metasedimentary terrane (the Ortega Group) from

GROUP	FORMATION	MEMBER	LITHOLOGY	THICKNESS[1] (m)	YOUNGING[2]
	Pegmatites		Coarse-grained granitic pegmatites.		
Embudo Granite	Penasco Quartz Monzonite Rana Quartz Monzonite Puntiagudo Gran. Porphyry Cerro Alto Metadacite		Variable, but generally coarse-grained, microcline-rich, quartz monzonites to granodiorites. (See Long, 1976).		
?	Hondo Schist[3]		Interbedded, banded, garnet-staurolite schists and metasiltstones; calc-silicates; amphibolites and black slates locally.		YOUNGING (↑)
—?—?—?	Pilar Phyllite		Black slate and fine-grained phyllite with iron-oxide stains and local white schist beds; Distinctive basal blue-black garnet quartzite. (see Montgomery, 1953).	84 (N)	
Ortega Group	Rinconada[4] (see Montgomery, 1953)	R_6 / Rs_3	Banded gray to red quartz-muscovite-biotite-garnet phyllite and schist. Strongly crenulated.	240–565 (M) 350 (N)	
		R_5 / Rq_2	Massive, micaceous, white to gray quartzite grading up to massive, blue crystalline quartzite.		
		R_4 / Rs_2	Reddish weathering gray schist with small euhedral garnets and staurolites. Local staurolite phyllite.		
		R_3 / Rq_1	Tan to white, micaceous, cross-bedded quartzite grading up to massive, crystalline, blue quartzite.		
		R_2 / Rs_1	Muscovite-staurolite schist with large twinned staurolites.		
		R_1 / Rs_1	Biotite-muscovite-andalusite "salt and pepper" schist.		
	Ortega Quartzite		Massive, clean, coarsely crystalline, white to gray to yellow brown, cross-bedded quartzite; local interlayered aluminous schists. (See Montgomery, 1953).	760 (M) 835 (N)	
Felsic metavolcanic-metasedimentary complex (includes the Rio Pueblo Schist).			Metarhyolite, metasediments, red-mica schist, metaconglomerate. (see Gresens and Stensrud, 1974).	Base Unexposed	
?	Piedra Lumbre[5]		Fine-grained, carbonaceous, muscovite-biotite phyllite with garnet porphyroblasts; local calc-silicates. Internally transposed. (See Montgomery, 1953).	183 (N) 369 (L)	YOUNGING (↑)
—?—?—?	Marquenas Quartzite (see Scott, 1980)	Mc_1	Conglomerate with deformed quartzite pebbles in a micaceous quartz matrix. "Flaser" structures.	610 (M) 410 (L)	
Vadito Group		Mrq	Cross-laminated, rippled, micaceous quartzite with isolated pebble layers.		
		Mq	Sparsely cross-bedded, vitreous, massive gray quartzite with pebble layers.		
		Mc_2	Quartzite, felsite, and mafic pebble conglomerate with micaceous quartzite matrix.		
	Vadito Schist		Various quartz-muscovite schists and phyllites with interbedded mafic metavolcanics. (See McCarty, 1983).	760 (M) 700 (L)	
	Vadito Amphibolite		Various mafic metavolcanics with interlayered metaclastics and metavolcaniclastics. (See McCarty, 1983).	Base Unexposed	

1 (M) = Montgomery, 1953; (N) = Nielsen, 1972; (L) = Long, 1976

2 Relative ages of Ortega and Vadito Groups based partially on criteria in Rio Mora area by Grambling and Codding, 1982; and Grambling, 1983.

3 The Hondo Schist is recognized in the northern Picuris only, and may be correlative with the Piedra Lumbre Formation.

4 According to Grambling (pers. commun., 1984) the R_1-R_2 mineralogical change is metamorphic rather than original sedimentary, thus the Rs-Rq system is preferred.

5 In the southern Picuris, the Piedra Lumbre - Pilar contact is a major fault (Holcombe and Callender, 1982). If the Hondo Schist is equivalent or partially equivalent to the Piedra Lumbre, then the Marquenas - Piedra Lumbre contact must also be a major fault.

FIGURE 2. Lithologic column for Precambrian rocks of the Picuris Range, New Mexico.

an underlying metavolcanic terrane (the Vadito Group). Grambling and others (1983) recognized a viridine-bearing (Mn-rich andalusite) horizon which defines the contact between the Ortega Group and the underlying Vadito Group in the Rio Mora area, the Truchas Peaks region, and the western Picuris Range near the town of Pilar. Williams and Wobus (1984) have identified the same marker horizon in basal Ortega Quartzite throughout the Tusas Range. Viridine has been found near the base of the Ortega Quartzite during recent mapping near U.S. Hill in the eastern Picuris Range. The presence of this unusual, perhaps time-stratigraphic, horizon establishes a firm stratigraphic correlation between uplifts across north-central New Mexico.

STRUCTURE

The dominant east–west fabric of the Picuris Range is due to a set of large-scale, east- to east–northeast-trending, tight to isoclinal folds overturned to the north (Montgomery, 1953). The major folds are the Copper Hill anticline to the south and the Hondo syncline to the north (Fig. 3). These folds are cut by three generations of faults (Nielsen and Scott, 1979), from oldest to youngest (1) shear zones subparallel to axial surfaces of major folds, (2) oblique-slip faults which trend either northeast–southwest or northwest–southeast (e.g., the Pilar–Vadito fault of Montgomery, 1953), and (3) normal faults of late Tertiary age.

Structural-fabric Elements

Deformational fabrics have been defined on the basis of detailed mapping and structural and thin-section analyses of rocks in the northern Picuris Range, in an area that includes both limbs of the Hondo syncline. The terms discussed below are defined in Holcombe and Callender (1982).

Folding events are denoted F_1 and F_2, with associated axial plane cleavages S_1 and S_2. L_{10} and L_{20} are intersection lineations of surfaces designated by subscripts. In most rocks compositional layering is thought to represent true stratigraphic bedding (S_0) due to well-preserved sedimentary structures within layering. S_1, the earliest cleavage, appears as a schistosity in micaceous rocks of the Vadito Group in the southern Picuris Range (Holcombe and Callender, 1982). S_1 is generally parallel to bedding (S_0), thus S_0 and S_1 will be included together as S_0/S_1. S_2 is the second-generation penetrative cleavage and appears to be dominant in both the southern and northern Picuris Range schistose rocks.

This study has identified a persistent schistosity generally parallel or subparallel to compositional layering (S_0). In most samples this schistosity is the only cleavage present. In several samples collected from fold noses, in which the schistosity is axial planar to the fold, it appears that only compositional layering is folded. This suggests that the schistosity is S_1 and cuts bedding in F_1 folds. However, if this fabric is S_1, one would expect to see the more dominant S_2 cleavage superimposed on it. In fact, only one thin section shows convincing evidence for both S_1 and S_2. In this sample, from a quartz–mica schist of the Rinconada Formation, S_1 appears as oriented micas in quartz-rich layers of S_2 differentiated layers (Fig. 4). The appearance of S_1–S_2 relationships such as this, in which a segregated S_2 crenulation-cleavage folds an earlier S_1 schistosity, is highly dependent on the quartz/mica ratio of the schist. A ratio close to 50/50 is most conducive to the visibility of both S_1 and S_2. This overprint geometry has commonly been described in Proterozoic rocks of northern New Mexico (Codding, 1983; McCarty, 1983). It is not known whether this sample is representative of the area and both S_1 and S_2 exist in these rocks, or whether the sample is atypical and there is only one penetrative cleavage. Holcombe and Callender (1982) noted that although both S_1 and S_2 were recognized in Piedra Lumbre and Vadito Group rocks, S_1 was not positively identifiable in the Rinconada Formation or Pilar Phyllite in their study area. It is possible that in most samples of Ortega Group rock both S_1 and S_2 do not appear together because either S_2 has obliterated S_1 or the quartz/mica ratio was not appropriate for the mutual development of S_1 and S_2. Alternatively, the sample may be a special case and either S_1 or S_2 is not developed in the northern Picuris Range.

Both S_0/S_1 and S_2 are locally overprinted by several less pervasive, weak crenulation cleavages (S_3). The various S_3 cleavages are difficult to separate and are not considered important in the deformational geometry.

It is generally impossible to distinguish S_1 and S_2 in outcrop. Compositional layering (S_0) is cut by a subparallel axial-plane cleavage (S_2).

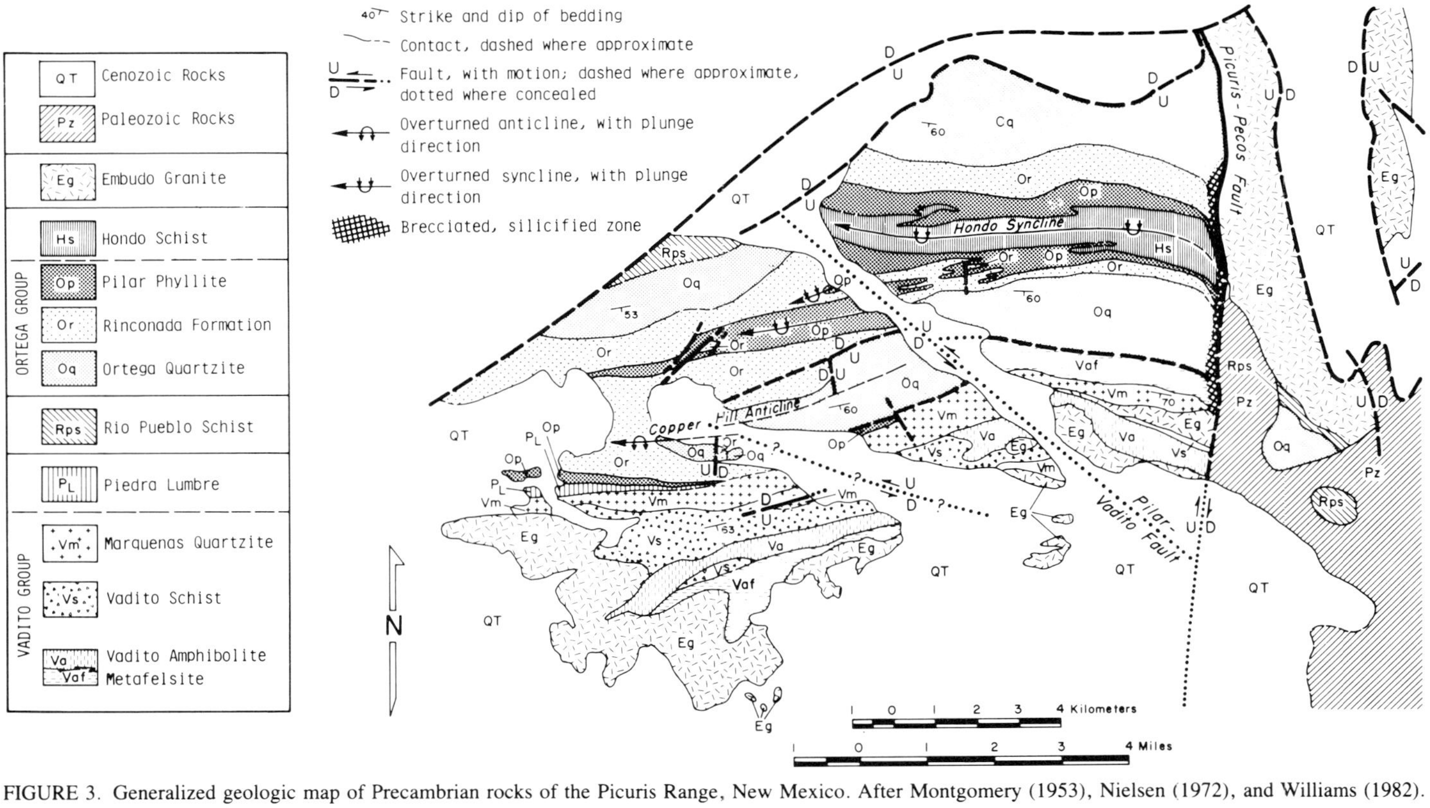

FIGURE 3. Generalized geologic map of Precambrian rocks of the Picuris Range, New Mexico. After Montgomery (1953), Nielsen (1972), and Williams (1982).

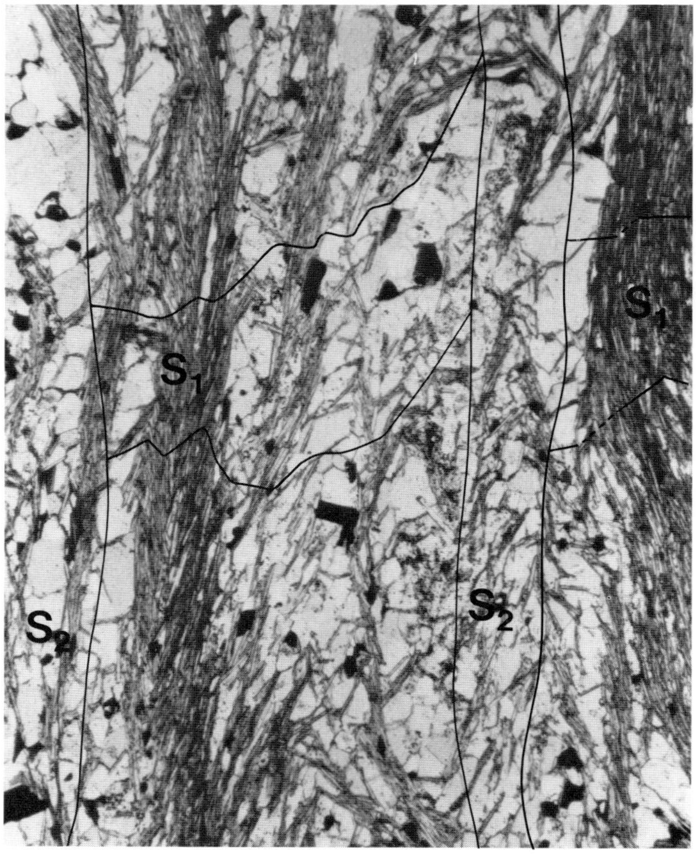

FIGURE 4. Photomicrograph of a schist of the Rinconada Formation, northern Picuris Range, showing the orientations and overprinting relationships of S_1 and S_2 cleavages. S_1 appears as scattered micas in quartz-rich S_2 layers. Field of view is 3 mm across.

There are two ways of approaching the problem of S_1–S_2. The first method is based on work by Holcombe (pers. comm. 1983) in which he has tentatively recognized a correlation between stratigraphy, deformation, and porphyroblast growth in schists of the Picuris Range. Holcombe found that not only does the appearance of axial-plane cleavages depend on stratigraphic/geographic position, but that the growth of certain metamorphic minerals can be related temporally to cleavage formation. The second approach is through a study of folding and refolding within the area.

Folds

Folds in the Picuris Range vary in size from the Copper Hill anticline–Hondo syncline with wavelengths of 4 or 5 km to folds only millimeters across. A complete description of various fold types was given by Nielsen (1972). Nielsen believed that extremely large isoclinal F_1 folds (so large that the entire Picuris Range may be on one limb) were in turn tightly folded by east–west-trending folds coaxial with F_1 (i.e., the Copper Hill anticline and Hondo syncline) whose axial planes dip steeply south. These F_2 folds were later broadly cross-folded by F_3, forming the present outcrop pattern of gentle, doubly plunging structures shown by Montgomery (1953) and Nielsen (1972).

This sequence of folding is consistent with descriptions of structures elsewhere in northern New Mexico. In the Rio Mora area, Grambling and Codding (1982; pers. comm. 1983) have described several map-scale isoclinal F_1 folds refolded by nearly upright, tight F_2 folds, plus rare S_3 crenulations. In the Tusas Mountains, M. L. Williams (pers. comm. 1984) has recognized a set of large, tight, upright F_2 folds with wavelengths of about 1 km, which locally fold a pre-existing layer-parallel cleavage (S_1); both are cut by several later F_3 crenulations.

In the northern Picuris Range both limbs of the Hondo syncline dip about 55° to the south. Because this and many minor folds are very tight, one cannot say whether S_1 or S_2 is axial planar to those folds. One way to approach this problem is to look at refolding patterns.

Refolding

Although most workers agree that the Picuris rocks have experienced multiple deformations, the absence of map-scale refolded fold patterns (other than those formed by minor F_3 structures) in the Ortega Group is puzzling. Montgomery (1953) mapped one very small type-III fold (Ramsay, 1967) closure on the southern limb of the Hondo syncline in the northern Picuris, but the structure does not appear on the map by Nielsen (1972). Exposure is very poor in this area. Recent mapping there yielded no convincing evidence of F_1 fold closure, although a detailed structural survey in the northern Picuris has revealed that refolding does occur on a scale of less than 100 m. In several places it was found that S_0/S_1–S_2 structural vergence and stratigraphic facing disagreed. That is, older units were found to core F_2 synforms and younger units cored antiforms, thus forming synformal anticlines and antiformal synclines (Fig. 5). The small distances between these reversals imply that at least one of the two folds involved has a wavelength on a scale of tens of meters. These small folds do not seem to interfere greatly with the larger-scale folds that form the Hondo syncline. The interference pattern is probably between minor F_1 folds and minor F_2 folds. If the major F_1 folds are enormous, it is not clear why two relatively small sizes of parasitic folds (centimeters and tens of meters in wavelength) but none of intermediate map-scale size folds would appear, unless either F_1 formed shear zones rather than large folds, or F_1 and F_2 were coaxial so that local reversals in structural vergence and stratigraphic facing are the only evidence of refolding.

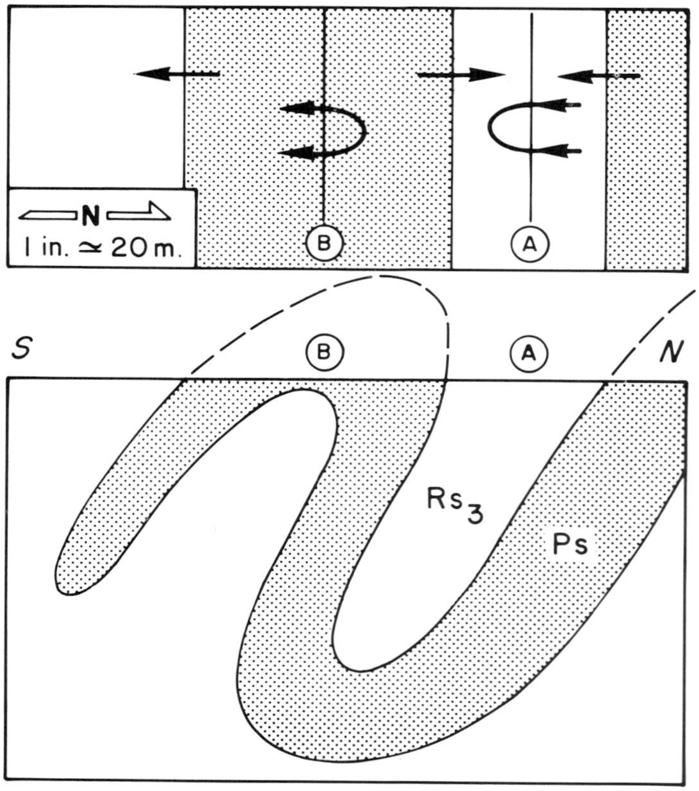

FIGURE 5. Idealized sketch map and cross section showing one possible refolded fold geometry for the Tierra Amarilla Canyon area, northern Picuris Range. Arrows on map point in the direction of synform. The older Rs_3 unit (Rinconada Schist) cores synforms; the younger Ps (Pilar Phyllite) cores antiforms, thus forming synformal anticlines and antiformal synclines.

Shear

Shear fabrics are common in quartzites of the Picuris Range. In the northern Picuris Range parallel elongate quartz grains commonly show a preferred crystallographic orientation along with sweeping extinction and strain ribbons. In one area on the north limb of the Hondo syncline the R_5 quartzite (Fig. 2) exhibits an extreme mylonitic texture (Fig. 6). It is not yet known if this outcrop represents a local zone or is part of a larger shear zone. Holcombe (pers. comm. 1983) has recognized a complex, anastomosing set of minor S_2-related ductile-shear zones (less than 1 m across) within the Embudo Granite in the southern Picuris Range. Callender and Holcombe (pers. comm. 1983) have speculated that thin, micaceous layers within the Ortega Quartzite and underlying metarhyolites in the Pilar area may be local areas of high-shear strain. It is not yet known whether ductile-shear zones make up a significant part of the deformation of the Picuris Range, nor whether major zones of ductile shear exist there at all. An interesting speculation is that F_1 deformation had a major component of subhorizontal shear and the S_1 fabric was well developed in local shear zones only. It is possible that this localized shearing was responsible for the destruction of all large F_1 fold noses.

STRUCTURAL ANALYSIS

I have made several preliminary structural analyses designed to detect both geographic and stratigraphic structural variations within rocks of the northern Picuris Range. In terms of geographic variations, no significant changes in S_0/S_1, S_2, and L_{20} appeared from the north to the south limb nor from the east to the west along the axis of the Hondo syncline. Local areas of anomalous fold axis and S-plane orientations are attributed to effects of broad F_3 folding and Phanerozoic faulting. However, in terms of stratigraphic variations of particular interest is an analysis of S_0/S_1, S_2, and L_{20} for Ortega Group rocks versus the schist of the Hondo Canyon. In the southern Picuris Range, Holcombe and Callender (1982) identified a strong S_1 in schists of the Piedra Lumbre Formation, but found no definite S_1 cleavage in Rinconada Schist and Pilar Phyllite. Furthermore, they documented a change in both vergence and L_{20} orientation between the Pilar and Piedra Lumbre, and established that the Piedra Lumbre is internally transposed. They proposed that the S_0/S_1–S_2 vergence change across the Pilar–Piedra Lumbre contact is probably due to a single S_2 having been superimposed on slightly discordant S_0 planes across a boundary fault. It is therefore crucial for an understanding of the Piedra Lumbre–Hondo Schist situation to examine S_0/S_1–S_2 relations between these rocks in the northern Picuris Range.

FIGURE 6. Photomicrograph of mylonite containing strained porphyroclasts, from R_5 quartzite north of the schist of Hondo Canyon. Field of view is 3 mm across.

Stereograms for the fabric elements S_0/S_1, S_2, and L_{20} in the Ortega Group (Ortega Quartzite, Rinconada Formation, Pilar Phyllite) and the Hondo Schist (Piedra Lumbre of Nielsen, 1972) are shown in Figure 7. Both F_3 deformation, and refraction and fanning of S_1 and S_2 cleavages are probably responsible for much of the scatter of points. S_0/S_1 and S_2 for both rock sequences have about the same east–west strike. The only significant variation in their dips is that S_0/S_1 for the Hondo Schist dips about 10° steeper than does the 55° dip of the other S_0/S_1 and S_2 planes. This 10° variation in S_0/S_1 may represent an original S_0/S_1 orientation difference between the two groups, but is not large enough to rule out errors in measurement or variations due to cleavage refraction. The most conspicuous difference between these data and those of Holcombe and Callender (1982) is in the orientations of L_{20}. Here both groups show similar southwest plunging statistical L_{20} maxima (Ortega at 239, 37°; Hondo at 247, 32°), although the Hondo Schists show a southeast plunging group at 120, 30° that is absent in Ortega rocks. Holcombe and Callender (1982) found that L_{20} for Ortega Group rocks plunged southwest, whereas L_{20} plunged southeast for Vadito Group rocks.

Although the data set is not shown, L_{30} and L_{32} intersection lineations define a fairly consistent orientation of 55° to the south, indicated by L_3 on the summary stereoplot of Figure 7.

Results suggest that if there is an early S_0/S_1 structural disparity between the Ortega and Hondo Schists, the disparity is too small to detect within the analytical error of measurements. S_2 fabrics appear to have been imposed identically over the two rock groups, suggesting that if their boundary is a fault, the major motion occurred prior to F_2 deformation. Both groups show similar southwest-plunging L_{20} patterns, although a southeast-plunging trend in the Hondo Schists may be related to an undocumented but possible S_0/S_1 disparity.

SUMMARY AND CONCLUSIONS

It is not yet clear whether there are two regional, penetrative, axial-plane cleavages (S_1 and S_2) in the Precambrian rocks of the northern Picuris Range. The existence of two cleavages is suggested by a single thin section showing an S_1–S_2 microfabric relationship, and by structural-vergence measurements in the field which imply refolded folding. The style of F_1 deformation is unclear; no map-scale F_1 folds have been recognized, but minor F_1 folds or shear zones do exist. One possibility is that F_1 deformation is concentrated along shear zones and S_1 is thus localized. F_2 has formed tight, gently plunging folds overturned to the north and a strong, penetrative axial-plane cleavage (S_2), which are the dominant structural features in the northern Picuris Range.

The Hondo Schist may or may not be stratigraphically equivalent to the Piedra Lumbre Formation of the southern Picuris. Although the Piedra Lumbre seems to have been deformed during F_1 along with the Vadito Group, it is likely that the Piedra Lumbre is in fault contact with both the Ortega and Vadito Groups and that the faulting predated, or was concurrent with, F_1 deformation. A preliminary structural analysis comparing the fabrics of the schist of Hondo Canyon with Ortega rocks has shown that the only structural difference between them is a 10° difference in the dips of modal S_0/S_1. This variation may be within analytical error, and, therefore, does not unequivocally demonstrate pre-F_2 difference between these two rock groups. If the contact between the Pilar Phyllite and the schist of Hondo Canyon is not a fault, and the schist of Hondo Canyon is equivalent to the Piedra Lumbre For-

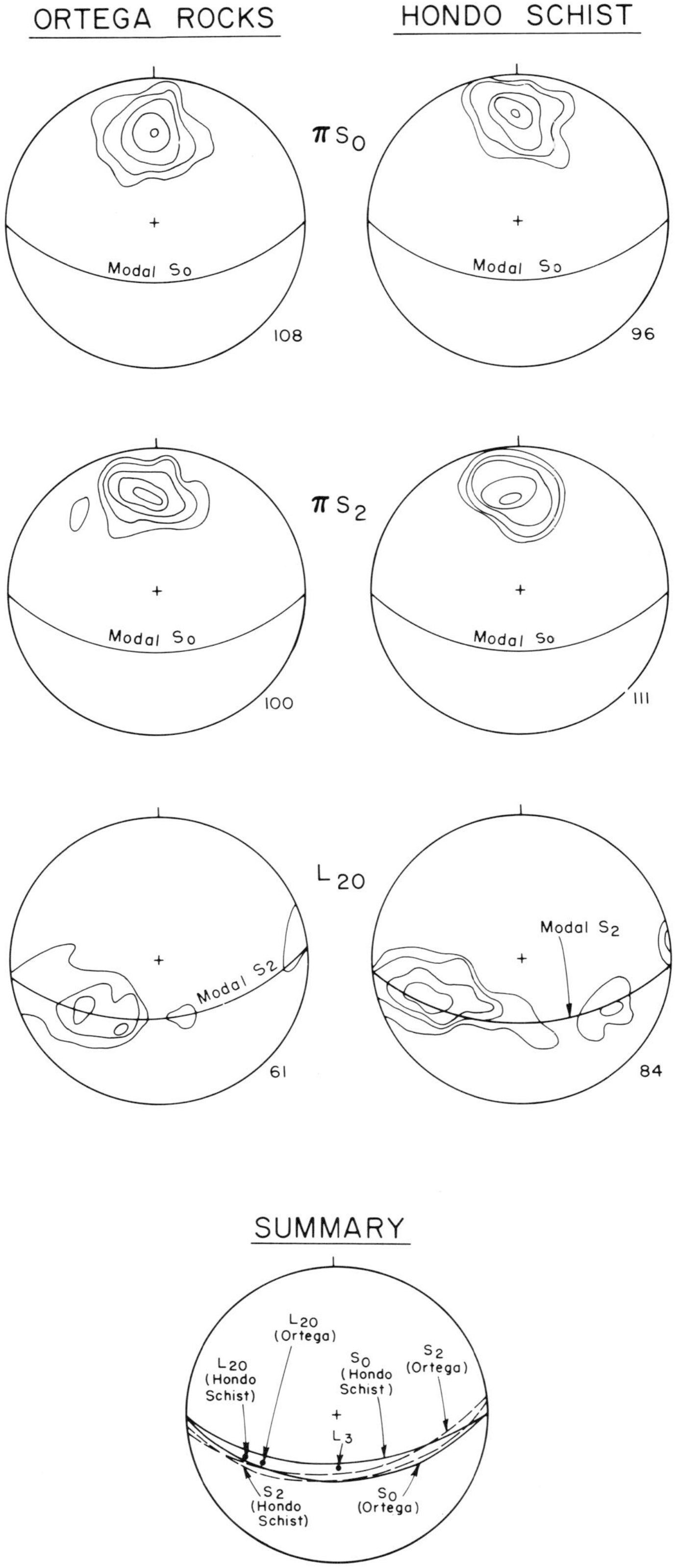

FIGURE 7. Stereograms showing the orientations of the major structural-fabric elements of the Ortega Group (Ortega Quartzite, Rinconada Formation, and Pilar Phyllite) and the Hondo Schist. Lower hemisphere, equal-area projections, tic to true north. Contours are at 2%, 4%, 8%, and 24% per 1% area, with the 1% contour added to both L_{20} plots. The number of data points is given at the lower right of each stereogram.

mation, then the schist of Hondo Canyon is a more complete section than the Piedra Lumbre in the southern Picuris Range. However, until more work is done in and around the schist of Hondo Canyon, it seems preferable to keep the schist of Hondo Canyon separate from the Piedra Lumbre Formation, the Vadito Group, and the Ortega Group.

ACKNOWLEDGMENTS

This paper stems from research begun in partial fulfillment of Ph.D. requirements at the University of New Mexico. Financial support was provided by the Department of Geology, University of New Mexico, and by National Science Foundation grant EAR-8018506 to J. F. Callender.

Special thanks go to J. F. Callender and J. A. Grambling for guidance and critical manuscript reviews. R. C. Holcombe and M. L. Williams provided much additional insight. Reviews by K. C. Condie and J. C. Reed, Jr. improved this paper considerably.

REFERENCES

Codding, D. B., 1983, Precambrian geology of the Rio Mora area, New Mexico: structural and stratigraphic relations: M.S. thesis, University of New Mexico, Albuquerque, 128 pp.

Grambling, J. A., and Codding, D. B., 1982, Stratigraphic and structural relationships of multiply deformed Precambrian metamorphic rocks in the Rio Mora area, New Mexico: Geological Society of America, Bulletin, v. 93, pp. 127–137.

———, Williams, M. L., and Codding, D. B., 1983, Mn- and Cr-rich marker horizons in multiply deformed Proterozoic metamorphic rocks, northern New Mexico (abstract): Geological Society of America, Abstracts with Programs, v. 15, no. 5, p. 424.

Gresens, R. L., and Stensrud, H. L., 1974, Recognition of more metarhyolite occurrences in northern New Mexico and a possible Precambrian stratigraphy: Mountain Geologist, v. 11, pp. 109–124.

Holcombe, R. J., and Callender, J. F., 1982, Structural analysis and stratigraphic problems of Precambrian rocks of the Picuris Range, New Mexico: Geological Society of America, Bulletin, v. 93, pp. 138–149.

Holdaway, M. J., 1978, Significance of chloritoid-bearing and staurolite-bearing rocks in the Picuris Range, New Mexico: Geological Society of America, Bulletin, v. 89, pp. 1404–1414.

Just, E., 1937, Geology and economic features of the pegmatites of Taos and Rio Arriba Counties, New Mexico: New Mexico Bureau of Mines and Mineral Resources, Bulletin 13, 73 pp.

Long, P. E., 1976, Precambrian granitic rocks of the Dixon–Peñasco area, northern New Mexico: Ph.D. thesis, Stanford University, 533 pp.

McCarty, R. M., 1983, Structural geology and petrography of part of the Vadito Group, Picuris Mountains, New Mexico: M.S. thesis, University of New Mexico, Albuquerque, 159 pp.

Miller, J. P., Montgomery, A., and Sutherland, P. K., 1963, Geology of part of the Sangre de Cristo Mountains, New Mexico: New Mexico Bureau of Mines and Mineral Resources, Memoir 11, 106 pp.

Montgomery, A., 1953, Precambrian geology of the Picuris Range, north-central New Mexico: New Mexico Bureau of Mines and Mineral Resources, Bulletin 30, 89 pp.

Nielsen, K. C., 1972, Structural evolution of the Picuris Mountains, New Mexico: M.S. thesis, University of North Carolina, Chapel Hill, 47 pp.

———, and Scott, T. E., 1979, Precambrian deformational history of the Picuris Mountains, New Mexico: New Mexico Geological Society, Guidebook 30, pp. 113–120.

Ramsay, J. G., 1967, Folding and fracturing of rocks: McGraw–Hill, New York, 568 pp.

Scott, T. E., Jr., 1980, A strain analysis of the Marquenas Quartzite and contact relationships of Ortega–Vadito Groups: M.S. thesis, University of Texas at Dallas, 173 pp.

Williams, M. L., and Wobus, R. A., 1984, Stratigraphic implications of Mn-andalusite in Proterozoic rocks of the Tusas Mountains, New Mexico (abstract): Geological Society of America, Abstracts with Programs, v. 16, n. 4, p. 260.

OCCURRENCES OF PRECIOUS METALS AND URANIUM ALONG THE RIO GRANDE RIFT IN NORTHERN NEW MEXICO

VIRGINIA T. McLEMORE and ROBERT M. NORTH
New Mexico Bureau of Mines and Mineral Resources, Socorro, New Mexico 87801

INTRODUCTION

Silver and gold occur in at least minor concentrations (above 1 ppm Au and 14 ppm Ag) in 25 mining districts within and adjacent to the Rio Grande rift in Colfax, Taos, Rio Arriba, Sandoval, Santa Fe, and Torrance Counties in northern New Mexico (Fig. 1, Table 1). No platinum or platinum-group metals are known to occur in the area. Uranium occurs in 15 of these districts and in four additional areas (Fig. 1, Table 1). Uranium also occurs in Precambrian pegmatites in the Tusas and Sangre de Cristo Mountains, and in Jurassic and Cretaceous sediments in the eastern San Juan Basin, Chama Basin, and near San Ysidro. These occurrences are not discussed in this report, but they are described elsewhere by Hilpert (1969), Chenoweth (1974a, 1974b, 1979), and McLemore (1983).

Gold, silver, and/or uranium occurrences are found in eight types of deposits (Table 1), many of which are genetically related to the formation of the Rio Grande rift. Gold occurs in late Tertiary to Recent placers in eight districts within basins of the Rio Grande rift and surrounding areas. Gold, silver, and uranium are associated with three types of Tertiary-age deposits: (1) hydrothermal gold–silver vein and breccia deposits, (2) base-metal veins, and (3) replacement deposits in limestones. Silver and rarely gold are found in barite–fluorite–lead vein

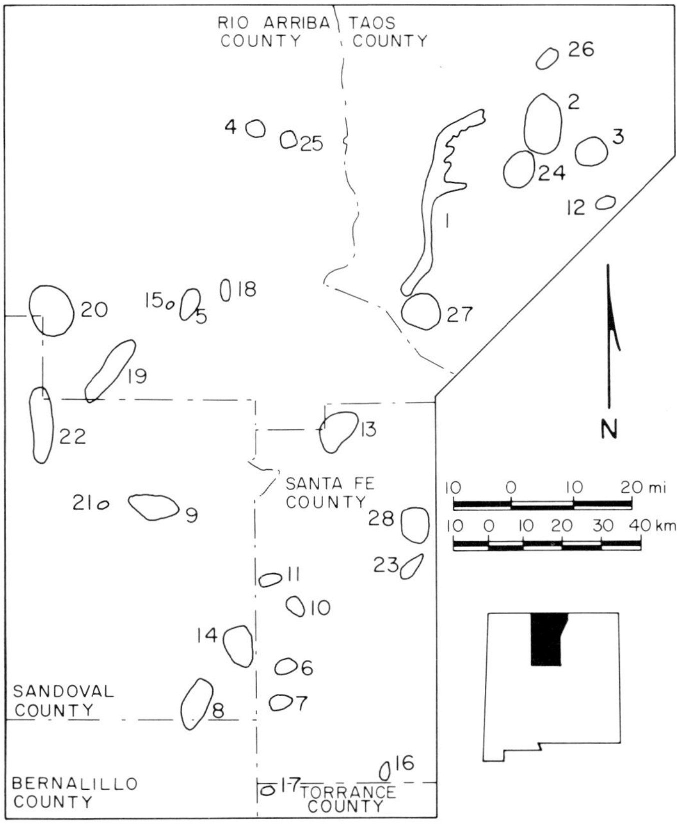

FIGURE 1. Gold, silver, and/or uranium districts or areas in northern New Mexico. Numbers represent localities given in Table 1.

TABLE 1. Precious metals and uranium districts along the Rio Grande rift and vicinity in northern New Mexico. [1]Commodities: Au—gold, Ag—silver, W—wolframite (tungsten), Cu—copper, Pb—lead, Zn—zinc, U—uranium, Ba—barite, Fl—fluorite. [2]Status of district: 1—active and producing presently, 2—inactive with past production, 3—inactive with no precious metal or uranium production. Production is listed in Table 2.

Map no. in Fig. 1	District	Commodities present[1]	Status of district[2]
	PLACER DEPOSITS		
1	Rio Grande valley	Au	2
2	Red River	Au	2
3	Elizabethtown-Baldy	Au	2
4	Hopewell	Au	2
5	Chama Placers	Au	2
6	Old Placers	Au	2
7	New Placers	Au	1
8	Placitas	Au	2
	HYDROTHERMAL Au-Ag VEIN AND BRECCIA DEPOSITS		
2	Red River	Cu, Pb, Zn, Au, Ag, Mo, U	2
3	Elizabethtown-Baldy	Au, Ag, W	2
6	Old Placers	Ag, Cu, Pb, W, Au	1
7	New Placers	Cu, Pb, Zn, Ag, Au	2
9	Cochiti	Au, Ag, Cu, Pb, Zn, U	2
	BASE-METAL VEIN DEPOSITS WITH GOLD AND/OR SILVER		
8	Placitas	Cu, Pb, Ag	2
10	Cerrillos	Cu, Pb, Zn, Au, Ag, U	2
11	La Bajada	Cu, U, Ag, Zn	2
	REPLACEMENT DEPOSITS IN LIMESTONES		
3	Elizabethtown-Baldy	Au, Ag, W	2
6	Old Placers		
7	New Placers	Cu, Pb, Zn, Ag, Au, W	2
12	Cimmaroncito	Au, Ag, Cu	2
	URANIUM DEPOSITS IN SEDIMENTARY ROCKS		
13	San Jose	(Tesuque Formation)	2
14	Hagan Basin	(Galisteo Formation)	2
15	Box Canyon	(Todilto Formation)	2
	BARITE-FLUORITE-LEAD (ZINC, COPPER, SILVER) VEIN DEPOSITS		
8	Placitas	Ba, Fl, Pb, Cu, Ag	2
16	Crow Butte	Ba, Fl, Pb, Ag	3
17	Edgewood	Ba, Fl, Pb, Ag	3
	STRATABOUND COPPER DEPOSITS (RED BEDS)		
18	Abiquiu	Cu, U, Ag	3
19	Coyote	U, Cu, Ag	3
20	Gallinas	Cu, U, Ag	3
21	Jemez Springs	Cu, Pb, Ag, Au, U	2
22	Nacimiento	Cu, Pb, U, Ag	2
23	Glorieta	Cu, Ag, U	3
	DEPOSITS IN PRECAMBRIAN ROCKS		
2	Red River	Cu, Pb, Zn, Au, Ag, U	2
3	Hopewell	Cu, Pb, Au, Ag	2
8	Placitas	Cu, Ag, Au	3
24	Twining	Cu, Ag, Au	2
25	Bromide #2	Cu, Au, Ag, U	2
22	Nacimiento	Cu, Au, Ag, U	3
26	Costilla Massif	U	3
27	Picuris	Cu, Ag, Au, W, U	2
28	Santa Fe	Cu, Pb, Zn, Ag, Au	3

TABLE 2. Reported gold, silver, uranium, and vanadium production from deposits in northern Rio Grande rift and adjacent areas. From North and McLemore (1984), Johnson (1972), Anderson (1957), Wells and Wootton (1940), U.S. Atomic Energy Commission ore production reports (1950–1970), and NMBMMR files. [a]—Estimated; [b]—Type of deposit: 1—placer gold deposit, 2—Au–Ag vein deposit, 3—base-metal vein deposit, 4—replacement deposit, 5—uranium deposit in sedimentary rocks, 6—stratabound copper deposit, 7—Ba–F–Pb (Zn, Cu, Ag) deposit, 8—deposit in Precambrian rocks.

Map No.	District	Au oz	Ag oz	U_3O_8 lbs	V_2O_5 lbs	Type of deposit[b]
	RIO ARRIBA					
18	Abiquiu	--	--	7	43	6
19	Coyote	--	--	182	206	6
25	Bromide #2	300[a]	4,500[a]	12	10	8
4	Hopewell	24,000[a]	10,000[a]	--	--	1, 8
15	Box Canyon	--	--	253	212	5
5	Chama placers	100[a]	--	--	--	1
	SANDOVAL					
20	Gallinas	--	--	12	24	6
21	Jemez Springs	1.0	159	--	--	6
22	Nacimiento	--	75,068	--	--	6
9	Cochiti	41,016	203,895[a]	--	--	2
8	Placitas	49	48	--	--	1, 3
	SANTA FE					
10	Cerrillos	930.68	27,864	--	--	3
11	La Bajada	--	52	27,116	42	3
7	New Placers	117,000[a]	304,625	--	--	1, 2, 4
6	Old Placers	200,000[a]	311	--	--	1, 2, 4
13	San Jose	--	--	12	--	5
	TAOS					
27	Picuris	14.75	1,351	--	--	8
2	Red River	364.89	8,051	3	--	2, 8
1	Rio Grande Valley	<1,000[a]	--	--	--	1
24	Twining	80[a]	1,000[a]	--	--	8
	COLFAX					
3	Elizabethtown-Baldy	471,400[a]	--	--	--	1, 2, 4
12	Cimmaroncito	100[a]	1,000[a]	--	--	4

deposits in three areas along or near the Rio Grande rift in northern New Mexico. Silver, uranium, vanadium, and rarely gold also occur in stratabound sedimentary copper deposits and probably formed prior to the Rio Grande rift. Mineral deposits in Precambrian rocks also contain gold, silver, and/or uranium mineralization and are occasionally associated with younger mineralization.

Production of gold and silver from placer deposits, deposits of Tertiary age, and deposits in Precambrian rocks has been moderate (Table 2). Except for La Bajada mine, uranium and vanadium production has been insignificant (Table 2).

PLACER-GOLD DEPOSITS

Placer gold is reported from eight localities in the vicinity of the northern Rio Grande rift. Most are adjacent to known lode-gold producing areas, where the eroded gold has been concentrated in Tertiary and Quaternary gravels formed by erosion of the Rio Grande valley.

Rio Grande Valley

Less than 1,000 oz (31,100 g) of placer gold (Table 2; Johnson, 1972) have been produced from the Rio Grande valley between Red River and Embudo (#1, Fig. 1) since Spanish Colonial times (Schilling, 1960). Early reports on the extent and grade of the gold-bearing gravels and sand bars of the present Rio Grande channel and Pliocene Servilleta Formation predicted great success for dredging operations (Silliman, 1880; Payne, unpubl. report 1932). Although large-scale operations were unsuccessful due to the presence of large basalt boulders in the gravels (Schilling, 1960), the area remains a favorite of recreational gold panners. The gold is fine- to medium-grained, averaging about pinhead size (Schilling, 1960), and is estimated to occur in deposits containing from 0.012 (Silliman, 1880) to 0.082 oz per cubic yard (Payne, unpubl. report 1932) running about 900 fine. The gold was probably derived from Precambrian and Tertiary lode deposits of the Taos Range to the east.

Red River

Less than 500 oz (15,550 g) of gold (Table 2; Johnson, 1972) have been produced from placer-gold deposits in the Red River district (#2, Fig. 1) in Bitter, Gold, Spring, and Placer Creeks (Johnson, 1972). These deposits were probably worked originally by Spanish miners and later in the 1800's by miners attracted to the rich Elizabethtown–Baldy placers (#3, Fig. 1). The Red River placers are predominantly in Quaternary gravels, although Tertiary gravel terraces have been prospected and possibly some gold was produced (Park and McKinlay, 1948). The gold was derived mainly from local Oligocene skarn and vein deposits and to a lesser extent from Precambrian vein deposits.

Elizabethtown–Baldy

The largest producing placer-gold district in the state was Elizabethtown–Baldy (#3, Fig. 1) where placer gold was found in nearly all the gulches draining the Baldy Mountain. Total placer production is estimated at 250,000 oz (7,775,000 g) of gold (Table 2; Anderson, 1957). Gold was discovered in 1866 in Willow Creek, and by 1868 several thousand people inhabited Elizabethtown (Clark and Read, 1972). Work declined in the 1870's but probably continued on a small scale well into the 20th century (Clark and Read, 1972). Placer gold, occasionally containing some fairly coarse nuggets, was mined extensively from Quaternary gravels in Willow, Ute, and South Ponil Creeks, Grouse and Hamburg Gulches, and the Moreno River. The placers were derived from the Oligocene vein and skarn deposits of the Baldy Mountain.

Hopewell District

Placer gold was discovered in the Hopewell district (#4, Fig. 1) in the 1880's, and $175,000 worth of gold was produced within only three years (Lindgren and others, 1910). Production slowed considerably as the high-grade Fairview placer was worked out by the late 1880's and work almost ceased by 1900 (Johnson, 1972). Nuggets up to 2 oz (62 g) were found during this period (Little, unpubl. report 1947). Total placer production is estimated as 16,000 oz (497,600 g; Table 2; Johnson, 1972). The placer deposits in the district are of two ages. The richest placer, the Fairview, is in Tertiary gravels exhumed by Placer Creek, whereas the lower-grade placers in Placer Creek are Quaternary (Bingler, 1968). The gold was derived from the Precambrian vein deposits of the district.

Chama Placers

Gold was found prior to 1848 in the sand and gravel deposits of the Chama River near the intersection of Canones Creek west of Abiquiu (#5, Fig. 1). Since then there has periodically been considerable interest in these placers, most recently in the late 1930's. Total production is unknown; Johnson (1972) estimated less than 100 oz (3,110 g), whereas Mitchell (unpubl. report 1938) claims $300,000 worth of production in only a few months. This appears exaggerated, but the district has probably produced more than 100 oz (3,110 g). We estimate the production of this district to be less than 1,000 oz (31,100 g).

According to Johnson (1972), the gold was derived from Precambrian rocks along the deep canyons of the Chama River; however, the Chama River does not cut Precambrian rocks north of Abiquiu; it cuts mainly Mesozoic sedimentary rocks (Bingler, 1968, pl. 1b). Another possible source of the gold is from Tertiary volcanic rocks to the south, drained by the Canones Creek, but no ore deposits are known to occur in these rocks. Currently much of the area is inundated by the Abiquiu reservoir.

Old Placers

Placer gold was discovered in the Old Placers district (#6, Fig. 1) in 1828 and was mined actively prior to 1900 and intermittently since that time. Total placer production is estimated to be greater than 100,000 oz (3,110,000 g; Table 2; Johnson, 1972). Gold occurs in both Quaternary and Tertiary gravels in the Dolores and Cunningham Gulches, and lower-grade material is found in the Lucas Canyon (Anderson, 1957). The richest gravels were mined from the upper part of a mesa-forming (Tertiary?) alluvial fan at the mouth of the Cunningham Gulch (Koschmann and Bergendal, 1968). The placers were derived from Oligocene vein and breccia deposits in the area.

New Placers

The New Placers district (#7, Fig. 1) was discovered in 1839 (Atkinson, 1961) and has produced until recently, although since 1900 on a small scale. The district is estimated to have produced about 97,000 oz (3,016,700 g) prior to 1900; since 1900 the production has been about 3,000 oz (93,300 g; Table 2; Johnson, 1972). The area is currently exploited on a small scale for the coarser material (>1 mm) that is sold chiefly as mineral specimens.

Gold was produced from unconsolidated Quaternary gravels surrounding the San Pedro Mountains and the best material was found in the Old Timer and San Lazarus Gulches. The cemented gravels (Tertiary?) in the vicinity of Golden contain gold, but all attempts to mine them have failed (Wells and Wootten, 1940). The placers were derived from the copper–silver–gold skarn and vein deposits of the San Pedro Mountains.

Placitas District

About 49 oz (1,520 g) of gold were produced from cemented gravels in the Placitas district at the north end of the Sandia Mountains (#8, Fig. 1; U.S. Bureau of Mines Mineral Yearbooks). Little is known about the history and total production of these placers, but apparently they were worked in the early part of this century when up to $3 worth (about 0.15 oz) of gold a day was recovered by dry-washing methods (Heikes, 1913).

HYDROTHERMAL GOLD–SILVER VEIN AND BRECCIA DEPOSITS

Hydrothermal vein and breccia deposits containing dominantly gold and silver and minor base metals occur in three districts in north-central New Mexico. The value of precious-metal production exceeds the value of base-metal production. These Oligocene-age deposits are associated with a period of igneous activity beginning around 33 m.y. and continuing until about 25 m.y. B.P., roughly coinciding with the earliest Tertiary rifting in the area (Chapin, 1979).

Red River District

This is a large district (#2, Fig. 1) which includes Red River, Cabresto Creek, lower Red River, Black Copper Canyon, and the Anchor–Midnight areas. Although many prospects and small mines occur in the district, reported production is small (Table 2).

Veins occur along fissures and shear zones in and adjacent to Miocene(?) soda granite and associated volcanic rocks, and in some places Precambrian rocks are mineralized (Schilling, 1960). The mineralogy of the veins is variable, but they generally contain auriferous pyrite or limonite pseudomorphs after pyrite. Some veins in the area were prospected for molybdenite (Schilling, 1960) and uranium (McLemore, 1983). The quartz-vein deposits have been dated from 21.0 ± 3.0 to 23.5 ± 0.8 m.y. B.P. (Ishihara, 1967; K–Ar on biotite) and hence are contemporaneous with rifting.

Elizabethtown–Baldy District

Lode deposits in the Elizabethtown–Baldy district (#3, Fig. 1) were discovered in 1866 and mined continuously until 1940. Since 1900 most mining has been on a small scale, except for operations of the Maxwell Land Grant Co. from 1910 to 1940 (Pettit, 1946). Total lode-gold production from the district is estimated at 221,400 oz (6,885,500 g; Koschmann and Bergendahl, 1968), most of which came from the vein deposits. The district also produced silver, copper, and small amounts of lead.

Mineralization fills fissures and major and minor fault zones that cut sills of quartz-diorite porphyry and Cretaceous Pierre Shale (Pettit, 1946; Clark and Read, 1972), and occurs along the unconformable contact between the Pierre Shale and the overlying Tertiary Raton Formation (Lee, 1916). The veins contain dominantly quartz and pyrite with minor calcite, chalcopyrite, pyrrhotite, and in places magnetite and molybdenite (Clark and Read, 1972). These deposits are thought to be the result of a hydrothermal system associated with the quartz-diorite porphyry (Lindgren and others, 1910). Similar sills in the Ute Park–Cimmaron Canyon Palisades have been dated at 33.8 m.y. B.P. (Armstrong, 1969; K–Ar on biotite) and mineralization at Elizabethtown–Baldy is probably of similar age.

Old Placers District

Vein and breccia deposits occur in a horst block northwest of the Tijeras fault zone in the Old Placers district (#6, Fig. 1). Mining began prior to 1850 (Jones, 1904) and continued intermittently until about 1930. Lode mining in the district was inactive until early 1980 when Gold Fields Mining Corp. opened a 3,000 short ton a day open-pit mine (Wright, 1983; Fig. 2). The ore averages 0.053 oz/ton (1.82 ppm) and is processed by cyanide heap-leaching methods (Hickson, 1982). The earliest mining was confined to quartz veins containing free gold, specular hematite, magnetite, pyrite, chalcopyrite, and, in the Live Oak vein, molybdenite (Lindgren and others, 1910). A breccia pipe composed of clasts of Cretaceous Mesaverde Group and intrusive quartz-latite porphyry is currently being mined (Wright, 1983).

The dominant igneous rock in the district is the nepheline-bearing augite-monzonite Ortiz stock which intruded the Cretaceous Mesaverde Group and was subsequently intruded by latite and quartz-latite dikes, sills, and laccoliths and overlain by an ash-flow tuff (Wright, 1983). The quartz veins cut the latite stock (Lindgren and others, 1910), indicating that mineralization occurred after emplacement of the latite. The breccia deposits contain only latite and Mesaverde clasts but border the ash-flow tuff (Wright, 1983), suggesting that mineralization occurred before the deposition of the tuff. A latite sill on the west side of the Ortiz intrusive complex has been dated at 34.0 ± 2.2 m.y. B.P. (Bachman and Mehnert, 1978; K–Ar on hornblende).

FIGURE 2. The open-pit Ortiz mine of Gold Fields Operating Co.–Ortiz, Old Placers district (#6, Fig. 1). Gold Fields mines about 3,000 tons of ore a day running about 0.05 oz/ton (1.8 ppm), making this the tenth largest gold producing mine in the U.S. in 1982. Photo by Susan Beattie. Published with permission and through the courtesy of Gold Fields Operating Co.–Ortiz.

New Placers District

Gold-bearing veins in the New Placers district (#7, Fig. 1) cut metasedimentary and intrusive igneous rocks around the margins of stocks in the San Pedro Mountains (Atkinson, 1961). Although production from these veins has been minor compared to the replacement (skarn) and placer deposits, they may represent an important source for the placer gold. Lead and zinc vein deposits occur at the Carnahan mine in the southern part of the district (Atkinson, 1961) and contain argentite (acanthite?), native silver, and chlorargyrite (Northrup, 1959). Precious-metal production from these veins is small.

Cochiti District

Early prospecting in the Cochiti district in the Jemez Mountains (#9, Fig. 1) began around 1880, but serious mining was prevented because of claims of Mexican land-grant owners. Federal Courts subsequently declared the lands open to location and by the 1890's the district was the site of considerable activity (Bundy, 1958). The most productive period was 1894–1904 when the Albemarle deposit was mined, with later flurries of activity in 1914–16 and 1932–40 (Koschmann and Bergendahl, 1968). Prospecting and possibly a small amount of production are currently underway.

The ore deposits are in veins filling faults in Tertiary andesite, volcanic breccia, and quartz diorite. The early vein filling was calcite which was subsequently replaced by quartz-sulfide mineralization. Sulfides present include pyrite, chalcopyrite, sphalerite, and argentite (Bundy, 1958). The gold is associated with pyrite and argentite. Uranium minerals are reported to occur at a few localities (Chenoweth, 1974a; Lindgren and others, 1910). Veins vary from a few millimeters to about 15 m at the surface and are reported to be as wide as 45 m at depth (Bundy, 1958).

The deposits of the district are best classified as epithermal deposits despite their replacement character, because they have many of the characteristics of epithermal deposits as discussed by Buchanan (1981). Fluid-inclusion data by Parkison and others (1984) indicate temperatures of mineralizing fluids of 240–315°C (see Wronkiewicz and others, this guidebook). The age of these deposits is given by Parkison and others as between 6.5 and 1.5 m.y. B.P., but no analytical data are given. Radiometric-age determinations in this geographic area must be subjected to close scrutiny due to possible resetting, or partial resetting, of radiometric ages by Pleistocene volcanic activity.

BASE-METAL VEIN DEPOSITS

Precious metals have been produced as byproducts of base-metal mining from vein deposits in three districts in north-central New Mexico. In all cases, silver is the most important byproduct, with subordinate gold and uranium.

Placitas District

Only one of the many base-metal veins in the Placitas district (#8, Fig. 1) has produced. The Montezuma mine produced 48 oz (1,493 g) of silver, and some copper and lead, in 1920 and 1926 (Elston, 1967). Numerous small fissure veins containing base-metal with silver and traces of gold occur in the vicinity of barite–fluorite and fluorite veins throughout the district (Kelley and Northrop, 1975; Hedlund and Kness, 1984).

Cerrillos District

Mining in the Cerrillos District began with prehistoric turquoise mining and continued with exploitation of lead oxides for glaze paint used on pottery (Warren and Weber, 1979). Metal mining for lead and possibly silver was initiated during the Spanish Colonial period prior to the Pueblo Revolt in 1680 (Warren and Weber, 1979). Very little mining activity is known to occur until around 1880; afterwards, activity was fairly continuous until 1960 (Elston, 1967). The district has been essentially idle since that time, with the exception of intensive exploration for copper mineralization in porphyry (Akright, 1979).

The important vein deposits of the district cut Oligocene augite–biotite-monzonite porphyry and hornblende-monzonite porphyry. Less important veins cut Oligocene Espinaso volcanics and underlying Eocene Galisteo Formation (Disbrow and Stoll, 1957). The veins are dominantly lead–zinc with minor copper, silver, uranium, and gold, and are probably related to the intrusion of the augite–biotite-monzonite porphyry. Primary ore minerals include sphalerite, galena, chalcopyrite, and pyrite, with silver associated with galena and gold with chalcopyrite (Disbrow and Stoll, 1957). An intrusive thought to be correlative with the augite–biotite-monzonite porphyry has been dated at 30.2 ± 0.7 m.y. B.P. (Baldridge and others, 1980; K–Ar on biotite), which is probably close to the age of mineralization in the district.

La Bajada Deposit

La Bajada (Lone Star) deposit is a unique, low-temperature, base-metal vein deposit that formed during the Oligocene or Miocene (Hilpert, 1969; Haji-Vassiliou and Kerr, 1972). Thin veins of uranium mineralization and base-metal sulfides occur along the footwall of a limburgite dike which was emplaced along a north-trending fault in the Oligocene Espinaso Formation. The uniqueness of the deposit is due to the presence of ore-controlling carbonaceous material (Haji-Vassiliou and Kerr, 1972) in volcanic rocks including intrusive limburgite that are intensely altered by hydrothermal solutions. The organic material is thought to have been derived from the underlying Cretaceous sediments (Haji-Vassiliou and Kerr, 1972). Hydrothermal alteration and mineralization probably occurred during the Oligocene or Miocene (Hilpert, 1969).

Lustig (1957) described 23 minerals from this deposit, including (in order of decreasing abundance) pyrite, sphalerite, marcasite, colusite, ($Cu_3(As,Sn,V,Fe)S_4$), chalcopyrite, and bornite. In addition, uranium, cobalt, nickel, molybdenum, germanium, and gold occur in appreciable amounts (Hilpert, 1969). Due to the complex association of uranium and carbonaceous material, specific uranium minerals have not been identified. One select sample assayed 0.09% U_3O_8, 1.51% Cu, 19 ppm Th, 0.06% Pb, 0.03% Zn, 0.54 oz/ton (19 ppm) Ag, and no gold.

Copper mineralization was first discovered in 1915 or 1916. La Bajada Mining Co. was formed in 1923 and by 1928 the American Smelting and Refining Co. controlled the deposit. In 1928 and 1929 the American Smelting and Refining Co. produced 5,345 lbs (2,424 kg) of copper and 52 oz (1,617 g) of silver (Table 3). The deposit was mined through two shafts. Uranium was discovered in 1950, and from 1956 to 1966 27,116 lbs (12,300 kg) of U_3O_8 and 42 lbs (19 kg) of V_2O_5 were produced (Table 2). In 1957 the underground workings were found to be unsafe and further development was by open pit (Chenoweth, 1979). The pit has since filled with water because the rim is only several feet above the river level.

In the late 1970's Bokum Resources and Union Carbide Corp. drilled exploratory holes in the vicinity of La Bajada mine, but their results are unknown. In 1979 the Lone Star Mining Co. lost the lease to the deposit for failure to submit an approved mining plan to the Bureau of Land Management. Reserves are probably present at the mine, but environmental problems will hamper future development. One additional uranium occurrence, the Hiser Moore #1, is known in the area (Hilpert, 1969).

TABLE 3. Production from La Bajada mine, Santa Fe County From Lustig (1957), U.S. Atomic Energy Commission ore production reports (1956–1966), and NMBMMR files (1979).

Year	Tons ore	Ag oz	Cu lbs	U_3O_8 lbs	V_2O_5 lbs	
1928	8	24	2,423	--	--	American Smelting and Refining Co.
1929	9	28	2,922	--	--	American Smelting and Refining Co.
1930-1955	no production					
1956	46	--	--	166	--	Lone Star Mining Co.
1957	51	--	--	204	42	Lone Star Mining Co.
1962	1,617	--	--	5,277	--	Lone Star Mining Co.
1963	5,465	--	--	14,482	--	Lone Star Mining Co.
1964	2,106	--	--	5,882	--	Lone Star Mining Co.
1966	364	--	--	1,105	--	Lone Star Mining Co.
TOTAL	--	52	5,345	27,116	42	

REPLACEMENT AND SKARN DEPOSITS

Skarn deposits account for the major production of the New Placers (#7, Fig. 1) and Cimmaroncito (#12, Fig. 1) districts. Small replacement and skarn bodies also occur in the Elizabethtown–Baldy and Old Placers districts.

New Placers District

The important lode deposits of the New Placers district are skarn deposits replacing limestone beds in the upper Madera Formation (Pennsylvanian) adjacent to an Oligocene monzonite-porphyry laccolith. The ore deposits are as much as 15 m from the contact of the laccolith with the Madera Formation and are localized by an overlying rhyolite sill (Atkinson, 1961). At the San Pedro mine, ore minerals were deposited in two stages in garnet tactite as a late phase of contact metasomatism. The first stage filled small cavities in the tactite and consists of chalcopyrite, bornite, pyrite, pyrrhotite, calcite, specular hematite, quartz, and chlorite. The second stage filled larger cavities in the tactite with quartz, pyrite, chalcopyrite, calcite, and minor siderite and scheelite (Atkinson, 1961). Gold and silver are associated with copper minerals in both stages. At the Carnahan mine, gold and silver occur with the lead–zinc deposits that have replaced favorable limestone beds (Atkinson, 1961). No age data are available for intrusive rocks in the New Placers district; however, based on geologic relations and on other dated rocks in the "porphyry belt," an Oligocene age is reasonable.

Cimmaroncito District

The ore deposits of the Cimmaroncito district (#12, Fig. 1) are confined to skarns in calcareous beds in the Pennsylvanian–Permian Sangre de Cristo Formation adjacent to quartz-monzonite-porphyry sills (Robinson and others, 1964). The deposits contain chalcopyrite, galena, and gold in a gangue of pyrite, quartz, garnet, epidote, calcite, magnetite, and specular hematite (Pettit, 1946). Small pockets of high-grade ore were mined and reported to assay as high as 0.5 oz/ton (17 ppm) gold, 4 oz/ton (137 ppm) silver, and 8–10% copper (Lindgren and others, 1910); however, total production has been small (Table 2). The ore deposits are related to the quartz-monzonite sills, which have been dated in the area at 33.8 m.y. B.P. (Armstrong, 1969; K–Ar on biotite).

URANIUM DEPOSITS IN SEDIMENTARY ROCKS

Uranium deposits occur in the Tesuque Formation (Tertiary) near San Jose, the Galisteo Formation (Eocene) in the Hagan Basin, and the Todilto Formation (Jurassic) at Box Canyon (Fig. 1). Small amounts of uranium and vanadium have been produced from the Tesuque and Todilto deposits (Table 2). All of these deposits are small, low-grade, and uneconomic, although the Galisteo deposits may have a potential should the uranium market improve.

Small and scattered uranium and vanadium occurrences are present in sandstones and conglomerates of the Tesuque Formation. The sediments of the Tesuque Formation were derived from Precambrian rocks in the Sangre de Cristo Mountains and volcanic rocks in the Jemez Mountains during later stages of rifting (Hilpert, 1969). Uranium typically occurs as coatings around opal and chert grains, with organic debris, and in clay zones, and probably represents accumulations of uraniferous ground waters. Uranium may have been derived from the Sangre de Cristo Mountains, the Jemez volcanics, or alteration of detritus in the host rocks. It is unlikely that any large, economic ore deposits could occur in these sediments because (1) past production was small and low-grade (Table 2), (2) known occurrences are low-grade (Hilpert, 1969), (3) organic material needed to trap uranium is not abundant in the Tesuque Formation, and (4) not enough time has elapsed to form economic deposits.

Uranium mineralization occurs in high-energy, braided-stream sediments of a complex alluvial-fan sequence in the Galisteo Formation in the Hagan Basin, which is a structural basin within the Rio Grande rift (Moore, 1979). The Galisteo Formation consists of fluvial–lacustrine sandstones, siltstones, conglomerates, and volcanic tuffs and rests unconformably on the Cretaceous Mancos Shale. It is overlain by the Espinaso Formation volcanics and intruded by uranium-bearing latite dikes and sills (McLemore, 1983; Hilpert, 1969). Uranium was probably derived from these latite dikes and the Espinaso volcanics, and was redistributed as uraninite and coffinite in roll-type ore deposits (Moore, 1979). One of these orebodies, the Diamond Tail, is estimated to contain 0.9 million lbs (0.4 million kg) of U_3O_8 at an average grade of 0.09% U_3O_8 at depths of 3–122 m. High production costs, low-grade ore, environmental constraints, and a declining uranium market forced Union Carbide Corp. to abandon development of uranium mineralization in this area.

A small uranium deposit occurs in the Todilto Limestone at Box Canyon (Hilpert, 1969; Chenoweth, 1974b). Uranium and vanadium are associated with intraformational folds in the limestones, and the deposit is similar to other limestone deposits in the Grants district (McLemore, 1983; Rawson, 1980). These deposits probably formed during the Jurassic and are not genetically related to the Rio Grande rift.

BARITE–FLUORITE–GALENA VEIN DEPOSITS

Silver has been produced as a byproduct from barite–fluorite–galena veins elsewhere in New Mexico (North, 1983; North and McLemore, 1984), but not from deposits along the northern Rio Grande rift. Barite–fluorite–galena and fluorite–galena veins occur in three districts or areas in northern New Mexico (Fig. 1) and silver ranges up to 0.46 oz/ton (16 ppm) in these veins (Table 4).

These deposits occur as veins and minor replacement bodies along fault zones primarily in Pennsylvanian and Permian limestones and rarely in sandstones and siltstones. Silver tends to occur with galena and an increase in lead concentration is accompanied by an increase in silver (Table 4). One variety of mineralization contains dominantly barite and minor amounts of fluorite, lead, zinc, copper, and calcite, whereas another variety contains dominantly fluorite with minor galena, zinc, and copper (Table 4). Uranium is rarely present in these deposits, although one deposit, the Shakespeare in Bernalillo County (not described in this report), contains 0.005% U_3O_8.

Most of the barite and fluorite deposits in New Mexico occur along, or in the vicinity of, the Rio Grande rift. However, no significant barite deposits occur north of the Placitas district in the Sandia Mountains. The barite and fluorite deposits were formed by basin brines (sedimentary hydrothermal) during the early stages of the Rio Grande rift.

TABLE 4. Chemical analyses of selected samples of barite–fluorite (± galena, silver) and fluorite (± galena, silver) veins in northern New Mexico. [1]—Trace of gold (<0.7 ppm); [2]—0.02 oz/ton Au (0.7 ppm).

Location	Name of Mine	Ag oz/ton (ppm)	% Pb	% Zn	% Cu	% $BaSO_4$	% $CaCO_3$	% CaF_2
Placitas district								
SW1/4 5 T12N R5E	Victo-Roco	0.20 (7)	0.21	0.0006	0.001	52.51	9.36	10.99
NW1/4 34 T13N R5E	Montezuma[1]	0.10 (3)	0.19	0.001	0.004	--	0.85	62.41
NW1/4 9 T12N R9E	Sec. 9	0.46 (16)	4.99	0.65	0.0007	18.3	0.40	38.88
SW1/4 6 T11N R5E	La Luz[2]	0.20 (7)	0.23	0.13	0.12	0.70	5.24	58.90
Edgewood								
NW1/4 5 T9N R7E	Shockley	0.20 (7)	0.74	0.001	0.002	67.72	0.97	16.46
NW1/4 5 T9N R7E	Shockley[1]	0.08 (3)	0.34	0.11	0.001	66.67	3.02	13.07
Crow Butte (El Cuervo)								
22 T10N R10E	Section 22	0.20 (7)	0.0006	0.0005	--	0.12	--	--
23 T10N R10E	Section 23[1]	0.22 (8)	2.31	0.001	--	27.23	--	--

1 Trace of gold (<0.7 ppm)
2 0.02 oz/ton Au (0.7 ppm)

STRATABOUND SEDIMENTARY COPPER DEPOSITS

Although stratabound sedimentary copper deposits (red-bed deposits) are not genetically related to the Rio Grande rift, several of them occur in the vicinity of the rift (localities 18–23, Fig. 1). In the Rio Grande rift area, stratabound copper deposits occur in the Permian Abo and Cutter Formations and the Triassic Chinle Formation. Minor occurrences are also present in the Pennsylvanian Madera and Permian Yeso Formations. In addition to copper mineralization, these deposits contain varying amounts of silver, uranium, vanadium, and rarely lead, zinc, and gold (Table 5). Silver contents average about 0.5 oz/ton (17 ppm), but may range in value from 0.03 to 20 oz/ton (1 to 600 ppm) Ag. Uranium values range from trace amounts up to 0.10% U_3O_8 (Table 5).

TABLE 5. Chemical analyses of selected samples from stratabound sedimentary copper deposits in northern New Mexico. No gold was detected in any of these samples.

Location	Name	% Cu	% U_3O_8	Ag oz/ton	(ppm)
Gallinas					
NW1/4 20 T23N R1E	Max Jacque	0.25	0.05	0.03	(10)
Coyote					
NW1/4 12 T21N R2E	Section 12	35.4	0.14	20.5	(703)
SE1/4 12 T21N R2E	Jarosa	0.22	0.06	0.3	(10)
Nacimiento					
NW1/4 35 T18N R1E	Deer Creek	5.92	0.144	0.88	(30)
Jemez Springs					
SW1/4 3 T17N R2E	Spanish Queen West	3.49	0.006	1.06	(36)
SE1/4 3 T17N R2E	Spanish Queen East	4.90	0.018	0.62	(21)

These mineral deposits were probably derived from: (1) erosion of Precambrian rocks enriched in these metals; (2) erosion of Precambrian mineral deposits; or (3) leaching of detritus in fluvial sediments (LaPoint, 1974).

DEPOSITS IN PRECAMBRIAN ROCKS

Twining and Red River Districts

Vein deposits of similar composition are restricted to Precambrian rocks in the Twining (#24, Fig. 1) and Red River districts. The veins contain quartz, chalcopyrite, malachite, azurite, pyrite, and minor gold and silver. They are probably of both Tertiary and Precambrian ages (Clark and Read, 1972, p. 109). Uranium occurs in some of these veins, and has been produced from the Black Copper mine (Table 2).

Hopewell and Bromide Districts

The lode deposits of the Hopewell (#4, Fig. 1) and Bromide (#25, Fig. 1) districts are similar and best considered together. The deposits are quartz fissure veins and replacement veins in schists of the Precambrian Moppin Metavolcanic Series. The replacement veins are from a few to about 70 cm wide and contain auriferous pyrite, chalcopyrite, galena, sphalerite, fluorite, and uranium. The quartz fissure veins are 1–100 cm wide and contain quartz, siderite, limonite after pyrite, and gold (Bingler, 1968).

The age of mineralization is uncertain. The replacement veins are structurally controlled by schistosity and hence the mineralization is post-metamorphic. No mineralization cuts Tertiary rocks, but mineralized clasts are found in the early Tertiary Ritito Conglomerate (Bingler, 1968). The mineralization is probably Precambrian, as deduced from the following lines of evidence: (1) the mineralization is restricted to Precambrian rocks; (2) there is no evidence of Tertiary intrusive-igneous activity in the area; and (3) the overall regional tectonic setting favors a Precambrian age over an early Tertiary age. In the Hopewell district uranium occurs with fluorite and is probably also of Precambrian age (Goodknight and Dexter, 1984a).

Placitas District

Minor base-metal and fluorite veins containing silver and traces of gold occur along faults in the Precambrian Sandia Granite (Kelley and Northrup, 1975). These veins, including the La Luz mine, are probably related to barite–fluorite, fluorite, and base-metal veins previously discussed.

Nacimiento Mountains

Although no production of silver, gold, or uranium has been reported from deposits in Precambrian rocks in the Nacimiento Mountains, several minor prospects and occurrences are found in these rocks. Copper mineralization, possibly with gold and silver, occurs in a schist xenolith in Precambrian granite at the Chalcocite prospect near the Nacimiento copper mine (Lindgren and others, 1910). Oxidized copper minerals and free gold occur in a 1 m wide vein along a shear zone in Precambrian granite at the Morningstar prospect south of the Nacimiento copper mine (R. H. Weber, field notes 1961; Woodward and others, 1972). Minor radioactive anomalies occur in Precambrian granite in the southern Nacimiento Mountains (Easton, 1955).

Costilla Massif

Uranium occurrences and geochemical anomalies in stream sediments occur in the Costilla massif (Precambrian) in northern Taos County (#26, Fig. 1). Uranium mineralization occurs along faults in the coarse-grained granite and in Precambrian pegmatites. Uranosilicates are common to many mineralized samples and chemical analyses up to 0.38% U_3O_8 are reported (Goodknight and Dexter, 1984). No sulfide mineralization is known to occur in the area.

The age of the mineralization is problematical. The Costilla Granite is anomalously high in uranium (Goodknight and Dexter, 1984), indicating a potential source. However, the massif may have been covered by Tertiary volcanics of the Amalia Tuff, which could have released uranium during weathering (Goodknight and Dexter, 1984b).

Picuris District

Mineralization in the Picuris district (#27, Fig. 1) consists of quartz veins, disseminated sulfides, and mineralization filling small fractures in quartz veins and quartzite. The mineralogy of all of these is similar, consisting of varying amounts of malachite, chrysocolla, and chalcocite with small amounts of gold, silver, and uranium (Williams, 1982). Tungsten (wolframite) has also been produced.

The ore deposits are restricted to the Precambrian Ortega Quartzite. The quartz veins and disseminations show some evidence of metamorphism. The mineralization is probably epigenetic and formed during the Precambrian (Williams, 1982).

Santa Fe District

In the Santa Fe district (#28, Fig. 1) the mineralization is disseminated in Precambrian biotite-rich phyllite and two-mica schist and consists of chalcopyrite, pyrite, bornite, pyrrhotite, and magnetite with rare sphalerite, tennantite, chalcocite, and covellite (Fulp, 1982). The formation of these deposit was associated with submarine-shelf volcanic activity (Fulp, 1982) during late Precambrian time.

ASSOCIATION OF GOLD, SILVER, AND URANIUM

The association of gold and silver with uranium is not common, although uranium is found in some of the world's largest gold deposits. Among the most important sources of gold in the world are the lower Proterozoic conglomerates, which produce a substantial amount of uranium as well (Pretorius, 1981). Platinum-group metals, silver, and thorium also occur in substantial quantities in some of these ancient conglomerates. Lower Proterozoic rocks are absent in New Mexico, and there is no potential for these gold–uranium deposits in the state. An unusual copper–uranium–gold deposit occurs in the hematite-rich sedimentary breccias of the Olympic Dam Formation (Precambrian) at Roxby Downs, Australia (Roberts and Hudson, 1983). A unique deposit at the Virgen de Guadalupe mine, Chihuahua, Mexico, contained radiating "sunbursts" of native gold in spheroidal masses of pitchblende (Leach, 1946). Uranium is also associated with a few of the Canadian gold deposits, but is rare in most of the Canadian precious-metal districts (Boyle, 1979).

In New Mexico, uranium is typically associated with copper deposits containing gold and silver. Some gold–quartz veins may also contain uranium minerals (McLemore, 1983). Most of the stratabound sedimentary copper deposits contain minor amounts of silver, uranium, and vanadium. Gold is rare in these deposits. Many of the copper deposits in Precambrian rocks, and vein deposits associated with Tertiary volcanic and intrusive rocks also contain varying amounts of gold, silver, and uranium (Table 1).

The association of copper, gold, silver, and uranium may be a result of a common source, mixing, and similar transport–deposition conditions. Much of the gold and silver is possibly derived from hydrothermal systems associated with Tertiary intrusives. Although some uranium may be derived from the same systems, most of uranium is probably in ground waters where it is leached from Precambrian rocks, older uranium deposits, and Tertiary volcanics. Copper may be derived from both hydrothermal systems and ground waters. The high heat flow associated with the Rio Grande rift and Tertiary intrusives would raise the temperature of the uranium- and copper-rich ground waters. Mixing of the heated ground waters with gold- and silver-bearing (with or without copper) hydrothermal fluids along faults produced by rifting could result in the copper, gold, silver, and uranium association.

ECONOMIC POTENTIAL

Precious Metals

Some of the gold deposits in northern New Mexico could have significant economic potential. Production continues at the Ortiz mine and from several small placer deposits. Gold and silver veins and base-metal deposits have a probable mineral-resource potential in the Red River district, Taos County (Ludington and Brown, 1984), and silver veins also have a probable mineral-resource potential in the Placitas district, Sandoval County (Hedlund and Kness, 1984). Both the districts are located in Federal Wilderness areas. The mineral-resource potential for gold and silver deposits in the Cochiti, Elizabethtown–Baldy, Old and New Placers, Cerrillos, and Hopewell districts is moderate to high; however, additional geologic work is needed in many of these areas.

Uranium

The mineral-resource potential for uranium in northern New Mexico is poor partly due to the present economic decline of the uranium industry. Most of the deposits in northern New Mexico are insignificant because of small size and low grade. However, should economic conditions improve, the uranium deposits in the Galisteo Formation in the Hagan Basin and at La Bajada could have a high potential. The U.S. Department of Energy (1980) estimates that probable resources of 185 tons (168 mt) of U_3O_8 at $30/lb exist in the Hagan Basin.

ACKNOWLEDGMENTS

This report is part of an ongoing evaluation of gold and silver occurrences in New Mexico. We are appreciative of discussions with various Bureau staff members. Robert Weber, Frank Kottlowski, and Robert Eveleth critically reviewed the manuscript. Mr. Jim Gourdie, Manager of Gold Fields Operating Company–Ortiz kindly granted permission to photograph their operation, and Mr. David Rubio took time from his busy schedule to provide a mine tour. G. Robert Osburn provided photo enhancement. Chemical analyses were provided by Lynn Brandvold and associates.

REFERENCES

Akright, R. L., 1979, Geology and mineralogy of the Cerrillos copper deposit, Santa Fe County, New Mexico: New Mexico Geological Society, Guidebook 30, pp. 257–260.

Anderson, E. C., 1957, The metal resources of New Mexico and their economic features through 1954: New Mexico Bureau of Mines and Mineral Resources, Bulletin 39, 183 pp.

Armstrong, R. L., 1969, K–Ar dating of laccolithic centers of the Colorado Plateau and vicinity: Geological Society of America, Bulletin, v. 80, pp. 2081–2086.

Atkinson, W. W., Jr., 1961, Geology of the San Pedro Mountains, Santa Fe County, New Mexico: New Mexico Bureau of Mines and Mineral Resources, Bulletin 77, 50 pp.

Bachman, G. O., and Mehnert, H. H., 1978, New K–Ar dates and the late Pliocene to Holocene geomorphic history of the central Rio Grande region, New Mexico: Geological Society of America, Bulletin, v. 89, pp. 283–292.

Baldridge, W. S., Damon, P. E., Shafiqullah, M., and Bridwell, R. S., 1980, Evolution of the central Rio Grande Rift, New Mexico; new potassium–argon ages: Earth and Planetary Science Letters, v. 51, pp. 309–321.

Bingler, E. C., 1968, Geology and mineral resources of Rio Arriba County, New Mexico: New Mexico Bureau of Mines and Mineral Resources, Bulletin 91, 158 pp.

Boyle, R. W., 1979, The geochemistry of gold and its deposits: Geological Survey of Canada, Bulletin 280, 584 pp.

Buchanan, L. J., 1981, Precious metal deposits associated with volcanic environments in the Southwest; *in* Dickinson, W. R., and Payne, W. D. (eds.), Relations of tectonics to ore deposits in the southern Cordillera: Arizona Geological Society Digest, Tucson, v. XIV, pp. 236–262.

Bundy, W. M., 1958, Wall-rock alteration in the Cochiti mining district, New Mexico: New Mexico Bureau of Mines and Mineral Resources, Bulletin 59, 71 pp.

Chapin, C. E., 1979, Evolution of the Rio Grande Rift; *in* Riecker, R. E. (ed.), Rio Grande Rift: tectonics and magmatism: American Geophysical Union, Washington, D.C., pp. 1–5.

Chenoweth, W. L., 1974a, Uranium in the Petaca, Ojo Caliente, and Bromide districts, Rio Arriba County: New Mexico Geological Society, Guidebook 25, p. 315.

———, 1974b, Uranium occurrences of the Nacimiento–Jemez region, Sandoval and Rio Arriba Counties, New Mexico: New Mexico Geological Society, Guidebook 25, pp. 309–313.

———, 1979, Uranium in the Santa Fe area, New Mexico: New Mexico Geological Society, Guidebook 30, pp. 261–264.

Clark, K. F., and Read, C. B., 1972, Geology and ore deposits of Eagle Nest area, New Mexico: New Mexico Bureau of Mines and Mineral Resources, Bulletin 94, 152 pp.

Disbrow, A. E., and Stoll, W. C., 1957, Geology of the Cerrillos area, Santa Fe County, New Mexico: New Mexico Bureau of Mines and Mineral Resources, Bulletin 48, 73 pp.

Easton, W. W., 1955, Airborne radiometric survey of the Nacimiento Mountains and adjacent areas, Rio Arriba and Sandoval Counties, New Mexico, 1954–1955: U.S. Atomic Energy Commission, Technical Memorandum TM-291, 40 pp.

Elston, W. E., 1967, Summary of the mineral resources of Bernalillo, Sandoval, and Santa Fe Counties, New Mexico: New Mexico Bureau of Mines and Mineral Resources, Bulletin 81, 81 pp.

Fulp, M. S., 1982, Precambrian geology and mineralization of the Dalton Canyon volcanic center, Santa Fe County, New Mexico: M.S. thesis, University of New Mexico, 199 pp.

Goodknight, C. S., and Dexter, J. J., 1984, Evaluation of uranium anomalies in the southwest part of the Costilla massif, Taos County, New Mexico; *in* Reports on field investigations of uranium anomalies: U.S. Department of Energy, Report GJBX–1(24), pp. IV 1-IV 28.

Haji-Vassiliou, A., and Kerr, P. F., 1972, Uranium–organic matter association at La Bajada, New Mexico: Economic Geology, v. 67, pp. 41–54.

Hedlund, D. C., and Kness, R. F., 1984, Sandia Mountain Wilderness, New Mexico: U.S. Geological Survey, Professional Paper 1300, v. 2, pp. 833–835.

Heikes, V. C., 1913, Dry placers in New Mexico; *in* Mineral resources of the United States, 1912: U.S. Geological Survey, Washington, D.C.

Hickson, R. J., 1982, Heap leaching practices at Ortiz gold mine, Santa Fe County, New Mexico; *in* Interfacing technologies in solution mining: American Institute of Mining Engineers, New York, pp. 209–222.

Hilpert, L. S., 1969, Uranium resources of northwestern New Mexico: U.S. Geological Survey, Professional Paper 603, 166 pp.

Ishihara, S., 1967, Molybdenum mineralization at Questa mine, New Mexico, U.S.A.: Geological Survey of Japan, Report 218, p. 5.

Johnson, M. G., 1972, Placer gold districts of New Mexico: U.S. Geological Survey, Bulletin 1348, 46 pp.

Jones, F. A., 1904, New Mexico mines and minerals: The New Mexico Printing Co., Santa Fe, 346 pp.

Kelley, V. C., and Northrop, S. A., 1975, Geology of Sandia Mountains and vicinity, New Mexico: New Mexico Bureau of Mines and Mineral Resources, Memoir 29, 136 pp.

Koschmann, A. H., and Bergendahl, M. H., 1968, Principal gold-producing districts of the United States: U.S. Geological Survey, Professional Paper 610, pp. 200–210.

LaPoint, D. J., 1974, Possible source areas for sandstone copper deposits in northern New Mexico: New Mexico Geological Society, Guidebook 25, pp. 305–308.

Leach, F. I., 1946, Mexico's richest uranium mine: The Mining Journal, Jan. 30.

Lee, W. T., 1916, The Aztec gold mine, Baldy, New Mexico: U.S. Geological Survey, Bulletin 620, pp. 325–330.

Lindgren, W., Graton, L. C., and Gordon, C. H., 1910, The ore deposits of New Mexico: U.S. Geological Survey, Professional Paper 68, 361 pp.

Ludington, S., and Brown, S. D., 1984, Columbine–Hondo Wilderness Study Area and Wheeler Peak Wilderness, New Mexico: U.S. Geological Survey, Professional Paper 1300, v. 2, pp. 807–809.

Lustig, L. K., 1957, The mineralogy and paragenesis of the Lone Star deposit, Santa Fe County, New Mexico: M.S. thesis, University of New Mexico, 55 pp.

McLemore, V. T., 1983, Uranium and thorium occurrences in New Mexico; distribution, geology, production, and resources, with selected bibliography: New Mexico Bureau of Mines and Mineral Resources, Open-file Report 183, 950 pp.

Moore, J. C., 1979, Uranium deposits in the Galisteo Formation of the Hagan Basin, Sandoval County, New Mexico: New Mexico Geological Society, Guidebook 30, pp. 265–267.

North, R. M., 1983, History and geology of the precious metal occurrences in Socorro County, New Mexico: New Mexico Geological Society, Guidebook 34, pp. 261–268.

———, and McLemore, V. T., 1984, Silver and gold occurrences in New Mexico: New Mexico Bureau of Mines and Mineral Resources, Open-file Report 191, 28 pp., 1:1,000,000 scale map

Northrop, S. A., 1959, Minerals of New Mexico, revised edition: University of New Mexico Press, Albuquerque, 665 pp.

Park, C. F., Jr., and McKinlay, P. F., 1948, Geology and ore deposits of Red River and Twining districts, Taos County, New Mexico: New Mexico Bureau of Mines and Mineral Resources, Circular 18, 35 pp.

Parkison, G. A., Emanuel, K. M., Wronkiewicz, D. J., and Norman, D. I., 1984, Geology of the Cochiti mining district, Sandoval County, New Mexico (abstract): New Mexico Geological Society, Spring Meeting, Abstracts and Programs.

Pettit, R. F., Jr., 1946, Mineral resources of Colfax County: New Mexico Bureau of Mines and Mineral Resources, Open-file Report 15, 50 pp.

Pretorius, D. A., 1981, Gold and uranium in quartz-pebble conglomerates: Economic Geology, 75th anniversary volume, pp. 117–138.

Rawson, R. R., 1980, Uranium in Todilto Limestone (Jurassic) of New Mexico: example of a sabkha-lake deposit; *in* Rautman, C. A. (compiler), Geology and technology of the Grants uranium region 1979: New Mexico Bureau of Mines and Mineral Resources, Memoir 38, pp. 304–312.

Roberts, D. E., and Hudson, G. R. T., 1983, The Olympic Dam copper–uranium–gold deposit, Roxby Downs, South Australia: Economic Geology, v. 78, pp. 799–822.

Robinson, G. D., Wanek, A. A., Hays, W. H., and McCallum, M. E., 1964, Philmont Country, the rocks and landscape of a famous New Mexico Ranch: U.S. Geological Survey, Professional Paper 505, 152 pp.

Schilling, J. H., 1960, Mineral resources of Taos County, New Mexico: New Mexico Bureau of Mines and Mineral Resources, Bulletin 71, 124 pp.

Silliman, B., Jr., 1880, Report on the newly discovered auriferous gravels of the upper Rio Grande del Norte in the counties of Taos and Rio Arriba, New Mexico: Herald Publishing House, Omaha, Nebraska, 34 pp.

U.S. Department of Energy, 1980, An assessment report on uranium in the United States of America: U.S. Department of Energy, Report GJO-111(80), 150 pp.

Warren, A. H., and Weber, R. H., 1979, Indian and Spanish mining in the Galisteo and Hagan Basins: New Mexico Geological Society, Special Publication no. 8, pp. 7–11.

Wells, E. H., and Wootton, T. P., 1940, Gold mining and gold deposits in New Mexico: New Mexico Bureau of Mines and Mineral Resources, Circular 5, 26 pp.

Williams, M. L., 1982, Geology of the copper occurrence of Copper Hill, Picuris Mountains, New Mexico: M.S. thesis, University of Arizona, 98 pp.

Woodward, L. A., and others, 1972, Geologic map of Cuba quadrangle, New Mexico: New Mexico Bureau of Mines and Mineral Resources, Geologic Map 25, scale 1:24,00.

Wright, A., 1983, The Ortiz gold deposit (Cunningham Hill)—geology and exploration; *in* Papers given at the precious-metals symposium, Sparks, Nevada, November 17–19, 1980: Nevada Bureau of Mines and Geology, Report 36, pp. 42–51.

GEOLOGY OF THE SOUTH FORK MOLYBDENUM OCCURRENCE, TAOS COUNTY, NEW MEXICO

DAVID M. JONES[1] and JAMES R. NORRIS[2]
[1]Department of Geosciences, University of Arizona, Tucson, Arizona 85721; [2]Noranda Exploration, Inc., P.O. Box 50326, Tucson, Arizona 85703

INTRODUCTION

The South Fork molybdenum occurrence is exposed along the South Fork of Rio Hondo in the Sangre de Cristo Mountains, Taos County, New Mexico (Fig. 1). South Fork is in the Rio Hondo (Twining) mining district approximately 16 km northeast of Taos and 16 km south of the Questa Molybdenum mine. The South Fork prospect lies entirely within the South Fork addition to the Wheeler Peak Wilderness Area in Carson National Forest.

History

Lindgren and others (1910) document the occurrence of molybdenite in quartz veins near the mouth of the Rio Hondo. In the same paper they mention reports of "claims showing galena and stibnite at the head of South Fork." Subsequent reports on the same area (Park and McKinlay, 1948; Schilling, 1960) do not mention molybdenite mineralization, though Clark and Read (1972) note pyrite mineralization and copper and zinc geochemical anomalies in the vicinity of South Fork Peak. Climax Molybdenum Co. apparently conducted limited geologic reconnaissance over exposed molybdenite mineralization at South Fork peak in the late 1960's (Harry Olson, oral comm. 1982). Intensive study of the area did not commence until 1981 when the U.S. Geological Survey published a report documenting a quartz–pyrite–molybdenite stockwork mineralization over a square kilometer near South Fork Peak (Ludington, 1981). Subsequent reconnaissance by Noranda Exploration, Inc. determined that molybdenite at South Fork is exposed over an area of roughly 4 km^2. In the fall of 1981 Noranda Exploration, Inc., as well as other groups, staked claims covering exposed veining. Reconnaissance geologic mapping and water, soil, and rock-chip geochemical sampling were also completed at this time.

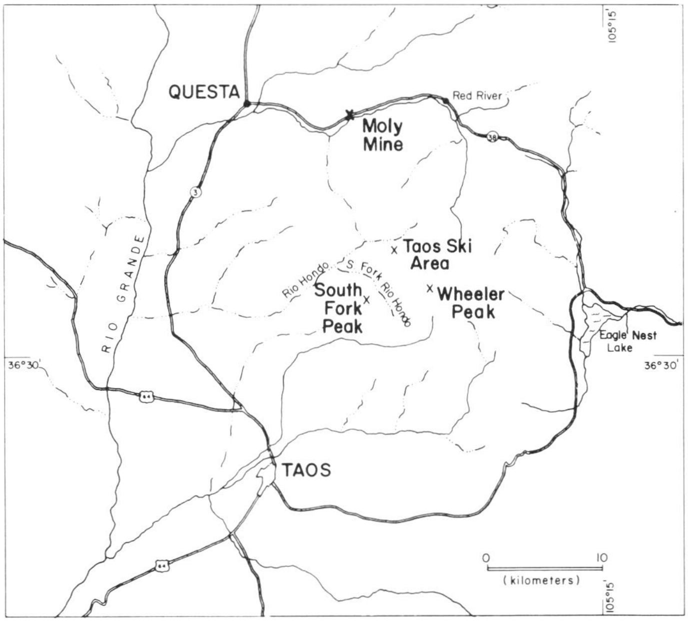

FIGURE 1. Location of map of the South Fork prospect.

In 1982 we carried our detailed (1:6,000) outcrop and float mapping at South Fork. Our objective was to evaluate the extent and controls of molybdenite mineralization and to characterize the geology and alteration. This paper presents the results of our studies.

REGIONAL GEOLOGY

Precambrian metaigneous and metasedimentary rocks overlain by mid-Tertiary volcanic rocks of the Latir volcanic field and intruded by mid-Tertiary batholithic granitic rocks constitute the lithologies in this part of the Sangre de Cristo Mountains (Fig. 2). Much of the Tertiary igneous activity is related to development of the Miocene Questa caldera, although post-caldera magmatism may be of a different genesis. The regional geology and tectonics of the Questa caldera are described by Reed and others (1983) and Lipman (1981, 1982). We will discuss only the salient features of the Questa caldera as they pertain to South Fork.

Intrusive Rocks

Quartz-monzonite to granite intrusions in the Rio Hondo area represent the southern extent of regionally extensive syn- and post-caldera magmatism in the Questa region (Fig. 2). Two major intrusive bodies

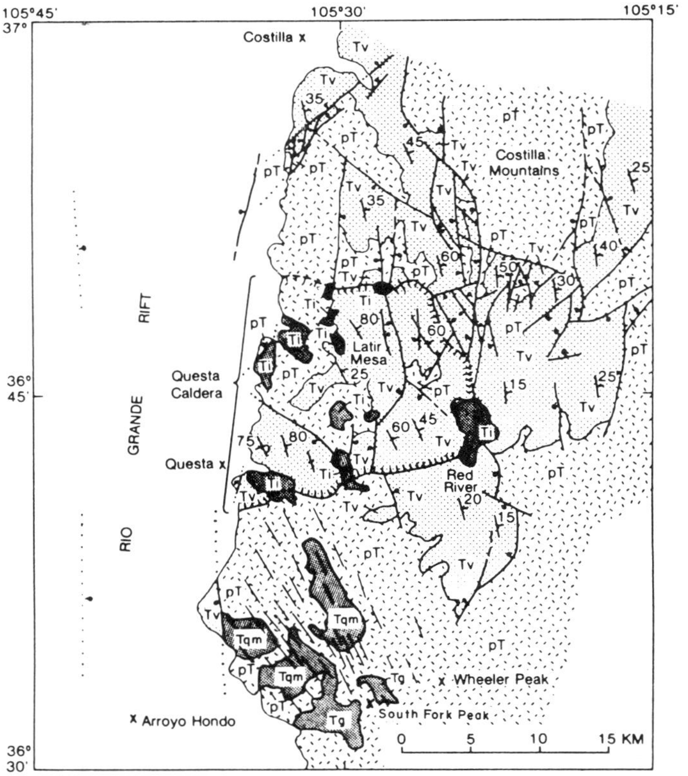

FIGURE 2. Generalized geologic map of the Questa region (from Lipman, 1981). Tg–Lucero Peak granite, Tqm–Rio Hondo quartz monzonite, Ti–undifferentiated Tertiary intrusive rocks, Tv–Tertiary volcanic rocks, pT–pre-Tertiary rocks.

are recognized, the Rio Hondo and Lucero Peak plutons. Quartz-monzonite of the Rio Hondo pluton is thought to be a southward extension of the more evolved, molybdenite-bearing Bear Canyon granite of the Bear Canyon pluton (Lipman, 1982). Granite of the Lucero Peak pluton occurs south of the Rio Hondo pluton and is exposed at South Fork. These batholithic rocks are considered cogenetic with the Latir volcanic field and the mineralized granites at Questa. Lipman (1982) reports K–Ar and fission-track age determinations of 20–26 m.y. for quartz-monzonite of Rio Hondo and 22 m.y. for granite of Lucero Peak.

A dike swarm of dominantly rhyolitic to quartz-latitic composition intrudes both the Rio Hondo and Lucero Peak plutons (Fig. 2). This dike swarm is a prominent feature at South Fork and in the region to the northwest. Characterized by northwest trends and local close spacing, the dike swarm can be tracked along strike for over 10 km. Lipman (1982) relates dike orientation to Rio Grande rift-related extension and emplacement of the batholithic rocks.

Mineralization

Molybdenum deposits in the Questa area are related to granites intruded along the southern structural margin of the Questa caldera. Molybdenum-ore zones occur within granite intrusions and their host volcanic rocks. Deposits are spatially associated with the upper contacts of causative intrusions. Both metallic mineralization and hydrothermal alteration are strongly controlled by local north–northwest fracture zones and east–northeast caldera-boundary structures (Leonardson and others, 1982).

LOCAL GEOLOGY

Lithologies

The South Fork molybdenum prospect occurs in regionally extensive Precambrian rocks intruded by mid-Tertiary biotite granite and variably porphyritic quartz-latite to rhyolite dikes. Figures 3 and 4 are a geologic map and cross section of the South Fork prospect.

Precambrian Rocks

Proterozoic amphibolite gneiss and quartz-monzonite gneiss comprise the high ridges and peaks along the south and east margins of the area. Similar Precambrian rocks are exposed at lower elevations along South Fork in the northern and western portions of the prospect. Gneissic layering and foliation in the Proterozoic rocks generally strike northeast, with moderate to steep dips to the northwest.

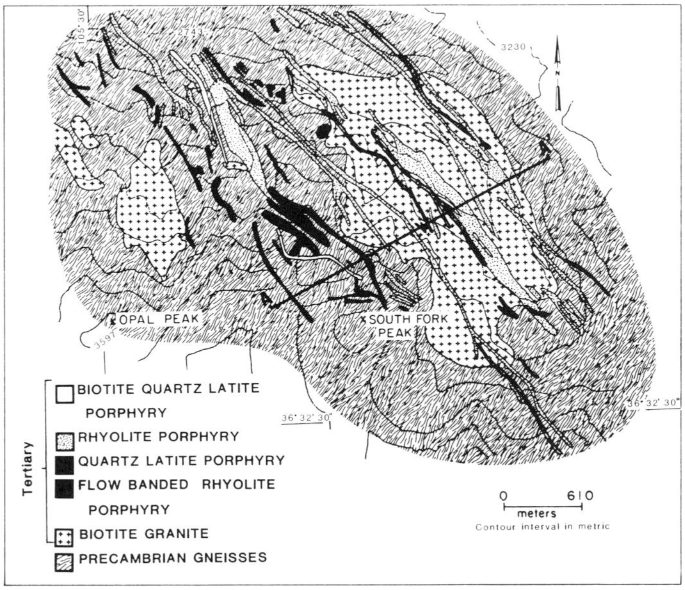

FIGURE 3. Simplified geologic map of the South Fork molybdenum prospect. (See Table 1 for lithologic descriptions of dike rocks.) The name Opal Peak is from Clark and Read (1972).

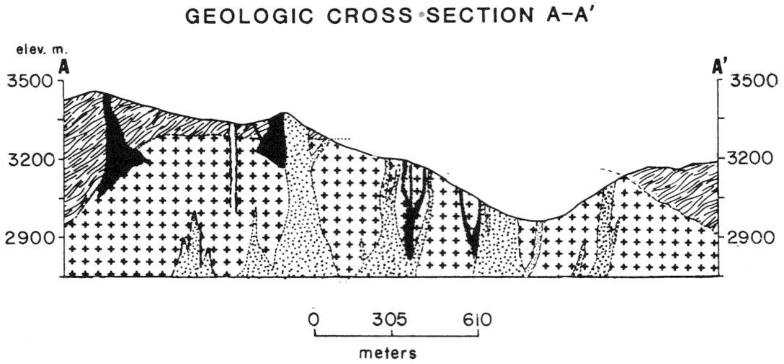

FIGURE 4. Generalized geologic cross section through South Fork. (Lithologic symbols are same as in Figure 3.)

Granite Intrusions

Biotite granite is exposed in two areas within the South Fork drainage (Fig. 3). In the eastern half of the project area a large (1.9 km^2) body of granite is exposed along upper South Fork and on the northeastern slopes of South Fork Peak. The apparent north–northwest orientation of this body may be due to the level of exposure revealed by the present drainage pattern. In the western half of the area a small (0.5 km^2) body of biotite granite crops out in the drainage immediately north of Opal Peak. Limited mapping indicates that biotite granite is also exposed at the western edge of the map area. The geometry of this westernmost intrusion is not known.

The southern contacts of the granite intrusions appear to be the topographic trace of a relatively flat roof. The eastern granite contact dips steeply to the north along its northern flank. Passive intrusion of the granites is indicated by the lack of contact breccias.

The biotite granite is texturally variable, ranging from porphyritic aplite and medium-grained porphyritic granite to medium- and coarse-grained granular granite. Systematic textural variations within the biotite granite were not delineated. Although aplite commonly occurs as a chilled rind at contacts with Precambrian rock, aplite bodies are also observed within granite far from contact zones. The megascopic mineralogy and texture of biotite granite in both outcrop areas are the same, suggesting they could be one intrusion.

Dike Rocks

The dominant trend of the dike swarm that transects South Fork is northwest (Fig. 3). Dikes in the northwest part of the prospect strike north to northeast. The dikes dip steeply, generally at angles greater than 60°. Dikes form up to 40% of the rock in some areas. Host rocks in these areas are present only as intradike slivers. Because they are more resistant to erosion than granite, dikes locally constitute more than 90% of the outcrop.

Our mapping delineated several dike types based upon modal mineralogy, texture, and crosscutting relations (Table 1). These dikes are depicted in simplified form in Figure 3. Flow-banded rhyolite forms the earliest dikes. These dikes intrude only Precambrian rocks above and proximal to granite contacts. They have not been observed cutting the granite. Flow-banded rhyolite is similar in composition and texture to aplitic phases of the granite. We therefore consider the flow-banded rhyolite dikes to be offshoots of the biotite granite. Quartz-latite porphyry, rhyolite porphyry, and biotite–quartz latite-porphyry dikes (Table 1) clearly cut the eastern exposure of biotite granite. Dikes are not well developed in the western outcrop of biotite granite. The western granite exposure is cut by a few thin dikes of quartz porphyry which exhibit a distinct porcellaneous texture. These dikes are not shown in Figure 3 due to their limited extent. Also not shown are two thin, northwest-trending dikes of andesite which cut the eastern granite.

Breccias

Several small breccias occur at South Fork, but are not shown in Figure 3. They occur as dike-like features or small pods which crosscut

TABLE 1. Lithologic description of Tertiary dikes at South Fork (based on megascopic examination only, rock names are field terms not based on petrographic or geochemical study; all dike rocks have an aphanitic groundmass). P–plagioclase, AF–alkali feldspar, Q–quartz, B–biotite, H–hornblende.

Unit	Distinguishing Features	Phenocrysts
Biotite-quartz latite porphyry	Biotite-rich (< 4%) character; dull gray-colored groundmass.	approximately 4% Q (1-5 mm) anhedral rounded, 9% AF (1-4 mm) subdedral to euhedral, 2%-4% B (0.5-2 mm) euhedral to subhedral.
Rhyolite porphyry	Pink to purple, but also exhibits cream-colored varieties. Q phenocrysts commonly rimmed by AF coronas. B poor.	8-20% AF (105 mm) subdedral to euhedral, 1-5% Q (1-4 mm) anhedral, 0-2% B subhedral.
Quartz-latite porphyry	Medium- to coarse-grained phenocrysts dominant, high Q:F ratio, and moderate B content; 'salt and pepper' texture.	2-6% Q (2-5 mm) anhedral, 6-9% AF (1-20 mm) subhedral to euhedral, 2-5% B+H (0.5-2 mm) euhedral to subhedral.
Flow-banded rhyolite	Flow-banding; Q phenocrysts are typically elongate, iron-oxide stained, and subhedral to glomeroporphyritic.	10-25% Q (1-7 mm) anhedral to euhedral, 10-15% AF (2-10 mm) subhedral to euhedral, 0-2% B (1-2 mm) subdedral to euhedral.

Precambrian rocks, biotite granite, flow-banded rhyolite, and rhyolite-porphyry dikes. Clasts are angular to subangular and range in width from millimeters to several centimeters. Breccia clasts are generally composed of the immediate host rocks. Drusy to massive quartz and fluorite are common matrix materials. Some clasts show quartz veining which does not cut the breccia matrix.

Alteration and Mineralization

Hydrothermal alteration is manifested as a variety of fracture-controlled vein assemblages. Pervasive, non-fracture-controlled alteration is not an extensive or volumetrically significant phenomenon.

Vein-controlled Alteration

The types and distribution of vein-controlled alteration are shown in Figure 5. The paragenetic sequence of veining was not determined due to the lack of exposed crosscutting relationships. Veinlet alteration is mostly restricted to Precambrian rocks, biotite granite, and flow-banded rhyolite dikes. Younger dike types possess only limited vein alteration consisting of fluorite ± gypsum veinlets and quartz veinlets with little or no pyrite.

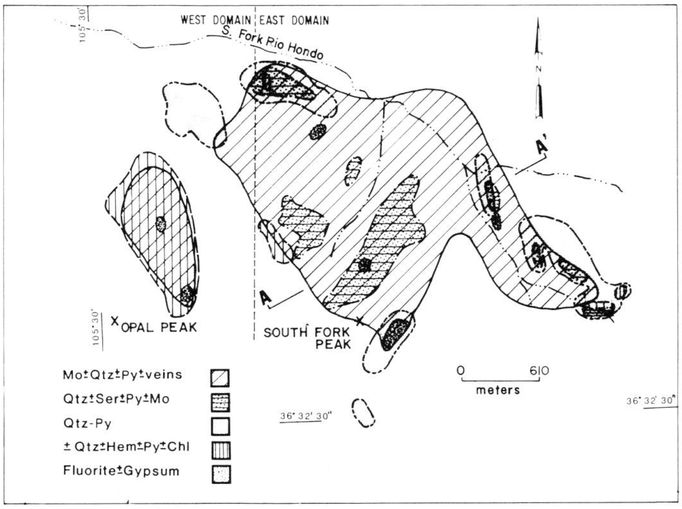

FIGURE 5. Generalized distribution of vein-controlled alteration at South Fork.

The bulk of vein-controlled alteration lies within the limits of exposed molybdenite veining (Fig. 5). Two broad areas of molybdenite-bearing veins are evident. These areas correspond with, but extend beyond, the east and west exposures of biotite granite.

Within the large eastern zone of molybdenite veining are isolated zones of quartz–sericite–pyrite ± molybdenite, quartz–pyrite, quartz–hematite–chlorite, and fluorite ± gypsum veins (Fig. 5). In the eastern granite, molybdenite-bearing veins are exposed over at least 183 m of relief within the granite intrusion. Orthoclase selvages on quartz–molybdenite veins are noted in only one area north of South Fork Creek at 2,957 m elevation.

The western zone of molybdenite-bearing veins is coextensive with well-developed quartz–hematite–pyrite veining (Fig. 5). Within the western granite, molybdenite veins occur over 213 vertical meters of exposure.

Veins are typically wide (>0.5 cm) and continuous (lengths >2 m). Quartz–molybdenite veins average 1–2 cm in width, but compound veins as much as 30 cm thick are present. Within the granite, rock between vein sets is commonly unaltered, spheroidally weathering granite. Only flow-banded rhyolite dikes and their host rocks exhibit many thin, discontinuous veins typical of "stockwork" deposits. Vein density in the granite is high (>2/m) in limited patches. The most extensive area of moderate to high vein density is in the western granite north of Opal Peak. Integrated vein densities greater than 1/1.5 m are exposed over large portions of the western granite and its Precambrian host rocks. The eastern granite exhibits overall low vein densities (<1/1.5 m). In both granite exposures, vein abundance does not exhibit any change with depth in the intrusion.

Pervasive Alteration

Pervasive argillization of feldspar along with associated bleaching and pyritization is best developed in the Precambrian rocks peripheral to flow-banded rhyolite dikes. This alteration is texturally destructive and typically results in 1–3% pyrite addition as disseminations and thin veinlets. The granite is not argillized to any significant extent.

In areas of closely spaced quartz–sericite–pyrite veins pervasive sericitization occurs locally as a result of overlapping sericite-vein selvages. However, some 1–3 m wide sericitized zones in the granite show no clear association with veining.

Sulfide Deposition

Molybdenite occurs dominantly as a constituent of quart–molybdenite–pyrite and quartz–sericite–pyrite–molybdenite veins. Molybdenite is also observed as medium-grained rosettes disseminated through aplite and fine-grained granite. However, this mode of occurrence accounts for little of the total molybdenite content of the rocks.

Pyrite-bearing veins account for most of the occurrence of pyrite. Disseminated and microveinlet-controlled pyrite occur in pervasively argillized rock and zones of highly veined flow-banded rhyoite, biotite granite, and Precambrian rock. Pyrite is locally found as disseminations and in quartz veinlets in rhyolite porphyry and quartz-latite porphyry dikes.

The average sulfide content of the surface rocks at South Fork is less than 1%. Sulfide (mainly pyrite) contents up to 5% are encountered in pervasively argillized zones and limited areas of closely spaced quartz–pyrite and quartz–pyrite–sericite veins.

Structure

The orientations of joints and veins at South Fork are shown in Figure 6. Vein orientations are shown for major vein sets regardless of mineral assemblage. Data from the east and west outcrops of biotite granite are plotted as separate domains to show the variation in structural fabric.

Veins in the east domain strike northwest to west–northwest. This same orientation is shown by a bimodal distribution of joint surfaces at N35W and N55W. The strike of joints and veins in the west domain is less systematic. Veins show a range of orientation from northwest through north–northeast. Joints exhibit a north–northeast maxima and less developed west–northwest and north–northwest directions.

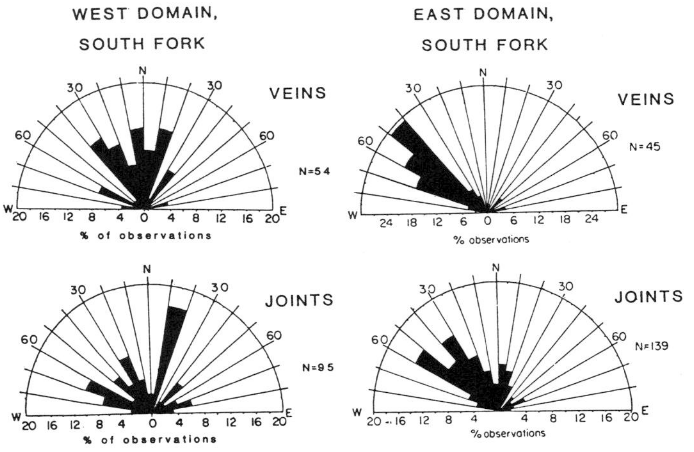

FIGURE 6. Strike rose diagrams for vein and joint orientations in the east and west domains. Figure 5 shows the boundaries for the east and west domains.

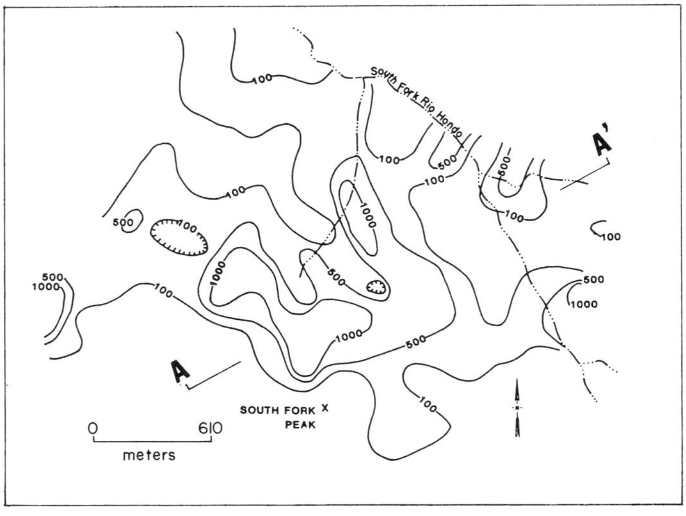

FIGURE 7. Molybdenum in oxalic-acid leachates of soil at South Fork. Concentrations in parts per million.

Veins and joint orientations along with dike orientations (Fig. 3) in the east and west domains indicate that two distinct structural fabrics exist in the granites at South Fork. The structural grain in Precambrian rocks does not appear to influence vein and joint orientations.

GEOCHEMISTRY

Soil Geochemistry

Soil samples were collected on 183-m centers from the area north of South Fork Peak and south of the South Fork Rio Hondo, covering only part of the area mapped. Soil samples were leached with oxalic acid as described by Alminas and Mosier (1975).

Elements exhibiting focused anomalies in the South Fork area include Mo, W, Bi, Cu, Pb, Be, and Y. As much as 5,000 ppm molybdenum is found in oxalic-acid leachates of soils at South Fork (Fig. 7). The highest anomalies center on the cirque immediately north of South Fork Peak. Q-mode factor analysis of soil-leachate geochemistry further substantiates the single-element anomalies. Factor analysis creates multi-element factors which can be interpreted as endmember components accounting for variability in the data. High-factor loadings indicate the degree to which samples contain the element suite of that factor. Figure 8 shows factor loadings for a tungsten–beryllium–bismuth–copper–molybdenum factor.

Hydrogeochemistry

Figure 9 shows the raw data for molybdenum in stream waters of the Rio Hondo drainage. This figure is based on more than 50 water samples from Rio Hondo, South Fork Rio Hondo, and primary and secondary tributaries of those streams. High concentrations of molybdenum (as much as 120 ppb) are present. For comparison, streams draining the area of the Log Cabin molybdenum orebody along the south margin of the Questa caldera contain an order of magnitude less molybdenum. The northwest trend of the anomaly correlates with dike-swarm, vein, and joint orientations in the eastern granite.

Q-mode factor analysis produced several elemental factors which account for variance in the hydrogeochemical data. Of these, two factors correlate with observed mineralization at South Fork and elsewhere in the Questa region. In part, these factors mimic the molybdenum distribution in Figure 9. They indicate high factor scores in the South Fork area and exhibit a northwest grain. Fluorine and uranium are important indicator elements of mineralization in the hydrogeochemistry of this region.

DISCUSSION

Interpretation of Alteration

Determining the origin of quartz–molybdenite, quartz–sericite–pyrite–molybdenite, and quartz–pyrite veins is critical in assessing the South Fork prospect. Much quartz–pyrite and some quartz–molybdenite veining appears genetically related to flow-banded rhyolite dikes and the granite, and apparently predates other igneous phases. However, some molybdenite veining has no obvious source. Molybdenite in the granite is not restricted to aplitic or fine-grained phases, or to its contact with Precambrian rocks. Molybdenite is found as much as 213 m below the granite contact in medium- to coarse-grained granite. The vertical continuity of the South Fork system is unknown. The unfractured and unaltered nature of the western granite between systematic vein sets is not typical of hydrofracturing related to a volatile-rich magma. However, a large volume of molybdenum-bearing fluid was required to produce the observed veining. A source other than the exposed biotite granite may be responsible for some of the molybdenite veining. While alteration of the post-granite dikes is not conspicuous, volatile content in quartz–latite–porphyry and rhyolite–porphyry dikes resulted in quartz ± pyrite and fluorite ± gypsum veins. This demonstrates that the later magmas exsolved silica and fluorine in hydrothermal fluids.

Interpretation of Structure

Joints, veins, and dikes in the east granite have northwest orientations indicating NE–SW extension (Fig. 6). This direction is compatible with the orientation of the regional stress field during Miocene time (Rehrig and Heidrick, 1976; Lipman, 1981). The orientation of joints, veins, and dikes in the western granite (Fig. 6) does not match those of the eastern domain. Although some northwest- to north–northwest-trending dikes are found, the west domain also shows a distinct north to north–northeast component. The northwesterly striking features may reflect northeasterly extension as exemplified by the east domain. The north to north–northeast fabric, however, is partly compatible with the orientation of Miocene and younger faults in the Latir volcanic field (Lipman, 1981, fig. 9A).

The different orientations of structural elements in the granites of the eastern and western domains indicate development in different stress regimes. Two possible reasons for variation in the stress regime are: (1) the east and west exposures of granite are temporally separate intrusions which crystallized under different regional stress conditions, and (2) the granites are essentially contemporaneous intrusions and the stress field at the time possessed local deviations.

The latter suggestion appears unlikely. The northwest-trending dike swarm in the eastern granite can be traced for 10 km with little variation. The dikes were emplaced in a regionally uniform stress field which allowed significant extension to accommodate the volume of dike rock intruded. Local perturbation of this stress field is not consistent with the areal extent of the structural fabric.

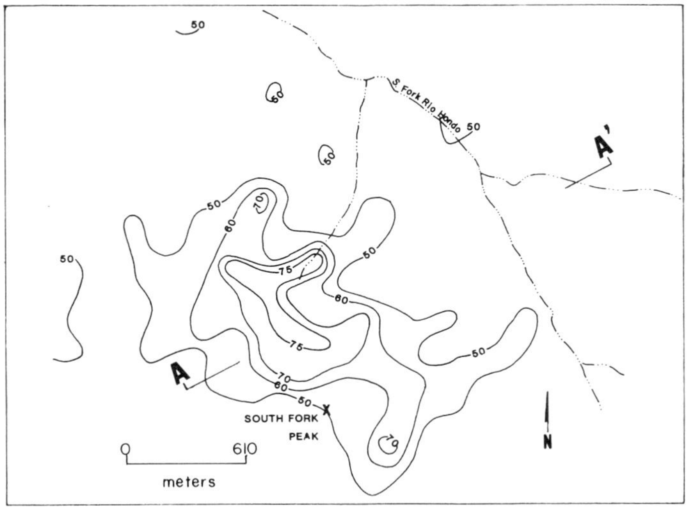

FIGURE 8. Tungsten–Beryllium–Bismuth–Copper–Molybdenum Factor for soil geochemistry at South Fork. High factor loadings indicate greater degrees of correlation between elements.

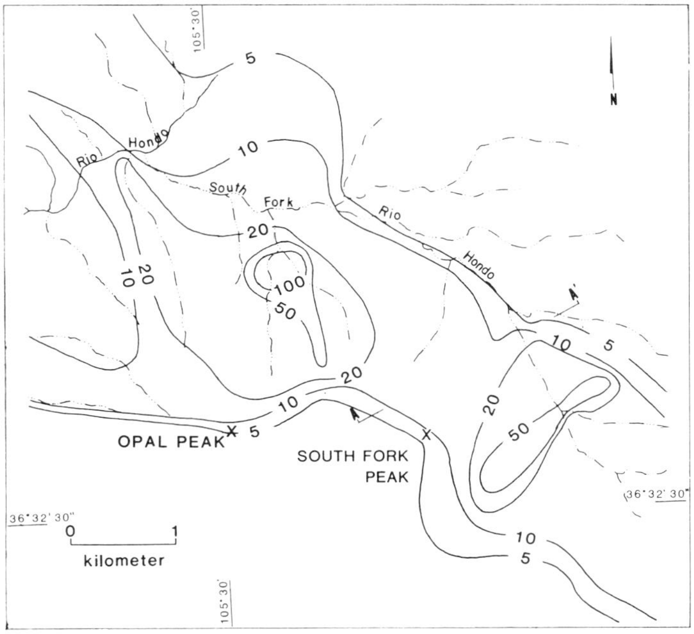

FIGURE 9. Molybdenum hydrogeochemistry of the Rio Hondo drainage. Concentrations in parts per billion.

The former suggestion bears consideration. The dike swarm which cuts the east granite also cuts, and is more widespread in, the Rio Hondo quartz monzonite (Fig. 2). These dikes do not intrude the main mass of Lucero Peak granite or the west granite at South Fork. This suggests a younger age for the west granite, possibly similar to the main mass of the Lucero Peak granite which lies further to the west.

Regional Controls

Unpublished mapping by Jones indicates that South Fork and other molybdenum prospects several kilometers to the northwest are localized along an arcuate structural zone. The concave northward, inward-dipping zone is defined by rhyolite dikes northwest of South Fork prospect and Tertiary faults, hydrothermal alteration, and intrusions northeast of the prospect. The senior author favors interpreting these features as being caldera-related. Additional work on this problem is presently in progress.

Genetic Model

We envision the following model for the South Fork molybdenum occurrence. The intrusions at South Fork are high-level parts of a composite batholith which rose to even higher levels in the Questa area. The first phase of the intrusive sequence is the biotite granite. During passive emplacement of the granite magma, volatile constituents (H_2O, Mo, F) concentrated in the upper portions of the intrusion. Brittle failure of the host rocks allowed injection of volatile-rich melt forming the flow-banded rhyolite dikes. As a consequence of rapid cooling and high volatile content, the flow-banded rhyolite possesses an aphanitic groundmass and closely spaced veins of quartz, molybdenite, and pyrite. Pressure drop and water loss within the magma caused portions of the granite melt to crystallize rapidly, forming aplites and porphyries with aplitic groundmass. Volatiles remaining in the melt after formation of the flow-banded rhyolite dikes produced disseminated molybdenite and at least some of the molybdenite-bearing veins in the aplites and granites. However, the remaining volatile content was not sufficient to induce extensive hydrofracturing. As a result, extensive stockwork veining did not occur at the magma–host rock content.

At least three distinct quartz-latite to rhyolite dike types intrude the east granite. This indication of ongoing magmatism suggests that granitic plutons were intruded at depth. While alteration of the post-granite dikes is limited, volatile content in the rhyolite-porphyry and quartz-latite-porphyry dikes is indicated by the presence of quartz ± pyrite and fluorite ± gypsum veins. Also, fluorite-bearing breccias of hydrothermal origin cut rhyolite-porphyry dikes, suggesting ongoing hydrothermal activity.

Though the age relationship between the eastern and western granites is speculative, the exclusion of the west granite from the structural trends and diking well developed in the eastern granite argues for a younger age. If so, multiple periods of intrusion and mineralization can be inferred. If the east and west granite are of the same age, veining in the granites may be of different ages as indicated by different structural fabrics.

Several potential modes of molybdenum deposition are provided by this model. First, molybdenite veining may be the result of fluids evolved from the biotite granite. This mechanism appears responsible for much of the molybdenite showing at South Fork. Second, molybdenum veining may persist at depth in association with the source(s) of the post-granite dikes or other undiscovered intrusions. The situation at South Fork may be analogous to the "intermineral" dikes at Climax which cut surface molybdenite veins but are mineralized at depth (White and others, 1981).

CONCLUSIONS

The following conclusions are based upon observations of the geology, alteration, and mineralization at South Fork.

(1) The South Fork prospect is a multiple intrusion system involving mid-Tertiary granitic magmatism.
(2) Subeconomic molybdenite occurrences are found at the surface over 4 km^2 of the prospect.
(3) Molybdenite veining is spatially associated with biotite granite and flow-banded rhyolite intrusions.
(4) Alteration is composed of a variety of vein-controlled assemblages.
(5) Dikes of quartz-latite porphyry and rhyolite porphyry cut molybdenite-bearing veins and most alteration zones. Quartz-latite-porphyry and rhyolite-porphyry dikes contain sparse quartz ± pyrite and fluorite ± gypsum veins suggesting derivation from a volatile-rich magma.
(6) The parent magmas of the post-biotite granite dikes and perhaps other unknown intrusions at depth may have produced some of the molybdenum mineralization we see on the surface.

ACKNOWLEDGMENTS

We thank Noranda Exploration, Inc. for allowing us to present these data, and Albert L. Lamarre for encouraging preparation of this paper.

Dave Jones particularly wishes to thank William D. Payne for support early in this work. Robert Peale contributed to the initial recognition of mineralization at South Fork.

REFERENCES

Alminas, H. V., and Mosier, E. L., 1975, Oxalic acid leaching of rock, soil and stream-sediment samples as an anomaly-accentuation technique: U.S. Geological Survey, Open-file Report 76-275, 25 pp.

Clark, K. F., and Read, C. B., 1972, Geology and ore deposits of Eagle Nest area, New Mexico: New Mexico Bureau Mines and Mineral Resources, Bulletin 94, 152 pp.

Leonardson, R. W., Dunlap, G., Starquist, V. L., Bratton, G. P., Meyer, J. W., Osborne, L. W., Atkin, S. A., Molling, P. A., Moore, R. F., and Olmore, S. D., 1982, Preliminary geology and molybdenum deposits at Questa, New Mexico; *in* The genesis of Rocky Mountain ore deposits: changes with time and tectonics: Exploration Geologists Society Symposium, Denver Region, November 1982, pp. 151–155.

Lindgren, W., Graton, L., and Gordon, C., 1910, The ore deposits of New Mexico: U.S. Geological Survey, Professional Paper 68, 361 pp.

Lipman, P. W., 1981, Volcano-tectonic setting of Tertiary ore deposits, southern Rocky Mountains; *in* Dickinson, W. R., and Payne, W. D. (eds.), Relations of tectonics to ore deposits in southern Cordillera: Arizona Geological Society Digest, v. 14, pp. 199–213.

———, 1982, The Miocene Questa caldera, northern New Mexico: relation to batholith emplacement and associated molybdenum mineralization; *in* The genesis of the Rocky Mountain ore deposits: changes with time and tectonics: Exploration Geologists Society Symposium, Denver Region, November 1982, pp. 133–148.

Ludington, S., 1981, Quartz–pyrite–molybdenite stockwork near South Fork Peak, Taos County, New Mexico: U.S. Geological Survey, Open-file Report 81-1080, 8 pp.

Park, C. F., Jr., and McKinlay, P. F., 1948, Geology and ore deposits of Red River and Twining districts, Taos County, New Mexico: New Mexico Bureau Mines and Mineral Resources, Circular 18, fig. 7 (scale 1:6,000).

Reed, J. C., Lipman, P. W., and Robertson, J. R., 1983, Geologic map of the Latir Peak and Wheeler Peak Wildernesses and Columbine–Hondo Wilderness Study Area, Taos County, New Mexico: U.S. Geological Survey, Map MF-1570B.

Rehrig, W. A., and Heidrick, T. L., 1976, Regional tectonic stress during Laramide and late Tertiary intrusive periods, Basin and Range province, Arizona: Arizona Geological Society Digest, v. 10, pp. 205–228.

Schilling, J. H., 1960, Mineral resources of Taos County, New Mexico: New Mexico Bureau of Mines and Mineral Resources, Bulletin 71, 124 pp.

White, W. H., Bookstrom, A. A., Kamilli, R. J., and Ganster, M. W., 1981, Character and origin of Climax-type molybdenum deposits: Economic Geology, 75th Anniversary Volume, pp. 270–316.

GEOLOGY OF THE COCHITI MINING DISTRICT, SANDOVAL COUNTY, NEW MEXICO

DAVID J. WRONKIEWICZ[1], DAVID I. NORMAN[1], GARY A. PARKISON[2], and KARL M. EMANUEL[2]
[1]Department of Geology, New Mexico Institute of Mining and Technology, Socorro, New Mexico 87801;
[2]NICOR Mineral Ventures, Inc., 2659-G Pan American Freeway, NE, Albuquerque, New Mexico 87107

INTRODUCTION

The Cochiti or Bland mining district is located in Sandoval County in north-central New Mexico. The district lies about 80 km (50 mi) due north of Albuquerque and 50 km (30 mi) due west of Santa Fe on the southeast flank of the Jemez Mountains (Fig. 1). Elevations in the district range from about 2,100 to 2,500 m. The topography in the area is very rough, with steep northwest–southeast-trending canyons and intervening ridges. Two main drainages occur in the area, Bland Canyon to the east and Colle Canyon to the west. The old townsite of Bland has only a few buildings still standing and about three permanent residents.

HISTORY

Earliest prospecting and mining activity in the Cochiti district dates back to 1893 when the first claims were located. Work at the two largest mines in the district, the Albermarle and the Lone Star, was concurrent and lasted from about 1897 until 1903. Some smaller mines in the area, notably the Iron King, Washington, and Crown Point, were also active at this time. It has been estimated that nearly 1,000 people inhabited the area of the Bland townsite with perhaps an additional 250 at the Albermarle townsite during this peak of activity. Prior to 1903, estimated production from the district was $695,000 in gold and $345,000 in silver. The best description of the geology is by L. C. Gration in Lindgren and others (1910). Other historical accounts of the district were published by Wynkoop (1900), Barbour (1908), and Statz (1912). More recent descriptions of the geology are by Bundy (1958), Elston (1967), and Stein (1983). The largest mine in the district is the Albermarle, where total past production may have been 105,000 tons (95,250 mt) at an average grade of about 0.22 oz/ton (6.8 g/t) gold and 4.0 oz/ton (124 g/t) silver. Total recovered value is estimated to have been $667,500. The mine was developed by an inclined shaft some 220 m deep and a total of about 1,500 m of drifts. One of the first cyanide mills in New Mexico was installed at the mine site, with a capacity of approximately 250 TPD.

Total past production from the Lone Star mine is estimated to have been 65,000 tons (59,100 mt) grading 0.20 oz/ton (6.2 g/t) gold and 4.0 oz/ton (124 g/t) silver. About 30,000 tons (27,250 mt) of this ore were produced between 1897 and 1903. The production from this period was either direct-shipped or processed at the Woodbury mill at the mouth of Bland Canyon. From 1914 to 1916 about 33,000 tons (30,000 mt) were mined and milled at the recently constructed Cossak mill. This 20-stamp cyanide leach facility was constructed adjacent to the Lone Star mine. Lesser activity from 1932 to 1936 resulted in about 2,000 tons (1,800 mt) of ore being direct-shipped to the El Paso smelter.

Total production from the Washington, Iron King, and Crown Point mines was typically 5,000 tons (4,500 mt) or less each. This ore was generally hand-sorted and direct-shipped. A small flotation mill was built near the Iron King mine in 1947. After processing a total of 12 tons of ore, the mill was sold for back taxes in 1951.

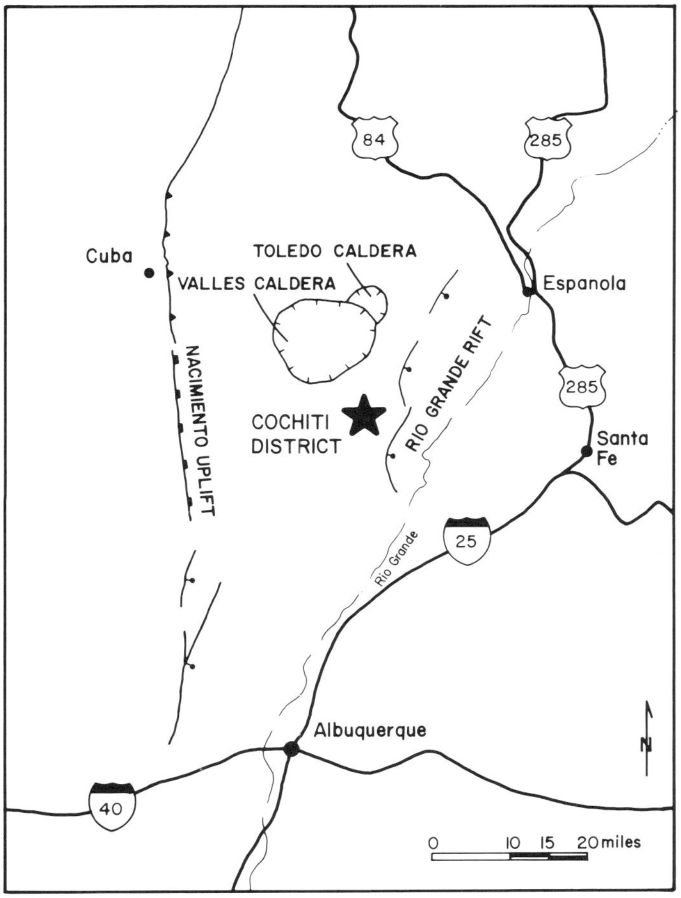

FIGURE 1. Map of north-central New Mexico showing location of the Cochiti district.

REGIONAL GEOLOGY

The Cochiti district is located on the southeast flank of the Jemez Mountains. This late Cenozoic volcanic landform owes its present topographic relief to the evolution of the Toledo and Valles calderas, as well as development of the Rio Grande rift and Nacimiento fault systems. Much of the volcanic stratigraphy underlying the Cochiti district was formerly interpreted to be Eocene and Oligocene in age (Bailey and others, 1969; Smith and others, 1970). Stein (1983) reports a single K–Ar date which suggests that the volcanic stratigraphy of the Cochiti district might actually be of Miocene age and thus significantly younger than previously estimated.

The informal Bland group (Stein, 1983) is composed of a predominantly hypabyssal and extrusive suite of gabbro, basalt, basaltic andesite, dacite, and volcanic breccias. This sequence has been intruded by a composite sequence of quartz monzodiorites consisting of an older granular intrusion which has been injected by a younger porphyritic dike swarm (Fig. 2). These rocks appear to reflect the emplacement history of a single intrusive event. Outcrop exposures of the Bland group are dominated by quartz-monzodiorite porphyries followed by intimately mixed basaltic andesite and dacite collectively referred to as andesite. Outcrops of arkosic sandstone have been interpreted to be older than any rocks of the Bland group and may represent xenoliths or roof pendants within the younger intrusive rocks.

Overlying and intruding the rocks of the Bland group are those of the Keres Group which has been subdivided into the Paliza Canyon Formation and the younger Bearhead Rhyolite (Smith and others, 1970). The Keres Group is petrogenetically similar to the Bland group, al-

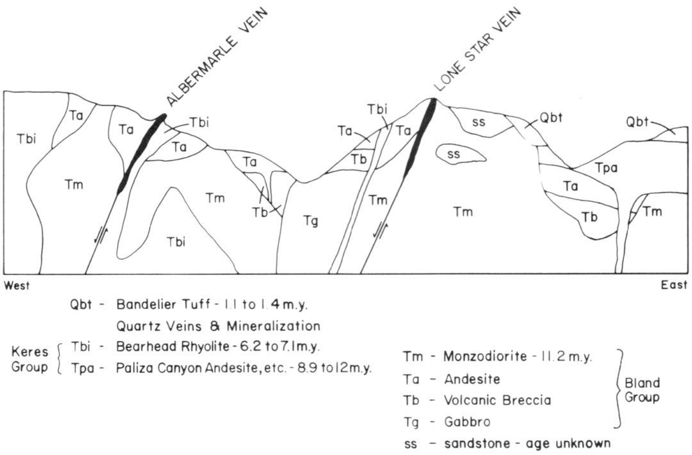

FIGURE 2. Diagrammatic cross section of the Cochiti district showing possible geologic relationships. Not to scale.

though stratigraphic continuity between the two groups has not been established as yet. The Paliza Canyon Formation generally consists of flow-banded rhyodacite, volcanic breccia, and massive andesite in about equal amounts. The Bearhead Rhyolite is generally seen to have intrusive contacts with older rocks. This unit is white to gray, flow-banded, and is generally restricted to the western part of the district.

The Tewa Group represents the latest episode of volcanism in the Jemez Mountains and is represented in the area by the Bandelier Tuff (Smith and others, 1970). The Bandelier has been subdivided into the basal air-fall and non-welded ash-flow deposits of the Otowi Member and the ovelrying welded ash-flow deposits of the Tshirege Member. The Tewa Group in the Cochiti district appears to occupy a paleovalley coincident with the present Bland Canyon. Quaternary surficial deposits cover large areas underlain by older units.

Geochronology of the rocks of the Cochiti district (Fig. 2) is based on correlations and radiometric-age data, and has recently been summarized by Stein (1983). From an extensive review of existing and new K–Ar dates, the Keres Group appears to range from 13 to 6 m.y. (Gardner and Goff, this guidebook). A single K–Ar date from quartz-monzodiorite porphyry (Bland group) of 11.2 m.y. suggests that the Bland group and Keres Group may, in part, be contemporaneous and could be locally equivalent (Stein, 1983; Gardner and Goff, this guidebook). K–Ar dates from 7.1 to 6.2 m.y. have been obtained from the Bearhead Rhyolite, establishing it as the youngest intrusive unit in the area (Bailey and Smith, 1978; Gardner and Goff, this guidebook).

The structure of the district is characterized by numerous generally north–south trending faults, fractures, and breccia zones. These dislocations may be related to the evolution of the Rio Grande rift, which has been active since 27 m.y.B.P. (Chapin, 1979) and currently forms the eastern margin of the Jemez Mountains. Areas of fractured and brecciated rock are ubiquitous throughout the district. The actual number of faults and the magnitude of fault displacements are difficult to determine due to the presence of widespread alteration, poor exposures, and the lack of reliable stratigraphic marker horizons. Brecciated margins between veins and wallrocks, and brecciated and locally resilicified quartz veins suggest that most veins have followed faults. Indeed, most faults have been noted by the presence of quartz veins. Faults are typically near-vertical or dip steeply to the west. Only rarely do faults cut through units of the Tewa Group; therefore, most faulting is older than 1.4 m.y.

In the Albermarle area, mineralized veins generally occupy a fault zone between andesite of the Bland group and the Bearhead Rhyolite (Figs. 2, 3). The observed relations suggest that the mineralization followed both the intrusion of the Bearhead Rhyolite (6.2 m.y.) and subsequent faulting. The volcanic rocks of the Tewa Group (1.4–1.1 m.y.) are not cut by the veins. These structural relationships suggest that mineralization occurred between 6.5 and 1.4 m.y. ago.

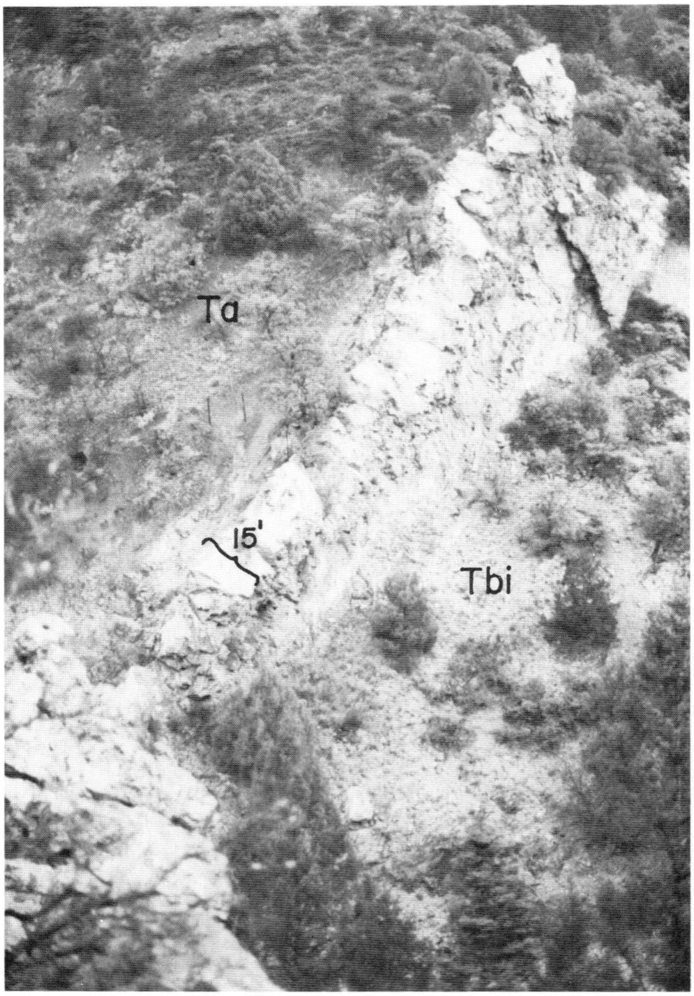

FIGURE 3. Photo of Albermarle vein just east of the Albermarle shaft. Ta – Andesite of the Bland group; Tbi – Bearhead Peak Rhyolite intrusion. (Photo by D. Wronkiewicz)

MINERALIZATION

Mineralization at the Cochiti district is fairly typical of epithermal quartz-vein-hosted gold–silver deposits (Buchanan, 1981). The mineralization is similar to better known epithermal districts such as Oatman (Arizona), Mogollon (New Mexico), and Creed (Colorado). Most of the quartz veins in the Cochiti district are quite similar, varying mostly in width and grade. Therefore, only a general description will be given for all veins of the district with variations noted where appropriate.

Veins generally strike north–south, but can vary, notably to a northeast trend at the Albermarle mine and a northwest trend at the Crown Point mine (Fig. 4). These flexures seem to localize ore shoots within dilatant zones, as do smaller bends at the Lone Star and Washington mines. Veins vary greatly in strike length, but the Lone Star vein system can be traced for over 3 km. This feature is the major system within the district and is locally composed of one or two subparallel veins, each up to 15 m in width, or a number of anastomosing smaller veins. Most mappable veins (more than 1 m wide) can be traced for at least 150 m. Veins generally dip steeply, ranging from about 65° W to 80° E. At least 13 separate significant veins or vein systems have been mapped for a total of 10 km of strike length.

Veins typically exceed 3 m in width and locally range up to 21 m or more when the silicified wallrock envelope is also included. Mining widths at the Albermarle mine were up to 18 m, at the Lone Star up to 14 m, and down to about 1 m at the Crown Point mine. Veins do not appear to be restricted to, or show a preferential development within,

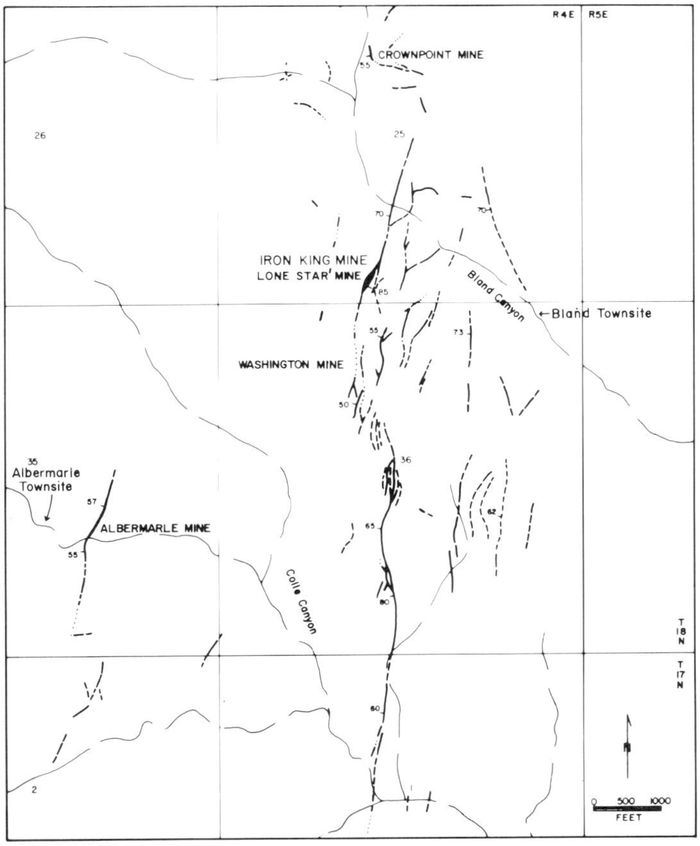

FIGURE 4. Map of veins in the central part of the Cochiti district.

specific type of wallrock. They are commonly brecciated, but are generally still cohesive. Multiple periods of quartz deposition followed by brecciation can be documented in many areas. Significant post-mineralization faulting has not affected the veins in most cases; when present, it has often been localized along the western wallrock contacts. Where hanging-wall stockwork zones are noted, precious-metal values appear to be restricted to quartz veinlets.

Mineralogy of the veins is quite simple. Most vein material averages 80–85% quartz and appears massive in outcrop. Coarse-grained calcite has been noted in drill holes. Quartz is generally white (although local amethyst has been locally noted), of a sugary texture and grain size, but can vary from chalcedonic to vug-lining crystals approaching 5 cm in length. Distinct banding in veins is uncommon; when present, it is generally manifested by changes in grain size of the quartz and by small, elongate vugs. Silicified wallrock horses and fragments are quite common and may locally exceed 50% by volume of some veins. Quartz-after-calcite textures have been noted, but are not common.

Metallic minerals in the veins are relatively sparse, rarely amounting to more than a few percent. Pyrite is the predominant sulfide phase and is often accompanied by related oxidation products. Only trace amounts of sphalerite, chalcopyrite, covellite, and galena have been noted. Native gold and electrum are generally not found to be hosted within, or directly associated with, pyrite. Silver minerals are predominantly argentite and proustite–pyrargyrite (ruby silver). Grain sizes of sulfides other than pyrite are generally less than 0.5 mm.

Paragenetic and cross-cutting relations suggest that a single mineralizing period occurred within the Cochiti district. While some veins show banded or layered quartz deposition, a consistent relation between varying pulses of quartz deposition and mineralization has not been demonstrated. Cross-cutting relationships within quartz veins cannot be definitely related to the timing of mineralization. Multiply brecciated and resilicified veins testify to at least three pulses of quartz deposition and brecciation. Paragenetic relations among the base-metal sulfides and the various precious-metal phases suggest near-simultaneous introduction and deposition, but with a slightly earlier base-metal stage followed by the introduction of gold and silver.

Higher grades of gold and silver mineralization are confined to irregular and vaguely defined ore shoots which generally appear to plunge vertically and have a larger vertical than horizontal dimension. Most production from the Lone Star and Albermarle mines came from single ore shoots. Higher grades are typically associated with greater vein widths. This suggests that ore shoots are related to dilatant zones along vein-filled faults which acted as conduits for ascending hydrothermal fluids.

It is difficult to find an outcrop within the district which has not been at least propylitically altered. More advanced alteration phases in wallrocks surrounding veins are generally present, but with quite variable widths (Bundy, 1958). Where alteration is most intense, an advanced argillic assemblage (dickite–kaolinite) occurs adjacent to veins and grades outward to an argillic zone (kaolinite–montmorillonite). Rarely present is a phyllic zone (sericite, quartz), which is found closest to the vein. Advanced argillic alteration is typically associated with the upper structural levels of veins systems, although some large areas of argillic alteration have been noted well away from known veins. In addition, certain rock types within the district are more susceptible to alteration than others; quartz monzodiorite is the most reactive, followed by andesite, dacite, and rhyolite.

FLUID-INCLUSION STUDIES

Three types of primary fluid inclusions were observed from quartz and calcite vein material. The most common type of inclusion contains a vapor-phase bubble occupying 2–30% of the inclusion volume. A second category of fluid inclusions, which has a higher vapor/liquid ratio (exceeding 90% by volume), is ubiquitous throughout the district and indicates boiling of the hydrothermal fluids (Roedder, 1979). A third type of inclusion is similar to the first type, but also contains an additional H_2S liquid phase. No true daughter minerals were observed in any of the inclusions.

Three hundred fifty filling temperatures and 100 freezing temperatures were determined from 34 doubly polished thick sections (Fig. 5). Overall homogenization temperatures ranged from 193 to 377°C, while salinities ranged from 0 to 5 wt% equivalent NaCl. Areas of higher-grade precious-metal values seem to correlate well with homogenization temperatures of 240–315°C.

Preliminary analyses of the vapor phase from all types of inclusions in vein material within previously mined areas show average compositions (in mole percent) of H_2S–3.50, organic compounds–3.30, CO_2–2.15, N_2–1.65, and H_2O–89.40.

SUMMARY AND ORE-DEPOSITION MODEL

A genetic model for ore deposition at the Cochiti district incorporates salient features from the above discussion as well as new information, and, hopefully, integrates them into a cohesive picture. Chronologically, the major events are:

(1) Deposition of the pre-Bland group arkosic sandstone which may have filled basins related to the early history of the Rio Grande rift, starting at 27 m.y.B.P. The sandstone is thought to have been later incorporated as xenoliths or roof pendants within the quartz-monzodiorite porphyry.

(2) The volcanic breccia and andesite, and hypabyssal intrusion of rhyodacite, andesite, and gabbro of the Bland group are similar to the rocks of the Paliza Canyon Formation of the Keres Group. Deposition of these units generally predated the intrusion of consanguineous quartz-monzodiorite stocks (11.7 m.y.), dikes, and sills within the district (Stein, 1983). The Bland group could, in part, be equivalent to the Paliza Canyon Formation (Gardner and Goff, this guidebook). These events may reflect the early stratovolcano-building stage of the Jemez Mountains volcanic field (Smith and others, 1970).

(3) Intrusion of the Bearhead Rhyolite (6.2–7.1 m.y.) mostly as subvolcanic stocks and dikes. These felsic intrusives acted as the heat source which drove a closed-cell ground-water convection system within the district (Berger and Eimon, 1982). The area affected by regional

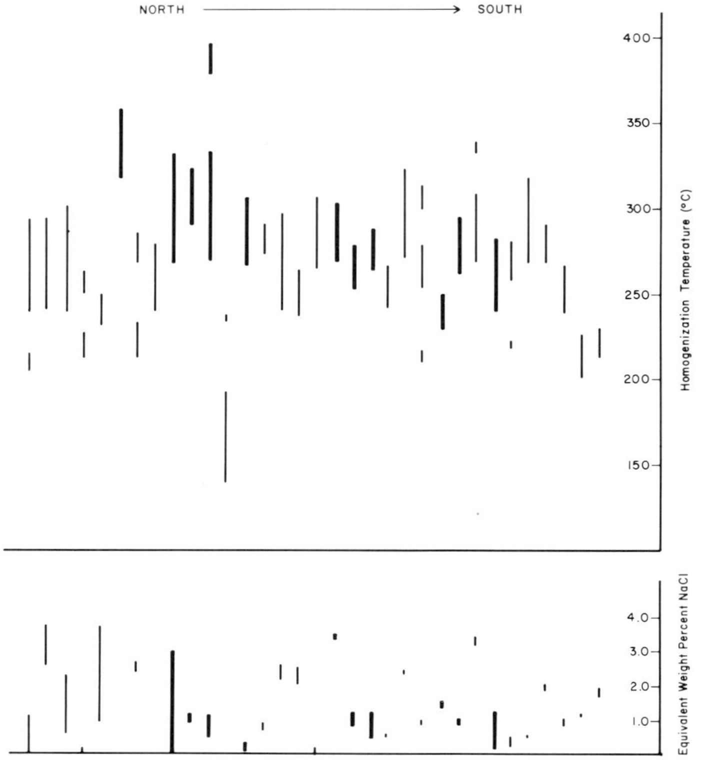

FIGURE 5. Fluid-inclusion homogenization temperatures and salinities from quartz and calcite vein material. Bars give ranges of temperatures and salinities observed. Thicker bars correspond to inclusions obtained from areas with higher-grade precious-metal mineralization.

propylitic alteration may mark the boundaries of this convection system. The main conduits for circulation were active tensional faults related to the Rio Grande rift. Scavenging of gold, silver, and quartz was accomplished by meteoric water as it circulated through the predominantly andesitic-composition stratigraphy.

(4) Hydrothermal fluids were typically quite dilute with an indicated 2% equivalent NaCl, which may reflect the concentration of organic compounds and H_2S alone. There may have been little NaCl in the solutions. Fluid inclusion studies suggest deposition of precious metals at or above the level of boiling at temperatures ranging from 240 to 315°C. Consequent effects of fluid boiling which may have caused deposition of metals and gangue minerals are oxidation, reduction of temperature and pressures, increase of pH, and the loss of H_2S (Buchanan, 1981). Gold and silver were probably transported as bisulfide complexes (Barnes, 1979).

(5) Episodic boiling of solutions resulted in the deposition of both gangue and ore minerals and, in conjunction with recurrent faulting, gave rise to multiply brecciated and banded veins as well as local silicification of wallrock. Dilatant zones or cymoid structures along faults localized extensive fluid flow, which resulted in the formation of ore shoots.

(6) Boiling of the hydrothermal fluids was associated with a decrease in pH above the level of ore deposition, which caused widespread argillic alteration peripheral to veins and particularly in hanging wall zones.

(7) Gradual cooling of the intrusive rhyolite caused a cessation of the Cochiti hydrothermal-covection system.

(8) This was followed by continued faulting and significant erosion interrupted by a second episode of volcanism related to the collapse of the Valles and Toledo calderas.

ACKNOWLEDGMENTS

The authors wish to thank NICOR Mineral Ventures, Inc. and Mr. and Mrs. J. L. Jones of Albuquerque for permission to publish this paper. We are also indebted to F. E. Goff, C. H. Maxwell, and W. S. Baldridge for constructive reviews of this manuscript. We are especially grateful to M. W. Selke for preparing the figures and manuscript.

REFERENCES

Bailey, R. A., and Smith, R. L., 1978, Volcanic geology of the Jemez Mountains, New Mexico: New Mexico Bureau of Mines and Mineral Resources, Circular 163, pp. 184–196.

———, ———, and Ross, C. S., 1969, Stratigraphic nomenclature of volcanic rocks in the Jemez Mountains, New Mexico: U.S. Geological Survey, Bulletin 1274-P, 29 pp.

Barbour, P. E., 1908, The Cochiti mining district, New Mexico: Engineering and Mining Journal, v. 86, pp. 173–175.

Barnes, H. L., 1979, Solubilities of ore minerals; *in* Barnes, H. L. (ed.), Geochemistry of hydrothermal ore deposits: John Wiley and Sons, New York, pp. 404–460.

Berger, B. R., and Eimon, P. I., 1982, Comparative models of epithermal silver–gold deposits: American Institute of Mining Engineers, Preprint no. 82-13, Society of Mining Engineers—American Institute of Mining Engineers Meeting, Dallas, Texas.

Buchanan, L. J., 1981, Precious-metal deposits associated with volcanic environments in the Southwest; *in* Dickinson, W. R., and Payne, W. D. (eds.), Relations of tectonics to ore deposits in the southern Cordillera: Arizona Geological Society Digest, v. 14, pp. 237–262.

Bundy, W. M., 1958, Wall-rock alteration in the Cochiti mining district, New Mexico: New Mexico Bureau of Mines and Mineral Resources, Bulletin 59, 71 pp.

Chapin, C. E., 1979, Evolution of the Rio Grande rift: a summary; *in* Riecker, R. E. (ed.), Rio Grande rift: tectonics and magmatism: American Geophysical Union, Washington, D.C., pp. 1–6.

Elston, W. E., 1967, Summary of the mineral resources of Bernalillo, Sandoval, and Santa Fe Counties, New Mexico: New Mexico Bureau of Mines and Mineral Resources, Bulletin 81, pp. 17–22.

Gardner, J. W., and Goff, F. E., 1984, Potassium-argon dates from the Jemez volcanic field: implications for tectonic activity in the north-central Rio Grande rift: this guidebook.

Lindgren, W., Graton, L. C., and Gordon, C. H., 1910, The ore deposits of New Mexico: U.S. Geological Survey, Professional Paper 68, pp. 150–162.

Roedder, E., 1979, Fluid inclusions as samples of ore fluids; *in* Barnes, H. L. (ed.), Geochemistry of hydrothermal ore deposits: John Wiley and Sons, New York, pp. 684–737.

Smith, R. L., Bailey, R. A., and Ross, C. S., 1970, Geologic map of the Jemez Mountains, New Mexico: U.S. Geological Survey, Miscellaneous Geological Investigations Map I-571, scale 1:125,000.

Statz, B. A., 1912, Geology of the Cochiti mining district, New Mexico: Mining Science, v. 6, pp. 276–277.

Stein, H. L., 1983, Geology of the Cochiti mining district, Jemez Mountains, New Mexico: M.S. thesis, University of New Mexico, Albuquerque, 122 pp.

Wyncoop, W. C., 1900, The Cochiti district, New Mexico: Engineering and Mining Journal, v. 70, pp. 215–216.

HYDROCARBON POTENTIAL OF THE ESPAÑOLA BASIN—A PROGRESS REPORT

BRUCE A. BLACK
Black Oil, Inc., P.O. Box 537, Farmington, New Mexico 87499

INTRODUCTION

In a series of earlier articles, Black (1979, 1982, 1984) discussed oil- and gas-exploration developments in the tectonic basins of the Rio Grande rift in central and north-central New Mexico. This article is a short progress report and an update on the most recent developments in the Espñola Basin portion of the rift.

DISCUSSION

The Española Basin, including the Hagan and Santa Fe embayments, is now the focus of active oil and gas exploration by at least two major and four smaller companies. Seismic programs and sporadic wildcat drilling are in progress, and a number of additional wells is planned for the summer of 1984.

Unlike the Albuquerque Basin to the south and the San Luis Basin to the north, the Española Basin lies directly on line with Cretaceous stratigraphic trends and with established production of the San Juan Basin in northwestern New Mexico. Figure 1 shows the major Cretaceous strandlines, as well as the producing areas of the San Juan Basin and their relationship to the Española Basin. It is obvious why the Española Basin is now the focus of a Cretaceous oil and gas play, a detailed discussion of which can be found in Black (1984).

Because of the somewhat different structural evolution of the Española Basin (Black, this guidebook), the basin also appears to have a different and perhaps greater potential for shallow oil and gas prospects in structures involving the Paleozoic and lower Mesozoic sections.

In virtually every well drilled to date in the Española Basin (Fig. 2) there have been oil and/or gas shows, the most recent of which were well-documented shows of live oil and gas. Several of the wildcat wells could have probably yielded commercial quantities of gas if there had been an existing pipeline system in the area. Up to 20 bbls of high-gravity oil per day, with no water, has reportedly been swabbed from the Cretaceous rocks in at least one recent well. There were collapsed casing problems with that hole, which now awaits a workover.

Only four of the 24 wells drilled since 1914 (Table 1) have penetrated

TABLE 1. Chronology of oil and gas tests in the Española Basin, including the Santa Fe and Hagan embayments.

Well	Section Township(N) Range	Total depth (ft)	Year completed	Deepest formation penetrated	Remarks
1. Tejon Oil & Develop. Co. No. 1	7-14-6	1850	1914	Santa Fe	Reported oil show at 1100 and 1087 ft.
2. Wood et al. Galisteo No. 1	7-12-10	3240	1921		Numerous oil and gas shows reported between 760 and 2715 ft.
3. Wood et al.	30-13-10	1540	1925		
4. E.M. Elliott No. 1	26-21-9	1760	1937	Penn.	Reported oil show at 775 and 1100-25 ft, gas shows at 1175-1200 ft.
5. Castle & Wigzell, Kelly Fed. No. 1	11-20-9	2703	1961	Penn.	Minor oil shows reported.
6. Eastland Oil Co., McKee No. 1	8-13-9	2092	1974	Morrison	Reported show of gas and oil.
7. CPGS Bicentennial No. 1 Federal	35-13-6	1416	1976	Morrison	Reported show of gas.
8. CPGS Bicentennial No. 2 Federal	35-13-6	1411	1976	Morrison	Reported show of gas.
9. CPGS Bicentennial No. 3 Federal	35-13-6	1112	1976	Morrison	Reported show of gas.
10. CPGS Bicentennial No. 4 Federal	35-13-6	1403	1976	Morrison	Reported show of gas.
11. Brent Exploration No. 1	29-13-6	1875	1978	Morrison	
12. Brent Exploration No. 3	19-13-6	2070	1978	Entrada	Reported show of oil.
13. Brent Exploration No. 5	19-13-6	769	1978	Dakota	Reported show of gas.
14. TransOcean Oil Co., McKee No. 1	4-13-9	8128	1978	Penn.	Reported show of gas; flared gas for Dakota test.
15. Whigham, Inc., Pinon-Davis No. 1	21-14-10	1514	1979	Dakota	Reported show of oil and gas.
16. Whigham, Inc., Pinon-Davis No. 2	21-14-10	1150	1979	Greenhorn	Reported show of oil and gas.
17. Pelto Oil Co., McKee No. 1	5-13-9	3100	1981	Morrison	Oil and gas shows in Cretaceous. Oil on ditch in Greenhorn. No tests.
18. Pelto Oil Co., Harmon No. 1	18-13-6	1398	1981	Entrada	Cored 50 ft of residual oil saturated sand in Dakota and spotty oil saturation in Entrada.
19. Pelto Oil Co., Ortiz No. 1	26-14-8	7450	1981	Morrison	Flared gas and recovered oil-cut mud on DST in Greenhorn. Ran casing and perforated Dakota - swabbed 20 bbls/day, 40° oil - no water. Plugged back to Greenhorn and perforated, swabbed 22 bbls/day 42° oil, no water. Now shut in.
20. Pelto Oil Co., Blackshare Federal No. 1	35-14-6	7018	1981	Chinle	Small gas shows in Mancos. Dakota faulted out. Cored spotty oil-stained sand in Entrada.
21. Pelto Oil Co., Pelto State No. 1	36-14-8	5504	1981	Morrison	Gas shows in Cretaceous.
22. CPGS Ferrill No. 1	7-13-9	3192	1981	Mancos	Large volumes of gas shows throughout Cretaceous Mancos, increasing to T.D. No DST's. Would probably be good gas well if pipeline were present.
23. CPGS Ferrill No. 2	5-13-9	1596	1982	Morrison	Numerous gas shows in Cretaceous. Large gas kick in Dakota. Well might be a small gas producer if pipeline were present.
24. CKS, Inc., Gianardi No. 1	22-15-8	7,000±	1983	Abo?	No data yet released.

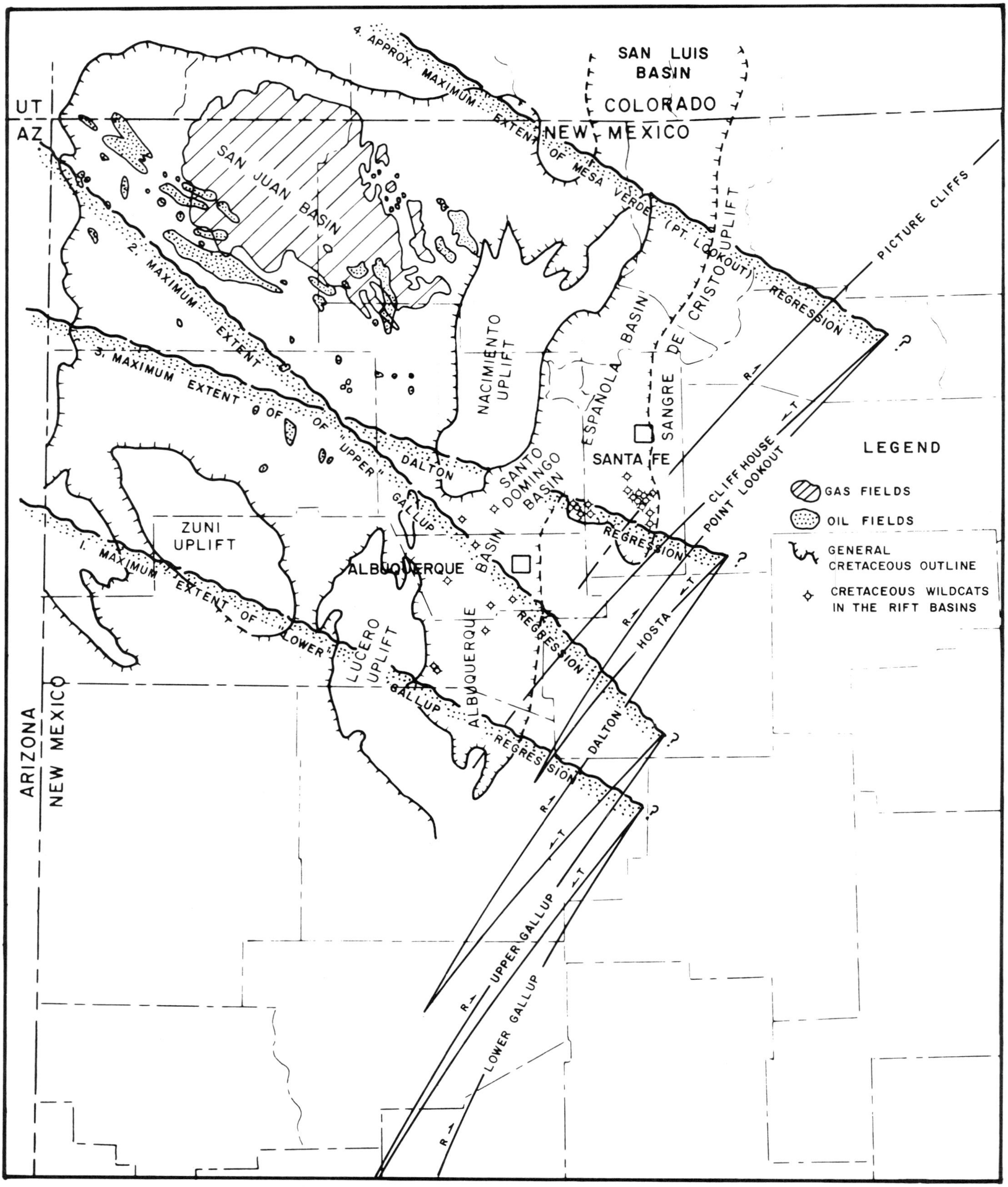

FIGURE 1. Strand lines in the San Juan Basin and in the Rio Grande rift basins.

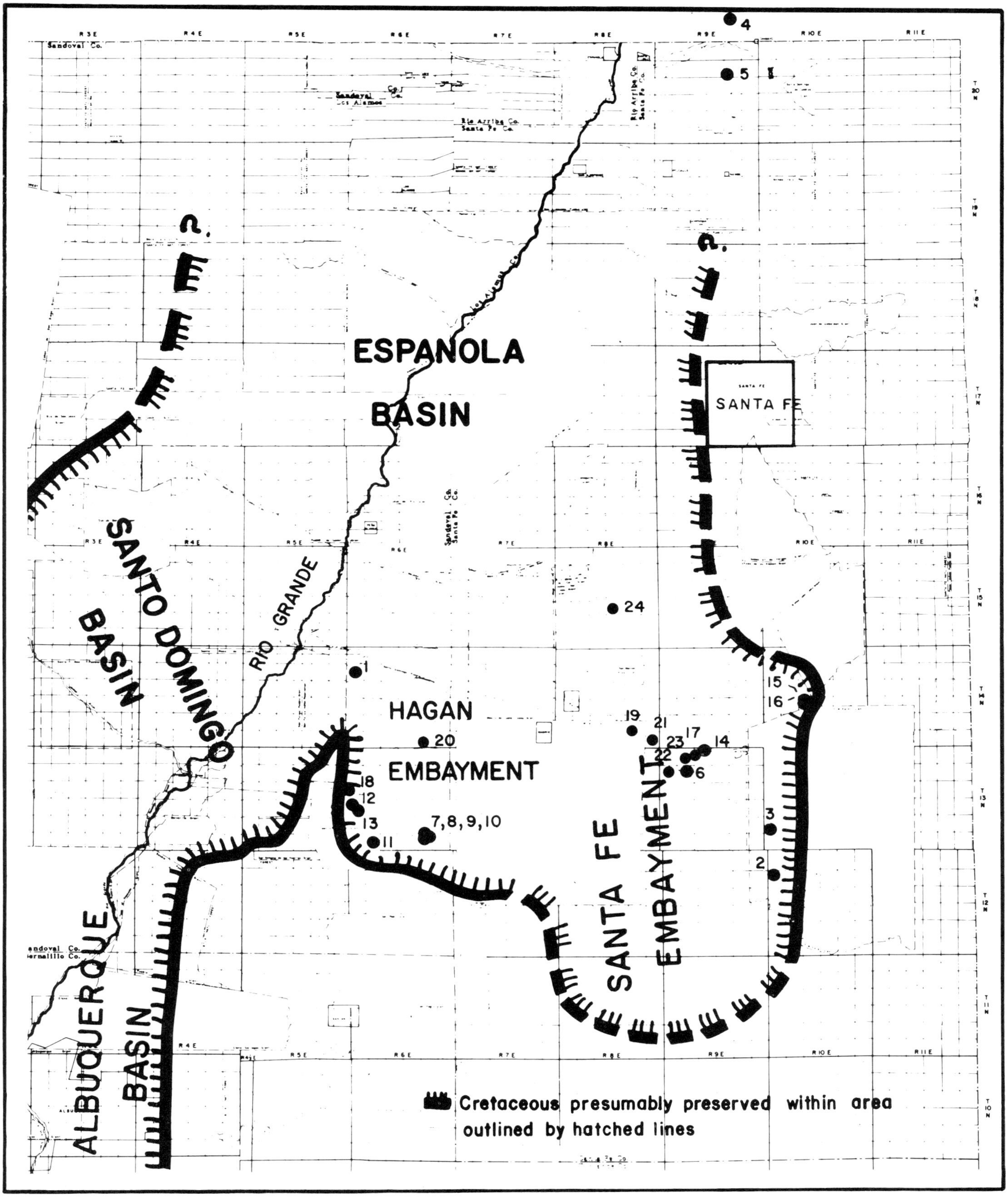

FIGURE 2. Location map showing oil and gas tests in the Española Basin. Numbers are keyed to Table 1.

Paleozoic rocks in the basin. Two of these are old wells and two are recent tests. Records from both old tests reported oil and gas shows from the Paleozoic section and at least one of the recent tests encountered gas shows.

CONCLUSIONS

Oil and gas exploration in the Española Basin will continue, and the primary objective will probably continue to be the Cretaceous section. Source-rock potential and maturation history for Paleozoic rocks—unlike those for the Cretaceous section—are still unknown; however, the potential for oil and gas structurally trapped in Paleozoic rocks appears to be significant and that concept will also be tested further.

REFERENCES

Black, B. A., 1979, Oil and gas exploration in the Santa Fe–Galisteo–Hagan area of New Mexico: New Mexico Geological Society, Guidebook 30, pp. 275–279.

———, 1982, Oil and gas exploration in the Albuquerque Basin: New Mexico Geological Society, Guidebook 33, pp. 313–324.

———, 1984, Oil and gas exploration in the Rio Grande rift; *in* Oil and gas fields of New Mexico: Four Corners Geological Society, v. 3.

EROSIONAL HISTORY AND SOIL DEVELOPMENT ON QUATERNARY SURFACES, NORTHWEST ESPAÑOLA BASIN, NEW MEXICO

DAVID P. DETHIER and KAREN A. DEMSEY
Department of Geology, Williams College, Williamstown, Massachusetts 01267

INTRODUCTION

Extensive late Cenozoic fans and pediments form surfaces that flank the Rio Grande River along much of its course in New Mexico. The surfaces record pauses during general downcutting by the Rio Grande, as well as episodes of aggradation. Surface formation is likely linked with tectonic control of base level, but climate change may also be important. Where ages can be determined for pediments and fans, these features can be used to help constrain local uplift and denudation histories. However, easily dated material is uncommon in the arid and semiarid regions of the West, and radiometrically dated Quaternary sequences are rare.

Arid and semiarid basins with axial drainage display a general sequence of landforms outward from flanking mountain ranges: mountain front, pediment or fan, coalescing pediments or fans, and stepped alluvial terraces along the major drainage. In most semiarid regions, pediments and adjacent alluvial fans have similar gradients and transport similar material by flow in one or more channels (Moss, 1977). Surfaces are more likely to be erosional near the mountain front and depositional near the axial drainage, but any surface may display evidence for contemporaneous erosion and deposition. Grain size in many active alluvial fans decreases away from the mountains (Bull, 1964), and preservation of stratigraphic relations and soils is more likely downslope. However, once a surface is abandoned by base-level lowering, lateral erosion may proceed more rapidly in downslope areas where surfaces are covered by finer material. Preservation of a complete stratigraphic record is thus unlikely at any point, but comparison of morphostratigraphic and soil relations on several surfaces aids in reconstruction of geomorphic history.

Soil Development

Available water and organic matter are sparse and rates of chemical weathering are slow in arid regions. Carbonate accumulates in dry soils because eolian sources, precipitation, and slow weathering of soil minerals result in high concentrations of calcium (or magnesium) carbonate in evaporating soil solutions. Dry fallout and precipitation constitute most of the calcium carbonate, which accumulates in arid soils at rates of 0.2–0.5 g cm^{-2}/1,000 years (Machette, 1982). The zone of carbonate precipitation is usually below the depth of minor erosional events that affect upper soil horizons. Thick accumulations of carbonate resist erosional stripping by rainsplash or sheetflow and tend to preserve surfaces. Total soil carbonate is thus likely to increase regularly with time, allowing soil ages to be estimated where there is some radiometric control (Bachman and Machette, 1977; Machette, 1978).

Carbonate accumulation in soils follows a regular sequence, first described by Gile and others (1966), who proposed a field classification for carbonate development. Stage I is marked by discontinuous grain coatings and suggests an age of several thousand years; stage II is characterized by thicker coatings, nodule development, and formation of minor cement in soils as young as 10–15,000 years (ka); in stage III cement is widespread as a slowly permeable, $CaCO_3$-rich K horizon develops in soils of late Pleistocene age; stage-IV morphology consists of laminar carbonate overlying a strongly cemented carbonate layer and is developed in soils over about 250 ka old. Careful field study of carbonate horizons can thus be used to provide age estimates for surfaces.

As arid soils increase in age, B horizons become increasingly clay-rich, and color hues become redder and chromas darker. The clay and iron content begin to provide some cementation in perhaps 10–20 ka, affording additional protection for underlying carbonate horizons. Argillic horizons have formed in southern New Mexico on soils of early Holocene age (Gile and others, 1981). However, surface erosion, limited flocculation due to concentrated soil solutions, and mechanical disruption by carbonate crystals make argillic horizons unstable in most older surface soils. Texture of the B horizon is thus a guide to the age of upper Pleistocene and Holocene soils, but sytematic increases in fine material are less common with increasing age.

Regional Geology

The central part of the Rio Grande rift consists of three en-echelon basins (Albuquerque, Española, and San Luis) flanked by mountain ranges and drained by the Rio Grande (Fig. 1). In Miocene through mid-Pliocene time the Española Basin was probably internally drained and accumulated sediment of the Santa Fe Group on fans constructed from upland areas to the west, north, and northeast of the basin (Manley, 1979). The Rio Grande–Rio Chama system first became integrated through the Española Basin in Pliocene time, probably between about 2.8 and 4.0 m.y. ago, and followed a course through the center of the Velarde graben. Santa Fe deposition ended about 3 m.y. ago, and the event is marked by an erosional unconformity and a layer of resistant cobble gravel in areas near the Rio Grande, and by erosion surfaces near the Sangre de Cristo Range and Ortiz Mountains (Manley, 1979). Kelley (1979) correlated all high-level pediments in the Española Basin with Bryan's (1938) Ortiz surface. Basin deposition near the Rio Grande resumed after cutting of the Ortiz pediment, covering that surface with the Ancha Formation east of the Rio Grande, and with the Puye Formation west of the river. General basin aggradation apparently continued until after the Puye fanglomerate was covered by local basalt flows and the Bandelier Tuff. Regional erosion resumed sometime after deposition of the youngest member of the Bandelier Tuff, some 1.1 m.y. ago, and a series of cut and fill surfaces formed as the Rio Grande eroded down to present levels (Fig. 2).

Surfaces of pediments, fans, and terraces cut during Quaternary time are prominent landforms of the northern Española Basin, particularly in the area along the east flank of the NE Jemez Mountains. Because the area includes active faults and major climatic shifts have marked the past million years, erosion and depositional episodes may represent a response to these two forces, as well as to regional base-level control by the Rio Grande. In this paper we report the results of detailed geomorphic mapping and soil sampling near the Rio del Oso, and reconnaissance studies elsewhere in the northern Española Basin.

SETTING

Topography and Climate

The Española Basin comprises a dissected lowland bounded by the volcanic Jemez Mountains to the west, the Sangre de Cristo Range to the north and northwest, the Brazos–Tusas Ranges to the northwest, and low volcanic plateaus to the north and south. Local relief in the badlands and dissected surfaces is generally 100–250 m, but reaches 300 m near the Picuris Mountains. The Rio Chama and Rio Grande are the main perennial drainages in the area of interest; numerous tributaries such as the Ojo Caliente, Santa Cruz, and Pojoaque Rivers, Santa Clara Creek, and Rio Embudo and Rio del Oso have some flow during most of the year.

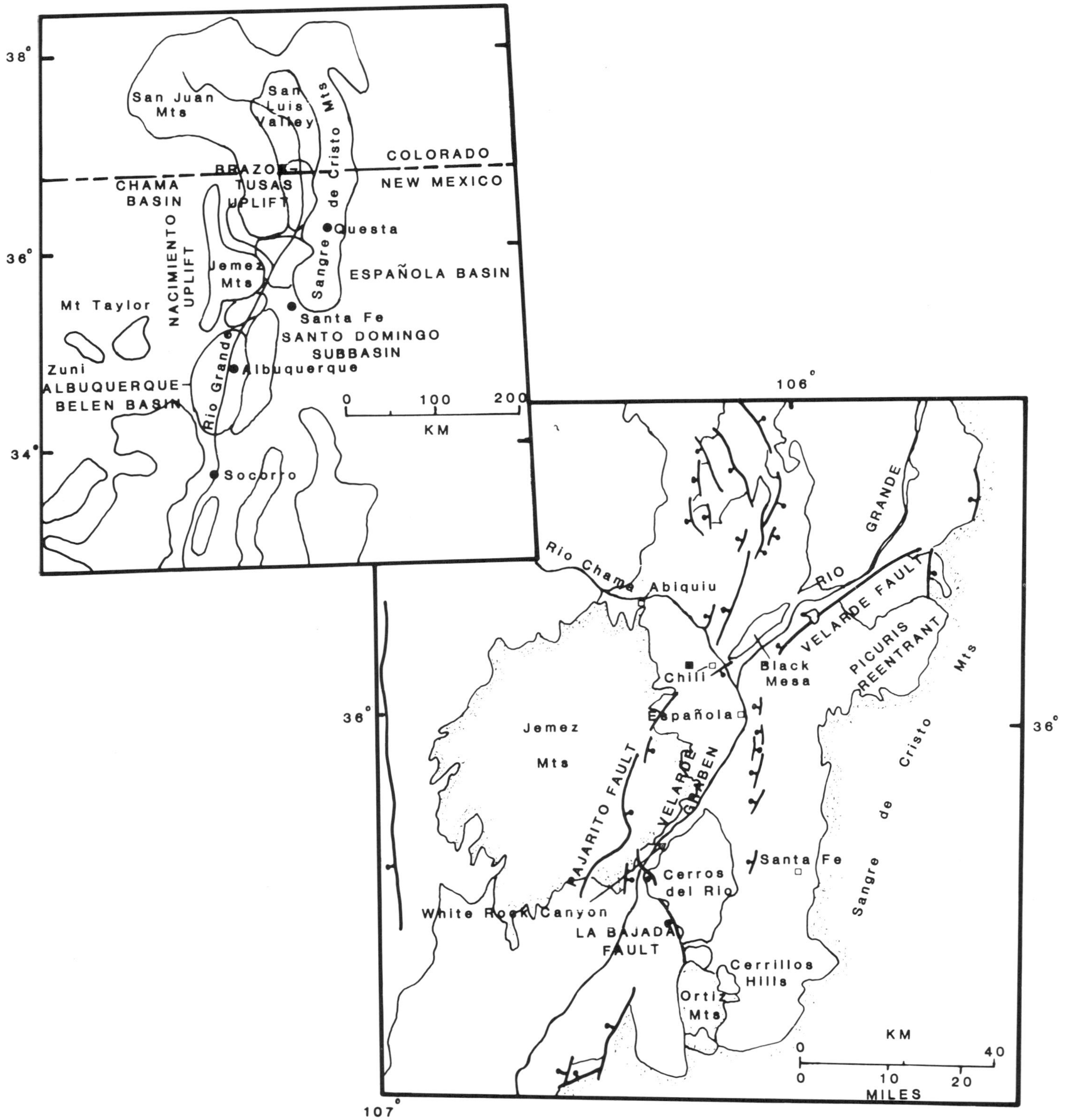

FIGURE 1. Map showing location of the Rio Grande rift and geographic features of the Española Basin (after Manley, 1979). Field area is shown as a square near Chili.

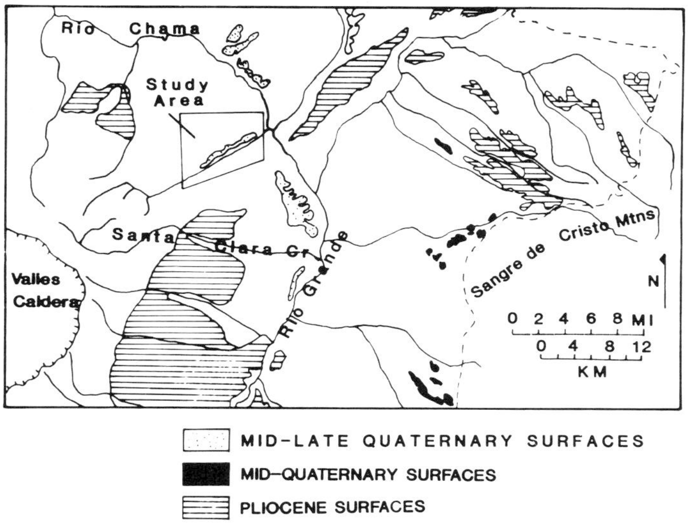

FIGURE 2. Map showing distribution of Pliocene and Quaternary erosion surfaces in the northern Española Basin (after Kelley, 1979).

Climate in most of the Española Basin is semiarid. Mean annual precipitation averages 25 cm at Española, 36 cm at Santa Fe (Spiegel and Baldwin, 1963), and 46 cm at Los Alamos (Griggs, 1964). Significant understory vegetation and soil organic matter are present only above 2,200 m. Detailed information about paleoclimate in the Española Basin is sparse, but general inferences about Quaternary climates can be drawn from studies in nearby areas (Van Devender and Spaulding, 1979). Full "glacial" climates in northern New Mexico were probably expressed by greater effective moisture. Cirque and small valley glaciers formed in the Sangre de Cristo Range and vegetative cover likely increased in lowland parts of the Española Basin. Interglacial climates, by analogy to the Holocene, were warmer and drier than glacial climates and the change to interglacial climates likely resulted in landscape instability, erosion, and sedimentation. Holocene climate was initially warm and dry (Martin and Mehringer, 1965), but in the past 4,000 years has become more variable (Leopold, 1951).

Pliocene and Quaternary Geology

High pediments correlative with the Ortiz are preserved in the Peñasco embayment and slope WNW toward the top of Black Mesa. The lowest of the high pediments (Truchas surface of Manley, 1976) predates deposition of the lower Bandelier Tuff (1.4 m.y.), which is preserved on a surface cut about 35 m below the Truchas level. A basalt flow that caps Black Mesa was dated at 2.8 m.y. (Manley, 1979); it overlies cobble gravel deposited by the ancestral Rio Grande, giving a close approximation for base level above the present channels at 3 m.y. ago. A lower surface, termed the Santa Cruz (Manley, 1976), is preserved as isolated remnants in the eastern Española Basin. Kelley (1979) suggested that the surface is a pediment near the mountains and a fan down-gradient. Manley (1976) noted upper Pleistocene terrace deposits at two levels as much as 50 m above present channels.

Surface remnants west and northwest of Española cannot be traced laterally to surfaces east of the Rio Grande. However, basal Puye Formation is a channel gravel deposited by the Rio Grande at grade with the Ortiz surface about 3 m.y. ago (Manley, 1979). Base level rose from 3 to 2 m.y. ago and probably 1.4 m.y. ago as the Puye, flows and maar deposits from the Cerros del Rio volcanic field, and finally the lower Bandelier Tuff periodically dammed the Rio Grande at White Rock Canyon. Base level lay about 180 m above the present level when a basalt flow entered an extensive lake in Los Alamos Canyon some 2 m.y. ago (Manley, 1979). We know of little specific evidence for relative base-level changes between 1.4 and 1.1 m.y. ago. Smith, Bailey, and Ross (1970) noted that deep canyons were cut into the lower Bandelier and filled by upper Bandelier Tuff, but published maps and reconnaissance fieldwork suggest that much of the distal upper Bandelier covered a surface of low relief. Upper Bandelier Tuff exposed at similar elevation east and west of White Rock Canyon strongly suggests that base level did not change dramatically until after 1.1 m.y. ago.

Late Quaternary base-level changes on the axial drainage must have been controlled by relative motion between the Santo Domingo Basin and the Española Basin along the La Bajada fault, and by resistance to erosion provided by the basalt flows and tuff in the 10 km upstream from the fault. After the volcanic rocks were removed, headward erosion doubtless occurred rapidly in the poorly consolidated Santa Fe Group. We also recognize the importance of deformation along late Quaternary faults to individual drainages, but concentrate here on the influence of Rio Grande base level.

METHOD

We mapped surfaces and deposits along Rio del Oso in detail, and described and sampled soils in this area from surfaces south of the Rio del Oso and from one exposure in the Santa Cruz surface. All soil samples were analyzed for bulk density (Chleborad and others, 1975), carbonate content, and gravel/sand ratios. We performed complete granulometric analyses on selected samples and determined extractable iron (Jackson, 1973) for all B horizons and for selected samples of parent material. Our carbonate analyses followed a CO_2 generation technique using Chittick apparatus as proposed by Dreimanis (1962) and modified by Machette (written comm. 1982). Precision of carbonate analyses was $\pm 6\%$, whereas replicate analyses for bulk density and extractable iron gave a precision of about $\pm 8\%$.

We calculated total carbonate content for each horizon (c_s) using methods suggested by Machette (1982). Initial carbonate content was estimated using parent material beneath soil profiles, or assuming an initial carbonate value of 1% (Gile and others, 1981) where unaltered parent material was absent. We calculated total secondary carbonate in a soil profile (cS) by summing c_s values for all horizons. For undisturbed soils and constant accumulation rates, cS should be directly related to age. Machette and others (1982) suggested that carbonate-accumulation rates have averaged 0.5 gm cm^{-2} ka^{-1} for the past 50 ka and 0.25 gm cm^{-2} ka^{-1} prior to 50 ka.

Two samples were submitted to Beta Analytic, Inc., for ^{14}C analyses. Sample KD-29 consisted of charcoal fragments disseminated in silty sand beneath an alluvial surface some 3 m above the modern Rio del Oso. The second sample (KD-8) consisted of an overthickened A horizon buried by some 5 m of alluvium and colluvium and overlying a poorly developed soil. The surface above the buried soil lies some 26 m above the Rio del Oso in an erosional zone dissected by many small gullies.

RESULTS

Surfaces and Deposits Along the Rio del Oso

Five surfaces could be consistently distinguished above the active channel of the Rio del Oso and in adjacent areas along the Arroyo del Palacio (Fig. 3). The highest surface, designated Q1, is exposed as isolated remnants in the western part of the Chili and eastern part of the Vallecitos quadrangles. The surface is generally cut into a stratigraphic sequence consisting of Santa Fe Group truncated and overlain by Puye deposits which underlie the lower Bandelier Tuff. Overlying the tuff are deposits of basaltic cobble and boulder gravel that display stripped stage-IV soils overlain by stage-III soils. The Q2 surface is more widely exposed at lower elevations and is cut across basaltic cobble and boulder gravel that fills channels in Santa Fe sediment near the mountain front. Deposits beneath the Q2 surface are finer and soils are better preserved east of the Vallecitos quadrangle. Surfaces Q3 and Q4 overlie arkosic sand and basaltic gravel from one to more than 20 m thick that fill channels cut into the Ojo Caliente Sandstone. The Q3 surface truncates an aggradational sequence near the Rio Chama. Deposits beneath the Q3 surface contain buried soils at a number of locations and surface-boulder cover is minimal. Deposits beneath the Q4 surface contain less gravel than those beneath the Q3 surface. Qal represents low, alluvial terraces near modern channels.

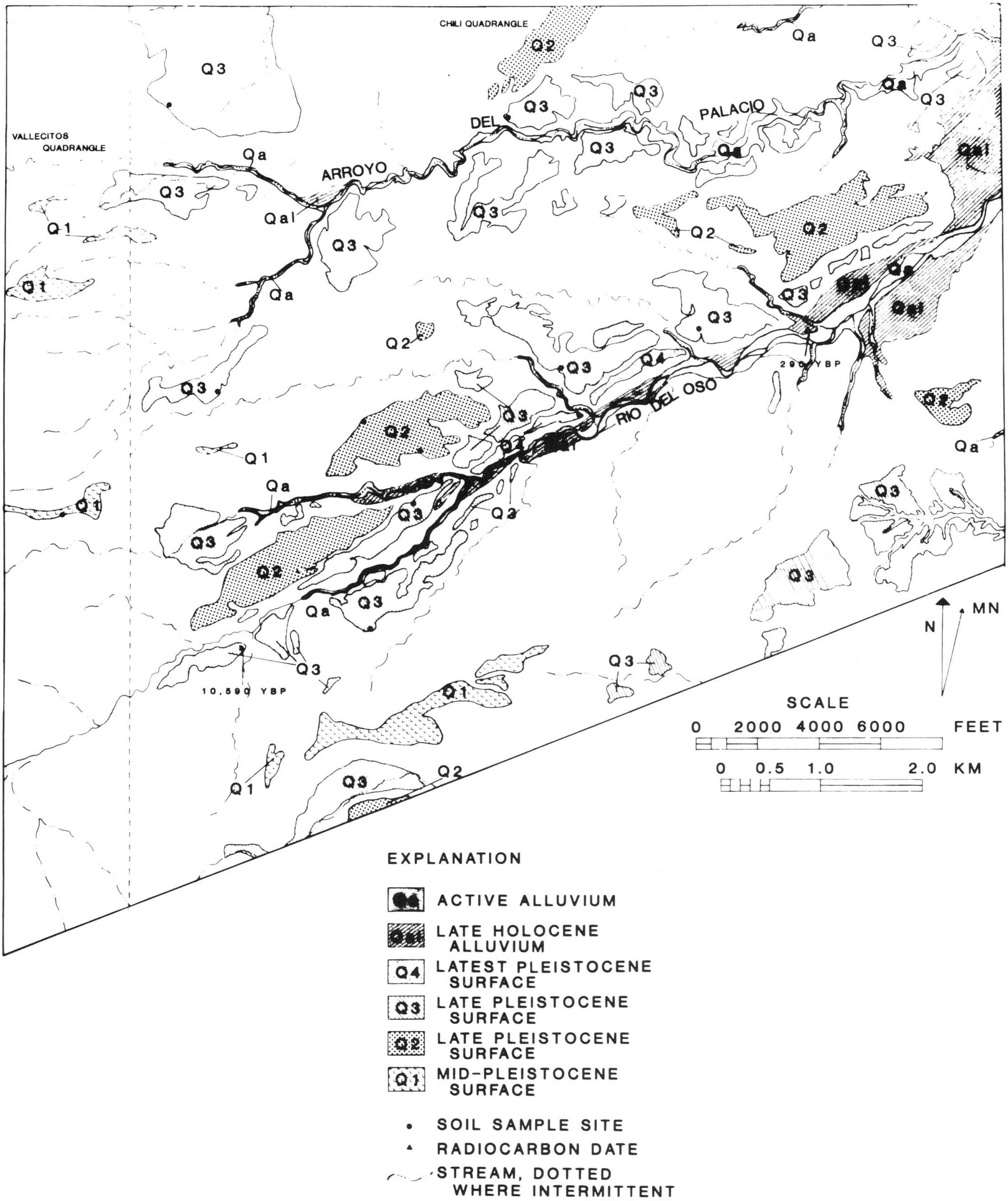

FIGURE 3. Map showing surfaces of the Rio del Oso area, northeast Jemez Mountains.

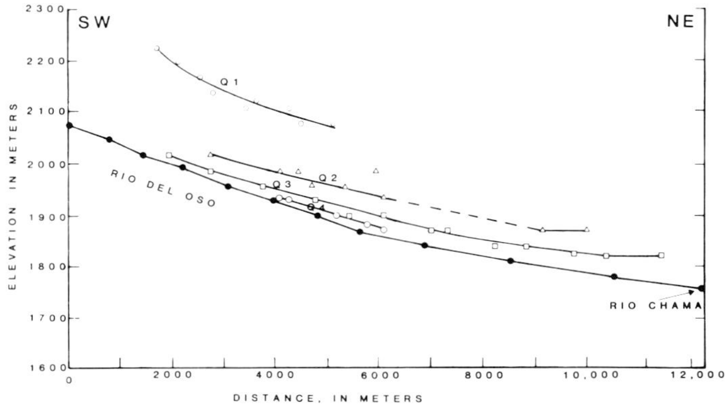

FIGURE 4. Gradients of surfaces in the Rio del Oso area. Distances measured from the west border of Figure 3.

Gradients on surface remnants along Rio del Oso (Fig. 4) are similar to the river with the exception of Q1, which appears to steepen toward the mountain front. Easterly segments of Q1 project to an elevation of ~2,000 m at the Rio Chama. The present channel of the Rio Chama at Black Mesa lies at an elevation of 1,735 m, whereas the Pliocene gravels beneath the capping flow on Black Mesa crop out at about 2,010 m.

Soils

Laboratory analyses of carbonate in 21 soil profiles generally support field descriptions of soils (Demsey, unpubl. data 1983) and clearly show the influence of erosion on the accumulation of soil carbonate. Soil ages calculated from carbonate content give minimum age for surface stabilization ranging from about 200 to 20 ka (Table 1). Bulk densities ranged from 1.2 to 2.1 g cm^{-3}, whereas dithionite-extractable iron ranged from 0.4 to 1.04% in B horizons and from 0.16 to 0.28% in parent material, but the data show no clear trend related to surface position or inferred age.

Carbonate content of typical profiles on each surface ranges from a maximum of 60% (Q1) to about 25% for a stage-II soil on the Q4 surface (Fig. 5). The strongest K horizon on the Q1 surface displays stage-IV morphology, and several soils elsewhere on the surface expose two or more stripped and buried K horizons of at least stage-III development. Maximum carbonate content of the Q2 soils is about 40%, typical of "strong" stage-III field descriptions. Soils on several Q2 surfaces expose weak Holocene B horizons overlying stronger argillic or carbonate-cemented buried B horizons. Soils on Q3 and Q4 surfaces along the Rio del Oso are more poorly developed than those on adjacent higher surfaces.

TABLE 1. Selected analyses and calculated ages for soils developed along the Rio del Oso and in nearby areas.

Surface	Carbonate stage	Soil carbonate in gm cm^{-2}	Age[1] of best developed soil on surface in ka	Iron content[2], in percent
Q-1	IV	59		0.24
Q-1[3]	IV	78	200	0.84
Q-2	Strong III	52	160	0.64
Q-3[4]	III	28	51	0.40
Q-3[5]	III	21		0.87
Q-3	I	2		0.6
	III	13		0.26
	I	~1		0.60
	III	23		0.67
	III	12		0.52
Q-4	II	17	35	0.39
Santa Cruz	Strong III	--		1.0

1 Calculated from carbonate accumulation rates listed by Machette, 1982.
2 Dithionite extractable in strongest B horizon.
3 Colluvial slope adjacent to Q-1 surface, Vallecitos Quadrangle.
4 Surface south of Arroyo del Palacio, Chili Quadrangle.
5 Surface north of Arroyo de la Plaza Larga, San Juan Pueblo Quadrangle.

Approximate minimum ages for Rio del Oso surfaces were estimated using the soil-carbonate stage (Gile and others, 1981) and values listed in Table 1. Soils on the Q1 and Q2 surfaces have carbonate contents that suggest ages of about 150–250 ka. However, the stripped stage-IV carbonate which underlies stage-III soils on the Q1 surface probably took at least 200 ka to develop, and the surface probably formed at least 400–500 ka ago. Comparable stages of surface soils on the Q1 and Q2 surfaces suggest that both stabilized at about 200 ka. Maximum profile stage suggests Q3 soils are slightly older than Q4 soils, but the surfaces are of broadly similar age. A buried A horizon beneath an eroding, low Q3 surface gave a radiocarbon date of 10,590 ± 150 yrs B.P. (Beta-8359), and soil development and carbonate content suggest it was about 5 ka when it was buried. This evidence shows that at least portions of the Q3 surface were active at about 15 ka, implying complex erosional and depositional events in latest Pleistocene time. Tree-ring counts on the Qal surface suggest it was abandoned at least 100 yrs ago; soil development is modern. Charcoal in deposits beneath the surface gave a date of 280 ± 90 yrs B.P. (Beta-7821).

DISCUSSION

Rio del Oso Surfaces

Our data and those of Manley (1976, 1979) indicate that the late Pliocene and Quaternary landscape northwest of Española was generally stable until after deposition of the upper Bandelier Tuff, and downcutting which isolated Q1 surfaces did not occur until mid- to late Pleistocene time. Once erosion began, denudation of areas east of Lobato Mesa occurred rapidly, dropping local arroyo levels at least 150 m and average landscape elevation more than 80 m. No evidence is preserved for periods of significant aggradation between at least 200 and 50 ka before present. Deposits beneath the Q3 surface near the Rio Chama record

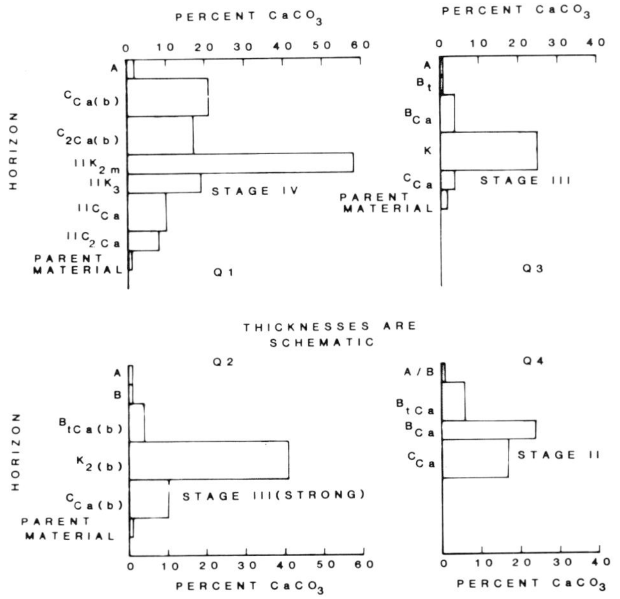

FIGURE 5. Carbonate content and stage of the best preserved soils on surfaces of the Rio del Oso area.

at least 10–20 m of net deposition during latest Pleistocene time. Soil carbonate gives reasonable minimum-age estimates for surfaces, but profiles that closely match surface age are fairly uncommon on Q4 and Q4 surfaces, rare on Q2 surfaces, and have not been observed on Q1 surfaces. Erosion rates in the study area were influenced by Rio Grande/ Rio Chama base-level control that depended on downcutting rates at White Rock Canyon and regional uplift. Similar flights of surfaces are present south of the Rio del Oso (Fig. 2), but their history was also influenced by the great thickness of Puye cover, and by late Pleistocene deformation (C. Harrington, written comm. 1984).

Surfaces along the Rio del Oso and in nearby areas stabilized between about 250 and 500 ka before present. The surface of pediment Q1 was an aggradational area until after 1.4 m.y. ago, and probably until after 1.1 m.y. ago, assuming surfaces are similar to those exposed 8 km to the south of Rio del Oso. Carbonate stage and total-profile carbonate suggest that colluvium which slopes from Q1 surfaces has been stable for more than 200 ka. After erosion removed most of the Q1 surface and cut through into underlying Miocene sediment, the Q2 surface formed some 40 m lower and was abandoned at about 200 ka. Although the Q1 surface formed more than 400 ka ago, soil development implies that erosion slowed on the Q1 and Q2 surfaces at similar times. Total soil carbonate suggests that parts of the Q3 surface stabilized at 25–40 ka, carbonate stage indicates latest Pleistocene, and a single radiocarbon date demonstrates that portions of the surface were active at about 10 ka. Carbonate stage on Q4 soils suggests that they are slightly younger than Q3 soils, and cS values and absolute elevations of the surfaces overlap. Both surfaces must have been abandoned in latest Pleistocene or earliest Holocene. Periods of aggradation and erosion probably characterized much of Holocene time, but the Rio del Oso area records evidence only for 3–5 m of aggradation followed by erosion in historical time.

For most of their length, surfaces in the study area are pediments carved by streams during high-energy, episodic transport in systems of shallow channels that sloped toward Black Mesa. Dethier and Martin (this guidebook) mapped a north-trending fault at the west edge of the Chili sheet that displays at least 5 m of Quaternary motion. Activity of this fault and others along the mountain front could have influenced deposits beneath the Q1 surface and local stream gradients, but our field studies give no positive evidence. None of the younger surfaces show gradient changes or offsets that can be attributed to faulting.

The depositional nature of the distal Q3 surface and latest Pleistocene aggradation along major rivers in northern New Mexico was probably a response to latest Pleistocene climatic changes or to slowing of regional uplift. Aggradation in Q3 time along the Arroyo del Palacio, Rio del Oso, and Arroyo de la Presa produced terraces at the mouths of these drainages and other terraces along the west side of the Rio Grande. Local evidence is not sufficient to decide whether the Rio Grande produced aggradation by decreasing gradients or whether the tributaries aggraded in response to excessive loads from their catchments.

Surfaces of the Northern Española Basin

Late Quaternary dissection is most complete and surfaces youngest in the northwest Española Basin. South of Española, canyons are narrow and steep-walled, and dissection of the surface cut across the Bandelier Tuff (the Pajarito Plateau) has apparently occurred more slowly than in the Rio del Oso area because unconsolidated Miocene sediment is covered with tuff and thick Puye gravels. Gradients for the high pediments (Q1) and lower surfaces are comparable north and south of Española (Fig. 6). However, lower surfaces are principally terrace deposits near the Rio Grande, and there are no low pediments parallel to stream channels. The high surfaces of Manley (1976) probably correlate with the Ortiz pediment and have gradients similar to that plotted in Figure 6. The late Pliocene surface that corresponds to the Ortiz is buried beneath the Puye Formation west of the Rio Grande and is displaced by north- and northeast-trending faults.

Our reconnaissance investigations and analyses of soils developed on the Santa Cruz surface are not sufficient to establish its age. Soil carbonate is at least stage III, and clay and iron accumulation are greater than in any of the soils we examined in the Rio del Oso area. Kelley's (1979) reconstruction of the surface from isolated remnants (see Fig. 6) shows that it is much lower than the Ortiz and grades toward a Rio Grande level close to, or lower than, the present level. Because soil development suggests that the Santa Cruz formed prior to latest Pleistocene time, and its gradients suggest modern base levels, the Santa Cruz event does not correlate with surfaces preserved in the northwest Española Basin.

We believe evolution of surfaces in the northwest Española Basin resulted primarily from base-level control by the Rio Grande, but apparent west-to-east asymmetry suggests at least some local tectonic influence. Once White Rock Canyon was cut through the resistant volcanic rocks into unconsolidated sediment, river grade probably responded to regional erosion and deposition controlled by tectonism and climatic change. We have sketched base-level changes from late Pleistocene to the present for an area near Española, assuming minimal tectonic influence over short intervals (Fig. 7). The graph shows inferred, relative local changes and should be used as a guide for discussion. Integration of the Rio Grande in late Pliocene resulted in cutting of widespread erosion surfaces graded to a base level some 35 m higher than the present level about 2 m.y. ago. Afterwards the base level in the NW Española Basin rose as a result of damming and aggradation by Jemez Mountain rocks, reaching levels some 350 m above the present level about 1 m.y. ago. Erosion followed, lowering the base level 200 m by at least 200 ka and perhaps by as much as 400 ka.

Bachman and Mehnert (1978) suggest that slowing of regional-uplift rates caused cutting of the Llano de Albuquerque surface at about 500 ka ago, and formation of the Llano de Manzano by coalescing fans some 200 or 300 ka ago. Surfaces Q1 and Q2 may broadly correlate with these Albuquerque Basin surfaces, which would imply regional control of Rio Grande base level in late Quaternary time. However,

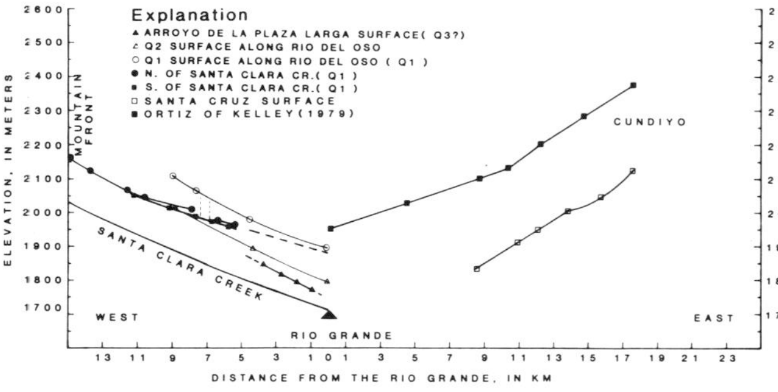

FIGURE 6. Gradients of the Ortiz, Santa Cruz, and mid- to late Pleistocene surfaces near Española, New Mexico. Vertical dashes are faults mapped along Santa Clara Creek.

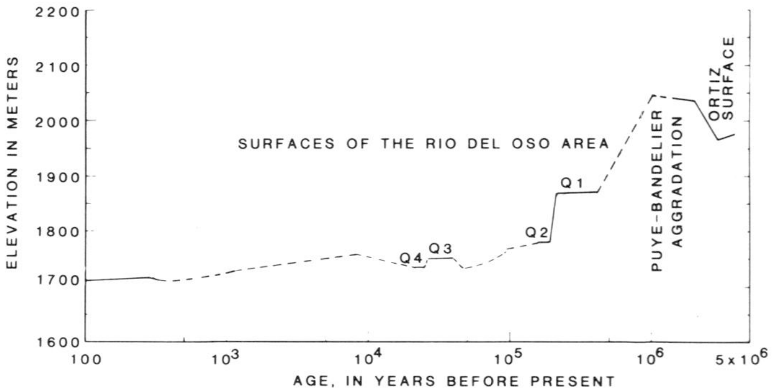

FIGURE 7. Schematic illustration of changes in base level near Española from late Pliocene to present.

base level during most of the 3–1 m.y.B.P. interval was controlled locally. The upper Pliocene to mid-Quaternary deposits in White Rock Canyon provided local-base level, and understanding their evolution is a key to advancing our knowledge of Quaternary geomorphic evolution of the Española Basin.

ACKNOWLEDGMENTS

Portions of this work were supported by Los Alamos National Laboratory and the Bronfman Science Center at Williams College. Chuck Harrington provided encouragement in the field and a thorough review of an earlier version of this manuscript, and John Hawley made a number of helpful suggestions.

REFERENCES

Bachman, G. O., and Machette, M. N., 1977, Calcic soils and calcretes in the southwestern United States: U.S. Geological Survey, Open-file Report 77-794, 163 pp.

———, and Mehnert, H. H., 1978, New K–Ar dates and the late Pliocene to Holocene geomorphic history of the central Rio Grande region, New Mexico: Geological Society of America, Bulletin, v. 89, pp. 283–292.

Bryan, K., 1938, Geology and ground-water conditions of the Rio Grande depression in Colorado and New Mexico; *in* Rio Grande joint investigation in the upper Rio Grande basin in Colorado, New Mexico, and Texas: National Resources Committee, Regional Planning, pt. 6, pp. 197–225.

Bull, W. B., 1964, Geomorphology of segmented alluvial fans, western Fresno County, California: U.S. Geological Survey, Professional Paper 352-E, pp. 89–128.

Chleborad, A. F., Powers, P. S., and Farrow, R. A., 1975, A technique for measuring bulk volume of rock materials: Association of Engineering Geologists, Bulletin, v. 12, no. 4, pp. 317–322.

Dethier, D. P., and Manley, K., 1984, Geology of the Chili quadrangle, Rio Arriba County, New Mexico: U.S. Geological Survey, Miscellaneous Field studies Map, in press.

Dreimanis, A., 1962, Quantitative determinations of calcite and dolomite by using Chittick apparatus: Journal of Sedimentary Petrology, v. 32, no. 3, pp. 520–529.

Gile, L. H., Hawley, J. W., and Grossman, R. B., 1981, Soils and geomorphology in the Basin and Range area of southern New Mexico—guidebook to the Desert Project: New Mexico Bureau of Mines and Mineral Resources, Memoir 39, 222 pp.

———, Peterson, F. F., and Grossman, R. B., 1966, Morphological and genetic sequences of carbonate accumulation in desert soils: Soil Science, v. 101, no. 5, pp. 347–360.

Griggs, R. L., 1964, Geology and ground-water resources of the Los Alamos area, New Mexico: U.S. Geological Survey, Water-supply Paper 1753, 107 pp.

Jackson, M. L., 1973, Soil chemical analysis—advanced course (2nd revised edition): Published by author, Madison, Wisconsin, 895 pp.

Kelley, V. C., 1979, Geomorphology of Española Basin: New Mexico Geological Society, Guidebook 30, pp. 281–288.

Leopold, L. B., 1951, Pleistocene climate in New Mexico: American Journal of Science, v. 249, pp. 152–168.

Machette, M. N., 1978, Dating Quaternary faults in the southwestern United States by using calcic paleosols: U.S. Geological Survey, Journal of Research, v. 6, no. 3, pp. 369–381.

———, 1982, Morphology, age, and rate of accumulation of pedogenic $CaCO_3$ in some calcareous soils and calcretes of the southwestern United States (abstract): Geologic Society of America, Abstracts with Programs, v. 14, pp. 182–183.

Manley, K., 1976, The late Cenozoic history of the Española Basin, New Mexico: Ph.D. thesis, University of Colorado, Boulder, 171 pp.

———, 1979, Stratigraphy and structure of the Española Basin, Rio Grande rift, New Mexico; *in* Riecker, R. E. (ed.), Rio Grande rift: tectonics and magmatism: American Geophysical Union, Washington, D.C., pp. 71–86.

Martin, P. S., and Mehringer, P. J., Jr., 1965, Pleistocene pollen analysis and biogeography of the Southwest; *in* Wright, H. E., and Frey, D. G. (eds.), The Quaternary of the United States: Princeton University Press, pp. 433–451.

Moss, J. H., 1977, The formation of pediments: scarp backwearing or surface downcutting? *in* Doehring, D. O. (ed.), Geomorphology in arid regions: Annual Geomorphological Symposium no. 8, State University of New York at Binghamton, Publications in Geography, pp. 51–78.

Smith, R. L., Bailey, R. A., and Ross, C. S., 1970, Geologic map of the Jemez Mountains, New Mexico: U.S. Geological Survey, Miscellaneous Geological Investigations Map I-571.

Spiegel, Z., and Baldwin, B., 1963, Geology and water resources of the Santa Fe area, New Mexico: U.S. Geological Survey, Water-supply Paper 1525, 258 pp.

Van Devender, T. R., and Spaulding, W. G., 1979, Development of vegetation and climate in the southwestern United States: Science, v. 204, pp. 701–710.

Narrow-gauge yard at Antonito. This is now the eastern terminus of the Cumbres and Toltec Scenic Railroad. It was formerly the point where the branch lines to Santa Fe and Durango separated.

DEVELOPMENT AND DEFORMATION OF QUATERNARY SURFACES ON THE NORTHEASTERN FLANK OF THE JEMEZ MOUNTAINS

CHARLES D. HARRINGTON[1] and M. J. ALDRICH, Jr.
Mail Stop D462, Los Alamos National Laboratory, Los Alamos, New Mexico 87545
([1]Permanent address: Department of Marine, Earth and Atmospheric Sciences, North Carolina State University, Raleigh, North Carolina 27650)

INTRODUCTION

The eastern Jemez Mountains volcanic pile overlies the western margin of the Española Basin, one of three en-echelon basins forming the central portion of the Rio Grande rift. The Jemez Mountains have served as a major source of volcanics (between approximately 13 and 0.13 m.y. ago; Gardner and Goff, this guidebook) and volcaniclastic materials (from approximately 10 m.y. ago to present) into this portion of the basin. The local basin dynamics therefore reflects not only regional tectonic events related to the development of the central rift, but also the more localized volcanic and tectonic pulses within the Jemez Mountains. During the Pliocene and Quaternary, a sequence of erosional surfaces developed along the northeastern flank of the Jemez Mountains. Their surfaces slope away from the mountains and grade toward the axial streams (Rio Chama or Rio Grande) of the Española Basin.

Studies of the Quaternary development of the Española Basin include investigations by Bryan (1938), Manley (1976), Kelley (1979), and Dethier and Manley (in press). In this paper we present a progress report on a tectonic–geomorphic study of an area west of Hernandez, New Mexico, in the northeastern Jemez Mountains–northwestern Española Basin (Fig. 1).

QUATERNARY EROSION SURFACES

Surface Characteristics

The primary topographic features within the study area (Fig. 1) are three erosional surfaces that developed across this portion of the northeastern Jemez Mountains during the Quaternary. The surfaces gradually

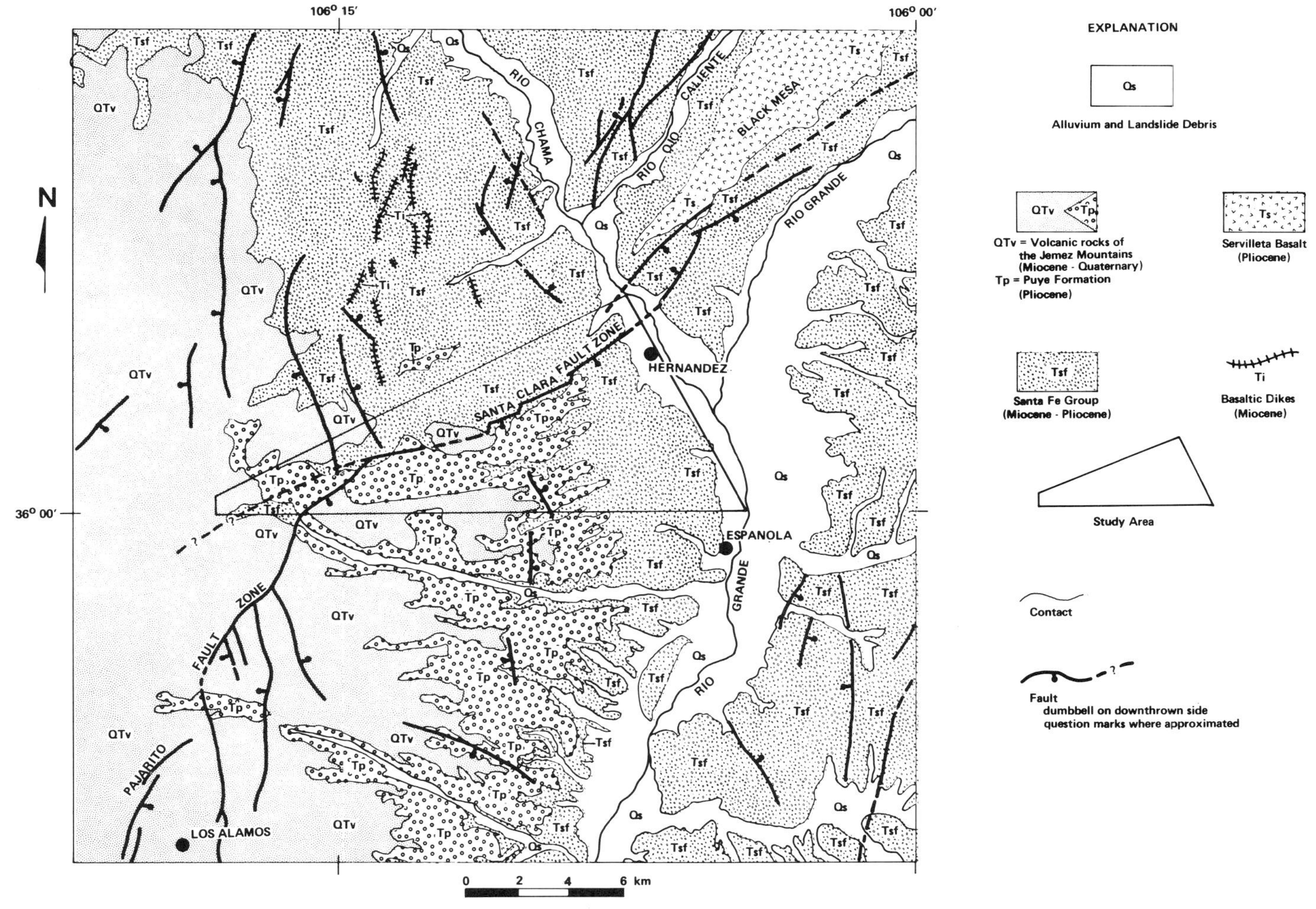

FIGURE 1. Location of field area along and south of the Santa Clara fault zone, northeast Jemez Mountains. Modified from Kelley (1978) and Dethier and Manley (in press).

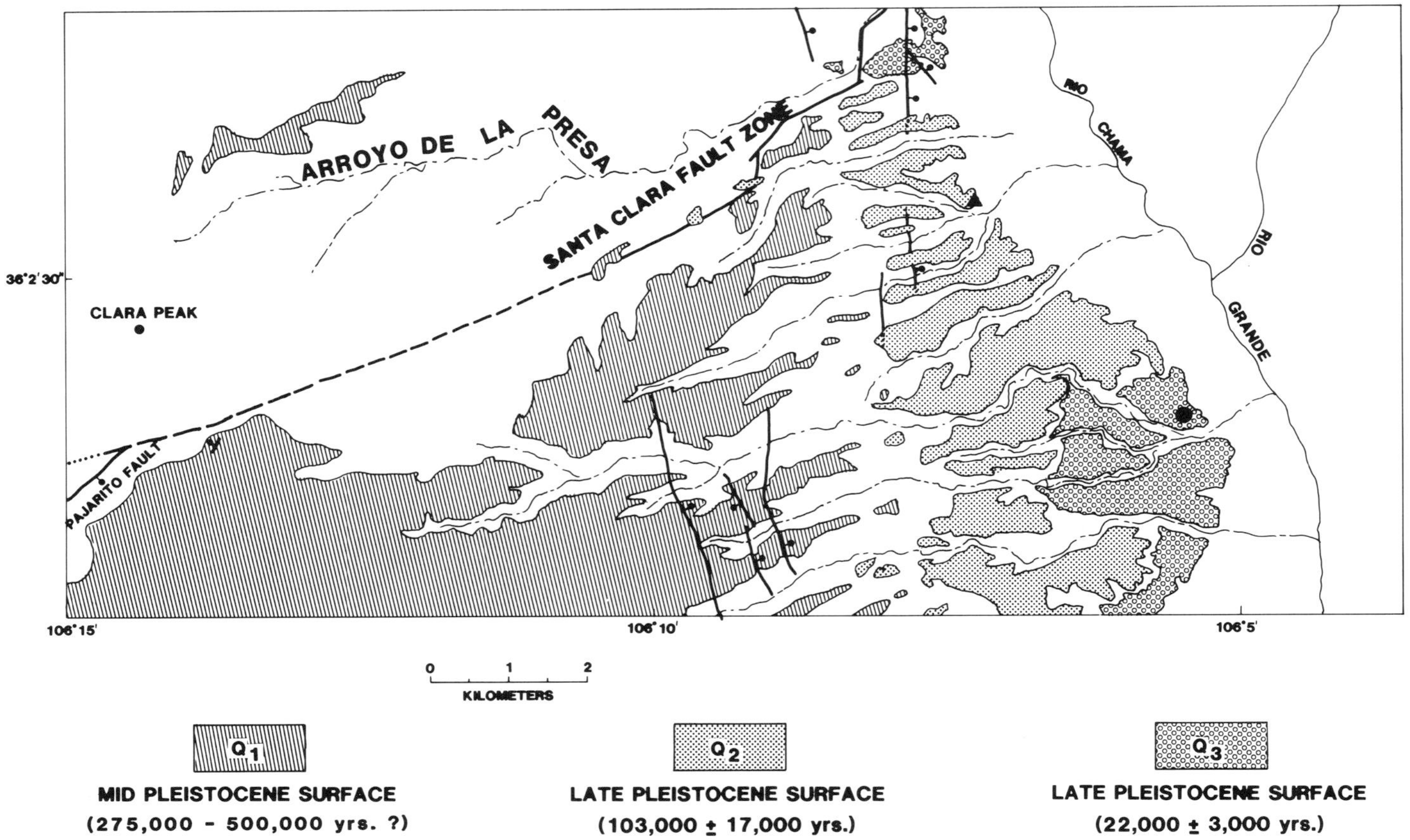

FIGURE 2. Quaternary surfaces within the study area west of Hernandez, New Mexico. Δ – Location of dated sample from Q_2 surface; o – location of dated sample from Q_3 surface.

descend in elevation, forming a series of three steps that slope northeast to east and cut across the upper Santa Fe Group (Chamita Formation and Ojo Caliente Member of the Tesuque Formation, Miocene), Puye Formation (Pliocene), and Bandelier Tuff (Quaternary). They are primarily erosional surfaces (including pediments) with a relatively thin gravel and sand veneer. The two lower surfaces also extend as terraces up major arroyos at elevations below the next higher (older) surface.

Remnants of the highest, and oldest, surface (Q_1) have the greatest areal extent, descending from elevations of 2,200 m near the mountains to 1,920 m along its eastern margin (Fig. 2). The surface, which cuts across the Bandelier Tuff, Puye Formation, and parts of the Santa Fe Group, is graded to the northeast with an average surface slope of 15 m/km (1.5%), which is significantly gentler than the slopes of the two younger surfaces (Fig. 3). Projection of this profile gradient east to the Rio Grande indicates base level on the axial stream at Q_1 time was 1,895 m, 185 m above the present Rio Grande elevation (1,710 m). Relief on the Q_1 surface is generally less than 10 m except along major arroyos and several scarps where it steps down to the east. The scarps are structural in origin and are related to a series of younger, north-trending faults. Capping deposits on the Q_1 surface consist mainly of a thin veneer of cobble gravel and sand. Although the Q_1 surface has undergone some regrading since formation, the primary type of its erosional modification has been arroyo incision and enlargement. Major arroyos that head on the Q_1 surface were incised as much as 100 m before they stabilized during Q_2 surface formation. These arroyos had wide valley floors (200–300 m) when the Q_2 surface formed. Remnants of these arroyo floors remain as terraces lying well above (>20 m) present channel elevations. These terrace segments are graded downstream onto remnants of the Q_2 surface.

The Q_2 surface extends basinward from the foot of the dissected scarp (elevation of 1,840 m), which separates this surface from the Q_1 surface (Fig. 2). The Q_2 surface is cut into the Chamita Formation and Ojo Caliente Member, has an average width of 4 km, an elevation of 1,780 m at its basinward margin, and a surface gradient of 29 m/km (3%). Projection of this erosional surface to the Rio Grande indicates base-level elevation was approximately 1,755 m during Q_2 time, 45 m above the present Rio Grande (Fig. 3). The Q_2 surface is capped by alluvial-fan and colluvial deposits along its western margin, while farther down-slope to the east the erosional surface is veneered with gravel and sand deposits generally 3–6 m thick, with thicknesses of 10 m near the Rio Grande (Dethier and Manley, in press). The surface is displaced in several places by north-trending faults producing a series of steps, up

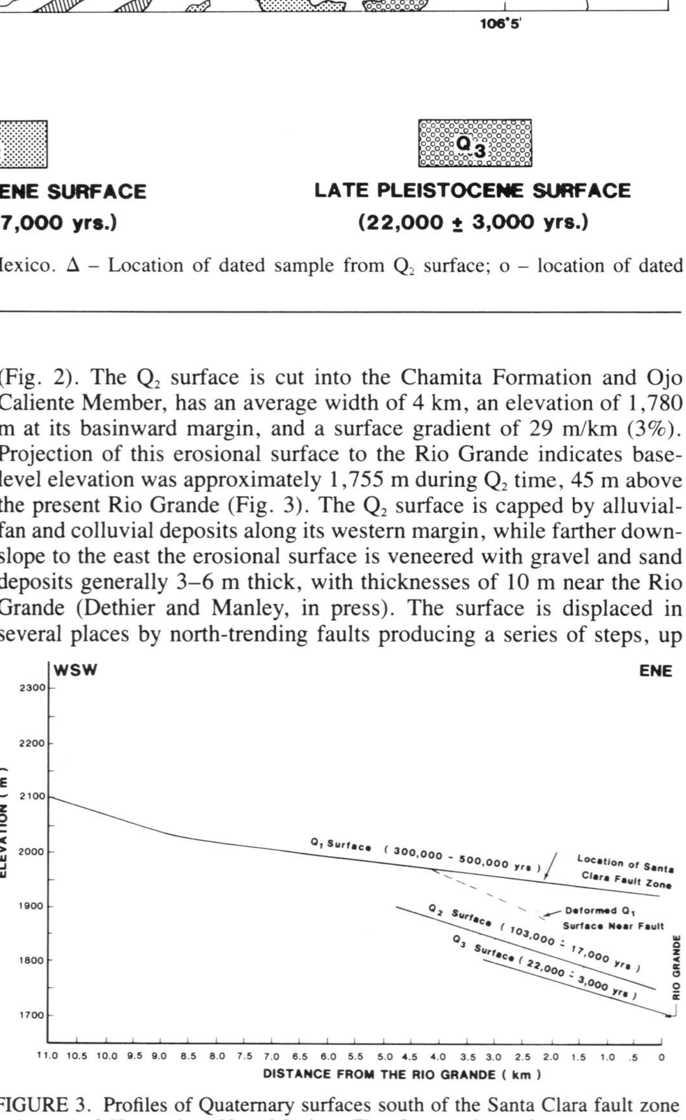

FIGURE 3. Profiles of Quaternary surfaces south of the Santa Clara fault zone and west of Hernandez, New Mexico. The Q_2 and Q_3 surfaces slope more eastwardly than the Q_1 surface and do not cross the Santa Clara fault zone.

to 30 m high, which step down to the east. Relief on the surface, not related to structure, is less than 10 m. Erosion by eastward-flowing streams incised the Q_2 surface by 43 m (2 km west of the Rio Grande) before reaching grade during formation of the Q_3 surface. Narrow terrace remnants, which flank arroyos cut into the Q_2 surface and grade to Q_3 remnants, are parts of these graded valley floors.

The Q_3 surface, cut primarily across the Chamita Formation, has the most limited areal extent of the three surfaces and generally lies basinward of the Q_2 surface (Fig. 2). The surface slopes eastward from maximum elevations of 1,795–1,735 m at its eastern margin. Surface gradients on the surface are 27 m/km and project to an elevation of 1,720 m at the Rio Grande, 10 m above present river level (Fig. 3). The Q_3 surface is erosional in its upper reaches and is generally capped by a veneer of sand and gravel less than 7 m thick. In its lower reaches the surface is aggradational, suggesting a rise in Rio Grande base level during its formation. Displacement of the Q_3 surface has occurred along various north-trending faults producing stepped topography similar to the Q_1 and Q_2 surfaces.

Time of Formation of Erosional Surfaces

The time of formation for the oldest erosional surface (Q_1) within the area can be estimated based on several lines of geologic evidence. For the two younger surfaces, ^{230}Th–^{234}U dating has yielded absolute minimum dates of formation (see Appendix).

The Q_1 surface truncates both members of the Bandelier Tuff, thus establishing its maximum age as less than 1.1 m.y. Soils on the Q_1 surface have a well-developed horizon of carbonate accumulation. Gile and others (1966) have found that such carbonate accumulation in soils develops through a particular sequence of morphological stages that are time dependent (see Dethier and Demsey, this guidebook). Based on correlation with soils on dated geomorphic surfaces bordering the Rio Grande valley in the Albuquerque Basin (Bachman and Machette, 1977), the presence of such soil-carbonate stages can be used to obtain an approximate time of surface stabilization and beginning of soil development. The oldest soils found on the Q_1 surface within the study area contain a laminar carbonate horizon and thus would be classified as stage IV (Gile and others, 1966). The presence of such advanced soil horizons suggests that the Q_1 soil is a minimum of 250,000 years old (Bachman and Machette, 1977). Additional evidence for the minimum age of Q_1-surface development can be derived from estimates of erosion rates. The Q_1 surface lies 90 m above the Q_2 surface that in turn lies 43 m above the Q_3 surface. Utilizing the ^{230}Th–^{234}U dates of formation for the two lower surfaces (discussed below), the 43 m of arroyo incision of the Q_2 surface occurred in approximately 80,000 years. If equal incision rates are assumed (the authors realize that such an assumption is tenuous, at best) during dissection of the Q_1 surface, then the Q_1 surface formed approximately 170,000 years before the Q_2 surface, or 275,000 years ago. We believe that an age for the Q_1 surface of 275,000–500,000 years is likely. Such an age estimate is broadly correlative with the highest (Q_1) surface of Dethier and Demsey (this guidebook) north of our area.

Soils on the Q_2 surface have a stage-III pedogenic carbonate horizon. Such a stage suggests a minimum age for this surface of approximately 100,000 years (Bachman and Machette, 1977). Carbonate coatings were extracted from pebbles from the K horizon of soils on this surface. Th–U dating of the carbonate yielded an age of $103{,}000 \pm 17{,}000$ years (see Appendix). This date for the Q_2 surface is consistent with other geologic data from the area, but does not correlate with any ages estimated by Dethier and Demsey (this guidebook) for surfaces to the north.

The youngest Quaternary erosional surface (Q_3) within the area is also capped by deposits with pedogenic horizons of carbonate accumulation. The more developed soils have carbonate-horizon morphologies similar to stage II of Gile and others (1966). Th–U dating of carbonate coatings from pebbles extracted from the Q_3 surface soil shows that these coatings are $22{,}000 \pm 3{,}000$ years old. This age is consistent with geologic and soil data related to this surface and indicates our Q_3 surface correlates with either the Q_3 or Q_4 surface of Dethier and Demsey (this guidebook).

QUATERNARY TECTONICS

Several faults and fault zones have exerted an influence on the development of the present-day topography (Fig. 2). These include the Pajarito fault zone, Santa Clara fault zone (new name), and a series of north-trending faults. The Pajarito fault zone bounds the area on the west, and has normal, east-side-down displacement. The Santa Clara fault zone trends northeast and has had a complex history of oblique-slip motion in which the south side has moved down relative to the north side. A number of generally north-trending down-to-the-east normal faults are also present in this area.

Quaternary movement on the Pajarito and Santa Clara fault zones resulted in rotation of the Q_1 surface, decreasing its gradient and changing its slope direction from east to northeast. Evidence that the Q_1 surface was initially graded to the east and later rotated is derived from analysis of drainage channels cut into it (Fig. 4). Major arroyos incised into this surface trend east across the slope of the surface rather than parallel to the slope dip. If this surface formed with an original northeast slope, major arroyos should trend parallel to the slope direction. Most of the tributaries to the major east-draining arroyos, which head within the area of the Q_1 surface, trend northeast, parallel to the present-surface slope. Only a few tributaries enter the north side of the larger east-trending arroyos. This asymmetry may be the result of formation of the tributaries after tilting of the Q_1 surface to the northeast. Alternatively, the tributaries may have formed prior to rotation of the surface, but only those with courses approximately parallel to the new slope were not abandoned and destroyed. Evidence for deformation of the Q_1 surface by movement along the Santa Clara fault zone is the marked increase in gradient along the surface near the fault. In sections 12 and 14 of T21N, R7E, the Q_1 surface is cut across beds of the Puye Formation. The base of the Puye at this location is at 1,890 m, while 0.5 km to the south its elevation is 1,920 m—a slope of 61 m/km. This slope on the base of the Puye differs in orientation and far exceeds its slope in other areas. The Q_1 surface along the same traverse has a surface gradient of 45 m/km, approximately three times its gradient away from the fault (Fig. 3). Projection of the lower-gradient portion of the Q_1 surface to the fault zone places the surface at an elevation approximately 50 m higher than that obtained by a similar projection of the portion with the steeper gradient. Thus, depression of the block southeast of the Santa Clara fault zone by about 40 m following formation of the Q_1 surface is indicated.

East-draining arroyos near (within 1 km) the Santa Clara fault zone, which head in the Puye, have extensive gravelly valley fills. These fills are preserved as terraces composed mostly of cobble-sized gravel in the upper reaches and pebble-sized gravel downstream. The surfaces of these terraces grade downstream to remnants of the Q_2 surface. Arroyos farther to the south of the fault zone do not contain such fills. These gravelly valley fills probably resulted from reduction of gradients along the east-trending arroyos by downwarping of the area during movement along the Santa Clara fault zone following formation of the Q_1 surface. Later, these fills were regraded to the Q_2 surface level as that surface formed. There apparently has been no appreciable movement on the Santa Clara fault zone since the development of the Q_2 surface.

Normal Faults

North-trending normal faults cut all three surfaces. Movements along these normal faults, which generally trend perpendicular to surface slope, have created a series of stepped surface segments, most of which are down-to-the-east. As none of the faults completely cross the area, the number and location of steps along east–west profiles differs at various locations. Most of these north-trending faults show Quaternary displacement, some in excess of 10 m. Recent displacement along several of these faults is indicated. One north–northwest-trending fault, showing several recurrent movements, has displaced a basalt boulder gravel formed on the Q_3 surface ($22{,}000 \pm 3{,}000$ years) a minimum distance of 10 m. Soil profiles across the fault are significantly different, suggesting latest movement was quite recent. Other north-trending faults within the area also indicate similar recent movements. The down-to-the-east movement on these faults probably increased gradients and

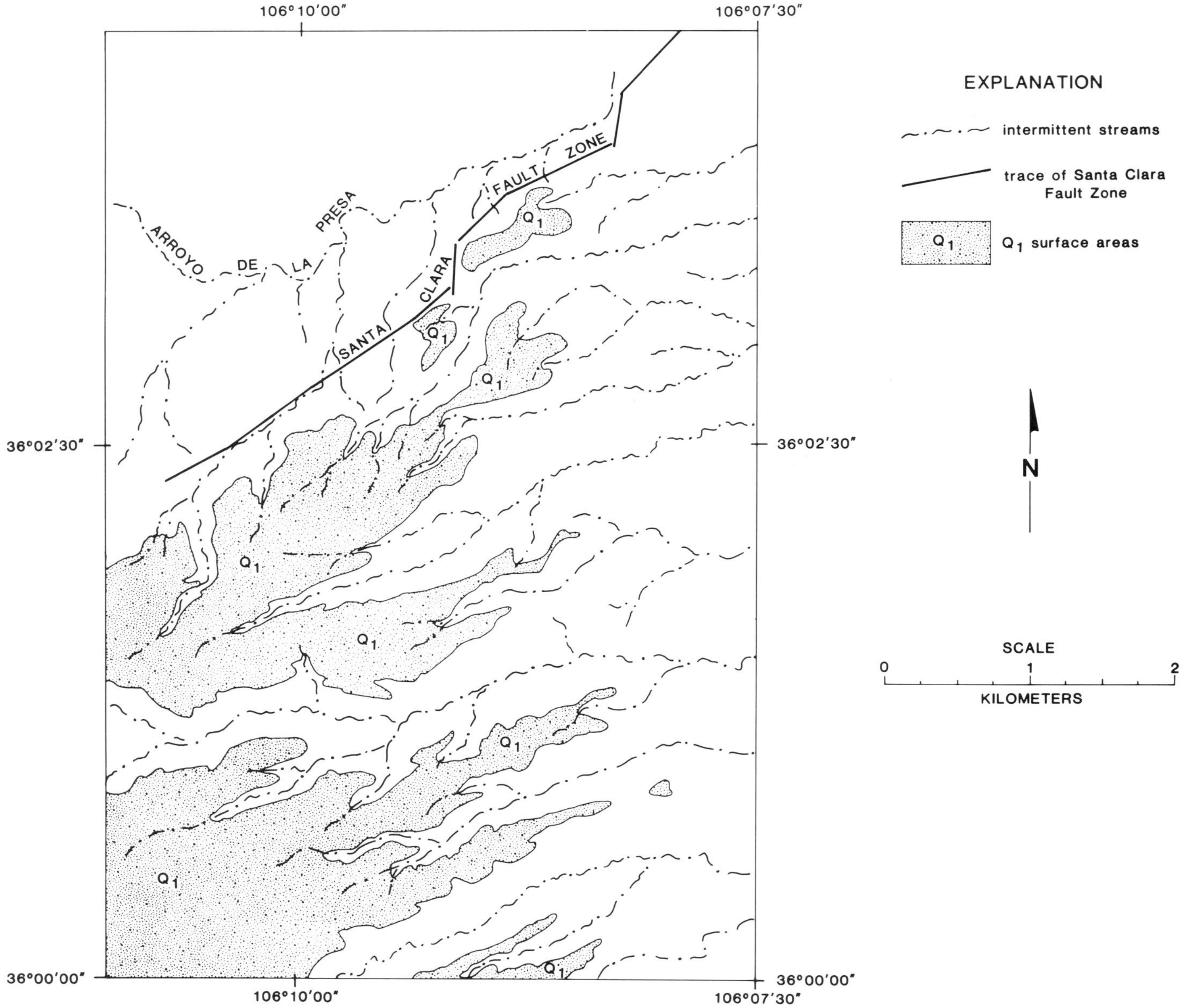

FIGURE 4. Drainage map of streams heading on or crossing the Q_1 surface within the study area.

erosive ability along the east-trending arroyos. The periods of arroyo incision that marked the termination of development of the Q_2 and Q_3 surfaces were significantly influenced by displacements on the north-trending faults.

DISCUSSION

Tectonic–geomorphic analysis along and south of the Santa Clara fault zone permits a general reconstruction of Quaternary geomorphic and tectonic events affecting this portion of the northeastern Jemez Mountains and central Rio Grande rift.

The oldest erosional surface (Q_1) is believed to have formed within the interval from 275,000 to 500,000 years B.P. Prior to the formation of this surface, significant movement had occurred along the Santa Clara fault zone. Analysis of slickensides indicates this movement was dominantly oblique slip. South of the fault, the Puye Formation overlies the Chamita Formation (upper Santa Fe Group); however, 2 km north of the fault a Puye(?) outcrop directly overlies the Ojo Caliente Member (middle Santa Fe) at an elevation of 2,040 m (Dethier and Manley, in press). Adjacent to the fault on the downdropped block the Chamita–Ojo Caliente contact is at 1,980 m, thus suggesting that a minimum of 60-m uplift with an unknown amount of lateral displacement occurred following deposition of Santa Fe Group sediments, but prior to deposition of the Puye. This time interval (~4.5–3.0 m.y.) represents an erosional episode leading to the development of an erosion surface called the Ortiz by Bryan (1938). Continued motion on the Santa Clara fault zone during and following deposition of the Puye Formation is indicated. The thick (>100 m) wedge of Puye beds south of the fault zone differs markedly from the few scattered outcrops to the north. The latter are elongated outcrops that occur adjacent to major arroyos and generally parallel their trend. These deposits may represent gravel fills in paleochannels cut in the erosional surface. If true, this suggests that either the wedge-like Puye deposits were never deposited north of the fault or, more likely, the upper Puye beds were so thin north of the fault zone that they eroded away between ~2.1 and 1.5 m.y. ago. In either case an elevated area must have existed north of the fault zone following formation of the pre-Puye erosion surface. Although the amount of uplift that occurred between the beginning of Puye deposition and formation of the Q_1 surface is not known, it was certainly of a much lesser magnitude than that which immediately followed deposition of the Santa Fe Group.

Both the Pajarito and Santa Clara fault zones were active following formation of the Q_1 surface (275,000–500,000 years B.P.). Deformation of the Q_1 surface along the Santa Clara fault zone indicates approximately 40 m of downwarping of this surface by movement on the fault.

The period of erosion that led to the dissection of the Q_1 surface terminated in the development of the Q_2 surface, culminating prior to 103,000 years ago. Deformation of this surface has been dominated by motion along the north-trending normal faults. Displacements on these faults have segmented the surface producing a sequence of steps (up to 25 m in height) generally down-to-the-east. No strong evidence exists for significant vertical motion on the Santa Clara fault zone during or following formation of the Q_2 surface. It is not clear if the initiation of dissection of the Q_2 surface was related to lowering of base level along the axial drainage by movement on faults near the southern margin of the Española Basin, but the possibility of this having happened seems reasonable. Certainly, dissection of the Q_2 surface within the study area was markedly influenced by movement along the north-trending faults.

There are indications that the period from 120,000 to 100,000 years ago was a time of extension within the central rift. Machette (1978) found that movement occurred along the County Dump fault in the north Albuquerque Basin 120,000 years ago. McCalpin (1982), based on analysis of offsets in various-age glacial deposits, reported that displacements occurred on a number of faults along the west flank of the Sangre de Cristo Mountains in Colorado (eastern margin of San Luis Basin) about 120,000 years ago. Apparently, movement on the north-trending faults west of Hernandez also occurred at this time.

Formation of the Q_3 surface was completed by about 22,000 years ago. Movement on several of the north-trending faults occurred during and subsequent to this time. During the formation of this surface the lower reaches of Arroyo de la Presa extended across the Santa Clara fault zone. The Arroyo de la Presa no longer follows this portion of its Q_3 channel, but instead turns north around an outcrop of a steeply dipping basalt flow, following a longer course to the Rio Grande than its Q_3 channel. We suggest that recurrent movement on the north- and northwest-trending faults increased gradients along the Q_3 paleochannel, causing rapid incision. The stream occupying the paleochannel was hindered by the presence of a resistant basalt flow cropping out across the channel. Increased gradient along an adjacent arroyo permitted rapid headward extension of the stream at an elevation below that of the Q_3 paleochannel. This stream eventually captured the Arroyo de la Presa stream, rerouting it north around the margin of the flow. Arroyo incision of the Q_3 surface has continued for the past 22,000 years at more rapid rates than those which existed earlier in the Quaternary, except for periods in the Holocene when alluvial terraces were formed (see Dethier and Demsey, this guidebook). Stratigraphic and geomorphic (soils) evidence indicates that recurrent movements are still occurring on the north-trending normal faults.

The ultimate cause of incision and abandonment of the Quaternary erosional surfaces is not certain. The role of climate in the development, or termination in development, of geomorphic surfaces is difficult to assess. Episodic climatic shifts which occurred during later Quaternary time, and the attendant changes in vegetation and discharge of water and sediment, must have produced a pronounced effect on surface development. Gile and others (1981) found that in the Las Cruces area of New Mexico times of climatic change from semiarid to arid conditions (associated with change from pluvial to interpluvial intervals) were marked by valley incision and a concomitant increase in landscape instability. The arid conditions were characterized by significant decreases in effectiveness of vegetative cover and increases in erosion and sedimentation. Climatic shifts from arid to semiarid climates were times of increasing landscape stabilization and more continuous vegetative cover (Gile and others, 1981).

Comparison of the timing of late Quaternary surface development and valley incision in the northeast Jemez Mountains to the Las Cruces area indicates that these areas have virtually the same landscape-evolution chronology. Valley incision in the Las Cruces area, which initiated dissection of the Tortugas and Picacho surfaces (<150,000 and <25,000 years, respectively), generally coincides with the beginning of valley incision into the Q_2 and Q_3 surfaces in the Española area. This coincidence in timing of late Quaternary landscape development suggests that climatic changes had a significant influence on the formation of surfaces within the northeast Jemez Mountains.

It is not clear whether climatic influences or basin-margin tectonics were the predominant cause of surface dissection. However, there is strong evidence that the terminations in development of the Q_2 and Q_3 surfaces were significantly influenced by recurrent movements along the north-trending faults.

APPENDIX

U–Th ages of pedogenic-carbonate rinds from the underside of pebbles were determined by T. L. Ku. Activity ratios of $^{234}U/^{238}U$ and $^{230}Th/^{234}U$ were calculated from the radiochemical data in Table A-1. The ages derived from these ratios indicate the time elapsed since carbonate illuviation began. Uncertainties given represent one standard deviation derived from counting statistics.

Sample 2-84-MJA, 36°03′05″N, 106°07′20″W. Q_2 surface. San Juan Pueblo 7½′ quadrangle, New Mexico. Activity ratio $^{234}U/^{238}U$ 1.31 ± .14, $^{230}Th/^{234}U$ .632 ± .065. Age 103 ± 17 Ka.

Sample 4-84-MJA, 36°03′13″N, 106°07′57″W. Q_3 surface. Chili 7½′ quadrangle, New Mexico. Activity ratio $^{234}U/^{238}U$ 1.19 ± .11, $^{230}Th/^{234}U$ .185 ± .019. Age 22 ± 3 Ka.

TABLE A-1. Radiochemical data on leachate (L) and residue (R).

Sample	f_L		Disintegrations per minute per gram of total sample			
			^{238}U	^{234}U	^{232}Th	^{230}Th
2-84-MJA	0.899	L	.662±.024	.856±.029	.045±.003	.553±.012
		R	.214±.010	.233±.011	.194±.010	.220±.010
4-84-MJA	0.828	L	3.05±.12	3.61±.14	.058±.002	.912±.015
		R	.599±.024	.690±.027	.097±.005	.529±.020

REFERENCES

Bachman, G. O., and Machette, M. N., 1977, Calcic soils and calcrete in the southwestern United States: U.S. Geological Survey, Open-file Report 77-794, 163 pp.

Bryan, K., 1938, Geology and ground-water conditions of the Rio Grande depression in Colorado and New Mexico; *in* Rio Grande joint investigation in the Upper Rio Grande Basin in Colorado, New Mexico and Texas: National Resources Committee, Regional Planning, pt. 6, pp. 197–225.

Dethier, D. P., and Manley, K. (in press), Geology of the Chili quadrangle, Rio Arriba County, New Mexico: U.S. Geological Survey, Miscellaneous Field Studies Map MF.

Gile, L. H., Peterson, F. F., and Grossman, R. B., 1966, Morphological and genetic sequences of carbonate accumulation in desert soils: Soil Science, v. 101, no. 5, pp. 345–360.

———, Hawley, J. W., and Grossman, R. B., 1981, Soils and geomorphology in the Basin and Range area of southern New Mexico—Guidebook to the Desert Project: New Mexico Bureau of Mines and Mineral Resources, Memoir 39, 222 pp.

Kelley, V. C., 1978, Geology of the Española Basin, New Mexico: New Mexico Bureau of Mines and Mineral Resources, Geologic Map 48.

———, 1979, Geomorphology of the Española Basin: New Mexico Geological Society, Guidebook 30, pp. 281–288.

Machette, M. N., 1978, Dating Quaternary faults in the southwestern United States by using calcic paleosoils: U.S. Geological Survey, Journal of Research, v. 6, no. 34, pp. 369–381.

Manley, K., 1976, The late Cenozoic history of the Española Basin, New Mexico: Ph.D. thesis, University of Colorado, Boulder, 171 pp.

McCalpin, J. P., 1982, Quaternary geology and neotectonics of the west flank of the northern Sangre de Cristo Mountains, south-central Colorado: Colorado School of Mines Quarterly, v. 77, no. 3, 97 pp.

The San Luis Valley Southern train at Jaroso in 1955. The grain elevator and the Conoco terminal in the background constitute the industrial district of Jaroso (photo by Bob Richardson).

Looking east at Jaroso today. The San Luis Valley Southern tracks ran across the street in front of the big tree. The railroad depot was located to the right, next to the Conoco station.

REGIONAL HYDROGEOLOGY AND THE EFFECT OF STRUCTURAL CONTROL ON THE FLOW OF GROUND WATER IN THE RIO GRANDE TROUGH, NORTHERN NEW MEXICO

L. M. COONS and T. E. KELLY
Geohydrology Associates, Inc., Albuquerque, New Mexico 87107

INTRODUCTION

Ground-water flow in any porous medium is influenced by various hydraulic properties of the medium, together with hydrologic and hydraulic boundaries surrounding it or within it. Two rather distinct regimes of ground-water flow exist in the Rio Grande structural trough in north-central New Mexico from the Colorado border south to Española; the structure and stratigraphy of the Rio Grande rift effectively control the movement of shallow ground water in the area, and also the movement of a deeper ground-water body contained in saturated sediments of the Santa Fe Group. Where the cross-sectional area of Santa Fe Group sediments has been constricted or narrowed, in places like the Embudo "constriction" between Taos and Española near Pilar, mixing of the regimes occurs. Velocity and direction of shallow ground-water flow change and become similar to those of the deeper ground water in transient storage. The significant features of the Rio Grande rift region in north-central New Mexico are shown in Figure 1.

REGIONAL GEOLOGY

Structural Setting

The Rio Grande rift is a structural feature extending the entire length of New Mexico. The dominant structural feature in the rift region is the broad sag, whose axis, in general, was followed by the Rio Grande. East of the axis, the beds dip generally westward. West of the axis, the dip of the beds is generally lower and less uniform in direction; south and east inclinations are most common. All of the faulting is normal and forms conjugate systems of fault sets dipping either east or west. The faults are relatively steep, dipping 50–80° from horizontal (Kelley, 1978).

All of the rocks underlying the Santa Fe Group deposits are considered in this study to form the basement complex. Definite hydrologic differences between the Santa Fe deposits and the pre-Santa Fe rocks make this distinction appropriate.

The Santa Fe Group was deposited in a series of grabens and fault structures which extends roughly from Leadville, Colorado, to the vicinity of Ft. Quitman, Texas. Bryan (1938, p. 197) identified eight structural basins which comprise the Rio Grande trough. According to this nomenclature, Taos is located near the southern end of the San Luis Basin, whereas Española is located in the Española Basin. According to Kelley (1952, p. 93), these two basins are separated by the "Embudo constriction" which was formed by the Brazos uplift and the Picuris salient (Fig. 1).

Basement Rocks

There are several distinctly different basement-rock complexes in the region. The basin fill rests predominantly on Precambrian-age crystalline basement. Crystalline rocks, mostly in the Sangre de Cristo range between Penasco and Santa Fe, consist of granite, gneiss, quartzite, various schists, and greenstones (Kelley, 1978). Metamorphic rocks (schists and quartzites) occur mostly in the Picuris salient and Brazos uplift (Fig. 1). The Chama Basin forms a wedge of Paleozoic, Mesozoic, and early- to middle-age Tertiary rocks. In the vicinity of Taos the crystalline-basement rocks range in age from Precambrian to Tertiary.

Santa Fe Group and Related Deposits

The Rio Grande trough or depression was filled by sediments during Miocene and early Pliocene time. During most of the Pliocene and earliest Quaternary time, tilting, faulting, erosion, and pedimentation took place. Since then, as much as 30% of the once-accumulated maximum volume of Santa Fe Group deposits has been removed (Kelley, 1978).

The name "Santa Fe marls" was first used by Hayden (1869, p. 60) to describe a sequence of sediments which had been deposited in the Rio Grande trough. Bryan (1938, p. 205) noted that "The main body

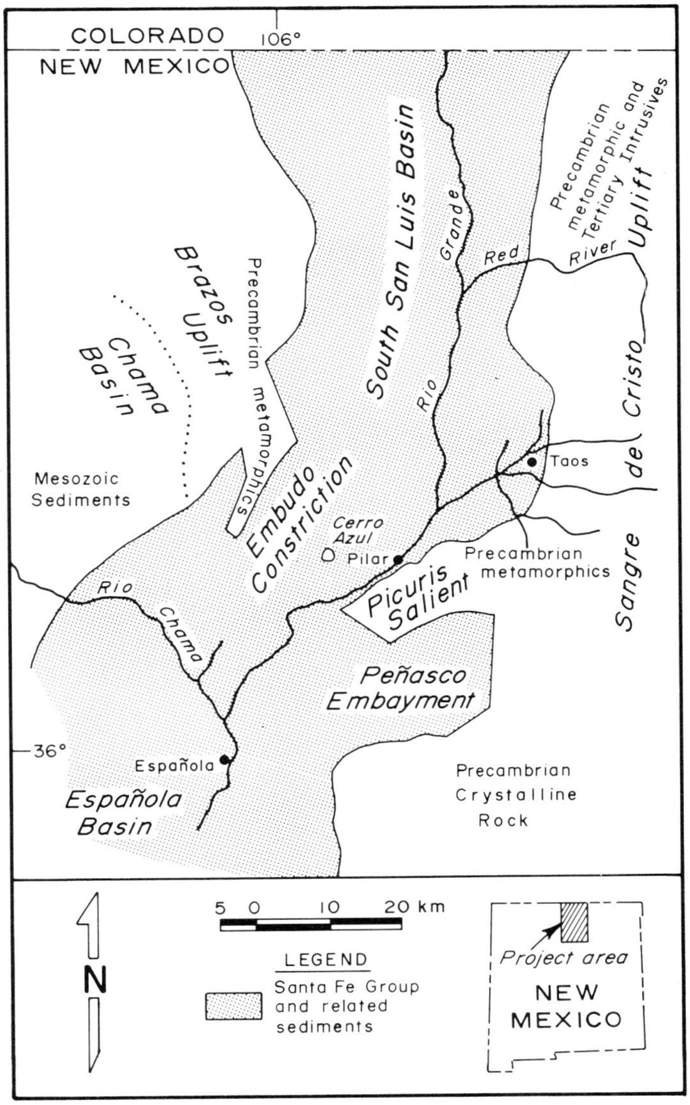

FIGURE 1. Map of north-central New Mexico showing major structural elements which form the Embudo constriction. Cerro Azul is a Precambrian block protruding through the late Tertiary Santa Fe sediments.

of sedimentary deposits of the Rio Grande depression, from the north end of the San Luis Valley, Colorado, to and beyond El Paso is considered to be the same general age and to belong to the Santa Fe Formation." Subsequently a great deal of stratigraphic nomenclature has been applied to various parts of the Santa Fe Group. In some cases the different terminology is based on lithologic characteristics (Smith, 1938; Spiegel and Baldwin, 1963; Bingler, 1968); in other instances the terminology is based on fossil content (Galusha and Blick, 1971). In other areas, different facies of the Santa Fe Group have been described without the application of additional nomenclature (Winograd, 1959; Wilson and others, 1978).

Spiegel and Baldwin (1963, pp. 34–35) first described the Tesuque Formation as a part of the Santa Fe Group. Where described south of Española, the formation ". . . consists of several thousand feet of pinkish-tan soft arkosic, silty sandstone and minor conglomerate and siltstone." The sorting and cementation are variable, but the beds are typically poorly sorted and unconsolidated. There are some interbedded basalt and ash deposits.

The Tesuque was deposited as a sequence of coalescing alluvial fans, and much of the material was reworked by the ancestral Rio Grande. Consequently, individual clastic beds probably are not continuous, and individual strata can seldom be traced for more than a mile or two (Miller and others, 1963, p. 50). Individual strata also are disrupted by faulting. The faults usually are less than 2 mi long (Manley, 1978, p. 12) and the displacement is usually less than 300 ft (Galusha and Blick, 1971, p. 101).

It is seldom possible to differentiate deposits of the Tesuque Formation in the subsurface. Also, it is often difficult to apply the surface nomenclature to subsurface deposits. Consequently, it has become a common practice to apply the name "Tesuque Formation" to any water-bearing strata in the Rio Grande basin which has a lithology similar to the type-Tesuque Formation.

The San Luis Basin may contain as much as 30,000 ft of sediments that are equivalent to the Santa Fe Group (Chapin, 1971, p. 191). Although this entire sequence may not be water-bearing, Emery and others (1971, p. 129) have estimated that the upper 6,000 ft are capable of producing water to wells. These thicknesses probably decrease to the south; McKinlay (1956, p. 21) estimated that the Santa Fe Group is about 3,000 ft thick in the Sunshine Valley north of Taos.

Due to the thickness of the volcanic beds on the Taos Plateau, very meager subsurface data from wells completed in the Santa Fe Group are available there. The thickness of the Santa Fe Group deposits has been estimated to be 3,000 ft or less; Kelley (1978) estimated a thickness of 8,000–9,000 ft south of Española.

Water-bearing Characteristics in the Southern San Luis Basin

The San Luis Basin is located in southern Colorado and extends as far south as the Embudo constriction; it includes the pediment slopes at the foot of the Sangre de Cristo Mountains and the basalt plateau west of the Rio Grande.

The four principal geologic units in the basin are the basement complex, the Santa Fe Group, the Cisneros Basalt, and fan–terrace deposits. The water-bearing characteristics of these units have been summarized by Winograd (1959) and Wilson and others (1978).

The basement complex in this area is non-water-bearing. The Precambrian and younger lithified strata of the basement complex are known to be highly fractured; however, there is very little intergranular permeability in these rocks.

According to Wilson and others (1978, pp. VI-11), drilling data from the Taos area indicated that the Santa Fe Group contains three distinct "sequences of rocks" or facies. These are: a lower sand facies, a middle facies of basalt deposits with interbedded lake and river deposits, and an upper facies of alluvial deposits. Winograd (1959, p. 15) recognized the same general sequence of strata in the Sunshine Valley; however, the basalt flows are much more prominent in that area.

Little is known of the hydrologic character of the lower facies. This is due principally to the fact that the lower unit generally is overlain by strata capable of producing adequate quantities of water for most needs. Wilson and others (1978, table 11) suggest that this lower unit may be a part of the Tesuque Formation. The maximum thickness probably does not exceed 3,000 ft and areally the unit pinches out against the mountain front. The water-bearing potential of this unit is unknown.

The middle facies of the Santa Fe Group at Taos is characterized by the presence of lake deposits and basalt flows (Cisneros Basalt of Kelley, 1978) interbedded with sediments more typical of the Santa Fe Group. The basalt flows thicken westward and interfinger with the relatively unconsolidated sand, silt, clay, and gravel of the Santa Fe Group toward the east. Ash beds are also present. Most individual basalt flows are less than 50 ft thick, but the cumulative thickness of basalt is as much as 670 ft near Taos Junction bridge (Winograd, 1959, p. 16). The basalt transmits water through the fractures and joints, as well as along bedding planes between flows. In addition, relatively thin deposits of unconsolidated sand and gravel which transmit additional water may be present between flows. Most well yields are 25 to 30 gpm (gallons per minute) (136 to 163 m^3/day).

The upper facies is the principal water-bearing zone in the Taos–Sunshine area where yields as high as 3,000 gpm have been reported. This aquifer consists of poorly sorted, unconsolidated deposits of clay, silt, sand, and gravel. Due to the variability of the deposits, few of the individual beds can be traced more than a short distance. Boulders as large as eight inches in diameter are present in exposures of these deposits. In most cases it is difficult to distinguish the surficial fan and terrace deposits from the underlying upper facies of the Santa Fe Group. Spiegel and Couse (1969) suggest that these alluvial deposits may have a greater water-producing capacity than the Santa Fe Group; however, they are thin, generally not exceeding 35 ft in thickness.

The Cisneros Basalt consists of highly fractured and jointed andesitic lava flows which have reported yields of 25–30 gpm (136–163 m^3/day). Most of the individual flows have zones of weathering on the upper surface. Thin deposits of alluvial and eolian sediments separate some of the flows. Owing to the fracture permeability in these basalts, there is very little runoff from the basalt plateau. Rather, the precipitation is either quickly lost by evapotranspiration or else it infiltrates the jointed Cisneros Basalt and moves laterally through joints and between individual flows. Much of this precipitation is discharged as springs along the wells of the Rio Grande gorge.

The cumulative thickness of the Cisneros Basalts is not known. On the west side of the plateau, near No Agua Mountain, a well was drilled to a depth of 700 ft without penetrating the Santa Fe deposits (Winograd, 1959, p. 14). The flows are about 670 ft thick in the gorge near Taos Junction bridge, and the total thickness decreases to less than 500 ft toward the north.

Water-bearing Characteristics in the Española Basin

The geologic units of the Española area are present as far north as the Embudo constriction. Along the west side of the basin, these units may be present beneath the lower Rio Chama in the vicinity of Abiquiu; however, they are not known to be present in the El Rito area.

Several geologic units have been identified in the Española Basin. These include the Tesuque Formation, Chama–El Rito Formation, Ojo Caliente Sandstone, the Cisneros Basalt, and alluvium. Galusha and Blick (1971, p. 64) and Kelley (1978) recognized that the Chama–El Rito Formation is a facies of the Tesuque, as is the Ojo Caliente Sandstone. Because of the similarity in lithology, it would be very difficult to differentiate these individual units in the subsurface.

Because of the similarity of these units, most studies have considered the entire geologic sequence to be a single hydrologic unit (Trauger, 1967; Koopman, 1975). In a comprehensive analytical study of the Pojoaque River basin, Hearne (1980) considered all the geologic units in the middle Rio Grande trough to be part of the Tesuque aquifer system. Yields of 500 gpm (2725 m^3/day) have been reported from wells drilled to sufficient depths.

The Cisneros Basalt which caps the Black Mesa north of Española is part of the extensive basalt-flow complex of the Taos Plateau. The basalt is highly fractured, and precipitation which falls on the Black Mesa is readily absorbed by these fractures and by unconsolidated alluvium which locally covers the flows. Much of the water is lost by evapotranspiration. When sufficient precipitation does occur, the water is discharged as intermittent springs along the edges of the mesa. Otherwise, the Cisneros Basalt is not considered to be water-bearing. Test holes drilled on top of Black Mesa are reported "dry" until they reach the level of the nearby Rio Chama and Rio Grande. This is a minimum depth of about 1,000 ft.

REGIONAL FLOW SYSTEMS

Shallow Ground Water and the Zone of Transient Storage

Emery (1971, p. 129) described two aquifer systems in the San Luis Valley of Colorado which contain at least two billion acre-feet (2.5 million cubic hectometers) of water in storage in the upper 6,000 ft. Although he described the deeper aquifer as a confined system, the discontinuity of the "clay series" resulted in varying degrees of hydraulic connection with the shallow aquifer. The shallow aquifer was found to be unconfined and therefore responsive to local conditions of recharge and discharge. The semi-confining "clay series" was generally present at depths of 50 to about 200 ft (15.2–61 m).

Underflow southward across the state line would be restricted by the San Luis Hills, a series of volcanic extrusives which form a constriction to ground-water flow approximately 5–10 mi (8.0–16.1 km) north of the state line. Nevertheless, McKinlay (1956, p. 21) estimated that there are about 3,000 ft (914 m) of saturated sediments near the state line, and the trough is approximately 21 mi (33.8 km) wide at this point. Assuming conservative aquifer parameters, there may be as much as 123 ft^3/sec (3.5 m^3/sec) of ground-water underflow from Colorado to New Mexico across the state line. This is compared to the long-term average discharge in the Rio Grande of 325 ft^3/sec (9.2 m^3/sec) at the state line (USGS, 1977, p. 85).

Emery's concept of two aquifer systems is applicable in New Mexico as well as Colorado. Although there is no evidence of a persistent confining layer, two different flow systems probably exist. The lower zone of transient storage would represent that ground water which flows through the deeper parts of the trough without being influenced by the shallow water-table system.

The zone of transient storage extends from the bed of the Rio Grande to the bottom of the unconsolidated sediments in the trough. The cross-sectional area would vary with the structural configuration of the Rio Grande basin (Fig. 1). At the state line the width of the zone would be about 21 mi (33.8 km), but this would be reduced to only 12 mi (19.3 km) in the Embudo constriction. The upper surface of the zone of transient storage would coincide with the bed of the Rio Grande; the gradient would generally be from north to south at the rate of about 21 ft/mi (4 m/km).

The shallow-aquifer system would be that zone of saturation between the water table and the zone of transient storage. This aquifer is influenced by various components of the hydrologic cycle, such as recharge by precipitation and snowmelt, surface runoff, depletion by shallow ground-water evaporation and transpiration, pumping from wells, and various discharges from springs. North of the Embudo constriction, the shallow aquifer is a major source of recharge to the Rio Grande.

Ground-water Flow Through the Embudo Constriction

The amount of aquifer material in the Embudo constriction is considerably less than in areas north and south. Not only has the cross-sectional area been reduced, but Cerro Azul, a Precambrian outlier, is located near the middle of the constriction. Consequently, the Santa Fe deposits probably are quite thin in the middle of the structural confinement. Inasmuch as the constriction has produced a vast reduction in the amount of permeable material in the trough, the ground-water flow regime would also be affected.

In order for the same quantity of water to flow through the constriction from north to south, the velocity of flow must increase. A water-table contour map of the area shows that the predominant direction of shallow ground-water flow in the constriction is from north to south, parallel with the river channel, in order to move the required quantity of ground

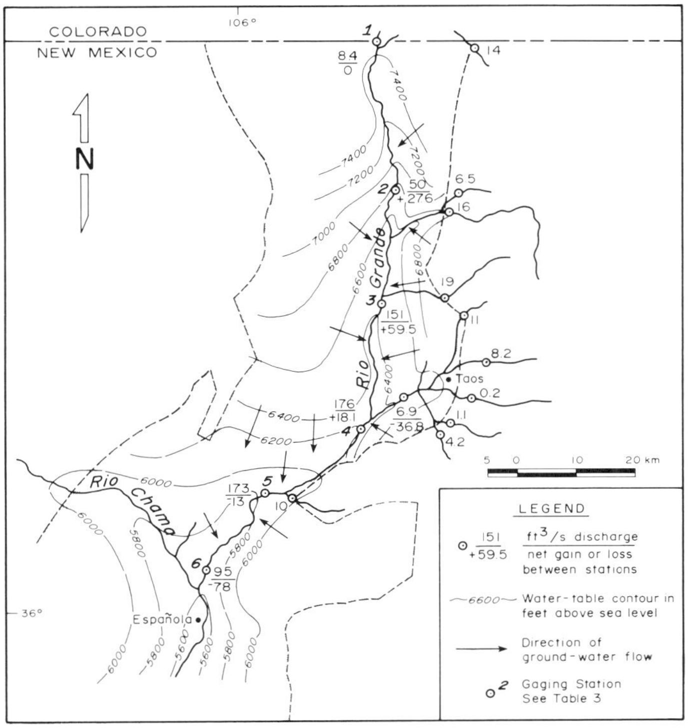

FIGURE 2. Water-table contour map and location of gaging stations in upper Rio Grande basin, New Mexico.

water through the constriction (Fig. 2). Elsewhere the flow direction is normal to the river, showing that there is ground-water discharge from the shallow aquifer to the Rio Grande.

An evaluation of stream-flow records shows that there is a significant increase in the flow of the Rio Grande north of the Embudo constriction. Through the constriction and southward, ground-water gains are minimal.

For the purposes of analysis, the date of September 10, 1977, was selected. It was on this date that the lowest discharge was recorded by the U.S. Geological Survey at the Rio Grande above San Juan Pueblo station. This was the southernmost station used in the analysis, and therefore indicated that low-flow conditions prevailed on that date. The stations used in the analysis are listed in Table 1. Data are from U.S. Geological Survey (1977).

The flow at the state line (station 1) on September 10, 1977, was 8.3 ft^3/sec (0.24 m/sec)—the amount of surface flow entering New Mexico from Colorado. At station 2, a distance of 20 river miles (32.2 km), the flow was 50 ft^3/sec (1.4 m^3/sec), or a gain of 27.6 ft^3/sec (0.78

TABLE 1. Discharge stations on the Rio Grande and its tributaries north of San Juan Pueblo.

	Name	Location		Discharge (ft^3/sec)
1.	Rio Grande at Colorado-NM State Line	37°00'03"	105°43'19"	8.4
	Costilla Creek near Costilla	36°58'01"	105°30'23"	14
2.	Rio Grande near Cerro	29.12.20		50
	Red River near Questa	29.13.32		16
	Cabresto Creek near Questa	29.13.21		6.5
	Rio Hondo near Valdez	36°32'30"	105°33'21"	19
3.	Rio Grande near Arroyo Hondo	27.12.31		151
	Rio Pueblo de Taos near Taos	26.13.36		8.2
	Rio Lucero near Arroyo Seco	36°30'30"	105°31'49"	11
	Rio Fernando de Taos near Taos	25.13.27		.24
	Rio Grande del Rancho near Talpa	36°17'52"	105°34'55"	4.2
	Rio Chiquita near Talpa	36°19'55"	105°34'42"	1.1
	Rio Pueblo de Taos below Los Cordovas	36°22'39"	105°40'05"	6.9
4.	Rio Grande below Taos Junction Bridge	24.11.15		176
	Embudo Creek at Dixon	23.10.19		10
5.	Rio Grande at Embudo	23.9.23		173
6.	Rio Grande above San Juan Pueblo	21.8.10		95

m^3/sec) when adjusted for inflow from upper Costillo Creek near Costillo. This represents a ground-water inflow of 618 gal/min (3,368 m^3/day) per river mile. The gain is even more significant between stations 2 and 3 which show an average gain of 1,666 gal/min (9,080 m^3/day) per river mile of channel. This reach of the Rio Grande is noted for the number of perennial springs along the channel.

The ground-water inflow diminishes to about 400 gal/min (2,180 m^3/day) per mile between stations 3 and 4, which includes the contributions from Rio Pueblo de Taos and its headwater streams. It should be noted that there is a net loss along this tributary system due to irrigation diversions in the Taos area.

Records for this date show that there was a net gain of 105.2 ft^3/sec (3.0 m^3/sec) to the Rio Grande from ground water between the state line and station 4 which is located at the upper end of the Embudo constriction. However, there was a net loss of 13 ft^3/sec (0.37 m^3/sec) between stations 4 and 5, which coincides with the length of the constriction.

The significant losses which occur between stations 5 and 6 can be attributed to diversion of streamflow for irrigation between Dixon and Española.

SUMMARY

Ground-water flow in the Rio Grande trough in northern New Mexico is affected by both the stratigraphy and structure of the region. Although as much as 8,000–9,000 ft (2,438–2,743 m) of basin-fill material of the Santa Fe Group fill the trough in many places, not all of this material is saturated. In the Embudo constriction, the amount of basin-fill material is decreased by structural controls. Shallow ground-water flow through the constriction changes direction in order to move the required amount of flow past that location. In the constriction, shallow and deep ground-water flow share a common direction, and probably a common gradient, from north to south. Ground-water budget calculations suggest a sharp decrease in the quantity of ground water draining into the Rio Grande through the constriction, thus indicating that a relatively larger quantity of water is moving in the subsurface through the constriction that is draining into the Rio Grande and then moving through.

REFERENCES

Bingler, E. D., 1968, Geology and mineral resources of Rio Arriba County, New Mexico: New Mexico Bureau of Mines and Mineral Resources, Bulletin 91, 158 pp.

Bryan, K., 1938, Geology and ground-water conditions of the Rio Grande depression in Colorado and New Mexico; *in* (U.S.) National Resources Planning Board, The Rio Grande joint investigation in the upper Rio Grande basin: U.S. Government Printing Office, Washington, D.C., v. 1, pt. 2, pp. 197–225.

Chapin, C. E., 1971, The Rio Grande rift, part I: modifications and additions: New Mexico Geological Society, Guidebook 22, pp. 191–201.

Emery, P. A,. Boettcher, A. J., Snipes, R. L., and McIntyre, H. J., Jr., 1971, Hydrology of the San Luis Valley, south-central Colorado: U.S. Geological Survey, Hydrologic Inventory Atlas HA-381.

Galusha, T., and Blick, J. C., 1971, Stratigraphy of the Santa Fe Group: American Museum of Natural History, Bulletin, v. 144, art. 1, 127 pp.

Geohydrology Associates, Inc., 1979, Preliminary assessment of water resources near Las Placitas and El Rito, New Mexico: Consultant report for Charles F. Horne & Associates, Inc. and Placitas Mutual Domestic Water Consumers Association (funding provided by Interstate Stream Commission), 67 pp.

Hayden, F. V., 1869, Preliminary field report of the U.S. Geological Survey of Colorado and New Mexico: U.S. Geological Survey, Third Annual Report, 155 pp.

Hearne, G. A., 1980, Mathematical model of the Tesuque aquifer system underlying Pojoaque River basin and vicinity, New Mexico: U.S. Geological Survey, Open-file Report 80-1023, 181 pp.

Kelley, V. C., 1952, Tectonics of the Rio Grande depression in central New Mexico: New Mexico Geological Society, Guidebook 3, pp. 93–105.

———, 1978, Geology of Española Basin, New Mexico: New Mexico Bureau of Mines and Mineral Resources, Geologic Map 48.

———, 1982, The right-relayed Rio Grande rift, Taos to Hatch, New Mexico: New Mexico Geological Society, Guidebook 33, pp. 147–151.

Koopman, F. C., 1975, Estimated ground-water flow, volume of water in storage, and potential yield of wells in the Pojoaque River drainage basin, Santa Fe County, New Mexico: U.S. Geological Survey, Open-file Report 74-159, 33 pp.

Manley, K., 1978, Structure and stratigraphy of the Española Basin, Rio Grande rift, New Mexico: U.S. Geological Survey, Open-file Report 78-667, 24 pp.

McKinlay, P. F., 1956, Mineral deposits; *in* Geology of Costilla and Latir Peak quadrangles, Taos County, New Mexico: New Mexico Bureau of Mines and Mineral Resources, Bulletin 42, 32 pp.

Miller, J. P., Montgomery, A., and Sutherland, P. K., 1963, Geology of part of the southern Sangre de Cristo Mountains, New Mexico: New Mexico Bureau of Mines and Mineral Resources, Memoir 11, 106 pp.

Roberson, J. A., and Crowe, C. T., 1975, Engineering fluid mechanics: Houghton Mifflin Company, Boston, 520 pp.

Smith, H. T. U., 1938, Tertiary geology of the Abiquiu quadrangle, New Mexico: Journal of Geology, v. 46, no. 7, pp. 933–965.

Spiegel, Z., 1954, Report on the possibilities of obtaining public water supplies at El Rito (Rio Arriba County), New Mexico: New Mexico State Land Office, Open-file Report, 8 pp.

———, and Baldwin, B. (with contributions by F. E. Kottlowski and E. L. Burrows, and a section on geophysics by H. S. Winkler), 1963, Geology and water resources of the Santa Fe area, New Mexico: U.S. Geological Survey, Water-supply Paper 1525, 258 pp.

———, and Couse, I. W., 1969, Availability of ground water for supplemental irrigation and municipal–industrial uses; *in* Taos Unit of U.S. Bureau of Reclamation, San Juan–Chama Project, Taos County, New Mexico: New Mexico State Engineer, Open-file Report, 22 pp.

Trauger, F. D., 1967, Hydrology and general geology of the Pojoaque area, Santa Fe County, New Mexico: U.S. Geological Survey, Open-file Report, 32 pp.

Wilson, L., Anderson, S. T., Jenkins, D. N., and Lovato, P., 1978, Water availability and water quality, Taos County, New Mexico: Lee Wilson and Associates, Inc., Santa Fe, New Mexico.

Winograd, I. J., 1959, Ground-water conditions and geology of Sunshine Valley, western Taos County, New Mexico: New Mexico State Engineer, Technical Report 12, 70 pp.

HYDROGEOLOGIC CROSS SECTION THROUGH SUNSHINE VALLEY, TAOS COUNTY, NEW MEXICO

W. K. SUMMERS and L. L. HARGIS
W. K. Summers and Associates, Inc., P.O. Box 684, Socorro, New Mexico 87801

INTRODUCTION

This paper presents a reinterpretation of hydrogeologic conditions in the Sunshine Valley area, east of the Rio Grande in Taos County. Winograd (1959) collected and studied an extensive suite of well records in this area and concluded that perched or partly perched aquifers are present, particularly in the large part of the area underlain by fine-grained deposits which he thought were lacustrine. Our interpretation is based on data from 145 wells (Fig. 1), including many of those studied by Winograd. We have also used the records of 16 additional wells just beyond the northern and southern boundaries of the area shown in Figure 1. Our analysis led us to doubt the lacustrine origin and hydraulic-confining function of the fine-grained sediment, and we concluded that perched conditions are much less extensive than Winograd thought. We therefore proceeded to investigate the form of the main or regional water table on the assumption that any perched aquifers are local phenomena.

We first describe our assumptions and procedures. Then we present an analysis of the vertical form of the flow field along a cross section normal to the river (Fig. 2), and discuss the difference between our interpretation and that of Winograd (Fig. 3).

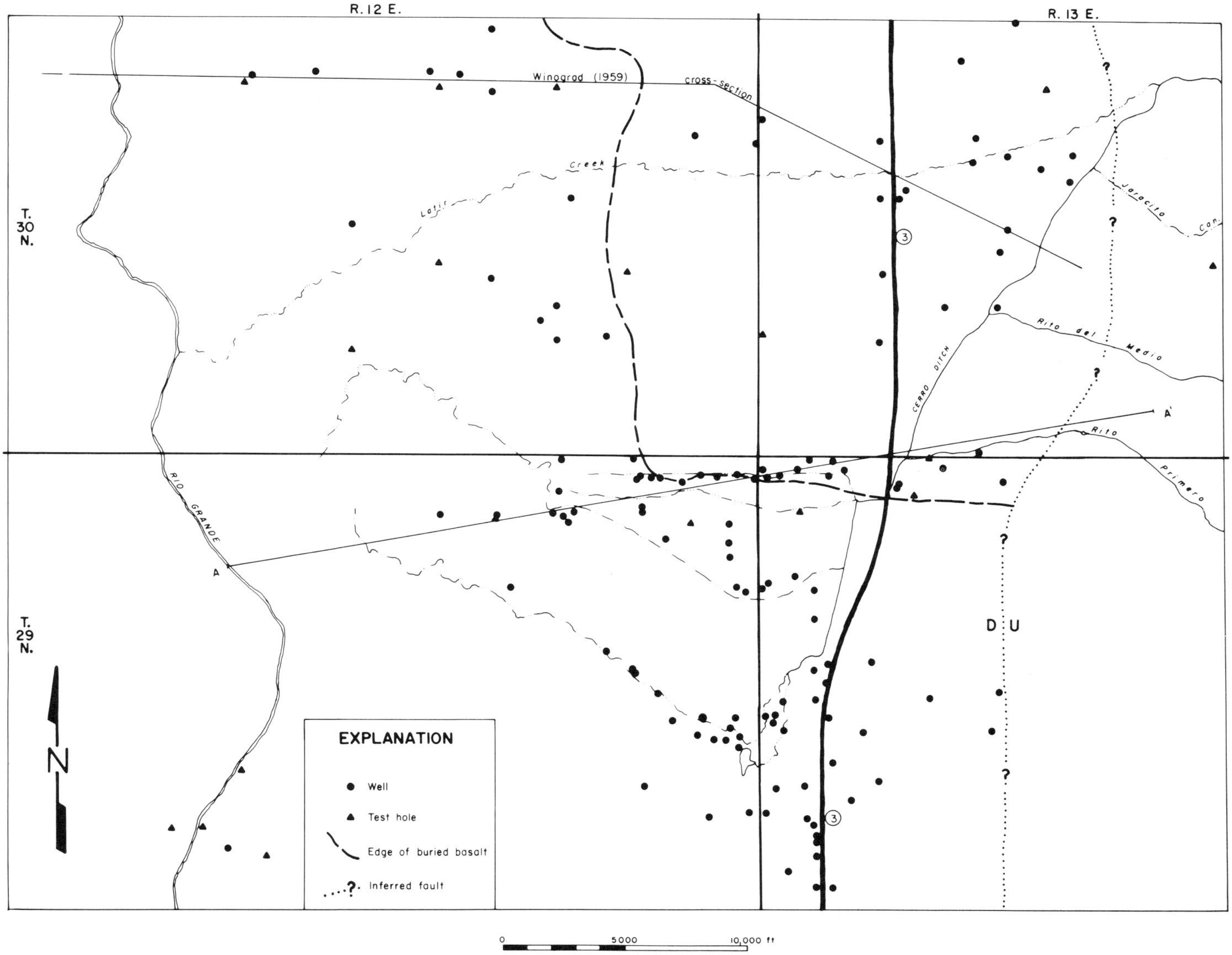

FIGURE 1. Reference map of part of Sunshine Valley, Taos County, New Mexico.

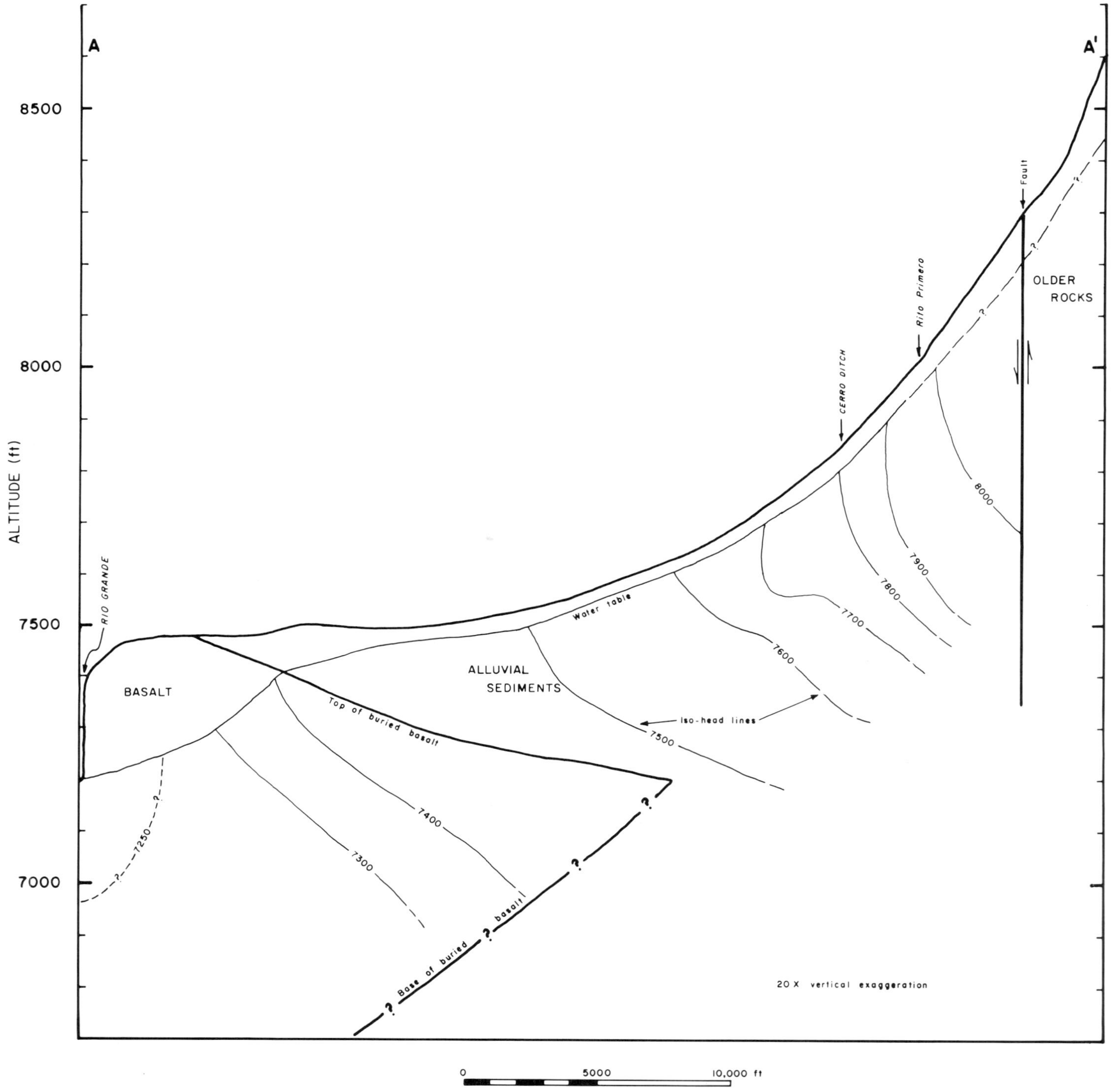

FIGURE 2. Cross section A–A′.

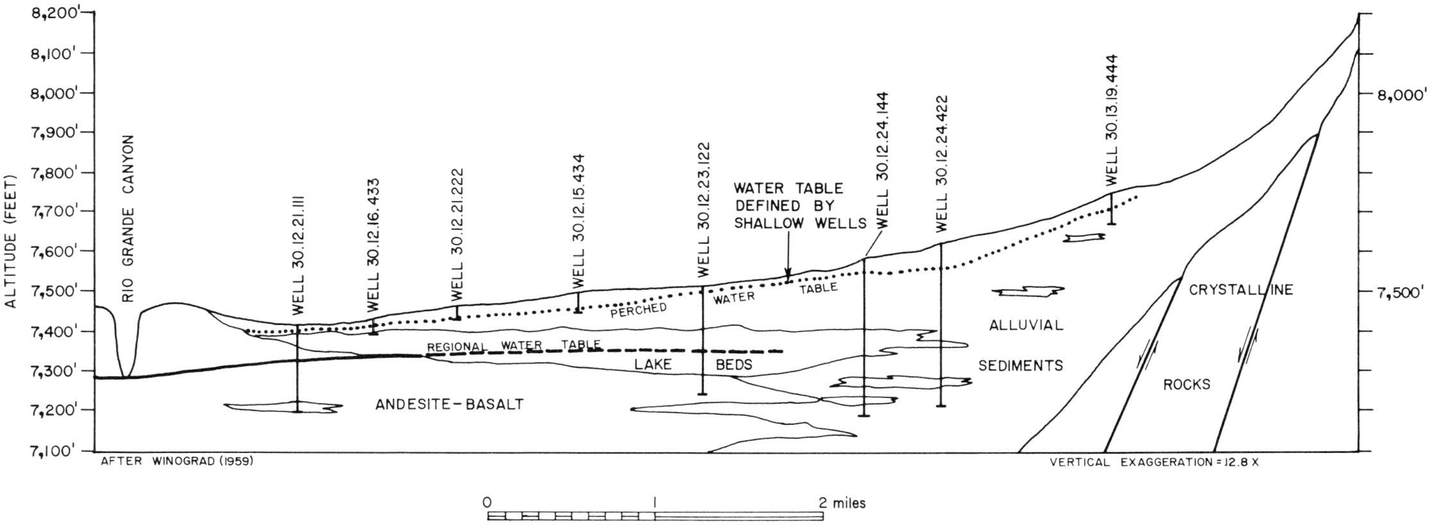

FIGURE 3. Winograd's cross section.

BACKGROUND

Winograd presented an east–west cross section through Township 30 North that showed his interpretation of hydrogeologic conditions in the Sunshine Valley, Taos County, New Mexico. Figure 3 of this paper shows the salient features of Winograd's cross section east of the Rio Grande. As the figure shows, he recognized two water tables; one he labeled as "Water table in alluvial sediments. Semiperched where underlain by saturated andesite–basalt. Perched in vicinity of test hole 30.12.21.111." The other he labeled as "Water table in andesite–basalt. Solid line indicates water table; dashes indicate piezometric surface of water confined in andesite–basalt under subnormal artesian pressure. Water in alluvium is semiperched from well 30.12.16.433 to a point east of well 30.12.23.122; perched from about well 30.12.16.433 to a point west of well 30.12.21.111."

Winograd came to this conclusion using data from 141 wells that he inventoried during 1955 and 1956 in an area of about 800 mi^2. He also came to the conclusion that clay beds in the western part of the area were lacustrine in origin and referred to them on his cross section as "lake beds."

Figure 1 shows the location of (1) wells in the area, (2) Winograd's cross section (our Fig. 3), (3) our cross section (Fig. 2), (4) the eastern edge of the basalt, and (5) the approximate location of the western bounding fault that separates rocks of Tertiary age from the rocks of Precambrian age.

METHODS

Figure 2 shows the land surface, the water table, the top of the basalt, and the approximate location of the fault. It also shows lines of equal total hydraulic head, which we call "iso-head contours."

The topographic profile came from U.S. Geological Survey topographic maps. We inferred the location of the fault from Winograd's geologic map and cross section. The water table, top of the basalt, and iso-head contours came from maps we drew using data from the wells shown in Figure 1.

We based the basalt map on the altitude of the basalt in outcrop, the depths to basalt reported in wells, and the altitudes of the bottoms of wells that did not reach basalt. The edge of the basalt shown in Figure 2 agrees fairly well with the edge of the basalt that Winograd shows.

To draw the water table map we made four assumptions:

(1) The water table exists everywhere and is a smooth, continuous surface that intersects the Rio Grande.
(2) Except where it intersects the Rio Grande, the water table is below the land surface.
(3) The altitude of the water level in a well whose depth is 100 ft or less equals the altitude of the water table.
(4) The water table in a well must be higher than the altitude of the bottom of the well.

We raised a basic question: If the water table is a smooth, continuous surface and flow occurs from recharge to discharge areas, how can we go about defining the hydraulic-head distribution beneath the water table using existing well data? To answer this question, we began by reasoning that:

(1) The water-level altitude in any well is some sort of a vertical average of the head in the rocks to which the well is open (that is of the slotted, screened, perforated, or open-hole interval).
(2) In areas such as the Sunshine Valley where we would expect the hydraulic gradient to have a large vertical component directed downward, the head at the bottom of the well must be lower than the head at any point above the bottom.

Therefore, the water-level altitude of a well is larger than the head at the bottom, but if we represent the water-level altitude of the well as the head at the bottom, we will be making a conservative estimate of the head at the bottom of the well.

To draw the iso-head contours on the cross section, we drew a series of four water-level altitude maps based on the depths of the wells. The water-table map was based on the depth interval from 1 to 100 ft, so the first iso-head map was based on wells ranging in depth from 101 to 150 ft. Other depth intervals used were 151–200 ft, 201–300 ft, and 301–500 ft.

To draw the iso-head contours in the plane of the cross section, we plotted the values of the contours at the bottom of their depth range below the land-surface profile. Then, starting at the water table, we drew lines through points of equal value. Making the iso-head contour interval 100 ft circumvented the problem of using water-level data obtained over several years. Because U.S. Geological Survey records show that seasonal water-level fluctuations and the variation in water level caused by pumping in the area are less than 50 ft, we estimate that the contours should be within approximately 50 ft of the mean annual condition.

The 7,250-ft iso-head contour line on the cross section reflects the existence downstream of the 7,200-ft iso-head contour on two of the iso-head maps.

DISCUSSION

Figure 2 shows a dynamic ground-water system. Water moves downward from the land surface to the Rio Grande. In the immediate vicinity of the river the vertical component of hydraulic gradient is directed upward.

The cross section differs from Winograd's in the following ways:

(1) It shows one continous water table.
(2) It shows a continuous uninterrupted zone of saturation.
(3) It does not discriminate between lake beds and other sediments of Tertiary age.

If we argue that the water table is the top of the zone of saturation, then the water table of Figure 2 marks the top of the zone in which the rocks are saturated. Winograd believed that beneath the water table in the alluvium are rocks in which the degree of saturation is less than 100%. Moreover, his water table implies movement of water toward the river through the sediments. However, no springs occur in the wall of the Rio Grande gorge, and we have no reason to believe that water evaporates or transpires from the water table, so the water must be moving downward. Figure 2 shows that the physics of the system need not require perching to explain the observed water levels. However, to resolve the question, several suites of piezometers would have to be installed.

Another problem posed by Winograd's interpretation is: What happens to the water table in the basalt at its eastern extremity? Winograd recognized that recharge from the streams that drain the Sangre de Cristo Range must move to the Rio Grande. The head distribution depicted in Figure 2 shows water moving from the water table at higher altitudes toward the river, taking some of the mystery from Winograd's interpretation of recharge, but in no way diminishing its significance.

The term "lake beds" implies that the rocks include clays that are less permeable than the alluvial clays elsewhere in the valley, that the clay beds are more continuous than elsewhere, and that as a result they should cause "perching" of ground water. After studying the well logs of the area, which includes geologists' lithologic descriptions, electric logs, and drillers' logs, we cannot justify the term "lake beds" for the rocks that overlie the basalt. Perhaps even more important, we could not establish criteria for defining the top of the lake beds so that we could map the top. The beds consist of clays plus variable amounts of sand and gravel, which are much like the alluvial deposits we see elsewhere in the Rio Grande valley. The rocks between the mountains and the Rio Grande consist of stream deposits that reflect the changes in stream velocity (or sediment-carrying capacity) as well as the sediment source and climatic variables. Ponds or playas may have developed locally, but they do not constitute the primary environment of deposition. These beds serve to "perch" ground water no better than do other fine-grained alluvial deposits in the valley.

ACKNOWLEDGMENTS

We thank the Wanda McGee Trust administrators, Robert Waltrip, and Robert Waltrip, Jr., for their support and for letting us publish data and maps originally prepared for them.

REFERENCES

Winograd, I. J., 1959, Ground-water conditions and geology of Sunshine Valley and western Taos County, New Mexico: New Mexico State Engineer, Technical Report no. 12, 70 pp.

THERMAL MINERAL SPRINGS IN CAÑON DE SAN DIEGO AS A WINDOW INTO VALLES CALDERA, NEW MEXICO

FRANK W. TRAINER
U.S. Geological Survey, 345 Middlefield Road, Menlo Park, California 94025
(Present address: P.O. Box 1735, Corrales, New Mexico 87048)

INTRODUCTION

Two groups of mineral springs are among the best-known surficial thermal features in the Jemez Mountains. Detailed chemical data for the spring waters were first collected in 1912. Despite accumulation of considerable information since then, the springs remain puzzling features. The objective of this report is a re-examination of the springs as a basis for inference of geohydrologic conditions up-canyon, toward and into Valles caldera.

The complex pile of volcanic rock that forms the Jemez Mountains straddles the western marginal fault zone of the Rio Grande rift (Fig. 1). Pre-volcanic rocks—Precambrian crystalline rocks and Paleozoic and Mesozoic sedimentary rocks outside the rift, and Tertiary and Quaternary valley-fill deposits within it—are covered in most of the Jemez Mountains. Volcanism began in late Tertiary time and culminated, about 1.4 and 1.0 m.y. ago, in explosive eruptions that led to the successive formation of two calderas and two extensive tuff sheets. The subsequent history of the younger (Valles) caldera included formation of a resurgent structural dome and of a series of rhyolite domes in the ring-fracture zone (Smith and Bailey, 1968); fumarolic and hot-spring activity that has occurred during at least part of the post-caldera period continues at present.

Part of the main aquifer in Valles caldera is known to be a hot-water geothermal reservoir beneath the resurgent dome. Vapor-dominated conditions are present locally. The reservoir rocks are fractured tuff, sedimentary and crystalline rocks, and semi-consolidated valley-fill sediments. The geothermal water, dominated by Na and Cl, contains about 2,500 mg/L Cl. Temperatures are on the order of 260°C and higher (Dondanville, 1978; Kerr, 1982). Recharge of the reservoir is believed to be by infiltration of meteoric water along fractures. Subsurface discharge from the reservoir is also controlled by fractures, and is largely through limestone beneath Cañon de San Diego, the canyon which carries surface drainage from the caldera.

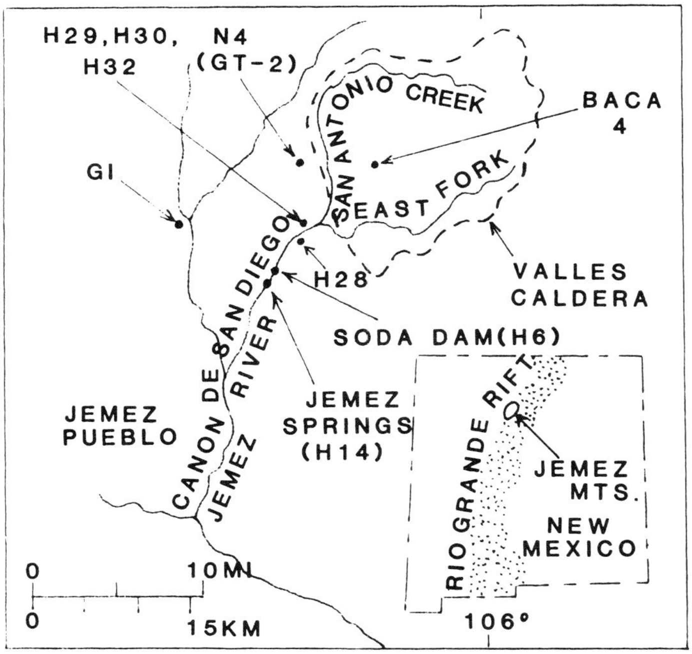

FIGURE 1. Map showing locations of features described; inset shows location of Jemez Mountains relative to Rio Grande rift.

The two groups of thermal mineral springs are at Soda Dam and Jemez Springs (Fig. 1) in the canyon, about 5.8 and 7.5 km, respectively, outside the rim of the caldera. Summers (1976) described the history and the geographic setting of the springs and presented representative chemical data. Goff and Kron (1980) have recently mapped the geology near the springs and logged a geothermal well at Jemez Springs. At Soda Dam the springs flow from fractured rock where a fault contact between sedimentary and crystalline rocks is exposed in the canyon floor. Jemez Springs issue from alluvium, but alignment of the springs indicates control by faults in the underlying bedrock. Observations to date have shown large, consistent, and puzzling differences between the two spring waters.

THE MINERAL WATERS

The waters at Soda Dam and Jemez Springs have Na and Cl as principal ions, high concentrations of Ca and HCO_3, and minor constituents that include B, Br, and Li (Table 1 and Fig. 2). The ratios Li/Na and B/Cl are in the ranges described by White (1957, pp. 1669, 1671) as typical of thermal waters in volcanic terranes. The Soda Dam water is nearly twice as mineralized as that at Jemez Springs, but the latter water is markedly hotter—75°C as against 50°C. Both groups of springs emit H_2S fumes.

Results of numerous studies show that the chemical character of these waters is unlikely to have formed under near-surface conditions. White and others (1963, p. F42) considered the water at Jemez Springs typical of thermal water from non-geyser areas associated with volcanism; Woltz (1972, pp. 42, 45–47) concluded from study of selected constituents that these waters, among others in the region, contain contributions from magmatic sources. After the discovery of subsurface geothermal conditions in Valles caldera in the 1960's during drilling for oil, the caldera was widely discussed as a possible source of the heat and mineral constituents in the spring waters. Dondanville (1971, p. 15) noted the similarity between geothermal fluid and the mineral water in Cañon de San Diego, although the relationship of the sample he cited to fluid in the main part of the reservoir as it has later become known has not been established. Trainer (1974) concluded that ion ratios in the spring waters are consistent with derivation from Valles caldera and (1975) suggested that mixing of geothermal and dilute ground waters could explain the mineral-spring waters. Trainer and Lyford (1979) and Goff and others (1981) used chloride-variation diagrams to show possible relationships between water types. Goff and others (1981, fig. 6) were the first to include data from the geothermal reservoir in their plots, and their diagrams provide the most convincing argument published to date for a caldera provenance for the mineral waters. New data from the reservoir (White and others, this volume; and BACA 4, Table 1 and Figs. 2, 3) make the case for a caldera source for the mineral waters even more convincing.

The mixing hypothesis assumes that geothermal fluids flow out of the caldera through the subsurface and mix with dilute ground water along discrete flow paths. So long as the dilute component of the mixture provides little or none of the diagnostic constituents, the ratios of those

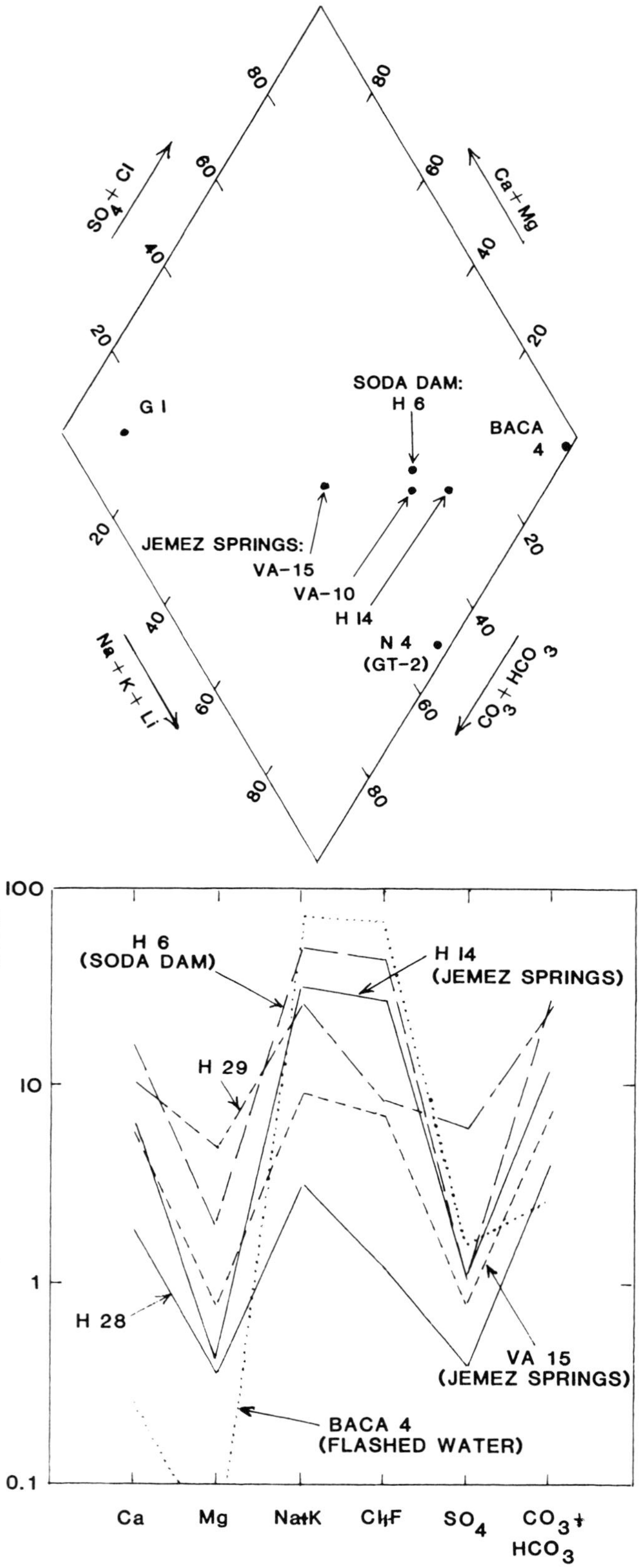

FIGURE 2. Chemical diagrams for waters. *Diamond field of Hill–Piper diagram* shows composition of geothermal and dilute waters, and of mixed waters at Soda Dam (H 6) and Jemez Springs (H 14, H 17, VA-15); analyses in Table 1 except VA-15 ("500-foot aquifer") from Goff and others (1981), and H 17 (78-foot well in Jemez Springs) from Trainer (1978). See Figure 1 for locations. *Schoeller (semilogarithmic) diagram* shows composition of waters and concentration of major constituents.

constituents remain constant over a wide range of concentrations; hence the linear relationship in the variation diagram (Fig. 3). The inferred course of change in composition is illustrated by the diamond diagram in Figure 2. Geothermal water from near the right (Na–Cl) vertex, on mixing with water in limestone (similar to G 1, near the $CaCO_3$ vertex) and with water from tuff or granitic rock (points for which fall near the southwest edge of the diamond), attains a composition similar to that of H 6 (Soda Dam), or of H 14 or VA-10 (Jemez Springs). Goff and others (1981, fig. 6) further illustrated a mixing line for samples from several springs at Jemez Springs which is represented in the diamond diagram by the shift from H 14 and VA-10 to VA-15. This further mixing appears to have occurred in the limestone near the springs, through dilution of mineral water by limestone water similar to G 1. (Data for VA-10 and VA-15, not given in Table 1, are from Goff and others, 1981, p. 234.) Even with all these changes, however, the Na–Cl composition of the geothermal water is preserved (see logarithmic plots in Figure 2).

The original mixing calculations were made with a simple mass-balance equation, assuming the geothermal water to be like sample N 4 (Table 1) from well GT-2 of the Los Alamos National Laboratory (Fig. 1); samples from the caldera were not then available. Recalculated mixing proportions based on average concentrations of Cl, Li, and B in analyses of geothermal fluid (White and others, this volume; and BACA 4, Table 1), taken as the parent water, are presented in Table 1.

A second plot by Goff and others (1981, fig. 4), based on stable-isotope data, reinforces the interpretation of mixing. Their isotopic line is replotted in Figure 4 as the continuous line through H 14 and H 15.

The isotopic data can be used to test the dilution factors in Table 1. Trainer and Lyford (1979) did this by assuming the parent geothermal fluid for the mineral water to be like sample N 4. In terms of Figure 4 the method is based on the argument that water H 6 has an isotopic composition determined by the combination of fractions having the isotopic compositions of N 4 and of the intercept on the meteoric-water line of the dashed line through N 4 and H 6. These fractions, from Table 1, are 1.0 part from N 4 and either 0.4 part or 0.8 part meteoric water (depending on whether data for flashed geothermal water or for water plus steam are used in calculation of the dilution factor).

Figure 4 shows the results of this test. The dilution factor based on flashed water gives a closer approximation of the observed value than

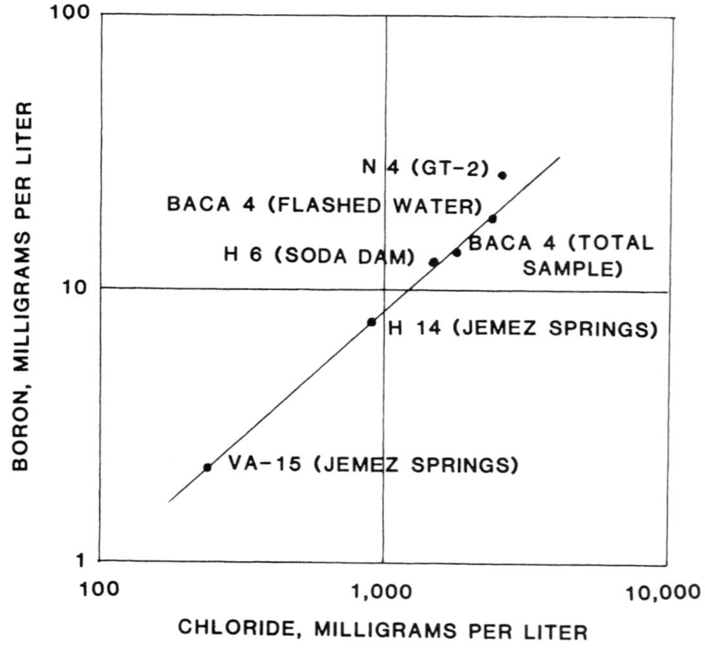

FIGURE 3. Covariation diagram for boron and chloride. Analyses in Table 1 except VA-15 (Goff and others, 1981).

TABLE 1. Chemical analyses, isotopic data, and calculated dilution factors for selected ground-water samples collected in and near Cañon de San Diego[0]. Concentration of constituents in milligrams per liter, specific conductance in micromhos per centimeter at 25°C, isotopes in per mil. Lesser dilution factor calculated from total sample of geothermal fluid, greater factor from flashed liquid; see text for discussion. Indicator constituents for dilution factor: Cl, 2,800 and 2,200 mg/L; Li, 23 and 18 mg/L; B, 23 and 28 mg/L. See Figure 1 for locations of sampled wells and springs.

Sample Number.....	Well BACA 4[1] (water)	(steam)	N 4 (GT-2)	G 1	H 6	H 14	H 28	H 29	H 30	H 32
Date of collection.	8-22-73		5-3-74	10-30-73	12-1-72	12-2-72	7-18-74	10-24-73	3-21-73	1-17-73
Silica (SiO_2)......	760	0.3	170	16	50	95[3]	87	37	37	45
Calcium (Ca).......	5	.6	7.3	100	330	130	38	210	130	140
Magnesium (Mg).....	0.6	.4	13	11	24	4.8	4.4	59	25	13
Sodium (Na)........	1,500	1	4,800	14	990	640	70	570	100	390
Potassium (K)......	300	.4	180	5.8	200	82	5.5	34	11	51
Lithium (Li).......	22		15	0.06	13	8.7	0.56	2.1	0.69	3.6[4]
Bicarbonate (HCO_3).	196	6	6,770	376	1,578	732	254	1,611	660	761
Carbonate (CO_3)....	0	0	0	0	0	1.5	0	0	0	0
Sulfate (SO_4)......	74	3	1,600	21	52	53	18	250	32	32
Chloride (Cl)......	2,500	1	2,600	13	1,500	920	40	300	61	470
Fluoride (F).......	7.1	.0	16	.3	3.3	4.8	1.4	1.6	1.0	4.6
Boron (B)..........	18		26	.06	12.5	7.4	.37	2.4	.51	3.3
Dissolved solids...	5,080	10	9,380	365	3,740	3,500	393	2,260	728	1,520
Specific conductance.........	8,000	50	14,200	640	6,160	3,930	540	3,200	1,100	2,200
pH.................	6.8	4.5	7.7	7.3	6.1	6.3	7.3	6.5	6.8	6.7
D..................			-78.9[2]		-84.8	-82.1				
^{18}O................			- 8.12[2]		-10.40	-10.52				
Dilution factor,										
from chloride....					0.5, 0.9	1.4, 2.0	54,69	6.3, 8.3	35, 45	3.7, 5.0
from lithium.....					.4 .8	1.1, 1.6	29,37	7.6, 10	25, 32	4.0, 5.4
from boron.......					.4 .8	1.4, 2.1	44,56	6.5, 8.6	35, 45	25, 32
average..........					.4 .8	1.3, 1.9	42,54	7, 9	32, 41	11, 14

[0] Analyses by U.S. Geological Survey.
[1] 25 percent of total flow flashed to stream.
[2] Sample collected 4-23-74.
[3] Sample collected 1-25-74.
[4] Sample collected 3-4-74.

does the factor based on geothermal water plus steam. Interestingly, and perhaps significantly, the original calculation in which the dilution factor was based on data from well N 4 produced a still better fit. These observations suggest that the fluid that flows out of the reservoir is of composition and concentration broadly similar to the samples now available from the reservoir but modified by boiling. The dilution factors in Table 1 appear to be of the correct general magnitudes.

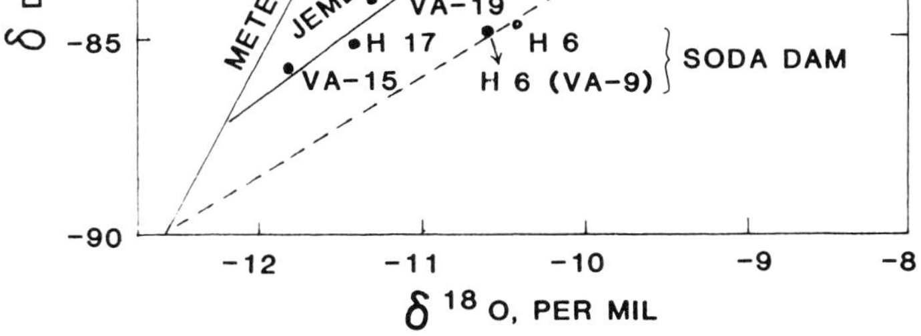

FIGURE 4. Plot of deuterium–oxygen-18 values for thermal mineral waters at Soda Dam, Jemez Springs, and well N 4 (GT-2); data in Table 1 except VA samples (Goff and others, 1981) and H 17 (Trainer, 1978). Also shown (x) are constructed positions of isotopic values for Soda Dam sample (H 6), calculated from calculated mixing proportions for water H 6 (Table 1) and from D, ^{18}O data for N 4 and for intercept of dashed line on meteoric-water line. See text for discussion.

Composition of the mineral waters has remained rather stable over time, but concentration of constituents has fluctuated. Figure 5A illustrates fluctuation in Cl and HCO_3 in one spring at Jemez Springs over a period of nearly a year. This record is too short and too fragmentary to suggest more than an orderly change in concentration.

In a second approach all available analyses for Cl were tabulated by month of sampling to determine whether fluctuations in concentration of Cl could be found within the year. Thirty-four analyses were available from springs H 14 and H 17 (VA-10 in Figure 2) at Jemez Springs, 58 analyses for Soda Dam. Two disadvantages were recognized in this procedure: (1) only one analysis was available for each of several months (none for one month at Soda Dam), and (2) if change in concentration occurred as a result of some seasonal effect, changes in timing of the seasons from one year to another could lead to poor definition of the change in concentration. In an attempt to address these two difficulties, the data were summarized as median values for successive two-month periods (December–January, January–February, etc.) to obtain 12 values for the year. Sizes of the two-month samples ranged from three to 10 analyses for Jemez Springs and from three to 18 analyses for Soda Dam. The two-month running median values are shown in Figure 5B.

Periods of relatively high concentration in the plots in Figure 5B, as in November–January, can be explained as representing times of higher discharge from the geothermal reservoir, lesser dilution of the thermal-mineral water, or both; and periods of low concentration, as in late spring (at Soda Dam) and in early autumn, can be explained as times of low flow from the reservoir, greater dilution, or both. Our knowledge of the geothermal reservoir is inadequate to support informed speculation about fluctuation in discharge; but fluctuation in head is typical of shallow aquifers, and the following discussion is based on the assumption that the troughs in the Cl hydrographs represent annual episodes of ground-water recharge, increase in ground-water head, and increase in the flow of dilute water into the conduits that contain the thermal-mineral water.

The principal "event" which leads to recharge of shallow aquifers in this region is melting of the snow cover during February–June (but

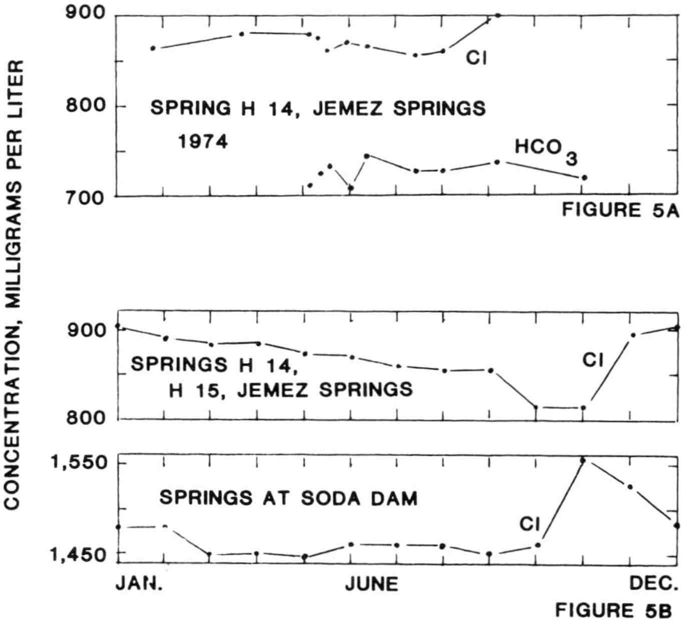

FIGURE 5. Chemical plots for mineral waters at Soda Dam and Jemez Springs. **A,** Concentration of Cl and HCO_3, spring H 14, Jemez Springs, 1974; data from Trainer (1978). **B,** Summary hydrographs of Cl concentration, Jemez Springs (H 14, H 15) and Soda Dam, based on two-month running median values for all available data. Data from Goff and Sayer (1980); Goff and others (1981); Purtymun and others (1974, 1975, 1976, 1978a, 1978b, 1981); Summers (1976); Trainer (1978); and files of the U.S. Geological Survey.

during a shorter period within that range, in a particular year). Given its presence in both hydrographs, the low concentration of Cl in the autumn is reasonably correlated with recharge and dilution following melting of the snow cover. A lesser pulse of ground-water recharge and dilution may occur during the late-summer thundershower season, and the writer correlates the spring trough at Soda Dam with the episode of rain.

The suggested correlations indicate lag times of five to six months which represent, in this hypothesis, the time required for mass transport of diluted mineral water from the sites of mixing to the springs. The geohydrologic setting of aquifer recharge is considered in a later section of this report, but comments concerning residence time of the shallow ground water and lag time for flow in the conduits should be added here.

Tritium data provide a basis for examination of likely residence time. The concentration of tritium in precipitation had not been determined before fusion testing during 1952–63 produced large quantities of artificial tritium that mask the natural concentration in the atmosphere. Thatcher (1962, p. 48) concluded that in the continental-interior region of the United States this natural concentration was about 8–10 TU (tritium units, 1 atom ^3H in 10^{18} atoms H). Because tritium has a half-life of 12.26 years, it can be used to investigate the residence time of ground water where that time is short enough for measurable tritium to remain, but long enough to predate artificial tritium.

In 1973 the mineral water at Jemez Springs (spring H 14) contained 6.4 ± 0.6 TU, and that at Soda Dam (H 6) 4.0 ± 0.4 TU (Trainer, 1978, p. 127). These values are thought to be of the correct relative magnitude because Jemez Springs water contains the larger component of dilute (presumably young) water.

Samples from deep wells in Valles caldera contained less than 1 TU in 1982 (A. F. White, pers. comm. 1984). If the Jemez Springs mineral water is assumed to contain 1.0 volume of geothermal water and 1.6 volumes of dilute water, and if the geothermal water is assumed to have contained no measurable tritium by the time it reached the mixing sites, then by calculation the tritium content of the dilute-water component was about 10 TU in 1973. By working back in time using the half-life of 12.26 years we find that the dilute water could have contained about 20 TU in 1960 and 40 TU in 1948. The 40-TU estimate is invalid if atmospheric concentration was on the order of 10 TU before fusion testing. The 20-TU estimate is credible if the water entered the ground early in the period of testing, before atmospheric tritium levels had reached more than about three times the prior natural level. Hence, the dilute-water component of the Jemez Springs water is thought to have entered the ground in 1952 or 1953, giving a residence time for the shallow ground water of about 20 years. According to the interpretation presented in this report, water from the shallow aquifer entered a geothermal-water conduit, near the end of the 20-year residence time, and mixed with the thermal-mineral water.

Similar calculation yields a 9-Tu value for the dilute-water component of the Soda Dam mineral water. Considering the degree of uncertainty inherent in the dilution factors used in the calculations, this value does not differ from that calculated for the Jemez Springs water.

Lag time between mixing of waters and discharge from the springs is inferred to be five to six months. So far as interpretation of the hydrographs can show, this lag time could as well be another period, one or more years longer than five to six months (that is, 17–18 months, or 29–30 months, for example). Later discussion notes that the limestone conduits under Cañon de San Diego probably are cavernous, and in such a setting the half-year flow time is believed to be more realistic than longer estimates.

Differences in the plots for Jemez Springs and Soda Dam in Figure 5B may be apparent rather than real. However, three types of evidence already mentioned are consistent with the conclusion that the two groups of springs drain conduit systems that are isolated from one another. These are: (1) the consistent differences in concentration of dissolved solids in the two spring waters; (2) the interrelations of D and ^{18}O; and (3) the inverse relationship between mineralization and water temperature. If the conduit systems *are* isolated, the differences in Figure 5B may be real.

SPRING DISCHARGE

Measurement of Cl and water discharged by the groups of mineral springs is another approach to characterization of the conduit systems. Figure 6 summarizes these measurements in terms of Cl load in the Jemez River. The lowest hydrograph is for the river at a gaging station of the U.S. Geological Survey just outside the rim of Valles caldera. There Cl load is adequately explained by contributions from precipitation and from small mixed-water springs in upper Cañon de San Diego (data in Trainer, 1978). Because very little Cl is added to Jemez River between the gaging station and Soda Dam, this hydrograph can be taken as background against which to judge Cl discharge of the mineral springs. The hydrographs are arranged from bottom to top in downstream order; the numerical difference between the lower pair of graphs is Cl discharge of the Soda Dam springs, and the difference between the upper pair is Cl discharge of Jemez Springs. The mineral springs discharge quantities of Cl substantially above background.

The most striking aspect of these hydrographs is their peakedness. Peaks in all the curves occur during streamflow peaks, notably during the melt-water flood in spring. This observation is expectable at the gaging station but surprising with respect to the mineral springs. Reference to the Cl curve in Figure 5A shows that these conspicuous peaks are not due in an appreciable way to change in Cl concentration in the mineral water.

Increase in Cl discharge of the springs during surface-water floods is accompanied by corresponding increase in springflow (not shown in Figure 6; data given in Trainer, 1978). The rapid transmission of head change in the mineral-water system necessary to explain such an increase in springflow is believed to have been by means of pressure waves in confined conduits, which account for the near-synchronism of cause and effect. Three explanations for such pressure waves can be considered: (1) changes of head in the reservoir, (2) repetitive external loading and unloading of the conduits by flood water in the canyon, and (3) fluctuations of head in the conduits, transmitted to them by diluting water at sites of mixing.

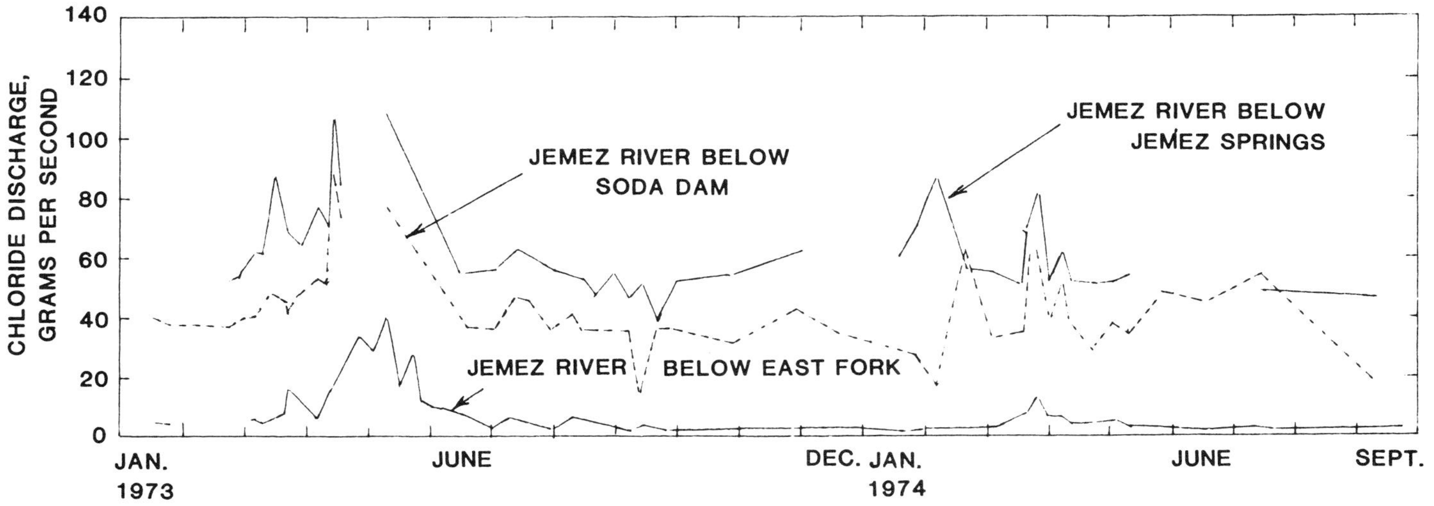

FIGURE 6. Chloride discharge at three stations on the Jemez River, 1973–74. See text for discussion.

The first hypothesis would explain increase in geothermal reservoir head and flood wave in the stream as due to the same episode of snow melt or to the same rain storm(s). Near-synchronism of flood wave and springflow peak would require almost instantaneous recharging of the reservoir. By any reasonable estimate the time required for water to percolate from the land surface to the reservoir would be many years. The tritium concentrations of less than 1 TU in deep water in Valles caldera suggest a travel time of 30 years or more.

Repetitive loading and unloading of confined aquifers by effects at the land surface is a widely recognized phenomenon capable of causing pressure waves that are practically synchronous with the loading and unloading. Such a process would accelerate the discharge of water already in the conduits near the springs but would not affect its dissolved load. Hence it does not account for even the small fluctuations in water chemistry noted earlier. Moreover, the most conspicuous transient loading effects in near-surface confined aquifers are caused by rather heavy and very localized loads such as railroad trains. In the local instance, loading is unlikely to amount to more than a meter or two of water on the canyon floor, believed to be more than 300 m above the confined conduit. For these reasons loading is thought to provide only part of the explanation of the springflow peaks, if that.

The third hypothesis, which explains the pressure pulse through temporary increase in head transmitted to the conduits at the sites of mixing, is believed to provide an adequate explanation for the observations. As noted in the last section, recharge is thought to occur where shallow aquifers receive infiltration from the stream, so that substantial increases in both recharge and ground-water head are highly seasonal.

In summary, near-coincidence of peaks in streamflow and springflow is seen as a transient pressure effect caused by pulses of recharge to the conduits. The fluctuations in water chemistry described earlier, on the other hand, are transport phenomena whose effect is seen months after the causative pulses of recharge and dilution.

Geothermal water discharged by the mineral springs at Soda Dam and Jemez Springs is one component of the fluid discharged from the caldera and traveling down Cañon de San Diego. Discharge of geothermal fluid down the upper canyon is believed to be almost entirely by flow in subsurface conduits. Beginning at Soda Dam, two additional flow components come into play: surface-water transport, and transport in ground water flowing through the canyon-floor alluvium. Geothermal water and solutes enter the river chiefly by springflow into the stream, largely at Soda Dam, Jemez Springs, and Jemez Pueblo (Fig. 1). Two mineral springs at the Pueblo show that conduit flow continues at least that far down the canyon, some 32 km outside the caldera as measured along major faults (Trainer, 1975). Some conduit flow very likely continues past the sites of leakage and discharges mineral water into the valley-fill aquifer in the Rio Grande rift. Geothermal water and solutes reach the alluvium by discharge directly from the bedrock and by infiltration from losing reaches of the river. Flow through the alluvium near Soda Dam and Jemez Springs is limited by the small cross section of the deposit in the narrow canyon floor. At Jemez Pueblo, where the canyon floor is wider and the alluvium may be thicker, ground water flowing through the alluvium must account for a significant part of the geothermal water and solutes being transported down Cañon de San Diego at any time.

The magnitude of that transport cannot be estimated from observations at the mineral springs alone because those observations do not cover the other modes of transport. The chief significance of the observations at Soda Dam and Jemez Springs, for the present discussion, lies in what they show of conduit flow. In brief, the annual hydrograph of either Cl or water discharge consists of two members: *peak flow* during a pronounced surface flood, with decay to a longer episode of *low flow*.

In 1973–74 low flow fluctuated within a relatively narrow range, peak flow was substantially greater, and total peak flow was considerably greater during the large flood of 1973 than during the small one of 1974. Median flow of the group at Soda Dam was 23 liters per second (Lps), or 365 gallons per minute (gpm); at Jemez Springs it was 21 Lps (333 gpm). However, more than half the measurements for each group of springs was made during low flow, and the median values therefore represent that flow regime. Average values include the effects of extreme values that may be in error, but have the advantage of representing the entire flow cycle. Average discharge for the entire period of record in 1973–74 was 25 Lps (397 gpm) at Soda Dam and 23 Lps (365 gpm) at Jemez Springs. Estimates of the long-term geothermal component of the spring discharge would require a longer period of record so that the magnitudes of discharge in the two flow regimes could be established more precisely.

GEOHYDROLOGIC SETTING AND THE CONDUIT SYSTEMS

The area surrounding the mineral springs, relatively well known as a result of geologic mapping by Wood and Northrop (1946), Smith and others (1970), and Goff and Kron (1980), lies within the complex, southwest-trending Jemez fault zone. The main fault, as it is now known, branches northeast of Soda Dam. The southern strand passes down-canyon beneath the village of Jemez Springs. The northern strand bifurcates to enclose a small horst of granitic rock at Soda Dam, southwest of which the branches merge into a single strand that passes through the mesa to the west. The tuff sheets on the mesa, cut by the near-vertical normal fault, are offset about 45 m. Thus the fault has been active for at least part of the time since formation of the caldera, estimated by Doell and others (1968, p. 211) to be about 1 m.y. Paleozoic rocks in the canyon are offset about 230 m across the fault, indicating a pre-caldera history of deformation (Goff and others, 1981, p. 231). According to Slemmons (1975, p. 10) the fault does not offset a young rhyolite flow, farther up the canyon, that is thought to be younger than 0.1 m.y. (Smith and Bailey, 1968, p. 641).

The springs at Soda Dam issue from the north side of the horst, where limestone is in fault contact with the granitic rock. These springs are at the lowest altitude at which the fault is exposed in the canyon. The modern dam is a linear travertine mound built along a line of springs across the canyon floor. Older, inactive spring deposits lie on higher ground to the northeast and southwest. They mantle the horst hill to a level nearly 215 m above the active springs; the highest travertine on the hill is cut by the horst-bounding fault (Goff and Kron, 1980). Still older travertine on the canyon walls is at levels as high as 300 m above the modern springs (Bailey and Smith, 1978, p. 186). These observations imply a considerable age for the springs at Soda Dam, or for earlier mineral springs there if discharge has not been continuous. Moreover, the vertical separation of old and modern travertines indicates a substantial change in hydraulic-head relationships in the mineral-spring system. Even if part of this separation were due to faulting, a large part of it appears to postdate the deformation and to be due to deepening of the canyon. Migration of the spring outlets to successively lower levels must have been accompanied by steepening of hydraulic gradient in the conduit system, with consequent increase in discharge, or, to the extent that supply of water to the reservoir was insufficient to maintain increased draft, by decline in reservoir pressure. M. L. Sorey (written comm. 1982) has suggested that lowering of the outlet at Soda Dam may have contributed to development of vapor-dominated conditions in the reservoir through such a decrease in pressure.

The Jemez Springs locality marks the downstream limit of limestone exposures in the canyon, beyond which the limestone is covered by younger sandstone and shale. The springs issue from alluvium in and near the Jemez River. Linear areal distribution of the springs and the fact that the outlets maintain their positions despite annual flooding suggest fault control.

Goff and others (1981) summarized and discussed geohydrologic information obtained from a test well drilled at Jemez Springs; the information includes driller's log, additional geologic and hydrologic data, and chemical and isotopic analyses of water samples. The hole penetrated about 70 ft (22 m) of gravel, 700 ft (215 m) of sedimentary rock, chiefly limestone, and 60 ft (18 m) of gneiss. Water was found at several levels in the gravel, and confined thermal water was found in an "80-foot aquifer" at the contact of alluvium and limestone and in a "500-foot aquifer" in fractured shale within the limestone. No major aquifer was found at the top of, or in, the crystalline rock.

Goff and others (1981, p. 235) concluded that the samples from springs H 14 and H 15 (which are the principal springs and are thought by several investigators to be those in this group most representative of the source water) are from the 80-foot aquifer. This conclusion is consistent with temperature data from the well and from nearby springs. Maximum water temperatures recorded during drilling were 68°C in the 80-foot aquifer and 60.5°C in the 500-foot aquifer. (The first of these values may be in error on the low side and the second on the high side; Goff and his colleagues concluded that the 80-foot and 500-foot aquifers had not been successfully isolated during sampling, because of problems of well construction, and that the shallower sample contained water from both levels.) In comparison, as noted earlier, the maximum temperature measured at the surface outlets (H 14) is 75°C. The writer believes the anomalous thermal gradient in the well to reflect a ground-water flow pattern developed locally in the fractured and solution-eroded limestone and in the alluvium (in which some strata have probably been selectively cemented by calcite). Major conduits in the limestone are most reasonably explained as steeply dipping, solution-widened fractures in the fault zone, but flow to individual springs is probably largely along a combination of vertical fractures and near-horizontal bedding-plane openings in the limestone and along porous layers in the gravel. In such a situation local hydraulic and thermal fields tend to be layered horizontally, and thermal anomalies such as that observed in the well can develop.

Flow through the gravel also provides opportunity for dilution of mineral water near the springs. Such dilution would explain the mixing line defined by the Jemez Springs samples as a local phenomenon. The waters of the two principal springs may also have been affected by local mixing, but data to test this possibility are lacking.

No direct information on the nature of the conduit systems is available; except for a single geophysical study, only inference can be used to postulate the character of the conduits. Using direct-current resistivity data from a surface survey, Pearson and Goff (1981) interpreted an electrically conductive zone beneath the village of Jemez Springs as a thermal aquifer. This zone, at a depth of 24 m and in alluvium just above bedrock, appears to be equivalent to the 80-foot aquifer in the test hole. Pearson and Goff traced this conductive zone for several kilometers up and down the canyon from Jemez Springs but found resistivity to increase with increasing distance from the village. The writer believes the thermal water in the 80-foot aquifer to have been isolated from water at shallower levels in the gravel, and hence probably to have been derived from a bedrock conduit, basing these inferences on the relatively high temperature of the spring water and on its low content of tritium. Both lines of evidence argue lack of interconnection with shallow aquifers, which are likely to have received recharge from the surface recently.

Because of the general absence of primary porosity in the limestone, fractures must form the continuous and hydraulically effective conduit systems implied by the data summarized in earlier sections. The areal pattern of faults, taken with the three observed occurrences of mineral water (the two groups of springs, and well GT-2) in limestone, suggets that fault zones modified by dissolution of wall rock provide a reasonable explanation for the conduits.

Given the enormous quantity of limestone that has been removed from the conduit systems (obvious at Soda Dam, inferred from the dissolved load at Jemez Springs), the conduits are likely to be far larger and more conductive than is required to transport the relatively modest quantities of water discharged by the springs. From this reasoning and from the inference, already described, of rapid pressure transients and slow mass transport, the conduits are believed to contain large quantities of confined water in transient storage, with fluctuations in discharge being governed by fluctuation in prevailing head.

One consequence of slow transport and large conduit volume (large wall area) is high conductive heat loss from fluid to wall rock. The Soda Dam conduit system must be significantly larger than the Jemez Springs system, inasmuch as the springs at Soda Dam discharge far more Ca, Mg, and HCO_3 than those at Jemez Springs in a roughly comparable volume of water. Greater heat loss in the conduits is thought to explain the lower temperature of the Soda Dam water. This general reasoning suggests that very hot water is not likely to be found beneath Cañon de San Diego very far outside the caldera.

Fumaroles just outside the rim of Valles caldera reveal the presence of water at or above the boiling temperature (about 93°C at that altitude) beneath the canyon there. The area of fumaroles extends along the axis of the canyon floor for about a kilometer, its course marked by patches of alteration of soil and rock. H_2S is emitted at one locality. The linear distribution of fumaroles suggests that the hot-water conduit is an unmapped strand of the Jemez fault.

This area near the caldera rim is of unusual geohydrologic interest because data from three springs and three shallow wells show the presence of six near-surface perched aquifers within an area of less than 0.5 km^2; one well taps alluvium, and two wells and three springs (including wells H 29 and H 30 and spring H 32, Table 1) are in limestone. Samples from H 29, H 30, and H 32 are mixtures of geothermal and dilute waters; H 29 also contains H_2S and therefore must have received fumarolic vapor. Tritium analyses for H 32 before and after the snowmelt flood of 1973 revealed 18.8 ± 2.1 TU and 75.0 ± 4.0 TU, respectively (Trainer, 1978, p. 127). The aquifer drained by spring H 32 evidently received recharge during the flood, and, because the aquifer is situated beside the river (and probably beneath it, farther upstream), flood water is thought to have moved directly into the limestone at some point near the spring.

These observations bear on the problem of how and where mixing of geothermal and dilute fluids may occur. The reservoir in Valles caldera and its outflow conduits just outside the caldera appear to be the regional aquifer in this part of the Jemez Mountains; other known aquifers are perched above the main zone of saturation. In such a situation, flow from a perched aquifer to the main aquifer can occur where two conditions are met: (1) the perched aquifer received recharge in excess of its storage capacity and loses water by leakage through its confining bed or through spillover; and (2) the rock that confines the geothermal conduit (which locally is the regional aquifer) is sufficiently fractured locally to permit flow into or out of the conduit. Under these circumstances a recharge mound is maintained above the conduit by percolation from the perched aquifer. Where the head in the mound exceeds the head in the conduit, water flows in and dilutes the stream of geothermal water. Where the conduit head is greater, geothermal water leaks out into the mound and creates a mixed water in the country rock. The latter process has evidently occurred near the caldera rim, where several variants of mixed water have been found in the limestone (H 29, H 30, H 32, Table 1). This general area may also contain sites of flow into the conduit inasmuch as the presence of fumaroles shows the conduit to be imperfectly confined here.

Upper Cañon de San Diego is thought to be a favorable general site for mixing because perched aquifers are present in the limestone, the rocks appear to be sufficiently fractured to facilitate rapid flow, and the river is a continuous but fluctuating source of recharge. The ring-fracture zone in the caldera may also satisfy the conditions required. The ring fracture seems, from areal geological relations, to be occupied by intrusives over much of its extent (Smith and others, 1970). However, geophysical data suggest that this zone contains water in an area in the southwestern part of the caldera (Jiracek and others, 1975). That area is beneath or near San Antonio Creek, one of the headwaters of the Jemez River, and thus from the consideration of a potential source of recharge it appears a favorable site for the mixing of waters.

Isolation of the two systems of conduits from one another, as required by the conceptual model developed in this report, is difficult to explain on the basis of the pattern of faults as it is now known. Future mapping may resolve this difficulty despite the obstacle of large covered areas. Bifurcation of the Soda Dam conduit, with subsequent mixing of water along only one branch to form the Jemez Springs water, may give an alternative explanation more consistent with the known structural setting.

ACKNOWLEDGMENTS

Appreciation is gratefully expressed to Union Oil Company of California for the opportunity of collecting and publishing chemical data for well BACA 4; and to the following individuals for helpful and stimulating discussion: R. L. Borton, F. E. Goff, R. H. Mariner, M. L. Sorey, F. G. West, and A. F. White.

REFERENCES

Bailey, R. A., and Smith, R. L., 1978, Volcanic geology of the Jemez Mountains, New Mexico; *in* Hawley, J. W. (compiler), Guidebook to Rio Grande rift in New Mexico and Colorado: New Mexico Bureau Mines & Mineral Resources, Circular 163, pp. 184–196.

Doell, R. R., Dalrymple, G. B., Smith, R. L., and Bailey, R. A., 1968, Paleomagnetism, potassium–argon ages, and geology of rhyolites and associated rocks of the Valles caldera, New Mexico; *in* Coates, R. R., and others (eds.), Studies in volcanology: Geological Society of America, Memoir 116, pp. 211–248.

Dondanville, R. F., 1971, The hydrothermal geology of the Valles caldera, Jemez Mountains, New Mexico: Consultant's report to Union Oil Company of California, 36 pp. (1978).

———, 1978, Geologic characteristics of the Valles caldera geothermal system, New Mexico: Geothermal Resources Council, Transactions, v. 2, sec. 1, pp. 157–160.

Goff, F. E. and Kron, A., 1980, In-progress geologic map of Cañon de San Diego, Jemez Springs, New Mexico and lithologic log of Jemez Springs geothermal well: University of California, Los Alamos Scientific Laboratory, Informal Map LA-8276-Map, scale 1:12,000.

———, and Sayer, S., 1980, A geothermal investigation of spring and well waters of the Los Alamos region, New Mexico: University of California, Los Alamos Scientific Laboratory, Informal Report LA-8326-MS, 21 pp.

———, Grigsby, C. O. Trujillo, P. E., Jr., Counce, D., and Kron, A., 1981, Geology, water geochemistry, and geothermal potential of the Jemez Springs area, Cañon de San Diego, New Mexico: Journal of Volcanology and Geothermal Research, v. 10, pp. 227–244.

Jiracek, G. R., Smith, C., and Dorn, G. A., 1975, Deep geothermal exploration in New Mexico using electrical resistivity: Second United Nations Symposium on the Development and Use of Geothermal Resources, Proceedings, v. 2, pp. 1095–1102.

Kerr, R. A., 1982, Extracting geothermal energy can be hard: Science, v. 218, pp. 668–669.

Pearson, C., and Goff, F. E., 1981, A Schlumberger resistivity study of the Jemez Springs region of northwestern New Mexico: Geothermal Resources Council, Transactions, v. 5, pp. 119–122.

Purtymun, W. D., Adams, W. H., and Owens, J. W., 1975, Water quality in vicinity of Fenton Hill site, 1974: University of California, Los Alamos Scientific Laboratory, Report LA-6093, 16 pp.

———, Adams, W. H., and Stoker, A. K., 1978a, Water quality in vicinity of Fenton Hill site, 1976: University of California, Los Alamos Scientific Laboratory, Report LA-7307-MS, 17 pp.

———, West, F. G., and Adams, W. H., 1974, Preliminary study of the quality of water in the drainage area of the Jemez River and Rio Guadalupe: University of California, Los Alamos Scientific Laboratory, Report LA-5595-MS, 26 pp.

———, Ferenbaugh, R. W., and Adams, W. H., 1981, Water quality in the vicinity of Fenton Hill, 1980: University of California, Los Alamos National Laboratory, Report LA-9007-PR, 16 pp.

———, Adams, W. H., Stoker, A. K., and West, F. G., 1976, Water quality in vicinity of Fenton Hill site, 1975: University of California, Los Alamos Scientific Laboratory, Report LA-6511-MS, 21 pp.

———, Stoker, A. K., Adams, W. H., and Owens, J. W., 1978b, Water quality in the vicinity of Fenton Hill site, 1977: University of California, Los Alamos Scientific Laboratory, Report LA-7468-PR, 16 pp.

Slemmons, D. B., 1975, Fault activity and seismicity near the Los Alamos Scientific Laboratory geothermal test site, Jemez Mountains, New Mexico: University of California, Los Alamos Scientific Laboratory, Report LA-5911-MS, 26 pp.

Smith, R. L., and Bailey, R. A., 1968, Resurgent cauldrons; *in* Coats, R. R. and others (eds.), Studies in volcanology: Geological Society of America, Memoir 116, pp. 613–662.

———, ———, and Ross, C. S., 1970, Geologic map of the Jemez Mountains, New Mexico: U.S. Geological Survey, Miscellaneous Geological Investigations Map I-571, scale 1:125,000.

Summers, W. K., 1976, Catalog of thermal waters in New Mexico: New Mexico Bureau Mines & Mineral Resources, Hydrologic Report 4, 80 pp.

Thatcher, L. L., 1962, The distribution of tritium fallout in precipitation over North America: International Association of Scientific Hydrology, Bulletin, v. 7, no. 2, pp. 48–58.

Trainer, F. W., 1974, Ground water in the southwestern part of the Jemez Mountains volcanic region, New Mexico: New Mexico Geological Society, Guidebook 25, pp. 337–345.

———, 1975, Mixing of thermal and nonthermal waters in the margin of the Rio Grande rift, Jemez Mountains, New Mexico: New Mexico Geological Society, Guidebook 26, pp. 213–218.

———, 1978, Geohydrologic data from the Jemez Mountains and vicinity, north-central New Mexico: U.S. Geological Survey, Water-resources Investigations 77–131, 146 pp.

———, and Lyford, F. P., 1979, Geothermal hydrology in the Rio Grande rift, north-central New Mexico: New Mexico Geological Society, Guidebook 30, pp. 229–306.

White, D. E., 1957, Magmatic, connate, and metamorphic waters: Geological Society of America, Bulletin, v. 68, pp. 1659–1682.

———, Hem, J. D., and Waring, G. A., 1963, Chemical composition of subsurface waters: *chapter F in* Data of geochemistry, 6th ed.: U.S. Geological Survey, Professional Paper 440-F, 67 pp.

Woltz, D., 1972, The chemistry of groundwaters in the Jemez area and a magnetic survey of a potential source of magmatic fluids: M.S. thesis, University of New Mexico, Albuquerque, 90 pp.

Wood, G. H., and Northrop, S. A., 1946, Geology of the Nacimiento Mountains, San Pedro Mountain, and adjacent plateaus in parts of Sandoval and Rio Arriba Counties, New Mexico: U.S. Geological Survey, Oil and Gas Investigations Map OM-57.

Taos Pueblo, ca 1920. Photo by Burt Harwood.

FLUID CHEMISTRY OF THE BACA GEOTHERMAL FIELD, VALLES CALDERA, NEW MEXICO

ART F. WHITE[1], JOAN M. DELANY[1], ALFRED TRUESDELL[2], KATHY JANIK[2], FRASER GOFF[3], and HARRISON CRECRAFT[4]
[1]Lawrence Berkeley Laboratory, Berkeley, California 94720; [2]U.S. Geological Survey, Menlo Park, California 94025;
[3]Los Alamos National Laboratory, Los Alamos, New Mexico 87545; [4]Union Oil Co. of California, Santa Rosa, California 95406

INTRODUCTION

The Baca geothermal field is located on the western flank of Redondo Peak, a resurgent-dome structure near the center of the Valles caldera in northern New Mexico. The geology of the Valles caldera region is described by Smith and Bailey (1968). The regional hydrology has been presented by Trainer (1974, 1975), Trainer and Lyford (1979), Laughlin (1981), Goff and others (1981), and Goff and Grigsby (1982). Production characteristics of the Baca geothermal field are discussed by Bodvarsson and others (1982) and Union Oil Co. (1982).

To date, 24 geothermal wells have been drilled into the Baca geothermal field in the Sulphur Creek area (Fig. 1) and along the Redondo Creek. The wells at Sulphur Creek penetrated high temperature (>260°C) but low-productivity zones. Wells completed along Redondo Creek, in a large northeast-trending central graben within the caldera, encountered similar high temperatures in a more permeable fracture- and stratigraphically controlled reservoir.

After the initial completion of five commercial wells in the Bandelier Tuff by Union Oil Co. of California (Union), a 50 MWe power plant was proposed in 1977 by the Public Service Co. of New Mexico in cooperation with the U.S. Department of Energy Geothermal Demonstration Power Plant Program. However, drilling of 13 additional wells aimed at completing the required steam capacity resulted in only two production wells. Subsequent attempts to find additional steam in deeper Paleozoic limestones and Precambrian granite, as well as steam derived from hydraulic fracturing in the tuff, failed to adequately increase production and the project was terminated in January 1982.

The fluids tapped by the Union wells represent the deepest and highest-temperature fluids encountered within the Valles caldera. Their chemistry is important in the context of understanding both the overall hydrothermal recharge and discharge system in the caldera and the potential reservoir contributions to thermal springs outside the caldera, which add significantly to surface flow in the Jemez River. In addition, the Valles caldera is a principal target for proposed deep continental drilling into a magma–hydrothermal regime. The geothermal fluids in the Baca reservoir are probably the best available indicators of potential interaction with magma-derived components at greater depth.

Although Union had previously collected geochemical data applicable to their exploration program, final flow tests at a number of the wells during the summer and fall of 1982 represented an excellent opportunity to collect a more comprehensive suite of chemical data. This paper presents preliminary major- and trace-element data and reinterprets some aspects of the regional hydrothermal system. Additional work is underway in interpretation of existing data for other wells, interpretation of isotopic data, and modeling water/rock interaction within and beyond the geothermal reservoir.

GEOCHEMICAL SAMPLING AND RESULTS

Five Baca wells were available in 1982 for chemical sampling (Table 1). Four of the wells, Baca 4, 13, 15, and 24, were classified as commercial based on production rates, and Baca 19 was considered subcommercial. Except for Baca 19, two sets of samples per well were collected during separate flow tests.

The sampling techniques employed were those of Nehring and Truesdell (1977), in which separate liquid water and steam fractions were collected from condensers attached to wellhead separators. The initial 7/23/82 sample from Baca 15 may have been contaminated due to prior injection of waste water from Baca 13. Table 1 lists the sampling conditions of the wells and separators during the flow tests.

TABLE 1. Well and separator characteristics.

Well	Depth T.D. (m)	Production at line pressure (kg/km)	Well head presure (psig)	Separator pressure (psig)	Water fraction (X^*fluid)
Baca 4					
6/11/82	1939	20,000	144	118	0.719
7/02/82			145	124	0.713
Baca 13					
6/04/82	2501	24,000	142	121	0.762
7/01/82			140	122	0.762
Baca 15					
7/23/82	1673	48,000	164	125	0.792
9/08/82			187	125	0.641
Baca 19					
9/08/82	1705	14,000	45	21	0.798
Baca 24					
6/18/82	3233	15,000	152	127	0.811
7/16/82			145	122	0.807

Unstable chemical species that could not be analyzed in the field were taken to the Fenton Hill Laboratory of Los Alamos National Laboratory (LANL) and analyzed soon after collection. The remaining major and trace constituents were analyzed independently at both Lawrence Berkeley Laboratory (LBL) and LANL. The two sets of results showed generally good agreement with charge balances within a maximum of ±4%. Table 2 lists a composite of field, Fenton Hill, and LBL chemical analyses.

CALCULATION OF RESERVOIR CHEMISTRY

Due to extensive steam flashing in the wellbore during production, chemical concentrations in Table 2 do not represent the chemical composition of the reservoir fluids.

TABLE 2. Chemical analyses of liquid water from the separators (mg l^{-1}).

	Baca 4		Baca 13		Baca 15		Baca 19	Baca 24	
	6/11/82	7/02/82	6/04/82	7/01/82	7/23/84	9/08/82	.0/08/82	6/18/82	7/16/82
pH*	7.28	7.20	7.30	-	7.21	7.12	8.00	-	7.42
Li	21.1	21.2	22.5	22.7	24.9	24.0	26.6	24.1	24.0
Na	563	1607	1504	1533	1904	1867	1970	1867	1822
K	336	336	320	310	391	407	369	281	286
Mg	0		.049	0	0	.03	.02	0	.02
Ca	4.9	3.5	4.8	4.4	16.1	19.4	18.7	17.5	24.9
Al**	.09	.09	.13	.12	.14	.06	.12	.16	.22
Al^{3+}**	<.001	<.001	.003	.001	.003	.002	.002	<.001	.006
B	22	20	22	20	29	27	29	26	28
Si	376	380	335	333	341	322	294	309	317
SiO_2	804	813	717	712	730	689	629	661	678
NH_3**	2.2	2.8	2.0	2.2	--	1.8	1.5	.09	1.2
F	5.5	4.8	9.4	6.2	8.6	8.6	8.6	5.0	4.8
Cl	2770	2750	2489	2650	3328	3266	3356	3082	3046
Br	5.0	5.7	5.4	6.2	9.8	10.3	10.3	8.3	10.3
HCO_3**	215	190	221	236	89	75	139	89	90
CO_3**	0	0	0	0	0	0	4.8	0	0
SO_4	50	50	56	51	42.6	45.6	48.5	50	46.2
S^{2-}**	.35	.34	.21	.22	.45	41	.23	.04	.14
PO_4	.09	.31	.12	.15	0			.49	
As	1.9	2.6	2.1	3.2	3.5	3.6	4.0	5.0	3.6
Cr	.0016	.0008	<.0001	.0006	.0036	.0022	.0014	.0006	.0012
Cs	-	.4	-	.3	3.2	3.2	3.3	.4	3.6
Hg	.0048	<.001	.0015	<.001	<.001	<.001	<.001	<.001	<.001
Fe	0.22	0	.006	0	0	0	0		0
Mn	.0039	-	.0086	-	.0086	.0050	.0050	-	.018
Pb	<.01	<.01	<.01	<.01	<.001	<.001	<.01	<.01	<.001
Se	<.001	<.01	<.001	<.01	-	-	-	<.01	-
Sr	.158	0	.184	0	.2	.2	.2		.35
Zn	.012	<.01	.24	<.01	<.01	<.01	<.01	<.01	<.01

* measured in field
** measured at Fenton Hill Laboratory (LANL)

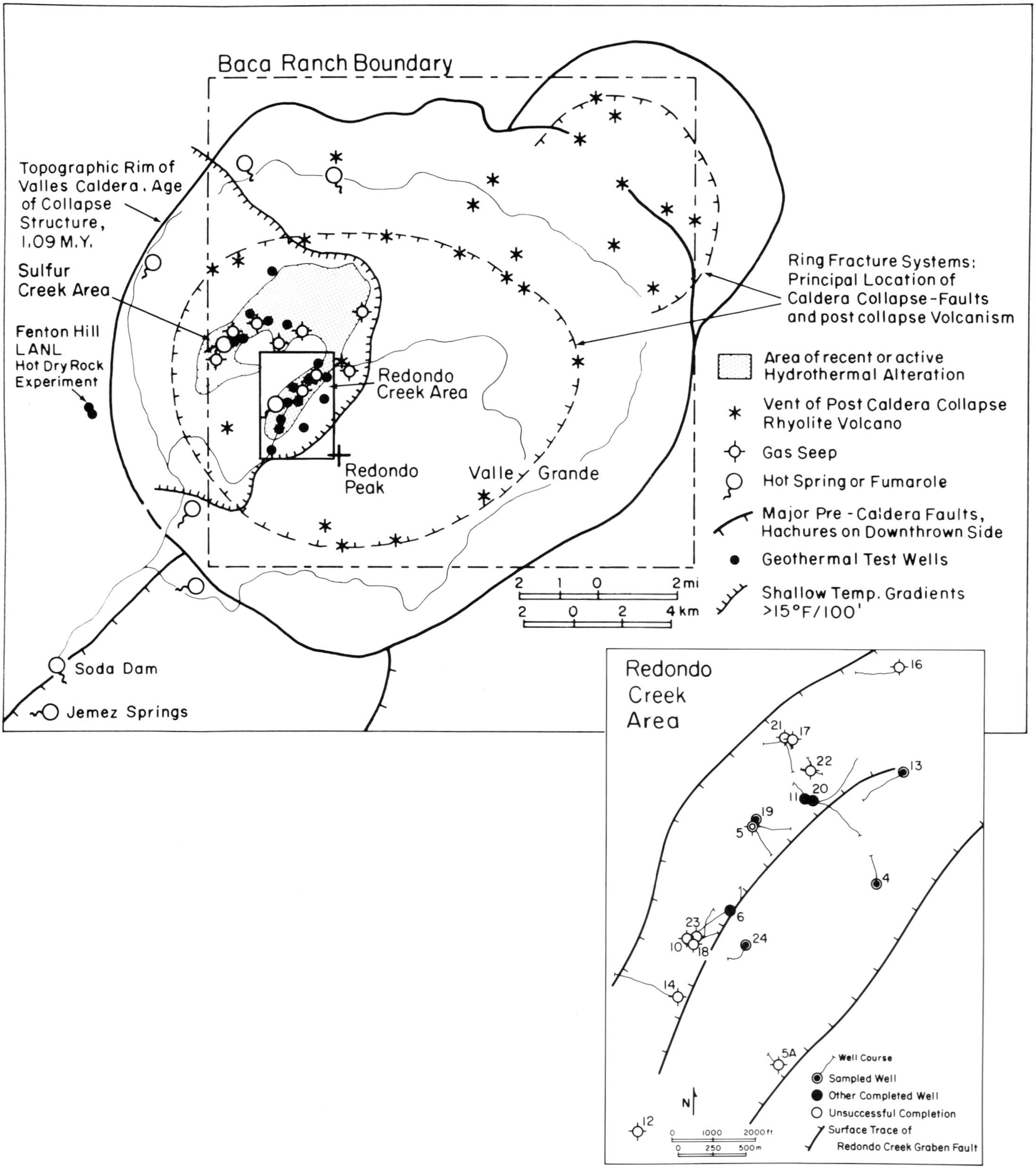

FIGURE 1. Map of the Valles caldera area and locations of the Baca wells and hot springs.

The weight fraction of liquid water, X^*_{liquid}, flashing at the separator pressure (Table 1) was calculated from the relationship

$$H = X^* \times H^*_{liquid} + (1 - X^*) \times H^*_{steam} \quad (1)$$

where H is the reservoir enthalpy (J gm^{-1}), and H^*_{liquid} and H^*_{steam} are the enthalpies of liquid water and steam at the separator pressure.

In addition to production from flashing, steam may also originate in the reservoir. Non-volatile chemical species remain in liquid water and volatile species are contained in the coexisting steam phase. The proportions of steam and liquid in the reservoir are related to the reservoir enthalpy by the expression

$$H = X \times H_{liquid} + (1 - X) \times H_{steam} \quad (2)$$

in which the liquid water and steam enthalpies are for the temperature and pressure of the reservoir. The reservoir enthalpy can be calculated from equation 1 using the measured separator liquid-water fraction (Table 1) and data for H^*_{liquid} and H^*_{steam} at the separator pressure (Keenan and others, 1969). The liquid fraction in the reservoir was calculated from equation (2) and H_{liquid} and H_{steam} values using reservoir temperatures based on the Na–K–Ca geothermometer (Table 3).

The concentration of a non-volatile chemical species, C, in the liquid-water phase of the reservoir can then be calculated from the relationship

$$C = C^* \times \frac{H^*_{H_2O}}{X_{H_2O}} \quad (3)$$

where C* is the concentration in the flashed liquid (Table 2). The relationship assumes no chemical precipitation during flashing and no corrosion in the wellbore.

Bodvarsson and others (1982) concluded that the Baca reservoir contains a single-flow system with hot water at depth and two-phase fluid in the regions of upflow. A comparison is shown in Figure 2 between the produced enthalpies determined from equation 1 and the calculated enthalpies from equation 2 by assuming the wells were fed by liquid water only at the Na–K–Ca-geothermometer temperature of the reservoir. The 9/7/82 sample from Baca 15 plots above the equal enthalpy line, indicating excess steam originating from a two-phase reservoir. All other samples lie either on or below the line, indicating a single-phase hot-water reservoir. The lower than ideal fluid enthalpies for many of the wells are probably due to conductive cooling or steam loss from the wellbore.

Union drilling records and previous flow tests indicate that Baca 4 and 15 had penetrated steam zones in the upper part of the reservoir. The absence of two-phase conditions at Baca 4 during sampling may be due to the effects of final well completion in which these steam zones were cased off or the effects of thermal equilibration in the wellbore. Earlier flow tests of Baca 15 in 1976 and 1977 produced very high enthalpies (>2,100 J gm^{-1}), indicating a large steam contribution. Part of this excess steam may have resulted from flashing in the reservoir during these previous long-term flow tests (Bodvarsson and others, 1980). The very low-production enthalpy for Baca 15 on 7/23/82 is probably related to prior injection of fluid from Baca 13. The later 9/7/82 Baca 15 sample is assumed to be more representative of the two-phase nature of the well.

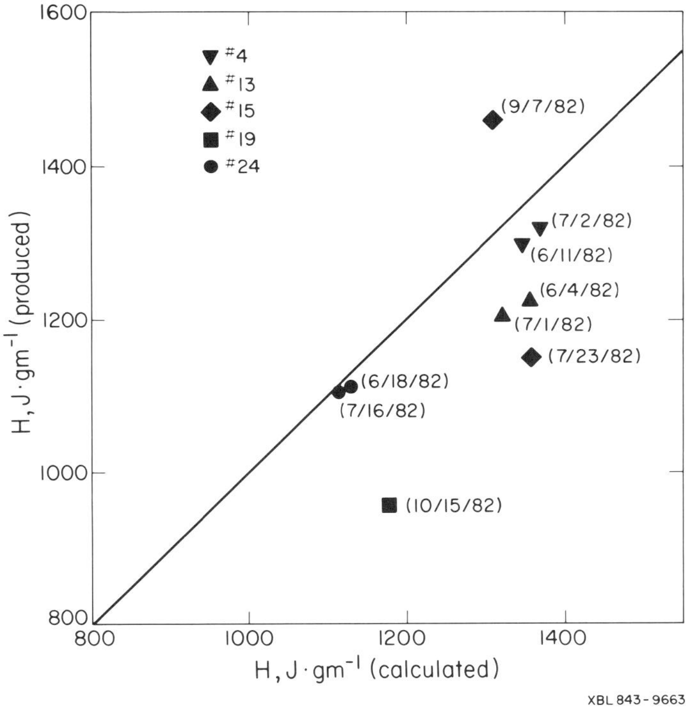

FIGURE 2. Comparison of produced enthalpies and calculated enthalpies assuming single-phase hot water at Na–K–Ca-geothermometer temperatures.

TABLE 3. Geothermometer temperatures (°C).

Location	Sample date	$T_{measured}$	T_{SiO_2}	T_{Na-K}	$T_{Na-K-Ca}$	T_{Na-Li}
Baca 4	6/11/82	295	263	288	300	305
	7/02/82	295	264	283	271	301
Baca 13	6/04/82	295	254	286	299	320
	7/01/82	295	254	278	294	319
Baca 15	7/23/82	270	255	281	285	301
	9/08/82	270	251	290	287	298
Baca 19	10/08/82	-	244	267	288	304
Bacca 24	6/18/82	247	248	236	258	299
	7/16/82	224	250	242	257	302
Soda Dam	12/01/72	48	102	277	230	310
	1/04/79	48	102	278	226	301
Main Jemez Sp.	11/19/79	35	128	195	194	300
Travertine Mound	1/19/79	72	127	202	196	308
Buddhist Sp.	1/19/79	50	120	205	192	291
Jemez Well "80 ft"	1/19/79	68	118	201	192	296
Jemez Well "500 ft"	1/19/79	60	71	245	192	294

ENTHALPY–CHLORIDE RELATIONSHIPS

Truesdell and Fournier (1976) demonstrated that plots of enthalpy–chloride delineate hydrologic characteristics of geothermal reservoirs and hot springs. In Figure 3, aqueous-reservoir chloride concentrations are determined from the produced enthalpies and calculated liquid-water enthalpies using equations 1–3. The chloride data generally cluster into two groups with low chloride concentrations in Baca 4 and 13 and higher chloride concentrations in Baca 15, 19, and 24. Baca 4 and 13, which are located in the northeastern section of the Redondo Creek graben, also have higher measured and calculated temperatures and larger enthalpies than do the other wells situated to the southwest. Baca 15 and 19 have exhibited intercommunication during flow and interference tests, which explains similar chloride concentrations. No direct interconnection has been demonstrated for the other sampled wells.

The chloride–enthalpy contents of the Baca wells in Figure 3 are bracketed by diagonal lines indicating trends resulting from boiling and steam loss. These straight lines assume a constant steam enthalpy of 2,700 J gm^{-1} as indicated by intersection with the pure-steam phase in the upper left of the figure. This enthalpy is consistent with the separator temperatures (±120°C), but can be used to describe higher temperature processes because the enthalpy of steam only varies ±4% to the critical point. The lower enthalpy–higher chloride fluids in wells 15, 19, and 24 can be derived from adiabatic cooling and later conductive reheating of fluids with a chloride composition of Baca 4 and 13. Localized boiling requires the separation and loss of steam from

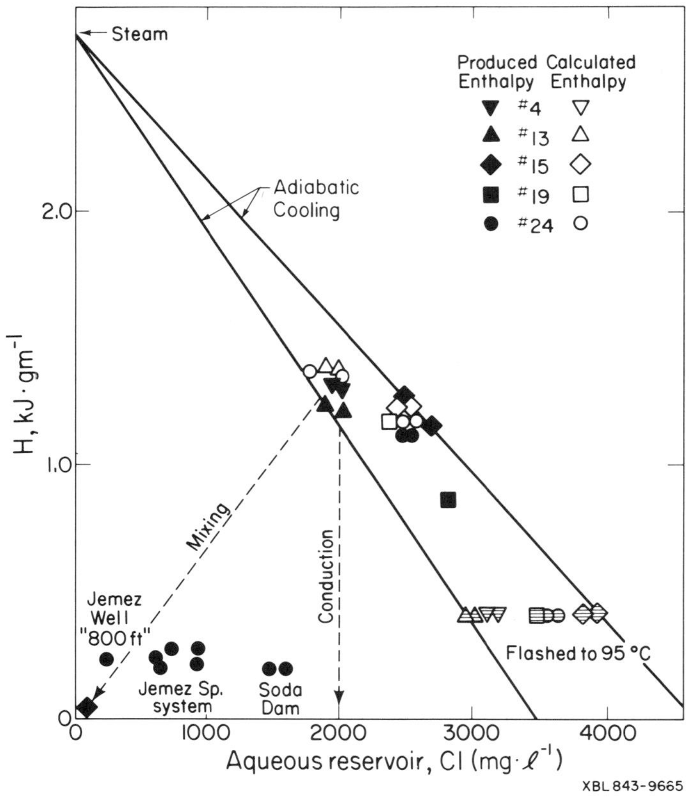

FIGURE 3. Enthalpy–chloride relationships for hot springs and Baca wells.

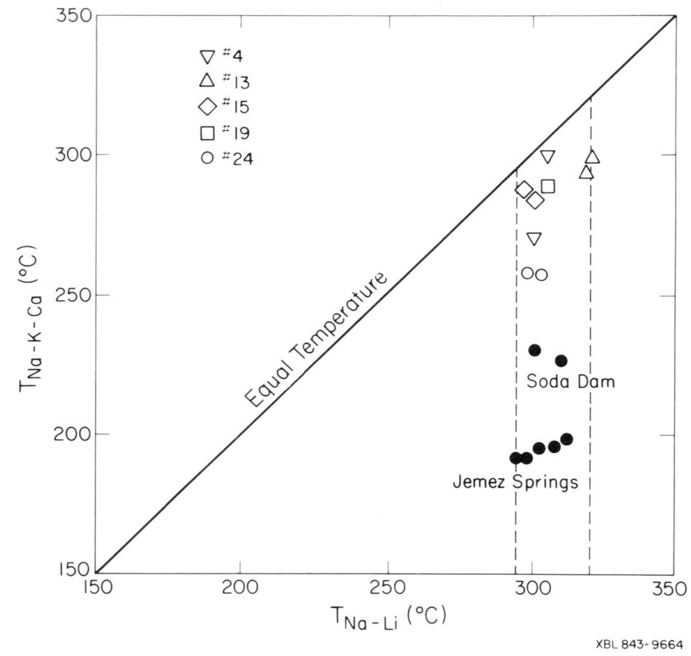

FIGURE 4. Comparison of temperatures of wells and hot springs calculated from Na–K–Ca and Na–Li geothermometers.

the reservoir fluid. The upper steam-producing zones in Baca 15 could result from such boiling. Although there are currently no surface manifestations of reservoir steam loss in the area of Redondo Creek, acid-sulfate springs indicative of steam separation do occur at Sulphur Creek.

Chloride–enthalpy data from the Jemez Springs system and Soda Dam (Fig. 4) are also plotted in Figure 3. Previous workers, including Trainer (1974) and Goff and others (1981), have proposed that these hot springs, which discharge from Paleozoic carbonates along fault systems defining the course of the Jemez River, represent a mixture of local meteoric waters and hot waters originating from a geothermal reservoir in the Valles caldera. Both the enthalpies and chloride concentrations of these springs are lower than those of the Baca fluids. The temperatures of the springs range from 70°C at Travertine Mound at Jemez Springs to 48°C at Soda Dam (Goff and others, 1981).

Hot-spring chloride–enthalpy relationships can evolve from deeper–hotter geothermal reservoirs such as the Baca system by three mechanisms: (1) adiabatic cooling, (2) conductive cooling, and (3) mixing with cold meteoric water. The chloride compositions plotted at the lower end of the adiabatic cooling trend in Figure 3 represent Baca fluid compositions flashed to atmospheric pressure at 95°C. Such adiabatic cooling would result in much higher chloride concentrations than are present in the hot springs without extensive mixing with meteoric water.

Changes in chloride concentrations and enthalpies as functions of conductive cooling and mixing with meteoric water are shown in Figure 3, assuming a fluid source comparable to the hottest, most dilute composition of the Baca reservoir. Mixing is assumed to occur with cold meteoric water of the composition of Agua Durme Spring (Goff and Grigsby, 1982). Except for the 500-ft aquifer intersected by the Jemez well (Goff and others, 1981), hot-spring enthalpy–chloride contents fall between the mixing and thermal conduction lines in Figure 3, suggesting both dilution with meteoric water and heat loss to cooler aquifers surrounding the geothermal reservoir. Thermal conduction clearly is more important at Soda Dam than at Jemez Springs.

CHEMICAL GEOTHERMOMETRY

Chemical geothermometers are widely used to estimate geothermal reservoir temperatures. Four such geothermometers were used to calculate reservoir temperatures for the Baca wells and the hot springs along the Jemez River. The temperature relationship in degrees centigrade are:

$$T_{SiO_2} = \frac{1535}{0.989 - \log SiO_2} - 273.15 \quad (4)$$

$$T_{Na\text{-}K} = \frac{855.6}{1.099 + \log(Na/K)} - 273.15 \quad (5)$$

$$T_{Na\text{-}Li} = \frac{1000}{0.38 + \log(Na/Li)} - 273.15 \quad (6)$$

$$T_{Na\text{-}K\text{-}Ca} = \frac{1647}{\log(Na/K) + 0.33 \log(\sqrt{Ca}/Na)} - 273.15 \quad (7)$$

where all concentrations are in moles l^{-1}.

The silica geothermometer (Truesdell, 1976) is used to calculate reservoir temperatures from discharged surface-silica concentrations assuming adiabatic isoenthalpic cooling in the wellbore. The Na–K (Truesdell, 1976), Na–Li (Fouillac and Michard, 1981), and Na–K–Ca (Fournier and Truesdell, 1973) geothermometer temperatures are calculated from elemental ratios which are essentially independent of the amount of reservoir flashing. As indicated for the Baca data in Table 3, calculated reservoir temperatures increase progressively in order $T_{SiO_2} < T_{Na\text{-}K} < T_{Na\text{-}K\text{-}Ca} T_{Na\text{-}Li}$.

Baca well 24 shows the closest agreement between all the geothermometer temperatures and the maximum measured downhole temperature.

The T_{SiO_2} estimated for fluids from the other wells is considerably lower than either the measured temperature or other geothermometer temperatures indicating precipitation of quartz during boiling in the wellbore or aquifer. The Na–K and Na–K–Ca geothermometers appear to reflect most closely both maximum measured reservoir temperatures and temperature differences between wells. This correspondence probably reflects local chemical-equilibrium conditions at different temperature and production zones within the fractured reservoir. The Na–Li temperature estimates are higher and more consistent than the other geothermometer temperatures. As indicated by Figure 4, maximum $T_{Na-K-Ca}$ values approach Na–Li temperatures in some of the wells. As discussed by Fouillac and Michard (1981), lithium is unlikely to re-equilibrate with secondary minerals with decreasing temperature due to its small ionic radius. Therefore, T_{Na-Li} probably reflects the maximum temperature achieved during the evolution of the reservoir. The fact that fluids in all the samples have comparable T_{Na-Li} values indicates a homogeneous, deep-seated heat source.

Comparison of different geothermometer temperature estimates for the hot springs along the Jemez River shows greater differences. The T_{SiO_2} and $T_{Na-K-Ca}$ temperatures are given in Table 3 as reported by Goff and others (1981), who concluded that silica in the springs either re-equilibrated at a lower temperature or was diluted by meteoric water. Figure 4 shows that although the $T_{Na-K-Ca}$ values of Jemez Springs and Soda Dam are much lower than those of the Baca wells, T_{Na-Li} values, which are possibly more indicative of maximum reservoir temperatures, are nearly identical for the springs and wells.

Goff and others (1981) suggested that lower Na–K–Ca temperatures are in part due to extensive dissolution of the carbonate aquifer from which the Soda Dam and Jemez Springs discharge. The resulting high calcium concentrations (>300 ppm) relative to the Baca well fluids (<25 ppm) tend to depress $T_{Na-K-Ca}$ values (equation 7). Therefore, the T_{Na-K} estimates are higher (Table 3) and probably more representative of maximum re-equilibration temperatures in the hot-springs system.

MIXING OF GEOTHERMAL WATERS

Some of the best indicators of a mixed geothermal component in hot springs are elemental or isotopic ratios which are unique to the geothermal reservoir and which do not re-equilibrate within the aquifer at lower temperatures. Common elemental ratios which may indicate geothermal origin are B/Cl, Br/Cl, Li/Cl, Cs/Cl, and As/Cl. Both Trainer and Lyford (1979) and Goff and Grigsby (1982) employed such ratios in concluding that waters at Soda Dam and Jemez Springs were derived from mixing of thermal waters of the Valles caldera with cold meteoric water.

Figure 5 shows a plot of boron to chloride concentrations for the Baca well waters and data for hot-spring water analyses from Trainer (1975) and Goff and others (1981, 1982). As indicated, a mixing line, similar to that determined by these workers, passes through the Baca well data. The Baca data are the estimated liquid-water concentrations corrected for wellhead flashing and two-phase reservoir conditions.

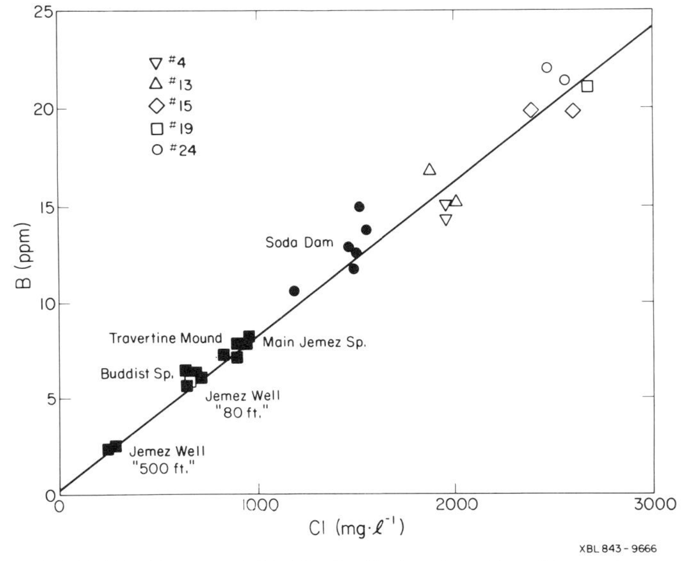

FIGURE 5. Boron and chloride distributions in wells and hot springs.

Based on distribution coefficients for boric acid between the liquid water and steam phases at the separator temperatures used at Baca ($K_d = 0.01$, Ellis and Mahon, 1977), total reservoir boron is assumed to be contained in the liquid-water fractions.

Two hydrologic processes can produce co-linear concentration charges; boiling which results in concentration increases upward along the line in Figure 5, and mixing with dilute meteoric water which results in decreases downward. Assuming that the highest-temperature wells, Baca 4 and 13, represent the primary hot-water component, boiling of this fluid will result in higher boron and chloride concentrations found in the cooler Baca 15, 19, and 24 wells. Dilution of geothermal fluids in the Valles caldera with meteoric water, as proposed initially by Trainer (1974), will result in the observed hot-spring boron and chloride concentrations. The mixing proportions in Baca 4 and 13 required to produce the spring-water compositions at Soda Dam, Jemez Springs, and the 500-ft aquifer intersected in the Jemez well (Goff and others, 1981) are 80, 50, and 10%, respectively. The proportion of geothermal water relative to meteoric water discharging from the springs would therefore be considerably higher than proposed by previous workers who overestimated chemical concentrations for the Baca geothermal fluids.

Elemental ratios of non-conservative chemical species can be used to estimate the extent of re-equilibration in a geothermal system. Figures 6a and 6b show respective Na/Cl and K/Cl ratios for the Baca wells and hot springs. Constant Na/Cl and K/Cl ratios for fluids in the wells suggest that differences in sodium, potassium, and chloride concentrations are due to boiling. The major exception is the significant decrease in the K/Cl ratios for Baca 24. This drop, coupled with lower production and geothermometer temperatures, indicates re-equilibration of potassium with reservoir rocks.

Na/Cl and K/Cl ratios for Soda Dam fluids plot close to the mixing line defined by the Baca wells, excluding the potassium values for Baca 24. Mixing lines drawn through the Jemez Springs waters are significantly offset from the Baca–Soda Dam mixing line with higher Na/Cl ratios and significantly lower K/Cl ratios.

DISCUSSION AND CONCLUSIONS

Based on the data presented in this paper and by previous workers, several interesting features of the geothermal system can be inferred.

Comparison of corrected liquid-reservoir compositions with measured temperatures and Na–K–Ca temperature estimates reveals that the fluids with the highest temperatures have the lowest dissolved chemical concentrations. This implies that the lower-temperature, chemically more concentrated fluids intersected in other wells have evolved by boiling. The general agreement between Na–K–Ca and measured temperatures indicates that local chemical equilibrium occurs in different thermal regimes in the reservoir. This is particularly well demonstrated for the low temperatures in Baca 24. The fact that Na–Li temperatures are identical in all the wells suggests that fluids in the reservoir had achieved the same maximum temperatures prior to re-equilibration. The highest calculated Na–K–Ca temperatures approach the Na–Li temperatures, suggesting maximum temperatures in the reservoir have not decreased significantly with time.

The B/Cl ratios show that the chemistry of the hot springs along the Jemez River can be explained by mixing of meteoric water and hot water in the Baca reservoir, as has been proposed by previous workers. Because the dissolved solids in the Baca fluids are lower than previously estimated, the water discharging at the hot springs requires even a greater proportion of thermal water to produce the observed chemistry. The Na–Li temperatures indicate that the fluids from the Baca wells and the hot springs have a common thermal source. These chemical data support the conclusion that the Baca geothermal reservoir is discharging through these springs via a fault system which extends into the Valles caldera.

The chemical re-equilibration data are the only evidence that suggests a more complex discharge pattern and possibly an additional, deeper geothermal reservoir. The Na/Cl and K/Cl ratios for the Soda Damn waters show no re-equilibration from the mixing line defined by the Baca fluids. This is also indicated by the high Na–K-geothermometer

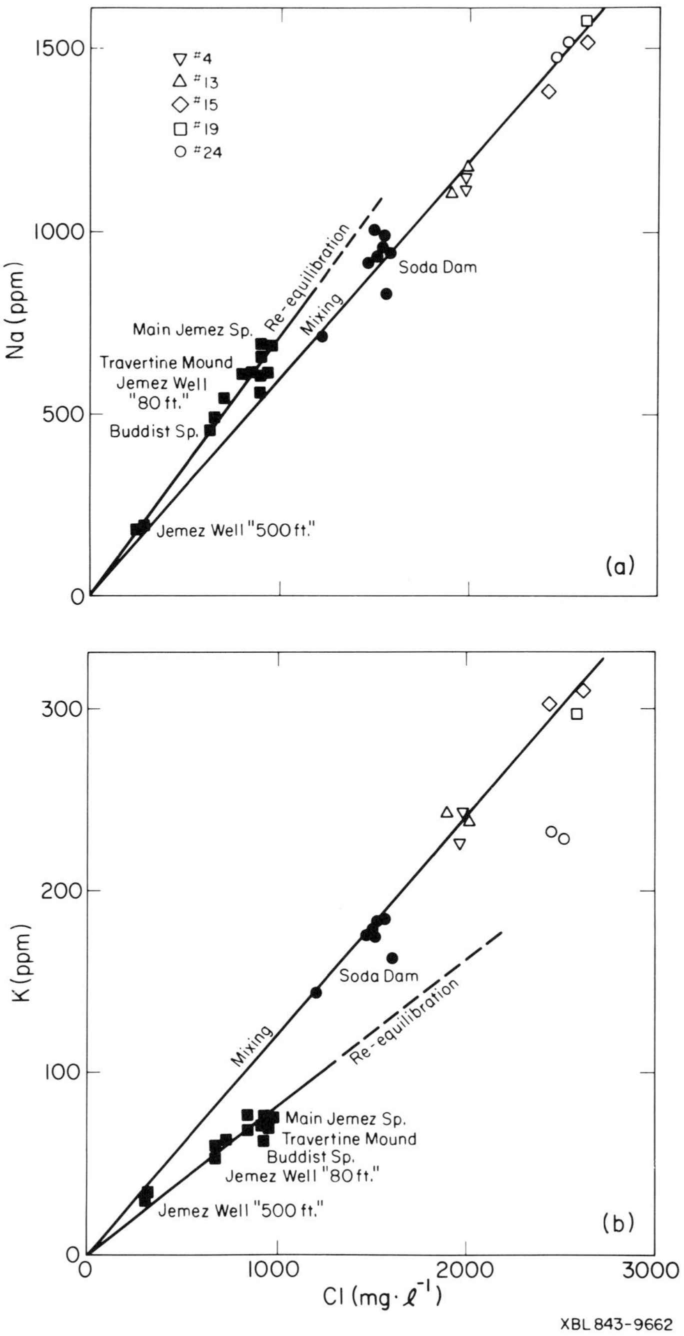

FIGURE 6. Sodium, potassium, and chloride distributions in wells and hot springs.

estimates of temperature for water from Soda Dam, which are in excess of 275°C. These temperatures are hotter than measured and calculated geothermometer temperatures for Baca 19 and 24.

The Jemez Springs waters show strong sodium and potassium re-equilibration at a calculated Na–K temperature of 200°C. Fluids from Baca 24 also show re-equilibration at a Na–K temperature of 240°C. Therefore, the question that must be answered is why thermal waters at Soda Dam failed to re-equilibrate with a cooler aquifer when waters at Jemez Springs and the cooler Baca wells did. The only apparent explanation is that thermal waters discharging at Soda Dam must have been confined in an aquifer or reservoir devoid of sodium- and potassium-silicate phases with which to re-equilibrate. The Paleozoic carbonates which underlie the Bandelier Tuff in the Baca field, and from which the springs at Soda Dam discharge, are the obvious choice. Reaction with a carbonate aquifer is also indicated by very high calcium and low SiO_2 concentrations at Soda Dam relative to Jemez Springs, and even relative to many meteoric springs in the Valles caldera region.

The Na–K equilibration temperature of 275°C indicates that the thermal waters discharging at Soda Dam must have entered the carbonate rocks at high temperatures from a source containing sodium and potassium silicates. This suggests discharge of the high-temperature reservoir waters, similar to fluids produced by wells 14 and 13 ($T_{Na-K} = 285°C$), from the Bandelier Tuff into the underlying Paleozoic carbonates. Because concentrations of elements such as Na, K, B, Br, and Li are low in such carbonates, original chemical signatures derived from the overlying tuff would tend to be preserved in the carbonate reservoir, as would the geothermometer temperatures. Deepening of Baca 22 and Baca 12 into the Paleozoic carbonate encountered high temperatures but low permeability. However, high permeabilities have been observed elsewhere in the Paleozoic carbonates at Fenton Hill.

Re-equilibration of the waters at Jemez Springs indicates that these waters have moved through sodium and potassium silicates at lower temperatures ($T_{Na-K} = 200°C$). These mineral phases could have been contained in lower-temperature zones of the Baca reservoir in the Bandelier Tuff as represented by Baca 24. Because the Jemez Spring waters discharge topographically and stratigraphically below Soda Dam, another possibility is that these waters have re-equilibrated with Precambrian granite which underlies both the Bandelier Tuff and the Paleozoic carbonate. Such granite was encountered at 250 m in the well drilled at Jemez Springs. Reactions with a silicate aquifer, either the Bandelier Tuff or Precambrian granite, is also implied by the higher silica and lower calcium concentrations relative to the Soda Dam waters. In summary, the avenues of transport from a high-temperature source must have been very different for thermal waters discharging at Jemez Spring and Soda Dam.

REFERENCES

Bodvarsson, G. S., O'Sullivan, M. J., and Tsang, C. F., 1980, The sensitivity of geothermal reservoir behavior to relative permeability parameter: Lawrence Berkeley Laboratory, Report 12040, p. 14.

———, Vonder Harr, S., Wilt, M., and Tsang, C. F., 1982, Preliminary studies of the reservoir capacity and generating potential of the Baca geothermal field, New Mexico: Water Resources Research, v. 18, pp. 1713–1723.

Ellis, A. J., and Mahon, A. J., 1977, Chemistry of geothermal systems: Academic Press, New York, p. 392.

Fouillac, C., and Michard, G., 1981, Sodium/lithium ratios in water applied to geothermometry of geothermal reservoirs: Geothermics, v. 10, pp. 55–70.

Fournier, R. O., and Truesdell, A. H., 1973, An empirical Na–K–Ca geothermometer for natural waters: Geochimica Cosmochimica Acta, v. 37, pp. 1255–1275.

Goff, F. E., and Grigsby, C. O., 1982, Valles caldera geothermal system, New Mexico, U.S.A.: Journal of Hydrology, v. 56, pp. 119–136.

———, ———, Trujillo, P. O., Jr., Counce, D., and Kron, A., 1981, Geology, water geochemistry and geothermal potential of the Jemez Springs area, Cañon de San Diego, New Mexico: Journal of Volcanology and Geothermal Research, v. 10, pp. 227–244.

———, McCormick, T., Trujillo, P. E., Counce, D., and Grigsby, C. O., 1982, Geochemical data for 95 thermal and nonthermal waters of the Valles caldera–Southern Jemez Mountains region, New Mexico: Los Alamos National Laboratory, Report LA-9367-OBES.

Keenan, J. H., Keyes, F. G., Hill, P. G., and Moore, J. G., 1969, Steam tables (International edition–metric units). Wiley, New York, p. 162.

Laughin, A. W., 1981, The geothermal system of the Jemez Mountains, New Mexico, and its exploration; *in* Ryback, L., and Muffler, L. J. P. (eds.), Geothermal systems: principles and case histories: Wiley, New York, pp. 295–320.

Nehring, N. L., and Truesdell, A. H., 1977, Collection of chemical, isotope and gas samples from geothermal wells: Proceedings of the Second Workshop on Sampling Geothermal Effluents, EPA-600/7-78-121, pp. 130–140.

Smith, R. L., and Bailey, R. A., 1968, Resurgent cauldrons; *in* Coats, R. R., Hay, R. L., and Anderson, C. A. (eds.), Geological Society of America, Memoir Series 116, pp. 613–662.

Trainer, F. W., 1974, Groundwater in the southwestern part of the Jemez Mountains volcanic region, New Mexico: New Mexico Geological Society, Guidebook 25, pp. 337–345.
———, 1975, Mixing of thermal and nonthermal waters in the margin of the Rio Grande rift, Jemez Mountains, New Mexico: New Mexico Geological Society, Guidebook 26, pp. 213–218.
———, and Lyford, F. P., 1979, Geothermal hydrology in the Rio Grande rift, north-central New Mexico: New Mexico Geological Society, Guidebook 30, pp. 299–306.
Truesdell, A. H., 1976, Summary of section III—geochemical techniques in exploration: Proceedings of the Second United Nations Symposium on the Development and Use of Geothermal Energy, San Francisco, pp. 1 iii–1 xxix.
———, and Fournier, R. O., 1976, Calculation of deep temperatures in geothermal systems from the chemistry of boiling water of mixed origins. Proceedings of the Second United Nations Symposium on Development and Use of Geothermal Energy, San Francisco, pp. 837–844.
Union Oil Company of California, 1982, Baca Project Geothermal Demonstration Power Plant: Final Report for the U.S. Department of Energy, ET-78-F-F-03-1717, p. 385.

The station at Servilleta was a boxcar, looking ready for collapse in 1941 when the 473 made a brief stop with its usual 2-car consist (photo by Bob Richardson).

This concrete basement is the most significant structure remaining at Servilleta. This desolate station, 10 miles south of Tres Piedras, was the transfer point for John Dunn's stage to Taos via the Arroyo Hondo bridge.

The rotting railroad tie in the foreground marks the location of the track at Servilleta. The site of the station is hidden in the sage brush.

GEOCHEMICAL BEHAVIOR OF A HOT DRY ROCK GEOTHERMAL RESERVOIR

C. O. GRIGSBY, FRASER GOFF, P. E. TRUJILLO, and D. A. COUNCE
Earth and Space Sciences Division, Los Alamos National Laboratory, Los Alamos, New Mexico 87545

INTRODUCTION

Technical feasibility of the Hot Dry Rock (HDR) concept of geothermal-energy extraction was proved in 1978 (Tester and Albright, 1978). Since that time a series of experiments has been conducted in the first (Phase I) reservoir to assess reservoir performance. The geochemistry of fluids obtained during these early experiments has been discussed by Grigsby and others (1983) and Goff and Grigsby (1982). These authors noted that the chemistry of fluids was subtly different from the natural hydrothermal fluids originating in the nearby Baca geothermal field, Valles caldera. After additional experiments, and after obtaining samples of pore fluid from deeper wells at Fenton Hill, the differences between these two fluids are even more noteworthy than previously thought. Because of the attention now being focused on the Valles caldera due to geothermal activities and the growing Continental Scientific Drilling Program, it is appropriate that the fluid chemistry of the Fenton Hill pore fluids be contrasted here with the Baca geothermal waters.

BACKGROUND

The HDR concept of geothermal-energy production involves extracting energy from hot, impermeable rocks by a man-made geothermal system. First, fracture-flow paths are created by hydraulically fracturing the rock between two wellbores (Fig. 1). Cold water is pumped down one wellbore, through the fracture system where it is heated, and up the second wellbore. At the surface, the hot water is maintained under sufficient pressure to prevent flashing and the attendant problems of scale formation. Energy is transferred from the hot water to a refrigerant that drives a turbine in a separate, closed loop. Fresh water is added to the cooled production water as required to make up the permeation losses in the reservoir before it is recirculated through the fracture system (Smith and others, 1975).

Field-scale investigations of the HDR concept in low-permeability Precambrian basement rock are being conducted at Fenton Hill, located on the west flank of the Valles caldera in the Jemez Mountains of northern New Mexico (Fig. 2). The initial reservoir design consists of two holes connected by a fractured heat-extraction system that ranges in depth from 2.5 to 3 km. The first borehole, Geothermal Test-2 (GT-2), was drilled in 1974 to a depth of 2.98 km, where the temperature was 197°C. A second borehole, Energy Extraction-1 (EE-1), was directionally drilled to 3.06 km in 1975. These two boreholes and the man-made hydraulic fractures connecting them constitute the Phase-I system. A third well, EE-2, was completed to 4.4 km (14,400 ft true vertical depth) in 1980, and a fourth well, EE-3, was completed to 4.1 km (13,399 ft) in 1981. This second pair of wellbores constitutes the Phase-II system that is being developed to demonstrate the reservoir

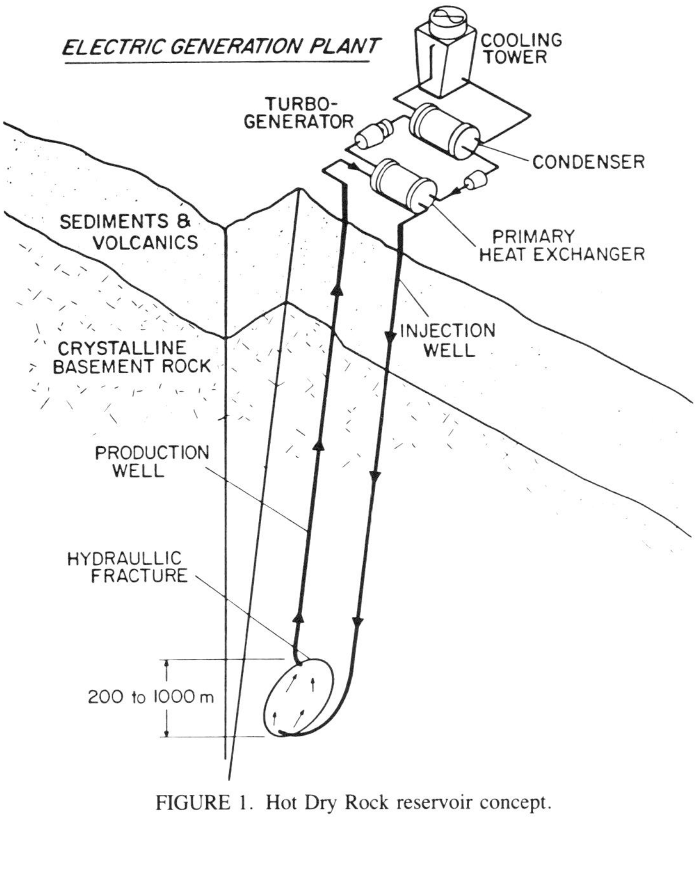

FIGURE 1. Hot Dry Rock reservoir concept.

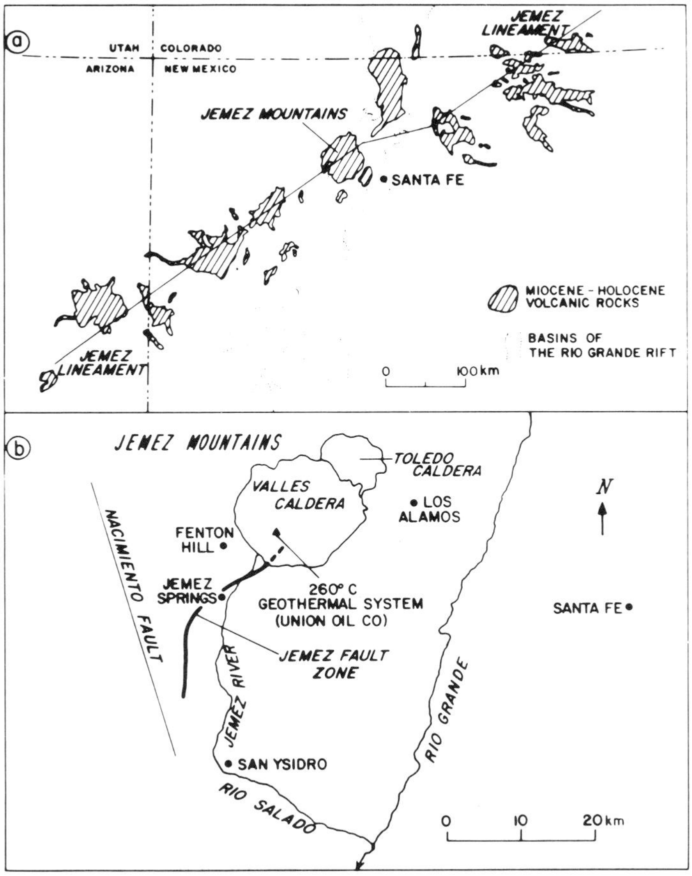

FIGURE 2. Map showing **(a)** location of the Jemez Mountains at the intersection of the Jemez lineament and the Rio Grande rift (from Chapin and others, 1978), and **(b)** location of Fenton Hill Hot Dry Rock site on the western flank of the Valles caldera in the Jemez Mountains.

lifetimes required for commercial heat extraction. As of the preparation of this paper, however, the two Phase-II wells have not been hydraulically connected. Figure 3 shows a simplified geologic column from cores and cuttings taken during the drilling of GT-2, EE-1, EE-2, and EE-3. Superimposed on this figure is an elevation view of the four wellbores drilled at Fenton Hill. The temperature scale from a background temperature log of EE-2 is also shown in Figure 3.

The Phase-I reservoir originally consisted of a set of near-vertical fractures that crossed the EE-1 wellbore at 2.75 km (behind the EE-1 casing). The performance of this original system was evaluated by a 75-day heat-extraction test which began on January 28, 1978 (Tester and Albright, 1978; Albright, 1978; Murphy, 1978; Grigsby and Tester, 1978). Following this test, the EE-1 casing was recemented and refractured, and a new, larger reservoir was formed with an injection point at 2.93 km in EE-1. The production zones for the enlarged reservoir were the same zones that were active during the 75-day test of the smaller reservoir. Preliminary evaluation of the enlarged reservoir was accomplished during a 24-day heat-extraction experiment that began on October 23, 1979 (Murphy, 1980; Murphy and others, 1981). More extensive testing of the enlarged reservoir ran for 280 days from March 10 to December 16, 1980. Results of this extended test are described by Zyvoloski (1981). A summary of the entire Phase-I experience is given by Dash and others (1983). Nominal values of various Phase-I operating conditions are given in Table 1.

Fluid samples from EE-2 and EE-3 have been collected during various experiments and workover operations. Three types of samples have been collected: (1) downhole samples collected at in-situ temperature and pressure using a one-liter high-temperature-fluid sampling tool developed by Los Alamos; (2) vent samples that result from the return of fluid that has been pumped downhole in a fracturing experiment; and (3) bottoms-up samples from reverse-circulating a wellbore to remove fluid that has been sitting in a wellbore prior to conducting workover operations (Fig. 4). Vent samples are typically collected at the liquid outlet of a steam separator, and they require correction for flashing. Bottoms-up samples have been cooled by counter-current flow of cold water as shown in Figure 4.

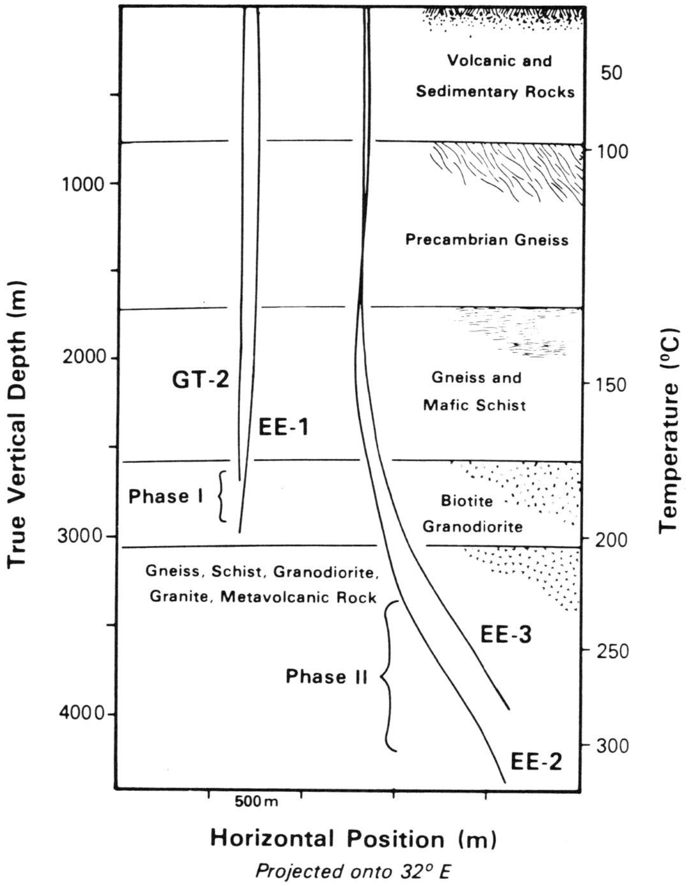

FIGURE 3. Elevation view of Phase-I wellbores (GT-2 and EE-1) and Phase-II wellbores (EE-2 and EE-3). Geologic column shows rock types interpreted from cuttings and core samples (Laughlin and others, 1983). Temperature scale at right is from a temperature log of EE-2.

TABLE 1. Summary of Phase-I operating conditions.

	Initial Reservoir	Enlarged Reservoir	
	75-day experiment	24-day experiment	280-day experiment
Dates	Jan. 28-Apr. 13, 1978	Oct. 23-Nov. 15, 1979	Mar. 3-Dec. 15, 1980
Duration (days)	75	23	286
Nominal Wellbore Separation Distance (m)	100	320	320
Injection Pressure (MPa)	9	9.5	8-9
Production Pressure (MPa)	1.2	1.1	1.4
Nominal Injection Flow Rate (m^3/s)	14×10^{-3}	7.6×10^{-3}	6×10^{-3}
Nominal Production Flow Rate (m^3/s)	12.5×10^{-3}	6×10^{-3}	5.7×10^{-3}
Impedance (GPa s/m^3)	1.7 - 0.4	1.9	1.75
Production Temperature (°C)	175-85	153	158-149
System Modal Volume (m^3)	11-27	120-130	155-180
System Integral Mean Volume (m^3)	35-56	260-280	400-1100
Average Thermal Power Extraction Rate (MW_t)	3.1	2.8	2.3
Total Power Extracted (MW-days)	240	60	625
Total Water Loss (m^3)	4900	4000	30,000

FLUID GEOCHEMISTRY

Fluid samples (except for the EE-3 reverse-circulation samples) were collected according to procedures described previously by Grigsby and others (1983). The EE-3 samples that were not collected according to those procedures are "grab" samples that were filtered and acidified or diluted in the laboratory after a 24–48-hour delay. Results of these various analyses are reported in Table 2.

The predominant ions in solution are Na^+, Cl^-, and HCO_3^-, with K^+, Ca^{+2}, and SO_4^- also present in significant concentrations. Important minor or trace elements are B, As, F, and Br. The high concentrations of all species, especially in the relatively undiluted EE-3 bottoms-up samples, are significant in that these are the most concentrated fluids ever observed in the Jemez Mountains (see Trainer, 1974; Goff and Grigsby, 1982; White and others, this guidebook). Because of the counterflow heat exchange that occurs in reversing out a well, this fluid arrives at the surface at temperatures substantially below the boiling point. There is no concentration effect due to steam loss or boiling. This is also evident from the low silica concentrations in the bottoms-up samples, although it is not clear at this time if some silica from these samples has polymerized before analysis.

Because of the addition of fresh make-up water of dilute chemical composition (i.e., Cl≤20 mg/*l*) to the Phase-I injected fluid and of similar fresh water to the Phase-II wellbores during drilling, workover, and fracturing operations, uncontaminated samples of formation fluids have never been obtained. Therefore, it is impossible to calculate the absolute concentration of species in the pore fluid at this time. Only the ratios of species can be defined with accuracy. The presence of a unique formation fluid or pore fluid in the Precambrian rocks is inferred from several lines of evidence:

(1) Oxygen- and hydrogen-isotope shifts suggest that long-term exchange has occurred between water and the rock;
(2) mixing relationships indicate the presence of a concentrated parent fluid;
(3) similar behavior of several species in response to changing operational conditions.

Shifts in oxygen- and hydrogen-isotope ratios are shown in Figure 5 for produced, injected, and make-up samples from the 24- and 280-

TABLE 2. Chemical and water-isotope analyses of Fenton Hill fluid samples. Elemental analyses by P. Trujillo and D. Counce, LANL; isotope analyses by L. Merlivat, Departement de Recherche et Analyse, Saclay, France.

Sample	Date	Time	pH	Conductivity µmho/cm	SiO_2	Na	K	Li	Ca	Mg	Fe	B	As	SO_4	HCO_3	Cl	F	Br	$\delta^{18}O$ (‰)	δD (‰)
GT-2 Production Fluid from 280-Day Experiment (Experiment 217)																				
2	10 Mar 80	15:00	6.57	7500	267	1490	110	n.d.	83	3.0	n.d.	42.3	n.d.	341	538	1750	8.7	11.2	-7.40	-74.7
3	11 Mar 80	00:30	5.95	5600	245	1100	90	17.0	61	2.3	5.55	33.1	n.d.	290	513	1220	10.6	12.0	n.d.	n.d.
4	11 Mar 80	07:00	6.06	4500	232	855	75	14.0	42	1.6	5.83	25.0	n.d.	248	493	921	15.3	3.7	-9.45	-81.4
8	14 Mar 80	10:00	6.33	3100	214	750	53	9.5	26	0.8	1.00	14.5	n.d.	227	400	583	16.5	2.1	-10.90	-84.6
15	21 Mar 80	10:30	7.33	2900	214	554	48	8.8	28	0.6	0.64	13.5	n.d.	260	471	528	11.2	5.3	-10.80	-85.0
22	28 Mar 80	10:30	6.78	2500	209	485	46	7.8	24	0.5	0.38	12.0	n.d.	241	455	422	14.1	1.2	n.d.	n.d.
29	04 Apr 80	10:30	7.31	2400	211	470	47	7.6	25	0.5	0.42	10.9	n.d.	222	483	387	14.4	1.5	-11.10	-85.3
36	11 Apr 80	14:00	6.15	2500	226	490	48	7.9	27	0.6	0.44	11.4	n.d.	245	477	281	15.8	2.6	n.d.	n.d.
50	25 Apr 80	10:00	5.13	2500	227	485	47	7.8	24	0.5	0.30	11.2	n.d.	258	454	412	14.6	2.1	-11.00	-85.0
63	09 May 80	06:00	6.27	2500	233	475	48	7.8	25	0.5	0.30	11.2	n.d.	256	448	420	13.3	1.7	n.d.	n.d.
76	30 May 80	10:30	6.14	2900	233	535	48	8.6	28	0.6	0.39	13.4	n.d.	286	450	485	17.0	2.3	-10.70	-84.6
94	20 Jun 80	10:00	6.28	2400	228	490	46	7.8	25	0.6	0.23	11.8	n.d.	229	490	380	13.0	1.6	-10.75	-81.5
104	11 Jul 80	10:00	6.37	2300	211	402	39	6.8	21	0.5	0.20	9.8	n.d.	190	471	314	12.6	1.1	-10.55	-82.7
114	25 Jul 80	11:10	6.37	2300	225	398	39	7.1	19	0.5	0.21	10.9	n.d.	205	467	351	13.0	<0.5	-10.75	-81.5
175	31 Oct 80	10:45	6.26	2000	204	360	35	6.0	19	0.5	0.38	8.7	n.d.	180	346	305	17.2	<0.5	n.d.	n.d.
183	21 Nov 80	11:45	6.26	2000	199	405	38	6.3	20	0.5	0.36	0.2	n.d.	208	434	302	20.3	<0.5	-11.05	-82.6
EE-2 Downhole Sample																				
V68	13 Jul 80	---	6.46	18000	435	3310	401	76.1	94	3.0	5.1	178	13.9	64	692	5870	6.1	42.3	n.d.	n.d.
EE-2 Vent Samples (Experiment 2032)																				
9	11 Dec 83	15:30	7.78	6100	440	1310	142	23.7	43	2.8	0.4	51	4.2	216	557	1859	7.8	8.5	n.d.	n.d.
10	12 Dec 83	10:30	6.85	7700	415	1360	154	25.4	98	3.8	3.6	59	4.9	176	800	2278	7.2	10.0	n.d.	n.d.
11	12 Dec 83	13:30	7.80	8000	430	1410	164	26.2	93	3.8	2.0	58	5.0	153	640	2297	7.0	10.1	n.d.	n.d.
15	14 Dec 83	09:30	6.15	7500	417	1410	157	26.8	129	4.0	10.0	58	4.2	179	787	2171	6.7	11.0	n.d.	n.d.
16	14 Dec 83	16:00	6.01	9100	420	1480	167	29.1	138	4.9	12.1	63	4.0	150	799	2557	6.6	11.6	n.d.	n.d.
EE-3 Bottoms-up Samples																				
V55	03 May 83	14:50	6.52	26200	191	4030	600	87	99	11.5	52	52	3.4	68	1498	7500	2.5	53	n.d.	n.d.
V56	03 May 83	15:23	6.67	30500	156	4830	730	106	140	9.8	0.16	272	18.3	51	1100	10500	2.3	71	n.d.	n.d.
V57	03 May 83	15:29	6.43	31400	148	4970	760	109	126	9.9	0.15	266	18.1	49	1018	9900	2.11	71	n.d.	n.d.
V60	03 May 83	15:42	6.43	33400	150	5150	760	113	126	10.1	0.14	278	20.3	50	978	9820	2.08	72	n.d.	n.d.
V62	03 May 83	15:49	6.47	32600	150	5270	800	119	124	10.8	0.16	266	19.6	49	939	9640	2.17	72	n.d.	n.d.
V64	03 May 83	16:00	6.32	33800	150	5240	810	120	133	11.2	0.26	271	20.9	52	937	9850	2.26	73	n.d.	n.d.

n.d. - not determined

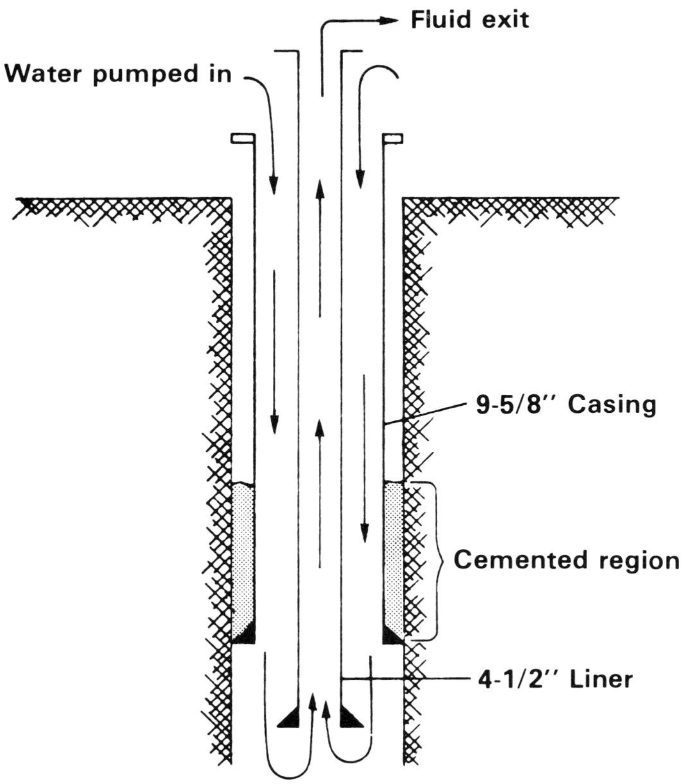

FIGURE 4. Reverse-circulating a well. "Bottoms-up" occurs when the fluid originally at the bottom of the well is displaced up the center pipe to the surface.

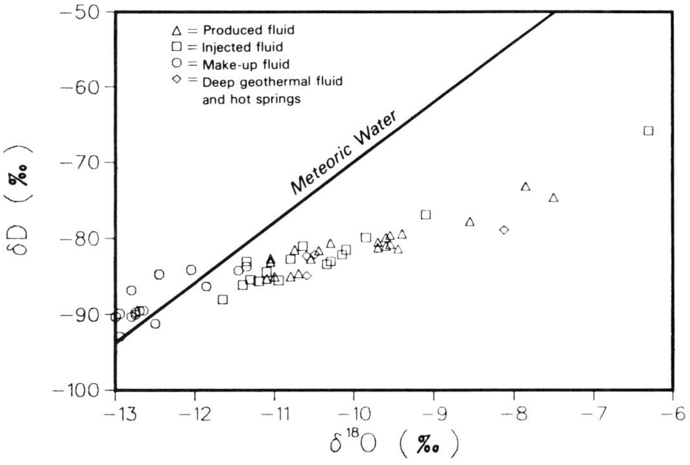

FIGURE 5. Oxygen- and hydrogen-isotope shifts for samples from the 280-day experiment.

day tests of the enlarged reservoir. Make-up samples consist of meteoric water that is pumped from the base of the volcanic section beneath Fenton Hill (approximate depth 780 m). These points cluster near the Craig meteoric-water line as they should (Craig, 1961). Injected fluid consists of produced fluid plus make-up water just before it is recycled through the system. Produced fluids (Fig. 5) represent a combination of pore fluid (formation fluid) that has resided in the reservoir rocks for extended periods of time and make-up fluid of meteoric composition. Early samples of the HDR produced fluids show enrichment in ^{18}O of about 6.5 per mil relative to the make-up fluid. As recirculation continues, the formation fluid is diluted with meteoric water so that the "steady-state" produced fluid shows only a 1.5–2.0 per mil shift. Likewise, there is a 15–20 per mil shift in hydrogen-isotope ratios in the early samples, but with time this shift drops to only 5–10 per mil.

Also shown in Figure 5 are isotope analyses of samples from the Baca wells, tapping the deep geothermal system in the Valles caldera (see paper by White and others, this guidebook, for description of the major element chemistry in the Baca wells) and from derivative hot springs which discharge in Cañon de San Diego (Goff and others, 1981; Goff and Grigsby, 1982). The similarity in oxygen- and hydrogen-isotope behavior between the Fenton Hill system and the Valles deep geothermal and derivative fluids, which appears in Figure 5, does not hold when ratios of other species are considered. Figure 6, for example, shows two unique mixing relationships between local, surface-meteoric water, Fenton Hill pore fluid, and Valles geothermal and derivative fluids. Notice that the ratios of B/Cl and Li/Cl in the Fenton Hill system are significantly different from the ratios in the Baca fluid. This mixing behavior is also observed in the Na/Cl and K/Cl ratios, but in these two cases the differences between the Fenton Hill samples and the Valles geothermal samples are not as striking. Possibly, Na and K are showing minor re-equilibration effects with silicate minerals. However, the time scale for recirculation is much shorter than the time scale for re-equilibration (Charles, 1979), and these re-equilibration effects are masked by recirculation.

Although the in-situ Fenton Hill pore fluid does not display unusual chemistry compared with representative fluids from other geothermal systems around the world (e.g., Fournier, 1981, table 4.2), the pore fluid is roughly three times more concentrated than the Baca geothermal fluids produced 8 km to the east (White and others, this guidebook). On the other hand, saline ground waters originating from Precambrian rocks have been recognized at other regions that do not have associated thermal anomalies (i.e., Fritz and Frape, 1980). Thus, at this time it is not entirely clear whether the saline Fenton Hill pore fluids would normally appear in the underlying Precambrian section or whether their appearance is somehow induced by heat from the Miocene through Quaternary volcanism that culminated in the Valles caldera (Gardner and Goff, this guidebook). We note that saline ground waters from the Precambrian of the Canadian shield do not show much enrichment in ^{18}O relative to the meteoric-water line (Fritz and Frape, 1980). In fact, they display a tendency toward D rather than ^{18}O enrichment. In contrast, the Fenton Hill pore fluid displays a pronounced ^{18}O enrichment of 6.5‰, suggesting to us rock–water interactions in a high-temperature environment. We contend, therefore, that both the Fenton Hill and Baca fluids are thermally driven by magmatic processes, but have different histories.

The similar behavior of ions in the various Fenton Hill fluids can be seen by defining a dimensionless-concentration scale such that the dimensionless concentration of each species in the pore fluid is 1 while the dimensionless concentration of each species in the make-up fluid is 0. The dimensionless concentration $<C>$ is given by

$$<C> = \frac{C(t) - C^m}{C^\infty - C^m} \qquad (1)$$

where C^∞ is the concentration of a species in the pore fluid, C(t) is the concentration of that species in the produced fluid at time t, and C^m is the concentration in the make-up fluid (Grigsby and others, 1983). A graph of $<C>$ for Na, K, Cl, and B is given in Figure 7 for the 280-day experiment. The values of C(t) and C^m are measured directly; C^∞ cannot be measured directly, but is approximated by initial values of C(t=0). Figure 7 shows that the dimensionless concentrations of all of these species are the same and that the response of each of the species to changes in operational conditions is identical. This is a clear indication of mixing of a concentrated fluid with a dilute fluid in the reservoir.

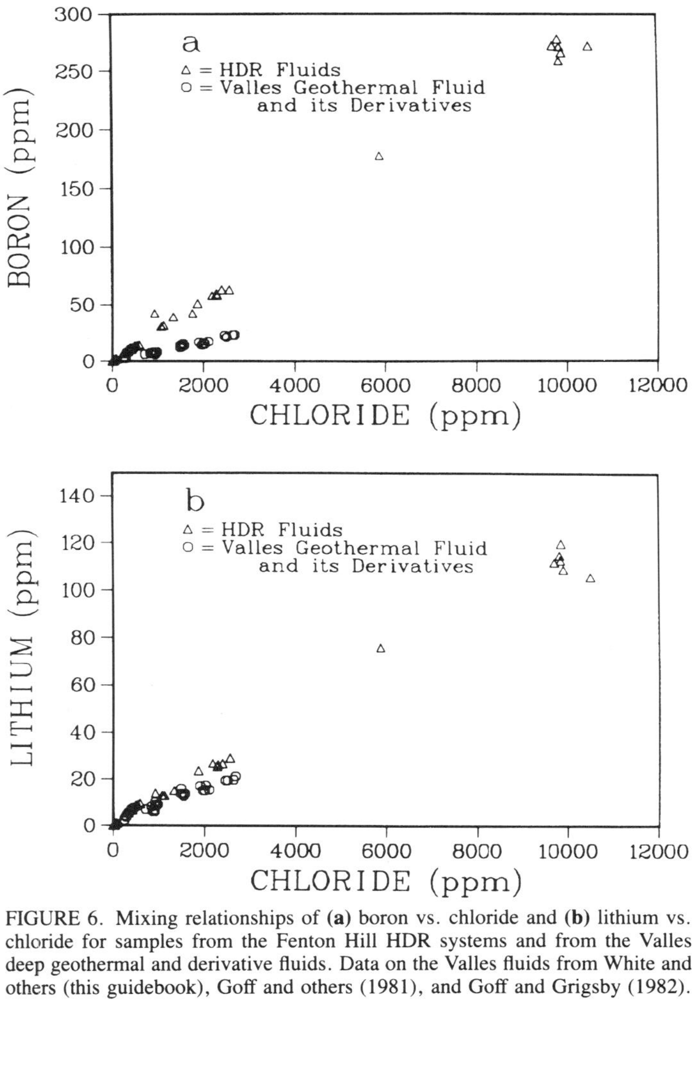

FIGURE 6. Mixing relationships of **(a)** boron vs. chloride and **(b)** lithium vs. chloride for samples from the Fenton Hill HDR systems and from the Valles deep geothermal and derivative fluids. Data on the Valles fluids from White and others (this guidebook), Goff and others (1981), and Goff and Grigsby (1982).

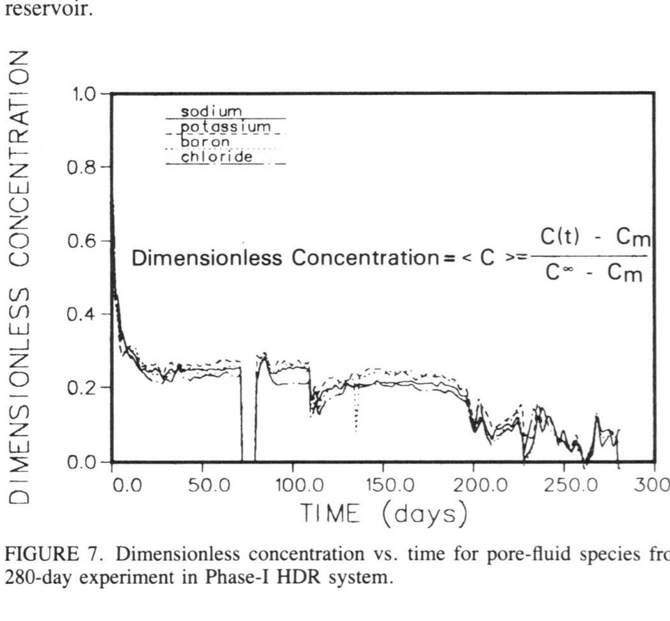

FIGURE 7. Dimensionless concentration vs. time for pore-fluid species from 280-day experiment in Phase-I HDR system.

GEOCHEMICAL MODELING

The evidence discussed in the previous section indicates that a unique aqueous phase containing dissolved salts is present in the Precambrian granitic rocks, and this phase is distinct from the Valles geothermal fluids. Furthermore, this fluid is apparently being added to the fluid recirculating through the Phase-I system for the entire duration of the 280-day experiment.

The rate at which this fluid is added can be roughly calculated by considering the increase in concentration of the pore-fluid species in the circulating fluid as it passes through the downhole system. If we assume that the pore-fluid composition is constant and that the pore fluid initially present in the main-fracture system has been completely flushed out so that no increase in concentration of Cl, B, etc. occurs in the system, then simple mixing of the fluid in the main-fracture flow path with the pore fluid can be used to determine the relative amounts of pore fluid and fresh fluid. This mixing relationship is simply expressed as a material balance

$$(1 - x)\, C_{in} + (x) C^{\infty} = C_{out} \qquad (2)$$

where x is the fraction of pore fluid in the produced fluid, C_{in} is the injected concentration, and C_{out} is the produced concentration. A graph of x vs. time for the 280-day experiment is given in Figure 8. By integrating this curve (after changing the flow fraction to a flow rate) we find that the cumulative volume of pore fluid displaced from the Precambrian rocks is of the order of 6,700 m^3 (1.8×10^6 gal).

GEOTHERMOMETRY

Quartz is by far the most reactive mineral under conditions of recirculating water flow past granodiorite samples from GT-2 at 200°C (Charles, 1979). While quartz is exposed to recirculating fluid, the aqueous-silica activity is almost completely controlled by quartz dissolution. The solubility of quartz as a function of temperature is well known (Morey and others, 1962; Kennedy, 1950). Aqueous-silica solutions become metastable as they are cooled and become supersaturated. They do not immediately precipitate silica when the tempertature is lowered, so the solubility vs. temperature relationship has been used as a means of estimating equilibration temperatures between quartz and water in wells and hot springs (Mahon, 1966; Fournier and Rowe, 1966). This quartz-geothermometer relationship is empirically expressed as

$$T(°C) = \frac{1309}{5.19 - \log_{10}[SiO_2]} - 273.15 \qquad (3)$$

where aqueous-silica concentration is in mg/*l*. This equation applies over the temperature range of 0–250°C, but works best for temperatures above 150°C.

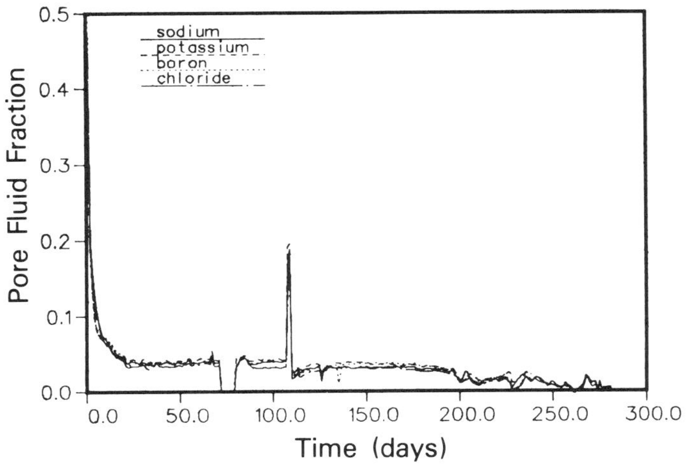

FIGURE 8. Fraction of produced fluid that was pore fluid in 280-day experiment.

Another chemical geothermometer that is often used to predict reservoir temperatures is the Na–K–Ca geothermometer (Fournier and Truesdell, 1973), which predicts the temperature of equilibration from the ratios of Na/K and Na/Ca. These ratios are set by cation exchange among the feldspars and feldspar-alteration minerals (particularly at temperatures above 100°C). The Na–K–Ca geothermometer is written as

$$T(°C) = \frac{1647}{\log_{10}(Na/K) + \beta[\log_{10}(\sqrt{Ca}/Na - 2.06] + 2.47} - 273.15 \qquad (4)$$

where Na, K, and Ca are concentrations in mg/kg, and

$$\beta = \begin{cases} 4/3, T < 100°C \\ 1/3, T > 100°C \end{cases}$$

Because the Na–K–Ca geothermometer depends upon the ratios among several dissolved species, dilution of a hot fluid containing large amounts of dissolved Na, K, and Ca with a dilute fluid should not significantly affect the ratios of the ions. Therefore, the predicted Na–K–Ca temperature should be relatively unaffected by dilution with fresh water. The calculated Na–K–Ca temperatures range from 190 to 195°C in samples collected during the 280-day experiment, and these temperatures correspond to the measured initial rock temperature.

Changes in calcium concentration within the system by dissolution or precipitation of calcite can affect the accuracy of the Na–K–Ca-geothermometer prediction (Fournier and Truesdell, 1973). If calcite is precipitating in the reservoir, the rate of precipitation of calcite would be limited by the rate at which calcium is supplied either by feldspar alteration or by the make-up fluid. Feldspar alteration is a slow process relative to the recirculation rate, and the calcium concentration in the pore fluid is higher than in the make-up fluid so that the calcium activity is low. It is therefore expected that the Na–K–Ca geothermometer is relatively unaffected by calcite precipitation in this system.

Comparisons of the silica and the Na–K–Ca-geothermometer temperatures with actual measured temperatures from the 280-day experiment are shown in Figure 9. In general, Na–K–Ca geothermometer temperatures are higher than silica-geothermometer temperatures and are close to the initial rock temperature. This conclusion is not surprising, since the Na/K ratio is set in the fluid that has been in long-term residence in the Precambrian rock.

The behavior of the quartz-geothermometer predicted temperatures is more difficult to explain. It was originally thought that quartz dissolution in the main-fracture flow path provided the source of additional

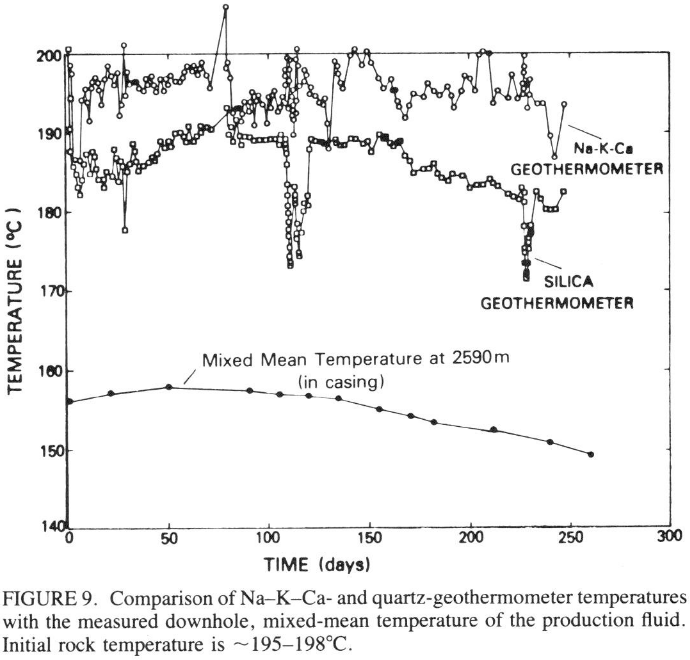

FIGURE 9. Comparison of Na–K–Ca- and quartz-geothermometer temperatures with the measured downhole, mixed-mean temperature of the production fluid. Initial rock temperature is ~195–198°C.

silica, which is observed in the production fluid. However, as shown by Grigsby (1983), the rate of quartz dissolution is much too slow and the residence time is much too short to account for the silica concentrations in the production fluid. Instead, some of the fluid injected into the reservoir must have sufficiently long residence time in regions that have not been thermally depleted so that the silica concentration can come to equilibrium at the initial rock temperature. This fluid is then mixed in the reservoir with water, which has a short residence time in cooler regions of the reservoir so that it contains only the silica concentration of the injected fluid. According to a simple mass-balance calculation, approximately 30% of the injected fluid must have a mean residence time greater than five days at temperatures of ~195°C to explain the silica behavior.

DISCUSSION AND CONCLUSIONS

A significant volume of a previously unknown geothermal fluid occurs in Precambrian basement rocks on the west flank of the Valles caldera. We suggest that this fluid has a different origin than the deep geothermal fluids inside the caldera. Moreover this in-situ fluid exerts a controlling influence over the observed fluid chemistry in the Fenton Hill HDR system. Detailed modeling of recirculated water in the deeper HDR reservoir may provide some clues about the origin of the in-situ fluid, but study and comparison of hydrothermal waters from additional deep wells drilled within and around the Valles caldera may ultimately be the key to addressing this problem.

ACKNOWLEDGMENTS

This project was sponsored by the U.S. Department of Energy, Division of Geothermal Energy. We thank Frank Trainer and Art White for reviewing this paper and thank Frank again for sharing with us the publication on saline waters from the Canadian shield. We also thank Marcia Jones for typing this manuscript and John Paskiewicz for drafting the figures.

REFERENCES

Albright, J. N., 1978, Evaluation of the Fenton Hill hot dry rock geothermal reservoir, part III. Reservoir characterization using acoustic techniques; *in* Proceedings of the Fourth Workshop on Geothermal Reservoir Engineering, Stanford University, pp. 256–263.

Chapin, C. E., Chamberlin, R. M., Osburn, G. R., and others, 1978, Exploration framework of the Socorro geothermal area, New Mexico: New Mexico Geological Society, Special Publication 7, pp. 114–129.

Charles, R. W., 1979, Experimental geothermal loop II: 200°C study: Los Alamos Scientific Laboratory, Report LA-7735-MS, 40 pp.

Craig, H., 1961, Isotopic variations in meteoric waters: Science, v. 133, pp. 1702–1703.

Dash, Z. V., Murphy, H. D., Aamodt, R. L., and others, 1981, Hot dry rock geothermal reservoir testing: 1978 to 1980: Journal of Volcanology and Geothermal Research, v. 15, pp. 59–99.

Fournier, R. O., 1981, Application of water geochemistry to geothermal exploration and reservoir engineering; *in* Rybach, L., and Muffler, L. J. P. (eds.), Geothermal systems: principles and case histories: J. Wiley and Sons, New York, pp. 109–143.

———, and Rowe, J. J., 1966, Estimation of underground temperatures from the silica content of waters from hot springs and wet-steam wells: American Journal of Science, v. 264, pp. 685–697.

———, and Truesdell, A. H., 1973, An empirical Na–K–Ca geothermometer for natural waters: Geochimica et Cosmochimica Acta, v. 37, pp. 1255–1275.

Fritz, P., and Frape, S. K., 1982, Comments on the ^{18}O, ^{2}H, and chemical composition of saline groundwaters on the Canadian shield; *in* Perry, E. C., Jr. (ed.), Isotope studies of hydrologic processes: Northern Illinois University Press, DeKalb, pp. 57–63.

Goff, F. E., and Grigsby, C. O., 1982, Valles caldera geothermal systems, New Mexico, USA: Journal of Hydrology, v. 56, pp. 119–136.

———, ———, Trujillo, P. E., and others, 1981, Geology, water geochemistry and geothermal potential of the Jemez Springs area, Cañon de San Diego, New Mexico: Journal of Volcanology and Geothermal Research, v. 10, pp. 227–244.

Grigsby, C. O., 1983, Rock–water interactions in hot dry rock geothermal systems: reservoir simulation and modeling: M.S. thesis, Massachusetts Institute of Technology, 134 pp.

———, and Tester, J. W., 1978, Evaluation of the Fenton Hill hot dry rock geothermal reservoir, part II. Flow characteristics and geochemistry; *in* Proceedings of the Fourth Workshop on Geothermal Reservoir Engineering, Stanford University, pp. 249–255.

———, ———, Trujillo, P. E., and others, 1983, Rock–water interactions in hot dry rock geothermal systems: field investigations of *in-situ* geochemical behavior: Journal of Volcanology and Geothermal Research, v. 15, pp. 101–136.

Kennedy, G. C., 1950, A portion of the system silica–water: Economic Geology, v. 45, pp. 629–653.

Laughlin, A. W., Eddy, A. C., Laney, R., and other, 1983, Geology of the Fenton Hill, New Mexico, hot dry rock site: Journal of Volcanology and Geothermal Research, v. 15, pp. 21–41.

Mahon, W. A. J., 1966, Silica in hot water discharged from drill holes at Wairakei, New Zealand: New Zealand Journal of Science, v. 9, pp. 135–144.

Morey, G. W., Fournier, R. O., and Rowe, J. J., 1962, The solubility of quartz in water in the temperature interval from 25°C to 300°C: Geochimica et Cosmochimica Acta, v. 26, pp. 1029–1043.

Murphy, H. D., 1978, Evaluation of the Fenton Hill hot dry rock geothermal reservoir, part I. Heat extraction performance and modeling; *in* Proceedings of the Fourth Workshop on Geothermal Reservoir Engineering, Stanford University, pp. 244–248.

———, (ed.), 1980, Preliminary evaluation of the second hot dry rock geothermal energy reservoir: results of Phase I, Run Segment 4: Los Alamos National Laboratory, Report LA-8354-MS, 88 pp.

———, Tester, J. W., Grigsby, C. O., and other, 1981, Energy extraction from fractured geothermal reservoirs in low-permeability crystalline rock: Journal of Geophysical Research, v. 86, pp. 7145–7158.

Smith, M. C., Aamodt, R. L., Potter, R. M., and other, 1975, Man-made geothermal reservoirs; *in* Proceedings of the Second United Nations Symposium on the Development and Use of Geothermal Resources, San Francisco, California, pp. 1781–1787.

Tester, J. W., and Albright, J. N. (eds.), 1979, Hot dry rock energy extraction field test: 75 days of operation of a prototype reservoir at Fenton Hill: Los Alamos Scientific Laboratory, Report LA-7771-MS, 104 pp.

Trainer, F. W., 1974, Ground water in the southwestern part of the Jemez Mountains volcanic region, New Mexico: New Mexico Geological Society, Guidebook 25, pp. 337–345.

Zyvoloski, G. A. (ed.), 1981, Evaluation of the second hot dry rock geothermal energy reservoir: results of Phase I, Run Segment 5: Los Alamos National Laboratory, Report LA-8940-HDR, 94 pp.

MAN ON THE RIO GRANDE: INTRODUCTION AND OVERVIEW

JOHN A. WARE
Laboratory of Anthropology, Museum of New Mexico, Santa Fe, New Mexico 87503

INTRODUCTION

A record of over 10,000 years of human occupation is preserved in the Rio Grande rift. The duration and character of that occupation was and remains intimately related to geological processes. The Rio Grande rift preceded the river of the same name. Rifting concentrated runoff, imposed a drainage corridor, and helped to create one of the most important resources in an arid environment: a perennial water source. The rift almost certainly influenced the seasonal movements of early hunters and gatherers in northern New Mexico. The region's first Paleoindian inhabitants were undoubtedly drawn to the water source, as were the herds of Pleistocene animals they hunted. As water became increasingly scarce during the millennia-long droughts that followed the close of the Pleistocene, the river exerted an important influence on settlement patterns and seasonal population movements.

The rift concentrated important resources other than water. Volcanism associated with rifting produced obsidian for stone tools, altered land forms, and created elevations, ecological diversity, and a greater variety of wild plant and animal resources. The floodplain of the Rio Grande eventually attracted farming settlements. When large areas of the Colorado Plateau were abandoned by prehistoric farmers during the first centuries of the present millennium, many of those who were displaced migrated to the rich bottomlands of the Rio Grande and its tributaries, where the certainty of a perennial stream flow and the advantages of a longer growing season encouraged population growth and agricultural intensification.

When Europeans colonized New Mexico in the sixteenth century, they encountered over a hundred farming communities along the Rio Grande and its tributaries (Fig. 1). In the next century the valley of the Rio Grande became a migration corridor for Spanish settlers probing deep into New Mexico from the mining towns of Chihuahua, and in 1680 it became an escape route for the descendants of immigrants fleeing the onslaught of the Pueblo Revolt. Two hundred years later the Rio Grande corridor of trade, migration, and communication helped to usher in a new era of trade with, and eventually political domination by, the United States.

The four papers that follow focus on the late prehistoric and early historic periods on the northern Rio Grande, from the twelfth-century Pueblo Coalition (Wendorf and Reed, 1955) to the seventeenth-century Pueblo Revolt, a span of five centuries that witnessed profound changes in human adaptations and cultural lifeways on the Rio Grande in northern New Mexico. Some of the more important changes are outlined and discussed in the first two articles by Peckham and Schroeder, which present concise overviews of the late prehistoric and early historic periods. The final two articles, by Cordell and by Bower and Snow, address specific research issues of the period. Cordell raises questions regarding some old assumptions about cultural continuity in the Rio Grande valley, assumptions that have influenced historical and archaeological research strategies in the northern Southwest since the turn of the century. Bower and Snow use x-ray fluorescence analysis of early historic Tewa pottery to investigate historic cultural interactions among Spanish and Pueblo populations in the Rio Grande valley.

The following papers make no attempt to cover the entire ca 10,000 years of documented human occupation of the northern Rio Grande valley. In introducing the papers, I will focus on some of the more obvious gaps, to provide some historical background for the more detailed archaeological and historical discussions that follow, and to provide a critical overview of some of the more important research issues that are raised. The sketch that follows relies primarily on secondary sources, recent syntheses, and discussions with colleagues. It touches on some, but by no means all, of the important issues. For more detailed introductions to Rio Grande prehistory, the reader should refer to the recent syntheses by Biella and Chapman (1977), Cordell (1978a, 1979b), and Stuart and Gauthier (1981).

EARLY HUNTERS AND GATHERERS

Much circumstantial evidence suggests that the Rio Grande rift supported intermittent human occupation for at least 10,000 years. For all but the last 1,500 years, humans lived in small groups as mobile hunters and gatherers, exploiting a diverse mix of wild animals and plants. Physical evidence of early Paleoindian hunters and gatherers (ca 12,000–6,000 years ago) is fragmentary, since only isolated Paleoindian artifacts have been recovered from the northern Rio Grande valley (Peckham, this guidebook). Subsequent Archaic hunting and gathering populations (ca 6,000–2,000 years ago) have been more thoroughly documented in the region, thanks to intensive archaeological surveys at Abiquiu and Cochiti Reservoirs (Biella and Chapman, 1977; Schaafsma, 1976). Nevertheless, data on pre-farming populations in the Rio Grande valley are extremely sketchy.

Arid-land hunters and gatherers typically live in small, highly mobile bands. Camp sites are often occupied for no more than two or three weeks before the band moves on to a new foraging area. The ephemeral remains associated with such short-term site occupations are difficult to identify archaeologically, and they are extremely susceptible to post-occupation disturbance, especially in a geologically active zone such as the Rio Grande rift. Site-occupation spans are often so brief that few formal artifacts are discarded. Consequently, unambiguous assignment of hunter–gatherer living sites to specific time periods or cultural groups

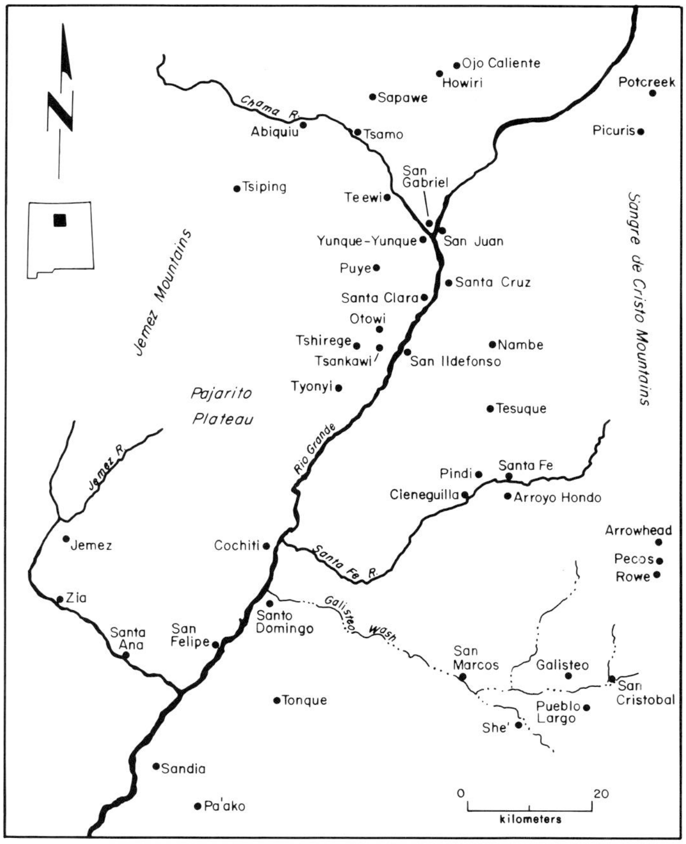

FIGURE 1. Late prehistoric and historic pueblo sites in the northern Rio Grande valley.

on the basis of temporally or culturally sensitive artifacts is difficult, and often impossible (Cordell, 1979a).

The paucity of physical remains and the long time spans that separate us from the earliest hunters and gatherers of the Rio Grande valley make it difficult to reconstruct dimensions of their environment, demography, settlement and subsistence patterns, and organization; archaeologists are just beginning to address these problems in a systematic way. Fundamental problems still exist in, for example, the areas of early hunter–gatherer chronology and culture history. The existence of a pre-Clovis (ca 12,000 years ago) occupation of the northern Rio Grande valley continues to be a major issue despite the widely discredited early occupation dates from Sandia Cave (Hibben, 1955; Stevens and Agogino, 1975). Other major points at issue include an apparent 1,400-year discontinuity between the Folsom (10,000 years ago) and the Cody Complex (8,600 years ago) (Irwin-Williams, 1979), an hypothesized discontinuity between Paleoindian and Archaic populations (Irwin-Williams, 1967, 1973), and an apparent truncation of the Oshara Archaic Tradition in the Galisteo Basin by the Chiricahua Cochise culture (Lang, 1977).

These discontinuities may be significant cultural–historical events, or they may be little more than failures of measurement, interpretation, and inference on the part of archaeologists who have recovered very little physical evidence and only a handful of unambiguous dates for early hunters and gatherers in the northern Rio Grande valley. Whatever the case, the experts appear far from a consensus, and there is a great deal of disagreement among archaeologists over some very basic issues.

Progress in understanding early hunter–gatherer adaptations will continue to be slow because of the paucity of material evidence. As dating techniques improve, paleoenvironmental reconstructions become more complete, and archaeologists refine their theories of early hunter–gatherer adaptations, research prospects will undoubtedly improve. What is most needed in the Rio Grande rift are more large-scale, intensive areal surveys, a more thorough understanding of contemporary hunter–gatherer adaptations and material correlates of those adaptations, and, perhaps most important, sufficient imagination to approach old problems in new ways.

FROM FORAGER TO FOOD PRODUCER

Corn and squash horticulture was probably introduced into the Rio Grande valley during the first millennium B.C. The centuries immediately following the introduction of agriculture witnessed important social, economic, and technological changes, including a shift from mobile hunting and gathering to food production, and the evolution of the first settled villages in the northern Rio Grande valley. These changes are not well documented in the upper Rio Grande valley, and we are not altogether sure why. As Cordell points out (this guidebook), the upper Rio Grande valley was a much less marginal environment for hunters, gatherers, and subsistence farmers than many other areas of the northern Southwest, and yet, examples of the most precocious Anasazi cultural development occur to the north and west of the Rio Grande valley, on the arid Colorado Plateau.

Cordell suggests several possible explanations. Alluviation and erosion on the floodplains of the Rio Grande and its tributaries may have destroyed much of the physical evidence of past human occupation. Recognition bias has almost certainly played a role as well. In focusing as they have on the large and spectacular remains of the late Classic pueblos, archaeologists may have systematically overlooked the more ephemeral remains of early hunters, gatherers, and farmers in the Rio Grande valley. Perhaps, as Cordell also suggests, the Rio Grande valley had such abundant wild resources that a mobile hunting and gathering adaptation persisted longer than in more marginal areas outside the valley.

This last point raises some important theoretical issues regarding the causes and consequences of food production. On the basis of recent studies of contemporary hunters and gatherers (e.g., Lee and DeVore, 1968; Lee, 1979), it appears that many of the "benefits" of agriculture are illusory. When wild foods are abundant and available year-round, hunting and gathering is the subsistence strategy of choice because it involves substantially less effort and risk than farming. It follows, then, that the benefits of agriculture might be more apparent in a marginal environment where hunting and gathering options are less attractive because of environmental uncertainty, resource scarcity, and competition in the food quest, or where wild foods are only available seasonally, creating long periods of food scarcity.

There are other advantages that agriculture affords in a "patchy" arid or semiarid environment where seasonal variability is often extreme and important wild-plant and animal resources tend to be localized and unpredictable (Yellen, 1976). In such an environment, farming not only provides a stable source of domestic calories, but the clearing of fields for agricultural crops provides new disturbed habitats for the concentration of economically important wild plants and animals which can be exploited along with the domestic crops (Emslie, 1981; Ford, 1982; Linares, 1976). In an important respect, field agriculture transformed the ecology of the semiarid northern Southwest by increasing the local density of edible wild plants and animals. These fundamental ecological changes perhaps contributed to the growth of population and the evolution of complex societies on the otherwise marginal Colorado Plateau.

In contrast to the western Colorado Plateau, the valley created by the Rio Grande rift was probably one of the more optimal foraging areas in the northern Southwest (Cordell, 1979b). This may explain why settled, village-farming economies developed so late in much of the northern Rio Grande valley, and why population exploded in the valley during the thirteenth and fourteenth centuries A.D. following the structural collapse of Anasazi systems in the San Juan Basin and Colorado Plateau to the west.

PUEBLOS AND SPANIARDS

From the twelfth through the seventeenth century A.D., the rate of culture change in the northern Rio Grande valley accelerated as Pueblo population increased and as European contact created new social and economic challenges for the indigenous Pueblo population. As population grew and became concentrated in larger communities, a wealth of physical archaeological evidence began to accumulate. It is in this period of growth and rapid change that Rio Grande archaeologists have concentrated their efforts for the past 100 years, and where Southwestern historians and archaeologists begin to work hand in hand. It is at this point also that the archaeology and history papers in this guidebook take up the story of culture change in the Rio Grande valley.

During the 500 years that elapsed between the Pueblo Coalition and the Pueblo Revolt, the population of the Rio Grande valley appears to have been very unstable. Significant population growth and settlement instability occurred in the valley during the thirteenth and early part of the fourteenth century A.D. (Peckham, this guidebook). Cordell (this guidebook) mentions significant population instability during the fourteenth through sixteenth centuries, as many large Rio Grande pueblos were abandoned and new pueblos established. A hundred years later, during the first century of Spanish occupation of the Rio Grande, there was a 90–95% decline in Pueblo population, from 60,000 to 10,000 individuals in A.D. 1600, to under 6,500 individuals in A.D. 1706 (Schroeder, this guidebook).

The causes and consequences of population growth and decline on the Rio Grande have been an issue of active discussion and debate for years (Cordell, 1979a). Of all the demographic episodes in question, the seventeenth century collapse of Pueblo population is undoubtedly the best documented because there are historical census data that verify the collapse. Schroeder (this guidebook) attributes seventeenth-century population decline to a combination of European diseases, famine, and warfare.

Much debate surrounds studies of population fluctuations during the thirteenth and fourteenth centuries, prior to European contact. Most investigators hypothesize an episode of explosive growth during the Pueblo Coalition and early Classic period, and most of the discussion has focused on the causes of such a rapid growth.

Immigration from the north and west is one of the most appealing theories, primarily because population growth on the Rio Grande follows large-scale depopulation of the northern Colorado Plateau. Others have cited the lack of physical evidence of large-scale migrations to argue for internal growth of an indigenous Rio Grande population during the thirteenth and fourteenth centuries A.D. Although the current consensus seems to be that large-scale migrations occurred (Peckham, this guidebook), the issue is, in fact, far from resolved. For one thing, the debate over immigration versus internal growth masks an important epistemological issue: archaeologists do not know how to measure accurately the size of prehistoric populations.

If adequate funerary data are available, archaeologists can say a great deal about such things as the age structure, morbidity patterns, and mortality rates of prehistoric populations. But with contemporary tools and techniques archaeologists can say practically nothing with certainty about the size of prehistoric populations, and, therefore, about flunctuations in the size of populations. Regional estimates of prehistoric populations are typically based on such measures as site frequency, site and artifact density, room counts, living-room-floor areas, and hearth frequencies. All of these measures, however, can change in response to variables other than population fluctuation. For example, the number of sites in a region can increase through time if there is an increase in population mobility.

Cordell (this guidebook) raises another important epistemological issue regarding culture change and continuity, and the use of the present to interpret the past. Traditionally, archaeologists have looked to ethnographic accounts of historic Pueblo Indians for help in interpreting Anasazi prehistory, on the assumption that Pueblo culture is basically conservative and, therefore, resistant to change. In emphasizing the effects of nearly 300 years of Spanish occupation of the northern Rio Grande, Cordell (this guidebook) argues persuasively for more skepticism in our use of ethnographic analogy in southwestern archaeology.

There is no question that ethnographic analogy has an important role to play in archaeological theory-building. But in the Southwest, where cultural continuity is too often assumed rather than tested, ethnographic resources are too often exploited as a source of answers about the past rather than as a source of questions. Ethnographic analogy is a valid tool for assessing the likelihood of alternative hypotheses and as a source of ideas and hypotheses about the past. But hypotheses generated by ethnographic analogy must ultimately be evaluated with independent archaeological data.

Perhaps nowhere is analogy employed more uncritically than in the study of prehistoric organization, which brings us to the final point in this discussion and, coincidentally, to the last paper in the set that follows. There is probably no greater challenge in prehistoric archaeology than to reconstruct aspects of prehistoric social organization. There are undoubtedly many important relationships between the structure and organization of a society and the material residue of a society. However, there are so many intervening variables between an extinct social system and the tangible archaeological remains of that social system that unambiguous correlations between the two classes of phenomena are difficult to construct.

Because of the difficulties inherent in identifying and measuring dimensions of prehistoric social organization, archaeologists have demonstrated a tendency to fall back on archaeological and historical analogy whenever confronted with an organizational problem. Without historical records, informant interviews, or eyewitness accounts, how does one go about measuring the kinds and degrees of social relationships and the extent of interaction among social groups in the distant past?

Methods of quantifying social interaction and organization on the basis of ceramic-design analysis developed in the late 1960's (Hill, 1970; Longacre, 1970) were based on many untested assumptions, and the validity of their conclusions has been questioned (Plog, 1977). In recent years, other approaches to studying social interaction have been developed, and some show a great deal of promise. One such approach, exemplified in the article by Bower and Snow (this guidebook), involves constituent analysis of ceramic raw materials and focuses on the movement of goods within and between communities.

Bower and Snow use x-ray fluorescence techniques to analyze ceramic samples from over a dozen historic pueblos in the Rio Grande valley and early Spanish settlements along the Santa Fe River. Although the results of the analysis must be considered preliminary, the fact that they are able to demonstrate significant differences in ceramic raw materials between the various pueblo sites suggests that they may be able eventually to discriminate between historic ceramic production centers and measure ceramic exchange and social interaction among early historic population centers in the northern Rio Grande.

These kinds of studies should eventually provide important information on prehistoric social and economic interaction in the northern Rio Grande. Such studies, for example, might help us to test the hypothesis proposed by Cordell (this guidebook) regarding the extent of pre-contact social interaction among the late prehistoric pueblos of the Rio Grande and the effects of Spanish intrusion on those pre-contact interaction patterns.

The history of human occupation of the Rio Grande valley of northern New Mexico is rich and extremely complex. The papers that follow reduce some of the inherent complexity. In the process, they contribute to the evolution of prehistoric research on one of America's most historic rivers.

ACKNOWLEDGMENTS

I would like to thank Scott Baldridge, Tim Maxwell, Yvonne Oakes, Stew Peckham, and Laura Ware for reading and commenting on preliminary drafts of this manuscript. Special thanks to Myra Ellen Jenkins for helping to coordinate the Man on the Rio Grande Colloquium at the College of Santa Fe in 1983 and 1984.

REFERENCES

Biella, J. V., and Chapman, R., 1977, Archaeological investigations in Cochiti Reservoir, New Mexico, vol. 1: Office of Contract Archaeology, University of New Mexico, Albuquerque.

Cordell, L. S., 1979a, Cultural resources overview of the middle Rio Grande valley, New Mexico: U.S. Government Printing Office, Washington, D.C.

———, 1979b, Eastern Anasazi; *in* Ortiz, A. (ed.), Handbook of North American Indians, vol. 9, Southwest: Smithsonian Institution, Washington, D.C.

Emslie, S. D., 1981, Birds and prehistoric agriculture: the New Mexico pueblos: Human Ecology, v. 9, no. 3, pp. 305–329.

Ford, R. I., 1982, People and plants on prehistoric Black Mesa: Paper presented at the School of American Research Advanced Seminar, Culture Change on Black Mesa, March 1982, 26 pp.

Hibben, F. C., 1955, Specimens from Sandia Cave and their possible significance: Science, v. 122, pp. 688–689.

Hill, J. N., 1970, Broken K Pueblo: prehistoric social organization in the American Southwest: Anthropological Papers of the University of Arizona, no. 18, Tucson.

Irwin-Williams, C., 1967, Picosa: the elementary southwestern culture: American Antiquity, v. 32, no. 4, pp. 441–457.

———, 1973, The Oshara tradition: origins of Anasazi culture: Eastern New Mexico University, Contributions to Anthropology, v. 5, no. 1, Portales.

———, 1979, Post-Pleistocene archaeology, 7000–2000 B.C.; *in* Ortiz, A. (ed.), Handbook of North American Indians, vol. 9, Southwest: Smithsonian Institution, Washington, D.C.

Lang, R. W., 1977, Archaeological survey of the upper San Cristobal Arroyo drainage, Galisteo Basin, Santa Fe County, New Mexico: The School of American Research, Contract Archaeology Program, Santa Fe.

Lee, R. B., 1979, The !Kung San: men, women, and work in a foraging society: Cambridge University Press.

———, and DeVore, I., 1968, Man the hunter: Aldine Publishing Co., Chicago.

Longacre, W. A., 1970, Archaeology as anthropology: a case study: Anthropological Papers of the University of Arizona, no. 17, Tucson.

Linares, O. F., 1976, "Garden Hunting" in the American tropics: Human Ecology, v. 4, no. 4, pp. 331–349.

Plog, S., 1977, A multivariate approach to the exploration of ceramic design variation: Ph.D. dissertation, University of Michigan, Ann Arbor.

Schaafsma, C. F., 1976, Archaeological survey of Maximum Pool and Navajo excavations at Abiquiu Reservoir, Rio Arriba County, New Mexico: The School of American Research, Santa Fe.

Stevens, D. E., and Agogino, G. A., 1975, Sandia Cave: a study in controversy: Eastern New Mexico University, Contributions in Anthropology, v. 7, no. 1, Portales.

Stuart, D. E., and Gauthier, R. P., 1981, Prehistoric New Mexico: background for survey: New Mexico Historic Preservation Bureau, Santa Fe.

Wendorf, F., and Reed, E. K., 1955, An alternative reconstruction of northern Rio Grande prehistory: El Palacio, v. 62, nos. 5–6, pp. 131–173.

Yellen, J. E., 1976, Long-term hunter–gatherer adaptation to desert environments: a biogeographical perspective: World Archaeology, v. 8, no. 3, pp. 262–274.

Man with horse (at what is now Kit Carson State Park), 1920's(?). Photographer unknown.

THE ANASAZI CULTURE OF THE NORTHERN RIO GRANDE RIFT

STEWART PECKHAM
Museum of Indian Arts and Laboratory of Anthropology, Museum of New Mexico, Santa Fe, New Mexico 87504

INTRODUCTION

It is not surprising that in the legends of origin of the modern Tewa-speaking Pueblo Indians, the extent of the "Tewa world" (Ortiz, 1969) very closely corresponds to the geological extent of the Española Basin of the northern Rio Grande rift area. (For this paper the terms "Tewa world" and "rift area" will be used synonymously to refer only to the Española Basin.) The Tewa world thus is comprised of the Rio Grande valley and its tributaries, the Tewa pueblos, their fields, their range for hunting and gathering, and their sacred areas. Beginning at a shrine in the center of the plaza in each of the Tewa pueblos, and extending outward in the cardinal directions to distant mountains that form the horizon, a series of other shrines serves to delineate the Tewa world in the minds of the Tewa people.

The Tewa world not only has areal extent, it has a time dimension; archaeological studies have shown that the Tewa world was a recognizable entity as early as A.D. 1325, and may be traceble back even further.

Today's Indian inhabitants of the Tewa world can easily refer to their legends of origin to demonstrate that their roots in the area lie in the distant past—"since time immemorial," as they sometimes put it. Their memory is based in part on the fact that the region abounds with ruins of ancient, long-abandoned villages still bearing Tewa names that are referred to in oral histories. But there are hundreds, or even thousands, of other ancient dwelling sites in the Tewa world that are apparently beyond memory or tradition.

RATIONALES OF ARCHAEOLOGICAL RESEARCH IN THE RIFT AREA

Beginning in the late 19th century, anthropologists sought to give scientific substance to the Tewa world through systematic studies of the cultures of the Rio Grande pueblos. As early as the 1880's, Adolf F. A. Bandelier, the Swiss anthropologist/naturalist, came to New Mexico to begin a study of the Indians of the living pueblos. In particular, he wanted to document the relationships between the modern pueblos and other pueblos that were occupied when the Spanish first came to the Southwest in the 16th century. Additionally, he sought to relate both the modern pueblos and those occupied in the early Historic period to still others whose construction and use was totally prior to the arrival of the Spanish and the writing of their chronicles. The latter pueblos were, therefore, prehistoric, and for them there could be no documents among the 19th century Tewa unless they were in the form of oral traditions—unwritten and passed on by word of mouth.

Bandelier's reports (1890–1892) intrigued scholars in the nation's museums, and for the next 40 years the northern Rio Grande region hummed with archaeological activity as pioneer archaeologists such as Edgar L. Hewett (1953), Alfred V. Kidder (1958), Nels C. Nelson (1914), and Jean A. Jeancon (1923) excavated thousands of prehistoric rooms in several dozen major pueblo ruins in the region.

Early 20th century excavations in the rift area focused on sites named in Tewa oral histories and reflected an evolutionary approach to the interpretation of Pueblo culture, i.e., the culture of the modern pueblos evolved from that of the pueblos occupied just before the colonization of the area by the Spanish in the late 1500's. What they hoped to discover was an earlier, less developed culture—how much earlier, they could not really say. Dendrochronology—tree-ring dating—was still in its infancy, and other methods, such as radiocarbon and obsidian-hydration dating, were still over a generation off in the future. Consequently, for the first 20 years of field research in the rift area, reference to chronology was conspicuously absent in archaeological reports.

One rationale of archaeology is that cultural identification and relationships must be based on repeated and reasonably consistent associations of such tangible items and attributes as settlement patterns, house types, layouts of architectural features, and categories of local and intrusive pottery, stone, bone, and other artifacts. Geological, botanical, and zoological data could permit the in-depth recognition of relationships between the cultural assemblages from the prehistoric sites and those that could be observed among the living pueblos, as well as the reconstruction of the prehistoric environment and its effects on the culture being studied. Thus, until as recently as the 1970's, archaeological research in the rift area strongly emphasized the evolutionary approach initiated by the first generation of archaeologists. It relied heavily on acquisition of collections for both museum exhibits and related research, and called for increasingly detailed description and classification of the fragmentary and incomplete attributes recovered from both excavations and archaeological surveys.

During the last decade or so, proponents of the "new archaeology" have questioned the goals and methods of "old archaeology." New archaeologists do not see simple description and classification of artifacts and the establishment of historical sequences as the principal purposes of archaeology. To them, those procedures too frequently inhibit research and are only preliminary steps in the explanation of the cultural processes.

In the rift area, there is comparatively little information that reflects the developing thought of the new generation of archaeologists. Still, the cultural–historical approach to archaeology has not been rendered obsolete, though what was formerly accepted as established facts is now viewed as "models" or postulates requiring substantiation through the application of more systematic methods of data acquisition and analysis. Nevertheless, a great body of data and archaeological specimens is available that can be used to provide relevant and palatable information to those who visit museums and national monuments and whose interests are primarily historical.

PECOS CLASSIFICATION VERSUS RIO GRANDE CLASSIFICATION

When a group of archaeologists convened at the ruined Pueblo of Pecos in 1927 to pool the results of their studies of the Anasazi Culture—the prehistoric sedentary agricultural Indians of the northern Southwest, they established an evolutionary sequence of numbered periods (Table 1) to which they could relate their archaeological discoveries. This was the "Pecos classification" familiar to all Southwestern archaeologists (Kidder, 1927). For those working in the Rio Grande region, excavations had concentrated on sites of the "Proto-Historic" period (or Pueblo IV) which was characterized by the "decline from the preceding cultural peak," of the Great Pueblo period (Pueblo III).

One archaeologist (Roberts, 1936) went so far as to re-name the periods of the Pecos classification and refer to Pueblo IV as the "Regressive Pueblo" period (Table 1). To some Rio Grande scholars this was an unkind cut, because it implied that the culture of the Rio Grande region was somehow inferior to that of the west. It was a matter of some controversy, and the problem was more than just terminological. The Pecos classification worked moderately well in the San Juan Basin of northwestern New Mexico where ruins were abundant and, with the notable exception of Chaco Canyon, fairly consistently reflected the scheme developed at Pecos. The Rio Grande region just didn't fit. No

TABLE 1. Stratigraphic-nomenclature chart comparing the principal Anasazi cultural-sequence designations for the northern Southwest.

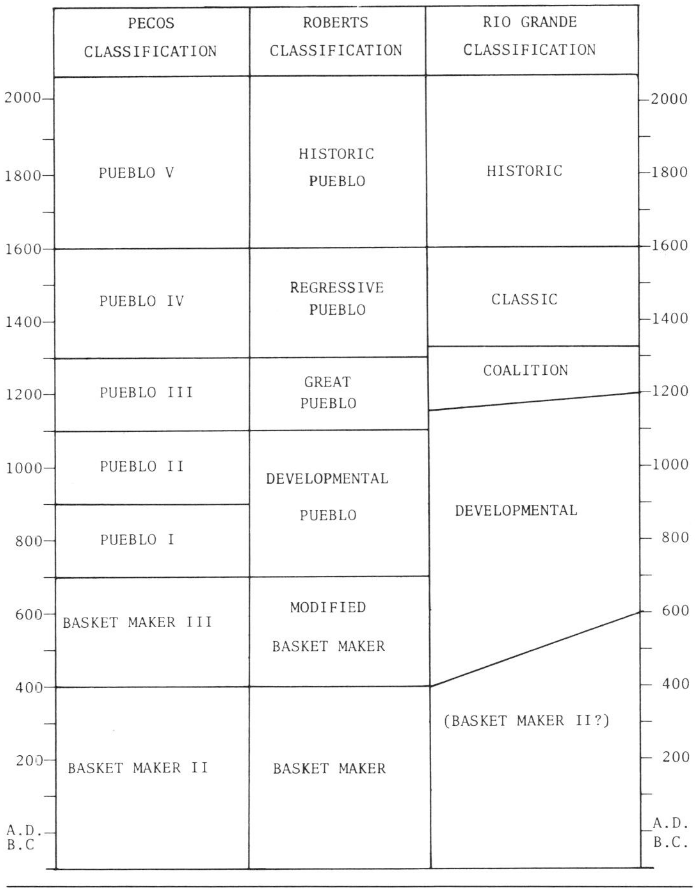

matter how hard Rio Grande archaeologists tried to adjust their interpretations to the Pecos classification, their field work suggested that prior to Pueblo IV evidence of cultural development was either missing, truncated, or inconsistent, and only occasionally corresponded to that in the west.

Between 1930 and 1950 archaeological surveys began to supplement data acquired from excavations, and it was found that there were indeed a few scattered prehistoric settlements in the Rio Grande region that at least superficially resembled their western contemporaries. These discoveries got further support from dendrochronology, but the region still yielded information that conflicted somewhat with that found in the San Juan Basin.

Several attempts were made to re-synthesize the cultural developments of the Rio Grande region, but it was not until 1955 that an acceptable alternative to the Pecos classification was developed (Wendorf and Reed, 1955). For convenience, we can call it the Rio Grande classification (Table 1). It was less detailed than the Pecos classification, but it nevertheless offered a framework on which cultural, demographic, and environmental change in the Rio Grande region could be reconstructed, compared, and, ultimately, explained.

THE RIFT AREA BEFORE A.D. 1325

No Paleoindian (late Pleistocene) campsites have, as yet, been found in the rift area, though isolated finds of diagnostic projectile points of that period have been recorded in and near Santa Fe and on the Pajarito Plateau. They may have been dropped by Paleoindians, but it is more likely that they were found elsewhere by later Indians and brought into the rift area.

Human use of the rift area appears to have begun during the Archaic period (ca 5500 B.C.–A.D. 400) when small, nomadic groups of hunters–gatherers explored the area and became familiar with its terrain, available resources, and climate. Sites of the Archaic period occur in both lowland and upland locations throughout the rift area, including the Tesuque Valley (Miller and Wendorf, 1958) and high-altitude sites in the Sangre de Cristo Mountains (Wendorf and Miller, 1959). Artifacts of that period are found throughout the rift area, notably in the vicinity of Cerro Pedernal (Bryan, 1939) near Abiquiu, where deposits of both Pedernal chert and Polvadera obsidian (Warren, 1974) were readily available to the Archaic and later Indians for quarrying and manufacturing chipped stone tools.

The Developmental Period, A.D. 400–600 to 1150–1200

Maize agriculture had been introduced into the rift area during the last centuries of the Archaic period (Skinner and others, 1980), but it would be several hundred years before the hallmarks of the Anasazi Culture—recognizable permanent settlements and pottery-making—were adopted by the inhabitants of the region. During this time, the Developmental period, the population of the rift area seems to have been small, dispersed, and largely settled along the lower portions of the permanent streams draining the western slopes of the Sangre de Cristo Mountains in the area between Santa Fe and Española (Fig. 1A).

At present, the earliest dwelling sites known for the rift area are those reported along the Rio Tesuque and Rio Nambe (McNutt, 1969; Skinner and others, 1980), which consisted of semi-subterranean pit houses (Fig. 2A) of irregular shape and generally non-systematic placement of a hearth and various pits, post holes, and other interior features. One of these structures in the Nambe drainage has been radiocarbon-dated at about A.D. 400.

LA 835 (a single-site number, but actually consists of a cluster of small, separate, not entirely contemporaneous dwelling sites), near Pojoaque and Tesuque Pueblos (Stubbs, 1954), may have begun as a scattering of single subterranean dwellings (Fig. 2B), but by A.D. 850 or 900 they were all accompanied by short contiguous alignments of surface dwellings and storage rooms with walls constructed of massive coursed adobe or jacal (vertical poles set into the ground and plastered with adobe). As surface-room units became the primary domiciles for nuclear or extended families, the subterranean rooms appear to have been converted to the almost exclusively ceremonial functions of a kiva, following the pattern established in the western Anasazi areas. Commonly, one or two of the largest surface rooms also had interior features arranged like those in the pit houses and later kivas (Fig. 2B, C).

Associated with these structures is Red Mesa Black-on-white, a mineral-painted pottery type of widespread occurrence that is a key indicator of the ascendance of Chaco Canyon, in northwestern New Mexico, as a major ceremonial and economic center during the 9th and 10th centuries (Marshall and others, 1979; Powers, Gillespie, and Lekson, 1983).

An isolated community ceremonial chamber, a Great Kiva (one of only two or three known for the entire Rio Grande region for this period), had been constructed at the Pojoaque site cluster (Stubbs, 1954; Peckham, 1979). Tree-ring dates are too few and incomplete for precise dating of the Great Kiva, but they suggest that it may have been constructed as early as the 10th century. However, later styles of western-derived, mineral-painted pottery [so-called "Chaco II Black-on-white" and, later, Kwahe'e Black-on-white (Mera, 1935)] associated with the structure indicate that its last use was in the 11th century, while there was still strong influence from the San Juan Basin area.

Unlike the massive house-blocks of Chaco Canyon, the greater LA 835 community occupied scattered locations through 5–10 km of the Nambe and Tesuque Valleys. However, the variety of non-local pottery types and artifacts associated with the site cluster suggests that it may have been a regional economic center, possibly approaching the status of a "Chaco outlier" (Powers, Gillespie, and Lekson, 1983; Marshall and others, 1979).

Some, but not all, dwelling units at LA 835 may have been vacated during a drought between A.D. 1125 and 1150. However, the demise of its Great Kiva at this time may be more significant, since its aban-

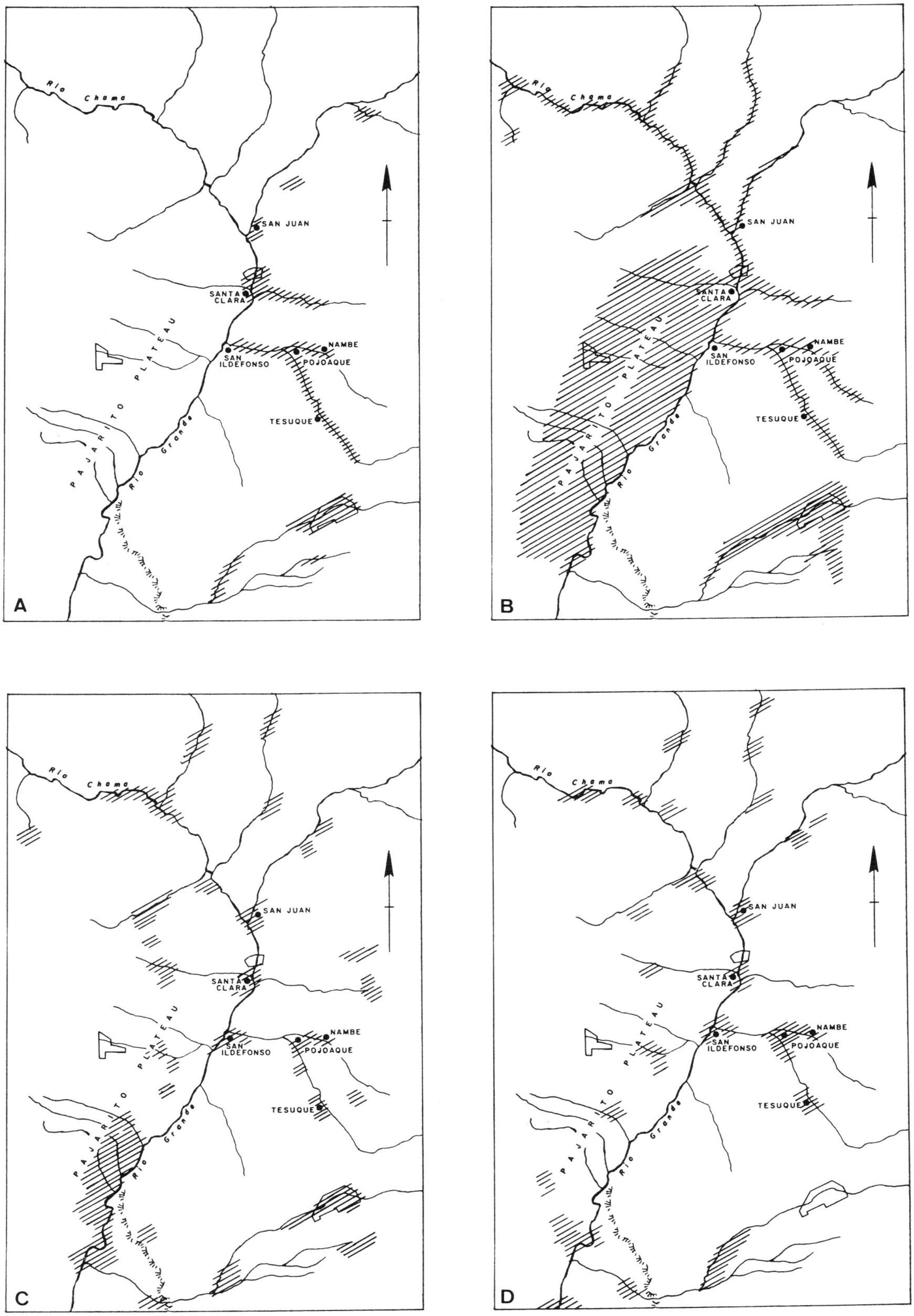

FIGURE 1. Changes in the settlement distribution in the Tewa world, circa A.D. 400–600 to A.D. 1600. **A,** Development period (A.D. 400–600 to 1150–1200); **B,** Coalition period (A.D. 1150–1200 to 1325); **C,** Early Classic period (A.D. 1325–1400); **D,** Late Classic period (A.D. 1400–1600).

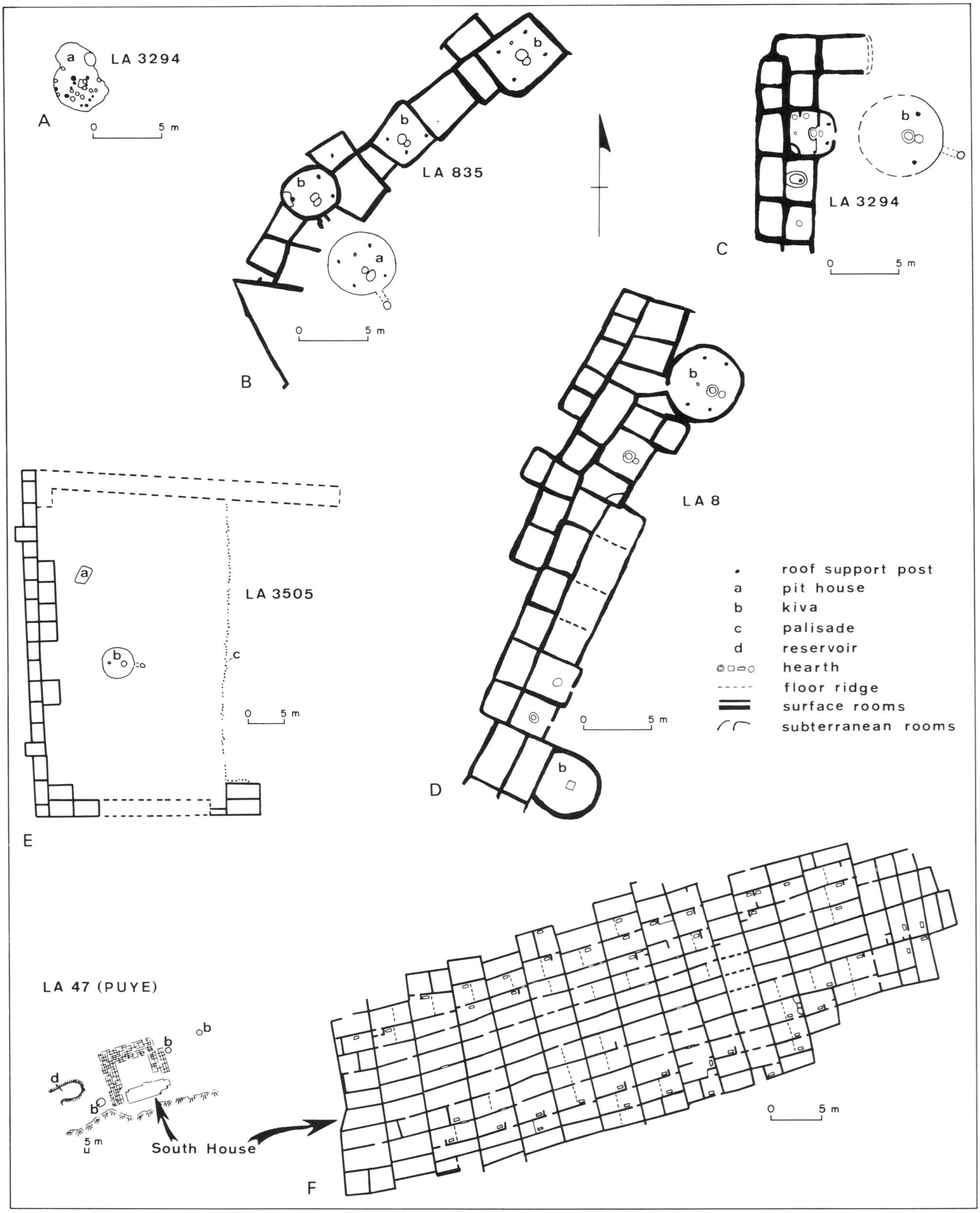

FIGURE 2. Sequence of development of dwelling ground plan in the Tewa world. **A,** Early Developmental (Tesuque area); **B,** Middle Developmental (Pojoaque area); **C,** Late Developmental (Tesuque area); **D,** Early Coalition (Santa Fe area); **E,** Late Coalition (Abiquiu area); **F,** Late Classic (Pajarito Plateau).

donment coincides with the collapse of the Chaco socio-economic system in the west. Reverberations of that event were felt throughout much of the Anasazi area, leading to the decline of many of the so-called "Chaco outliers."

Throughout the Developmental period, the rift-area population appears to have grown slowly, and mainly in lowland areas, but archaeological-survey data show that there was movement toward some locations along the extreme eastern perimeter of the Pajarito Plateau, the upland mesas that form the eastern flanks of the Valles caldera. Still, much of the rift area was largely vacant when the Coalition period began in the late 12th century.

The Coalition Period, A.D. 1150–1200 to 1325

Although not appearing to be a momentous event, the change from mineral to vegetal pigment on rift-area painted pottery announces the advent of the Coalition period. A similar change had taken place in a number of areas in northwestern New Mexico 50–75 years earlier and may have served as the model followed later in the rift area.

In the rift area at least two diagnostic pottery types (Mera, 1935) are known for this period: Santa Fe Black-on-white (ca A.D. 1150 or 1175 to 1300) which, though in vegetal paint, seems to continue design trends begun during the Developmental period; and Wiyo Black-on-white (A.D. 1275 or 1300 to 1350), whose brownish color, increased thickness, and heavy-handed decoration in vegetal paint seem to have no antecedents in the rift area.

The changes in ceramics more or less coincide with the beginnings of a population explosion in the rift area (Fig. 1B) that can only be explained by the arrival of migrants from the west or south. Beginning in the second quarter of the 12th century, areas of the San Juan Basin began to depopulate, and at least some of those leaving that area began to drift eastward, stopping for an undetermined period along the upper Rio Puerco before moving into the Rio Grande region. Subsequently, during the last quarter of the 13th century, the San Juan Basin, the Mesa Verde area, and much of southwestern New Mexico experienced many years of drought, leading the remaining inhabitants to giving up and moving into areas where water supplies were more reliable and where they could join those who had emigrated almost a century earlier.

During this period, hundreds of dwelling sites were established on the largely vacant Pajarito Plateau (Worman, 1967; Steen, 1977, 1982), especially in the southeastern part of the plateau (Biella, 1979). Early Coalition settlements east of the Rio Grande were either fewer in number, or were subsequently absorbed and obscured by the growth of substantial pueblos late in the period. Occupants of many of the latter villages continued the Developmental pattern of settling along the permanent streams draining the western Sangre de Cristo Mountains (Fig. 2D), but larger villages were frequently situated on mesa tops.

At least six large villages were located along the Santa Fe River, including Pindi Pueblo (Stubbs and Stallings, 1953) and the School House Site near Agua Fria, and an almost continuous village that underlies the Municipal Building, in downtown Santa Fe, and extends southwestward to St. Francis Drive (U.S. 84/285). Other settlements were established southeast of Santa Fe: Arroyo Hondo Pueblo (Schwartz and Lang, 1973), Pecos, Forked Lightning, Dick's, and Rowe ruins (Kidder, 1958), and northeastward along the western fringes of the New Mexico Plains near Watrous, north of Las Vegas. Still other major settlements were being founded along the lower Rio Chama (Fig. 2E) and its tributaries (Hibben, 1937; Leubben, 1953; Peckham, 1981), and, apparently for the first time, the lands along the Rio Grande north of Española and San Juan Pueblo to near Velarde were permanently settled.

Although the Anasazi are generally thought of as being very sedentary, the Coalition illustrates how mobile they really were. Early in the period, they seem to have had a pioneering spirit in spite of the stress they experienced in having to abandon their former homes. Once in the rift area, they had to explore it, locate areas of potential settlement and cultivation, as well as usable natural resources. They seem not to have been reluctant to experiment with new areas of settlement, expanding their development of some localities while abandoning others.

The Coalition period was certainly the time of establishment of all of the existing Tewa pueblos (Ellis, 1964). Tewa legends of origin say that they emerged from the underworld at a lake near Alamosa, Colorado, and split into two groups as they moved southward down both sides of the Rio Grande. Because places of "emergence" were often changed to accommodate the changing locations of migrating groups, this has been interpreted as indicating that the Tewa had originally lived in the upper San Juan River drainage, in the Mesa Verde region and in the vicinity of the Navajo Reservoir (Ellis, 1967), though neither the intervening area between the upper San Juan and the rift area nor the southern San Luis Valley have been systematically investigated for corroborative archaeological data.

Coalition dwellings on the Pajarito Plateau seem to continue the Developmental architectural tradition of one or two large rooms with kiva-like interior features and with as many as a dozen smaller contiguous rooms for storage (Fig. 2C). Unless migrants arriving from the west had adopted the prevailing rift-area dwelling layout, this room configuration may indicate that many of those moving to occupy the Pajarito Plateau at this time were the descendants of the Developmental occupants of the rift-area lowlands. On the other hand, Ellis (1967) raises the possibility that the indigenous rift-area people may have held the arriving migrants at bay, forcing them to settle in upland areas until their peaceful intentions were verified. However, both situations may have occurred, but at different times.

Many of the small early Coalition period sites on the Pajarito Plateau seem not to have been associated with subterranean kivas, possibly because of the shallowness of the soil capping the massive volcanic tuff that forms many of the mesas. However, subterranean kivas have been recorded at Coalition sites in the Santa Fe and Tesuque Valleys (Stubbs and Stallings, 1953; McNutt, 1969).

The merging of populations, the abandonment of old communities in the rift area, and the establishment of new ones appear to have continued into the second decade of the 14th century. It is generally felt that during this period both the original rift people and the arriving migrants experienced considerable stress as each competed for arable land, water, and other resources, while at the same time experiencing the vagaries of a not totally familiar environment. The development of pueblos with enclosed plazas may have been due to increased need for defense from attack by either unfriendly neighbors or even nomadic Indians from the plains east of the Sangre de Cristo Range. By A.D. 1325 the major period of unrest and resettlement passed, relative stability prevailed, and the Anasazi were again able to devote their efforts to less stressful and more cooperative undertakings as the Rio Grande Classic period began.

THE CLASSIC PERIOD, A.D. 1325–1600

In some respects, the Classic period continues events and developments that began during the Coalition period. The coalescence of the population (Fig. 1C, D) continued on the Pajarito Plateau as major Coalition communities, such as Puye (Morely, 1910), Shufinne, Tsankawi, Potsuwi (Otowi), Tyuonyi (Hendron, 1940), and Tsirege, grew even larger and many now-nameless settlements were abandoned. In the lower drainages of the Rio Chama, the huge villages of Teewi (Wendorf, 1953), Poshu-ouinge (Jeancon, 1923), Sapawe, Tsama, Howiri, Hupobi, Pose, and Abechiu (Abiquiu) (Hewett, 1906) had hundreds of rooms each, often surrounding multiple plazas. Other ancestral villages of the modern Tewa pueblos were all in these areas, but in most cases these were abandoned before A.D. 1600 (Fig. 2F).

The modern communities along the Rio Santa Cruz have all but obscured any prehistoric remains in its bottomland areas, but early archaeological surveys show that much of this drainage, though settled during the Coalition period, lay inexplicably vacant during all but the very early Classic period. Curiously, the extensive Coalition villages within the limits of Santa Fe were also vacant by A.D. 1400, and there is no evidence of any Classic settlement in the Santa Fe Valley, except Cieneguilla—a large, non-Tewa pueblo south of the Santa Fe Airport (incidentally giving credence to the early Spanish reports that the capital of New Mexico was moved to its present location because there were no Indians living there in A.D. 1609).

By A.D. 1400, the long-standing architectural tradition of ground-level kiva-like rooms being attached to houseblocks seems to have been discarded. Kivas were again subterranean. Those on the Pajarito Plateau were carved vertically into volcanic tuff, while those in the talus cliff-dwelling areas were often cut back into the faces of the tuff cliffs. Almost invariably, the ceremonial chambers had the same interior features as the Developmental-period pit houses. Added to these, however, were rows of holes in the floor, against the walls or on the north and south sides of the hearth, which held anchors for the bottoms of vertical looms. Perhaps sometime during the Coalition growing of cotton had been introduced to the lowland parts of rift area, and by the Classic period virtually every kiva, and some pueblo rooms, were equipped with looms, for it was mainly in the kivas that the men, rather than the women, did the weaving.

Most Classic pueblos had at least one circular "big kiva" measuring as much as 14 m in diameter, thought to have served as a community ceremonial chamber—as opposed to small kivas which may have been used by religious societies. Big kivas were equipped with loom sockets in the floors and with so-called "floor drums," usually pairs of rectangular sub-floor vaults whose functions are not totally understood. Covered with planks, the features may have held fetishes and may have been stamped upon during rituals to communicate with dieties in the underworld.

Big kivas may have also functioned in ritualized aspects of warfare (Hawley, 1950). Following the stressful times of the Coalition, the Classic-period Anasazi of the Rio Grande region show evidence of considerable preoccupation with warfare, particularly in the defensive layouts of villages and the depiction of shield- and weapon-bearing figures on rock art—petroglyph—panels found near some Classic sites. Formal warrior societies persisted among the Historic pueblos until the 19th century, and it is likely that these organizations had their beginnings in the Classic period, if not during the Coalition. Rift-area rock art also portrays many masked figures or katsinas, horned serpents, and other elements that suggest that the Katsina Cult had been introduced into the area from the south-central New Mexico (Schaafsma and Schaafsma, 1974), homeland of the Jornada Branch of the Mogollon Culture, by the beginning of the Classic period.

Archaeologists frequently note that the concentration of the population into large villages may have permitted the establishment of more authoritarian control over the populace, thereby enabling the Anasazi to accomplish major "public works" projects that lesser groups of people could not have undertaken. Foremost among these would have been projects that would have conserved, or improved the utilization of, scarce water.

Modern weather records in the Española Basin show that much of the area lies in a "rain shadow" caused by the proximity of the Jemez Mountains. Weather conditions of the past, as reconstructed from dendrochronological data (Dean and Robinson, 1977), show that the rift area has repeatedly experienced periods of extreme drought as well as years of surplus moisture. Upland areas of the Pajarito Plateau and the foothills of the Sangre de Cristo Range receive more precipitation than the lowland areas, and it is likely that some Coalition and Classic settlements in the rift area were located to use this advantage. Even so, water was a precious commodity, and the Anasazi developed a number of ingenious ways to maximize its use.

After building check dams and gridded gardens during the Coalition, the Anasazi drew on their awareness of the environment to reclaim thousands of hectares of marginal land for farming. Prairies on the bluffs overlooking Ojo Caliente Creek were divided up into extensive grid systems, with garden plots outlined with cobbles, covered with gravel mulches, and linked by what appear to be ditches or water catchments. A mesa top near Abiquiu, almost 300 m above the Chama Valley, is covered with literally hundreds of garden plots in grids laid out to trap heavy rainfall, preventing it from running off unproductively. Large reservoirs were constructed adjacent to villages at Puye, Shufinne, Tsirege, and Potsuwi. Whether for drinking purposes, agriculture, or construction, these structures reflect emergent engineering skills as well as the ability to organize village labor resouces for the common good.

Besides having the labor supply to simply clear the bottomlands, construction of ditch systems and diversion of water from major rivers were other achievements of the Classic period. Early Spanish chroniclers reported seeing fields under irrigation when they first visited the rift area. Unfortunately, modern irrigation farming has obliterated all but a few possible examples of prehistoric ditches, though some of the modern systems may still use ditch alignments established during the Classic period.

The Classic was the time of the so-called "Biscuit Wares" (Harlow, 1973). The soft, thick, porous, painted pottery of this period in the rift area reminded early archaeologists of the "bisque" or "biscuit" stage in the manufacture of modern glazed pottery. They named and described a series of pottery types, and conveniently alphabetized some of them in the order of their occurrence, as follows:

Type Name	Abbreviated Name	Time Range
Abiquiu Black-on-gray	Biscuit A	A.D. 1350–1425
Bandelier Black-on-gray	Biscuit B	A.D. 1425–1475
Cuyamungue Black-on-tan	Biscuit C	A.D. 1475–1600
Sankawi Black-on-cream	— —	A.D. 1550–1650

All of these types were decorated with vegetal pigment, showing continuity with Wiyo Black-on-white of the Coalition period. The earliest of the types was slipped, polished, and painted only on the interior of bowls (bowls were almost the only form of vessel for Biscuit A and Biscuit B); Biscuit B was slipped, polished, and painted on both interior and exterior; and Biscuit C and Sankawi Black-on-cream showed progressive thinning of vessel walls, thinner-lined decoration, and production of storage jars as well as bowls. Notable in the painted designs of the earlier types is the depiction of the Awanyu, the Tewa conception of the keeper of springs and other forms of water.

Produced in the rift area, and coincident with part of the period of manufacture of Biscuit C, was a unique incised and polished gray ware, Potsuwi'i Incised. Although some incised decorated pottery may have been made in the rift area during the Developmental period, there is no indication of any relationship between it and the Classic period type—unless it is viewed as a stylistic revival. It has been suggested that Potsuwi'i Incised was introduced from the Plains or lower Mississippi areas where incised pottery was being made at about the same time (Mera, 1932; Harlow, 1973). An alternative source may have been the Casas Grandes area in northern Chihuahua where similar incised pottery was being made only slightly earlier than Potsuwi'i Incised. Additionally, the Casas Grandes area may have been the stimulus for the over-all smudged black or gray ware which began to be made in the rift area by about A.D. 1500. In some variations, the black ware has been made continuously to the present day, and its production and sale by the potters of San Ildefonso and Santa Clara constitutes a major source of income for many creative craftspeople in those pueblos.

Concomitant with the manufacture of the Biscuit Wares in the rift area, Pueblo Indians in the middle Rio Grande region and the Galisteo Basin were making totally different pottery decorated with a lead glaze and other pigments and slip colors. As was the case throughout the prehistoric period, there was much reciprocal trade of desirable objects and materials, especially pottery, and the middle Rio Grande glaze wares are commonly found at Classic-period sites in the rift area. Sometime during the 1520's, glaze-decorated pottery began to be produced at Puye, west of Santa Clara Pueblo, and in the vicinity of modern San Juan Pueblo. By coincidence, a glazed-pottery producing village, Pueblo del Encierro (LA 70), approximately at Cochiti Dam, was abandoned very shortly after A.D. 1520, as determined by tree-ring dating. The people of Cochiti Pueblo say that Pueblo del Encierro was not one of their ancestral villages, whereas members of a clan at Santo Domingo Pueblo (whose Tano-speaking ancestors had moved into that Keres pueblo in the late 1700's) state that the pueblos at Cochiti Dam were occupied by some of their earlier ancestors. Thus, it is possible that a group of Tano people, on abandoning Pueblo del Encierro, moved north to Puye, introduced glaze-decorated pottery to the rift area and, by their presence, influenced the language of the original Tewa occupants of

Puye. Now extinct, Tano is considered by some linguists to have been a southern Tewa dialect. Although its influence on the Tewa at Puye may have been slight, it may have been enough to cause today's people of Santa Clara Pueblo (who claim Puye as an ancestral village) speak a dialect noticeably different from that spoken by the other Tewa pueblos (so different that the people of the latter pueblos make fun of the way the Santa Claras speak Tewa).

THE HISTORIC PERIOD, A.D. 1600 TO THE PRESENT

The first Spanish colony and capital were established, in A.D. 1598, at the Tewa pueblo of Yuque-Yunque, on the west side of the Rio Grande opposite San Juan Pueblo. The impact of Spanish culture on the Tewa was substantial. It introduced new domesticated food plants, domesticated animals, firearms, the Roman Catholic religion, social, political, and economic controls, and many other features that are still evident in both Pueblo and non-Pueblo societies that make the multicultural Southwest so distinctive. The Tewa have certainly lost many aspects of the culture of their Classic-period ancestors. The language, though still spoken, is changing. Aspects of the Tewa religion have changed. The warrior society is extinct. Tewa pueblos have produced leaders in the fields of education, anthropology, medicine, and in other professions as well. They are well aware of the changes, and some probably lament them. At the same time, they retain a measure of many of the cultural advances that archaeological studies have recorded for the periods covered by the Rio Grande classification. In many respects they are still in the Classic period.

REFERENCES

Bandelier, A. F., 1890–1892, Final report of investigations among the Indians of the southwestern United States, carried on mainly in the years from 1880 to 1885. 2 vols.: Papers of the Archaeological Institute of America, American Series 3 and 4, Cambridge, Mass.

Bryan, K., 1939, Stone cultures near Cerro Pedernal and their geologic antiquity: Texas Archaeological and Paleontological Society, Bulletin, v. 11, pp. 9–42.

Biella, J. V., 1979, Changing residential patterns among the Anasazi, A.D. 750–1515; *in* Archaeological investigations in Cochiti reservoir, New Mexico, vol. 4: Adaptive change in the northern Rio Grande valley: University of New Mexico, Office of Contract Archeology, pp. 103–144.

Dean, J. S., and Robinson, W. R., 1977, Dendroclimatic variability in the American Southwest: A.D. 680 to 1970: National Technical Information Service Publication PB 266 340, Springfield, Virginia, 147 pp.

Ellis, F. H, 1964, Archaeological history of Nambe Pueblo, 14th century to the present: American Antiquity, v. 30, no. 1, pp. 34–42.

———, 1967, Where did the Pueblo people come from?: El Palacio, v. 74, no. 3, pp. 35–43.

Harlow, F. H., 1973, Matte-paint pottery of the Tewa, Keres and Zuni pueblos: Museum of New Mexico Press, Santa Fe, 271 pp.

Hawley, F., 1950, Big kivas, little kivas, and moiety houses in historical reconstruction: Southwestern Journal of Anthropology, v. 6, no. 3, pp. 286–302.

Hendron, J. W., 1940, Prehistory of El Rito de los Frijoles, Bandelier National Monument: Southwestern Monuments Association (Coolidge, Arizona), Technical Series 1, 73 pp.

Hewett, E. L., 1906, Antiquities of the Jemez Plateau, New Mexico: Bureau of American Ethnology, Bulletin 32, 55 pp.

———, 1953, The Pajarito Plateau and its ancient people (Revised by Bertha P. Dutton): University of New Mexico Press, Albuquerque, 174 pp.

Hibben, F. C., 1937, Excavation of the Riana Ruin and Chama Valley survey: University of New Mexico Bulletin, Anthropological Series, v. 2, no. 1, 74 pp.

Jeancon, J. A., 1923, Excavations in the Chama Valley, New Mexico: Bureau of American Ethnology, Bulletin 81, 145 pp.

Kidder, A. V., 1927, Southwestern Archeological Conference: Science, v. 66, no. 1716, pp. 489–491.

———, 1958, Pecos, New Mexico: archaeological notes: Papers of the Robert S. Peabody Foundation for Archaeology (Andover, Mass.), 5, 360 pp.

Leubben, R. A., 1953, "Leaf Water Site," in salvage archaeology in the Chama Valley, New Mexico. Assembled by F. Wendorf: School of American Research, Santa Fe, Monograph 17, pp. 9–33.

Marshall, M. P., Stein, J. R., Loose, R. W., and Novotny, J. E., 1979, Anasazi communities of the San Juan Basin: Public Service Company of New Mexico and Historic Preservation Bureau, Planning Division, Department of Finance and Administration of the State of New Mexico, 363 pp.

McNutt, C. H., 1969, Early Puebloan occupations at Tesuque by-pass and in the upper Rio Grande Valley: University of Michigan, Museum of Anthropology, Anthropological Papers, 40, 144 pp.

Mera, H. P., 1932, Wares ancestral to Tewa polychrome: Laboratory of Anthropology, Santa Fe, Technical Series Bulletin 4, 18 pp.

———, 1934, A survey of the biscuit ware area in northern New Mexico: Laboratory of Anthropology, Santa Fe, Technical Series Bulletin 6, 25 pp.

———, 1935, Ceramic clues to the prehistory of north-central New Mexico: Laboratory of Anthropology, Santa Fe, Technical Series Bulletin 8, 64 pp.

Miller, J. P., and Wendorf, F., 1958, Alluvial chronology of the Tesuque Valley, New Mexico: Journal of Geology, v. 66, no. 2, pp. 177–194.

Morley, S. G., 1910, The south house at Puye: Papers of the School of American Research, Santa Fe, Old Series 7, 15 pp.

Nelson, N. C., 1914, Pueblo ruins of the Galisteo Basin, New Mexico: American Museum of Natural History, Anthropological Papers, v. 15, pt. 1, 124 pp.

Ortiz, A., 1969, The Tewa world: space, time, being, and becoming in a Pueblo society: University of Chicago Press, 340 pp.

Peckham, S., 1979, When is a Rio Grande kiva?: Papers of the Archaeological Society of New Mexico, Albuquerque, 4, pp. 55–86.

———, 1981, The Palisade Ruin (LA 3505): A coalition period pueblo near Abiquiu, New Mexico: Papers of the Archaeological Society of New Mexico, Albuquerque, 6, pp. 113–147.

Powers, R. P., Gillespie, W. B., and Lekson, S. H., 1983, The outlier survey: a regional view of the settlement of the San Juan Basin: Reports of the Chaco Center 3, Division of Cultural Research, National Park Service, Albuquerque, 437 pp.

Roberts, F. H. H., Jr., 1936, A survey of southwestern archeology: Smithsonian Institution, Annual Report for 1935, pp. 507–533.

Schaafsma, P., and Schaafsma, C. F., 1974, Evidence for the origins of the Pueblo Katchina Cult as suggested by southwestern rock art: American Antiquity, v. 39, no. 4, pp. 535–545.

Schwartz, D. W., and Lang, R. W., 1973, Archaeological investigations at the Arroyo Hondo Site, third field report—1972: School of American Research, Santa Fe, 47 pp.

Skinner, S. A., and others, 1980, Archaeological investigations at Nambe Falls: Southern Methodist University, Department of Anthropology, Archaeology Research Program, 231 pp.

Steen, C. R., 1977, Pajarito Plateau: survey and excavations: Los Alamos Scientific Laboratory, Report LASL 77-4, National Technical Information Service (Springfield, Virginia), 70 pp.

———, 1982, Pajarito Plateau: archaeological surveys and excavations, II: Los Alamos National Laboratory, Report LA-8860-NERP, National Technical Information Service (Springfield, Virginia), 60 pp.

Stubbs, S. A., 1954, Summary report of an early Pueblo site in the Tesuque Valley, New Mexico: El Palacio, v. 61, no. 2, pp. 43–45.

Stubbs, S. A., and Stallings, W. S., Jr., 1953, The excavation of Pindi Pueblo, New Mexico: School of American Research, Santa Fe, Monograph 18, 165 pp.

Warren, A. H., 1974, The ancient mineral industries of Cerro Pedernal, Rio Arriba County, New Mexico: New Mexico Geological Society, Guidebook 25, pp. 87–93.

Wendorf, F., 1953, Excavations of Te'ewi; *in* Salvage archaeology in the Chama Valley, New Mexico. Assembled by F. Wendorf: School of American Research, Santa Fe, Monograph 17, pp. 34–93.

———, and Miller, J. P., 1959, Artifacts from high mountain sites in the Sangre de Cristo Range: El Palacio, v. 66, no. 2, pp. 37–52.

———, and Reed, E. K., 1955, An alternative reconstruction of northern Rio Grande prehistory: El Palacio, v. 62, no. 5–6, pp. 131–173.

Worman, F. C. V., 1967, Archeological salvage excavations on the Mesita del Buey, Los Alamos County, New Mexico: Los Alamos Scientific Laboratory, Report LA-3636, National Technical Information Service (Springfield, Virginia), 39 pp.

Taos Pueblo, ca 1920. Photo by Burt Harwood.

THE TEWA INDIANS OF THE RIO GRANDE AND THEIR NEIGHBORS—A.D. 1450–1680

ALBERT H. SCHROEDER
Archaeological and Historical Societies of New Mexico, 1108 Barcelona Lane, Santa Fe, New Mexico 87501

INTRODUCTION

The Tewa country, flanked on the west side of the Rio Grande by the Jemez Mountains and Pajarito Plateau and on the east side by the Sangre de Cristo Range, extended along the river throughout the Española Basin of the Rio Grande rift, roughly from Bandelier National Monument north to the Abiquiu–El Rito area. The Tewa story, briefly reviewed herein and covering the transition from late prehistoric times into the early historic period, takes place within this region. However, interaction with neighboring people, both Indian and Spanish, provides a wider cultural environment within which the Tewas operated.

LATE PREHISTORIC TIMES

Between A.D. 1450 and 1540, more than 100 pueblos existed between the Chama Valley on the north and the Socorro area on the south, and from the Pecos, Galisteo Basin, and Salinas area on the east to the Acoma, Zuni, and Hopi villages on the west. These people had much in common. However, they differed regionally from one another in certain crafts, tool kits, architectural features, and cultural practices. Generally, their homes, small to large pueblos of masonry or coursed adobe, ranged from one to several stories high (seven or more at Picuris in early historic times) and were built in a linear or rectangular plan with a plaza in front or enclosed by the house blocks (as at Tesuque today). In some cases the pueblo was a large compact mass (like Taos). These prehistoric puebloans farmed, hunted, gathered wild foods, herded turkeys, and pursued a number of crafts which produced a variety of items.

Cultural variations that had developed prior to A.D. 1450 survived into late prehistoric times as distinct sets of characteristics in different regions of the Pueblo world. These regional clusters of pueblos with the same characteristics suggest that each cluster represented an identifiable ethnic unit as was the case in early historic times. For example, certain groups of closely associated pueblos, dating before Spanish entry, exhibit big kivas, either one (like the historic-period Tewas) or two (like the Keres), while other villages had little kivas alone (as among the Zunis) or in combination with a big kiva. These occurrences suggest differing socio-religious concepts occurred among different prehistoric regional groups as does the practice of cremation of the dead by some as opposed to interment unburned by others. Color combinations and decorative motifs of ceramics as well as use or non-use of glaze paint also reveal definite regional variations.

Other differences between specific areas in late prehistoric times appear to be environmentally rather than culturally determined. Irrigation farming, such as reported among the Tewas in the late 1500's, was practiced earlier in locales where stream runoff was dependable. Farming dependent on rainfall was more common and included water-control devices such as check dams, terraces, and reservoirs. Variations in growing seasons limited the types of crops in some areas. While cotton could be raised in the lower elevations south of the Tewas, pueblos to the north of Española probably found it necessary to harvest green corn on many occasions. Proximity to natural resources which were mined or quarried (lead, turquoise, obsidian, salt, jet, malachite, hematite, etc.) provided commodities to nearby pueblos for barter with other less fortunately located groups and with Indians on the plains. Identifiable late prehistoric clusters of pueblos, each with a set of characteristics in common as opposed to a set of another contemporary group, provide the archaeologist with leads by which to link them to early historic period ethnic or pueblo units recognized by the Spaniards. This task is simplified when a set of prehistoric regional traits occurs in the same region of a historic group that shares similar traits and customs.

Early Spanish journals of the late 1500's described the way of life of Pueblo Indians before they adopted any western European cultural items or practices. In this sense, the Pueblo culture of late prehistoric times had not yet been altered by contact with the Spaniards. As described in the journals, the material culture was similar to that found by archaeologists in late prehistoric sites. Socio-economic factors, unlikely to be gleaned in any detail from excavations in late prehistoric ruins, can be postulated from these early records.

According to the early journals, Pueblo Indians spoke several mutually unintelligible languages. By 1601, 130 pueblos were recorded as "surrounded entirely by uninhabited areas." The Spaniards also noted that the pueblos had contact with wandering Indians who lived in tents, obviously a reference to Indians on the plains. Reports covering this period from 1540 to 1608 remark on the trade of the eastern-frontier pueblos of Taos, Picuris, and Pecos with tribes on the plains (Hammond and Rey, 1940, p. 259; Hammon and Rey, 1953, pp. 629, 633, 640).

Relations with these people to the east were not always cordial. According to information obtained from Pecos Indians in 1540 (Hammond and Rey, 1940, p. 258), the Teyas, who may have been ancestral Wichita (Schroeder, 1962, p. 8), attacked Pecos and other pueblos in 1526. Similar hostilities existed between Plains Indians and Tano pueblos of the Galisteo Basin in the early 1580's (Hammond and Rey, 1927, p. 29), and with or between pueblos near the north end of the Sandia Mountains in 1591 (Schroeder and Matson, 1965, p. 160). In 1540, Pecos was attempting to obtain farmland by force from the Southern Tiwas of the Bernalillo area (Hammond and Rey, 1940, p. 328). In the 1580's, these Tiwas were at war with the Piros on their south boundary south of Isleta (Hammond and Rey, 1966, p. 303). It appears that some, if not all, of these enmities may well have originated prior to the entry of the Spaniards. The Tewas are not known to have been involved in hostilities before A.D. 1600, yet a few years later some of the pueblos were engaged in similar activities, particularly the four along the Chama drainage to the north of the other ten pueblos from San Juan Pueblo to Tesuque along the Rio Grande (Schroeder and Matson, 1965, pp. 112ff, 131).

In addition to these conflicts, we know that the pueblo people in prehistoric times faced disruptive extremes in their natural environment, such as the several years of cold winters in the 1540's (Hammond and Rey, 1940, p. 234; Schroeder, 1972, pp. 43–44). Also, because of experiences with droughts in the past, the Pueblo Indians of the early historic period practiced the storage of corn for three, four, and even up to seven years to provide insurance against crop failures (Hammond and Rey, 1953, pp. 653, 660, 693, 698, 863; Schroeder and Matson, 1965, p. 100). In fact, the droughts of the 1560's to the 1590's probably played a major role in the abandonment of the pueblos in the Bandelier National Monument area and on the Pajarito Plateau by the 1590's (Schulman, 1956; Fritts, 1965, fig. 3; Schroeder, 1968, p. 18).

The above, combined with known data recovered from sites occupied in late prehistoric times, suggests that the people of the latter period developed regional patterns and customs that survived and closely conformed to those of early historic times. The pueblo people of the transitional period were the sole occupants of their territory, interacted among themselves, traded with neighbors on the plains, and maintained their regional cultural identities.

CONTACT PERIOD

During the period of Spanish exploration, between 1540 and 1598, the pueblo way of life and material culture was little affected. It was, to all intents and purposes, nothing more than a time of contact between two vastly different cultures. The pueblos received gifts of pearl and glass beads, jingle bells, some metal artifacts, and a vague knowledge of the cross. They were introduced to, but did not receive any, wheeled vehicles, cattle, or other stock. Even the gifts were similar to items in use among Indians, the only difference being the material from which they were made. The Spaniards, however, gained considerable knowledge of the pueblo country, its resources, and its people and customs, all of which provided an incentive to colonize and exploit New Mexico's resources and souls—the latter variously stated to be some 20,000 men in the 1540's or 30,000 men plus numerous women and children in 1602 (Hammond and Rey, 1953, pp. 483, 863).

Only two exploratory entries visited the Tewas. In 1540, some of Coronado's men departed from the Jemez pueblos, traveled north over the Pajarito Plateau, and dropped down into Rio Grande–Chama River junction area. Here they visited two Tewa pueblos, San Juan and Yuqueyungue, the occupants of which had fled to four pueblos in the sierras (Hammond and Rey, 1940). In 1591, Gaspar Castaño de Sosa was the first to visit the Tewa pueblos below San Juan as well as nearby Picuris (Schroeder and Matson, 1965).

In 1598, Juan de Oñate's advance party hardly paused as it came up the Rio Grande until it arrived at San Juan Pueblo. Perhaps Oñate selected this area for settlement of his colonists because of Castaño de Sosa's report which noted good supplies of food among the Tewas derived from irrigated fields (Schroeder and Matson, 1965, pp. 117–118). He and those of previous expeditions had reported in their journals that the Tiwas of the Albuquerque–Bernalillo area were unfriendly. Oñate's temporary settlement or camp, as it is referred to in the journals, was established across the river from San Juan Pueblo at Yuqueyungue. He named his camp of 129 men, plus women and children, San Gabriel. While most of the Indian occupants of this 400-room pueblo moved to San Juan, some remained with the colonists.

As a protective measure (against enemies or to protect stored crops from rodents), pueblos were built without doors on the ground floor. Entry was through a hatchway in the roof or from the second story. When the Spaniards moved in, they added doors and windows, indicating that they used the ground-floor rooms for their homes. Within a month, a church large enough to accommodate the colonists was built. These changes, a flour mill, new crops and stock, and other Spanish features marked the Spanish occupation as different from that of the Tewa and other Indians (Hammond and Rey, 1953, pp. 323, 619, 639, 652, 653). At the same time, this frontier Spanish settlement bore little resemblance to towns in Mexico.

The day after the 1598 dedication of the church at San Gabriel, Oñate assigned eight priests to the pueblo provinces. A little more than a week after these assignments were made, the friars left for their new posts, probably accompanied by a few soldiers. Within three to five years, churches were erected among the Towa-speaking Jemez, the Keresan-speakers of Santo Domingo, the Tewas at San Ildefonso, and perhaps among the Tiwa-speakers of Picuris (Hammond and Rey, 1953, pp. 324, 630–631, 709, 711; Hodge, Hammond, and Rey, 1945, p. 260). With these padres came Spanish culture.

Oñate's report to the viceroy in March of 1599, written after exploration of some of the region, emphasized the economic potential of the area. He remarked on the returns expected in revenues from mineral lodes, pearls, salt and sulphur deposits, and buffalo skins, as well as from Indian tributes. The treasurer of New Mexico, Alonso Sanchez, had written a letter a few days prior to Oñate's report, indicating that he expected excellent returns because the Pueblos were such good farmers who also were "much given to commerce, taking from one province to another the fruits of their land and the products of their labor." This suggests that the exchange of food items among the Tewas and others played a large role in trade. In an effort to expand the potential of the local labor market, Oñate also requested that Indians be sent from Mexico, skilled in all trades, to teach the Pueblos various crafts (Hammond and Rey, 1953, pp. 425, 486, 489).

The experiences of the first few months included some hostilities, but none mentioned involved the Tewas. In spite of the difference in numbers between the colonists and the Pueblos, initially the Spaniards seem to have had the upper hand. Their horses provided mobility and rapid movement, their edged metal weapons and armor were a decided advantage in close combat, and the arquebus, though not as efficient as the bow and arrow, provided enough bluster to scare the Indians, as observed in 1581 (Hammond and Rey, 1927).

During this period of contact, there was an additional intrusion into New Mexico—the Apaches, who were referred to as masters of the plains in 1601 (Hammond and Rey, 1953, p. 749). In spite of statements made by Spanish explorers in the 1540's and 1601 that none other than the Pueblo Indians lived in the Pueblo country, some doubt has been expressed as to the date of the entry and settlement of Apaches in the state. Most investigators agree that Apaches were on the plains east of New Mexico by the middle 1400's and entered New Mexico in late prehistoric times for purposes of trade or to spend winters close to some pueblos. However, two statements suggest that some, at least, spent part of the growing season in New Mexico. The Antonio de Espejo expedition journals of March 1583 refer to "peaceful mountaineers" near the locale of present Laguna Pueblo. In June, they were reported living in shacks and raising corn and were referred to as "Corechos" (Hammond and Rey, 1929, pp. 86, 110–112), a term applied to nomads, usually Plains Apaches, prior to the early 1600's. This rancheria could have been ancestral Gila Apaches who lived south of, and were friendly with, Acoma and were first reported by that name in the 1620's. In addition, during the Vicente de Zaldivar inquiry of 1602, one witness reported that Zaldivar, on his 1599 exploratory trip west of the Rio Grande, met Apaches (perhaps in the locale of those noted in 1583), Cruzados (who were the Yavapais of central Arizona), and Tepeguanes (probably Pimas) (Hammond and Rey, 1953, p. 829).

The only other early references to Apaches concern groups in more northern locales, such as the Quinia Apaches west of Taos in the first decade of the 1600's and Navajo Apaches who in the 1620's lived within one day north of Santa Clara (Hodge, Hammond, and Rey, 1945, pp. 85, 89, 306). This also is the locale of early Navajo sites (near Abiquiu) dating from the early 1600's (Schaafsma, 1979). Oñate's report in 1598 of Cocoyes, who lived in the "neighboring sierras" of Jemez, evidently to the north since the following year he described these people as being numerous, living in jacales, and farming in the upper Rio Grande (Hammond and Rey, 1953, pp. 345, 485). The described locale seems to apply to the same area as that of the Quinias in the early 1600's, perhaps a band of those who later became known as Navajo Apaches. Navajos are not reported near the San Jose River north of Acoma until after the middle 1700's (Schroeder, 1963, pp. 7–11). In any case, it does appear that some Apaches were living west of the Rio Grande in the early 1600's—one group on the San Jose and another one, which evolved into the group known as Navajos, north of the Tewas, in the Chama drainage.

The Apaches on the north, perhaps just moving in at the time of Spanish entry in the late 1500's and perhaps identical with those reported as allied with Taos, Picuris, and Pecos, played a part in the events of the first decade of the 1600's, particularly in the Tewa area along the lower Chama River. Four pueblos in this drainage area, first noted by the Coronado journals of 1540–1542, seem to have survived into the very early 1600's. One of them, Tsama Pueblo, is designated as an existing pueblo on a map of 1602 relative to the Oñate colonization venture (Hammond and Rey, 1966, end papers). The coincidence of the previously mentioned long drought of the late 1500's, the establishment of the first Spanish settlement near the junction of the Chama and Rio Grande, and the first recorded Apache attacks on the Spaniards and pueblos prior to 1608 (Hammond and Rey, 1953, p. 1059) suggest that these events and the stress of the times led to the abandonment of the pueblos along the Chama. Excavations in some of the ruins of the latter have yielded evidence of violent deaths (Jeancon, 1912, pp. 29–30; 1923, p. 3; Wendorf, 1953, p. 46). These pueblos also may well have been those the Spaniards referred to in 1608 as having been destroyed and burned by Apaches (Hammond and Rey, 1953, p. 1059). This situation probably played a large part in forming the viceroy's decision to move the colony out of the Tewa country to an area unoccupied by Indians, in conformance with the 1573 laws of settlement (Hammond and Rey, 1953, pp.975–976, 1076, 1089).

During the first twelve years of residence among the Tewas, the Spaniards maintained their own, but definitely frontier, type of life. They also depended on their hosts for various needs, and the Tewas helped them build the irrigation ditch at San Gabriel. The colonists also exacted tribute to make up for some of their needs, such as blankets and food, the latter when their own crops of wheat, barley, chili, and peppers failed to provide sufficiently for their needs. Many of the colonists were unhappy with the venture after the first three years, but some felt otherwise (Hammond and Rey, 1953, pp. 619ff, 629, 655–656).

While two patterns of material culture representing pueblo and frontier-Spanish people existed side by side in Yuqueyungue, the rest of the Pueblo world away from the Spanish settlement remained unchanged except for the churches in nearby pueblos. The Apache entry from the plains, though it seems not to have brought about any significant material-culture change, marks the introduction of a nomadic group that generated among its sedentary neighbors both friendly and hostile reactions, the latter of which were to continue for more than 200 years.

SPANISH SETTLEMENT

When the colonists moved to Santa Fe in 1610, differences between pueblo and Spanish settlement patterns became obvious. The seat of government was a villa with a street pattern. According to the instructions, each household was to be assigned land for a house, garden, and grazing needs, and the villa divided into six districts with a square block for public buildings (Hammond and Rey, 1953, pp. 1087–1088). The pueblo communally arranged multistoried dwellings and farmlands stood in marked contrast. By the 1620's, Santa Fe had some 250 residents with about 1,000 servants (Hodge, Hammond, and Rey, 1945, p. 68). Headquarters for the Franciscan Order was at Santo Domingo.

The governor was instructed to congregate the dispersed or outlying pueblos to more defensible locations and to arrange for the young people to be instructed in the Spanish language so as to facilitate administration (Hammond and Rey, 1953, p. 1090). By the 1620's, only seven of the Tewa pueblos were occupied. How many Tewa and other pueblos were consolidated is not known, since possible pertinent documents were destroyed in the Pueblo Uprising of 1680. However, there was a drastic decrease in the number of pueblos and population in the 1600's possible also due to Spanish-introduced diseases—from about 130 recorded pueblos with perhaps 60,000–100,000 people in the first decade of the 1600's to 64 pueblos with 69,000 in the 1620's and only 40,000 in 1638, to 46 pueblos with 17,000 in 1679, and by 1706, after the uprising, to only 18 pueblos with 6,440 people, exclusive of the Hopis (Schroeder, 1979, p. 254).

Such a rapid decline suggests that in addition to disease other factors also were at work, such as reported famines, attacks on, or enslavement of, the pueblo people, and flights of pueblo refugees to plains tribes. Drought and resulting famine caused 3,000 deaths in 1640 (Scholes, 1936–1937, p. 324), and similar events in the 1660's resulted in an untold number of fatalities. Smallpox obviously took its toll among the Indians in the 1600's, if the number of pock-marked faces of Spaniards (colonists who fled in 1680) returning to Santa Fe in 1693 can be used as a gauge (Chavez, 1954).

The core of Spanish settlement was in the Santa Fe area, but small family rancherias were set up along the Rio Grande, mostly between Albuquerque and Española. They depended on nearby pueblos for some of their needs, such as pottery. The vessels for table and kitchen needs, made by the Tewas, were plain or decorated in black, white, and red-matte paint. Those made by potters to the south of the Tewas were decorated with glaze paint on a yellowish to reddish background (Snow, 1973). Though caravans from Mexico brought containers and dishes of majolica, olive jars, and some metal items of Mexican manufacture, these came in small quantities and only every few years, perhaps reaching only a minor sector of the population. The friars also depended on Indian potters for candlestick holders and plates, thus increasing the number of new shapes and forms in pueblo ceramics.

Trade with Plains Indians, a pueblo monopoly before the 1640's, was gradually taken over by Spanish governors who dealt directly with the suppliers, the latter often offering captives as items of barter (Scholes, 1940, p. 262). This interruption of Pueblo–Plains traffic in goods no doubt had an effect on the economy of the pueblos. It also brought about a significant change among the Apaches. By the mid-1600's they were acquiring horses not only to replace dogs (used to drag Apache belongings packed on tent poles), but also to be able to retaliate with high-speed vengeance against the Spaniards for taking their people as slaves (Hodge, Hammond, and Rey, 1945, p. 171; Scholes, 1930, p. 193; 1935, p. 109; 1936, p. 300). The expansion of Spanish settlement, decrease in pueblo population, shift in trade patterns, Apache mobility, and grazing of horses and sheep over larger and larger areas took place gradually in the 1600's.

During this period, interaction between the pueblos and Spaniards became increasingly strained. For administrative purposes, the Spaniards introduced a set of officers into the pueblos, to be elected by local occupants. These officers were to deal directly with the governor in Santa Fe (Bloom, 1928, p. 363), an obvious attempt to weaken the authority of the local caciques. The effect, however, was the reverse in that the new Indian officers acted as buffers between the Spanish authorities and pueblo leaders. Also, to further missionary efforts, caravans from Mexico brought additional friars to expand the mission system (Scholes, 1930; Hodge, Hammond, and Rey, 1945, pp. 109ff, 174). In pursuing their aims to convert the Indians, these friars became a major element in the spread of Spanish culture. One immediate effect of missionization was the need for the Pueblo Indians who spoke different languages and dialects to learn Spanish. This resulted in better communication and interaction with the Spaniards as well as among pueblo groups speaking different languages. It became the lingua franca.

In the 1600's, all but two of the Tewa pueblos after the 1620's had a resident padre for varying lengths of time. From a mission, the padre also attended to neighboring pueblos that lacked a resident friar. He also set up classes and shops at his mission to teach religion and music. New crafts introduced in these shops, such as carpentry, leather working, and smithing, provided the Indians with a livelihood by which to provide certain needs of the Spaniards (Hodge, Hammond, and Rey, 1945, pp. 100–102; Scholes, 1936–1937, p. 21). The priest's garden and cattle, tended by the Indians, introduced new elements to the pueblos.

The presence of the friars and their opposition to pueblo practices of worship and other customs obviously developed antagonism between the padres and the pueblo religious leaders. Spanish efforts to stamp out masked kachina dances and the use of kivas, as well as to weaken the authority of the ceremonial leaders (NMSRC, 1968, p. 274), succeeded only in driving pueblo rites into secrecy. This, along with the elected Indian officers, probably was responsible to a considerable measure for the preservation of much of pueblo ceremony and culture over the remaining years of the Spanish period. However, conversion to christianity was in part successful, to a great extent probably due to the presence of elements in catholicism that are similar to those in pueblo religious practices—the use of altars, singing, specialized instruction for neophytes, ornamentation and painting, formalized ceremonies and rituals, care of religious objects, and a religious calendar.

Friction between padres and pueblos, plus continuing disagreements between Spanish church and state on the handling and exploitation of the Pueblo Indians, led to a breakdown of the mission system to some extent in the 1600's. In addition, encomiendas were another burden that complicated relations. These were granted to favored Spanish individuals who were to provide protection for a pueblo and give aid to its local church in return for which tribute of blankets and corn could be collected from the pueblo. The Spaniards often went beyond this, illegally conscripting Indian labor to work their farms without compensation. Some of the governors went so far as to set up sweat shops in Santa Fe manned by Indians taken on one pretext or another, or led slaving expeditions and used the captives as miners in Chihuahua (Scholes, 1940; Hodge, Hammond, and Rey, 1945, pp. 168–173).

This treatment of, and demands on, the pueblo people drew many away from their routine and obligations in the village, undoubtedly leading to breakdowns in the local society. Converted Indians in the pueblo became outsiders, further depleting ritual representation in ceremonies and causing factional splits. These situations in turn led to some outward opposition against the Spaniards. In the early 1640's an unknown number fled from Taos, taking refuge among their Apache friends in Cuartelejo located in present Scott County, southwest Kansas. They came back to Taos in 1662. This is the same locale to which many Picuris fled in the 1690's, not returning until 1706 (Thomas, 1935; Schroeder, 1974, pp. 4, 6).

The combination of cultural and environmental stresses that smoldered among the Indians between 1640 and 1670's finally broke out in a general uprising in 1680, an event in which Popé, a Tewa from San Juan, played an important part. The Spaniards were driven out of the Pueblo world to El Paso del Norte, where they remained for 13 years. While the pueblo leaders attempted to stamp out everything Spanish during this period, they were unsuccessful in ridding themselves of Spanish crops, horses, cattle, sheep, and metal implements. Thus, like the Romans, the Spaniards in the 1600's came and saw, but, unlike the Romans, did not conquer, though they left a definite imprint of Spanish culture on the pueblos of the late 1600's.

REFERENCES

Bloom, L. B., 1928, A glimpse of New Mexico in 1620: New Mexico Historical Review, v. 3, pp. 357–380.

Chavez, A., 1954, Origins of New Mexico families in the Spanish colonial period: Historical Society of New Mexico, Santa Fe.

Fritts, H. C., 1965, Tree ring evidence for climatic changes in western North America: Monthly Weather Review, v. 93, pp. 421–443.

Hammond, G. P., and Rey, A. (eds. & transl.), 1927, The Gallegos relation of the Rodriguez expedition to New Mexico: Historical Society of New Mexico, Santa Fe, Publications in History, v. 4.

———, and ———, 1929, Expedition into New Mexico made by Antonio de Espejo, 1582–1583, as revealed in the journal of Diego Perez de Lujan: The Quivira Society (Los Angeles), v. 1.

———, and ———, 1940, Narratives of the Coronado expedition, 1540–1542: University of New Mexico Press, Albuquerque.

———, and ———, 1953, Don Juan de Oñate, colonizer of New Mexico, 1598–1628: University of New Mexico Press, Albuquerque.

———, and ———, 1966, The rediscovery of New Mexico, 1580–1594: University of New Mexico Press, Albuquerque.

Hodge, F. W., Hammond, G. P., and Rey, A., 1945, Fray Alonso de Benavides' revised memorial of 1634: University of New Mexico Press, Albuquerque.

Jeancon, J. A., 1912, Ruins at Pesedeuingue: Records of the Past (Washington, D.C.), v. 11, pp. 27–37.

———, 1923, Excavations in the Chama Valley, New Mexico: Bureau of American Ethnology (Washington, D.C.), Bulletin 81.

New Mexico State Records Center, 1968, Calendar of the Spanish archives of New Mexico, 1621–1821: Santa Fe.

Schaafsma, C. F., 1979, The Cerrito site (AR-4), A Piedra Lumbre phase settlement at Abiquiu Reservoir: School of American Research, Santa Fe.

Scholes, F. V., 1930, The supply services of the New Mexican missions in the seventeenth century: New Mexico Historical Review, v. 5, pp. 186–210.

———, 1935, Civil government and society in New Mexico in the seventeenth century: New Mexico Historical Review, v. 10, pp. 71–111.

———, 1936–1937, Church and state in New Mexico, 1610–1650: New Mexico Historical Review, v. 11, pp. 9–76, 297–349.

———, 1940, Troublous times in New Mexico, 1659–1670: New Mexico Historical Review, v. 15, pp. 249–268.

Schroeder, A. H., 1962, A reanalysis of the routes of Coronado and Oñate into the plains in 1541 and 1601: Plains Anthropologist, v. 7, pp. 2–23.

———, 1963, Navajo and Apache relationships west of the Rio Grande: El Palacio, v. 7, no. 3, pp. 5–23.

———, 1968, Tentative ecological and cultural forces and effects on Southwestern Indian farmers; *in* Contributions to Southwestern prehistory, v. 4: Proceedings, VII Congress of International Association of Quaternary Research, Eastern New Mexico University Contributions, v. 1, no. 1.

———, 1972, Rio Grande ethnohistory; *in* Ortiz, A. (ed.), New perspectives on the pueblos: University of New Mexico Press, Albuquerque, pp. 41–70.

———, 1974, A brief history of Picuris Pueblo: Adams State College (Alamosa), Series in Anthropology, no. 2.

———, 1979, Pueblos abandoned in historic times; *in* Handbook of North American Indians, v. 9: Smithsonian Institution, pp. 236–254.

———, and Matson, D. S. (eds. & transl.), 1965, A colony on the move, Gaspar Castaño de Sosa's journal: School of American Research, Santa Fe.

Schulman, E., 1956, Three ring hydrology of the Colorado River basin: University of Arizona, Bulletin, v. 16, pp. 45–48, 69.

Snow, D. H., 1973, Some economic considerations of historic Rio Grande Pueblo pottery; Schroeder, A. H. (ed.), *in* The changing ways of Southwestern Indians: Rio Grande Press, Inc., Glorieta, pp. 55–72.

Thomas, A. B., 1935, After Coronado, Spanish exploration northeast of New Mexico, 1696–1727: University of Oklahoma Press, Norman.

Wendorf, F., 1953, Salvage archaeology in the Chama Valley, New Mexico: School of American Research, Santa Fe, Monograph 17.

RIO GRANDE PREHISTORY: PRELUDE TO CONTACT

LINDA S. CORDELL
Department of Anthropology, University of New Mexico, Albuquerque, New Mexico 87131

INTRODUCTION

The basic descriptive and chronological framework for Rio Grande prehistory was established early in the 20th century (e.g., Kidder, 1915, 1916, 1924; Hewett, 1906; Nelson, 1914). Before the advent of various chronometric dating techniques the Rio Grande area was especially attractive to prehistorians, because it offers the ruins of villages abandoned during the historic period. Archaeologists could then begin from a known abandonment date and, as their excavations proceeded down through various strata, work their way back into prehistory. Not only were documents crucial for chronology, but historic accounts and observations of the modern Pueblo Indian villages provided the context for interpreting archaeological finds. The Rio Grande Pueblo Indians were considered living analogs for their ancestors by generations of archaeologists.

New data have been assembled since the early days and syntheses have been prepared periodically (e.g., Cordell, 1979; Lang, 1977; Wendorf and Reed, 1955). Yet only within the past five years have southwestern prehistorians begun to evaluate in a systematic manner the discrepancies between the late prehistoric and historic periods and the kinds of changes within Pueblo culture begun by initial contact with Europeans (e.g., Cordell and Plog, 1979; Cordell, 1984; Upham, 1982). Once the changes are fully appreciated and understood, a clearer picture will emerge of the culture encountered by the Spaniards in 1540.

ENVIRONMENT

Natural Setting

Compared to the San Juan Basin, the Hopi Buttes, and other regions of prehistoric Pueblo Indian development, the Rio Grande valley is relatively lush. The Rio Grande and some of its tributaries provide abundant sources of water. The mountains on both sides of the river serve as large catchment areas for precipitation, and the growing season over much of the valley proper is adequate for maize.

Nevertheless, prehistoric farming was not without risk. Floods have been catastrophic during the historic period and would have been beyond the technological abilities of the prehistoric Indians to control (Cordell and Earls, 1983). Poorly drained soils were damaged by mineralization (Fosberg and Husler, 1979). Dense stands of hardwood along the river were difficult to clear with stone tools, and crops planted in fields on the floodplain were subject to insect infestations and disease (Ford, 1972).

Relying on observations of agricultural practices among the modern Pueblo Indians, most prehistorians have underestimated the risks and assumed that irrigation or floodwater farming provided the bulk of agricultural produce during the late prehistoric period. However, recent archaeological surveys have documented hundreds of thousands of square meters of prehistoric upland-water and soil-control features throughout the Rio Grande area (Cordell and Earls, 1983; Cordell, 1984). The features, consisting of rock and earth terraces, rock grids, and raised and bordered fields, suggest that there was far more dependence on rainfall farming than is usually credited.

Given what appear to be considerable risks to agriculture, gathering and hunting must have constituted important elements in the prehistoric Pueblo Indian economies. Although few quantitative data are available, most studies support the importance of game (e.g., Emslie, 1981; Snow, 1974) and wild-plant food (Toll, 1983).

Cultural Setting

The archaeological record of prehistoric occupation of the Rio Grande area is as long as it is elsewhere in the Southwest. Sites dating to the Paleoindian Period (ca 11000–7000 B.C.) are known through excavation and reconnaissance survey (Hibben, 1955; Judge, 1973; Weber and Agogino, 1968). Sites of the Archaic Period (ca 6000 B.C. to A.D. 100) are not abundant, nor are they unusually scarce (e.g., Campbell and Ellis, 1952; Schaafsma, 1976).

Nevertheless, the record of prehistoric Pueblo occupation before A.D. 1200 is scant compared to areas such as the San Juan River drainage, the San Juan Basin, and portions of southeastern Utah. It is possible that many earlier sites are deeply alluviated, or have long since been eroded into the river, or lie under modern villages, towns, and cities. It is also possible that the relative abundance of game and wild-plant foods permitted Rio Grande populations to rely substantially on hunting and gathering, maintaining a mobile-residence pattern.

The major prehistoric increase in Rio Grande area settlement occurred between about A.D. 1200 and 1400. These dates follow the collapse of the structured social system centered in Chaco Canyon (Judge, 1983) and the abandonment of large areas of the Colorado Plateau. Some of the people from the San Juan Basin and Mesa Verde areas may well have moved into the upper Chama and Rio Grande valleys, accounting for the population increase. At the same time, however, other densely populated areas also included the area around Zuni, the upper Little Colorado and Winslow/Chavez Pass area of central Arizona, and the Tonto Basin and adjacent areas of Arizona below the Mogollon Rim (Cordell, 1984; Upham, 1982).

Most of the large and better known sites of the Rio Grande area date to the period between 1300 and 1600. These include Te'ewi, Tsiping, Howiri, Tsama, and Sapawe of the Chama Valley; Tsankawi, Tshirege, Otowi, Tyouny, and Puye of the Pajarito Plateau; Pot Creek Pueblo, Old Picuris, and the ancestral Taos village "Cornfield Taos" near Taos; Arroyo Hondo, Cieneguilla, Pindi, Pueblo Lumbre, Pueblo Largo, San Cristobal, San Marcos, and Las Madres of the Santa Fe and Galisteo Basin areas; Rowe Ruin, Loma Lothrop, Arrowhead, and Pecos Pueblo on the upper Pecos River; Kuaua, Alameda Pueblo, Paa-ko, Tijeras Pueblo, and San Antonio in the vicinity of Albuquerque; Tenabo, Gran Quivira, and Chilili east of the Sandias; and Pottery Mound, Senecu, and Teypama south of Albuquerque.

The number of very large late prehistoric sites of the Rio Grande area is impressive in contrast to the number of late prehistoric sites in central Arizona. In addition, the Rio Grande sites were occupied longer, many into the historic period, again in contrast to the upper Little Colorado and Tonto Basin sites that were abandoned in prehistoric times. The relative stability of the Rio Grande sites suggests that natural resources were sufficient to support large settlements and the social mechanisms of village integration were well developed.

CONTINUITIES AND DISCONTINUITIES

It is easy to recognize continuities between late prehistoric pueblo villages and the modern pueblos—a fact that has long encouraged the archaeological practice of using the modern villages as direct sources of analogy. Both prehistoric and modern villages consist of a massed series of roomblocks, one or more open, community-plaza areas, and one or more semisubterranean, ceremonial rooms or kivas. The cotton mantas (dresses), kilts, sashes, and tablitas (headdresses) worn by dancers at ceremonies are depicted in rock art and in murals dating to the

14th century. Many symbolic representations found painted on late prehistoric ceramics, rock art, and kiva murals are clearly recognized and understood by modern pueblo people (e.g., Dutton, 1963; Smith, 1952). These include representations of awanyu (plumed serpents), stepped clouds or kiva stairs, and lightning, as well as particular ceremonies and cosmological personages. On a more prosaic level, modern pueblo women make pottery, know how to grind corn using manos and metates, and know the use of many wild plants found in archaeological contexts, just as modern pueblo men know the rituals and techniques of the hunt and farming, the regulation of ritual events, and how young men are to be educated for positions of leadership in Pueblo society. On the most obvious levels, Pueblo Indians maintain their native languages and value their traditions.

Perhaps one of the more common paths of human reasoning is to emphasize similarities and continuities, thereby minimizing or overlooking the unfamiliar and the different. Archaeologists traditionally employ this mode of thought. This has lead to numerous attempts to assign linguistic or tribal identity to pueblo archaeological remains, especially ceramic styles (e.g., Mera, 1935; Ford and others, 1972) and to interpretations of prehistoric patterns of kinship and post-marital residence rules (e.g., Longacre, 1964, 1970). Emphasizing similarities has also lead to assuming that prehistoric pueblo villages were politically and economically independent, as the modern villages are, that each prehistoric village produced its own ceramics, and that agriculture provided a greater portion of food than did hunting and gathering.

Despite the undeniable similarities, however, there are discontinuities between the present and the prehistoric archaeological remains. For example, some symbolic representations (e.g., Dutton, 1963; Cordell, 1980) are not identifiable today. Petrographic analyses of rock used to temper ceramics, as well as analyses of clays and pigment minerals (e.g., Warren, 1980; Kidder and Shepard, 1936) indicate that not only were large quantities of ceramic vessels exchanged or traded among villages, but residents of some sites produced no ceramics of their own (Plog, 1980), and some sites, or clusters of sites, supplied certain kinds of vessels for an entire region.

There are other contrasts between the late prehistoric and historic periods which suggest that prehistoric organizational patterns were unlike those of today. The similarities between the Rio Grande and Western Pueblo stylistic treatment of ceramics and in kiva art indicate considerable interaction, oriented along an east–west axis, over the enormous distance from Pecos to the Hopi villages. A well-known example is the Sityatki style, a complex decorative treatment produced at Hopi, that also occurs on jars made at Pottery Mound. The murals from Kawaika-a and Awatovi, in the Jeddito area of Arizona, are similar in execution, pictorial treatment, and motif to those from Kuaua, at Bernalillo, and Pottery Mound (Cordell, 1984). It seems likely that at least some of the motifs relate to the Kachina cult, and it is possible that Kachina ceremonialism entered the Western Pueblos from the Jornada and Rio Grande areas, as the Schaafsmas (1974) propose. Probably most important though is that the similarities reflect a strength of interaction among villages that is not documented historically.

Two additional lines of evidence suggest that late prehistoric villages may not have been completely autonomous. First, within clusters of contemporary sites, some large sites seem to lack kivas whereas other sites within the cluster have several. This situation seems to characterize the late prehistoric upper Pecos Valley sites of Rowe Pueblo, which lack kivas, and Arrowhead, which has several. Second, modern pueblos maintain shrines and sacred areas well beyond their reservation boundaries, sometimes on other reservations. For example, the Zuni maintain shrines in the Sandia Mountains, and the Hopi make pilgrimages to the Zuni salt lakes. Retreats to shrines and sacred areas should not be interferred with. Today, arrangements for undisturbed pilgrimages are made by telephone (E. Ladd, oral comm. 1981). It is likely that in the late prehistoric period communication among villages relied on some coordination of ritual events by the religious hierarchies of different villages. Upham (1982) discusses modes of alliance systems within the 14th century Western Pueblo that may have been similar and have not continued into historic times.

FACTORS BEHIND THE DISCONTINUITIES

The brief discussion of differences between the late prehistoric and historic periods focuses attention on the systemic changes that must have accompanied contact with Europeans. Only three of these are discussed here: introduced disease, domestic livestock, and laws regulating colonial settlement.

Native-American populations had no immunities to the diseases introduced by Europeans (i.e., smallpox, tuberculosis, measles, etc.). When diseases are introduced in such situations, mortality rates of between 20 and 100% are documented. A severe smallpox epidemic struck the First Mesa Hopi villages as late as in 1853, reducing the population from 1200 to 650 persons by 1862 (Adams, 1981). Schroeder's (1979) review of various archival records lists 61 pueblos known to have been abandoned since 1540. Although reasons other than epidemic disease are given for some of these abandonments, disease is an all too common factor in most cases.

Heavy population losses must have had profound effects on pueblo economy and political organization. For example, some villages may not have been able to perform traditional ceremonies, because there were not enough people to carry out important roles or the leadership positions of various societies could not be filled. It is also possible that there were not enough able-bodied people to plant, tend, and harvest crops. Villages that could not continue to survive socially may have joined other villages, as the inhabitants of Pecos did when they went to Jemez in the 19th century (Cordell, 1984; Lycett, 1984).

Wild-plant and animal foods were not only important to the pueblos prehistorically, but often may have made the difference between starvation and survival. European domestic livestock grazed on land that had provided the wild resources essential to the pueblos. At first, the diminishing quantities of wild resources must have been alarming. However, as pueblos adopted domestic livestock and European crops, former hunting, trading, and foraging expeditions must have become less common, and the social ties that facilitated access to resources in the vicinity of distant villages declined.

Finally, Spanish law required colonists to establish their communities on land not already occupied by the Indians. Thus, in most cases, Spanish settlements were established in between the pueblos. While it is true that the general absence of Spanish colonists within Indian villages enabled the Indians to preserve many of their cultural traditions, the rule would have disrupted some aspects of the native system. The presence of Spanish settlements in between the pueblo villages further inhibited economic, ceremonial, and political interaction among the pueblo villages. The end result was the village autonomy and independence we are familiar with today.

REFERENCES

Adams, E. C., 1981, The view from the Hopi Mesas; *in* Wilcox, D. R., and Masse, W. B. (eds.), The Protohistoric Period in the North American Southwest, A.D. 1450–1700: Arizona State University, Anthropological Papers, no. 24, pp. 321–335.

Campbell, J. M., and Ellis, F. H., 1952, The Atrisco sites: Cochise manifestations in the middle Rio Grande valley: American Antiquity, v. 17, pp. 211–221.

Cordell, L. S, 1979, Cultural resources overview of the middle Rio Grande valley, New Mexico: U.S. Government Printing Office, Washington, D.C., 198 pp.

——— (ed.), 1980, Tijeras Canyon: analyses of the past: University of New Mexico Press, Albuquerque, 200 pp.

———, 1984, Prehistory of the Southwest: Academic Press, New York, 450 pp.

———, and Earls, A., 1983, Mountains and rivers: resource use at three sites: Paper presented at the Anasazi Symposium, Farmington, New Mexico.

———, and Plog, F., 1979, Escaping the confines of normative thought: a reevaluation of Puebloan prehistory: American Antiquity, v. 44, pp. 405–429.

Dutton, B. P., 1963, Sun Father's Way, the kiva murals of Kuaua, a pueblo ruin, Coronado State Monument, New Mexico: University of New Mexico Press, Albuquerque, and School of American Research, Santa Fe, 237 pp.

Emslie, S. D., 1981, Prehistoric agricultural ecosystems: avifauna from Pottery Mound, New Mexico: American Antiquity, v. 46, pp. 853–860.

Ford, R. I., 1972, An ecological perspective on the Eastern Pueblos; *in* Ortiz, A. A. (ed.), New perspectives on the pueblos: School of American Research, Santa Fe, and University of New Mexico Press, Albuquerque, pp. 1—18.

———, Schroeder, A., and Peckham, S., 1972, Three perspectives on Puebloan prehistory; *in* Ortiz, A. A. (ed.), New perspectives on the pueblos: School of American Research, Santa Fe, and University of New Mexico Press, Albuquerque, pp. 22–40.

Fosberg, S., and Husler, J., 1979, Pedology in the service of archeology: soil testing at LA 13086; *in* Biella, J. V., and Chapman, R. C. (eds.), Archeological investigations in Cochiti Reservoir, New Mexico, vol. 4: Office of Contract Archeology, University of New Mexico, pp. 307–318.

Hewett, E. L., 1906, Antiquities of the Jemez Plateau, New Mexico: Bureau of American Ethnology, Bulletin 2, Smithsonian Institution, p. 187.

Hibben, F. C., 1955, Specimens from Sandia Cave and their possible significance: Science, v. 122, pp. 688–689.

Judge, W. J., 1973, The Paleo-Indian occupation of the central Rio Grande valley, New Mexico: University of New Mexico Press, Albuquerque, 130 pp.

———, 1983, The Chaco system: A.D. 900–1200: a trial reconstruction: Paper delivered at the American Anthropological Association Annual Meeting, Chicago.

Kidder, A. V., 1915, Pottery of the Pajarito Plateau and of some adjacent regions in New Mexico: American Anthropological Association, Memoir, v. 2, pp. 407–462.

———, 1916, The Pueblo of Pecos: Archaeological Institute of America and School of American Research, Santa Fe, no. 33, 86 pp.

———, 1924, An introduction to the study of southwestern archaeology, with a preliminary account of the excavations at Pecos: Papers of the R. S. Peabody Foundation Southwest Expedition 1, Phillips Academy Andover, Yale University Press, New Haven, 377 pp.

———, and Shepard, A. O., 1936, The pottery of Pecos, vol. II: the glaze paint, culinary, and other wares: Papers of the R. S. Peabody Southwest Expedition 7, Phillips Academy Andover, Yale University Press, New Haven, 418 pp.

Lang, R. W., 1977, The prehistoric pueblo cultural sequence in the northern Rio Grande: Paper delivered at the 50th Pecos Conference, Pecos, New Mexico.

Longacre, W. A., 1964, Archaeology as anthropology: a case study: Science, v. 144, pp. 1454–1455.

——— (ed.), 1970, Reconstructing prehistoric Pueblo societies: University of New Mexico Press, Albuquerque, and School of American Research, Santa Fe, 247 pp.

Lycett, M., 1984, Social and economic consequences of aboriginal population decline from introduced diseases: Paper presented at the Society for American Archaeology 49th Annual Meeting, Portland.

Mera, H. P., 1935, Ceramic clues to the prehistory of north-central New Mexico: Laboratory of Anthropology (Santa Fe), Technical Series, 8, 109 pp.

Nelson, N. C., 1914, Pueblo ruins of the Galisteo Basin, New Mexico: American Museum of Natural History, Anthropological Papers, 15, pt. 1, 38 pp.

Plog, S., 1980, Stylistic variation in prehistoric ceramics: University of Cambridge Press, 160 pp.

Schaafsma, C. F., 1976, Archaeological survey of maximum pool and Navajo excavations at Abiquiu Reservoir, Rio Arriba County, New Mexico: School of American Research, Santa Fe, Contract Program, 143 pp.

Schaafsma, P., and Schaafsma, C. F., 1974, Evidence for the origins of Pueblo katchina cult as suggested by southwestern rock art: American Antiquity, v. 39, pp. 535–545.

Schroeder, A. H., 1979, Pueblos abandoned in historic times; *in* Ortiz, A. A. (ed.), Handbook of North American Indians, vol. 9, Southwest: Smithsonian Institution, Washington, D.C., pp. 236–254.

Smith, W., 1952, Kiva mural decorations at Awatovi and Kawaika-a with a survey of other wall paintings in the Pueblo Southwest: Harvard University: Papers of the Peabody Museum of American Archaeology and Ethnology, vol. 37, 363 pp.

Snow, D. H., 1974, The excavation of Saltbush Pueblo, Bandelier National Monument, New Mexico, 1971: Museum of New Mexico (Santa Fe), Laboratory of Anthropology, Note 97, 74 pp.

Toll, M. S., 1983, Wild plant use in the Rio Abajo, some deviations from the expected pattern throughout the central and northern Southwest: Paper delivered at the Rio Abajo Conference, "A seminar on archaeology and history of the Socorro District," Socorro, New Mexico.

Upham, C. S., 1982, Policies and power, an economic and political history of the Western Pueblo: Academic Press, New York, 225 pp.

Warren, A. H., 1980, Prehistoric pottery of Tijeras Canyon; *in* Cordell, L. S. (ed.), Tijeras Canyon: analyses of the past: University of New Mexico Press, Albuquerque, pp. 149–168.

Weber, R. H., and Agogino, G. A., 1968, Mockingbird Gap Paleo-Indian site: excavations in 1967: Paper presented at the Society for American Archaeology 33rd Annual Meeting, Santa Fe.

Wendorf, F., and Reed, E. K., 1955, An alternative reconstruction of northern Rio Grande prehistory: El Palacio, v. 62, pp. 131–173.

Taos Pueblo, ca 1920. Photo by Burt Harwood.

A COMPARATIVE STUDY OF EARLY HISTORIC "TEWA" POTTERY

NATHAN W. BOWER[1] and DAVID H. SNOW[2]
[1]Colorado College, Colorado Springs, Colorado 80903; [2]Cross-cultural Research Systems, Santa Fe, New Mexico 87501

INTRODUCTION

Pottery has long been used by archaeologists to trace trade routes and to determine the geographic extent of a given culture. Various archaeometric methods have supplemented the traditional stylistic analyses, particularly when only small, unidentifiable sherds of culinary ware are available (Bower and others, 1974; Hall and others, 1973; Isphording, 1974; Snow and Fullbright, 1976). The parametric data generated by an elemental-analysis method such as neutron activation, atomic absorption, or x-ray fluorescence are particularly useful when combined with computer-generated pattern recognition (Ward, 1974).

In this study pueblo pottery of the late protohistoric or early historic period was analyzed by x-ray methods in order to provide information about the number and locations of the historic pueblos trading with the early Spanish settlers in the Santa Fe area. After 1000 A.D., the Española Basin was primarily the home of the Tewa-speaking pueblos (Fig. 1), though a linguistically similar group known as the Tano (or southern Tewa) lived south of Santa Fe in the Galisteo River basin. This southern group may have been split off from the northern Tewa when the Keresans settled the area around the confluence of the Jemez and Rio Grande Rivers between 1200 and 1300 A.D. The Towas, who settled along the Jemez River about 1250 A.D., may also have been split at some point, as some authors feel that Pecos was also Towa, while others hold that it was probably Tewa (Ford and others, 1972). The Tiwas, who are generally believed to have developed in the Rio Grande valley originally, were displaced by the later pueblos to the Taos and Albuquerque areas.

Primarily Tewa Red culinary ware (which might be gray or black if it is fired in a reducing atmosphere) and Tewa Polychrome ware (Fig. 2) were used for the analysis. The Tewa Polychrome and Red wares were introduced at about the same time as the first Spanish settlement, and their popularity may have been influenced by the Spanish or the Mexican Indians who came with them. The Tewa Polychrome jars and bowls are distinctive because of their design band, which is white with thin, black line decorations made with carbon-based pigment instead of the mineral-based pigment generally used on pottery elsewhere. A red slip was usually applied both above and below this belly band (Fig. 2), though in the late 18th or early 19th century the red was used only in a band below the design (Harlow, 1967), and the original Tewa Red and Polychrome wares faded from production. Some Biscuit wares which were made by the northern Tewa, and some Glaze wares made by the southern Tewa or the Keresans, were also analyzed. These latter wares date from 1400 to 1600 A.D. (Dittert and Plog, 1980).

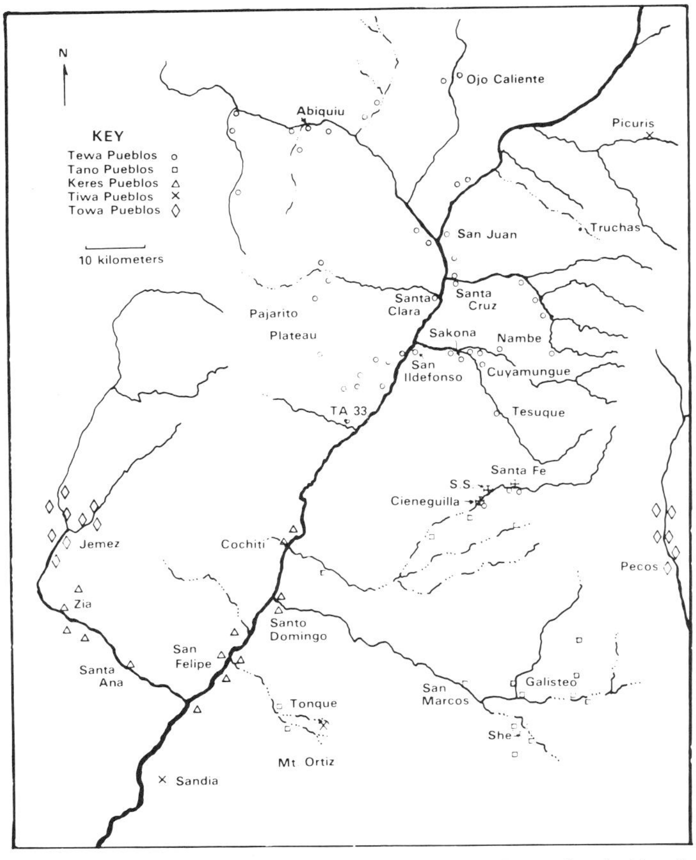

FIGURE 1. Location of protohistoric (ca 1400–1540 A.D.) and early historic (ca 1540–1750 A.D.) pueblos and Spanish settlements of the upper Rio Grande valley (Swanton, 1952).

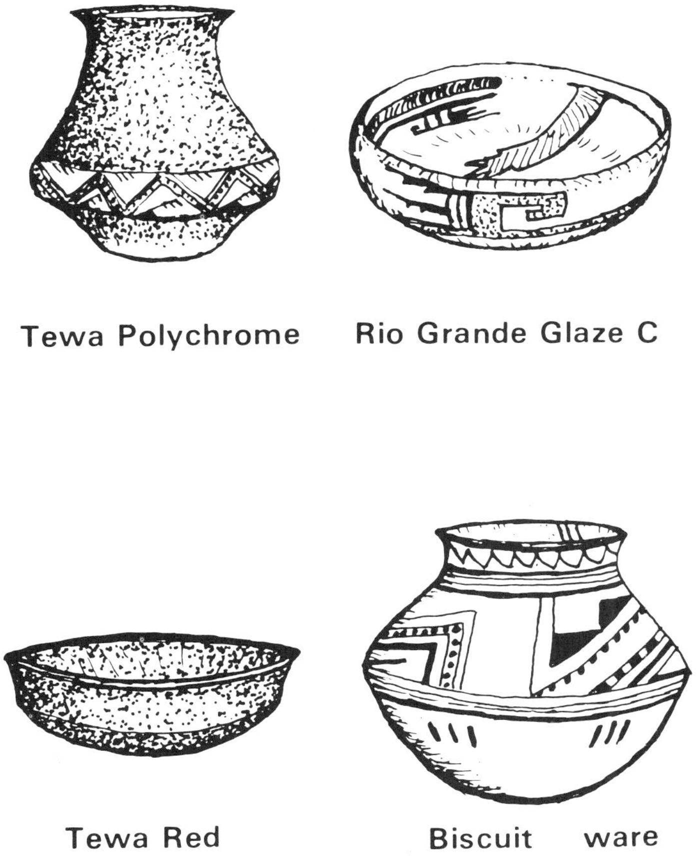

FIGURE 2. Typical pottery of the upper Rio Grande valley, 1450–1730 A.D. Stipling indicates red slip.

Besides providing useful information for understanding cultural interactions between the Spanish settlers and the native populations, a work such as this should provide archaeometrically useful data. Although the efficacy of XRF for archaeometric analyses is well known, the minimum number of samples and parameters to be measured on those samples is not known for the pottery of this area. This study will provide knowledge of what constitutes a diagnostic set of parameters as well as initiating a library of data for comparison with further analyses of pottery from this region. By extending these analyses into other historic, ceramic, geographic, and cultural groups, questions about the origins of the various pueblos and their interactions through time can eventually be addressed as well.

METHOD

Samples from 14 pueblos in the northern Rio Grande valley were collected by snipping off 1–2 g pieces from undecorated portions of the wares. For simplicity, the red slip or glaze was not removed from the analyzed pieces. This presented few problems in the analysis or the conclusions, and it actually increased separation between pottery groups. Generally, however, analyzing the surface separate from the body would be the best procedure. Most of the samples were taken from the collection of sherds at the Museum of New Mexico Laboratory. The 1–2 g weight was selected for these relatively homogenous samples as adequate to give analyses which would be within 5% precision with 95% confidence (Bromund and others, 1976) for each sherd. Of course, the variance between sherds at a given site might well be greater than this, especially if they came from different sources. Thirty samples of red ware from a Spanish settlement located along the Santa Fe River just west of the Santa Fe airport (site 16/T), and 12 samples from two other Spanish settlements (LA 16 and LA 930) located on either side of the main site were analyzed with the pueblo samples. An additional 23 prehistoric samples from three sites (Cochiti, Truchas, and TA 33) were analyzed so that more complete comparisons could be made.

A preliminary study of 20 sherds demonstrated that sample sizes of three to five sherds would be satisfactory for discerning differences between pueblos. The number of samples needed can be estimated from n = (Student's t statistic × the standard deviation / the difference in the means of two pueblos to be discerned)^2. For example, for separating Cuyamungue and Tesuque using only the strontium data, $n = (3.2 \times 53/(271-161))^2 = 3$ samples of each pueblo. If other elements were also included, even fewer samples could be used, though the chance of getting an imported pot without recognizing it as such becomes too great to warrant the savings.

After grinding the samples in a mullite mortar and pestle, 1 g of the powder and 9 g of lithium tetraborate were fused at 1150°C for 20 minutes with constant vigorous agitation. Glass disks 3 cm in diameter were poured from the melt and these were analyzed for 10 major (Na, Mg, Al, Si, P, K, Ca, Ti, Mn, and Fe) and 12 trace (Ba, La, Ce, V, Cr, Ni, Co, Zn, Sr, Rb, Y, and Zr) elements using a Rigaku model 3064 x-ray fluorescence (XRF) instrument.

X-ray powder-diffraction (XRD) spectra were collected on selected sherds in order to identify the different minerals present in the samples from the Spanish settlement. A Siemens model D500 was used to scan in 0.02-degree increments from a two theta of 2 to 36 degrees, counting for two seconds at each increment. A binocular microscope was also used to visually sort the sherds based upon the appearance of their temper.

The cluster analysis was calculated using a program based on the distance coefficient presented in Davis (1973) after normalizing the data to z-scores. The major oxides (except CaO) and Sr were used as a diagnostic set for the statistical analyses. CaO was not used because of Ca contamination from the ground water in this area. If it were included, the composition would not accurately reflect the composition the pot had when it was made. The remaining trace elements were not used in the diagnostic set as they did not have greater variances between pueblos than they had within pueblos, implying they would not add significantly to the discerning ability of the diagnostic parameters while they would increase the complexity of the calculations. Since the calculations were done with a small computer, except for Sr, the trace elements were dropped for simplicity.

RESULTS AND DISCUSSION

The sample sizes, means, and standard deviations of 10 major oxides and Sr for five historic pueblos are presented in Table 1. Tewa red pottery collected from three Spanish settlements located along the Santa Fe River, Biscuit ware from four northern Tewa pueblos, and prehistoric pottery from the surrounding region are presented in Tables 2, 3, and 4, respectively. A typical sherd from the Española Basin would fall within the following ranges of composition: 1.0–2.0% Na_2O, 1.5–2.5% MgO, 14.0–16.0% Al_2O_3, 62.0–65.5% SiO_2, 0.1–0.3% P_2O_5, 3.0–4.0% K_2O, 2.0–8.0% CaO, 0.4–0.8% TiO_2, 0.02–0.2% MnO, and 3.5–5.5% Fe_2O_3. For the trace elements it would have 300–1000 ppm Ba, 20–80 ppm La, 50–180 ppm Ce, 10–60 ppm V, 10–40 ppm Cr, 0–50 ppm Ni, 0–20 ppm Co, 70–130 ppm Zn, 100–400 ppm Sr, 100–170 ppm Rb, 20–80 ppm Y, and 150–400 ppm Zr. For La, Ce, Ni, and Co these concentrations are at or below the detection limits.

From the data in Tables 1, 3, and 4 it is apparent that the sherds from the northern portion of the Española Basin contain higher levels of SiO_2 and K_2O and lower levels of Al_2O_3 and Sr when compared to the sherds from south of Santa Fe, making it possible to discern differences between the pueblos and to approximately locate the sources of the Spanish settlement pottery. This difference in the ratios of the elements suggests that more alkaline feldspars are present in the northern samples and more calcic feldspars, which concentrate Sr, are favored in the southern samples. The higher concentration of Ca in the SS 2 samples compared to SS 1 supports this hypothesis, as any groundwater contamination should be fairly similar for these two groups of sherds.

For most of the pueblos there is enough of a distinction, so an unknown sherd can be "fingerprinted" using pattern-recognition methods. No temporal trend is apparent. Rather, pottery made at a given locale tends to have the same composition throughout history, though this has not been studied over a very large time span. The differences in the mean compositions of the Biscuit wares (Table 3) from Ojo Caliente, Pioge, Tesuque, and San Ildefonso imply different centers of production for these pueblos. The differences in the Glaze wares from Tonque and She, and in the Tewa wares from Santa Clara, Sakona, Cuyamungue Tesuque, and Pecos also imply different centers. The compositions for Truchas, TA 33, and Cochiti, which contain pottery

TABLE 1. Major oxide (%) and trace-element (ppm) composition of Tewa Red and Polychrome wares from five pueblos.

Sample Size	Santa Clara (LA 925) 4	Sakona (LA 1065) 7	Cuyamungue (LA 38) 16	Tesuque (LA 1064) 8	Pecos (LA 625) 5
Na_2O	1.48	1.42	1.32	1.34	1.49
	0.17	0.15	0.20	0.16	0.21
MgO	2.32	2.22	2.39	2.07	2.15
	0.56	0.14	0.17	0.36	0.33
Al_2O_3	15.46	14.54	15.00	15.04	15.15
	0.83	0.63	0.45	1.71	0.88
SiO_2	63.96	64.08	62.56	64.86	64.28
	3.14	1.82	1.94	0.92	1.83
P_2O_5	0.15	0.15	0.17	0.13	0.23
	0.03	0.05	0.06	0.04	0.08
K_2O	3.91	3.93	3.95	3.48	3.73
	0.35	0.10	0.20	0.22	0.64
CaO	4.39	4.41	5.05	4.47	4.27
	0.95	0.57	0.90	1.81	1.64
TiO_2	0.62	0.60	0.61	0.56	0.55
	0.10	0.06	0.06	0.06	0.08
MnO	0.15	0.10	0.13	0.09	0.08
	0.08	0.01	0.03	0.02	0.02
Fe_2O_3	4.38	4.69	4.69	4.45	4.80
	0.93	0.35	0.25	0.74	0.70
Sr	280	318	271	161	192
	71	40	50	56	72

TABLE 2. Major oxide (%) and trace-element (ppm) composition of pottery from selected early historic Spanish settlements.

	SS 1 (16/T)[1]	SS 2 (16/T)[1]	Santa Fe[2]	Cieneguilla[3]
	-	-	(LA 930)	(LA 16)
Sample Size	10	20	7	5
Na_2O	1.40	1.66	1.41	1.46
	0.40	0.26	0.22	0.18
MgO	1.85	2.28	1.75	2.03
	0.27	0.18	0.22	0.33
Al_2O_3	14.09	17.12	14.46	14.44
	2.01	0.76	0.86	0.93
SiO_2	65.48	61.08	66.65	65.61
	2.08	1.07	2.47	1.90
P_2O_5	0.21	0.20	0.36	0.17
	0.06	0.04	0.33	0.02
K_2O	3.21	2.82	3.60	3.84
	0.61	0.12	0.28	0.36
CaO	4.77	5.46	4.45	4.30
	1.69	0.63	1.59	1.06
TiO_2	0.55	0.83	0.54	0.47
	0.11	0.11	0.03	0.03
MnO	0.08	0.13	0.10	0.08
	0.03	0.03	0.02	0.01
Fe_2O_3	4.26	6.54	4.19	3.99
	0.95	0.68	0.44	0.34
Sr	216	519	126	197
	55	58	66	52

[1]All pottery appeared to be culinary red ware.
[2]Pottery consisted of a mixture of red wares.
[3]Sample was composed entirely of Tewa Polychrome plates, which were made exclusively for the Spanish.

TABLE 3. Major oxide (%) and trace-element (ppm) composition of protohistoric pottery of the northern Rio Grande valley.

	Bisquitware B	Bisquitware B	Bisquitware A	Bisquitware B
	Ojo Caliente	Pioge	San Ildefonso	Tesuque
	-	(LA 144)	(LA 122)	(LA 1064)
Sample Size	4	4	6	4
Na_2O	2.14	1.88	1.69	1.52
	0.19	0.34	0.38	0.34
MgO	1.81	1.55	1.58	1.86
	0.19	0.29	0.36	0.26
Al_2O_3	14.03	15.49	14.28	14.80
	0.96	0.58	1.19	0.34
SiO_2	69.03	67.78	68.45	65.07
	0.99	0.92	3.84	0.78
P_2O_5	0.06	0.09	0.14	0.10
	0.01	0.03	0.07	0.02
K_2O	3.79	4.06	3.94	3.50
	0.72	0.52	0.26	0.37
CaO	2.02	2.32	3.16	4.60
	0.53	0.90	1.61	2.44
TiO_2	0.26	0.42	0.43	0.57
	0.07	0.11	0.13	0.04
MnO	0.03	0.06	0.06	0.08
	0.01	0.01	0.01	0.03
Fe_2O_3	2.63	3.33	3.22	4.20
	0.52	0.23	0.83	0.08
Sr	178	163	172	202
	24	69	50	39

TABLE 4. Major oxide (%) and trace-element (ppm) composition of prehistoric and protohistoric pottery of the northern Rio Grande valley.

	Truchas[1]	TA-33[1]	Cochitil	Gallisteo[2]	Tonque[3]	She[4]
	(LA 321)	-	(LA 543)	(LA 26)	(LA 240)	(LA 239)
Sample Size	5	14	4	5	6	3
Na_2O	1.37	1.64	1.44	1.73	2.20	1.17
	0.20	0.61	0.42	0.30	0.32	0.14
MgO	2.33	1.80	2.25	1.77	1.90	2.50
	0.17	0.75	0.52	0.49	0.36	0.24
Al_2O_3	15.27	17.40	15.58	14.44	18.60	18.16
	0.74	1.50	1.41	1.22	1.10	0.45
SiO_2	64.85	65.35	63.50	67.35	61.32	60.91
	0.95	2.77	6.10	2.24	1.69	2.95
P_2O_5	0.19	0.23	0.22	0.14	0.24	0.26
	0.07	0.14	0.07	0.05	0.04	0.05
K_2O	3.79	2.89	2.79	3.83	2.15	2.84
	0.30	0.47	0.74	0.44	0.52	1.10
CaO	3.33	2.38	6.20	3.62	4.69	4.75
	1.35	1.31	2.53	1.47	0.95	0.63
TiO_2	0.57	0.63	0.78	0.44	0.83	0.73
	0.03	0.12	0.33	0.08	0.06	0.05
MnO	0.07	0.06	0.08	0.08	0.11	0.11
	0.02	0.02	0.03	0.03	0.02	0.03
Fe_2O_3	4.60	4.89	5.11	3.89	5.07	6.03
	0.38	0.97	1.97	0.69	0.79	1.11
Sr	236	252	320	207	748	622
	62	155	117	97	98	134

[1]Pottery from these pueblos consisted of a mixture of many types and from many periods.
[2]Pottery consisted of a mixture of Bisquit (2), Tewa Polychrome (2), and Powhage Polychrome (1) wares.
[3]Pottery consisted of a mixture of Glaze A(1) and Glaze C(5) wares.
[4]Pottery consisted of a mixture of Glaze A(1) and Glaze C(2) wares.

from widely varying cultures, traditions, and periods of history, show a greater heterogeneity than the samples from the other pueblos, as expected.

By comparing the magnitude of the variances for each of the oxides or elements within each pueblo with that between the pueblos, those elements which show the greatest ability to discern one pueblo from another can be selected. An analysis of this sort shows Sr is the most useful and K is second, followed closely by Al and Si. The other major elements contribute to a much lesser degree, and the remaining trace elements provide almost no additional separation. Although clustering could be done with just Sr or any subset, it is generally wise to include as much of the data available as possible. This is especially true when the variables are highly correlated, as Sr, K, Al, and Si are in this case, since the increase in information as variables are added to the data matrix is not as great as expected. For the maximum discerning ability a more sophisticated treatment such as principal-component analysis should be employed.

Figure 3 is a cluster plot of the most useful diagnostic variables, Sr and K_2O, for the Tewa Red and Polychrome wares of some pueblos and Spanish settlements (Glazed ware was used for Tonque). Clearly, with just these two variables many of the pueblos can be differentiated, though they cannot be sorted stylistically or by temper. SS 16/T can also readily be divided into two distinct groups, SS 1 and SS 2. Though Sr accomplishes most of the separation, K_2O was significantly lower (95% confidence, t test) in the Tewa Polychrome compared to the Tewa Red ware (3.85% versus 4.05%) for Cuyamungue, where eight samples of each were available and could be clearly identified. This is probably due to differences in the surface slips rather than any difference in the main body, and none of the other elements showed a difference. The variability between the Polychrome and Red ware of a given pueblo is also much less than that between pueblos, so for the cluster analyses these two stylistic types were combined.

The results of a pattern-recognition cluster analysis of selected pueblos are presented in Figure 4. Four major clusters are apparent in this diagram: the Española Basin, with Santa Clara, Sakona, Cuyamungue,

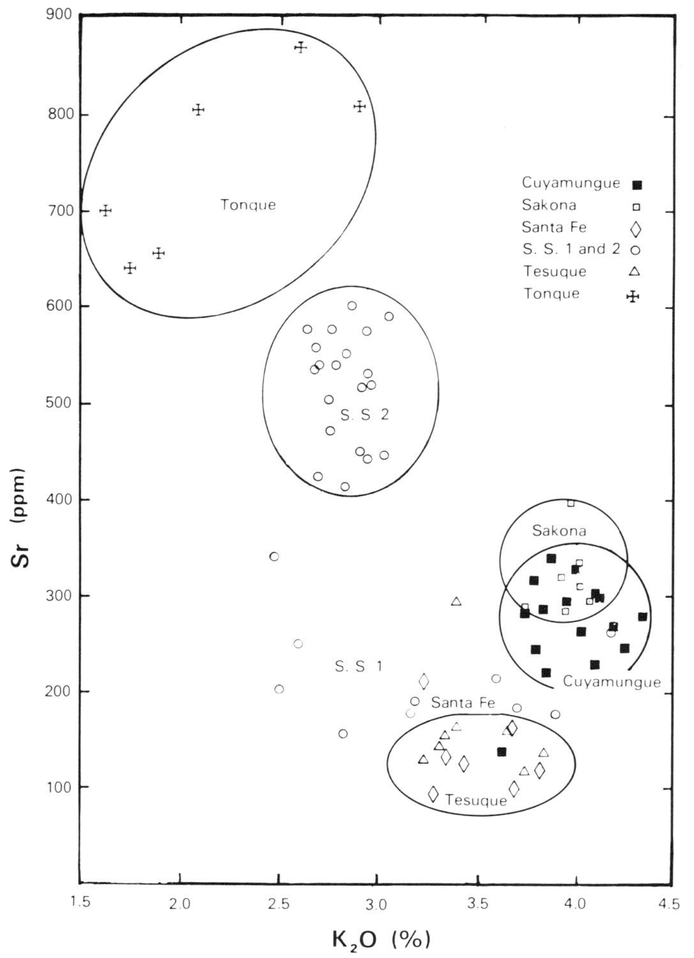

FIGURE 3. Cluster diagram using two variables for comparison of four pueblos and two Spanish settlements.

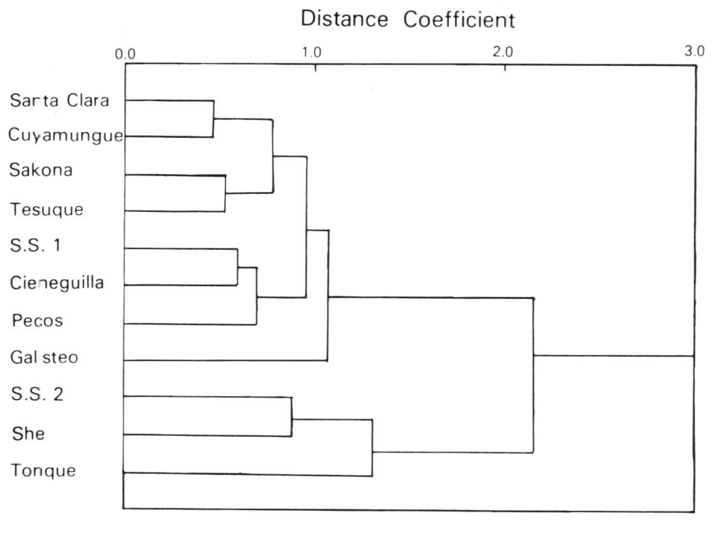

FIGURE 4. Cluster-analysis dendrogram using ten variables to compare eight pueblos and two Spanish settlements (SS 1 and 2, and Cieneguilla).

and Tesuque; the Santa Fe group, including part of the Spanish settlement's pottery, Cieneguilla (which was both a Spanish settlement and a pueblo), and Pecos; the Galisteo Basin north of the Galisteo River; and the region around the Ortiz Mountains south of the Galisteo River.

The samples of Glaze ware from She and Tonque which were analyzed contained hornblende that could be readily identified by XRD (Fig. 5). The hornblende was also found in two-thirds of the Spanish settlement samples. The differences in the Na_2O and Al_2O_3 between SS 2, Tonque, and She could be due, in part, to the large differences between the slipped surface of the Tewa Red and Polychrome ware compared to the glazed surface of the Glazed ware, though actual decorations were avoided. It is also possible, however, that the SS 2 samples came from yet another source somewhere near the Ortiz Mountains.

It is worth noting that the XRD spectra (Fig. 5) are also useful for making distinctions between the pueblos of the Española Basin, clearly pointing out the differences in the varieties of mica and feldspars present. For example, illite has a broader spectral line at 8.8 degrees than muscovite and some of the more crystalline micas, and is present in the samples from Tesuque, but not from Sakona. On the other hand, the samples from Sakona show a broad background due to glassy tuff centered at 24 degrees, which is less apparent in the Tesuque samples and nearly absent in the samples from south of Santa Fe. The feldspars are in different ratios in the samples from below Santa Fe compared to those of the Española Basin. The Spanish settlement group 1 samples could not be differentiated from the Tesuque Pueblo samples by XRD alone. Because XRD does not provide quantitative data with the same degree of precision as XRF, it is less useful for parametric pattern-recognition techniques. However, XRD measures a different physical parameter than XRF, so it can discern differences not apparent with XRF.

By comparing the compositions of the Spanish settlement group 2 pottery with those from She, it is possible to conclude that a common clay or temper source may have been used for their manufacture. The higher concentration of Sr in the She pottery is caused by a single Glaze A piece, pointing out the major problem with an analysis such as this. The sherd may not be of local manufacture, as its Sr content matches those from Tonque. There is little guarantee that all the pottery samples from a given pueblo were locally made. However, with an appropriate cluster analysis the number of distinct sources can be identified, and then the only problem is to match them with the correct pueblos.

If the SS 2 samples came from near She, the presence of a source of "Tewa" pottery around She strengthens the idea that the Tano were related to the northern Tewa. Possibly the SS 2 sherds actually predate the SS 1 sherds, coming from trade which took place prior to the revolt of 1680–1696. After the revolt many of the Tano left the Galisteo Valley, moving first to the Santa Cruz River and then in 1706 to Hano Pueblo located among the Hopi of northeastern Arizona. Possibly trade between the Spanish settlement and the Tano ended at that time and the Spanish obtained their pottery from a more local source, as the pottery labeled SS 1 is very similar to the Santa Fe and Pecos pottery. If this is so, the southern Tewa were involved in the production of "Tewa" ware from its inception.

CONCLUSIONS

XRF and XRD demonstrate that differences can be detected between the various pueblos as production centers even during a period when exchange of ideas made stylistic differences difficult to discern. In addition, the parametric nature of the data obtained from XRF readily lends itself to cluster analysis, helping to make clear the pattern of exchange taking place between the early Spanish settlements and the pueblos. An unidentified center of production, probably located south of Galisteo, was important during the early historic period, presumably being responsible for two-thirds of the red wares unearthed at site 16/T. A search for this site should provide an interesting piece of archaeological detective work, and, if found, could give important new information about trade in the area.

The archaeometric study demontrated that a single element (Sr) may be used as a diagnostic tool for locating this site, though a more complete

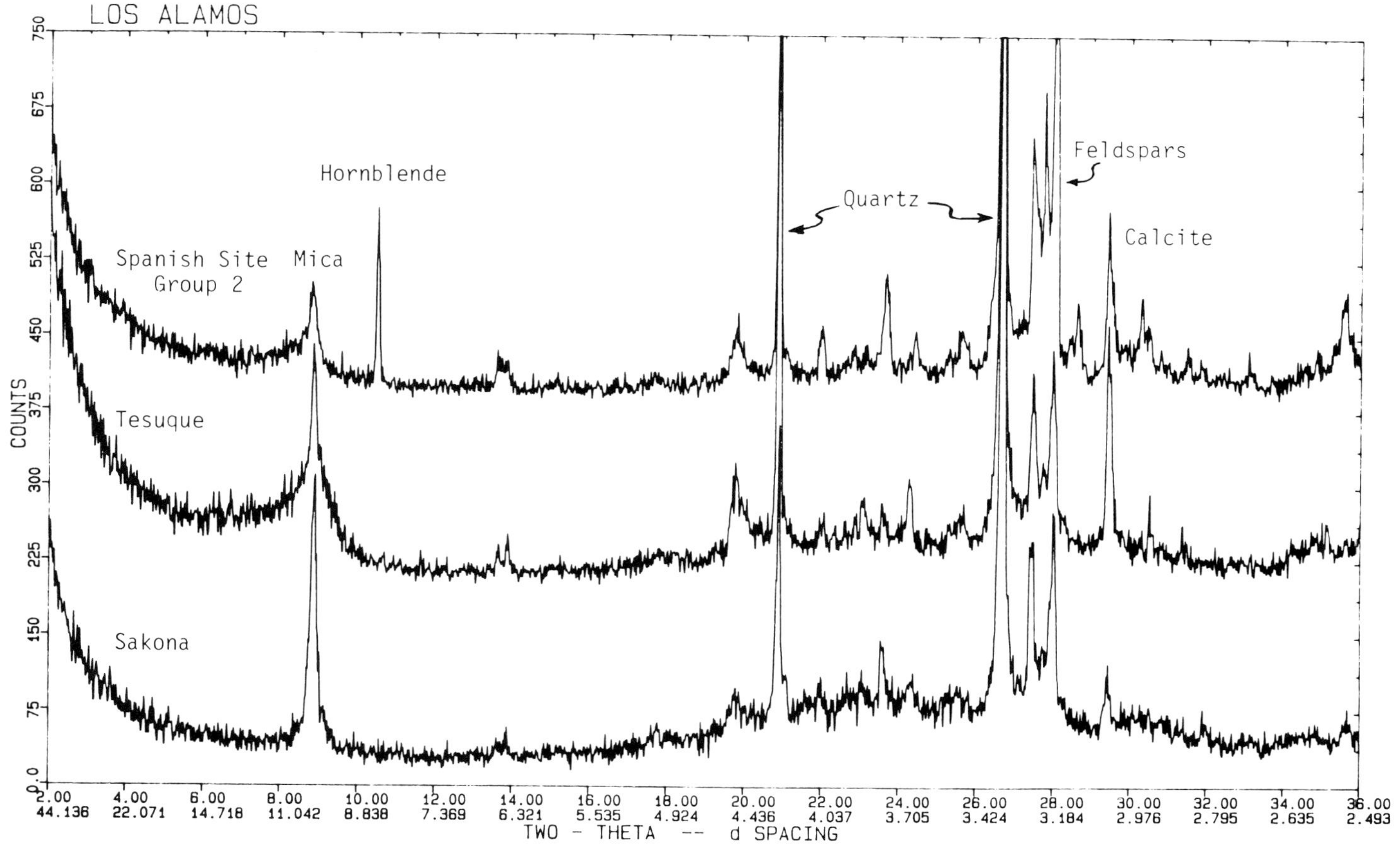

FIGURE 5. XRD spectra of three typical sherds from two pueblos and SS 16/T.

analysis used in conjunction with a petrographic study or XRD analysis for hornblende would be more conclusive. The study also demonstrated that the number of sherds analyzed from a given site need not be large if enough parameters (elements in this case) are measured and if the number of imported sherds at that site is not great. It is important to use as few samples as possible not only as a saving in labor, but also because the analysis is destructive of valuable artifacts. By using cluster analysis on parametric data, even imported pottery can be identified as such and the number and locations of centers of production can be estimated.

ACKNOWLEDGMENTS

This work was in part supported by a grant from Colorado College and the Los Alamos National Laboratory. We would like to thank Stewart Peckham, Museum of New Mexico, for his help with this work by making freely available his time, knowledge, and the museum sherd collection for analysis. Without his patience and generous support the significance of this work would be greatly reduced or non-existent.

REFERENCES

Bower, N. W., Bromund, R. H., and Smith, R. H., 1974, Atomic absorption for the archaeologist: an application to pottery from Pella of the Decapolis: Field Archaeology, v. 2, pp. 389–398.

Bromund, R. H., Bower, N. W., and Smith, R. H., 1976, Inclusions in ancient ceramics: an approach to the problem of sampling for chemical analysis: Archaeometry, v. 18, pp. 218–221.

Davis, J. C., 1973, Statistics and data analysis in geology: John Wiley and Sons, Inc., New York, p. 457.

Dittert, A. E., Jr., and Plog, F., 1980, Generations in clay: Northland Press, Flagstaff, Arizona, pp. 108–134.

Ford, R. I., Schroeder, A. H., and Peckham, S. L., 1972, Three perspectives on Puebloan prehistory; *in* Ortiz, A. (ed.), New perspectives on the pueblos: University of New Mexico Press, Albuquerque, pp. 19–40.

Hall, E. T., Schweitzer, F., and Toller, P. A., 1973, X-ray fluorescence analysis of museum objects: a new instrument: Archaeometry, 15, pt. 1, pp. 53–78.

Harlow, F. H., 1967, Historic Pueblo Indian pottery: The Monitor Press, Los Alamos, pp. 6–11.

Isphording, W. C., 1974, Combined thermal and x-ray diffraction technique for identification of ceramic-ware temper and paste minerals: American Antiquity, v. 39, pp. 477–483.

Snow, D. H., and Fullbright, H. J., 1976, Strontium/rubidium occurrence patterns in Apachean sherds: Pottery Southwest, v. 3, pp. 3–6.

Swanton, J. R., 1952, The Indian tribes of North America: U.S. Government Printing Office, Smithsonian Institution, Bureau of American Ethnology, Bulletin 145, pp. 330–346.

Ward, G. K., 1974, Comparison of source and artifact characterization data using a generalized distance measure: American Antiquity, v. 39, pp. 473–477.

The southbound "mixed train daily" stops at Embudo in July 1941. The stone veneer depot was the work of station agent Henry Wallace (photo by Bob Richardson).

RAILROADS OF THE SAN LUIS VALLEY

ALLEN L. BOWMAN
Los Alamos National Laboratory, Los Alamos, New Mexico 87545

The Denver and Rio Grande Railway (D&RG) pushed its 3-foot-gauge track over La Veta pass in the spring of 1877, reaching the new town of Alamosa on July 10, 1878 (Athearn, 1962). The isolated residents of the San Luis Valley (Fig. 1) could now travel the 130 miles to Pueblo in eight hours, or 250 miles to Denver in only 14 hours (Anderson, 1963, p. 405). More significantly, the convenience of rail transportation quickly made Alamosa the business and mercantile center of the San Luis Valley, a position that it still holds today. The position of Alamosa at end-of-track also made it the terminus for stage lines and wagon freight to the mining districts of the San Juan Mountains, supplanting the previous terminal towns of Garland City (4 miles east of Fort Garland) and La Veta. The railroad moved on to Chama in 1880, to Durango in 1881, and to Silverton in 1882, but the growth of the San Luis Valley was sufficient to support Alamosa, while Garland City became a ghost town and La Veta became a small farm and ranch center (Athearn, 1962).

The Denver and Rio Grande Railway was incorporated on October 27, 1870, by General William Jackson Palmer and several associates. The route of the main line was described as: from Denver via the South Platte, Plum Creek, Monument Creek, and Fontaine qui Bouille to the Arkansas River at or above Pueblo, thence to Cañon City and through the Big Cañon of the Arkansas (Royal Gorge) to the mouth of the South Arkansas, thence via Poncha Pass and the San Luis Valley to the Rio Grande and along it to El Paso in the state of Chihuahua, with authority to connect with and operate a railroad from that point to the city of Mexico (Anderson, 1963). Seven branch lines were also described. This description was fairly typical of railroad incorporation papers of the period, starting with a detailed description of a short line (Denver to the site of present-day Colorado Springs in this case) and then becoming progressively more vague. However, this plan was more specific than many, with a definite goal of El Paso. El Paso in 1870 could mean either the growing town in Texas or El Paso del Norte, Chihuahua, which became Ciudad Juarez in 1888 (Lister, 1966).

General Palmer was a Civil War veteran who went west at the end of the war to seek his fortune (the Rocky Mountain states were largely settled by such men, seeking adventure and opportunity). Palmer had been with the Pennsylvania Railroad before the war as a private secretary to the president, and thus looked to the new transcontinental railroad for employment. He joined the Union Pacific Eastern Division in 1865 as treasurer, and in 1869 became manager of construction for the newly renamed Kansas Pacific Railway, completing the connection from Kansas City to Denver in August 1870. While acting in this capacity, he either personally surveyed or directed the survey by others of most of the early route of the D&RG. Palmer thus used the D&RG to complete some of the unfulfilled plans that he had developed with the Kansas Pacific. The drastic difference was in the adoption by Palmer of 3-foot-gauge track for his new railroad, as contrasted to the 4-foot 8½-inch "standard" gauge. This choice of "narrow gauge" was made to allow for easier and, most importantly, cheaper construction in the Colorado Mountains. It was based at least partially on the experience of the 2-foot-gauge Festiniog Railway of Wales (Anderson, 1963).

The financial arrangements of the D&RG were typical of 19th-century railroad building. Palmer and his associates sold stock in the railroad company, being careful to hold enough stock to maintain control. The company then sold bonds to obtain money for construction. The construction job was then contracted to a separate company, the Union Contract Company, wholly owned by Palmer and his associates. The actual work was then normally done by subcontractors. The construction company operated with an enormous profit margin, and Palmer and his friends soon become millionaires. It is not surprising that the bonded indebtedness of the Railway Company soon outstripped its earning capacity. The first default on bond interest payments occurred in 1877 (Athearn, 1962). Another lucrative source of profit came from the practice of creating new towns on the railroad, on land conveniently owned by Palmer's group. These townsites were then sold as town lots to businessmen wishing to benefit from the new railroad. If a town

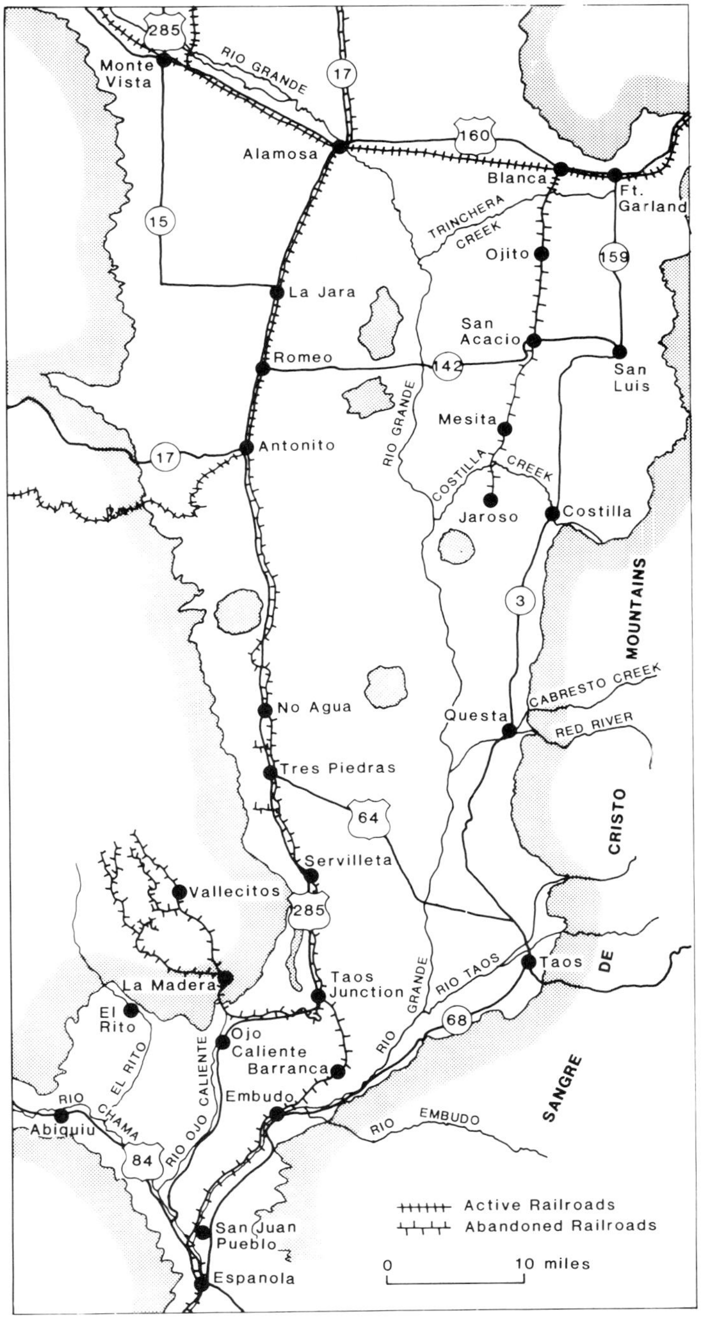

FIGURE 1. Map of active and abandoned railroads of the San Luis Valley.

already existed in the vicinity, it was simply bypassed, and it either declined or disappeared. Alamosa was such a company town, rising at the expense of Del Norte (Elston, 1954), which recovered only partially when a branch line arrived in 1881. Similarly, the railway bypassed Conejos to build the town of Antonito, but Conejos, which was an old Mexican farm village, managed to survive. La Veta and Garland City were also company towns that were not allowed to decline until after the town lots had been sold.

With the arrival of the railroad at Alamosa, on the bank of the Rio Grande, the corporate name of the Denver and Rio Grande had been fulfilled. This was an unusual accomplishment during a period of fanciful railroad names, most of which included "Pacific," even though they were often planned as short feeder lines. But Palmer wanted his railroad to go to El Paso, and he now had two possible routes available, one going south from Alamosa and the other crossing Raton Pass from El Moro (the company town near Trinidad, reached in 1876). When the Santa Fe railroad won the race to occupy Raton Pass in February 1878, however, this route was effectively closed to the D&RG (Athearn, 1962), and Palmer returned to his originally proposed main-line route from Alamosa to Albuquerque. Grading was completed to about 10 miles south of Antonito during the summer and fall of 1878, while the surveyors pushed on, looking for a good route off the Mesa Prieta to the Rio Grande (Chappell, 1981).

The Royal Gorge war with the Santa Fe railroad and the subsequent pressure to build the line up the Arkansas River to Leadville delayed the work south of Alamosa. Construction was started early in 1880 by the Rio Grande Extension Company, a new construction company owned by Palmer and friends. A separate line over Cumbres Pass towards Durango was being graded at the same time. The track was completed to Antonito on March 30, 1880, bypassing Conejos. Meanwhile, the grading contractors were working in separate sections to the south and by June were grading into White Rock Canyon. However, a legal agreement or, alternatively, a peace treaty, had been reached in Boston on March 27 between the D&RG, Santa Fe, and Union Pacific railroads that in effect banned any competitive overlap between the three roads. This has been described as the Tripartite Agreement or as the Treaty of Boston. As a part of this agreement, the D&RG was not permitted to build south into New Mexico more than about 90 miles for a period of 10 years. This ended any hope of a main-line connection to Albuquerque and El Paso. Thus, after a significant delay, all work south of San Juan Pueblo was stopped and the remaining line was ordered to be built as cheaply as possible. This meant taking the most direct route to the Rio Grande at Embudo, with the 4% grade on Barranca Hill, and abandoning a more expensive route to the east with a gentler grade. The track reached the new company town of Española at the end of 1880, and ended there (Chappell, 1981). The Albuquerque extension had been reduced to the Española branch.

The city of Santa Fe had expected to be on the new D&RG route, although it would probably have been on a branch much like the Santa Fe railroad connection. But when the rail line stopped at Española, a 30-odd-mile gap was left to be bridged by Barlow and Sanderson stages. The merchants and businessmen of Santa Fe incorporated the Texas and Santa Fe Northern Railroad in December 1880 to build a connecting road from Santa Fe to San Juan or Santa Cruz, and incidentally to other, more faraway places. After years of delays, the connection with the D&RG at Española was made in January 1887. This short-line railroad was reorganized as the Santa Fe Railroad in 1889, and again as the Rio Grande and Santa Fe Railroad in 1895. Joint operation of through trains between Antonito and Santa Fe was begun in 1893, and the D&RG leased the shorter line from 1896 to 1908, when the short line was made a part of the D&RG (Chappell, 1981). It is interesting to note that the 1893 schedule of the Santa Fe Southern shows a 40-mile trip from Española to Santa Fe (Allen, 1893), while in 1916 the D&RG required only 34 miles to make the trip over the same tracks (Chappell, 1981). The scheduled time was approximately two hours over both distances.

The D&RG rail net in the San Luis Valley slowly grew over the two decades after the completion of the Española branch. A line was built from Mears Junction (on the Marshall Pass line) over Poncha Pass to Villa Grove and Orient in 1881 as a mining spur (Athearn, 1962). The line from Villa Grove to Alamosa was completed in 1890 (Wilkins, 1974), and through traffic from Denver to Santa Fe and the San Juan was then routed via Pueblo and Salida over the new line (Hauck, 1974, pp. 14–18). A mining spur from Moffatt to Cottonwood was added in 1901. The line west from Alamosa was completed to Del Norte and South Fork in 1881 and to Wagon Wheel Gap in 1883 (Wilkins, 1974). The mining boom in Creede brought an extension to North Creede, built by the Rio Grande Gunnison Railway Company in 1891, but operated by the D&RG, with transfer of ownership in 1908 (Hauck, 1979). Meanwhile, conversion to standard gauge had begun with the addition of a third rail between Denver and Pueblo in 1881 (Athearn, 1962). This had become absolutely necessary to facilitate interchange of traffic with the rail lines from the east. The 3-rail or dual-gauge trackage was extended first to Trinidad and then up the Arkansas River to Leadville. As the through trains became standard gauge, the conversion of main lines to standard gauge was speeded up. All track south of Pueblo was converted to standard gauge by 1891, including the line from Cuchara Junction to La Veta. The line over the Culebra Range was relocated in 1899 to a slightly lower pass seven miles south, which was renamed Veta Pass, and the line was made standard gauge to Alamosa. The line from Alamosa to Antonito was converted to dual gauge in 1901, and the line to Creede was converted to standard gauge in 1902 (Wilkins, 1974).

The names of the passes over the Sangre de Cristo Mountains give rise to some confusion. If we follow the names on the Army Map Service map, Trinidad section, we see that the original D&RG narrow-gauge route was over La Veta Pass. The standard-gauge realignment in 1899 used Veta Pass (formerly Wagon Creek Pass), and the old route became first a wagon road and then a paved highway which still exists, although it is badly deteriorated. The highway was relocated in the 1960's to the present route over the North La Veta Pass. The old wagon road over Sangre de Cristo Pass was about two miles further north (Koch, 1980). The D&RG timetables, however, seem to consistently show the narrow-gauge route crossing at Veta Pass, and the present standard-gauge route as La Veta Pass.

The completed D&RG San Luis Valley system thus had a north–south narrow-gauge line from Poncha Pass to Antonito, with extensions to Santa Fe and Durango, and an east–west standard-gauge line from Veta Pass to Creede. Two more short-line, standard-gauge roads were added as farm-produce feeder lines. The San Luis Southern Railway was built from Blanca (5 miles west of Fort Garland) south 31 miles to the New Mexico border in 1910 (Griswold, 1980). The San Luis Central Railroad was built in 1913 from Sugar Junction (2 miles east of Monte Vista) north to Center, a distance of 13 miles (Ormes, 1963). Standard-gauge lines were projected to Durango, over either Weminuche Pass or Wolf Creek Pass, and routes to Taos were projected from the end of the SLS and from points on the "Chili Line" (Athearn, 1962). But the period of great railroad expansion was over, and Taos never got a railroad and Durango never got a connecting standard-gauge line.

The early D&RG passenger service was run as through trains from Denver to Durango with a section for Española split off at Antonito. With the completion of the Valley line in 1890, these trains were rerouted via Salida and Poncha Pass. The 1893 timetable shows the narrow-gauge Colorado and New Mexico Express leaving Denver at 8:00 p.m. and arriving at Española at 2:05 p.m. the next day, and at Santa Fe over the Santa Fe Southern at 4:05 p.m. The alternatives for a Denver to Santa Fe traveler were to go to Trinidad via the Fort Worth and Denver City Railway or to La Junta via the Santa Fe, stop overnight, and take the Santa Fe train the next day. Soon afterward, the D&RG cut their train back to a daily from Salida (Allen, 1893). After the conversion to standard gauge at Veta Pass, the service became an overnight standard-gauge train from Denver to Alamosa, with a connecting narrow-gauge train leaving in the morning for the daytime run to Durango and Santa Fe (Hauck, 1974). The Alamosa to Santa Fe schedule fluctuated between seven and ten hours at different times over the years (Chappell, 1981).

The railroad promoters of the 19th century would typically build a road, making their money from the construction as described above, and then get out, leaving for others the job of trying to operate with a profit. Revenues could be derived from passengers, still a significant factor at the turn of the century, from freight, and from mail service. The mail contracts were an important source of revenue, and also a compelling reason for running daily trains on a regular schedule. The standard argument for building a railroad into an unpopulated region was that the presence of a good transportion system would encourage and support development of the region, with an ensuing rise in business for the railroad. It was also often necessary in the west to traverse empty regions in order to reach good sources of business. The D&RG extension into the San Luis Valley was based on the prospects of reaching the San Juan mining districts, of extending a through line to Albuquerque and El Paso, and finally of benefiting from the development of the valley. The San Juan extension was profitable for several years, but with the repeal of the Sherman Silver Purchase Act in 1893 this business began to decline. The Cumbres Pass crossing also made this line an expensive one to operate in the winter. The diversion of the New Mexico line to Santa Fe reduced its value as a through line, and it soon became an unimportant branch line. Only the development of the San Luis Valley led to a stable, long-term railroad operation, which is reflected in the trackage that remains today.

The "Chili Line," as the route from Antonito to Santa Fe came to be called, traversed a particularly empty region. There were small Mexican farming villages such as Ojo Caliente, La Madera, and Petaca along the Rio Ojo Caliente and its tributaries. These were groups of isolated subsistence farmers, clustered where irrigation was feasible. There was also some farming, both by Pueblo Indians and by Mexicans, in the Española Valley (Swadesh, 1974). The largest farming community in the area was Taos, located some 20 miles off the rail line, and it still had a population of less than 1,000 people in 1912 (Williams and McAllister, 1981). Some sheep were grazed on the Mesa Prieta, but mostly on a small scale. A stage line ran from Santa Fe to the San Luis Valley starting in 1858, following the Rios Ojo Caliente and Tusas (Williams and McAllister, 1981). This stage line was of course eliminated by the coming of the railroad. Taos was connected by stage line, first to the Barlow & Sanderson Overland Stage at Rayado, then to the D&RG at Fort Garland, and then by various routes to different points along the "Chili Line" (Williams and McAllister, 1981). John Dunn ran the Taos stage from 1902 to 1938, first from Servilleta and then from Taos Junction, eventually using automobiles (Evans, 1959; Gjevre, 1969).

The only type of major freight shipment originating on the "Chili Line" was lumber (Chappell, 1971a). R. W. Stewart and Company set up a sawmill at Tres Piedras in the mid-1880's, and a spur line was built up Cañon de Tio Gordito (4 miles south of Tres Piedras) in 1888 to haul logs to the mill. The spur was moved to a point 3 miles north of Tres Piedras in 1890, and lasted for two more years. Both switch points were called Stewart Junction. The Hallack and Howard Lumber Company set up a large sawmill and lumber camp at La Madera in 1914, and paid the D&RG to build a branch line from Taos Junction. This line followed one of the old survey routes for the main line to the Rio Ojo Caliente, then ran upstream to La Madera. The lumber company then built its own logging railroad up into the timber. This line ran first up Cañon La Madera with short spurs into the timber, for a total of 20 to 30 miles. The logging spurs were repeatedly torn up and moved to new cutting areas. The central line was moved in 1922 to the Rio Vallecitos to tap new timber stands near Vallecitos and Cañon Plaza. By 1926, however, the supply of good, cheap timber was gone, and the lumber company moved out in 1927. La Madera reverted to being a sleepy village, and business on the branch line dropped to nearly nothing. The track was removed in 1932.

Although the branch-line track to La Madera was removed over 50 years ago, much of the grade is still identifiable. Looking south from Taos Junction along the main-line grade, the branch line can be seen curving off to the west just south of State Highway 96. It then crosses U.S. Highway 285 twice, swings to the south about 1 1/2 miles, and then curves to the west, crossing the highway three more times as it loops back and forth to avoid steep grades. Some of this portion of the grade can be driven by car, but only with care. The route leaves the highway on Forest Road 97 and then swings west toward the Cañada de los Comanches, crossing the arroyo and meeting State Highway 96 about a mile north of the intersection with U.S. Highway 285. Much of the grade from there to La Madera has been obliterated, but it can probably be found in places. It is fairly easy to recognize the grade when you are on it—usually a narrow, poor-surface dirt road with cuts and fills. The major difficulties in driving on the grade arise at culvert and trestle sites. The timbers are often either rotted out or completely gone. It should be an interesting exercise to look for traces of the lumber-company railroad northwest of La Madera. Logging railroads were usually laid with a minimum of grading and in random directions as they headed for the timber-cutting areas. At least some of the Hallack and Howard Lumber Company grade was used later for wagon roads.

The trains ran over the "Chili Line" for 60 years with minimal difficulties (Gjevre, 1969). Snow was rarely a problem and the grades were gentle, with the exception of the 4% climb from Embudo on the Rio Grande up Comanche Cañon to Barranca. The steam locomotives became bigger and more powerful, and heavier rail was laid to accommodate them. The 90-mile run from Antonito to Española was scheduled to take from four to six hours, with no obvious pattern. The D&RG management obviously decided that the Santa Fe branch was not important, for all the major lines were converted to standard gauge by 1910. After the Hallack and Howard Lumber Company shut down, freight business on the entire line dropped severely. With the company, by then reorganized as the Denver and Rio Grande Western Railroad (D&RGW), in serious financial difficulties (at least partly from the Depression), normal maintenance on the "Chili Line" was curtailed. This policy of "deferred maintenance" was a standard predecessor to abandonment. The line was shut down on September 1, 1941, with scrapping commencing almost immediately. It has been speculated that the work at Los Alamos, starting only one and half year later, would have provided enough business to make the line profitable again. I am inclined to doubt that this would have happened. In my opinion, it would not have been worthwhile to transfer freight from standard to narrow gauge in Santa Fe for the 20-mile run to Otowi and then load it onto trucks for the climb up the hill. It was cheaper and faster to make a single transfer to trucks in Santa Fe. Furthermore, only a limited amount of freight would have moved south from Alamosa over a route that would have required extensive repairs.

The decline in business for the railroads began in the 1910–1920 period with growing competition from autos and trucks. The railroads owned and maintained their right-of-way, and usually paid substantial taxes as well. The county and state governments then used some of this same tax money to build and improve the road system for the competition. The end result was preordained; gradual decline in business and eventual abandonment of the marginal rail lines. The railroads often contributed to their own difficulties with ridiculous rate schedules and indifferent service over poorly maintained track. The lines that survived were mostly the ones with long hauls of bulky freight.

The San Luis Southern Railway (SLS) was incorporated on July 3, 1909 (Griswold, 1980). Construction started at once, and a 31 miles long, standard-gauge line from Blanca to the New Mexico line at Jaroso was completed on November 6, 1910. This road was built by the Costilla Estates Development Company (CED) to further a large land-development scheme. The CED proposed to build reservoirs and then sell a portion of the Sangre de Cristo land grant as irrigated farm land. The railroad was intended to supply transportation and supplies for the farmers and to haul their produce to market. The CED created the towns of San Acacio, Mesita, and Jaroso along the railroad as farm-supply and shipping centers. The Mormon farming community of Eastdale, near Jaroso, had been settled years before. It soon faded and disappeared, as did the older Mexican villages of Old San Acacio and Plaza de los Manzanares. The town of Blanca, founded in 1908, was also the result of another land-development scheme. The railroad was easy to construct, with nine gentle curves, a maximum grade of 0.8%, and eight bridges of which the one over Trinchera Creek, or Rattlesnake Canyon, was the only large structure, 190 feet long and 46 feet high. The straight, level track, coupled with only light snowfall, made operations on the SLS very easy, especially in comparison with most railroads of the Colorado high country.

The SLS started operations bravely with two locomotives, two passenger cars, and 10 freight cars, all purchased second-hand, with the standard short-line schedule of one mixed train daily, running on a two-hour schedule between Blanca and Jaroso (sometimes listed as Jarosa). Land sales were slow, business was slower, and the railroad showed an annual operating loss. Bankruptcy was inevitable under these conditions, and the road was reorganized in 1927 as the San Luis Valley Southern Railway (SLVS). The new company made definite plans for an 18-mile extension to Questa, with an implied continuation to Taos. The railroad was heralded for several years as "The Taos Route," but the Great Depression intervened, shutting off all possibility of financing the Taos extension that had been discussed periodically since the inauguration of the SLS. The SLS had purchased a new Baldwin locomotive in 1919, and bought four antique locomotives from the D&RGW between 1919 and 1947. But the irrigation system for the CED slowly declined, and farming operations decreased rather than showing the expansion that the railroad needed so desperately. The stage line from Jaroso to Taos provided some business, but by 1949 the SLVS was petitioning for abandonment.

Two local businessmen, William McClintock and George Oringdulph, took over the railroad and reorganized it in 1954 as the Southern San Luis Valley Railroad (SSLV). The operation was continued primarily to serve the Mizokami vegetable sheds and the Colorado Aggregate Company (CAC), both located at McClintock, a railroad siding 1 1/2 mile southwest of Blanca. The railroad had started gasoline railbus operation in 1924 for normal daily service, and discontinued regular daily service in 1946. The SSLV then built its own diesel locomotive (named the D-500) from various odds and ends, and soon discontinued all steam operations. The railroad continued to haul scoria from the CAC quarry at Mesita and some occasional farm produce until 1958, when the entire road south of McClintock was abandoned and removed. The remaining 1 1/2-mile section of track remains today as one of the nation's shortest operating railroads, moving loaded freight cars from CAC, and occasionally from Mizokami, over to the D&RGW siding at Blanca. Some of the excess side track is also being utilized as an off-season parking lot for San Luis Central boxcars.

There remains today a modest railroad freight system in the San Luis Valley. The D&RGW maintains the Creede branch, from Veta Pass to Wasson, just short of Creede, with a branch from Alamosa to Antonito. The San Luis Central continues its shipping of farm produce from the region north of Monte Vista, and the SSLV is still functioning at Blanca. But the narrow-gauge operations of the D&RGW are long gone. The Cottonwood branch from Moffatt was abandoned in 1926, and the branch from Villa Grove to Orient went in 1942. The northern part of the Valley Line, from Villa Grove to Hooper, was abandoned in 1950, and the remainder from Hooper to Alamosa was removed in 1959 (Wilkins, 1974). The "Chili Line" south of Antonito was abandoned in 1941, as mentioned before. The passenger train on the San Juan extension (San Juan Express) was discontinued in 1951, and the line from Antonito to Durango and Farmington was abandoned in 1968 (Chappell, 1971b). The three-rail operation between Antonito and Alamosa included mixed-gauge freight trains, which were an astonishing sight for me. The narrow-gauge line from Antonito to Chama was retained as the Cumbres and Toltec Scenic Railroad (CATS), which still is in operation for summer excursions. Many of the grades of the abandoned lines are still visible today, but very little else remains. The buildings are almost all gone, and the ties have been removed for firewood. The water tanks at Embudo and Tres Piedras are still standing, but I doubt if they will hold water. The "Chili Line" stations of Servilleta, Barranca, and Taos Junction have vanished, while the SLS towns have become virtually ghost towns. But there are still enough remnants to show us a little of the extensive railroad structure of a half-century in the past, and the CATS trains are living relics of that period.

ACKNOWLEDGMENTS

The author gratefully acknowledges the assistance of Margaret Bowman and Beverly Gonzales for editing and typing the manuscript; Bob Richardson of the Colorado Railroad Museum for providing pictures and stories of train operations on the "Chili Line" and the Southern San Luis Valley Railroad; and Manuel Griego of La Madera, New Mexico, and Major Oringdulph of Blanca, Colorado, for sharing their personal knowledge of the railroads and general history of the region.

REFERENCES

Allen, W. F. (ed.), 1893, Travelers' official railway guide for the United States and Canada: The National Railway Publication Company, New York, 896 pp. (reprinted in 1972 by Sheridan Printing Company, Alpha, New Jersey).

Anderson, G. L., 1963, Kansas West: Golden West Books, San Marino, California, 269 pp.

Athearn, R. G., 1962, The Denver and Rio Grande Western Railroad: University of Nebraska Press, Lincoln, 395 pp.

Chappell, G. S., 1971a, Logging along the Denver & Rio Grande; narrow gauge logging railroads of southwestern Colorado and northern New Mexico: The Colorado Railroad Museum, Golden, 190 pp.

———, 1971b, Steam revival, a tourist excursion railroad over Cumbres: The Colorado Railroad Museum, Golden, Colorado Rail Annual, no. 9, pp. 114–126.

———, 1981, To Santa Fe by narrow gauge: The D&RG's "Chili Line": The Colorado Railroad Museum, Golden, 56 pp. (originally printed in Colorado Rail Annual, no. 7, 1969).

Elston, A. V., 1954, Wagon wheel gap: J. B. Lippincott Company, New York (Pocket Book edition published in 1955, 181 pp.).

Evans, M., 1959, Long John Dunn of Taos: Westernlore Press, Los Angeles, California, 174 pp.

Gjevre, J. A., 1969, "Chili Line," the narrow rail trail to Santa Fe: The Rio Grande Sun Press, Española, New Mexico, 82 pp.

Griswold, P. R., 1980, Colorado's loneliest railroad, the San Luis Southern: Pruett Publishing Company, Boulder, Colorado, 190 pp.

Hauck, C. W. (ed.), 1974, The Collected Colorado Rail Annual, A Journal of Railroad History in the Rocky Mountain West: The Colorado Railroad Museum, Golden, 136 pp.

———, 1979, CREEDE!: The Colorado Railroad Museum, Golden, Colorado Rail Annual, no. 14, pp. 111–232.

Koch, D. W., 1980, The Colorado pass book, a guide to Colorado's backroad mountain passes: Pruett Publishing Company, Boulder, Colorado, 125 pp.

Lister, F. C., and Lister, R. H., 1966, Chihuahua; storehouse of storms: University of New Mexico Press, Albuquerque, 360 pp.

Ormes, R. M., 1963, Railroads and the Rockies, a record of lines in and near Colorado: Sage Books, Denver, Colorado, 406 pp.

Swadesh, F. L., 1974, Los Primeros Pobladores; Hispanic Americans of the Ute Frontier: University of Notre Dame Press, Notre Dame, Indiana, 262 pp.

Wilkins, T. E., 1974, Colorado railroads; chronological development: Pruett Publishing Company, Boulder, Colorado, 309 pp.

Williams, J. L., and McAllister, P. E. (eds.), 1981, New Mexico in maps: University of New Mexico Press, Albuquerque, 177 pp.

FIELD-CONFERENCE SCHEDULE
AND
ROAD LOGS

1984
FIELD-CONFERENCE SCHEDULE

Wednesday, October 10
Registration Day

3:00–10:00 p.m. Registration: Lobby, Kachina Lodge and Convention Center, Taos, New Mexico.

6:00–9:00 p.m. Cocktail Party and Poster Session: Patio-Pool, Kachina Lodge.

Thursday, October 11
First Day

7:30 a.m. Rendezvous, parking lot at Kachina Lodge.

8:00 a.m. Buses depart for touring the Picuris Range (lunch provided).

5:30 p.m. Arrival at Fort Burgwin Research Center for an "old time" dinner.

8:00 p.m. Return to Kachina Lodge.

Friday, October 12
Second Day

7:00 a.m. Rendezvous, parking lot at Kachina Lodge.

7:30 a.m. Buses depart for tour of northern Sangre de Cristo Mountains to Amalia (lunch provided).

6:00 p.m. Return to Kachina Lodge.

7:00 p.m. Banquet at Kachina Lodge.

8:00 p.m. Speaker: Dr. Harrison (Jack) Schmitt.

Saturday, October 13
Final Day

7:30 a.m. Rendezvous, parking lot at Kachina Lodge.

8:00 a.m. Private-vehicle caravan departs west across Rio Grande gorge and south to Ojo Caliente Mineral Springs (lunch provided).

5:00 p.m. End of 35th New Mexico Geological Society Field Conference.

FIRST-DAY ROAD LOG
FROM TAOS TO EMBUDO, DIXON, PEÑASCO, VADITO, TRES RITOS, HOLMAN HILL, AND BACK TO TAOS VIA U.S. HILL AND TALPA

KIM MANLEY

THURSDAY, OCTOBER 11, 1984

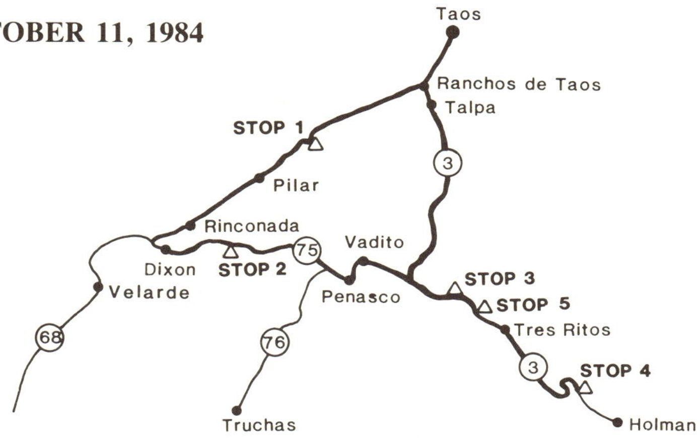

Assembly Point: Kachina Lodge
Departure Time: 8:00 a.m.
Distance: 91 miles
Stops: 5

SUMMARY

Today's route circles the Picuris Range. The Picuris Range is an east–west-trending spur on the west side of the Sangre de Cristo Mountains. It forms part of the division between the San Luis Valley and Española Basin segments of the Rio Grande rift, which is offset right-laterally here. The time of uplift of the Picuris Range is important to rift history; evidence suggests that the uplift took place in the Pliocene. The route follows the Embudo fault zone. Structural complexities of the zone will be viewed and discussed at Stop 1.

After turning eastward into the Dixon sub-basin of the Española Basin, where the Tertiary Santa Fe Group rocks are well exposed, the route progresses into the Precambrian rocks of the southern Picuris Range. A second stop is near the contact of the Vadito and Ortega Groups. The nature of the Precambrian units and their contacts will be discussed.

The route then passes through the type area of the Picuris Formation near Peñasco and makes an excursion into the Paleozoic rocks of the Sangre de Cristo Mountains in the vicinity of Rio Pueblo. The focus will be on paleoenvironmental interpretation of these rocks.

Return to Taos is via U.S. Hill and Ft. Burgwin. This segment of road crosses the saddle between the Picuris Range and the Sangre de Cristo Mountains. Both Tertiary and Pennsylvanian rocks are exposed. The former have not been studied; the latter represent several lithofacies deposited on the eastern margin of the ancestral Uncompahgre uplift.

Mileage

0.0 **Traffic light** at intersection of U.S. 68 and Cervantes Street on west side of Taos. Holy Cross Hospital on right; Taos High School on left. Proceed ahead on U.S. 68. **1.1**

1.1 U.S. 64 to Raton on left. View of Picuris Range. Continue southwest on U.S. 68. **1.3**

2.4 N.M. 3 to Talpa on left. Continue on U.S. 68. **0.4**

2.8 St. Francis of Assisi Church on left. Built in 1772, it is contemporary with the missions in California and an example of Spanish mission architecture. **0.5**

3.3 **Enter Llano Quemado.** Ascend to piedmont alluvial surface of Quaternary gravels deposited northward from the Picuris Range. Overlying fine, brown silts may be equivalent to Española Formation of Galusha and Blick (1971). **1.1**

4.4 N.M. 96 on right to Rio Grande Gorge State Park. **4.3**

8.7 View to right of Rio Grande gorge and Taos Plateau volcanic field (Fig. 1-8.7). **1.8**

FIGURE 1-8.7. View of the Rio Grande gorge west of Taos.

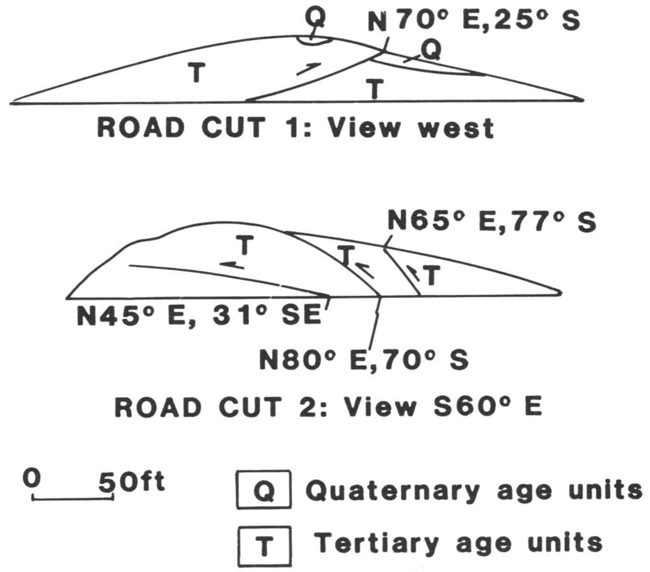

FIGURE 1-11.2a. Sketches of roadcuts at STOP 1 showing low-angle thrust faults along the Embudo fault zone. Low-angle fault in roadcut 2 experienced oblique motion upward to the northeast (modified from Muehlberger, 1979).

10.5 **STOP 1.** Picnic area north of Hondo Canyon. To the north, the Rio Grande gorge has been cut into the tholeiitic-basalt flows of the Servilleta Basalt whose K–Ar ages range from 4.5 to 3.6 m.y. (Ozima and others, 1967). The basalt is coarse-grained and diktytaxitic, with some interstitial glass; dominant minerals are plagioclase, augite, and olivine. To the southwest, the Rio Grande flows primarily in a canyon cut between the Taos Plateau volcanic field and the northwest flank of the Picuris Range.

Walk on left side of highway to roadcut at 11.2. **Watch for traffic.** Discussion of range-front faulting and Embudo fault zone. **0.7**

11.2 Roadcut on left is in Santa Fe Group sediments that have been faulted by a steeply northwest-dipping reverse fault (Fig. 1-11.2a). **0.4**

THE EMBUDO FAULT ZONE

The Embudo fault or fault zone forms the northwestern edge of the Picuris Range from near Rinconada to northeast of Hondo Canyon (Fig. 1-11.2b). Strike is generally N60°E. The fault zone is part of a larger

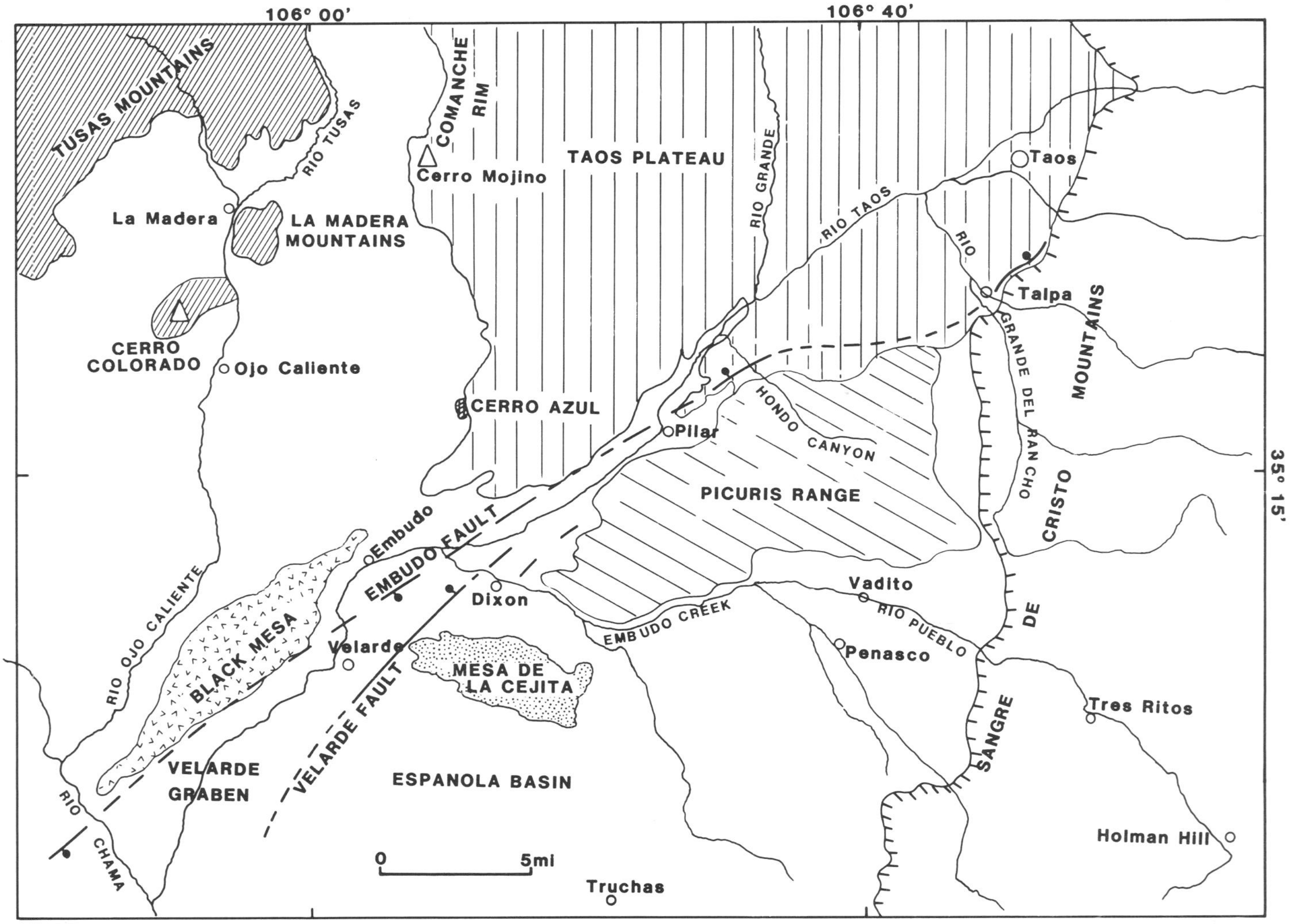

FIGURE 1-11.2b. Map of the Embudo fault zone and Velarde fault. (Data from Manley, 1977; Muehlberger, 1979; Steinpress, 1981; Personius and Machette, 1984).

lineament, the Jemez lineament, that trends northeast from the Springerville, Arizona, volcanic field to the Raton, New Mexico, volcanic field—a distance of roughly 400 mi (Chapin and others, 1978). The lineament appears to cross the Rio Grande rift from the Jemez volcanic field to the Taos area and is coincident with the Embudo fault zone. Along this zone, the Española and San Luis Basins are offset right-laterally; this suggests that the Embudo fault may act as a transform fault, although no significant amount of strike–slip motion has yet been recognized (Muehlberger, 1979).

Muehlberger (1979) describes the faulting in the Embudo zone as dominantly of: (1) north to west strike, averaging 45° southwest dip with reverse-slip motion; or (2) northeast strike with steep dips and left strike–slip motion. The evidence of reverse slip seen in Hondo Canyon contrasts with the northwest-dipping normal faults observed near Pilar and implies that the pivot point is over the Precambrian basement ridge which extends from the Picuris Range to the Tusas Mountains (Cordell, 1976; Muehlberger, 1979).

To the northeast of Hondo Canyon, the Embudo fault is recognized by displacement of Pleistocene deposits along several splits; the fault is inferred to continue across the piedmont slope at the base of the Picuris Range and to curve northward east of Rio Grande del Rancho to join the eastern boundary faults of the Taos Plateau, along which early Holocene or latest Pleistocene motion has been documented (Personius and Machette, this guidebook). This possible controlling effect of the Embudo zone, as a part of the Jemez lineament, on the structures in the Sangre de Cristo Mountains east of Taos is not known.

To the southwest, the Embudo fault appears to bifurcate. The eastern side, called the Velarde fault, strikes N40°E and forms the eastern boundary of the Velarde graben. The western segment continues, with a strike of roughly N60°E, but is hidden beneath the Servilleta Basalt of Black Mesa and the numerous landslide blocks in the area. Its presence is recognized in one location by steeply dipping beds of the Chamita Formation beneath the southeast edge of Black Mesa. This fault continues on the south side of the Chama River as the fault zone in Arroyo de la Presa (Manley, 1981).

The earliest age of motion of the Embudo fault is not known. Muehlberger (1979) concludes that it was active by at least late Miocene because it has offset eolian sandstone similar to the Ojo Caliente Sandstone Member of the Tesuque Formation of Galusha and Blick (1971) that has an estimated age greater than 7 m.y. (Manley, 1978). However, the following facts make this correlation questionable: (1) the presence of younger eolian sandstones within the Velarde graben to the southwest; and (2) the presence of Picuris Precambrian clasts in the conglomerates interbedded with the faulted eolian deposits, but the lack of similar clasts in the Ojo Caliente Sandstone south of the Picuris Range. The earliest activity along the Embudo fault probably is equivalent to the earliest movement along its southwestern extension. That faulting has offset a 9.9-m.y.-old basalt flow in Arroyo de la Presa (Manley, 1981) and has downdropped the Chamita Formation whose ash fission-track age is about 5 m.y. (Manley and Naeser, 1977). The implication is that the initiation of Embudo fault motion was more recent than 4 m.y. ago and that the Picuris Range is correspondingly young. Further study of the sedimentary rocks of the area, especially units currently lumped into the Picuris Formation, may shed light on the answer.

11.6 Cross fault. Santa Fe Group deposits overlain by coarse Quaternary gravel. **0.2**

11.8 Picnic area. **0.2**

12.0 Road crosses Embudo fault again. The Santa Fe Group sediments here contain eolian sandstone, fluvial conglomerate from the Picuris Range, and dark red-brown lacustrine mudstone. The age of these sediments and their relationship to the Santa Fe Group in the Española Basin are not known. **0.4**

12.4 Roadcut on right in Santa Fe Group overlain by Quaternary terrace gravels. Left-lateral strike–slip faults here dip steeply and strike about N15°E. **0.3**

12.7 Roadcut on right in conglomerate and sandstone of Santa Fe Group. **0.1**

12.8 View at 3:00 of Servilleta Basalt flow capping mesa. To the north, as many as four flows are present. **0.8**

13.6 Deeply incised Quaternary arroyo fill on left. South of arroyo are terrace gravels unconformably overlying Santa Fe Group sandstones and siltstones. **1.0**

14.6 Cross Rito Piedra Lumbre. **0.3**

14.9 On right, tilted tan sandstone and gray conglomerate of the Santa Fe Group are overlain by a nearly flat-lying layer of rounded Quaternary gravels derived from the north, a Servilleta Basalt flow, and Pliocene–Pleistocene gravel from the Picuris Range (Fig. 1-14.9). **0.6**

15.5 **Enter Pilar.** N.M. 96 on right to Rio Grande Gorge State Park. The road (N.M. 68) enters the Rio Grande gorge here and follows the river along the contact of the Tertiary and Precambrian rocks. On the north, Servilleta Basalt caps the Santa Fe Group sedimentary rocks and along the edge of the canyon these units are exposed in large landslide blocks.

To the south, the Precambrian rocks of the Picuris Range were mapped as Ortega Quartzite by Montgomery (1953). However, Gresens and Stensrud (1974) describe the lower part (a quartz–muscovite schist) as a metarhyolite and, therefore, a part of the underlying Vadito Group. **1.3**

THE PRECAMBRIAN ROCKS NEAR PILAR

J. A. Grambling
Department of Geology, University of New Mexico, Albuquerque, New Mexico 87131

The metarhyolite in the Vadito Group near Pilar contains abundant, conspicuous quartz-eyes that are thought to be relict phenocrysts. Whereas Ortega Quartzite is generally 98% modal quartz, this schist has only

FIGURE 1-14.9. Outcrop near Pilar shows northwest-dipping Santa Fe Group rocks unconformably overlain by horizontal gravel from the Picuris Range, Servilleta Formation basalt, and Quaternary gravel.

50–75% quartz and 25–50% muscovite. The rhyolite is not only metamorphosed but also severely deformed. Flattening probably exceeds 100%, and textures suggest significant shearing.

A few hundred feet above the highway, the metavolcanics are pink due to the reddish muscovite they contain. This coloration is from ferric iron (up to 6%) and trace manganese in the muscovite. The deep-red prismatic mineral appearing locally in the pink schist is piemontite, a manganese epidote.

Approximately 500 ft above the highway is the Vadito/Ortega contact. The contact is sharp and well exposed. At the contact is a viridine–muscovite–quartz schist about 2 ft thick. (Viridine is a bright-green variety of andalusite containing ferric iron and manganese.) This same thin viridine-schist unit has been noted at the Vadito/Ortega contact in about 30 or 40 localities in the Rio Mora area about 40 km southeast of here. It may be evidence for weathering of the volcanic section prior to deposition of the Ortega Quartzite, or, alternatively, it may represent a manganous chemical sediment deposited prior to the deposition of the Ortega Quartzite. This is the first reported occurrence of viridine schist in the Picuris Range.

Above the viridine schist is Ortega Quartzite. The quartzite here is typical: very pure quartzite with abundant and well-preserved crossbeds. Near the contact some good crossbeds show that the section is not overturned here.

The quartzite looks much less flattened and sheared than do the underlying metavolcanics—especially the viridine schist just beneath the contact. This is probably due to competency differences between units during major deformation, and it may explain the apparent concentration of shearing along the contact and in the uppermost part of the metavolcanic section.

16.8 An exposure of Precambrian rock at river level on the north side of river indicates that the river is now superposed over this spur of more resistant rocks. **0.8**

17.6 To north is an area of numerous springs among landslide blocks. **0.6**

18.2 Another exposure of Precambrian rock on the north side of the river (Fig. 1-18.2). **2.1**

20.3 Roadcuts in Pleistocene debris-flow deposit derived from the Picuris Range to the south. **0.2**

20.5 View ahead of basalt-capped La Mesita. The basalt is part of a 2.8-m.y.-old flow in the Servilleta Basalt and has apparently been displaced by movement along the Velarde fault that extends southward from the west end of the Picuris Range. To the left is the distal end of the Oso surface (Fig. 1-20.5).

On the right (northwest) is Black Mesa. Underlain by the Chamita Formation (Galusha and Blick, 1971), it is capped by a basalt flow of the Servilleta Basalt. **1.4**

21.9 **Enter Rinconada.** Santa Fe Group on left shows upward gradation from volcaniclastic conglomerates to nonvolcanic conglomerates. These deposits are at the west end of the Picuris Range; however, Precambrian rocks are continuous at depth along a northwest-trending ridge that connects with the Precambrian exposures in the southern Tusas Mountains near La Madera (Cordell, 1976). This ridge has a single surface expression at Cerro Azul about 6 mi northwest of here (Fig. 1-11.2b). The structural boundary between the San Luis and Es-

FIGURE 1-20.5. Aerial view of the Oso surface and Sangre de Cristo Mountains.

FIGURE 1-18.2. View northeast along the Rio Grande south of Pilar. In the Precambrian rocks on the right, both the Vadito Group and the Ortega Quartzite are present. The contact is about 500 ft above the highway.

FIGURE 1-22.2. Sandstone cliffs near Rinconada. Age is uncertain but important to the structural history of this area.

FIGURE 1-24.8a. Velarde fault exposed in hillside on right. View north to La Mesita on left and Taos Plateau in distance.

pañola Basins is this basement ridge. Cerro Azul itself is composed of Ortega Quartzite with structural trends, deformational fabrics, and pressure–temperature conditions similar to those of the Ortega Quartzite in the Tusas Mountains and Picuris Range (Kepes, 1984). **0.3**

22.2 Roadcut on right in crossbedded sandstone of uncertain age (Fig. 1-22.2). It is included in the Tesuque Formation (Ojo Caliente Sandstone

FIGURE 1-24.8b. Velarde fault south of La Mesita. View northeast, motion down to the northwest.

Member) by Galusha and Blick (1971), but is preserved within the younger Velarde graben of Manley (1978) and, therefore, is probably younger. **1.6**

23.8 Bridge over Rio Embudo. **Junction** of N.M. 68 with N.M. 75 to left. **Turn left** onto N.M. 75 to Dixon. **0.6**

24.4 Contact of steeply northwest-dipping Santa Fe Group and Precambrian Ortega Formation. **0.4**

24.8 Crossing a major fault zone that forms the eastern side of the Velarde graben. This Velarde fault zone appears to join with another zone to the north of the river to become the Embudo fault zone. The graben formed between 5 and 3 m.y. ago and has as much as 1,000 ft of offset (Manley, 1978). (See Figs. 1-11.2b and 1-24.8a, b.) **0.1**

24.9 View to left of three Quaternary stream-terrace levels along the Rio Embudo. **1.0**

25.9 **Dixon Post Office** on right. **1.2**

27.1 Forest Road 29 on right to Ojo Sarco. Cross Rio Embudo. **2.2**

NORTHEASTERN ESPAÑOLA BASIN

The Picuris Range marks the northeastern boundary of the Española Basin. To the south is the Picuris re-entrant which contains deposits of the Miocene Santa Fe Group and several Pliocene–Pleistocene surface gravels (Fig. 1-27.1a). Topographically, the re-entrant contains a large, fan-shaped plateau, with an apex near Jicarilla Peak, that extends from the Sangre de Cristo Mountains to the Rio Grande. On the south and north, the plateau is bounded by two erosional basins: that of the Rio Quemado and Rio Santa Cruz to the south, and of the Rio Embudo on the north.

Along the northern edge of the plateau are sedimentary rocks assigned to the Picuris Tuff (Miller and others, 1963), the Dixon Member (Steinpress, 1980), the Chama–El Rito and Ojo Caliente Sandstone Members (Galusha and Blick, 1971), and Cejita Member (Manley, 1977, 1979b) of the Tesuque Formation. The Picuris Tuff will be discussed later. The eolian Ojo Caliente Sandstone was so named because of its lithologic and paleontologic similarity to the more extensive deposits of that unit west and southwest of Black Mesa.

The Chama–El Rito is shown by Galusha and Blick (1971) and Steinpress (1980, 1981) to underlie the Ojo Caliente Sandstone. It

FIGURE 1-25.9. Stone building in Dixon, the old "Dixon Cash Store."

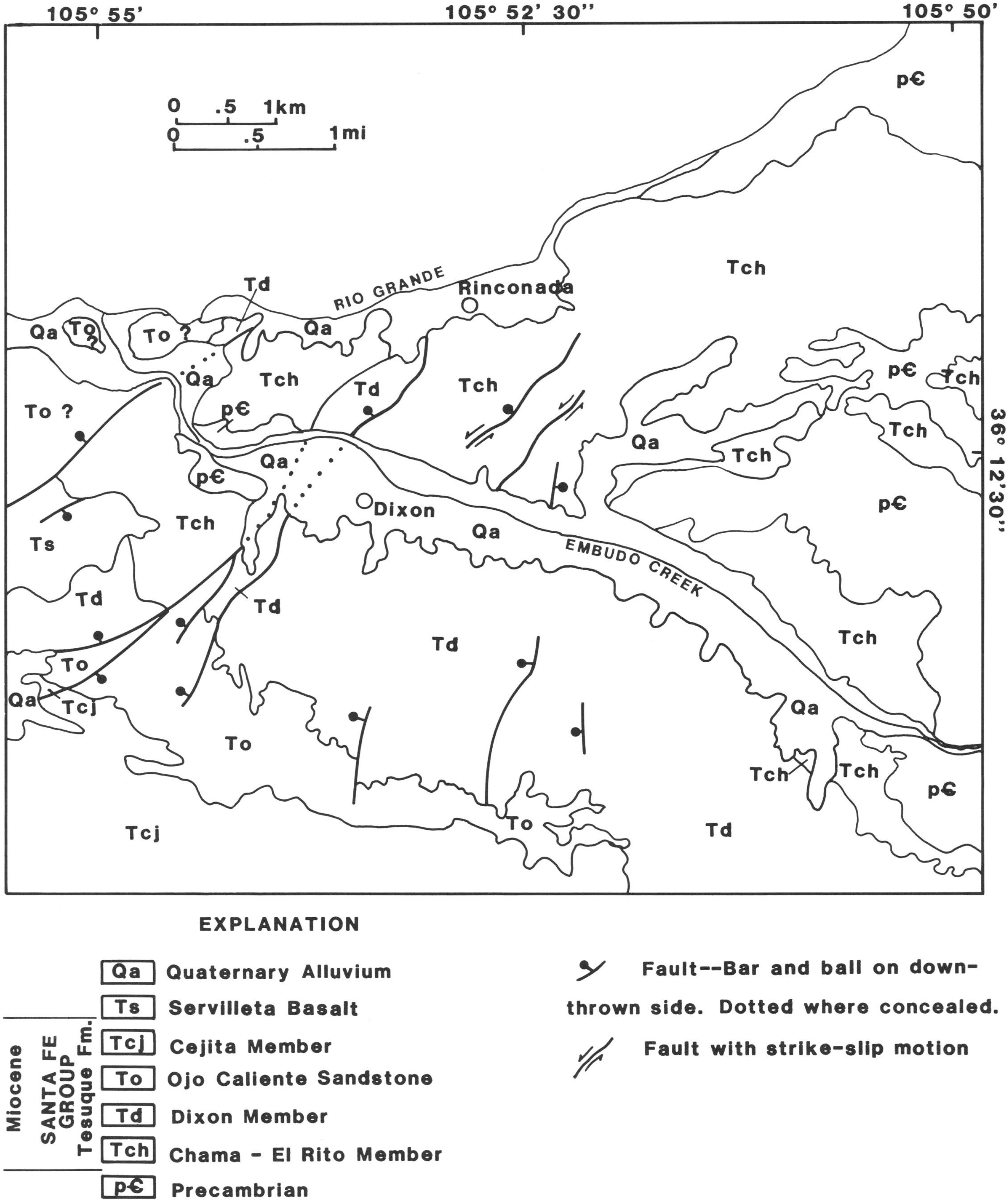

FIGURE 1-27.1a. Geology of the Dixon sub-basin of the Española Basin. (Modified from Steinpress, 1981.)

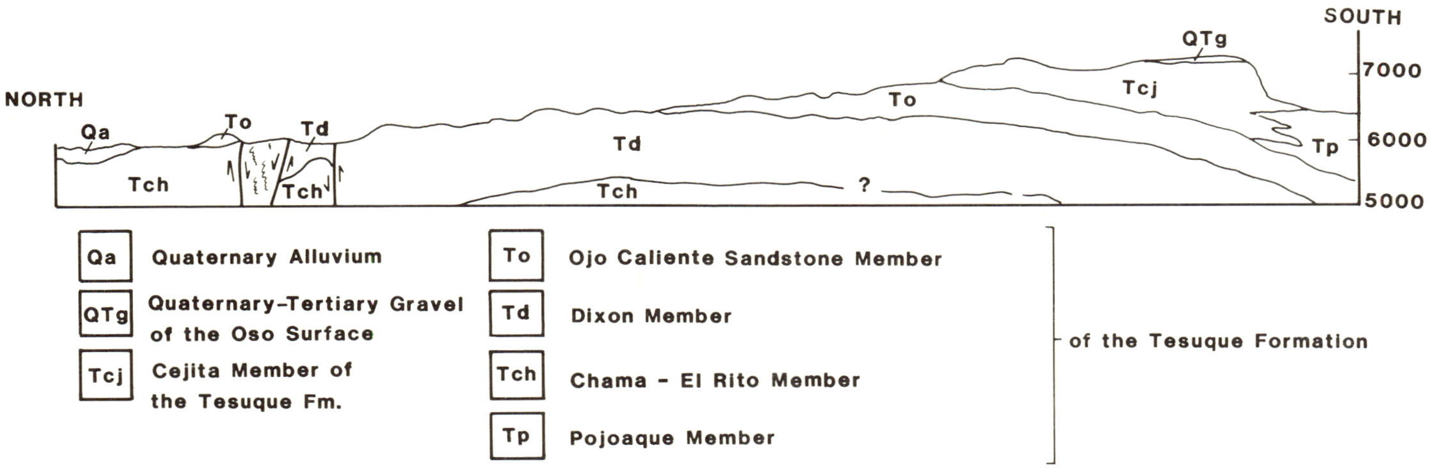

FIGURE 1-27.1b. Diagrammatic north–south cross section of the Dixon sub-basin and Mesa de la Cejita. (Modified from Manley, 1979b; Steinpress, 1980.)

consists of conglomerates and sandstones, and rare air-fall tuffs. Clasts consist of Tertiary volcanic rocks derived, perhaps secondarily, from the early Taos Plateau and Latir volcanic fields, and Precambrian rocks. Paleocurrent measurements indicate a southwesterly flow direction (Steinpress, 1981). A new unit, introduced by Steinpress (1980) as the Dixon Member, overlies the Chama–El Rito and underlies the Ojo Caliente Sandstone. It is similar to the Chama–El Rito, but contains additional clasts of Paleozoic sandstones and limestones as well as higher percentages of Precambrian clasts relative to volcanic ones.

The Cejita Member consists of well-rounded pebble to cobble conglomerate of Paleozoic clasts with interbedded sandstones. It is as much as 120 m thick and commonly forms vertical cliffs around the northern margin of Mesa de la Cejita (Fig. 1-27.1d). The Cejita Member is overlain by Precambrian clast-rich gravels of the Oso erosional surface (Fig. 1-27.1b).

Approximate ages for these units are given in Fig. 1-27.1c. They are inferred from paleontological and radiometric ages available in other parts of the basin. The only nearby age control is on the south side of Mesa de la Cejita, where the Cejita Member interfingers with piedmont deposits of the Pojoaque Member of the Tesuque Formation containing an ash with a fission-track age of 10.8 m.y. (Manley, 1979b).

The Oso erosional surface is the highest of three such surfaces on the northeast plateau. The surface projects (assuming no recent movement on the Velarde fault) to that beneath Black Mesa and is therefore close in age to the 2.8-m.y.-old Servilleta Basalt capping the mesa (Manley, 1979).

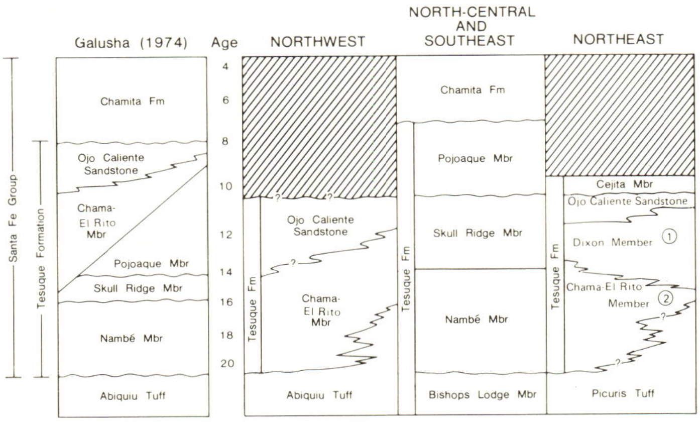

FIGURE 1-27.1c. Nomenclature and age for Santa Fe Group rocks in the Española Basin. (1) Dixon Member from Steinpress (1981). (2) Chama–El Rito Member mapped by Steinpress near Dixon and correlated with deposits in the northeast.

The Santa Fe Group units exposed in this sub-basin record the mid- to late Miocene tectonism of the rift. Source areas changed from northeast to east as the Sangre de Cristo Mountains were uplifted.

North–south-trending faults indicate down-to-the-west motion related to the formation of the Velarde graben; however, they also have a component of left-lateral motion that may have developed in response to a similar movement along the Embudo fault zone (Steinpress, 1981).

FIGURE 1-27.1d. Cliff of Cejita Member of the Tesuque Formation overlying older Santa Fe Group rocks in the southern Dixon sub-basin.

29.3 Contact of Santa Fe Group with Precambrian rocks. **0.6**

29.9 Precambrian Pilar phyllite on left. **1.6**

31.5 Precambrian metaconglomerate of the Vadito Group on left. Foliation strikes about N80°E and dips steeply to the south. Quartz dikes crosscut the formation (Figs. 1-31.5a, b). **0.6**

32.1 **STOP 2.** Road to Harding mine on right. **Turn right and park.** This stop will afford an opportunity to discuss and visit several parts of the Precambrian section. The topics will include the Marquenas Quartzite, the Vadito/Ortega contact, and the Harding mine.

The three major rock units defined by Montgomery (1953) for the Precambrian rocks of the Picuris Range were raised to group status by Long (1976). Until recently, the age relationship of the Vadito and Ortega Groups was in question. Recent work (Grambling and Codding, 1982; Holcombe and Callender, 1982) has presented structural and stratigraphic evidence for the Vadito being the older. Preliminary geochronological work also supports this relationship. Grambling, Williams, and Codding (1983) have determined that a mappable Mn-rich zone, with piemontite and Mn andalusite, exists near the contact of these two groups in the Rio Mora and Truchas areas of the southern Sangre de Cristo Mountains as well as in the Picuris Range, and a Cr-rich horizon is present 165–330 ft below the top of the Ortega Group. The Vadito to Ortega transition defines a shift from dominantly volcanogenic to cratonic deposition. Similar studies are underway in the Tusas Mountains. (See Third Day, Stop 2, mile 41.0.)

Precambrian rocks have undergone three episodes of deformation and similar mid-Proterozoic metamorphism of 4 kb and 525°C (approximately 13 km depth) (Grambling, Williams, and Codding, 1983; Grambling, 1984). The deformation resulted in foliation parallel to compositional layering, spaced cleavage cross-cutting foliation, and a crenulation cleavage (McCarty, 1983).

The Harding mine, developed in a pegmatite that contains large amounts of lepidolite, spodumene, tantalite–columbite, and beryl, was mined from 1919 to 1924, and again from 1927 to 1930, for lepidolite which was used by the ceramic industry. For the latter period of activity a mill was

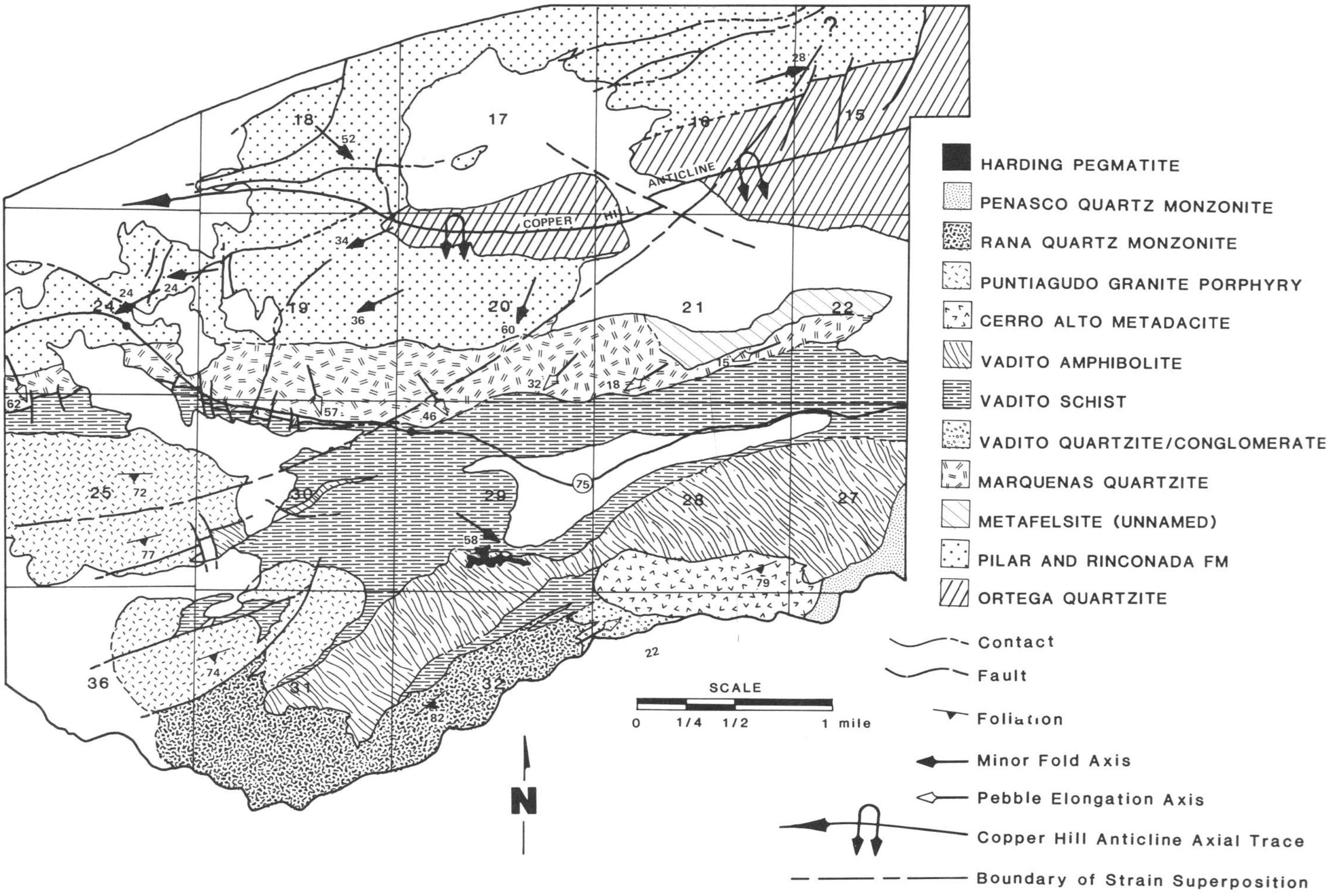

FIGURE 1-29.9. Geologic map of southwest Picuris Range (by K. C. Nielsen).

FIGURE 1-31.5a. Exposure of Precambrian Vadito Group conglomerate cut by quartz dikes.

FIGURE 1-31.5b. Close-up of texture of Vadito Group conglomerate. Broken clast in lower center is about 8 in. in length.

built at Embudo. From 1942 to 1947, microlite (calcium tantalate) was mined; it was processed at several locations until 1945, when a mill was built near the Rio Grande at Rinconada. A minor amount of spodumene and lepidolite was mined during this period also. The last phase of mining occurred between 1950 and 1958 when beryl was extracted.

The Harding pegmatites trend east–west near the boundary between quartz–muscovite schists on the north and amphibolitic rocks on the south. They are part of numerous bodies of pegmatite occurring in the Precambrian rocks of the Picuris Range, but differ from most others by their great thickness and unusual abundances of Be, Li, and Ta–Nb minerals. Radiometric ages from the mine are 1,260 m.y. and 1,350 m.y. using K–Ar and Rb–Sr (Aldrich and others, 1958). The pegmatites appear to be youngest of several intrusions (Long, 1974), although relationships are still unclear. (Data from Jahns and Ewing, 1976.) **1.4**

33.5 View of the Sangre de Cristo Mountains and valley of the Rio Santa Barbara. View ahead on skyline is of Jicarita Peak (Figs. 1-33.5a, b). In middle distance is remnant of valley floor before drainage was captured by stream to the south. Rio Santa Barbara now exits the valley through a gorge in Precambrian rocks onto which the captured stream was superposed.

Two terrace levels parallel the Rio Santa Barbara from the mountain front to the town of Peñasco. The terrace on the northeast, the higher of the two, is approximately 160 ft (50 m) above the valley floor and is underlain by 80 ft (25 m) of gravel. The lower terrace, 80 ft (25 m) above present floodplain, is underlain by 15 ft (5 m) of gravel. Both terrace deposits are predominantly well-rounded cobbles of Paleozoic rocks overlying the more quartzite-rich pebble conglomerates and sandstones of the Tesuque Formation. The terraces are cut into the Tesuque Formation below the elevation of the Oso surface and have more gentle gradients than the Oso surface at equivalent dis-

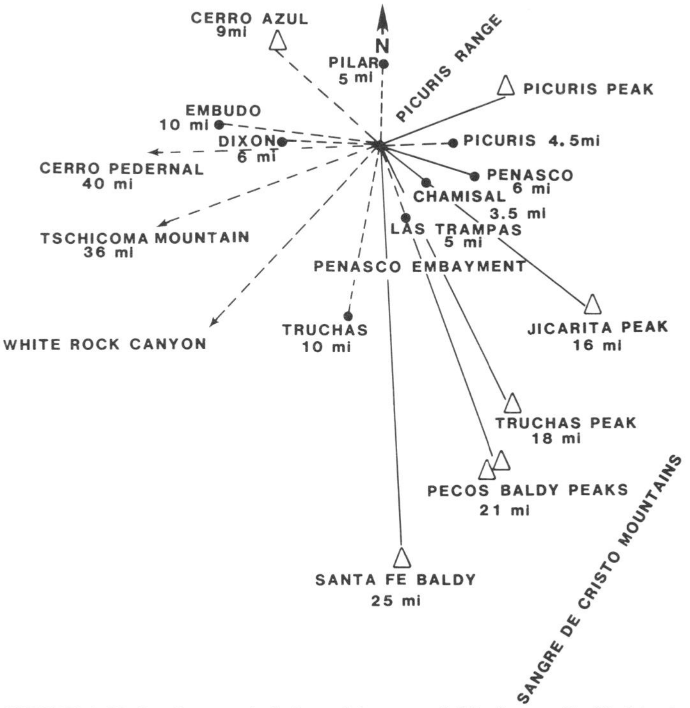

FIGURE 1-33.5a. Panoramic index of features visible from mile 33.5 in the southern Picuris Range.

tances from the mountain front (Manley, 1979b). They are probably outwash terraces of Wisconsin age. **1.6**

35.1 View ahead of the Picuris Range on the skyline; rounded hills of the Picuris Formation(?) in the middle distance. **0.3**

THE PICURIS FORMATION

The Picuris Tuff, described initially by Cabot (1938), has received little study. It is not an ash-flow tuff but rather a sedimentary rock consisting of conglomerates, sandstones (some tuffaceous), and possible air-fall tephra beds. The rocks included in the Picuris vary considerably. Preliminary study indicates at least two formations of greatly different age within the unit.

FIGURE 1-33.5b. View east from southern Picuris Range. Jicarita Peak on skyline; erosional surfaces paralleling the Rio Santa Barbara in middle distance.

FIGURE 1-36.8. View of southeastern Picuris Range with Picuris Formation in middle distance.

The exposures in the middle distance at mileage 35.1 consist of pale-yellow, very fine sand interbedded with pebble to cobble gravels containing clasts of Precambrian quartzites, phyllite, and schist. Toward the top, clasts of intermediate volcanic rocks become increasingly common, with a trace of vesicular basalt and welded rhyolitic tuff.

35.4 Road on left to Picuris Pueblo (west entrance). **0.5**

35.9 On left are well-rounded, coarse, upper Quaternary gravels. **0.9**

36.8 Road on left to Picuris Pueblo (east entrance); see Figure 1-36.8. **0.6**

37.4 N.M. 4 on right to Chamisal. Continue ahead on N.M. 75 to Peñasco. **0.2**

37.6 Left is Precambrian metagranite containing large porphyroblasts of K feldspar. **0.4**

38.0 **Peñasco Post Office** on right. **0.9**

38.9 N.M. 73 on right to Santa Barbara Camp Ground. **0.4**

39.3 Peñasco Ranger Station on left. Hill to right exposes Santa Fe Group. **0.6**

39.9 Large hill in middle distance exposes light-colored Picuris Formation. Picuris here is composed of conglomerates, sandstones, and debris flows. Conglomerate clasts include: Precambrian granite, quartzite, and some phyllite; numerous Tertiary volcanic rocks of intermediate composition; and minor rhyolitic, welded tuff. Debris flows are predominantly pumice (Fig. 1-39.9). **0.8**

40.7 **Enter Vadito. 0.3**

41.0 Bridge over Rio Pueblo. **0.2**

41.2 Forest Road 469 to left up Telephone Canyon to Picuris Peak Lookout. **0.9**

42.1 Roadcuts on left in Picuris Formation. **0.4**

42.5 View ahead on skyline of basanite flow capping Picuris Formation in the saddle between the east end of the Picuris Range and the Sangre de Cristo Mountains (Fig. 1-42.5). The flow is fine-grained and sparsely vesicular; it contains microphenocrysts of altered olivine (iddingsite), corroded feldspar, and euhedral pyroxene in a matrix of

FIGURE 1-39.9. Hills of Picuris Formation near Vadito.

FIGURE 1-42.5. Basanite flow near Vadito. Overlies Picuris Formation; K–Ar age is 5.1 m.y. (Manley, 1976).

olivine, pyroxene, plagioclase, opaque oxides, alkali feldspar, and analcite. Its K–Ar age is 5.1 m.y. (Manley, 1976). **0.3**

42.8 **Enter Placita. 1.4**

44.2 **Junction** of N.M. 3 with N.M. 75. **Turn right** on N.M. 3 to Tres Ritos and Mora. Roadcut ahead exposes tilted section of middle Desmoinesian Pennsylvanian shales and calcareous siltstones overlain by Quaternary(?) gravels. These Pennsylvanian rocks were deposited in shallow, deltaic-shelf areas between fan-deltas as the latter built eastward from the Uncompahgre uplift into the Taos trough.

The Pennsylvanian rocks of the Sangre de Cristo Mountains are more than 6,560 ft thick and span the entirety of Pennsylvanian time. Many depositional environments are represented; marine units are generally highly fossiliferous. Mollusc-dominated fauna is presently under study (Kues, this guidebook).

Subdivision and correlation of Pennsylvanian strata have been difficult because of abundant lateral facies changes. The terminology used in the Sandia Mountains (the Sandia, Madera, and Sangre de Cristo Formations) was replaced in this area by Sutherland (1963), who proposed three different formations: La Pasada, Flechado, and Alamitos (Fig. 1-44.2a). La Pasada, a predominantly carbonate deposit, extends northward to Jicarilla Ridge where it interfingers with the northward-thickening Flechado Formation, a mainly clastic unit that unconformably overlies the Mississippian Tererro Formation. The Alamitos Formation, consisting mostly of arkosic sandstones, overlies both the Flechado and La Pasada Formations.

The complex facies changes in the Pennsylvanian rocks indicate that the area was part of a north–south-trending trough between the Uncompahgre uplift on the west and the Sierra Grande uplift on the east (Fig. 1-44.2b).

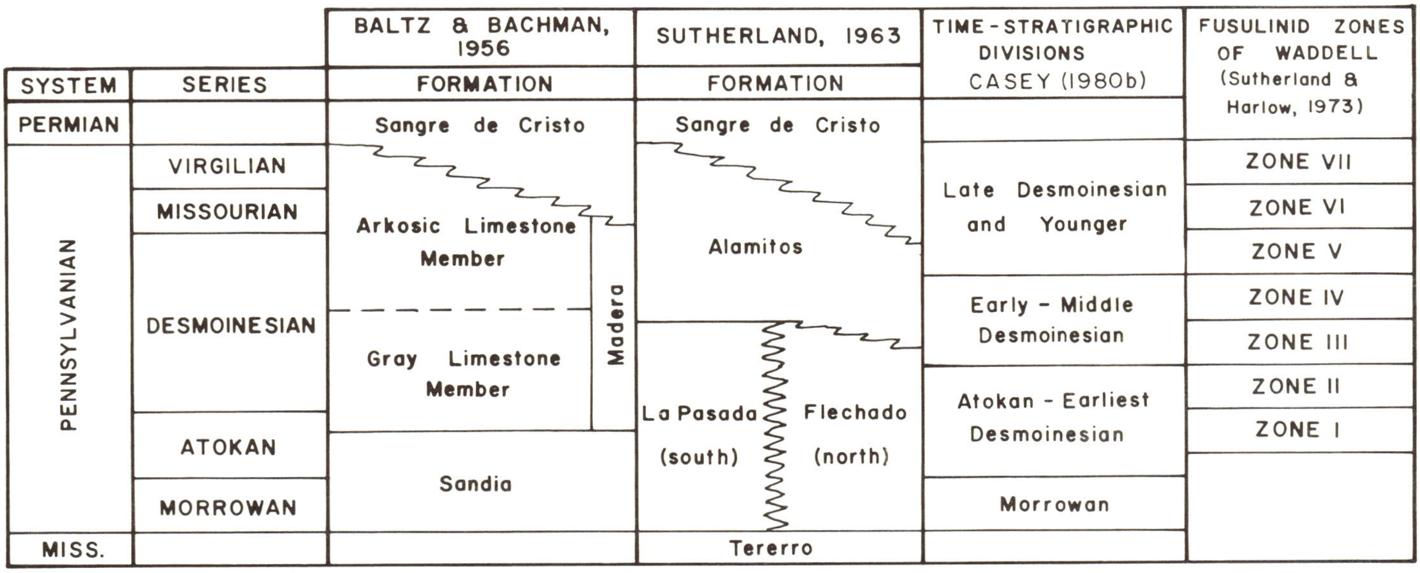

FIGURE 1-44.2a. Stratigraphic nomenclature of the Taos–Las Vegas area and informal time-stratigraphic divisions used by Casey (1980b).

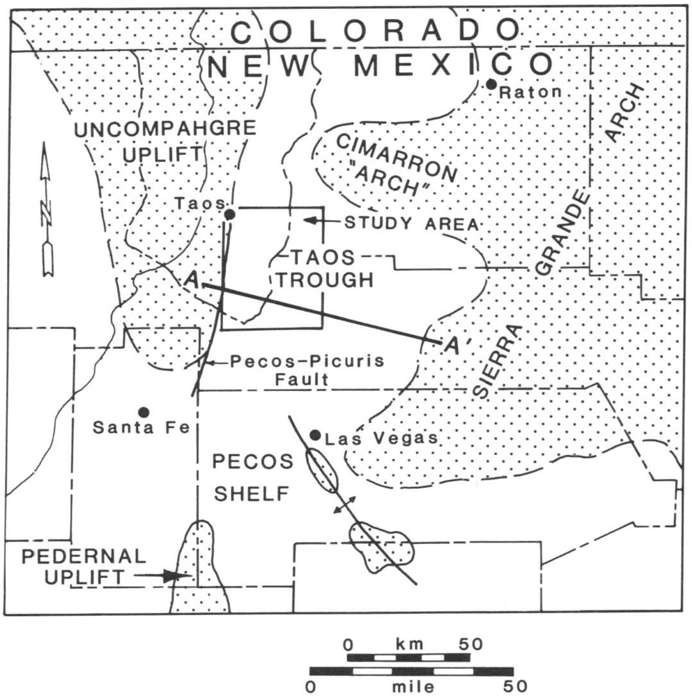

FIGURE 1-44.2b. Pennsylvanian tectonic setting of north-central New Mexico (after Casey, 1980b). Section line A–A′ gives location for Figure 1-44.2c.

Depositional environments included marine, shelf, deltaic, alluvial fan, and stream. The paleogeography changed with the structural evolution of the surrounding uplifts and intervening trough (Fig. 1-44.2c). (Data from Casey and Scott, 1979; Casey, 1980b; and Kues, this guidebook.) **0.3**

44.5 Roadcut on left in folded and fractured limestone (lower member of Madera Formation) which has been downdropped by a covered fault just ahead so that slopes ahead on the left are in younger rocks (Fig. 1-44.5). This outcrop contains abundant fusulinids. **0.7**

45.2 Roadcut in west-dipping Pennsylvanian sandstones. **0.4**

45.6 Forest Trail 20 to Osha Canyon on right. Roadcut in oldest Pennsylvanian rocks (Morrowan) which were deposited as a repetitive sequence of coarsening-upward progradational sediments on a muddy strand plain. Just east of the roadcut is the trace of a fault that places Pennsylvanian rocks against Precambrian quartzite, which is exposed along the crest of a major faulted anticline in the western part of the Sangre de Cristo Mountains. **0.2**

45.8 Comales Camp Ground on right. **0.3**

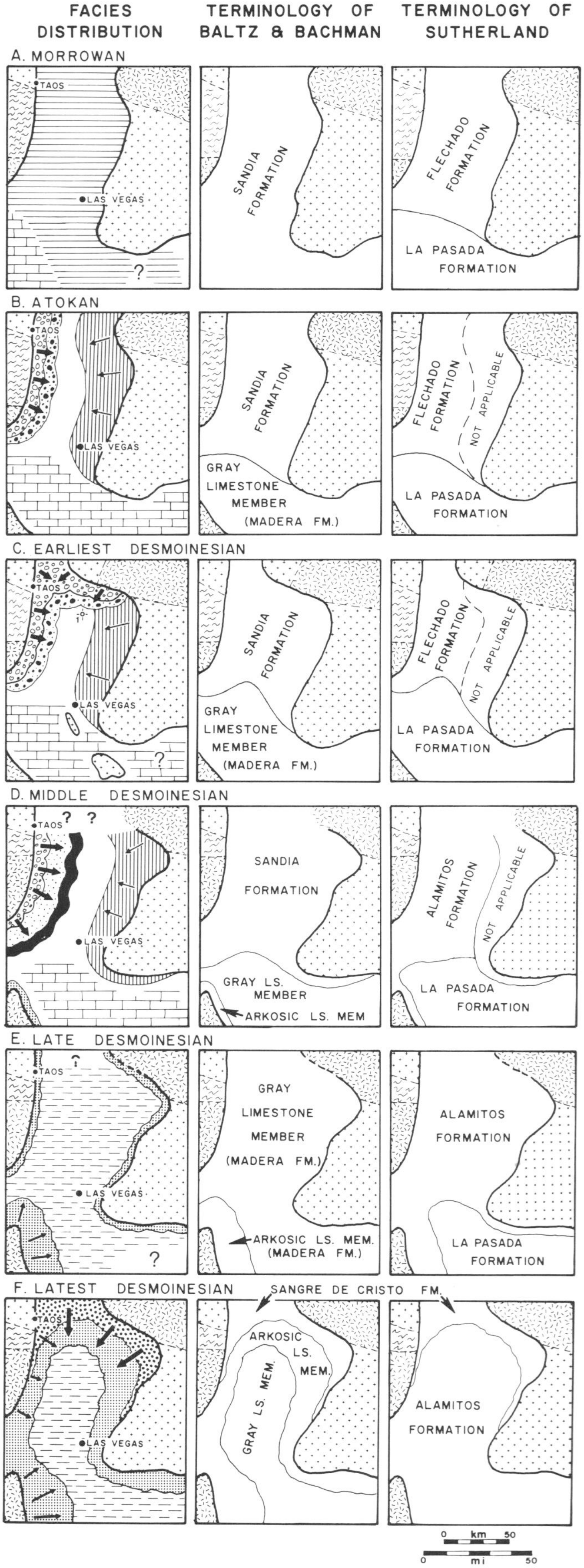

FIGURE 1-44.2c. Spatial distribution of the major depositional facies of the Taos trough and Pecos shelf and their relation to the stratigraphic nomenclature of Baltz and Bachman (1956) and Sutherland (1963). Facies distribution in the western part of the trough based on work by Roberts and others (1976). Facies distribution on the Pecos shelf based on work by Brill (1952), Sidwell and Warn (1953), Sutherland (1963), and Sutherland and Harlow (1973). Well labeled #1 in C is CONOCO No. 1 Mares–Duran. See Fig. 1-59.4 for explanation of symbols (from Casey, 1980a).

FIGURE 1-44.5 Deformation in Pennsylvanian limestone near junction of N.M. 3 and N.M. 75.

46.1 Approximate crest of the anticline. Precambrian quartzite with numerous fractures well exposed along west side of Rio Pueblo for about one mile in vicinity of Llanito Frio Camp Ground (Fig. 1-46.1). **0.4**

46.5 Forest Trail 22 to Comales Canyon on right. **0.6**

47.1 **STOP 3. Lunch.** Cross bridge over Rio Pueblo and park on left. Forest Trail 5 to Canyon Tio Maes. East side of anticline. Exposures include Devonian conglomeratic, quartzitic sandstone and Mississippian limestone overlying Precambrian quartzites.

Prominent cliffs on the hillsides up the valley are laterally persistent, coarse Pennsylvanian sandstones deposited as part of early Desmoinesian fluvial–deltaic complexes prograding into the Taos trough. A progression eastward along these units shows changes from alluvial fans with braided streams to low-sinuosity rivers, to lobate deltas. Between these sandstones, both vertically and laterally, are associated marine facies that progress from shelf to slope to basin deposits in an eastward direction (Casey, 1980a). Route climbs generally up-section from here to Holman Hill as strata dip eastward on flank of anticline.

The Mississippian rocks of the Sangre de Cristo Mountains comprise the Arroyo Peñasco Group which contains two formations: the Espiritu Santo and the Tererro. The Espiritu Santo Formation consists of a basal sandstone deposited over an erosional surface on Precambrian rocks (Del Padre Sandstone Member) and overlying carbonate rocks that represent marine transgression, followed by tidal flat and sabkha conditions. These carbonates have undergone extensive diagenetic alteration (see Ulmer and Laury, this guidebook). The Tererro Formation contains a collapse breccia at its base (Macho Member) that was present as an exposed karst landscape during the subsequent transgression and deposition of conglomerates and mudstones (Turquillo Member). Overlying that are oolitic and pelletoidal deposits (Manuelitas Member), and shales, mudstones, and limestones (Cowles Member). The Mississippian rocks are unconformably overlain by Pennsylvanian rocks. (Data from Armstrong and Mamet, 1979.) **0.1**

FIGURE 1-46.1. Fractured Precambrian quartzite, near Llanito Frio Camp Ground.

47.2 **Enter Las Mochas. 1.0**

48.2 Forest Trail 4 to Gallegos Canyon on left. **0.6**

48.8 Sipapu Ski Area on right. On left ahead, at road level, is one of the early Desmoinesian sandstone units seen from Stop 3. **0.5**

49.3 Forest Trail 7 to Flechado Canyon on left. **0.2**

49.5 Large cliff in Pennsylvanian (Desmoinesian) fan-delta complex. Location of Stop 5. See discussion under mileage 69.2. **1.0**

50.5 Areally persistent, transgressive limestone (lower Madera Formation) on hillside to left is offset 55 m down to the east by a fault. Still visible east of the fault, the limestone shows microfaults and large folds (Fig. 1-50.5). Strata are nearly flat-lying here. **0.6**

FIGURE 1-50.5. Deformation in major marine-transgressive Pennsylvanian limestone (lower Madera Formation near Agua Piedra Camp Ground).

51.1 Agua Piedra Camp Ground on right. **0.3**

51.4 Route continues in sandstones and shales of middle Desmoinesian age representing low-sinuosity, bed-load fluvial and upper delta-plain facies (Fig. 1-51.4). **0.3**

51.7 **Enter Tres Ritos.** Forest Trail 8 to La Cueva Canyon on left. **0.6**

52.3 Forest Road 76 to La Junta on left. Several miles up La Junta Canyon are excellent exposures of the lower Madera Formation. **0.3**

52.6 Exposures are of delta-plain deposits. **1.0**

53.6 Forest Trail 16 to Agua Sarca Canyon on left (Fig. 1-53.6). **1.2**

54.8 Outcrop on left of massive deltaic sandstone overlain by marine shale. **1.1**

55.9 Angostura Camp Ground on right. Exposures here and ahead are in a coarsening-upward marine deltaic sequence showing limestone at base, prodelta black shale, delta-front sandstone overlain by abundant distributary-channel sandstones, and capped locally by a large fluvial-channel sandstone cut into the delta front (Casey, 1980a). **1.5**

57.4 Outcrop on left of crinoidal limestone lens with weathering-accentuated trough crossbeds. **0.2**

57.6 Shales deposited on shallow upper shelf marginal to deltas. **0.4**

58.0 Fault near here cuts out delta sequence. Exposures are of marine calcareous shales and limestones. Phylloid algal mounds are locally present. **0.4**

58.4 Top of Holman Hill. Exposures are in calcareous shales and limestones with local, sandy, crinoidal grainstones—all shelf deposits.

The Permian–Pennsylvanian Sangre de Cristo Formation overlies or interfingers with the Alamitos Formation (or Madera Formation of older literature). It was deposited during the final stages of the Taos trough (Rowe–Mora Basin of some

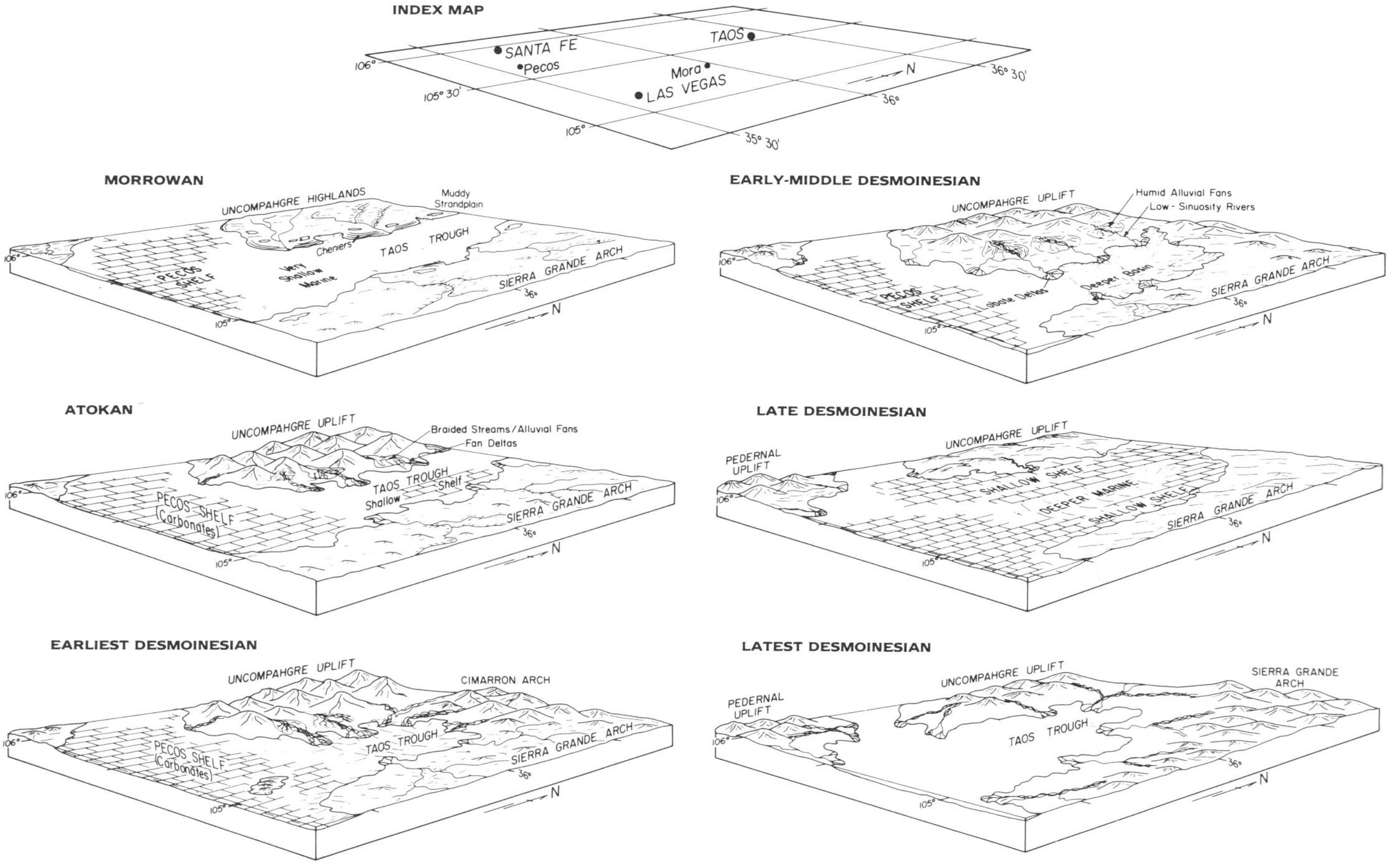

FIGURE 1-51.4. Block diagrams of the paleogeography of the Taos trough area during the Pennsylvanian (from Casey, 1980b).

FIGURE 1-53.6. Old ranch buildings along Rio Pueblo near Agua Sarca Canyon.

authors). Recent paleoenvironmental study (Brown, this guidebook) has divided the formation into three depositional systems in the defunct copper-mining district of Coyote Creek, 11 mi northeast of Holman Hill. Recognition of these systems has led to the construction of a depositional history for the formation. **0.2**

58.6 Black calcareous shales and siltstones of the middle Desmoinesian Taos trough (Fig. 1-58.6). Beds dip 5–20° southwest. **0.8**

59.4 **STOP 4.** This exposure is of calcareous shale and thin sandstones. The latter exhibit structures characteristic of turbidites. The penecontemporaneous folding is interpreted as slump folds forming in semiconsolidated slope deposits (Casey, 1980a). See Figure 1-59.4.

Return west on N.M. 3. **9.8**

69.2 **STOP 5.** Pullout on side road to right. The exposures in this area are of lens-shaped sandstone bodies that were probably deposited by low-sinuousity rivers on the Pennsylvanian (Desmoinesian) coastal plain. The observed sigmoidal cross-bedding indicates side-bar and sheet-splay deposits with fluvial channels migrating laterally in a preferred direction for a long period of time. **5.4**

74.6 **Junction** of N.M. 3 with N.M. 75 on left. **Continue** stright on N.M. 3 to Taos. **0.4**

75.0 Outcrop of Picuris Formation on right. **0.4**

75.4 Coarse conglomerate of unknown origin, probably Pliocene–Pleistocene and related to that seen at mile 74.6. **0.2**

75.6 Flat-topped ridges ahead on both sides of road are capped by the 5.1-m.y.-old basanite flow discussed at mile 42.5. **0.7**

76.3 Forest Trail 4 to Amole Canyon on right. **0.2**

76.5 Outcrop on left of Pennyslvanian sandstone, thin coal, and crinoidal limestone related to the marine transgression and deposition of the lower Madera Formation. **0.5**

FIGURE 1-58.6. View east from Holman Hill.

77.0 Outcrop on left of Pennsylvanian limestone. **1.1**

78.1 Forest Road 114 to Picuris Peak Lookout on left. Forest Road 442 to Gallegos Peak on right. **0.2**

78.3 Top of U.S. Hill. View of the Taos Range of the Sangre de Cristo Mountains to the north, including Wheeler Peak (13,161 ft), highest point in New Mexico. The roadcut immediately north of the overlook exposes a small, coarse-grained deltaic sequence in Pennsylvanian beds that dip gently to the east and southeast. Structural complexities make it difficult to determine the exact stratigraphic position of this and other sequences on the north slope of U.S. Hill; however, this exposure is very similar to the lower Desmoinesian deltaic deposits in nearby areas (Casey and Scott, 1979). **0.1**

78.4 Highly deformed, carbonaceous mudstones and sandstones of the lower Pennsylvanian in roadcuts for next mile. Structural complexity is due to a combination of tectonic deformation and penecontemporaneous, soft-sediment deformation. **2.1**

80.5 Roadcuts on both sides of the road are in Picuris Formation. **1.3**

81.8 Forest Road 476 on left to Vallecitos. **0.4**

82.2 Roadcuts on left expose Pliocene–Pleistocene gravel overlying the Picuris Formation. **0.2**

82.4 Outcrop on left exposes well-cemented boulder to cobble conglomerate of the Picuris Formation. Clasts include Precambrian metamorphic, Pennsylvanian sedimentary, and Tertiary volcanic rocks. **0.1**

82.5 Forest Road 439 to Rio Grande del Rancho on right. **1.1**

83.6 Forest Road 438 to Rito Olla on right. **0.4**

84.0 Fort Burgwin Research Center. Site of restored army post, Cantonment Burgwin, which was used from 1852 to 1860. Fort is now part of an academic campus operated by Southern Methodist University. During the summer months, the center offers courses in a variety of programs including archaeology, anthropology, biology, and geology. It

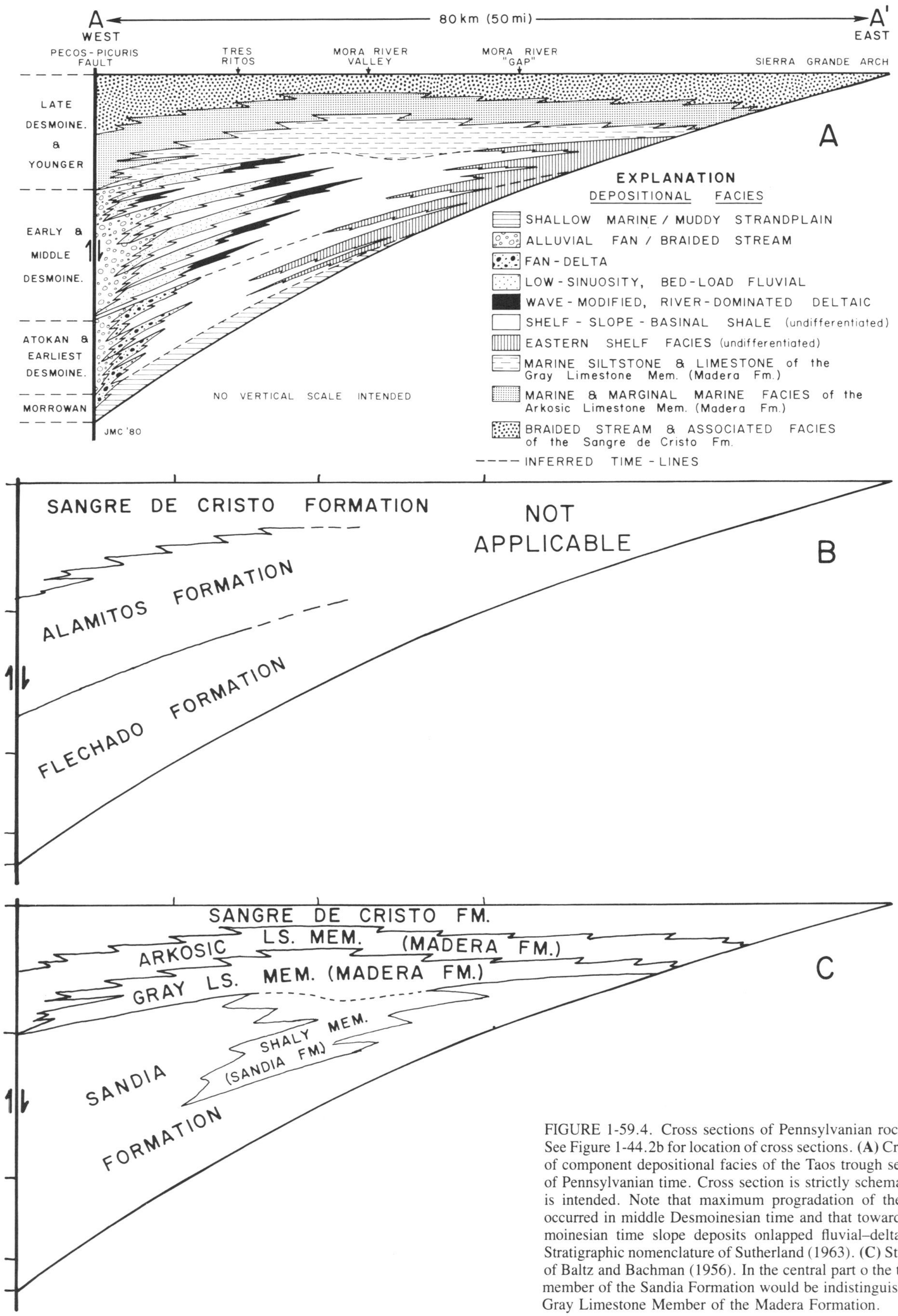

FIGURE 1-59.4. Cross sections of Pennsylvanian rocks (from Casey, 1980a). See Figure 1-44.2b for location of cross sections. (**A**) Cross-sectional distribution of component depositional facies of the Taos trough sedimentary fill at the end of Pennsylvanian time. Cross section is strictly schematic and no vertical scale is intended. Note that maximum progradation of the western clastic wedge occurred in middle Desmoinesian time and that toward the end of middle Desmoinesian time slope deposits onlapped fluvial–deltaic sandstone units. (**B**) Stratigraphic nomenclature of Sutherland (1963). (**C**) Stratigraphic nomenclature of Baltz and Bachman (1956). In the central part o the trough the upper or shaly member of the Sandia Formation would be indistinguishable from the overlying Gray Limestone Member of the Madera Formation.

FIGURE 1-75.2. View westward down Rio Pueblo valley.

also serves as a base for geology field camps for several universities. **0.5**

84.5 Outcrops of Picuris Formation. **1.0**

85.5 Roadcut in Picuris Formation on right. Picnic area on left. **0.2**

85.7 Forest Road 441 to Bear Wallow on right. **0.2**

85.9 Roadcut and hillside expose oblique-dip section of several small Pennsylvanian fan-delta lobes (Casey and Scott, 1979). Limestone bed at top of hill contains early Desmoinesian *Fusulina distenta*.

Fan deltas of this age are characteristically smaller than those that developed in middle Desmoinesian. They are commonly conglomeratic and vertically stacked in thick, overlapping, coarsening-upward sequences that reflect the first major uplift of the Uncompahgre. Large-scale, foreset-bedded sandstones are also common. **0.3**

86.2 Canyon to east exposes Pennsylvanian fan-delta deposits in three dimensions. To west of road, Pennsylvanian strata dip steeply to the east due to complex faulting and folding. High on the ridge, the Devonian(?)–Mississippian units are exposed overlying the Precambrian rocks of the Picuris Range. **1.0**

87.2 Leaving Carson National Forest. **1.2**

88.4 Coarse, partly cemented conglomerates of Tertiary age overlie lower Pennsylvanian rocks. **0.6**

89.0 **Enter Talpa. 1.5**

90.5 **Junction** of N.M. 3 with U.S. 68. **Turn right** onto U.S. 68 to Taos.

End of road log for First Day.

Taos Pueblo, ca 1920. Photo by Burt Harwood.

SECOND-DAY ROAD LOG
FROM TAOS TO HONDO CANYON, QUESTA, RED RIVER, AMALIA, AND BACK TO TAOS

P. W. LIPMAN and J. C. REED, Jr.

FRIDAY, OCTOBER 12, 1984

Assembly Point: Kachina Lodge
Departure Time: 7:30 a.m.
Distance: 144 miles
Stops: 6

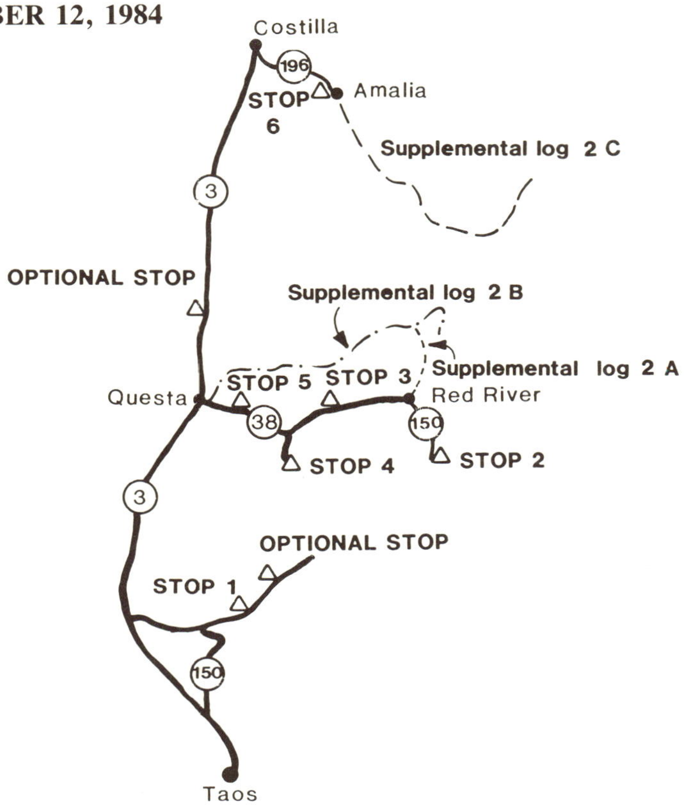

SUMMARY

The Sangre de Cristo Mountains are a narrow and rugged mountain uplift that extends for more than 300 km north and northwestward from Glorieta Pass east of Santa Fe, New Mexico, to Poncha Pass southwest of Salida, Colorado. The range is bounded on the west and southwest by a series of structural depressions that together form part of the Rio Grande rift (Fig. 2-0.0). On the east and northeast, the range is flanked by a number of lower and less continuous uplands and mountain blocks that separate it from the High Plains. The Sangre de Cristo Mountains owe their present relief and topographic configuration largely to Neogene and post-Neogene uplift, but they also follow closely the eastern flank of a now largely foundered Laramide (Late Cretaceous–early Tertiary) uplift (Tweto, 1975, 1979). The present range also lies at or near the eastern flank of a large uplift of late Paleozoic age (Mallory, 1972).

In northern New Mexico, the Sangre de Cristo Mountains are flanked on the west by the San Luis Valley, one of the component depressions of the Rio Grande rift, and on the east by the Raton Basin. The range is split longitudinally by the Moreno Valley and its northern extension, the valley of Comanche Creek, which occupy a complex north–south-trending structural depression probably related to the Rio Grande rift. The parts of the Sangre de Cristo Mountains east of the depression are the Cimarron Mountains; those west of the depression are the Taos Range. The Taos Range includes some of the most spectacular alpine topography in New Mexico, including Wheeler Peak (13,161 ft), the highest peak in the state. In the vicinity of the Questa caldera, the focus of the field trip on Day 2, the Taos Range is composed largely of Precambrian rocks partly mantled by mid-Tertiary volcanic rocks of the Latir volcanic field and invaded by, in part, coeval and cogenetic intrusive rocks related to the caldera (Reed and others, 1983). Topographic relief between the highest summits and the floors of major canyons where they debouch into the San Luis Valley is more than 1 mi; from the peaks to the floor of the Moreno Valley the relief is nearly 5,000 ft. Structural relief on the Precambrian surface between the crest of the range and the deepest part of the rift graben near Questa may be as much as 4–5 mi (Lipman and Mehnert, 1979); that between the range crest and the floor of the Moreno Valley may be 2 mi or more as inferred

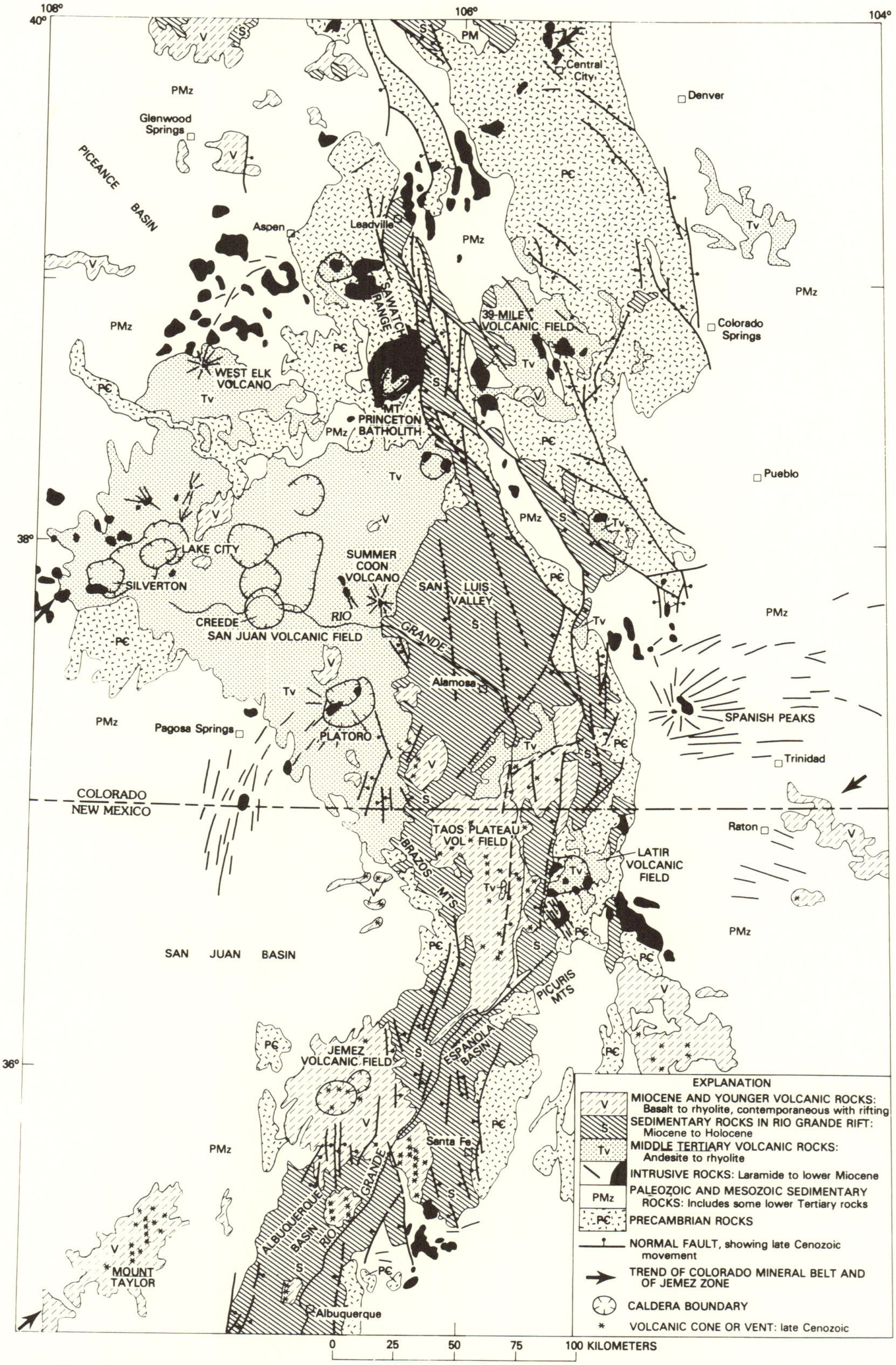

FIGURE 2-0.0. Index map of southern Rocky Mountains, showing location of Questa caldera and associated Latir volcanic field in relation to extent of major Laramide and Tertiary intrusions, Tertiary volcanic rocks, and late Cenozoic extensional faults (modified from Tweto, 1978; Woodward and others, 1978).

from cross sections and stratigraphic thicknesses given by Clarke and Read (1972).

Pre-Tertiary Rocks

Precambrian rocks that form the core of the range comprise supracrustal rocks of volcanic, volcaniclastic, and sedimentary origins, and a variety of mafic, intermediate, and felsic plutonic rocks, all of which have undergone amphibolite-grade regional metamorphism (Condie, 1980; Condie and McCrink, 1982; Reed, this guidebook). Most of these rocks have conspicuous foliations and many have distinct lineations produced by shearing and recrystallization during and after metamorphism. In many rocks shearing and recrystallization have outlasted the peak of regional metamorphism, so that the rocks display retrograde mineral assemblages (albite, epidote, chlorite, calcite) and cataclastic or mylonitic textures. These textures are most obvious in the granitic rocks and are best developed in a 5- to 6-mi-wide, northeast-trending zone centered approximately on Hondo Canyon. It is possible that this belt of cataclastic rocks is a manifestation of the Jemez zone, a northeast-trending zone of crustal weakness of Precambrian ancestry that has been an important control on late Cenozoic volcanism and tectonism (Mayo, 1956; Lipman and Mehnert, 1979).

Preliminary radiometric dates from the Precambrian rocks (Bowring and others, 1984) indicate that the volcanic rocks are about 1.75 b.y. old and that all the intrusive rocks were emplaced within an interval of about 120 m.y., between 1.750 and 1.644 b.y. ago.

Dikes of unsheared but locally altered diabase are widespread, cutting Precambrian rocks south of Red River. Dikes of diabase and hornblende diorite cut Precambrian rocks along Costilla Creek. All of the these dikes postdate Proterozoic regional metamorphism and the diabases south of Red River are truncated by the unconformity at the base of Mississippian strata. They resemble the dikes described by Hansen and Peterman (1971) in the Black Canyon of the Gunnison and dated by them as about 510 m.y. old (Cambrian or Ordovician). Near Tertiary intrusive bodies, pyroxene in the dikes is replaced by hornblende, sphene, leucoxene, and carbonates, and plagioclase is in part altered and sericitized. The alteration of the dikes suggests that some of the retrogressive alteration of the enclosing metamorphic rocks may have been associated with Tertiary igneous activity.

Tertiary Rocks

Structural and topographic relief in the segment of the Taos Range between Questa and the Colorado border provide a remarkable cross section through the recently recognized 23- to 26-m.y.-old Questa caldera and cogenetic volcanic and plutonic rocks (Lipman, 1981, 1983). The precaldera setting was a cluster of dominantly intermediate-composition stratocones (Fig. 2R-15.5a). Although largely eroded, remnants of an ash-flow sheet of silicic alkalic rhyolite (Amalia Tuff) and associated more mafic lavas of the Latir volcanic field are preserved as far as 28 mi beyond the source caldera. Within the caldera, the tuff ponded to a thickness of 1–2 mi and enclosed chaotic megabreccia fragments that slumped from the caldera walls. At the time of ash-flow eruption, the volcanic field was severely deformed along northwest-trending faults related to the Miocene Rio Grande rift, with the most intense deformation occurring within the caldera (Fig. 2R-12.8). Strata were steeply tilted and locally overturned along presumably listric faults (Fig. 2R-15.5b). Ash-flow tuffs locally underwent secondary flowage over concurrently developing fault scarps and accumulated within structural basins as rheoignimbritic lava flows.

Cogenetic batholithic granitic rocks, exposed over an area of 12 × 22 mi, range from mesozonal quartz monzonite to epizonal porphyritic granite and aplite, with the shallower and more silicic phases most abundant within the caldera. Compositionally and texturally distinct granitic phases define a highly evolved resurgent intrusion within the caldera, an incomplete ring dike along its southern margin, and a large mass of less fractionated quartz monzonite south of the caldera (Fig. 2R-0.0b). A negative Bouguer gravity anomaly (Cordell, 1978) is closely confined to the area of exposed granitic rocks; it also reflects boundaries of the Questa caldera (Fig. 2R-11.9). The gravity anomaly is interpreted as defining the extent of the underlying batholith, emplaced into lower parts of the volcanic sequence and underlying Precambrian rocks. Paleomagnetic determinations indicate that the granitic rocks were above Curie temperatures at the time of caldera formation and regional listric faulting (Hagstrum and others, 1982), yet most are little different in radiometric age from the intensely deformed volcanic rocks. These relations indicate that the batholithic complex represents the source magma body for the volcanic rocks and that it was largely liquid at the time of regional tectonic disruption.

Compositions of the volcanic and plutonic phases changed from early calc-alkaline metaluminous rocks to weakly peralkaline silicic rhyolite and equivalent aegerine–riebeckite granite at the time of the caldera formation, then back to postcaldera calc-alkaline granitic rocks. Maximum concentrations of alkalis and minor elements such as rubidium, thorium, uranium, niobium, zirconium, and yttrium were reached at the time the caldera formed about 25–26 m.y. ago, but the major molybdenum mineralization of the Questa district (Carpenter, 1968; Ishihara, 1967) took place when the late calc-alkaline granitic ring intrusions were emplaced about 23 m.y. ago along the south margin of the caldera.

The field-trip guide for Day 2 is organized in three segments that permit examination of progressively higher structural and stratigraphic levels in the Taos Range: 2H, northeast to Rio Hondo to examine Precambrian rocks and deep levels of Tertiary batholithic granitic rocks south of the Questa caldera; 2R, north of Questa and up Red River along the south caldera margin; and 2Q, north from Questa to Amalia to examine the Tertiary volcanic stratigraphy beyond the north rim of the Questa caldera. Three optional additional segments, 2A–2C, not suitable for travel by bus, provide guidance for further examination of the Tertiary and Precambrian rocks within and northeast of the caldera.

SEGMENT 2H: TAOS TO RIO HONDO TO QUESTA

This segment heads northeast from Taos, along the east margin of the Rio Grande rift zone, to the mouth of Hondo Canyon; then it progresses up Rio Hondo to examine Precambrian rocks south of the Questa caldera and the structurally deeper, less-silicic parts of the subvolcanic Miocene batholith.

Mileage

0.0 **Junction** of U.S. 64 and N.M. 3 with road to Taos Pueblo on right (southern route). **0.8**

0.8 Road to Taos Pueblo on right (northern route). **0.3**

1.1 **Enter El Prado. 0.7**

1.8 View to right (east) of large alluvial fans around the base of Pueblo Peak that show Quaternary faulting (see Personius and Machette, this guidebook). **1.7**

3.5 **Junction:** Three-way intersection of N.M. 3, N.M. 150, and U.S. 64. **Turn right** (northeast) onto N.M. 150 to Arroyo Seco and Taos Ski Valley. **2.6**

6.1 N.M. 230 on left to Valdez. **1.9**

8.0 **Enter Arroyo Seco.** Straight ahead center, the rounded rugged mountain is Lucero Peak, with outcrops of Miocene granite (Figs. 2H-8.0a, b). In distance on skyline is Vallecito Mountain (12,641 ft) consisting of Precambrian rocks, largely layered amphibolite and biotite gneisses. Ahead on right is Pueblo Peak (12,303 ft). This is the sacred mountain of the Taos Pueblo; it has not been examined geologically since the work of Gruner (1920). Presumably, it is composed largely of Precambrian rocks, with an unknown volume of Tertiary plutonic rocks and late Paleozoic sedimentary rocks. Just to the south there must be a major structural discontinuity, because the crest of the range for more than 31 mi to the south consists almost entirely of Paleozoic sedimentary rocks.

Between Pueblo Peak and Lucero Peak are drainages of Arroyo Seco and Rio Lucero. Prominent valley north of Lucero Peak is Hondo Canyon, our destination. North of Hondo Canyon, the mountain range consists of a thin roof of Precambrian rocks overlying Tertiary quartz monzonite on the lower hills. Conspicuous peaks visible include Gallinas Peak to the south and Lobo Peak in the center. Flag Mountain, south of Questa, is faintly visible to the north. **1.3**

FIGURE 2H-8.0a. Lucero Peak (10,831 ft), as viewed from Arroyo Seco. Massive outcrops are silicic granite of the Lucero pluton (22 m.y.), the youngest south of the Questa caldera.

FIGURE 2H-8.0b. Silicic granite of the Lucero pluton on Lucero Peak.

9.3 **Junction: Turn right** (east) at "T" intersection. Continue past the Buena Vista Chevron station on the left (north) and descend to the mouth of Hondo Canyon (Fig. 2H-9.3). Sharp ridge ahead is Cuchillo del Medio. **0.7**

10.0 Contact between sheared Precambrian amphibolite and Pleistocene alluvial-fan deposits (Fig. 2H-10.0). This contact shows signs of recent fault movement. **0.1**

10.1 Cross Rio Hondo and enter Hondo Canyon. Outcrops just beyond are layered amphibolite and felsic gneiss cut by a steep, west-dipping shear zone perhaps related to the range-front fault.

Dark rock at west end of first outcrop is a 16-ft-thick, northwest-trending dike that cuts the amphibolite sequence. The dike rock preserves a diabasic texture, but its original pyroxene is now entirely replaced by hornblende and biotite. Presumably, it is one of the swarm of pre-Mississippian dikes in the Precambrian terrane south of the caldera that has been thermally metamorphosed by the Tertiary plutons. **0.2**

10.3 U.S. Geological Survey gauging station on right.

FIGURE 2H-9.3. View east into mouth of Hondo Canyon.

STOP 1. Watch for heavy traffic! Medium- and coarse-grained amphibolite and amphibole gneiss interleaved with fine-grained, well-foliated, gneissic biotite granodiorite. The amphibolite sequence displays complex isoclines with axial planes parallel to foliation which strikes N20°W and dips 35°SW. Lineation plunges 15° S30°W. The granodiorite is apparently intrusive into the amphibolite, but its foliation is parallel to that in the enclosing rocks and most contacts are concordant, showing that the principal deformation postdates intrusion. A preliminary zircon age of 1,699 m.y. (Bowring and others, 1984) suggests that this rock is slightly younger than the quartz monzonite of Columbine Creek and similar in age to granitic rocks exposed a few kilometers to the south and southeast. The amphibolite is presumably of volcanic origin, part of the general supracrustal sequence dated as about 1,750 m.y. (Bowring and others, 1984). **0.4**

FIGURE 2H-10.0. Quaternary gravels overlying Precambrian amphibolite at mouth of Hondo Canyon.

10.7 Network of 1/3- to 4-in.-thick dikes of aplitic rhyolite cutting amphibolite are the harbingers of the enormous Tertiary dike swarm to the east (Fig. 2R-0.0b). **0.2**

10.9 Lower Hondo campground on right (south): approximate location of contact between Precambrian amphibolite and Tertiary quartz monzonite of the Rio Hondo pluton. The contact is not well exposed at road level. **0.7**

11.6 Outcrops of quartz monzonite of the Rio Hondo pluton. Roadcuts for next several miles are in similar rocks. **2.5**

14.1 Taos East Apartments at mouth of South Fork of Rio Hondo on right.

On South Fork Peak, at the head of the canyon, surface exposures of stockwork molybdenum mineralization have been recognized (Luddington,

TABLE 1. Generalized Tertiary igneous and stratigraphic sequence of the Latir volcanic field

[Ages estimated from unpublished K-Ar and fission-track determinations by H. H. Mehnert and C. W. Naeser (written commun., 1981, 1982)]

Unit	*Age (m.y.)*	*SiO_2 content*	*Composition and distinctive features*
Rift-related sedimentary and volcanic rocks			
Santa Fe Group			
Sediments from Precambrian sources	5-15		Silt to fanglomerate
Sediments from volcanic sources	16-22		Mostly conglomerate
Basaltic lava flows	15-16	44-51	Basanite to silicic alkalic basalt
Intrusive rocks of the Questa magmatic system			
Late mineralized plutons			
Intrusions along the Red River	22-23	76-77	Aplite and granite porphyry
Lucero pluton	22	76	Medium grained granite
Evolved intracaldera plutons			
Rito del Medio pluton	26	76	Medium grained granite
Canada Pinabete pluton	26	76	Medium- to fine-grained granite
Virgin Canyon intrusion and related rocks	26	76	Acmite-arfvedsonite granite and rhyolite porphyry
Cabresto Lake pluton	25-26*	72	Biotite-hornblende granite
Southern batholithic rocks			
Rio Hondo plutons	25-26*	64-76	Porphyritic quartz monzonite, locally grading into granite near roof
Amalia Tuff and cogenetic lava flows	26*	77-80	Rhyolitic ash-flow tuff and lava flows; phenccrysts of quartz, chatoyant sanidine
Precaldera volcanic rocks			
Comendite of Ortiz Peak	about 26	72	Forms thick flows and domes; large blocky K-feldspar phenocrysts
Latir Peak Quartz Latite	26-27	65-66	Forms thick flows and domes; phenocrysts of plagioclase, biotite, and hornblende
Andesitic and rhyodacitic flows	26-35	57-64	Variable in composition and texture; from many clustered central volcanos
Volcaniclastic sedimentary rocks	26-35		Mainly mudflow deposits, forming alluvial aprons around andesitic volcanoes
Tuff of Tetilla Peak	27	72	Weakly welded lithic-rich ash-flow tuff; highly variable in thickness
Early Tertiary sedimentary rocks (Vallejo Fm.)	Eocene(?)		Discontinuous post-Laramide sediments; weakly indurated; locally characterized by greenish cobbles of quartzite in reddish matrix

*Estimated age, discordant radiometric ages interpreted as variably reset by later igneous events

1982) and are, as of 1983, under study by industry. Outcrops of Precambrian amphibolite at bend in road on left define a narrow septum between two lobes of the Rio Hondo pluton; the septum is only about 220 yds wide at road level. Areas of weak copper mineralization have been prospected along contacts of the Rio Hondo pluton with Precambrian country rocks. The molybdenum mineralization up the South Fork is associated with late dikes cutting an eastern lobe of granite of Lucero Peak. **0.7**

14.8 Rhyolitic dikes make conspicuous fins on both sides of road; many late dikes cutting quartz monzonite of the Rio Hondo pluton (Fig. 2R-0.0b) are visible on both sides of the road for the next half mile. **0.5**

15.3 Turnoff on left (north) for Italianos Canyon.

OPTIONAL STOP 1. A 0.5-mi hike up the canyon permits examination of dike types, ranging from basalt to rhyolite porphyry, that cut the Rio Hondo pluton (Figs. 2H-15.3a, b). We can also see exceptionally clean exposures of the quartz monzonite along the stream bottom, displaying weak mineral foliation and abundant elongate mafic inclusions. **0.3**

15.6 **Turn around** at stores on left. Retrace route downstream **3.6**

19.2 **OPTIONAL STOP 2.** On left (north) side of road are both typical and mafic phases of the Rio Hondo pluton, mafic schlieren, and mafic inclusions. Note pervasive epidote coating of joints in the Rio Hondo pluton. Radiometric ages, presence of low-grade metamorphic alteration, and the distribution of dikes indicate that the Rio Hondo intrusion is older than the Lucero pluton to the south. **3.8**

23.0 **Junction: Continue straight ahead** (west) on gravel road (paved road makes sharp turn to left to Arroyo Seco). **1.3**

24.3 **Junction:** Join with paved road from right. Immediately after, **bear left** (south) on pavement. To right, along paved ramp that descends northeast to Rio Hondo and village of Valdez, are good exposures of dissected Pleistocene alluvial-fan deposits. **0.6**

24.9 **Turn right** (west) on unnamed paved road. **0.8**

25.7 After descending slope down dissected Pleistocene alluvial fan, enter outskirts of village of Arroyo Hondo. **0.3**

26.0 Sharp turn to left (west) in Arroyo Hondo. **1.0**

27.0 **Junction** of N.M. 3 and N.M. 150 at Rio Hondo bridge. **Turn right** (north) onto N.M. 3; cross Rio Hondo and ascend slope underlain by rhyodacite lava flow from Cerro Negro, a Pliocene volcano of the Taos Plateau field (Lipman and Mehnert, 1979), now largely buried beneath dissected alluvial-fan material.

In the distance behind us is the Picuris Range, extending westward as a prong into the Rio Grande rift. It is composed of stratified metasedimentary rocks of Proterozoic age somewhat resembling those

FIGURE 2H-15.3a. Large dike of porphyritic rhyolite exposed as fin cutting quartz monzonite of Rio Hondo pluton. Rio Hondo road, south of Italianos Canyon.

FIGURE 2H-15.3b. Basaltic dike (Tb), cutting quartz monzonite of Rio Hondo pluton (Tgm). Head of Italianos Canyon.

exposed within the Questa caldera. These mountains define the southern margin of the San Luis Valley, where the Rio Grande rift is offset en echelon to the west into the Española Basin. **2.9**

29.9 On left is outcrop of exceptionally fresh, nearly glassy rhyodacite from Cerro Negro volcano, to the east. **0.4**

30.3 On right is road to D. H. Lawrence Ranch, now owned by the University of New Mexico. **1.4**

31.7 On right is Forest Road 493 to San Cristobal. Road continues through thick, dissected Pleistocene alluvial-fan deposits from the Sangre de Cristo Mountains. **3.2**

34.9 Carson National Forest sign at Garrapata (Spanish for tick) Ridge.

The high ridge of the Sangre de Cristo Mountains to the east consists mostly of Precambrian quartz monzonite, amphibolite, and felsic volcanic rocks between Red River Canyon to the north and Rio Hondo on the south. A narrow belt of quartzite, similar to that more widespread to the north, extends from near the mouth of San Cristobal Canyon to the vicinity of Lobo Peak. The Precambrian rocks are underlain at shallow depth by Tertiary batholithic granitic rocks related to the Latir volcanic field.

Behind us is a good view of the most southerly volcano of the Taos Plateau field: Tres Orejas, meaning "three ears" in Spanish. The center point is a dike within an eroded cinder cone; the other two points mark high points on the rim of the cinder cone. **2.2**

37.1 N.M. 515 on left to Red River Fish Hatchery. This road provides access to the lower Red River Canyon. We are crossing the large, dissected alluvial fan at the mouth of the Red River. The scale of this alluvial fan can be appreciated by the view to the north across the valley. The size of the fan presumably attests to recency of uplift along the mountain front. **2.5**

39.6 Cross Red River. To the west, the low hills are Guadalupe Mountain, a Pliocene shield volcano of rhyodacite, surrounded by lavas of the Servilleta Basalt that make up the bulk of the Taos Plateau. The Servilleta flows are fleetingly visible along the flank of the Red River gorge as we proceed north. In a general way, the Servilleta Basalt constitutes the boundary between Pliocene and older fill of the Rio Grande rift, assigned to the Santa Fe Group, and the younger Pleistocene fan deposits through which we have been descending. **0.5**

40.1 **Enter** village of **Questa. 0.8**

40.9 **Junction** of N.M. 3 and N.M. 38 on right to Red River. **Turn right** for continuation of road log (Segment 2R).

End of Segment 2H of road log for Second Day.

SEGMENT 2R: QUESTA TO RED RIVER

This segment proceeds up the Red River valley from the town of Questa, providing a cross section through the Questa caldera (Fig. 2R-0.0a). The major focus is on structural relations along the south margin of the caldera, including mineralized granitic intrusions and the Molycorp mining operation. Above the town of Red River, the strati-

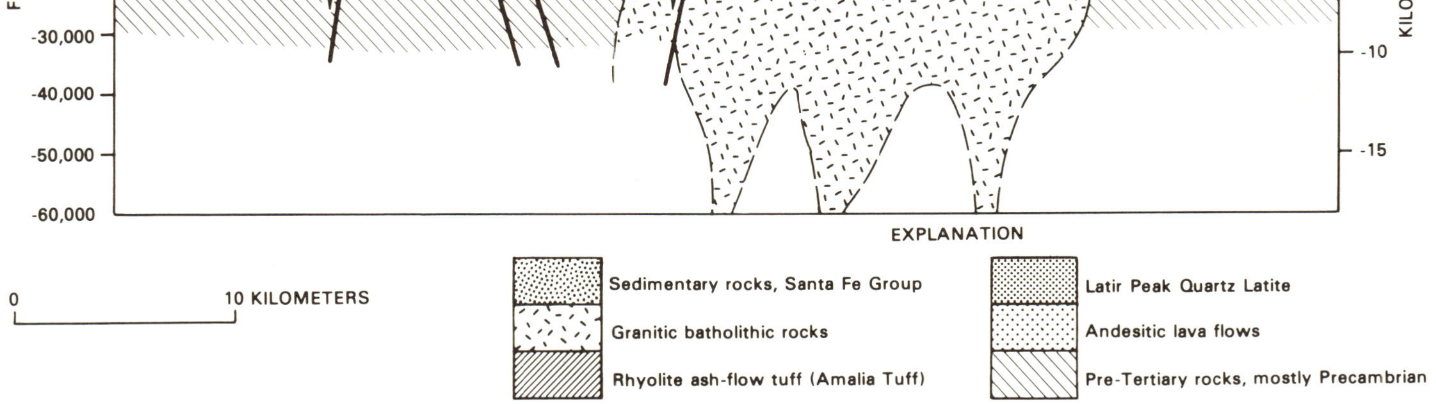

FIGURE 2R-0.0a. East–west cross section through the Questa caldera, showing steep chevron-style tilting of caldera-floor sequence away from caldera walls along listric faults.

FIGURE 2R-0.0b. Generalized distribution of Tertiary intermediate and silicic intrusions associated with the Latir volcanic field (basalt and andesite not shown). Dark stipple pattern – granitic intrusions; diagonal-line pattern (ql) – large masses of quartz latite; dashed pattern (r) – large rhyolite intrusions; thin lines – dikes of rhyolite and quartz latite. Intrusions of Moreno Creek area modified from Clark and Read (1972, pl. 1). Dashed lines indicate margins of mountainous areas; hachured line – topographic wall of the Questa caldera (dotted, where concealed in Rio Grande rift).

graphically coherent volcanic sequence southeast of the caldera can be examined in roadcuts. We will drive directly to the upper Red River site, making additional stops on the return trip.

Mileage

0.0 **Junction** of N.M. 3 and N.M. 38 in center of Questa: **Turn right** (east) on N.M. 38. This area offers a commanding view of the south margin of the Questa caldera, dissected by two major east–west canyons—Red River to the south and Cabresto Creek farther north.

Describing features from south to north: the high mountain at 2:00 is Flag Mountain (11,942 ft), consisting of Precambrian quartz monzonite with 3 ft- to 1/2-mi-sized inclusions of Precambrian felsic-volcanic rocks. The north shoulder of Flag Mountain at 1:00 is occupied by a Tertiary granitic to aplitic intrusion, part of an east–west-trending alignment of ring intrusions along the south margin of the Questa caldera, which extend approximately 50 mi up the Red River (Fig. 2R-0.0b). The conspicuous roads zigzagging up the west face of this intrusion provide access to drill pads that were used in blocking out a large, low-grade molybdenum deposit near the crest of the north ridge of Flag Mountain. This deposit is currently marginally economic and is not being developed because better ore is found upstream to the east.

Looking up the Red River Canyon, the prominent hill with an encircling road near its top is Goat Hill (Fig. 2R-0.0c). Approximately beneath this hill is the site of the new large Molycorp underground mine (Leonardson and others, 1983).

FIGURE 2R-0.0c. View up Red River Canyon from town of Questa. Slopes to left and conical Goat Hill, in middle distance (with road making horizontal line near summit), consist of thick intracaldera Amalia Tuff with lenses of caldera-collapse megabreccia. Dark, heavily timbered slope on right is underlain largely by silicic granite of the Bear Gulch pluton, the westernmost of the mineralized 23-m.y.-old intrusions along the southern margin of the Questa caldera.

Goat Hill consists mostly of volcanic rocks that are part of the caldera fill of the Questa caldera—mostly welded tuff and interlayered large megabreccia blocks of precaldera lava flows that slid in from the wall. At the crest of Goat Hill is an east–northeast-trending dike of altered, coarsely porphyritic rhyolite, which also represents part of the ring-intrusion complex along the south margin of the Questa caldera. In the distance up the Red River Canyon south of the caldera is a shoulder of Gold Hill (11,622 ft) above treeline, consisting of Precambrian volcanic rocks, quartz monzonite, and gabbro.

At 11:00, the high timbered ridge, between Cabresto Creek to the north and Red River to the south, consists entirely of densely welded Amalia Tuff, rotated to nearly vertical (Fig. 2R-15.5b), with an aggregate thickness measured normal to its foliation of at least 1.2 mi. This represents part of the caldera-filling welded tuff. Rotation of this block occurred late during caldera collapse, or during resurgent uplift of the caldera interior, as indicated by unrotated magnetic-pole positions of intracaldera granitic rocks that intrude the tilted welded tuff.

North of Cabresto Creek at 10:00, the high point is Pinabete Peak (11,946 ft) consisting of volcanic rocks of the caldera floor that have been uplifted in the resurgent core. The lower slopes are Precambrian rocks of the floor, and (difficult to see) on the north skyline are granitic rocks of the resurgent pluton within the core of the caldera.

Thus, from south to north we see a cross section through a caldera, including the caldera wall of Precambrian rock, thick welded tuff filling the moat, and the caldera floor rising up on the resurgent dome. **0.4**

0.4 Questa Post Office on left. Straight ahead is Goat Hill; on the right is Bear Gulch stock; on the left is massive ridge between Red River and Cabresto Creek, consisting of mainly vertically dipping intracaldera welded tuff. **1.3**

1.7 Questa District Forest Service Ranger Station on right. Roadcuts on left expose andesitic debris of caldera fill, underlying and interleaved with welded tuff. **0.2**

1.9 Andesite is well exposed in roadcuts on left. **0.2**

2.1 Eagle Rock Campground; cliffs on both sides of road are granite of Bear Canyon pluton. **0.5**

2.6 Pipeline road on left (north) heads into mouth of Capulin Canyon (see **Stop 5**). Continue up Red River road, which generally follows the southern margin of the Questa caldera. At this point we are just inside the topographic wall of the caldera, and we gradually angle from intracaldera volcanic fill out into Precambrian wall rocks in the next few miles. **0.7**

FIGURE 2R-3.5. Open-pit mine of Union Molycorp Questa operation (currently inactive); actual pit in the former site of Sulphur Gulch is barely visible in upper right. View is to northwest, from south of the Red River. The Molycorp mill and crusher are in left center, and the large waste dumps in upper left are covering the southern margin of the Questa caldera, with Precambrian rocks to the left of the Red River Canyon, and silicic granitic rocks of the Moly Mine pluton (23 m.y.) to the right.

FIGURE 2R-5.8. Contact between granite of Moly Mine pluton and andesitic country rocks within the Questa caldera as seen in the Union Molycorp open pit.

3.3 Roadcuts of altered andesite along tailings pipeline; see **Optional Stop** on return trip. **0.2**

3.5 Turnoff to Molycorp offices on left (Fig. 2R-3.5); Goat Hill Campground is on right (continue straight). High cliffs at 4:00 are Bear Canyon pluton; on left are intermixed volcanic debris and welded tuff of intracaldera fill. **0.4**

3.9 Good exposures of altered, chaotic andesite of caldera fill on right (south) across river. **0.6**

4.5 Headframe for access shaft of Goat Hill mine visible on left (north). **0.7**

5.2 Cottonwood Park on left; Columbine Canyon Campground on right (see **Stop 4**). **0.4**

5.6 Roadcut on left is Precambrian quartz monzonite cut by Tertiary dike swarm. **0.2**

5.8 At 10:00 ahead is spoils dump of Molycorp openpit mine. On rib within spoils dump, Precambrian quartz monzonite extends up to highest trees, and at this level are a few outcrops of andesitic rock on the southern wall of the Questa caldera (Fig. 2R-5.8). **0.2**

6.0 Main portal of old underground workings. Just to west about 250 ft above the road is 50-ft-thick dike of pre-Mississippian diabase cutting quartz monzonite. **0.8**

6.8 Ahead is Union Molycorp mill. On left (north) is spoils dump, filling former mouth of Sulphur Gulch. In between, tan, rugged cliff consists of Tertiary granite and aplite porphyry of Moly Mine pluton. **0.5**

7.3 Entrance to Molycorp mill and open pit on left. Straight ahead are east–west-striking dikes of light-tan rhyolite and aplite porphyry, cutting gray andesite within caldera fill. Precambrian rocks of the caldera wall are low on the slope north of the road and occupy the entire south slope. **0.8**

8.1 Rib of brown, iron-stained Precambrian quartz monzonite, interpreted as slump block just within caldera (see **Stop 3**). Metagabbro intruded by quartz monzonite makes up most of the steep slopes to the south. Iron staining is due to heavy pyritization. From here to Red River, rocks in great alteration scars are pyritized volcanics within the caldera. Alteration scars have been sources of recurrent mudflows during torrential summer rains that have intermittently interfered with, and closed road traffic to, Red River. In 1982, there was one fatality along this road during such a mudflow. **1.3**

9.4 Turnoff on left to Fawn Lakes Campground. **0.4**

9.8 Turnoff on right to Elephant Rock Campground. Elephant Rock is a local landmark on the ridge crest to the north, consisting of quartz-latite-breccia landslide debris within the caldera fill. **0.7**

10.5 Turnoff on right is Junebug Campground. Altered rocks on both sides of road are mainly intracaldera andesite cut by extensive semiconcordant sills and laccoliths of intrusive quartz latite. Apparent alteration is largely supergene effect of weathering of pyritized rock. **0.7**

11.2 **Enter Red River. 0.2**

11.4 Woodland Ski Basin on right. Large outcrop on left is dike of coarsely porphyritic quartz latite, containing K-feldspar megacrysts as much as 2 in. across. It cuts andesitic rock within the caldera. **0.5**

11.9 Turn to the right and enter main street of Red River. Rocks in the vicinity of the town are a complex assemblage of compositionally diverse intrusive rocks of the Red River pluton (Fig. 2R-11.9). **0.2**

FIGURE 2R-11.9. Bouguer gravity field in relation to Tertiary granitic intrusions associated with the Latir volcanic field. Contour interval 4 milligals; hachures point toward gravity lows. Gravity data from Cordell (1978 and unpublished). Granitic intrusions (dark pattern), from north to south: RM – Rito del Medio; VC – Virgin Canyon; CB – Canada Pinabeta; CL – Cabresto Lake; RR – Red River; MM – Moly Mine; BC – Bear Canyon; RH – Rio Hondo; LP – Lucero Peak. Dashed lines indicate margins of mountainous areas (light stipple); heavy hachured line – topographic wall of the Questa caldera (dotted where concealed in Rio Grande rift).

12.1 Forest Road 597 to Mallette Creek on left. **0.7**

12.8 **Junction** of N.M. 38 and N.M. 150. **Bear right** to N.M. 150. Large outcrops along road to left, down low, are intrusive rocks of the Red River pluton; large outcrops up the road ahead are sheared and altered Precambrian quartz monzonite. Leaving Red River, we are following approximate trend of major late Tertiary extensional fault (Fig. 2R-12.8), dropping volcanic rocks to the southwest (on our right) against Precambrian rocks to the northeast (on our left). **1.1**

13.9 On left, large varicolored outcrop of sheared and altered Precambrian quartz monzonite and Tertiary quartz latite along the fault zone. **0.1**

14.0 Red River Pass road on left. This road approximately follows the fault zone to the crest of the ridge to the east (left). **0.3**

14.3 On the left are more outcrops of sheared and altered Precambrian quartz monzonite and similarly sheared and altered Tertiary quartz latite and andesite. **0.3**

14.6 Poor exposures of grayish-brown volcaniclastic sediments on downthrown side of fault zone. Exposures of these sediments and interlayered lava flows continue to the next stop. **0.5**

15.1 Forest Trail 65 to Goose Creek on right. **0.4**

15.5 **STOP 2.** Rhyolite dike (Fig. 2R-15.5c) cutting typical andesitic volcaniclastic sediments and mudflows of lower part of precaldera volcanic sequence (Fig. 2R-15.5a). Note gentle easterly dips on these sediments, in striking contrast to near-vertical dips within the Questa caldera just to the north. These northwest strikes and gentle dips are characteristic of regional Tertiary structural trends outside the caldera (Fig. 2R-15.5b).

Retrace route to Red River. On return drive, high timbered hill above Red River (north) is Sawmill Mountain; it is underlain by thick intracaldera Amalia Tuff, and the eastern rim of the Questa caldera extends through the low pass on the far east side of Sawmill Mountain (Fig. 2R-16.9). **2.7**

18.2 **Junction** of N.M. 150 and N.M. 38. **Turn left** onto N.M. 38 into Red River.

OPTIONAL STOP: Turn right and immediately make left turn by sign for Eagle Nest and Raton and park. Relatively mafic quartz-monzonitic phase of the Tertiary intrusive complex near Red River.

After stop, turn right and retrace route on N.M. 38 into central Red River. **0.7**

18.9 Forest Road 597 (Mallette Creek) on right (see Supplemental Log 2A for route up Mallette Creek to Cabresto Creek). **4.0**

22.9 **STOP 3.** Rib composed of brown, iron-stained Precambrian quartz monzonite, metagabbro, and amphibolite blocks tens of meters across that are intermixed with lower Tertiary sedimentary rocks and juxtaposed against Tertiary volcanics within the caldera. The altered Precambrian rocks are interpreted as slump blocks along the south caldera wall. Iron staining is due to heavy pyritization localized by brecciation. Both peralkaline and metaluminous rhyolite dikes cut entire suite of caldera-filling rocks. **0.8**

23.7 Molycorp mill entrance. **0.5**

24.2 **OPTIONAL STOP.** Examine Moly Mine (Sulphur Gulch) granite pluton and discuss Molycorp mill and mining operation. **1.6**

25.8 **Junction: Turn left** into Columbine Canyon campground entrance. Follow one-way loop road to Campsite #10.

STOP 4. Lunch. Rugged cliffs across Red River Canyon to the north are a massive swarm of Tertiary dikes trending east–west and cutting Precambrian quartz-monzonite gneiss on caldera wall. Individual dikes of rhyolite porphyry, as much as 65–100 ft thick, stand as great fins. Dark band two-thirds of the way up to right skyline is a dike of pre-Mississippian diabase that is unsheared, but has been thermally metamorphosed by Tertiary igneous activity.

Outcrop east of campsite is medium- to coarse-grained, strongly foliated biotite–quartz monzonite. Strong cataclastic foliation trends about N70°E and is nearly vertical. Lineation plunges 10°W. This rock, which we refer to as the quartz monzonite of Columbine Creek, is part of an extensive body that makes up much of the Precambrian terrane on the south wall of the caldera; it has yielded a zircon age of 1,730 m.y. (Bowring and others, 1984). Part-way up the outcrop the quartz monzonite is cut by a 1/3–3/4-in. dike of Tertiary andesite.

Continue around campground loop road. **0.6**

26.4 **Turn left** and continue downstream on N.M. 38. **1.7**

28.1 Turnoff to Molycorp offices on right (north). **0.2**

28.3 **OPTIONAL STOP.** Roadcuts along Molycorp tailings pipeline display altered andesite of caldera fill intruded by coarsely porphyritic quartz latite. At westernmost outcrop, indurated but weakly welded Amalia Tuff containing lithic fragments, in contact with andesite, constitutes matrix of caldera-collapse megabreccia. The tuff becomes more welded away from andesitic blocks, which acted as heat sinks. **0.7**

29.0 **STOP 5.** Bear Canyon pluton and intracaldera Amalia Tuff. On north side of road, large outcrops are the north end of the pluton, which ends just above the cliffs on left (north) and intrudes andesite and welded tuff.

After examining granite, walk up pipeline road to mouth of Capulin Canyon (this can be driven by small vehicles but not by buses). Continue past

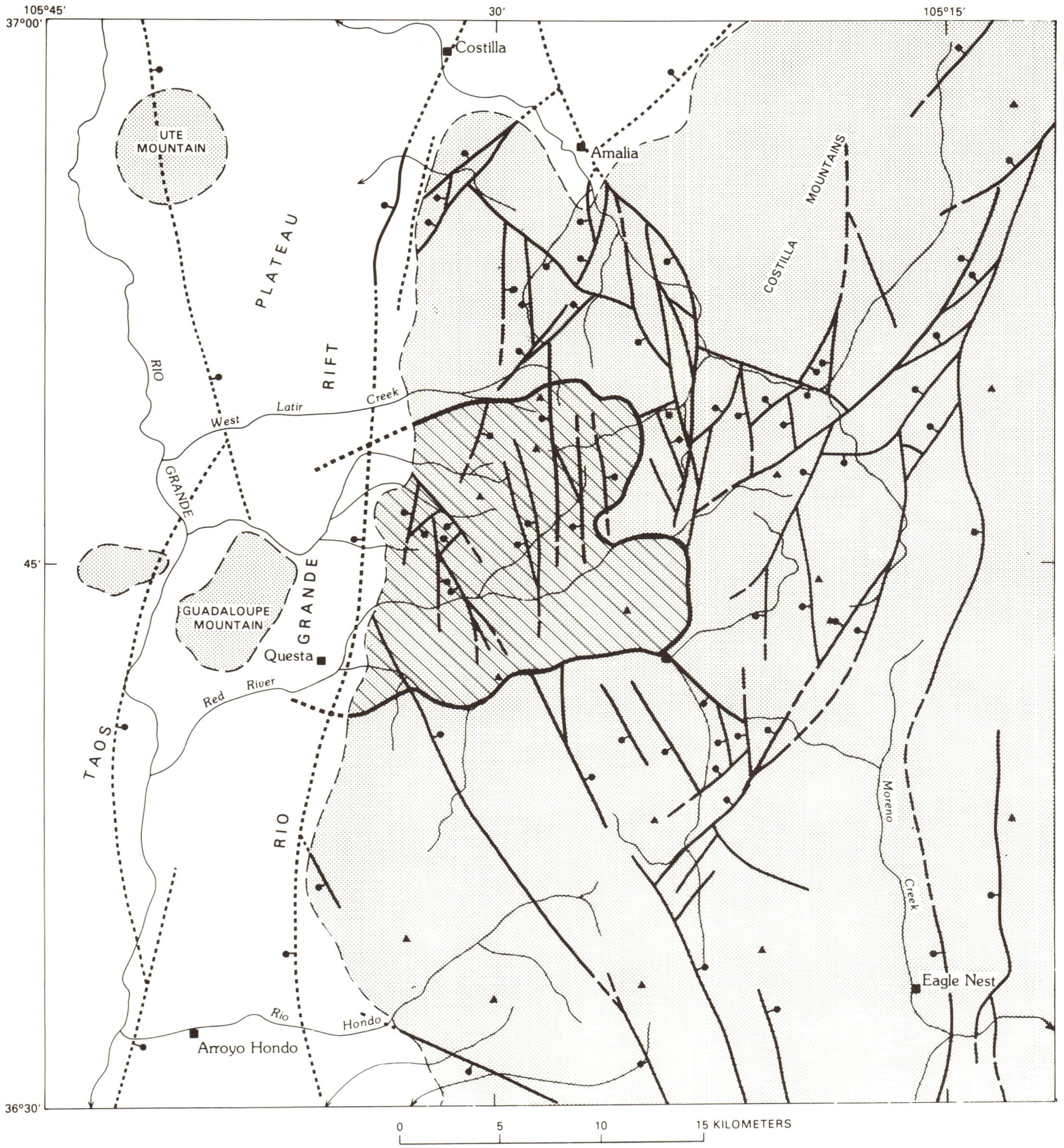

FIGURE 2R-12.8. Miocene and younger normal faults in relation to the Questa caldera (diagonal-line pattern). Faults are dashed where approximately located, dotted where concealed; bar and ball indicate downdropped side. Dashed line indicates margins of mountainous areas (light stipple).

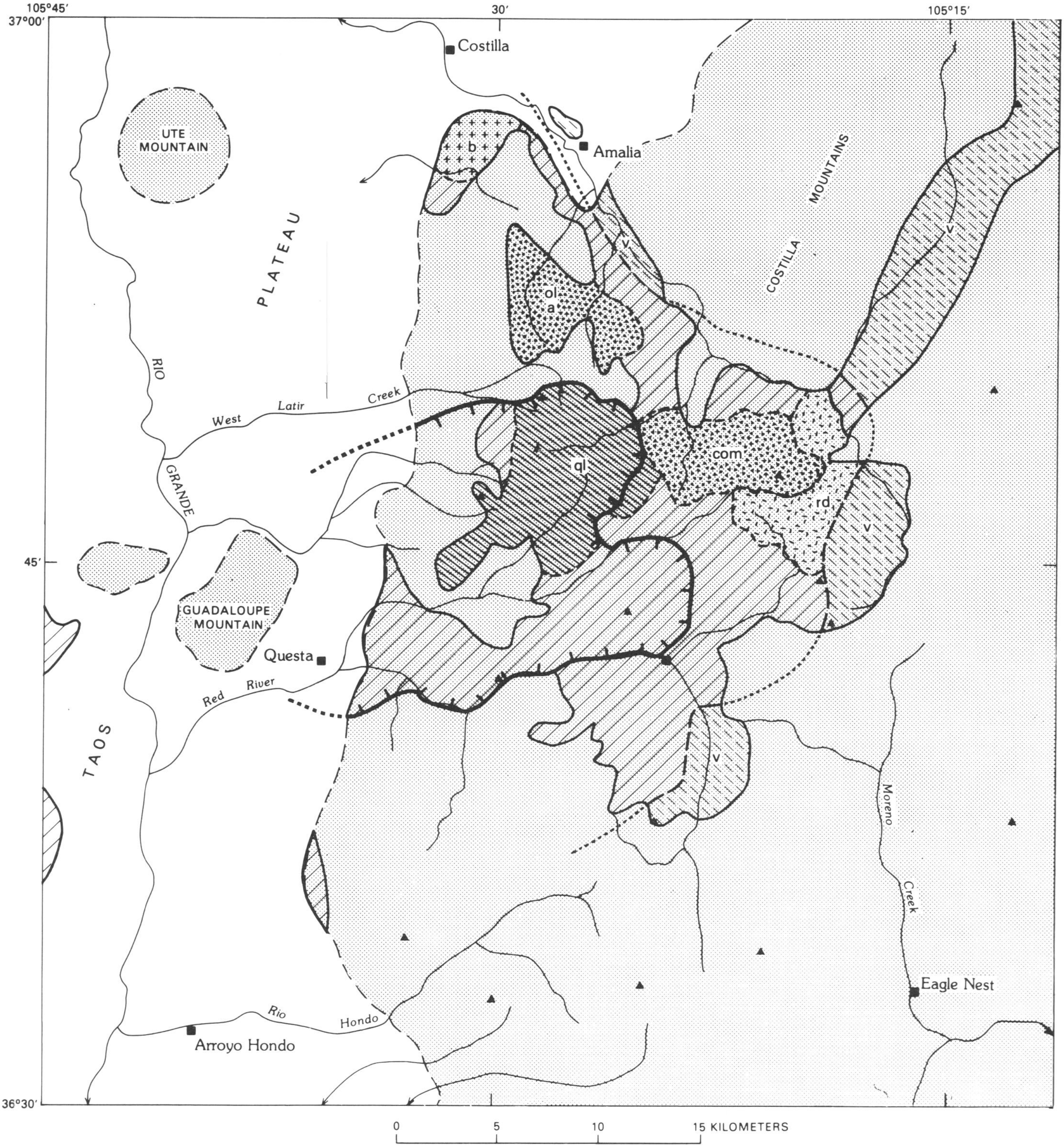

FIGURE 2R-15.5a. Generalized area underlain by dominantly intermediate-composition precaldera volcanic rocks of the Latir volcanic field. Pre-Tertiary rocks – light stipple pattern. Diagonal-dash pattern (v) – volcaniclastic sedimentary rocks (mainly mudflow deposits) that form alluvial aprons around central volcanoes. Dark patterns – near-source lavas and breccias erupted from central volcanoes: com – comendite of Ortiz Peak; ql – Latir Peak Quartz Latite; rd – hornblende-rich rhyodacite; ola – olivine–augite andesite; b – olivine basalt; diagonal-line pattern – undifferentiated andesite. Distribution of near-source and alluvial deposits demonstrates that the Questa caldera developed within a pre-existing volcanic highland of clustered central volcanoes; inferred boundaries between near-source and alluvial deposits indicated by dotted line. Dashed lines indicate margins of mountainous areas; hachured line – topographic wall of the Questa caldera (dotted where concealed in Rio Grande rift).

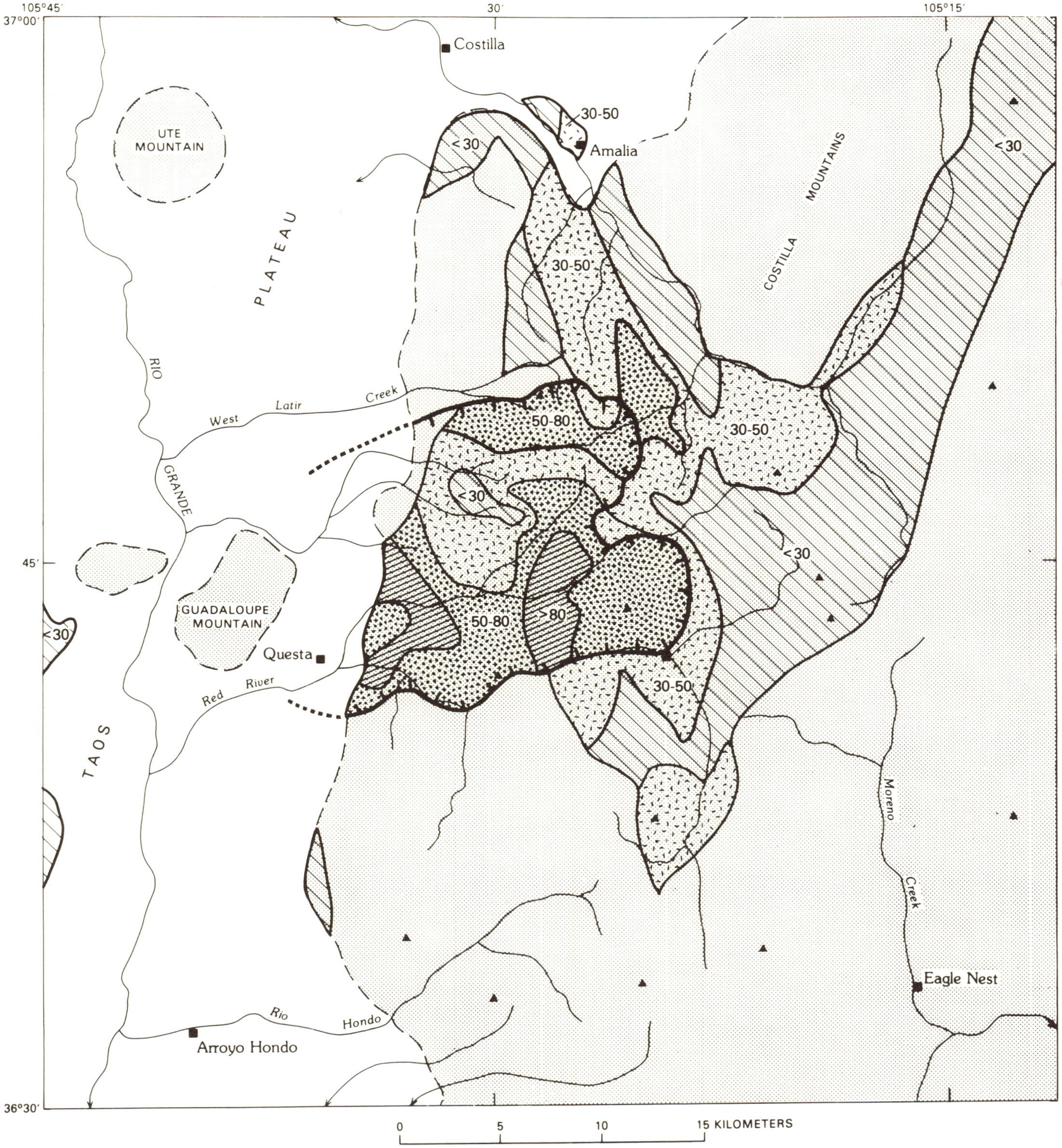

FIGURE 2R-15.5b. Averaged dips of Tertiary bedded and foliated rocks of the Latir volcanic field. Area was divided into 1-km^2 blocks, all attitudes within each block were averaged, and the averages were contoured. The steepest dips are confocal with the Questa caldera. Gentle dips within the north-central part of the caldera are interpreted as marking the crest of a resurgent dome. Steep dips just beyond the northeast caldera wall are associated with a major northwest-trending fault zone (Fig. 2R-11.7). Dashed lines indicate margins of mountainous areas; hachured line – topographic wall of the Questa caldera (dotted where concealed in Rio Grande rift).

FIGURE 2R-15.5c. View north along a prominent rhyolite dike exposed along upper Red River road. Margins of dike are columnar-jointed. Dike intrudes precaldera andesitic sedimentary rocks, poorly exposed on each side of outcrop.

dump about 220 yds up the road to first large outcrop.

This outcrop is typical intracaldera welded tuff in degree of welding and attitude. Look carefully to determine attitude: the strike is about N20°W and the dip averages vertical (Fig. 2R-15.5b).

As an option, continue up Capulin Canyon road to its end; park at highest part of road along stream. Walk along road above switchback to examine large megabreccia blocks in alteration scar visible from lower-dump parking area.

From the lower-dump parking area, a good view can be had of the north contact of Bear Canyon stock, where the bold outcrops of intrusive rock end in the scree and brushy slope just to the north.

Retrace route from dump parking area to Red River road; continue downstream. **0.9**

29.9 Questa Forest Service Ranger Station on left. **1.4**

FIGURE 2R-15.5d. Some unanticipated field activities. Black Copper Canyon (tributary to the upper Red River).

FIGURE 2R-16.9 View of town of Red River from the east, on Bobcat Pass Road. Darkly timbered slopes in foreground consist mainly of intrusive quartz latite along the southeast margin of the Questa caldera. Lighter-colored slopes behind the town are mostly thick, densely welded Amalia Tuff within the caldera. High point in right center is Sawmill Mountain (10,962 ft).

31.3 Forest Road 134 up Cabresto Creek to right. (See Supplemental Log 2B.) **0.3**

31.6 **Junction** of N.M. 38 with N.M. 3. **Turn right** for continuation of road log (Segment 2Q).

End of Segment 2R of road log for Second Day.

SEGMENT 2Q: QUESTA NORTH TO AMALIA

This segment follows the east margin of the modern Rio Grande rift north from Questa to Costilla, providing views of the Tertiary volcanic sequence resting on Precambrian rocks both within and outside of the Questa caldera. From the town of Costilla, the route follows Rio Costilla upstream along a series of Miocene fault blocks, related to early extension of the Rio Grande rift system, that tilt the volcanic sequence to the east and repeat it.

FIGURE 2R-24.2a. Crusher and mill of Union Molycorp Questa operation along the Red River at south margin of the Questa caldera. Rocks behind mill are silicic granite of the Moly Mine pluton (23 m.y., a structurally high part of a composite ring intrusion along the south caldera margin. The now-inactive open-pit molybdenum mine lies just behind these hills; current production is from underground workings to the west (left of photo).

FIGURE 2R-24.2b. Drilling oriented samples for paleomagnetic study—granite of Moly Mine pluton. Open-pit spoils dump in background. Along N.M. 38, 0.3 mi west of Molycorp mill.

Mileage

0.0 **Junction** of N.M. 3 and N.M. 38 in Questa. **Proceed north** on N.M. 3. Major drainages to east are Cabresto Creek on the north and Red River on the south. The high peak north of Cabresto Creek is Pinabete Peak (11,946 ft); south of the Red River is Flag Mountain (11,942 ft). The ridge between Cabresto Creek and Red River as seen from this perspective is welded tuff from bottom to top. Note the big alteration scar on the northwest side of the ridge; such scars mark sites of especially intense pyritization, causing supergene acid-sulfate alteration. **1.3**

1.3 Turnoff to Questa High School is on left; this is location of the most extreme Bouguer gravity low within the Rio Grande rift zone (Fig. 2R-11.9), reflecting downdropped western portion of the Questa caldera.

At 9:00 to 10:00 are the two summits of Guadalupe Mountain, double rhyodacitic volcanoes (K–Ar age: 4.8 m.y.) of the Taos Plateau volcanic

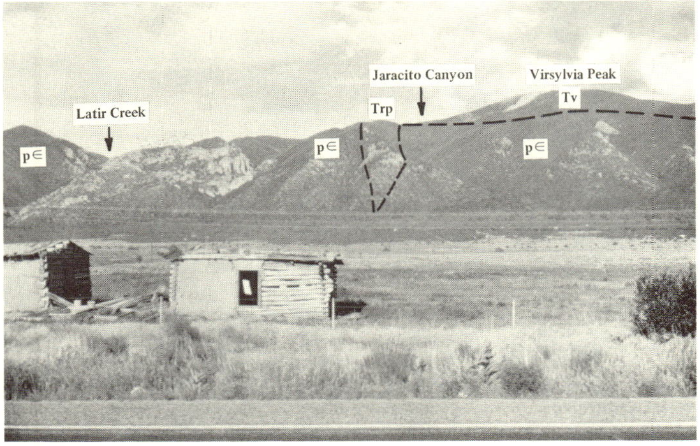

FIGURE 2Q-7.1a. North margin of the Questa caldera is marked by a large ring dike of peralkaline rhyolite–granite porphyry (trp), exposed on west-facing front of the Sangre de Cristo Mountains, just north of El Rito turnoff. The ring dike cuts Precambrian rocks (pЄ) of the caldera floor and wall; Tertiary volcanic rocks (Tv) are in depositional contact with the Precambrian within the caldera.

field (Lipman and Mehnert, 1979). At 11:00 in the distance is Ute Mountain, another Pliocene rhyodacitic volcano. **1.4**

2.7 N.M 378 to Cerro on left (road to Rio Grande Wild River Recreation area). **1.1**

3.8 New Mexico Port of Entry. **1.7**

5.5 Road to Buena Vista on left. **1.1**

6.6 Road to El Rito on right. **0.5**

7.1 **OPTIONAL STOP.** Crest of large alluvial fan affords view of cross-sectional structure of the Questa caldera along the mountain front and of late Tertiary volcanoes of the Taos Plateau (Figs. 2Q-7.1a, b).

The mountain front to the east exposes a cross section through the interior of the Questa caldera.

FIGURE 2Q-7.1b. Rio Grande gorge, cut into flat-lying flows of Servilleta Basalt (cliffs) and interlayered seimentary rocks of the Santa Fe Group (talus-covered slopes). View is from Big Arsenic Spring BLM campground northwest toward prominent shield volcano of Cerro de la Olla, an olivine-andesite shield (4 m.y.) of the Taos Plateau volcanic field. Small light-colored hills, silhouetted along base of Cerro de la Olla, are lower Miocene rocks of Brushy Mountain area (22–26 m.y.) on the major horst that defines the western structural margin of the deepest portion of the Rio Grande rift.

The high areas consist of caldera-floor stratigraphy (Fig. 2R-15.5a) resurgently uplifted within the core of the caldera. The highest peaks, such as Cabresto Peak at 3:30 (12,444 ft) and Pinabete Peak at 4:00 (11,946 ft), are mostly quartz-latitic and andesitic lava flows, locally underlain by volcaniclastic sediments, resting on Precambrian rocks in the middle timbered slopes. The Precambrian sequence inside the caldera is almost entirely multiply folded quartzite and pelitic schist of sillimanite grade. Wooded slopes of Pinebete Peak to the east are underlain by quartzite that forms white outcrops and scree slopes.

The large cliffs in the canyon at 3:00 are the granite of Rito del Medio (Fig. 2R-11.9) that is interpreted as representing resurgent magma within the caldera. The next canyon north is Jaracito Canyon, which marks the approximate position of the northern structural boundary of the caldera as it crosses the mountain front. Light rocks on nose of ridge just north of Jaracito Canyon are rhyolite porphyry in the ring dike along the caldera boundary. The trend of the Precambrian units is about north–south and dips are steep, so offset along the north caldera wall is not conspicuous.

The large canyon at 2:00 is West Latir Creek. Cliffs on the mountain front between Latir Creek and Jaracito Canyon are large Precambrian pegmatites in granodiorite gneiss. The caldera wall swings into the upper part of West Latir Creek from Jaracito Canyon, but both sides of the lower canyon are Precambrian rocks outside the caldera. At these lower exposure levels, the caldera boundary is marked by a large ring dike of peralkaline granite porphyry.

Straight ahead in the distance are the Sangre de Cristo Mountains in Colorado, with the peaks of Sierra Blanca (13,120 ft) at 12:30. Within the Rio Grande rift in the distance, the irregular, low San Luis Hills at 11:30 are composed of eroded, faulted, late Oligocene and Miocene lava flows exposed on a horst within the rift. This horst extends southward just west of the position of the present river, almost as far south as Taos, as marked by local erosional exposures of Miocene rocks (Lipman and Mehnert, 1979), and extends all the way to the Picuris embayment as indicated by gravity data.

In the farthest distance to the northwest, the low slopes are the southern flanks of the San Juan volcanic field, mainly outflow volcanic rocks from the Platoro caldera (Lipman, 1975a). The general area of this caldera is marked on days of good visibility by high, snow-covered peaks on the skyline at about 10:00.

To the west, coming into view behind Guadalupe Mountain, are three other major volcanoes of the Taos Plateau. Clockwise from due west these are Cerro Chiflo, a 10.3-m.y.-old lava dome of quartz latite whose slightly jagged summit is just visible; more gentle slopes of Cerro del Aire, a 3.5-m.y.-old shield of olivine and andesite consisting of compositionally monotonous flows of about 57% SiO_2; and in the distance, the large rhyodacite volcano of San Antonio Mountain, that has yielded a K–Ar age of about 3.2 m.y. (Lipman and Mehnert, 1975). **3.1**

10.2 Sunshine Valley turnoff on left. Low, irregular volcano between Cerro del Aire and San Antonio Mountain is the No Agua lava dome of silicic alkalic rhyolite (3.9 m.y.). On the right (north) side in morning light, the obvious quarry scars are the Johns Mansville perlite quarry. **4.2**

14.4 Urraca Ranch Wildlife Refuge turnoff on right. Low mesa now visible straight ahead is San Pedro Mesa (also known as Wildhorse Mesa), capped by Servilleta Basalt, consisting of flows of olivine tholeiite 3–4 m.y. old. This is the same lava sequence as exposed in the Rio Grande gorge, and cross-section interpretations indicate about 1,650 ft of subsequent uplift along the mountain-front fault system. Coming into view in front of, and just to the right, of San Pedro Mesa is a beheaded fan coming out of Cedro Canyon at 2:00. This fan is cut by a late Pleistocene fault scarp.

In the mountains to the east, all rocks up to the skyline are Precambrian rocks outside the caldera, chiefly granitic and fine-grained felsic gneisses and amphibolite. Precambrian quartzite caps the ridge at 2:00. Low orange-brown knobs along the mountain front are Precambrian pegmatite. The volcanic sequence is tilted to the east and rests depositionally on Precambrian rocks on the eastern slopes, just out of sight. **1.1**

15.5 Milepost 40. Low scarp about 220 yds east is fault scarp along which the most recent episode of movement was probably latest Pleistocene (10,000–25,000 yrs ago: Machette and Personius, 1983). **1.1**

16.6 Turnoff to Cedro Canyon on right. In Cedro Canyon is one of the best exposed sections of Tertiary volcanic rocks of the Latir field (Fig. 2Q-16.6). On the north side of the canyon, the big cliff is welded Amalia Tuff, from the Questa caldera. The welded tuff has been modified by hydrothermal alteration involving intense potassium metasomatism. Although they look fresh in hand sample, some samples have doubled in K_2O content and have lost almost all of their original Na_2O.

The Amalia Tuff is underlain by precaldera basaltic andesite and andesitic sediments (Fig. 2R-15.5a). It is overlain conformably by tilted volcaniclastic sediments that were eroded from the Latir field (volcaniclastic lower part of the Santa Fe Group), which in turn are overlain unconform-

FIGURE 2Q-16.6. Tertiary volcanic sequence exposed on west-facing front of Sangre de Cristo Mountains, just south of Cedro Canyon. East-dipping volcanic rocks rest depositionally on Precambrian granite (p€); in ascending order, they are: tuffaceous and volcanic sedimentary rocks (vS), xenocrystic-andesite flow (xA), olivine-bearing basaltic andesite (oA), alkalic andesite characterized by fissile jointing (FA), and Amalia Tuff (AT). Alluvial fan in foreground has been beheaded at Pleistocene scarp of the Costilla fault (CF).

ably by weakly consolidated cobbly gravels consisting of Precambrian clasts (upper part of the Santa Fe Group). The sequence is capped by Servilleta Basalt. South of the canyon, the slopes are underlain by additional andesitic lava flows fingering in between the Precambrian rocks and the welded tuff that thinly caps the ridge crest. The lowest flow is xenocrystic basaltic andesite; it is overlain by an olivine-bearing basaltic andesite that correlates with the sequence north of the canyon, and here is capped by a distinctive slabby, jointed alkalic andesite. Light-colored rocks in hills at head of canyon are Precambrian quartzite resting on layered amphibolite. **1.8**

18.4 "Highway B," formerly the "Highway Bar." At 3:00 is the mouth of Costilla Canyon, with the Costilla Mountains (entirely of Precambrian rocks) in the distance. At mouth of the canyon, the mesas on both sides are capped by Servilleta Basalt, underlain by Santa Fe sediments—all extensively involved in slumps and landslides. **1.0**

19.4 **Enter Costilla. Junction** of N.M. 3 and N.M. 196. **Turn right** onto N.M. 196 to Amalia. **1.5**

20.9 Enter Costilla Valley with landslides of Santa Fe Group sedimentary rocks and Servilleta Basalt on both sides. Costilla Mountains straight ahead are composed chiefly of Precambrian granite. **2.7**

23.6 End of landslide area. Mesas capped by Servilleta Basalt are underlain by poorly exposed andesitic lavas and interlayered volcaniclastic sediments below the Amalia Tuff that comes down to road level at the bend in road ahead. The slopes across valley to southwest are also Amalia Tuff. **0.2**

23.8 **STOP 6.** Roadcut exposure of Amalia Tuff, underlain by precaldera andesitic mudflows and overlain at bend in road by rhyolitic volcaniclastic sediments that are essentially basal Santa Fe Group (Fig. 2Q-23.8a–d). **0.4**

24.2 Poorly exposed on left, surrounded by talus below the capping basalt, is an obscure, dark outcrop of basanitic basalt, dated at about 16 m.y. (whole-rock K–Ar). This is interlayered within the volcaniclastic sediments. Straight ahead, high hills are Precambrian on left, with rhyolitic lava dome—the rhyolite of Gonzales Ranch—plastered against the bounding fault scarp that trends up the Costilla Valley. **0.7**

24.9 Scenic downtown Amalia; pavement ends. Heavily timbered north slopes of mountains on right are underlain by complexly faulted volcanic sequence north of Questa caldera, and locally by prevolcanic Tertiary sediments and Precambrian rocks.

End of Segment 2Q of road log for Second Day (see Supplemental Log 2C for continuation to east). **Turn around;** retrace route to Costilla, Questa, and finally Taos. Skeletal road log continues below. **4.2**

29.1 Road bends left, leaving valley of Rio Costilla and entering the San Luis Valley. Straight ahead on left is symmetrical shield volcano of Ute Mountain. Farther to right are low, irregular profiles of the San Luis Hills—Oligocene and lower Miocene volcanics on a horst within the Rio Grande rift. In the far distance is the profile of the southern San Juan Mountains. **1.3**

30.4 **Enter Costilla. Junction** of N.M. 196 with N.M. 3. **Turn left** onto N.M. 3. After turn, visible on right and ahead are several volcanoes of the Taos Plateau volcanic field, including the rhyodacitic

FIGURE 2Q-23.8a. Type locality of Amalia Tuff (ridge-forming unit at left center). View is toward Costilla Mountains (dark hills in distance on right), looking up valley of Rio Costilla along N.M. 196. Small village of Amalia is just ahead, around bend in road.

FIGURE 2Q-23.8b. Typical welded, devitrified Amalia Tuff, showing moderately drawn-out white-pumice lenses in reddish-purple groundmass of ash shards and phenocrysts. This area has been subjected to intense low-temperature hydrothermal alteration (potassium metasomatism) which is not obvious in outcrop, but which has removed most of the sodium and approximately doubled the potassium content of the rock. At type of locality of Amalia Tuff, along N.M. 196.

FIGURE 2Q-23.8c. Well-bedded, east-dipping volcaniclastic facies of the Santa Fe Group overlying the Amalia Tuff near the town of Amalia. These sediments represent material eroded from high central areas of the Latir volcanic field shortly after formation of the Questa caldera; they are overlain conformably by 16–17-m.y.-old basalt. View is to the north, along N.M. 196.

Ute Mountain at about 2:00, somewhat irregular jagged profile of Cerro Chiflo at 12:30, and Cerro Montoso at 1:00. At 12:00 is Guadalupe Mountain, just west of Questa, a double volcano of rhyodacite. **2.8**

33.2 To left (east) is jeep road to Cedro Canyon. Exposed in the prominent cliffs is the most southerly outflow volcanic sequence north of the caldera rim. The mountain barely above treeline at 10:00 is just inside the caldera—Virsylvia Peak (12,625 ft). It consists of precaldera andesitic lava flows, resting at about the level of the cirque basin on Precambrian. In the farthest distance, south along the Sangre de Cristo Mountains, is Flag Mountain, also Precambrian, just south of the southern margin of the caldera along the Red River. **9.5**

42.7 Crest of alluvial fan **(OPTIONAL STOP):** range-front view of structures of Questa caldera to east; panorama of Pliocene volcanoes of the Taos Plateau to the west. **0.5**

43.2 El Rito turnoff on left. At 9:00 are large cliffs of Rito del Medio granite, the resurgent granite within the Questa caldera. **1.2**

44.4 Milepost 30. Brush-covered slopes at 9:00 to 10:00 along the base of the mountain front are a southerly continuation of the resurgent pluton (Fig. 2Q-44.4). At 10:30, a conspicuous alteration scar at the front of an east–west-trending ridge marks the maximum thickness of the caldera moat fill, here entirely welded tuff from the base of the mountain to the top of the mountain front. The high peak at 10:30 is Flag Mountain (Precambrian) south of the caldera, and the switchback mining roads on the lower slopes toward us are mainly in the Bear Gulch pluton. This is one of the ring intrusions along the southern margin of the caldera. **1.7**

46.1 New Mexico Port of Entry. **1.1**

47.2 N.M. 378 to Cerro on right. **2.7**

49.9 Center of Questa; N.M. 38 to Red River on left. **14.0**

63.9 Rio Hondo bridge. **6.0**

69.9 Three-way intersection of N.M. 3, N.M. 150, and U.S. 64 at blinker light.

End of road log for Second Day.

DAY 2: SUPPLEMENTAL ROAD LOG 2A, FROM RED RIVER OVER SAWMILL PASS TO CABRESTO CREEK

This segment traverses the eastern margin of the caldera, thick, caldera-filling Amalia Tuff, and the intrusive complex of Red River. It connects the Red River and Cabresto Creek field-trip segments via Forest Road 597.

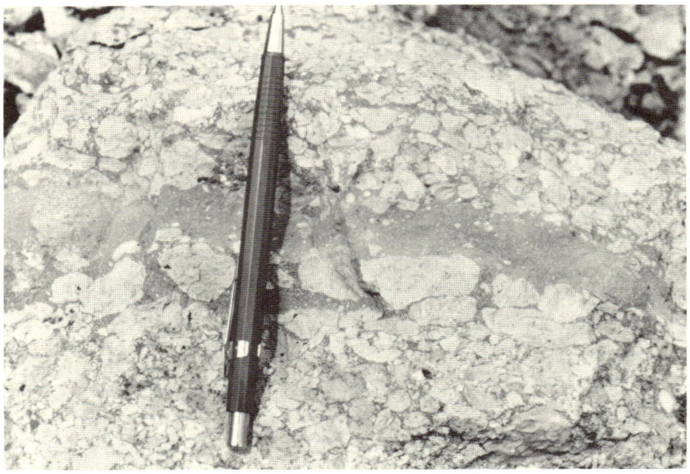

FIGURE 2Q-23.8d. Detail of volcaniclastic facies of the Santa Fe Group. Angular clasts of Amalia Tuff and genetically related rhyolitic lava are in red–brown sandy matrix.

FIGURE 2Q-44.4. Structurally high resurgent intrusion within the Questa caldera. The granite of Virgin Canyon (Tg) intrudes precaldera andesite and quartz latite (Tv) on the north face of Cabresto Peak (12,443 ft). Granite of Virgin Canyon is compositionally zoned from margins of peralkaline granite porphyry to metaluminous equigranular biotite granite in the interior.

Mileage

0.0 Main Street, Red River (N.M. 38); **turn north** onto Forest Road 597 up Mallette Creek road. **0.1**

0.1 Stop sign at High Street; continue straight ahead. **0.1**

0.2 **STOP 1.** Outcrops on left, at edge of town, are aplitic granite of Red River intrusion. **0.2**

0.4 Mallette Canyon Park. Cliffs on left are Precambrian quartzite intruded by aplitic granite of the Red River complex. Altered exposures along road to right ahead are Precambrian rocks complexly intruded by rhyolite porphyry and quartz latite. **0.8**

1.2 **STOP 2.** Outcrops of fine-grained monzonite or diorite of Red River complex on left. This is the most mafic phase of intrusive complex, yet is complexly intergradational with rhyolite and granite porphyry, with textures suggesting that both were melts. **0.1**

1.3 Fork in road; **bear left** and leave bottom of Mallette Canyon. **0.2**

1.5 Good exposures, in roadcuts on right, of heterogeneous, fine-grained, equigranular monzonite of the Red River complex. **0.1**

1.6 Crossing Trail Creek. Ahead are rhyolitic and aplitic phases of the Red River complex. **0.9**

2.5 Altered Precambrian country rocks of the Red River intrusions. **0.1**

2.6 At tight switchback, quartz-latite dike is intrusive into the Precambrian rocks. **0.5**

3.1 Indefinite contact between predominantly Precambrian rocks and biotite-bearing metaluminous rhyolite porphyry of the Red River complex. These are probably largely, but not entirely, postcaldera intrusions. **0.6**

3.7 Contact between rhyolite porphyry and poorly exposed intracaldera Amalia Tuff. The tuff is bleached and somewhat altered, but collapsed-pumice textures (eutaxitic foliation) are locally conspicuous. The road continues mainly in thick intracaldera tuff from this point to nearly all the way down the north side of Sawmill Pass. **1.0**

4.7 At tributary valley: obscure contact between intracaldera tuff and intrusive peralkaline rhyolite that is compositionally identical to the Amalia Tuff. See discussion for STOP 3.

Up the east side of Sawmill Pass ahead, the road passes first through the intrusive rhyolite, into welded tuff on the main nose, then back into the rhyolite to the pass. **0.5**

5.2 Sawmill Pass. The pass is along the contact between intrusive rhyolite to the south and a narrow zone of Amalia Tuff against the caldera wall to the north. The caldera wall is poorly exposed along the side road leading north from the pass, where float of welded tuff ends within about 325 ft against precaldera andesite. **0.5**

5.7 **STOP 3.** Road makes a tight turn around a pronounced nose on the ridge. The entire ridge is underlain by intrusive peralkaline rhyolite. The intrusive can be distinguished from the tuff by sparser, larger, unbroken phenocrysts of quartz and alkali feldspar, by a lack of eutaxitic foliation, and by a more rectilinear jointing. Intrusions such as this occur irregularly around margins of the caldera and may represent lower parts of ash-flow vents.

The deep canyon to the north, Bonito Canyon, marks the approximate trend of the caldera wall. Altered rocks on the ridge north of this canyon are mostly andesite overlying Precambrian rocks outside the caldera. The highest point on the ridge is capped by outflow Amalia Tuff containing a basal vitrophyre. **1.0**

6.7 A large mass of quartz latite is poorly exposed in roadcuts; this is interpreted as a megabreccia block enclosed within intracaldera Amalia Tuff. **0.5**

7.2 Smaller lens of quartz-latite megabreccia in roadcut. **0.5**

7.7 Good views to the northwest of Pinabete Peak and Latir Mesa, the above-timberline country in the resurgent core of the Questa caldera, that is underlain by quartz latite of the caldera floor. **0.5**

8.2 Last well exposed intracaldera Amalia Tuff, with well developed eutaxitic foliation. **0.3**

8.5 **STOP 4.** Precambrian quartzite at caldera wall, in poorly exposed contact with altered intracaldera Amalia Tuff. Intrusive rhyolite and quartz latite are present along the contact. **0.5**

9.0 Large dike of peralkaline rhyolite along caldera margin. **0.1**

9.1 Junction of Forest Road 597 (Mallette Creek) and Forest Road 134 (Cabresto Creek). Join Supplemental Log 2B at mile 9.3.

End of supplemental road log 2A for Second Day.

DAY 2: SUPPLEMENTAL ROAD LOG 2B, FROM QUESTA TO GREENIE PEAK, VIA CABRESTO CREEK

The road up Cabresto Canyon provides a west-to-east cross section through structurally lower parts of the Questa caldera and the east caldera wall; it continues structurally upward through the extracaldera sequence, including Precambrian rocks, precaldera intermediate-composition lavas and sediments of the Latir field, and the Amalia Tuff as an outflow unit.

Continuation of the road to the summit of Greenie Peak provides panoramic views of the southern part of the Latir field and the Questa caldera, as well as Wheeler Peak Wilderness and Moreno Valley to the south and east.

A passenger car is adequate to drive to the head of Cabresto Creek; to continue to the summit of Greenie Peak, a high-clearance vehicle (preferably with four-wheel drive) is recommended.

Mileage

0.0 **Junction** of N.M. 3 with N.M. 38 in center of Questa. **Turn right** on N.M. 38 to Red River. **0.3**

0.3 **Junction: Turn left** (north) onto Forest Road 134 to Cabresto Creek at bottom of hill at sign to Singing River Ranch. Follow "paved" road through residential Questa. Entire visible part of ridge between Cabresto Creek and Red River consists of intracaldera Amalia Tuff, which strikes north–northwest and dips nearly vertically. Whether the section is measured perpendicular to foliation or from bottom to top of topographic exposures, it is several thousand feet thick. **1.5**

1.8 Good view of the conspicuous alteration scar south of Cabresto Creek at 3:00. The scar exposes only altered Amalia Tuff; no intrusion has been found. **0.8**

2.6 Cattleguard; pavement ends at mouth of Cabresto Canyon. Rocks ahead on both sides of the valley are Questa caldera fill and upper parts of the caldera-floor sequence. Chaotic stratigraphic relations between precaldera units and the Amalia Tuff in the low ridges ahead are due to combined effects of: (1) complex, rift-related faulting along the mountain front; (2) faulting related to collapse of the caldera; and (3) intermixing of great slide blocks of megabreccia with lower parts of the intracaldera tuff. **0.4**

3.0 Up gully, to left (north), is altered white to light-tan intrusive peralkaline rhyolite that is compositionally similar to the Amalia Tuff. It cuts quartz-latitic and andesitic lava flows of the caldera floor. **0.2**

3.2 Prominent outcrops above road on left are tuff of Tetilla Peak, much pyritized, silicified, and iron-stained, and otherwise propylitically altered. This is the lowermost major volcanic unit of Latir field, consisting of local nonwelded to weakly welded, quartz-bearing, abundantly lithic ash-flow tuffs of low-silica rhyolite related to several rhyolitic lava domes scattered through the central part of the volcanic field. Here, it is structurally complex within the caldera-floor sequence, resting on Precambrian metamorphic and lower Tertiary sedimentary rocks. Strikes of weak eutaxitic foliation in the tuff are about N20°W and dips are nearly vertical. **0.2**

3.4 Rugged cliffs on north side of canyon are porphyritic, plagioclase-rich andesite that appears to form an irregular laccolithic intrusion. The andesite is intruded roughly along the lower part of the Tertiary section, which throughout this area strikes northwesterly and dips steeply. The andesitic intrusion is probably part of the precaldera sequence, but this relation cannot be determined with confidence. **0.1**

3.5 Gauging station and headgate of ditch. **0.2**

3.7 Low cliffs on north side of road are Precambrian muscovite quartzite and schist of the caldera floor. These rocks are tilted to the west and complexly faulted, probably related to resurgence of the caldera. The general stratigraphic sequence of the overlying Tertiary rocks of this area is: (1) discontinuous lenses of lower Tertiary conglomeratic sediments containing Precambrian clasts, interlayered with reddish shale and sandstone; (2) tuff of Tetilla Peak; (3) andesitic flows and breccias, mostly sparsely porphyritic in this area; and (4) biotite–plagioclase porphyritic Latir Peak Quartz Latite. **0.2**

3.9 The large, dark, prominent rib straight ahead is

another intrusive porphyritic andesite that was intruded along the basal Tertiary sedimentary horizon. The yellowish-tan cliffs to the left are Precambrian quartzite. **0.6**

4.5 Singing River Ranch. Prominent outcrops on both sides of road are Precambrian quartzite. Poorly exposed in the trees above is the Tertiary volcanic sequence just described, striking northwesterly and dipping steeply southwesterly. Rocks at road level are Precambrian for the next 1.5 mi. **0.2**

4.7 **STOP 1.** Outcrop of white granular quartzite, a typical Precambrian lithology within the caldera. Obscure layering (bedding?) is marked by thin micaceous partings that strike northwest and dip moderately northeast. Minor fold axes plunge about 10°SE. Close-spaced fracture cleavage in parts of the outcrop strikes about north–south and dips steeply west. This is parallel to axial planes of late folds in this area, but not seen in this outcrop. Some 30–50 ft to the west and 16–30 ft above road level layering steepens to near-vertical and can be traced around the nose of a large southeast-plunging fold whose axis is apparently parallel to minor fold axes and crenulations. **0.3**

5.0 **STOP 2.** Convenient parking on small dead-end road to right. Roadcut in crudely layered sillimanite–biotite–quartz gneiss. This is one of the principal lithologies interleaved with quartzite in the Precambrian sequence within the caldera. Here foliation and layering strike northwest and dip gently northeast; mineral lineation marked by elongated clots of sillimanite and biotite plunges northeast, downdip. Minor crenulations in some layers plunge 15°NE, parallel to fold axes seen at Stop 1. A few $1/3$–$2/3$-in. pods of white quartz–feldspar pegmatite are sheared and deformed. **1.0**

6.0 Cabresto Lake Road on left. Small exposures of the Cabresto Lake granite pluton are obscure in shrubs, just left of the junction. Much better exposures are about 0.5 mi up the Cabresto Lake Road or along the main fork of Cabresto Creek ahead. The Cabresto Lake pluton is a low-silica granite containing two feldspars, biotite, and hornblende. **0.3**

6.3 Large exposures of the Cabresto Lake pluton. Even though no exposures are visible at road level for the next mile, the road passes through the eastern portion of the Cabresto Lake pluton and Precambrian rocks that represent a structurally high portion of the caldera floor uplifted by the granitic intrusion. **1.7**

8.0 At crest of rise in road, ridges on left with light-colored outcrops are granitic rocks of the Weary Willie pluton, representing an isolated lobe of the Cabresto Lake granite pluton. **0.5**

8.5 Knob on left is altered quartz-latite lava along southeast contact of Weary Willie pluton, where floor of the caldera drops to river level. **0.1**

8.6 Cross mouth of Italian Canyon. **0.1**

8.7 Road crosses Cabresto Creek. Intense alteration in this area is associated with basal, weakly welded intracaldera Amalia Tuff and associated rocks in the structurally low eastern portion of the caldera. **0.4**

9.1 Bold white outcrops across stream to north are Precambrian quartzite on the east wall of the caldera. **0.2**

9.3 Forest Road 597 to Sawmill Pass and Creek on right (see Supplemental Log 2A). **0.1**

9.4 Precambrian quartzite in outcrops on right. Outcrops on left are Precambrian in lower part; weakly welded, much-altered Amalia Tuff within the caldera is exposed on the upper part of the ridge. The eastern caldera wall slopes gently, presumably because of subsequent tilting to the east along faults related to the Rio Grande rift. The road ahead, up Cabresto Creek, continues through Precambrian rocks for the next 3 mi. The Precambrian is mostly quartzite and pelitic schist, locally cut by irregular Tertiary intrusions of porphyritic quartz latite. **2.4**

11.8 Rito Claro on right; still entirely in Precambrian outcrops. **0.5**

12.3 Contact between Precambrian rocks and andesitic lava flows: a northwest-trending fault, down to the east. The andesitic flows strike northwest and dip 20–40° to the northeast, as indicated by obscurely interlayered, lighter-colored volcaniclastic units. Such andesitic rocks are exposed in roadcuts for the next 1.5 mi. **1.0**

13.3 Change in color of andesitic rocks from green–gray to purple-brown marks irregular, broad, northeast limit of zone of intense propylitic alteration. **0.5**

13.8 Official notice on right. Poorly exposed in grassy slopes on left is basal vitrophyre of outflow Amalia Tuff, one of few localities where the Amalia Tuff has such a basal vitrophyre. The Amalia Tuff forms a subhorizontal cap on the high ridges ahead and to the north, and also is plastered irregularly against lower slopes. This irregular contact in part perhaps represents a valley fill, but obscure structural complications are also likely present. **0.2**

14.0 Cattleguard. **0.4**

14.4 Outcrop on north side of road exposes margins of a thick lava dome, the Latir Peak Quartz Latite, that overlies the andesite sequence and underlies the Amalia Tuff in this area. **0.2**

14.6 Large outcrop of quartz latite on south side of road. **0.6**

15.2 Road crosses Cabresto Creek. **0.1**

15.3 Large outcrop of quartz-latite dike cutting andesitic-lava-flow sequence. **0.4**

15.7 Small outcrop on left is Precambrian quartzite, on

upthrown side of a normal fault, down to the west, that follows stream in this area. **0.7**

16.4 **Turn right** onto Greenie Peak road (four-wheel drive). Low hills ahead are andesite and quartz latite, discontinuously capped by flat-lying outflow Amalia Tuff. For the next couple of miles, the road winds through the trees, crossing numerous dikes and irregular intrusions of quartz latite cutting andesitic lavas. **0.4**

16.8 Obscure outcrop in grass to left, at bend in road, is a coarse-grained dioritic andesite that may represent an intrusive core of an andesitic volcano. **0.1**

16.9 Back into normal fine-grained andesite of the area. **0.4**

17.3 Road ahead largely follows poorly exposed intrusive quartz latite forming irregular large dikes and laccolithic bodies. **0.4**

17.7 Road forks; **turn right** to Greenie Peak. **0.2**

17.9 Road forks again; **turn right. 0.7**

18.6 Road crosses obscure contact into basal, weakly welded and altered Amalia Tuff, which caps Greenie Peak. **0.3**

18.9 Summit of Greenie Peak (11,249 ft). Float of lower part of Amalia Tuff, scattered around the summit area, is typically abundantly lithic and weakly welded; locally, it shows strong rheomorphic flow layering.

The panoramic view, counterclockwise, includes the upper Red River and the Red River Ski Area to the southwest, with Gold Hill (12,711 ft) behind it. To the right of Gold Hill and the ski area, the pointed timbered mountain in the near distance, with a road on its lower slope, is Sawmill Mountain underlain by thick intracaldera Amalia Tuff. The eastern caldera wall lies where the road goes through the pass just this side of Sawmill Mountain. The ridge trending west from Sawmill Mountain is the divide between the Red River to the south and Cabresto Creek to the north; it consists mostly of intracaldera welded tuff. The north-facing far slope of the Red River valley approximates the south caldera wall and consists entirely of Precambrian rocks. To the left of Gold Hill, the cluster of high mountains includes Wheeler Peak, the highest point in New Mexico (elevation 13,161 ft). These high peaks are all underlain by Precambrian rocks. Due south is the Moreno Valley, a Tertiary graben, flanked on the east by Baldy Mountain, a sill complex of Tertiary quartz latite intruding Cretaceous sedimentary rocks. Due east, the relatively low timbered ridges are volcanic rocks of the Latir field, dipping eastward and repeated several times by northwest-trending faults down to the southwest. In the far distance to the northwest, the highest mountain above treeline is Costilla Peak, consisting of light-tan Precambrian rocks resting on reddish Pennsylvanian and Permian arkosic sediments of the Sangre de Cristo Formation along a Laramide thrust-fault contact. To the north through the trees, the rounded ridges above timberline are the Costilla Mountains, consisting largely of Precambrian granite.

End of supplemental road log 2B for Second Day.

DAY 2: SUPPLEMENTAL ROAD LOG 2C, FROM AMALIA TO COSTILLA RESERVOIR

A continuation of road-log segment 2Q, this option permits further examination of Precambrian rocks, Tertiary volcanic stratigraphy northeast of the Questa caldera, and regional Miocene extensional faults. The route follows Rio Costilla upstream from the town of Amalia, across several eastward-tilted Miocene fault blocks related to early extension of the Rio Grande rift.

Mileage

0.0 N.M. 196 in scenic downtown Amalia; pavement ends; continue upstream. Heavily timbered north slopes of mountains on right are complexly faulted volcanic sequence north of Questa caldera, with underlying prevolcanic Tertiary sediments and Precambrian rocks locally exposed. **1.8**

1.8 Crossing Costilla Creek. Box canyon on east exposes chaotic vent structures of rhyolite of Gonzales Ranch. **0.3**

2.1 On right, exposure of basanitic basalt interlayered with volcaniclastic sediments. **0.3**

2.4 Sign to Rio Costilla Ski Area. Ridge on right is Amalia Tuff, as are low cliffs across river ahead at 11:00. In general, the road goes obliquely downsection in this area. Hills ahead on right are underlain by andesitic lavas. The major fault that bounds the volcanic block lies to the southeast, at the base of the Costilla Mountains. **1.0**

3.4 Mouth of Cordova Canyon along which a northeast-trending fault drops volcanics down against Precambrian and starts a repetition of the section. **0.3**

3.7 Volcaniclastic sandstones interlayered with andesite, dipping approximately 40° to the northeast. **0.3**

4.0 **STOP 1.** Propylitically altered olivine–augite andesite that is typical of the area. Two pointed hills on the right consist, respectively, of welded tuff on the lower hill to the west (right) and overlying andesite and basaltic andesite on the high point to the east (left). **1.1**

5.1 Entry gate to Costilla Park. This is the entry to private land of the Rio Costilla Cooperative Livestock Association (RCCLA), which encompasses much of the old Sangre de Cristo Grant. **0.2**

5.3 Cross a major bounding fault of the volcanic block and enter into Precambrian that makes up the narrow gorge ahead along Costilla Creek. **1.1**

6.4 Crossing Costilla Creek. **0.4**

6.8 Layered Precambrian amphibolite and pegmatite in roadcut on left. Pass into foliated Precambrian granitic rocks. **0.4**

7.2 Crossing Venado Creek (unnamed on Latir Peak 7.5-min quadrangle map).

Stream sediments from this drainage have uranium values of more than 200 ppm, highest of any sampled during the National Uranium Resource Evaluation survey of the Raton 2° sheet (Morgan and Broxton, 1978). Follow-up studies suggest that the uranium is in shear zones in the granitic rocks (C. S. Goodknight, pers. comm. 1983). **0.9**

8.1 Sign "Single lane next 1000 feet." **Watch for logging trucks! 0.4**

8.5 **STOP 2.** Quartz monzonite of Costilla Creek. Strongly foliated, interleaved, fine- to medium-grained quartz-monzonite gneiss and medium- to coarse-grained augen gneiss with potassium feldspar augen 1/3–2/3 in. in diameter. These rock types intergrade complexly, but locally the finer-grained rock seems to cut the coarser. Both are cut by slightly discordant boudined pegmatites consisting chiefly of potassium feldspar, plagioclase, and quartz and locally containing a few flakes of biotite.

Foliation strikes N70°E and dips 35°NW; lineation, marked by elongate feldspar augen and biotite streaks, plunges 10° N50°E. Both foliation and lineation are cataclastic structures. The pegmatites are interpreted as pre-cataclastic, now boudined and partly transposed into the foliation.

A coarse phase of the augen gneiss from a few hundred yards west along the road has yielded a zircon age of 1,644 m.y. (Bowring and others, 1984). **0.4**

8.9 Dike of dark post-metamorphic but pre-Tertiary biotite–hornblende diorite crosses road. Dike is about 80 yds wide and is part of a swarm that can be traced for at least 3 mi southwest along the valley of Costilla Creek. Dike is conspicuous in proper lighting in the lower of the cliffs on the south side of the creek, but is poorly exposed on the north side, although excellent samples are available in talus blocks. **0.1**

9.0 Mouth of Latir Creek. Leaving the Box Canyon. Outcrops still in Precambrian. **0.1**

9.1 Outcrops of gneissic granite and pegmatite. **0.2**

9.3 Irregular dike of diabase, 50–65 ft thick, with chilled margins, cuts gneissic quartz monzonite and pegmatite in outcrop north of road. Dike is part of the same swarm as diorite dike near mouth of Latir Creek. **0.2**

9.5 Outcrops of strongly foliated, gneissic quartz monzonite with pegmatite lenses on north side of road. Foliation makes vertical faces subparallel to road. Transition from quartz monzonite outcrops to smooth slopes ahead marks contact with layered amphibolite–felsic-gneiss sequence. **0.2**

9.7 Valley swings back into the fault zone bounding the volcanic block. For several miles ahead, this fault zone is occupied by quartz-latitic intrusions emplaced along the fault and dated at about 27 m.y. Ahead on right, light-colored rocks are nonwelded, lithic, rhyolitic ash-flow tuff (tuff of Tetilla Peak), a local prominent marker low in the precaldera section that delineates several fault repetitions on ridges ahead. Rocks overlying the lithic tuff on conspicuous cliffs are andesitic-lava flows and interlayered volcaniclastic mudflows between the lithic tuff and the overlying caldera-related Amalia Tuff. **0.8**

10.5 Broken exposures for several hundred yards along the road are in toe of large landslide from north wall of the valley. **0.2**

10.7 Obscure ford across river to mouth of Gate Creek. Small rocky exposure on west side is intrusive quartz latite. East of Gate Creek, rocky exposures with much talus on ridge and along Costilla Creek are of rhyolitic lava flows that overlie, and are cogenetic with, Amalia Tuff. **0.1**

10.8 Chocolate-brown exposures at mouth of gulley on left are in diorite dike. Outhouse on right, courtesy of RCCLA, may be welcome. **0.5**

11.3 Vermejo Gate. Enter the part of the Vemejo Land Grant that was recently donated to the Federal government and is now being managed by the U.S. Forest Service as the Vermejo Unit of the Carson National Forest. **0.1**

11.4 Good exposures of intrusive quartz latite. Intrusive quartz latite continues in outcrops ahead south of creek, with Precambrian quartzite and metamorphic rocks on north. **0.3**

11.7 Irregular body of diorite cuts highly fractured Precambrian quartzite and contains 7–10-ft blocks of quartzite. This body marks the east end of the dike swarm. Outcrops of quartzite for several hundred yards east. **0.3**

12.0 Green outcrops on slopes to north are amphibolite in the quartzite sequence, probably in a fold nose. **0.1**

12.1 Fold nose in quartzite visible in outcrops across creek to the south. **0.6**

12.7 A 30-ft-long pod of muscovite pegmatite in quartzite–mica-gneiss sequence. **0.3**

13.0 **STOP 3.** Junction of Comanche and Costilla Creeks. Small outcrop on north side of road is thinly laminated, muscovitic quartzite displaying at least two fold generations. Folds with north- or northeast-trending axes are warped by later folds with axes

FIGURE 2C-13.0a. Comanche Point, a 25-m.y.-old intrusion of low-silica biotite rhyolite at the junction of Costilla and Comanche Creeks. Irregular shape of intrusion is reflected by geometrically complex columnar ("cordwood") joints that formed perpendicular to cooling surfaces.

FIGURE 2C-13.0c. Dated hornblende-andesite lava flow (27 m.y.) in Comanche Creek, 1 km southeast of Comanche Point. Northeast-dipping andesite flow (Ta) and its basal flow breccia (Tab) overlie andesitic mudflows (Tac).

plunging northwest. A local fracture cleavage strikes northeast and dips gently to moderately northwest, parallel to axial planes of the younger folds. In spite of evidence for multiple deformations, the general trend of layering in the area is fairly consistent with northeast strikes and gentle to moderate southeast dips. To the east, the quartzite contains a 16–30-ft semiconcordant lens of crudely zoned, muscovite-bearing pegmatite.

About 55 yds west along the road are large blocks of lustrous muscovite schist with two lineations crossing at about 30°. The schist contains ovoid quartz–muscovite knots as much as 2 in. in diameter, possibly replacing cordierite. This lithology is interlayered with quartzite on the slopes above.

Just to the southeast is Comanche Point, a 25-m.y.-old intrusion of biotite rhyolite (Fig. 2C-13.0a). This is part of an irregular-laccolith complex, also containing quartz latite, that was emplaced along the bounding fault zone between the volcanic rocks to the south and Precambrian rocks to the north. Note irregular cordwood, columnar-cooling joints in the rhyolite of Comanche Point. High ridge behind Comanche Point is capped by silicic alkalic-basalt flows interlayered with weakly consolidated cobbly conglomerates of the Santa Fe Group (Figs. 2C-13.0b, c).

Follow main road straight ahead up Costilla Creek. **0.4**

13.4 **STOP 4.** North end of outcrops of rhyolite of Comanche Point. Contact with early Tertiary sediments and marginal vitrophyre zone are exposed up gully to west. The vitrophyre at this locality has yielded K–Ar dates of 25.3 m.y. (biotite) and 24.5 m.y. (sanidine). **0.3**

FIGURE 2C-13.0b. Typical exposure of weakly indurated, prevolcanic sediments (Eocene–Oligocene), containing distinctive greenish-quartzite cobbles in a clay-rich, reddish-brown sandy matrix. East of Comanche Point.

FIGURE 2C-14.1 Herd of elk in upper Comanche Creek.

FIGURE 2C-17.2a. Outcrop of partly welded, massive, glassy tuff, lower part of Amalia Tuff. First Creek area, upper Costilla Valley above Costilla Reservoir. View is to west; Costilla Mountains in distance.

13.7 **STOP 5.** Good exposures of intrusive quartz latite, dated by K–Ar at 26.9 (biotite) and 22.3 (hornblende) m.y. This is part of the Comanche Point laccolith emplaced into the lower Tertiary sedimentary rocks. The hill straight ahead is Precambrian and the right skyline is approximately the dip-slope on the basal Tertiary section. **0.4**

14.1 **STOP 6.** La Cueva camp. The poor exposures along side road to left are in lower Tertiary sedimentary rocks. These contain cobbles of a distinctive pale jade-green Precambrian quartzite and some intermixed volcanic material in this area. On right is the terminal moraine of the Costilla Valley glacier. Hill ahead is Precambrian quartzite and muscovite–quartz schist. **1.1**

15.2 Near-vertical fin of quartzite in spectacular exposure to the west (left). Quartzite is in contact with layered amphibolite to the north. Contact can be located within a few feet, but is not exposed. Layering in amphibolite is concordant with layering (bedding?) in quartzite. **0.7**

15.9 Fork in road; **bear left. 0.6**

16.5 Roadcut exposures of the east-dipping volcaniclastic sediments (precaldera) that contain clasts of Precambrian rocks and are overlain by a thin vitrophyric quartz-latitic welded tuff. **0.1**

16.6 **STOP 7.** Exposures of basal part of a thick basaltic-andesite sill that also makes large outcrops below road on next ridge to north. **0.6**

FIGURE 2C-17.2b. Partly welded, glassy Amalia Tuff, showing collapsed dark-pumice lenses in light-gray groundmass of ash shards and phenocrysts. First Creek area, in upper Costilla Valley above Costilla Reservoir.

17.2 View of Costilla Reservoir. Below road ahead, the pinkish roadcut exposure to left is weakly welded Amalia Tuff (Figs. 2C-17.2a, b), oxidized by underlying sill of basaltic andesite and overlain by weakly indurated cobble conglomerates of the Santa Fe Group. At 3:00 the high point is Costilla Peak, consisting of Precambrian resting on upper Pennsylvanian and Permian Sangre de Cristo Formation along a gently dipping Laramide thrust fault roughly at treeline.

End of supplemental road log 2C for Second Day.

Taos Pueblo, ca 1920. Photo by Burt Harwood.

THIRD-DAY ROAD LOG FROM TAOS TO TRES PIEDRAS, TUSAS, SPRING CANYON, TRES PIEDRAS, LAS TABLAS, PETACA, LA MADERA, AND OJO CALIENTE

KIM MANLEY and S. J. MAY

SATURDAY, OCTOBER 13, 1984

Assembly Point: Kachina Lodge
Departure Time: 8:00 a.m.
Distance: 102 miles
Stops: 4

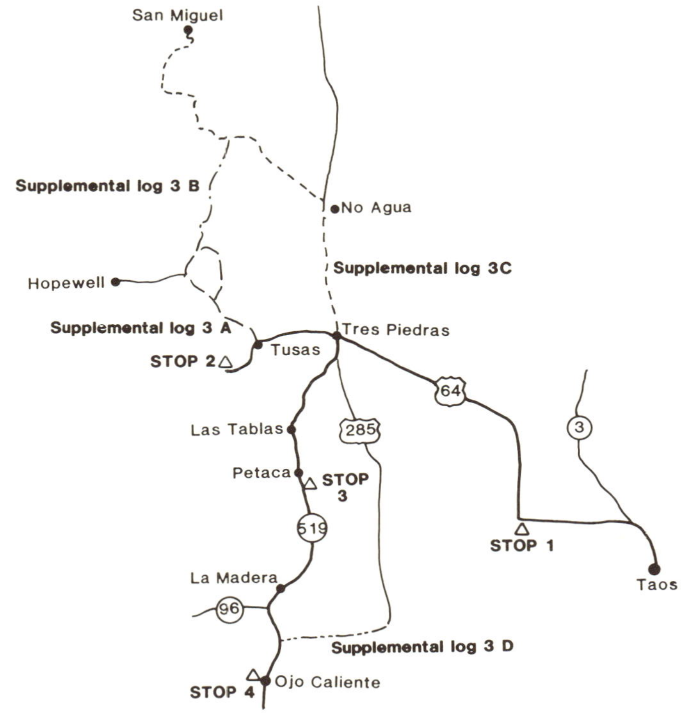

SUMMARY

This day the route will first cross the Rio Grande rift from Taos to Tres Piedras. The Rio Grande gorge (Stop 1) presents a close look at the Servilleta Basalt and discussion of present-day deformation in the Taos area. West of the gorge is a variety of volcanic centers from early (25–22 m.y.) to latest (3 m.y.) rift age. At Tres Piedras the route leaves the Taos Plateau and enters the Tusas Mountains where Precambrian rocks are partially covered by primarily Tertiary sedimentary deposits of the El Rito and Los Pinos Formations. Throughout the Tusas Mountains there are also exposures of pre-rift volcanic rocks of intermediate composition, similar to those of the San Juan and Latir volcanic fields.

The stratigraphy of the Precambrian rocks appears to loosely parallel that of the Picuris Range, with a prominent lower unit of metavolcanic and metasedimentary rocks overlain by voluminous quartzite. Near the boundary, a manganese-rich zone is currently being mapped. This stratigraphy will be viewed and discussed at the edge of Spring Canyon, at Stop 2.

From Spring Canyon the route returns to Tres Piedras and turns south to re-enter the Tusas Mountains where the Tertiary units are well exposed. Between the Tusas Ridge and Stop 3, south of Petaca, the complexity of the interrelationships of the Tertiary rocks can be seen. At Stop 3 are exposures of several of these rocks, including the Amalia Tuff. Volcanic and sedimentary stratigraphy as well as the nature of the western side of the Rio Grande rift will be discussed. Southward, the route enters the northwestern Española Basin. At Ojo Caliente, Stop 4, there will be a discussion of the zone of interfingering of several Santa Fe Group units and the Los Pinos Formation.

Four supplemental road logs accompany this log. The first (3A) follows a circular route north from U.S. 64, up Biscara Canyon, and south again along the Little Tusas Creek. Outcrops of two ash-flow tuffs (Treasure Mountain and Masonic Park) from southern Colorado can be seen on this route. The second (3B) connects the first and third logs through Cisneros Park and the Lucero Lakes. Supplemental log 3C begins at Tres Piedras and covers part of the northern Tusas Mountains via No Agua, the Rio San

Antonio, and high country east of Lagunitas campground to the Rio Los Pinos near the Colorado state line. The route is rough and not recommended for low-clearance vehicles. It passes young rhyolite domes and cinder cones of the Taos Plateau, older Hinsdale Formation basalts, and good exposures of the Esquibel Member of the Los Pinos Formation and the Masonic Park Tuff. Mileages are approximate. The fourth supplemental log section (3D) is a short extension eastward on U.S. 285 from the junction of U.S. 285 and N.M. 96 north of Ojo Caliente. The basaltic plug of Mesa Vibora is discussed.

Mileage

0.0 **Junction** of U.S. 64 and N.M. 3 with Taos Pueblo road (southern route). The pueblo, located 3 mi to the northeast, has a population of 1,516 (1972) and a reservation area of 95,333 acres. According to the U.S. Department of Commerce (1974): The Spanish, under Hernando de Alvarado, discovered Taos Pueblo in 1540 much as it is today, with two large communal houses facing each other across Taos Creek. Colonists from Don Juan de Oñate's community soon settled nearby. A series of Indian revolts against the settlers culminated in the successful Pueblo Revolt of 1680 spearheaded from Taos and led by Popé, a San Juan Indian. The Spanish were driven from New Mexico, not to return for 13 years. The Spanish returned in 1693, and in 1696, with the order by de Vargas that all pueblo governors be shot, effective resistance ceased. After the conquest of New Mexico by the U.S. in 1847, Taos again revolted against the occupation troops and murdered the American governor, Charles Bent, among others. The revolt was crushed and the leaders were hanged.

The Indians of Taos have an ancient and rich cultural past, much of which continues to survive. They live in large communal houses, the upper stories of which are reached by ladders. Round outdoor baking ovens and strings of chili drying in the sun are typical. The underground kivas serve as meeting places for the men, with women admitted only on certain occasions. The Deer, Turtle, and Sun-down Dances are unique and noted for their beauty and precision. The latter, given on September 30 every year, is the most important ceremony and expresses thanksgiving for the harvest. Most of the Taos people are farmers, many are artists, and others find employment in the nearby town of Taos. The Taos are the least typical of the Pueblo Indians because they share many characteristics with the Plains Indians, particularly the Kiowa with whom they have a linguistic relationship. **1.1**

1.1 **Enter El Prado. 2.4**

3.5 **Junction:** Three-way intersection of N.M. 3, N.M. 150, and U.S. 64. **Turn left** onto U.S. 64 to Tres Piedras. **0.2**

3.7 Roadcut in Pliocene–Pleistocene sandy gravel. The clasts are pebbles and cobbles of Precambrian igneous and metamorphic rocks and Tertiary volcanic rocks derived from the Taos Range (Clark and others, 1966). **0.5**

4.2 Crossing valley of Arroyo Seco. **2.0**

6.2 Low roadcut in Pliocene–Pleistocene alluvium. Note extensive soil development. **1.0**

7.2 Side road on left to Taos Municipal Airport. At 2:30 is Picuris Peak, elevation 10,801 ft (about 3,325 m), bald side toward us, which is composed of Precambrian quartzite. Right-lateral-slip Picuris–Pecos fault (Montgomery, 1953; Miller and others, 1963) lies along the north-trending valley to the left (east) of Picuris Peak. Jicarita Peak, elevation 12,750 ft (3,886 m) at 2:00 is on distant skyline (usually has snow patches all year). Jicarita Peak is underlain by gently dipping Pennsylvanian rocks of the southern continuation of the Sangre de Cristo Mountains. (See Fig. 3-7.2) **1.1**

8.3 View north of Rio Grande gorge. **1.8**

10.1 Roadcut in yellowish sandy gravel overlain by sandy alluvium with well-developed soil profile. Soils with prominent horizons of clay and carbonate accumulation are commonly present in the sandy alluvium that caps the piedmont slopes (Clark and others, 1966). **0.6**

10.7 Sandy gravel (Lambert, 1966) in roadcut. From here to valley of Arroyo Seco, the roadcuts are in sandy gravel capped with brown sandy alluvium or deposits of arroyo–terrace gravel. **0.4**

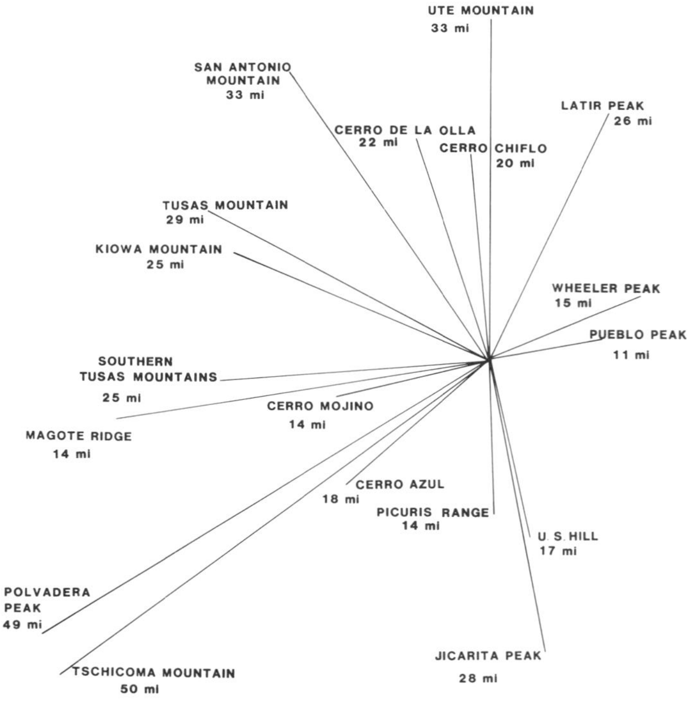

FIGURE 3-7.2. Panoramic index of features seen in the southern San Luis Valley from near the Taos airport.

11.1 East bridge abutment. "Rio Grande Gorge Bridge" historic marker ahead on left. **0.4**

11.5 **Junction: Turn left** to Rio Grande Gorge Overlook on west rim of gorge. **0.4**

11.9 **STOP 1.** Rio Grande Gorge Bridge. The gorge here is about 1,200 ft (366 m) wide, and the river is entrenched about 600 ft (185 m) below the Taos Plateau. The gorge is cut in tholeiitic flows of the Servilleta Basalt and interlayered sediments. K–Ar ages of flows in this area range from 4.5 to 3.6 m.y. (Ozima and others, 1967). Two main sedimentary units, comprising fluvial sand and gravel, are present in the gorge section, but they are largely covered by basaltic talus. A typical flow of the Servilleta Basalt can be examined at the bridge abutment. The basalt is relatively coarse-grained, diktytaxitic (vuggy), and contains nearly equigranular plagioclase, olivine, augite, and a few percent of interstitial glass. This basalt was erupted as highly fluid pahoehoe flows that traveled many miles, yet were only a few feet thick. (See Fig. 3-11.9.) For further information see Dungan and others (this guidebook). **0.3**

12.2 Exit from overlook area; **turn left** (west). **0.2**

12.4 Volcanic center of Tres Orejas at 10:00. This rhyodacite center is the most southerly volcano of the Taos Plateau volcanic field (Fig. 3-12.4). **0.2**

12.6 View on right of large Pliocene–Pleistocene alluvial fans against the Sangre de Cristo Mountains. **0.6**

13.2 Low roadcuts in Servilleta Basalt. **0.8**

14.0 Roadcuts in Servilleta Basalt overlain by brown sand with soil. **3.0**

17.0 Cerros de Taos on left (west). These hills are composed of two overlapping shield volcanoes of olivine andesite. **0.2**

17.2 On skyline to right is Cerro Chiflo, a 10.3-m.y.-old quartz-latite dome (Lipman and Mehnert, 1979). Irregularly shaped hills in near distance are eroded, structurally complex, early-rift volcanics, ranging in composition from andesite to rhyolite (K–Ar ages of 22–25 m.y.). These rocks are preserved in a horst that is structurally continuous with the San Luis Hills in southern Colorado; the structurally deepest part of the rift valley is confined to a relatively narrow zone (1.5–2 mi wide) between this horst and the Sangre de Cristo Mountains to the east. The symmetrical hill farther to the north is Cerro Montoso, a shield volcano of olivine andesite. **5.5**

22.7 View from 10:00 to 12:00 on skyline of west edge of Taos Plateau volcanic field called the Comanche Rim. Peak to south is Cerro Mojino, next to north is Petaca Peak. North of Comanche Rim are foothills of the Tusas Mountains (Fig. 3-22.7), which are composed of hills of Precambrian rock primarily overlain by two Tertiary formations, the El Rito and the Los Pinos. At 2:00 on skyline is San Antonio Mountain.

Although the terms Tusas and Brazos are often used interchangeably, preferred usage describes the Tusas Mountains as the present topographic expression of the structural Brazos uplift.

The 20-mi-wide mountain range, which contains many peaks over 10,000 ft and two (Brazos Peak and Grouse Mesa) over 11,000 ft, trends northwest–southeast from the San Juan Mountains of southwestern Colorado to the Rio Grande rift in north-central New Mexico. Around their southern terminus, the rift (Española and San Luis Basins) is offset to the east. **3.4**

26.1 Low roadcuts in Servilleta Basalt. Low mesa to northeast consists of gently dipping olivine andesite similar to that on Cerro del Aire; but the primary volcano morphology is more obscure, and these andesites may be older. **1.0**

27.1 North of road the 3.5-m.y.-old olivine-andesite shield volcano of Cerro del Aire is conspicuous, with a small spatter cone on its left (west) side. Obscurely silhouetted against the base of Cerro del Aire are low, timbered slopes that define two very low-relief basaltic shield volcanoes that are the largest exposed sources of Servilleta Basalt (Lipman and Mehnert, 1979). **3.0**

30.1 Crossing narrow bridge over Arroyo Aguaje de la Petaca. Good roadcut exposures of Servilleta Ba-

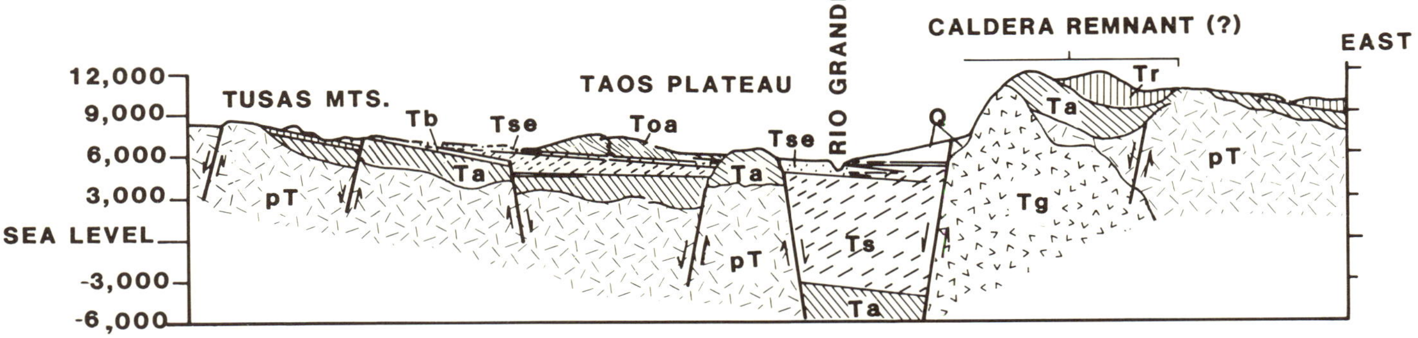

FIGURE 3-11.9. Diagrammatic cross section through the Taos Plateau volcanic field showing inferred structure of the Rio Grande rift based on general geologic relations in the vicinity of latitude 36° 30′ N. Symbols as in Figure 3-12.4. From Lipman and Mehnert (1979).

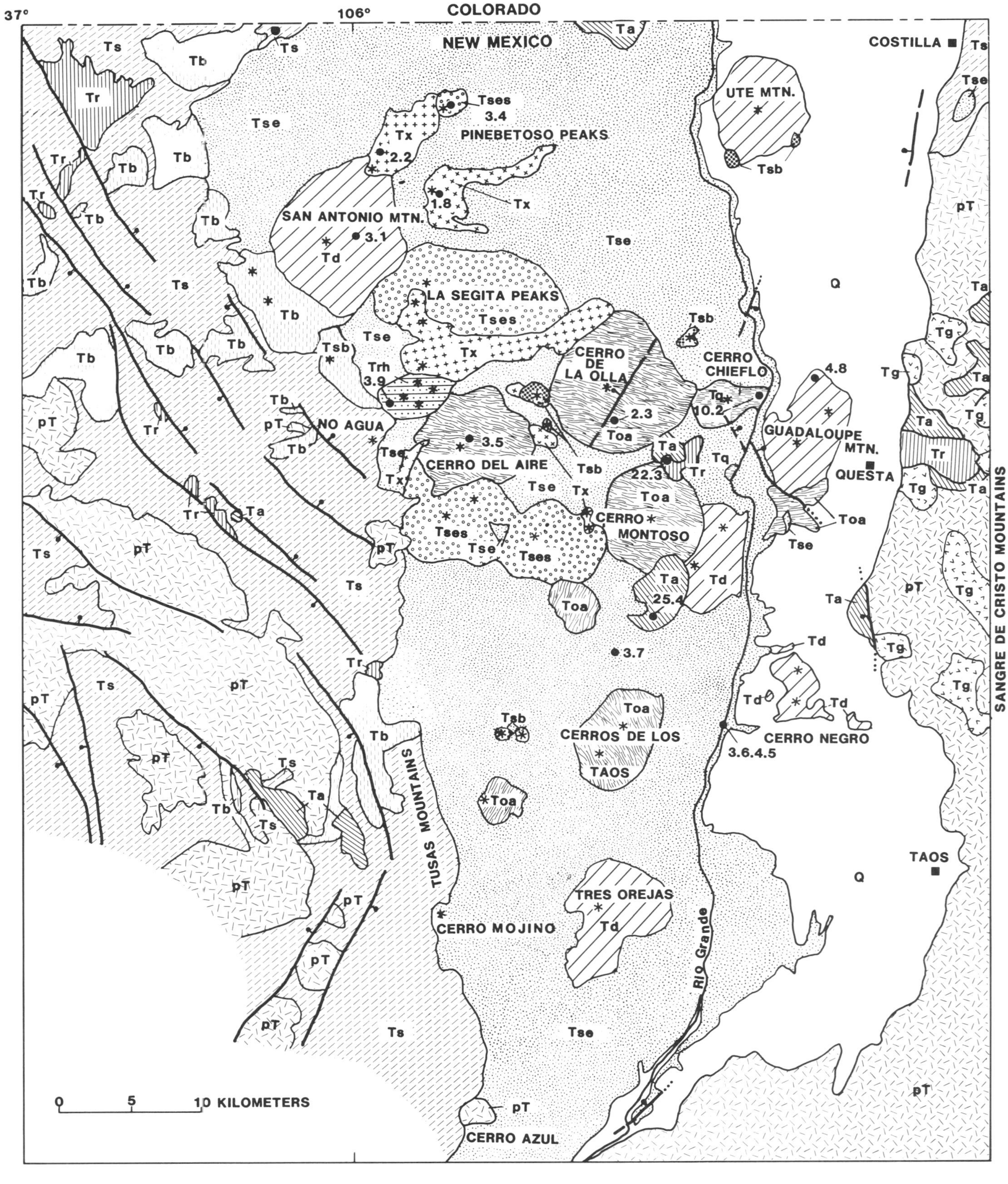
37°
106°
COLORADO
NEW MEXICO
COSTILLA
UTE MTN.
PINEBETOSO PEAKS
SAN ANTONIO MTN.
LA SEGITA PEAKS
CERRO DE LA OLLA
CERRO CHIEFLO
NO AGUA
CERRO DEL AIRE
CERRO MONTOSO
GUADALOUPE MTN.
QUESTA
CERRO NEGRO
CERROS DE LOS TAOS
TUSAS MOUNTAINS
TRES OREJAS
CERRO MOJINO
CERRO AZUL
TAOS
Rio Grande
SANGRE DE CRISTO MOUNTAINS
0 5 10 KILOMETERS

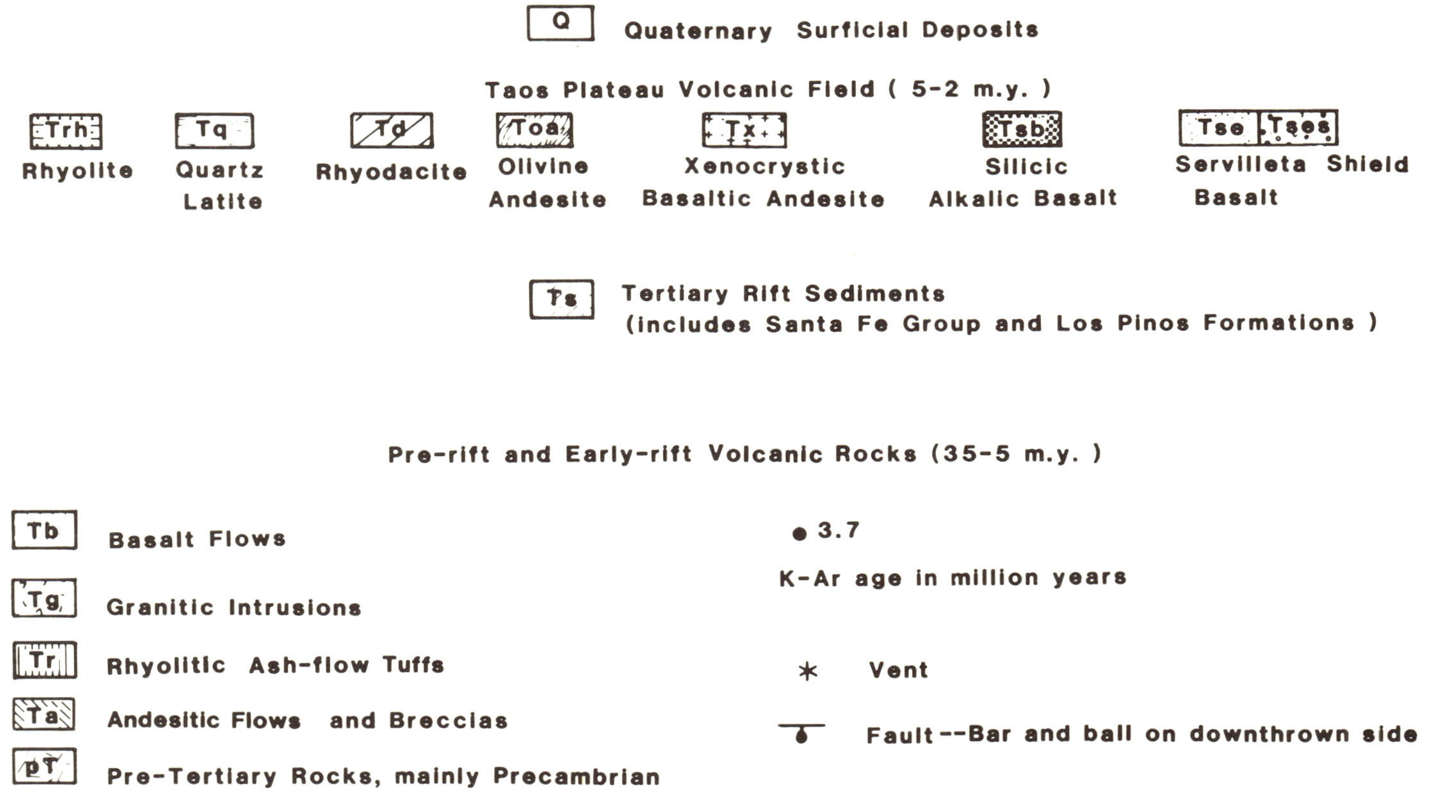

←FIGURE 3-12.4. Generalized geologic map of the Taos Plateau volcanic field based mainly on reconnaissance mapping by P. W. Lipman in 1967–1975. Geology of Tertiary rocks along the west side of the map (Tusas Mountains) is from Manley and others (in press). Figure modified from Lipman and Mehnert (1979).

salt. Roadcuts west of the bridge have good examples of segregations, veins, and pipe vesicles. The uppermost exposures show an obscure but interesting irregular contact between two pahoehoe flows; the top of the lower flow is characterized by tumuli (pressure ridges) with local relief of 3–7 ft. **0.7**

30.8 **Junction** of U.S. 64 with U.S. 285 in Tres Piedras. **Continue straight ahead** (west) on U.S. 64. **0.2**

31.0 On left, Tres Piedras Canyon cut in Precambrian Tres Piedras Granite. **0.5**

TRES PIEDRAS GRANITE

R. A. Wobus
Department of Geology, Williams College, Williamstown, Massachusetts 01267

The foliated Proterozoic granite that forms the conspicuous rounded outcrops (more than three) for which this village was named is representative of the older (ca 1,700 m.y.) plutonic episode of the Tusas Range. The Tres Piedras Granite seen here, and its slightly more mafic precursor, the Maquinita Granodiorite, bear the structural imprint of the same period of regional metamorphism that deformed their metasedimentary and metavolcanic wall rocks. The Tres Piedras is a high-silica (about 76% SiO_2), low-calcium granite (quartz monzonite in pre-IUGS terminology) with $K_2O : Na_2O$ of about 5 : 3. It is the same rock that underlies most of the Tusas Box, but is not the same as the granite stock underlying Tusas Mountain, with which it was formerly correlated. The latter rock, named the Tusas Mountain Granite (Wobus and Hedge, 1982), is a post-tectonic granite locally enriched in fluorine and a suite of trace elements indicative of late vapor-phase activity. Rb–Sr whole-rock analyses and U–Pb dating of zircon from the Tusas Mountain Granite indicate an age between 1,500 and 1,420 m.y.

31.5 Taos County line. **0.3**

31.8 Enter Carson National Forest. Outcrops of Tres Piedras Granite seen through trees on right. **0.9**

32.7 Approximate contact of Tres Piedras Granite with the Cordito Member of the Los Pinos Formation. **2.0**

STRATIGRAPHIC TERMINOLOGY

The Los Pinos Formation was named by Atwood and Mather (1932) and described later by Larsen and Cross (1956), who assigned it a Miocene and Pliocene(?) age. As defined, it includes sedimentary and volcanic rocks that overlie the Conejos Formation and Treasure Mountain Tuff, and underlie the Hinsdale Formation.

Butler (1946) subdivided the Los Pinos Formation south of Broke Off Mountain into three clastic members (Biscara, Esquibel, and Cordito) and a member of interbedded basalt flows (Jarita). North of Broke Off Mountain, Butler was unable to distinguish the three clastic units, but assumed the conglomerate beds present to be correlative to the two lower members (Biscara and Esquibel). Butler's terminology was confined to his unpublished dissertation until incorporated in a report pub-

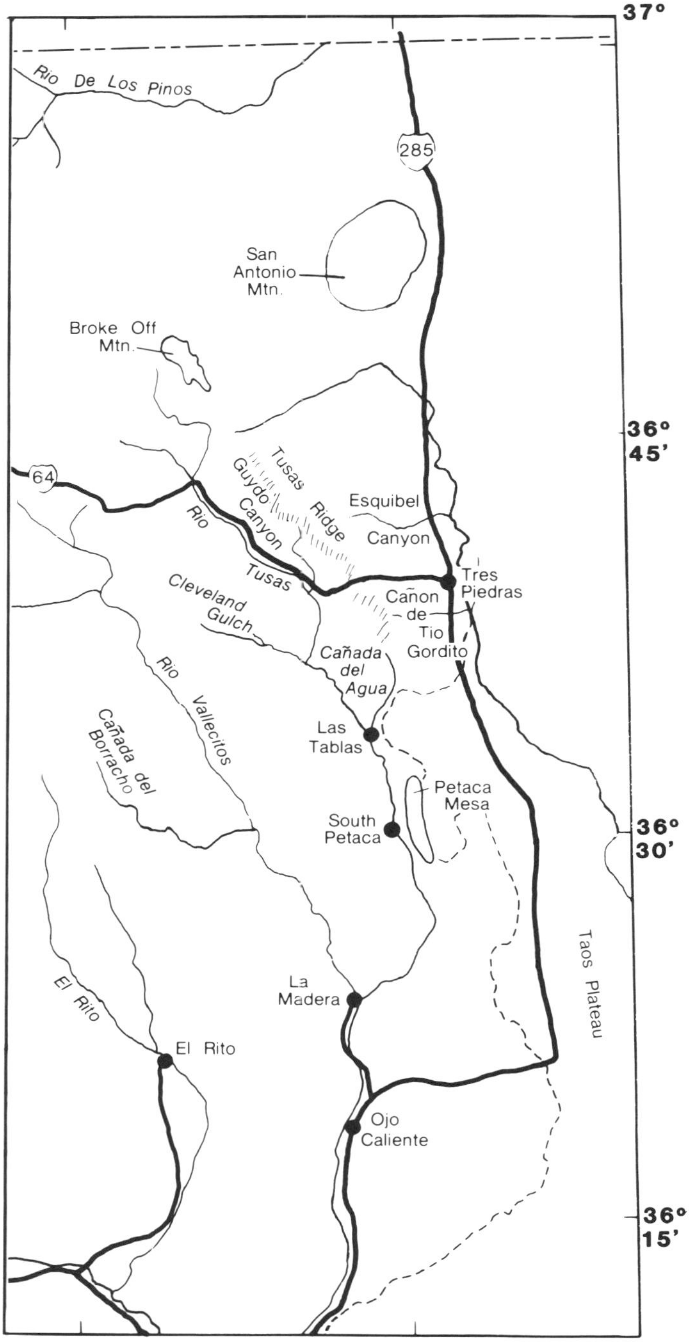

FIGURE 3-22.7. Geographic place names for eastern Tusas Mountains.

FIGURE 3-30.8a. Old water tower for narrow-gauge railroad, Tres Piedras.

lished by the New Mexico Bureau of Mines and Mineral Resources (Barker, 1958).

Mapping of the entire Tusas Mountains (Manley and others, in press) has led to the abandonment of the Biscara Member and minor revisions of the Esquibel and Cordito Members (Manley, 1981), which are now recognized as contemporaneous alluvial-fan deposits derived from the San Juan Mountains and the Sangre de Cristo Mountains, respectively. These members interfinger in the central Tusas Mountains (Fig. 3-32.7). The Jarita Basalt Member and all basalt flows interbedded in the Los Pinos Formation are now included within the Hinsdale Formation. This revision was suggested by Butler (1971) and formalized by Lipman and Mehnert (1975).

34.7 Forest Road 167 to right. **0.3**

35.0 Summit of Tusas Ridge, elevation 8,819 ft. Approximate contact of Cordito Member of the Los Pinos Formation with the Esquibel Member. Descend into valley of Rio Tusas in Esquibel Member. **1.6**

36.6 Cross Tusas Valley fault zone, which ranges from a single fault to a zone about 3 mi wide paralleling the north side of the Tusas Valley with southwest side down. Here it has two segments between which is the Esquibel Member, but southwest of both is a block of rotated Santa Fe Group (Manley and Wobus, 1982b).

A third fault brings the Esquibel Member to road

FIGURE 3-30.0. Outcrops of Tres Piedras Granite at Tres Piedras.

FIGURE 3-30.8b. Prospector's cabin in Tres Piedras.

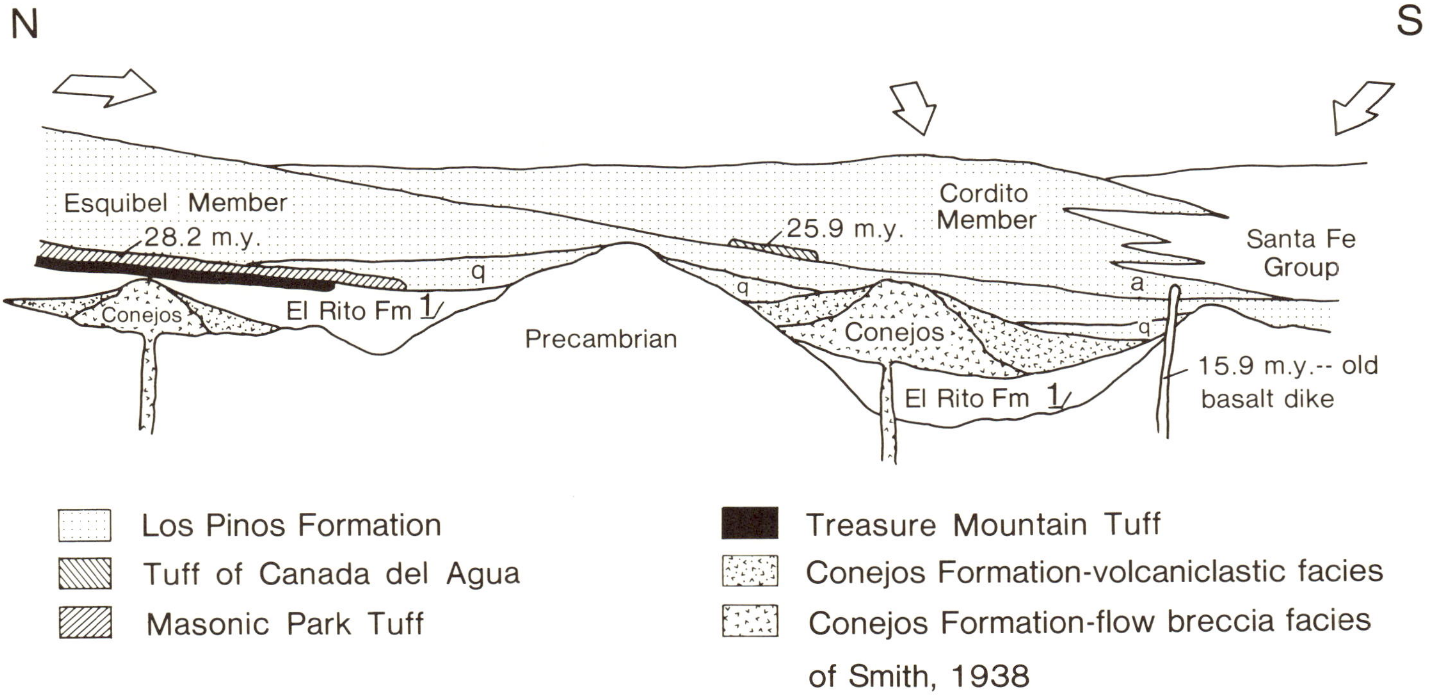

FIGURE 3-32.7. Diagrammatic longitudinal cross section of the Tusas Mountains, New Mexico. Precambrian-clast-rich bed in the Esquibel Member indicated by q; possible position of Abiquiu of Smith (1938) indicated by a. Arrows indicate direction of transport from different source areas.

level in the hillside to the left. The ridge to the left is partially capped by a Hinsdale basalt flow. **0.6**

37.2 At 11:30, Kiowa Mountain (spelled Kiawa in most literature, but Kiowa on the USGS topographic maps and in *New Mexico Place Names*) formed by Precambrian quartzite. **0.5**

37.7 **Junction: Turn left** onto N.M. 111. Cross Rio Tusas into Tusas, New Mexico (Tusas is Spanish for prairie dogs), a former trading point. Post office was here from 1898 to 1930.

At 11:00 is Tusas Mountain, elevation 10,143 ft, formed on Precambrian granite–gneiss (Tusas Mountain Granite) with hornblende schist (Moppin Metavolcanic Series of Barker) along the south, west, and northwest sides.

Supplemental road log 3A continues west on U.S. 64. **1.3**

39.0 On left the Rio Tusas exits the valley through a narrow canyon that leads into the Tusas Box. The outcrops exposed on the east, near river level, include the Tres Piedras Granite and the El Rito Formation. Road climbs through Cordito Member. **2.0**

41.0 **STOP 2.** Edge of Spring Creek Canyon (Fig. 3-41.0a). Park along road. The confluence of Spring Creek and the Rio Tusas define the beginning of the Tusas Box which terminates at the town of Las Tablas. The southwest side of the box is formed by Precambrian rocks; the northeast here is also Precambrian, but 2 mi downstream several Tertiary units create the canyon wall. A fault zone parallels the drainage (Manley and Wobus, 1982a).

Exposures along the road into the canyon show the Cordito Member unconformably overlying both the El Rito Formation and Precambrian silicic metavolcanics and metasediments (Figs. 3-41.0b and d).

Near the base of the Ortega quartzite and of Kiowa Mountain, the peak south of the canyon, is a manganese-rich zone similar to that found in the Picuris Range. It is being currently mapped by M. L. Williams. The presence of this zone supports the conclusion that only one quartzite unit is present in these mountains, that it correlates with the Ortega Quartzite, and that the underlying metasedimentary and metavolcanic rocks may be equivalent to the Vadito Group (Williams, 1984).

FIGURE 3-37.7. View northwest up the Rio Tusas valley from the old townsite of Tusas. Jawbone Mountain on skyline.

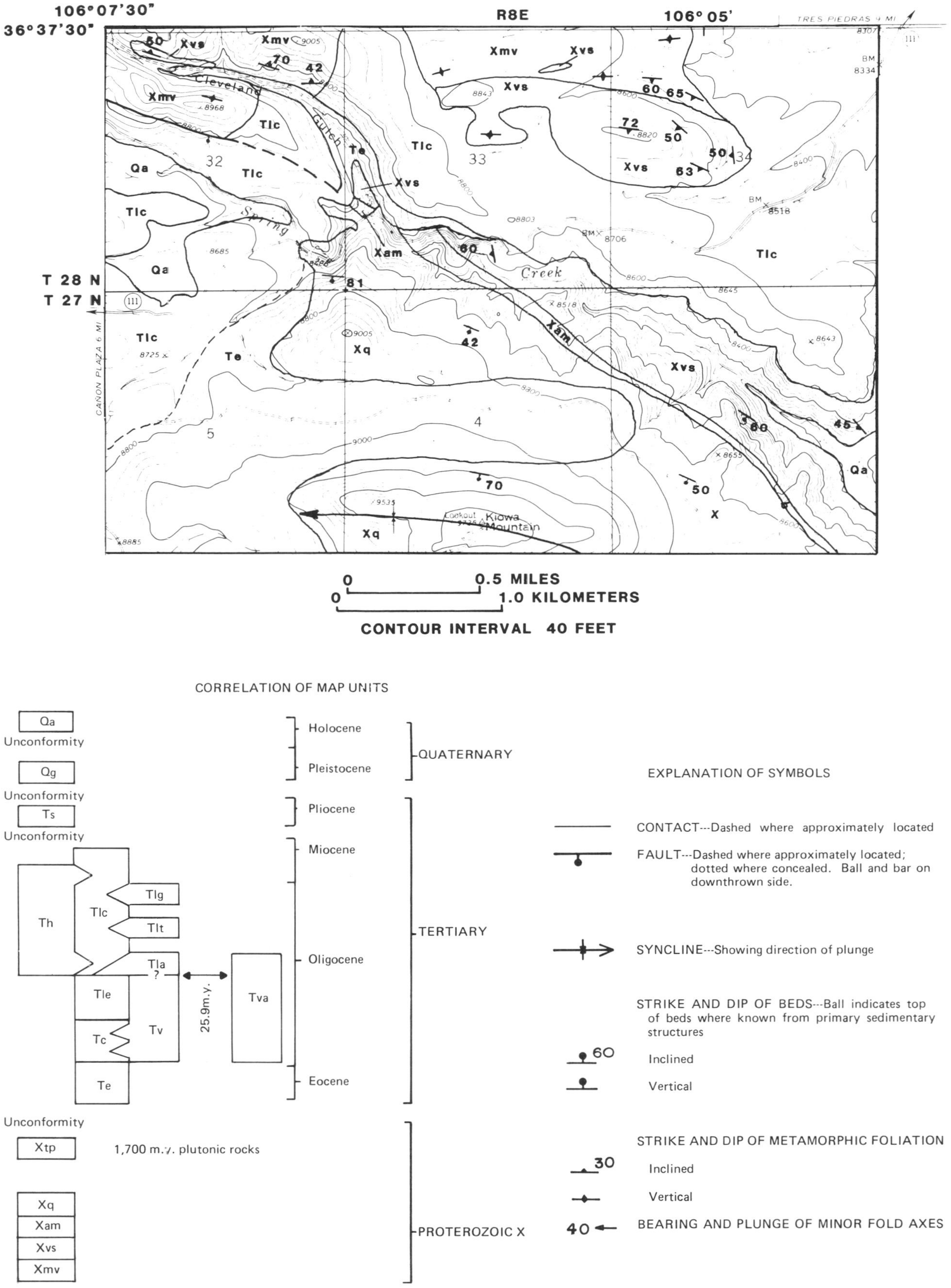

FIGURE 3-41.0a. Geologic map of the Spring Canyon part of the Tusas Box. From Manley and Wobus (1982a).

DESCRIPTION OF MAP UNITS IN FIGURE 3-41.0a

Qa **Alluvium** (Holocene)—Clay, silt, sand, and gravel deposited on lowest terraces and present floodplain. Locally includes colluvium. Variable thickness.

Tsf **Upper part of Santa Fe(?) Group** (Pliocene)—Fine to coarse, light-brown sand and silt at present preserved along Rio Tusas drainage. Thickness 5 m.

Qg **Gravel** (Pleistocene)—Poorly sorted boulder gravel on gently sloping surfaces. Thickness about 10 m.

Ts **Servilleta Basalt** (Pliocene)—Grayish-black tholeiitic basalt containing olivine phenocrysts and diktytaxitic texture. K–Ar ages near Taos are 3.6–4.5 m.y. (Ozima and others, 1967). Thickness about 20 m.

Th **Hinsdale Formation** (Miocene and Oligocene)—Grayish-black, fine-grained basalt flows interbedded with the Los Piños Formation (Tlc, Tle). Similar flows to the north have yielded K–Ar ages from 25 to 5 m.y. (Lipman and Mehnert, 1975, p. 130). Thickness 1–5 m.

Tl **Los Pinos Formation**

Tlc **Cordito Member** (Miocene and Oligocene)—Gray tuffaceous sandstone and moderately well-sorted conglomerate. Clasts, usually subrounded to well rounded, consist of several different rhyolitic ash-flow tuffs, gray rhyolite, intermediate and, more rarely, mafic volcanic rocks, and Precambrian rocks of several lithologies. Boulders are as much as 2 m long. Includes sandstones with abundant rhyolitic-pumice fragments of several compositions, mudflow units, and the Amalia Tuff. Interfingers with basalts of the Hinsdale Formation (Th). Conformably overlies Esquibel Member (Tle). Thickness at least 200 m.

Tle **Esquibel Member** (Oligocene)—White tuffaceous shale, mudflows, and moderately well-sorted conglomerate. Clasts, subrounded to well rounded, consist of intermediate volcanic rocks with large feldspar phenocrysts, vitrophyre, pumice, Precambrian rocks, and a few mafic volcanic rocks. Pumice clasts and shale especially common in upper part; Precambrian clasts dominant in lower part. Lower part gradational with the El Rito Formation (Te). Thickness at least 80 m.

Tlg **Gray rhyolite breccia** (upper Oligocene?)—Volcanic-flow breccia of granule- to pebble-sized gray rhyolite clasts. Contains rare clasts of basalt, Precambrian rocks, welded tuff, and intermediate volcanic rocks. Part of Cordito Member of the Los Pinos Formation (Tlc). Many small outcrops not mapped. Closely associated with tuff of Las Tablas (Tlt). Overlies Esquibel Member of the Los Pinos Formation (Tle), volcanic breccia (Tv), tuff of Las Tablas (Tlt), and El Rito Formation (Te). Thickness at least 7 m.

Tlt **Tuff of Las Tablas** (upper Oligocene)—White, nonwelded ash-flow tuff and reworked tuff. Phenocrysts include plagioclase and sparse biotite; many varieties of fibrous pumice are present. Lithic fragments include volcanic rocks, quartzite, and tuffs. Pebble-sized clasts of several rhyolitic ash-flow tuffs in reworked units suggest that the tuff of Las Tablas is contemporaneous with the Cordito Member of the Los Pinos Formation (Tlc). Thickness at least 40 m.

Tla **Tuff of Cañada del Agua** (Oligocene)—Light-gray, rhyolitic, welded ash-flow tuff containing about 40% phenocrysts, including quartz, sanidine and plagioclase, and sparse hornblende and biotite. Interfingers with lower Cordito Member of the Los Pinos Formation (Tlc). Now known to be part of the Amalia Tuff. K–Ar age of 25.9 m.y. from quadrangle to the south (Bingler, 1968, p. 36). Thickness about 7 m.

Tc **Conejos(?) Formation** (Oligocene)—Gray to greenish-gray, poorly sorted, volcaniclastic conglomerate. Clasts are subangular, intermediate- to mafic-composition volcanic rocks commonly containing phenocrysts of plagioclase, augite, hornblende, or biotite. Derived in part from local outcrops of volcanic breccia (tv). Lithology and stratigraphic position appear comparable to those of volcaniclastic facies of the Oligocene Conejos Formation in Colorado (Lipman and others, 1970). Includes part of Biscara Member of Los Pinos Formation of Barker (1958). Thickness at least 135 m.

Tv **Volcanic-flow breccias** (Oligocene)—Greenish-gray to maroon-red intermediate-composition breccia. Generally porphyritic with phenocrysts of plagioclase, pyroxene, and less commonly hornblende. Lithology and stratigraphic position appear comparable to those of volcanic breccias of the Oligocene Conejos Formation of Colorado (Lipman and others, 1970). May include some flow breccias younger than the 29.8 m.y. for uppermost Conejos Formation. Includes part of Biscara Member of Los Pinos Formation of Barker (1958). Thickness at least 35 m.

Te **El Rito Formation of Smith (1938)** (Oligocene and·Eocene)—Gray to red, medium-grained arkosic sandstone to poorly sorted boulder conglomerate. Sandstone is crudely bedded. Clasts in conglomerates angular to well rounded, consist of Precambrian quartzite, schist, and metavolcanic rocks with some quartz, amphibolite, and granite. Where formation rests against the Precambrian source rocks, it has the appearance of talus and may include some younger talus clasts. As described here, includes much of, and is equivalent to, the Ritito Conglomerate of Barker (1958); may also locally include part of the lower Esquibel Member of the Los Pinos Formation (Tle). Thickness of least 250 m.

PROTEROZOIC INTRUSIVE ROCKS

Xtp **Tres Piedras Granite** (Proterozoic X)—Pink to buff, fine- to medium-grained biotite–quartz monzonite, generally well foliated. Forms rounded outcrops where weathered; best exposed in Tusas Box, where contacts with layered metamorphic rocks are generally concordant. Named for prominent exposures in and near village of Tres Piedras, 14 km north of this quadrangle. Probably correlative with Maquinita Granodiorite of Barker (1958) to northwest, dated at 1,750–1,700 m.y. by L. T. Silver (*in* Barker and Friedman, 1974).[1]

PROTEROZOIC METASEDIMENTARY AND METAVOLCANIC ROCKS

Xq **Quartzite** (Proterozoic X)—Light-gray to bluish-gray, vitreous to pebbly quartzite with dark laminations and minor hematite, kyanite, and muscovite. Trough-type crossbedding locally well preserved. Amphibolite interlayers near base of section. Secondary muscovite increases in abundance to the south and southeast; quartzite grades into quartz–muscovite schist, with preserved crossbedding, along west side of Mesa de Jarita. Primary sedimentary structures in this unit are unrecognizable in extreme southwestern corner of map area because of tectonic deformation. Includes upper quartzite member of Kiowa Mountain Formation of Barker (1958) and unit previously mapped by Barker (1958) as Ortega Quartzite in southwestern part of quadrangle.

Xam **Amphibolite** (Proterozoic X)—Dark-gray to black, fine-grained amphibolite (metabasalt) in layers as much as 15 m thick, intercalated with vitreous to muscovitic quartzite. Retrograde metamorphism of amphibolite is indicated by chlorite and epidote replacing hornblende and by sodic (oligoclase) composition of plagioclase. Probably correlative with amphibolite member of Kiowa Mountain Formation of Barker (1958). Chemical analyses indicate a tholeiitic composition (Barker and Friedman, 1974).

Xvs **Silicic metavolcanics and associated metasediments** (Proterozoic X)—Thick section of metamorphosed silicic tuffs and flows, rhyodacite to rhyolite in composition. Interlayered with crossbedded muscovitic quartzite and discontinuous conglomerate beds (Xc). Metavolcanics occur in three textural varieties of similar chemical composition (Hovorka, 1978): dark-red to gray, fine-grained, porphyritic meta-tuff containing abundant reddish-quartz phenocrysts and devitrified flattened-pumice fragments; dark-gray (weathering to brown), fine-grained porphyry with pink phenocrysts of subhedral feldspar and rare pumice fragments; and light-gray, fine-grained, thinly laminated meta-rhyodacite with rare quartz phenocrysts and, locally, a tightly crenulated structure. Textural varieties occur together at outcrop scale and are not mapped separately. All silicic metavolcanic rocks were named Burned Mountain Metarhyolite by Barker (1958) and are approximately 1,750–1,800 m.y. old based on U–Pb dating (L. T. Silver, *in* Barker and Friedman, 1974).[2] Interlayers of muscovitic quartzite (Xc) represent parts of the lower quartzite member of Kiowa Mountain Formation (Barker, 1958).

Xmv **Moppin Metavolcanic Series of Barker (1958)** (Proterozoic X)—Greenish-gray phyllite, green-chlorite schist, silver-gray-sericite schist and sericitic meta-graywacke, and dark-gray to dark-green, equigranular to porphyritic amphibolite (metabasalt). Rare interbeds of sericitic meta-conglomerate with flattened or rod-like cobbles of sericite schist and phyllite, and of purplish, magnetite-rich schist. Unit is thought to represent metamorphosed mafic volcanic rocks and associated sedimentary rocks.

FOOTNOTE: [1]Recent changes in decay constants for U–Pb ages would alter these ages to 1,675–1,725 m.y. (Steiger and Jäger, 1977).
[2]Recent changes in decay constants for U–Pb ages would alter these ages to 1,725–1,775 m.y. (Steiger and Jäger, 1977).

FIGURE 3-41.0b. Cordito Member of the Los Pinos Formation unconformably overlying Precambrian rocks at Spring Creek.

PRECAMBRIAN ROCKS OF THE TUSAS MOUNTAINS

R. A. Wobus
Department of Geology, Williams College, Williamstown, Massachusetts 01267

The stratigraphic interpretation of the Proterozoic metavolcanic rocks and the quartzite–conglomerate sequence in the Tusas Mountains has been the subject of considerable controversy. For a summary of these interpretations see Wobus (this guidebook); the ensuing discussion follows the interpretation of Burns and Wobus (1983), Gresens and Stensrud (1974), and Just (1937), with an alternative interpretation to be found in Barker (1958).

FIGURE 3-41.0c. Conglomerate in basal Cordito Member of the Los Pinos Formation. Dog is standing on underlying El Rito Formation. Exposure in Cleveland Gulch near junction with Spring Creek.

Kiowa Mountain, to the south of this stop, is the southeastern end of a conspicuous ridge underlain by Proterozoic quartzite and quartz-pebble conglomerate that extends northwest into the Brazos Peak 15-min quadrangle, where the same rock underlies Brazos Box. These quartzites, and correlatives trending northwest from the Ortega Mountains and from Jawbone Mountain, are preserved in the troughs of major synclines, as recognized by Gresens and Stensrud (1974).

The quartzites are underlain by interlayered, metamorphosed, bimodal volcanic rocks and their associated volcaniclastic sediments. The mafic units of this assemblage (tholeiitic metabasalts, chlorite–muscovite phyllite, and schist) were named the Moppin Metavolcanic Series by Barker (1958), and the felsic components (metarhyolite sills and ash-flow tuffs, quartz–muscovite schist, and minor fluviatile conglomerate) constitute the Burned Mountain Metarhyolite, a name expanded from Barker's original (1958) definition.

Return via N.M. 111 to Tusas and Junction with U.S. 64, **turn right,** and continue east to Tres Piedras. **10.2**

FIELDWORK IN THE TUSAS MOUNTAINS IN THE 1930's

A. P. Butler, Jr.
9625 West 36th Ave., Wheat Ridge, Colorado 80033

Conditions for doing geologic fieldwork within a distance of 10–25 mi from Tres Piedras were more primitive in the 1930's than they are today in two main aspects. These were the kind of accommodations and the shape of the roads and highways, especially N.M. State Highway 74, now U.S. 285, that runs north–south through Tres Piedras.

A wall tent 10 ft long served as shelter and living quarters for a field party. We slept in bedrolls spread on the ground in the tent. Meals were cooked on a two-burner gasoline stove placed on an orange crate set on its side. Such crates were two-compartment affairs made of wood, and the space under the stove was used for storing utensils and some packaged food. Another crate and a wooden box provided additional storage space. Some kind of table was contrived for serving meals, but I do not recall any details about it. A gasoline lantern suspended from the ridge pole of the tent provided light in the evening.

FIGURE 3-41.0d. El Rito Formation overlying Precambrian rocks with nearly vertical foliation. Cleveland Gulch near junction with Spring Creek.

Cured meat such as bacon and ham and some fresh foods were placed in a burlap bag and hung by a rope from a branch on the shady side of a tree. That way it was out of reach of non-climbing animals, and no climbers were near our camp. Wetting down the bag with water daily helped to keep the contents cool during the middle of the day.

My fieldwork was done mostly in the eastern half of the Tusas 30-min quadrangle and was carried out from three main field camps. The first was in Colorado, about 1 mi north of the edge of the quadrangle and the state line. There our water had to be hauled in a bucket from a spring 25 or 30 ft lower than the campsite. The other camps were on homesteads of local residents who kindly allowed us to get water from their domestic wells. At a camp on the Hubert Hested homestead about 2.5 mi south of Tres Piedras, the boys in the family hauled water for us.

The condition of the state highway (now U.S. 285), particularly north of Tres Piedras, made travel to Colorado, where we obtained most of our supplies, slow—about 25 mi an hour. The highway was a graded dirt road where stretches of soft, potholed roadbed with protruding blocks of basalt alternated with stretches of almost unmodified bare rock. South of Tres Piedras the roadbed was somewhat better. Roads that extended west from U.S. 285 and provided access to Las Tablas, Petaca, Vallecitos, the upper Tusas Valley, and the valley of the Rio San Antonio were generally as good as, and in places better than, the highway. None was gravel-surfaced, however, and no maintained road extended from Tres Piedras to Tierra Amarilla as paved U.S. 64 does today. These subsidiary roads made much of the eastern half of the Tusas quadrangle reasonably accessible by a 1934 Ford sedan that served as a field car and means of travel from and to our academic base in New England.

51.2 **Junction** of U.S. 64 and U.S. 285 in Tres Piedras. **Turn right** onto U.S. 285. **1.3**

52.5 **Junction: Turn right** onto N.M. 519 and Forest Road 222. **0.7**

53.2 **Junction: Turn left** on N.M. 519 and Forest Road 222. Forest Road 675 bears right. **0.3**

53.5 On left is Precambrian Tres Piedras Granite with basalt of the Hinsdale Formation along its west and south sides and Servilleta Basalt along its east side. Road on Cordito Member of Los Pinos Formation. **1.1**

54.6 Exposure on right of light-gray, rhyolitic, moderately welded ash-flow tuff informally called the tuff of Cañada del Agua (Manley and Wobus, 1982a), but now recognized as the Amalia Tuff (see discussion at Stop 3, mile 66.2). The tuff contains about 40% phenocrysts, including quartz, sanidine, and plagioclase, and sparse hornblende and biotite. It defines the basal Cordito Member of the Los Pinos Formation and has a K–Ar age of 25.9 m.y. (Bingler, 1968), which would be 26.4 m.y. with current constants for K–Ar dating (Fig. 3-54.6). **0.1**

54.7 Cordito Member in roadcuts on right. **0.8**

55.5 Summit of Tusas Ridge (8,600 ft). Forest Road 578 crosses Forest Road 222. This ridge extends northwest–southeast for 17 mi from Cisneros Park to Petaca Peak. It is composed of Los Pinos Formation sedimentary rocks and Hinsdale Formation

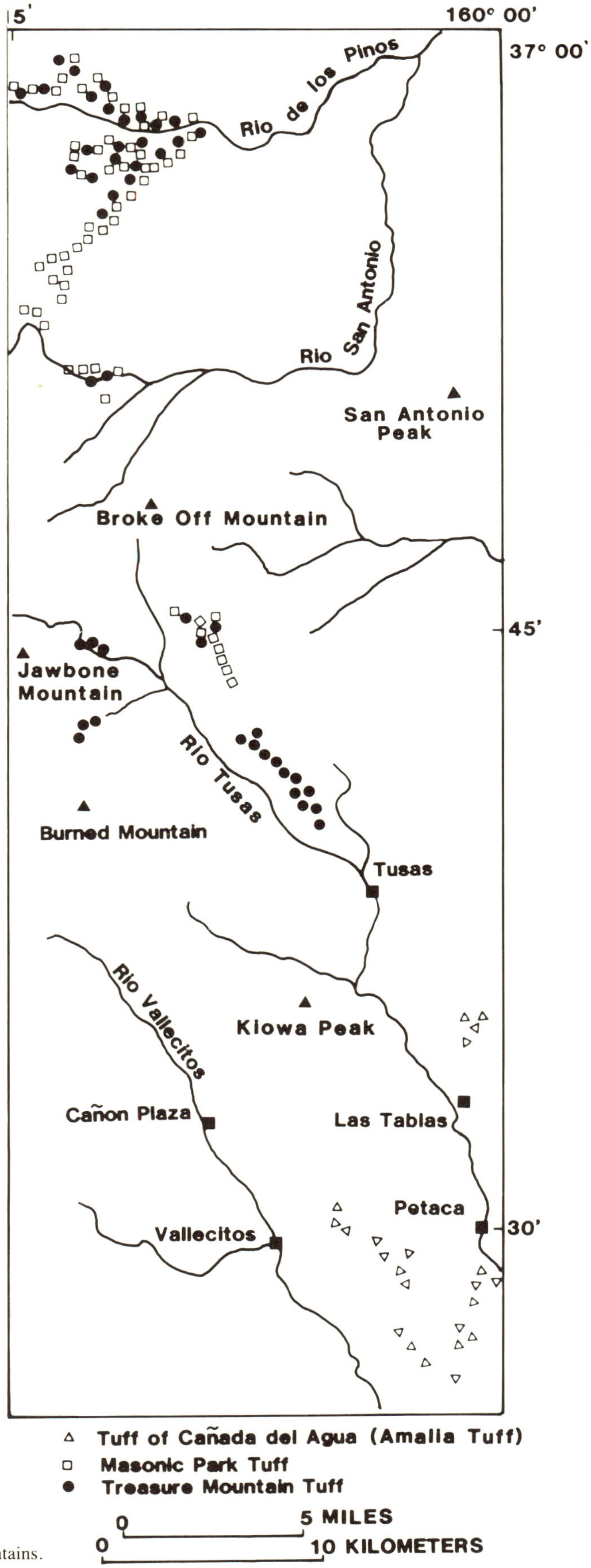

FIGURE 3-54.6. Distribution of outcrops of ash-flow tuff in the eastern Tusas Mountains.

basalt flows in an eastward-tilted fault block. Remnants of horizontal Servilleta Basalt(?) flows locally cap the ridge a few miles to the north and south of here. **0.2**

55.7 Amalia Tuff on right. Deposits beneath the tuff are in the Esquibel Member of the Los Pinos Formation, although mantled here by colluvium from the Cordito on hillside above. **0.6**

56.3 Cross fault bounding the west side of Tusas Ridge and continue in Cordito Member of Los Pinos Formation; this fault is continuous with the one crossing U.S. 64 at mileage 36.6 (Fig. 3-56.3). **0.6**

56.9 Large exposure of Tuff of Cañada del Agua (Amalia Tuff) overlying a flow of the Hinsdale Formation (Figs. 3-56.9a and b). **0.8**

57.7 Andesite-flow breccia exposed on both sides of the road. In this area, the Hinsdale Formation and Cordito Member of the Los Pinos Formation overlap a local volcanic field and its associated volcaniclastic sedimentary rocks. The andesite is generally porphyritic, with phenocrysts of plagioclase, pyroxene, and, less commonly, hornblende. Lithology and stratigraphic position appear comparable to those of the volcanic breccias in the northern Tusas Mountains that are part of the Oligocene Conejos Formation (Lipman and others, 1970). The volcaniclastic facies includes part of the Biscara Member of the Los Pinos Formation of Barker (1958). **0.5**

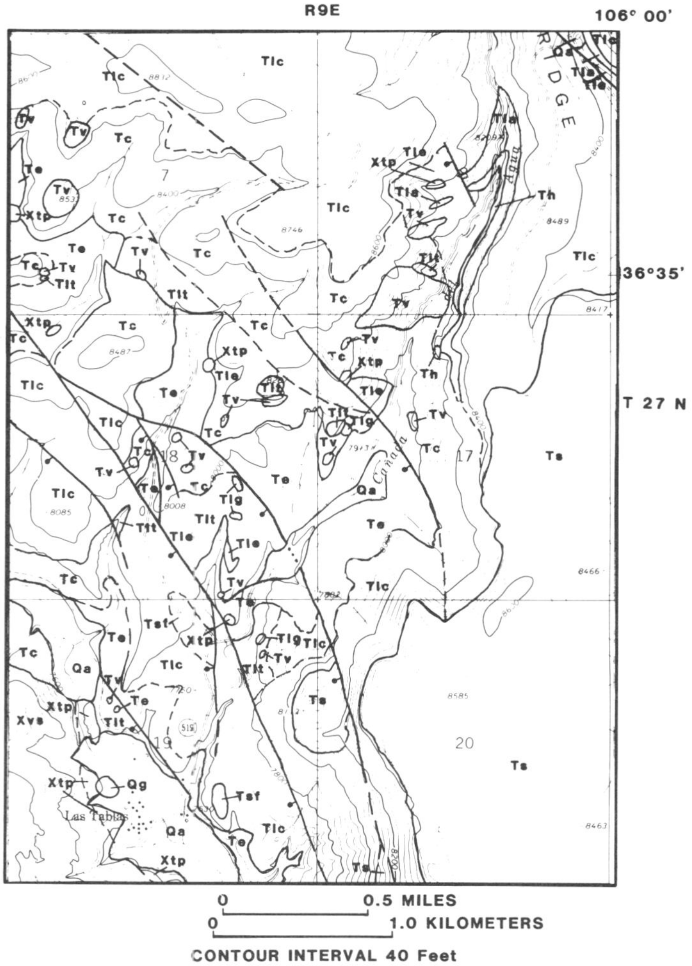

FIGURE 3-56.3. Geologic map of Cañada del Agua and part of the Tusas Box. From Manley and Wobus (1982a). See Fig. 3-41.0a for explanation of symbols and descriptions of map units.

FIGURE 3-56.9a. Amalia Tuff in Cañada del Agua.

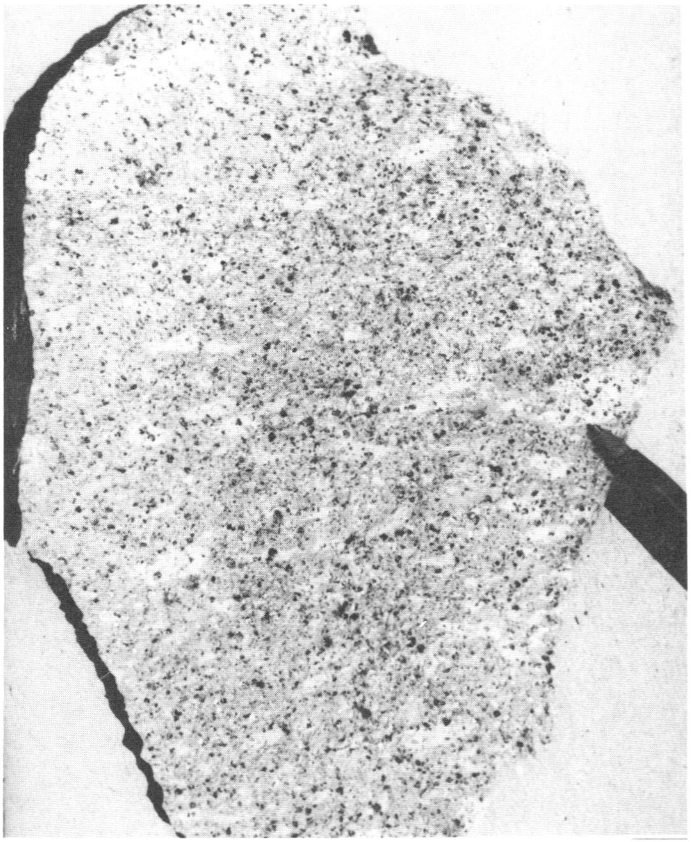
FIGURE 3-56.9b. Clast of Amalia Tuff showing phenocrysts and collapsed pumice. Sample from Cañada del Agua.

FIGURE 3-57.7a. Conejos(?) Formation andesitic-flow breccia, northeast side of the Tusas Box.

58.2 Crossing a down-to-the-southwest fault. Exposures are in Cordito Member overlying a paleohill of El Rito Formation. **0.2**

58.4 Low roadcut exposures are Esquibel Member. View left (east) of Red Mesa capped by Servilleta Basalt(?). Basalts have not been studied in this area; however, this flow overlies the older flow, crossed at road level (mile 57.2), with angular unconformity. **0.5**

FIGURE 3-57.7b. Texture of Conejos(?) Formation flow breccia.

58.9 Road crosses another down-to-the-southwest fault. Tuff of Las Tablas, a white, nonwelded ash-flow tuff, and associated reworked tuffaceous sedimentary rocks underlie most of the hillside on the right (Fig. 3-58.9). The tuff contains lithic fragments and fibrous pumice. Phenocrysts are plagioclase and sparse biotite. The tuff is contemporaneous with the Cordito Member. **0.1**

59.0 Contact of Tuff of Las Tablas with the underlying El Rito Formation. Tuff is well exposed ahead on left, across creek. **0.3**

59.3 Cordito Member on right. **0.3**

59.6 Road crosses down-to-the-southwest fault into the Cordito Member. **0.2**

59.8 Cross contact of Cordito Member with underlying El Rito Formation. **0.2**

60.0 **Enter Las Tablas.** Post office was established here in 1911. This is the downstream terminus of the Tusas Box. Precambrian rocks constitute the hillside across the river: the lower slopes are the Tres Piedras Granite; above it are silicic metavolcanic rocks and associated metasediments (Manley and Wobus, 1982b). These exposures are part of the

FIGURE 3-58.9. Exposure of tuff of Las Tablas on north side of Tusas Box.

prominent ridge of Precambrian rocks seen and discussed at Spring Canyon (mileage 41.0). The western edge of the ridge in this area is known as Mesa de la Jarita. There, the Precambrian rocks are overlain by El Rito Formation, Conejos Formation, Hinsdale Formation, the Amalia Tuff, and the Cordito Member of the Los Pinos Formation.

Mines in the Precambrian rocks 3 mi west–southwest (at Big Rock) produced kyanite in 1928 (Corey, 1960). The kyanite occurs in pods of quartz–kyanite and kyanite schist surrounded by sericite schist. **0.1**

60.1 Cliffs on left in El Rito Formation. **0.7**

60.8 View ahead of Petaca Mesa capped by Hinsdale Formation basalt. To left is Red Mesa capped by Servilleta Basalt(?). In the saddle between is the southeast continuation of the faults crossed in lower Cañada de Agua. Also in the saddle is a large exposure of a gray, rhyolite-flow breccia of granule- to pebble-sized fragments, which is found in several places along the west side of the Tusas Box. **1.0**

61.8 Hinsdale basalt to right of road in fault contact with Cordito Member to south. Flow is encountered again at mile 62.1. **0.3**

FIGURE 3-60.0a. View west into the Tusas Box. Far slope is of Precambrian rock.

62.1 Road crosses, then continues on, Hinsdale Formation basalt flow for 0.9 mi. The same flow is exposed in the low hills to the west (right) where it overlies the Precambrian rocks and is overlain by Cordito Member. **0.8**

62.9 Santa Fe Group deposits in roadcuts. **0.8**

63.7 Cross Rio Tusas. **0.2**

63.9 **Enter Petaca.** Post office established in 1900. **Bear left.** Forest Road 45 continues straight to Mesa de la Jarita, Kiowa Mountain, and Vallecitos.

The Petaca mining district extends from Kiowa Peak almost to La Madera. The mines are in Precambrian pegmatites of quartz, microcline, plagioclase, and muscovite. As many as 50 accessory minerals have been recorded. About 1870, sheet mica was mined for stove doors. Scrap-mica production began about 1900. Mining expanded about 1912, with a demand for mica for electrical uses. After World War I production declined, but peaked later in 1923 to 1930. Mica and pegmatite by-products (columbite–tantalite, beryl, and quartz) were produced during the 1950's and 1960's. (Data from Bingler, 1968.) **0.6**

64.5 Hinsdale basalt exposed on right for next one-half mile. **0.3**

64.8 **Enter** village of **South Petaca. 0.2**

65.0 Exposures overlying basalt are Cordito Member. **0.5**

65.5 Road is primarily on Conejos(?) Formation flow breccias for next 1.2 mi. These breccias include some flows younger than the ages for the Amalia Tuff and are therefore also younger than the 29.8-m.y. age for the uppermost Conejos Formation in Colorado (Lipman and others, 1970). **0.5**

66.0 Road on right to several mines in the Petaca district. **0.2**

66.2 **STOP 3. Lunch.** Discussion of geology and walk

FIGURE 3-60.0b. View of Kiowa Mountain and a cliff of Cordito Member of the Los Pinos Formation. The Tusas Box lies between.

up Cañada de Abrevadero (Fig. 3-66.2a). Ruins of a mill built in 1954 for dry bulk processing of pegmatite from the Capitan and other mines in the Cribbenville area about 2 mi to the west. It was well designed, constructed to standards unique for the Petaca district, and produced excellent grades of mica. In general an operational success, it was a financial failure mainly because the feed rarely contained more than 5% of muscovite instead of the 15–25% anticipated by the owners and their advisors. Operation ceased in 1957. (Data from Jahns, 1946; Bingler, 1968.)

A variety of volcanic rocks is exposed in this area. Near the entrance road to the old mill site are outcrops of a dark red-brown, welded ash-flow tuff that has been correlated with the Amalia Tuff from the Questa caldera on the basis of mineralogy, radiometric age (26.3 ± 0.9 m.y.), and trace- and major-element chemistry (P. W. Lipman, pers. comm. 1984). On the hill to the south of the mill are faulted exposures of vitrophyre and flows and flow breccias of intermediate composition. On the north side of Cañada de Abrevadero, one-fourth to one-half mile west of the road, are exposures of weathered vitrophyre, Hinsdale basalt, tuff of Cañada del Agua, and a quartz-latite flow with large feldspar phenocrysts. Exposures here and in Cañon de la Paloma establish that flow breccias (Conejos? Formation) underlie and overlie the tuff of Cañada de Agua (K–Ar age = 25.9 m.y.; 26.4 with current constants), and both overlie the Hinsdale basalt.

The tuff of Cañada del Agua is now known to be a less-welded part of the Amalia Tuff because of its similar mineralogy, radiometric age, and trace- and major-element chemistry (P. W. Lipman, pers. comm. 1984). Exposures of this tuff on Mesa de la Jarita are 35 mi from the source. **1.4**

67.6 Road on right to La Paloma mine and the Pineapple Group. Both these mines were active in 1953. **0.4**

68.0 Road crosses one of several faults trending north–south from Las Tablas to La Madera. **0.2**

68.2 Forest Road 376 to Servilleta Plaza on left (north entrance). **0.6**

68.8 Precambrian muscovitic quartzite on left side of road and again on both sides at 69.9 mi. **0.2**

70.0 Cross La Paloma drainage. **0.1**

70.1 Road on right leads to mines in the upper Cañon de los Alamos, including the Globe mine. **0.5**

70.6 Iron-stained Cordito on right. **0.5**

71.1 Forest Road 376 to Servilleta Plaza on left (south entrance). Miniature badlands have developed in

FIGURE 3-60.0c. Cordito Member conglomerate (Tc) overlying Precambrian rocks (p€), northeast side of Tusas Box.

FIGURE 3-60.8. Gray rhyolitic-flow breccia near Las Tablas.

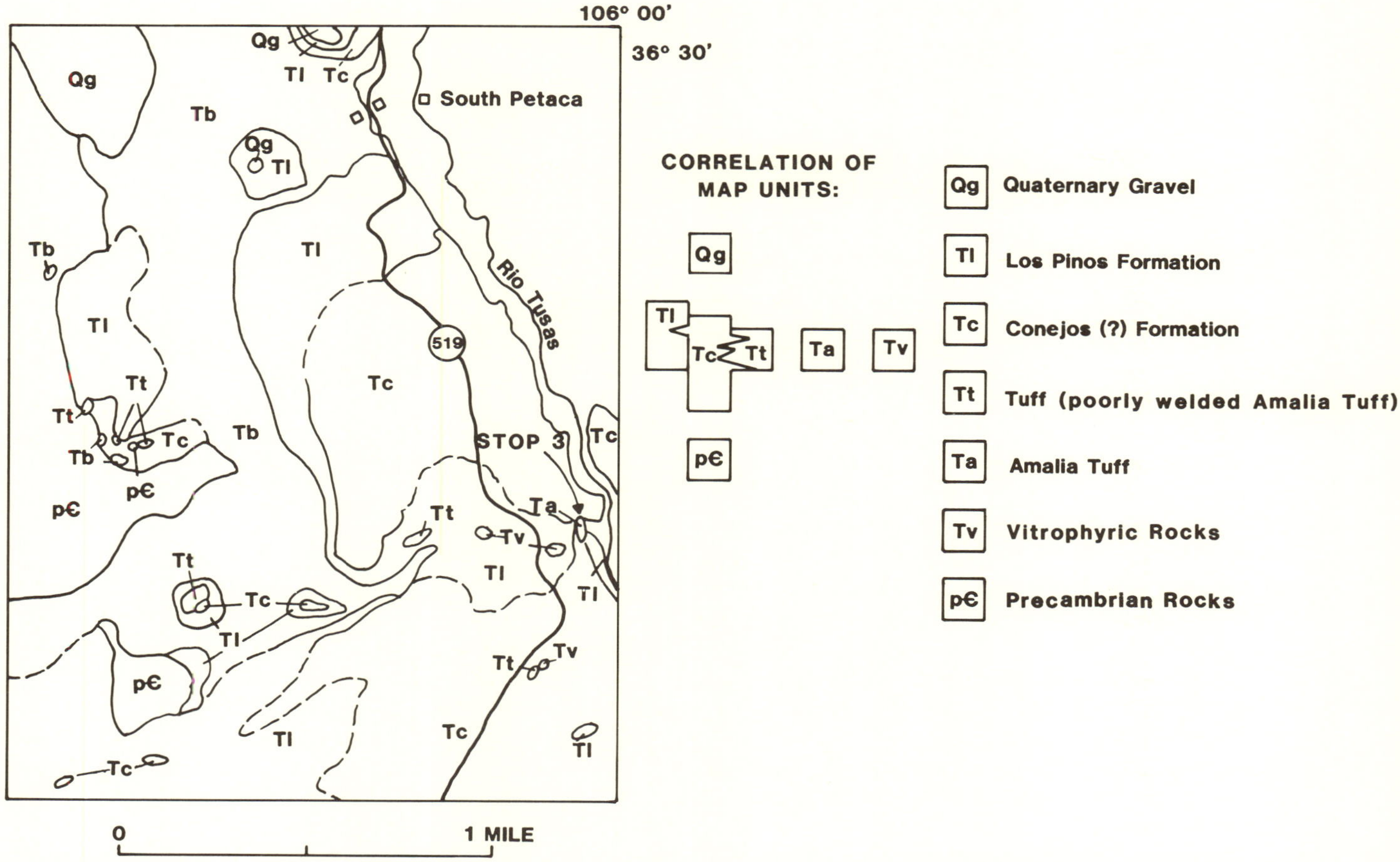

FIGURE 3-66.2a. Geologic map of the northeast corner of the La Madera quadrangle (from Manley, unpubl. mapping 1981). See Figure 3-41.0a for description of map units.

this valley from the erosion of Quaternary valley-fill sediments. Salt Lick Spring is at the distal end of these badlands. The Salt Lick pegmatite deposit is just south of the spring. The dike consists of coarsely crystalline microcline and quartz, with some albite, mica, and garnet; it crosscuts Precambrian muscovite quartzite. **0.5**

FIGURE 3-66.2b. Ruins of a mill built in 1954 near Petaca for bulk processing of pegmatite muscovite.

71.6 Fractured and faulted Cordito Member on right. **0.1**

71.7 View of Cañon de los Alamos on right. Five north–south-trending faults cut the canyon here. Roadcut on left exposes another two faults that bring the Santa Fe Group (on west) against the Cordito (on east). To the west, Los Pinos Formation onlaps the Precambrian rocks. **0.2**

71.9 Cross the western of two faults seen in last roadcut; outcrops in Santa Fe west of fault, Cordito on east. Basalt dike has intruded along fault. **0.3**

72.2 Two pegmatite dikes on left crosscut Precambrian rocks. **0.1**

72.3 Hinsdale Formation basalt onlaps the Precambrian rocks. Paloma Canyon prospect is located ahead in Cañon de la Paloma near its junction with the Rio Tusas. The prospect is in a N85°W-striking elongate pegmatite crosscutting Precambrian hornblende–chlorite schist. The pegmatite has a quartz core with a wall zone of feldspathic material. Rosettes of clevelandite are common, with garnet, fluorite, monazite, and columbite present. **0.4**

72.7 Cordito Member of Los Pinos Formation in fault

contact with Hinsdale Formation basalt and Precambrian rocks. **0.2**

72.9 Cordito Member of Los Pinos Formation in roadcuts. **0.8**

73.7 South of the Rio Tusas are low exposures of arkosic sand in the Santa Fe Group. They are overlain by Quaternary terrace gravels. Numerous exposures of Cordito Member on right. **1.2**

74.9 **Enter La Madera. Junction** of N.M. 519 with N.M. 111. **Turn left** onto N.M. 111 toward Ojo Caliente. **0.2**

75.1 Eolian Santa Fe Group sandstones exposed in next several roadcuts. **1.1**

76.2 La Madera Mountains on the left are Precambrian rocks fault-bounded on west side. This fault and several others, both east and west, provide control for the several hot springs in the area. These faults can be mapped from Petaca to the Chama River. The location of the hot springs may also be controlled in part by shear zones in the Precambrian rocks (Stix and others, 1982).

An electrical-resistivity survey (Stix and others, 1982; Vuataz and others, in press) shows that fluids emerge from the fault zone and flow as a plume of thermal water into cold aquifers of the valley alluvium east of the fault. Geochemistry of fluids indicates a maximum reservoir temperature of 80°C, with no evidence of high-temperature isotopic exchange between water and reservoir rocks. From these data it appears that the Ojo Caliente system is suitable only for small-scale, direct-use geothermal applications. (Data courtesy of F. E. Goff.)

Light-colored deposits on left are travertine from one set of springs along the fault zone. The hill ahead to the right is also capped by travertine. A Pleistocene age was established there based upon paleontological studies. **0.6**

76.8 N.M. 96 on right to El Rito. **0.8**

77.6 Enter narrow canyon in Precambrian rocks. **0.2**

77.8 Fault exposed on left. **1.2**

79.0 View to left of interfingering deposits of Los Pinos Formation and Santa Fe Group. **1.0**

FIGURE 3-76.2. Travertine north of Ojo Caliente.

80.0 **Junction** of N.M. 96 and U.S. 285. **Turn right** (south) on U.S. 285 to Ojo Caliente. Highway parallels Rio Ojo Caliente. **0.1**

80.1 Jemez volcanic field on skyline ahead. To left (east), vegetated hills expose Ojo Caliente and underlying Chama–El Rito Members of Miocene Tesuque Formation. Regional dip is 7°ESE. **0.1**

80.2 For the next 2 mi, high, dark-brown hills to right (west) of river expose Precambrian metarhyolite mapped and described by Treiman (1977). Lower flanks of hills are unconformably overlain by gray volcaniclastic conglomerate of Miocene Los Pinos Formation (Cordito Member). Low benches above river expose Rio Ojo Caliente terrace gravels capping the Los Pinos. The mature topography on the Precambrian rocks here is currently being re-exposed by erosion of the Los Pinos Formation and indicates the deeply dissected nature of the Laramide Brazos uplift by early Miocene time (Fig. 3-80.2). The present high elevation and regional dip of the Los Pinos Formation record the tectonism associated with formation of the Rio Grande rift. **0.3**

80.5 For the next mile, roadcuts on left expose uppermost Los Pinos Formation. Note continued abundance of gray volcaniclastic conglomerate and tuffaceous sandstone as in the Los Pinos farther north. Three miles south of Ojo Caliente, the abundance of the volcanic sandstone decreases as the Los Pinos grades into the more arkosic Chama–El Rito beds of the Española Basin (see May, this guidebook; Ingersoll and Ekas, this guidebook). **0.4**

80.9 To right, the highest dark-brown hill is Cerro Colorado, the expression of a metamorphosed Precambrian rhyolite stock described by Treiman (1977). **0.3**

81.2 Low benches across river expose Rio Ojo Caliente terrace gravels overlying Los Pinos Formation. **0.6**

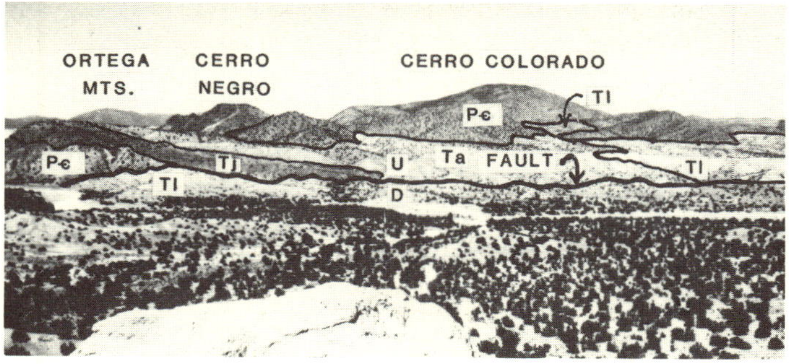

FIGURE 3-80.2. View north showing Precambrian and overlying Tertiary rocks exposed west of Ojo Caliente. Rocks dip to right (southeast) into Española Basin. Tertiary basalt flows and sedimentary rocks overlie mature Miocene paleotopography formed on Precambrian rocks. Photo taken from 2 mi southwest of Ojo Caliente, which is out of view to the right (east). Tl – Los Pinos Formation; Ta – Abiquiu Formation; Tj – Jarita(?) Basalt, Hinsdale Formation; P€ – Precambrian metamorphic rocks.

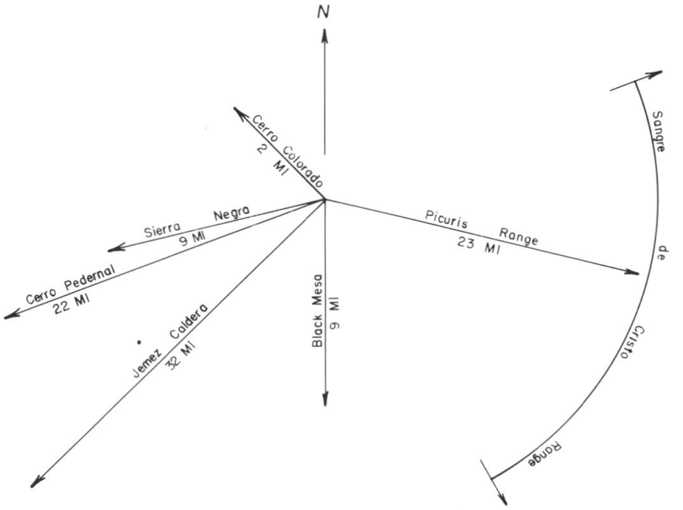

FIGURE 3-82.3a. Panoramic index for features seen from ridge crest at Stop 4 in Ojo Caliente.

81.8 **Junction** of U.S. 285 and N.M. 414. **Turn right** onto N.M. 414 at Ojo Caliente Mineral Springs sign. Historical records indicate a small settlement at Ojo Caliente by 1735 (Boyd, 1957). However, attacks by Comanche and Ute Indians caused recurrent abandonment of the settlement until the 1790's. The town is briefly described in Zebulon Pike's journal (March, 1807) during a trip south to Mexico while in custody of Spanish soldiers, following his arrest in southern Colorado. **0.1**

81.9 Old adobe church on left (Santa Cruz de Ojo Caliente) was built by the Spanish possibly in the early 1700's. The church and a plaza are mentioned in a correspondence by Governor Mendinueta, dated June 20, 1769, following a personal inspection of the settlement. The present corrugated pitched roof of the church does not appear in an 1880 photograph of the structure. **0.3**

82.2 Cross bridge over Rio Ojo Caliente and bear left. **0.1**

82.3 **STOP 4.** Welcome to Ojo Caliente Mineral Springs. Park vehicles on left (inside) shoulder of oval driveway or on far-right shoulder of road to allow room for other traffic. We will first briefly observe an outcrop of the Los Pinos Formation behind the bath houses and then climb the ridge of Precambrian rocks, 200 yds to the northwest, for a regional geologic perspective of the northwestern Española Basin. Ascent may be strenuous. Those preferring to stay with the vehicles may tour the facilities with Mr. George Mauro, the manager.

A small mineral-springs resort was first built here in 1862 by Antonio Joseph, a territorial representative. Kit Carson bathed here in the 1860's. The land was incorporated in 1929 and has been owned by the Mauro family since the 1930's.

From the top of the ridge, Black Mesa is seen forming the flat skyline to the south and approximately marks the southwest-trending synclinal axis of the Española Basin (Fig. 3-82.3a). The Jemez volcanic field is the highland area 32 mi to the southwest. This field straddles the western fault boundary of the basin. Cerro Pedernal, 22 mi to the west–southwest, is west of the basin fault margin, but contains the Abiquiu and Tesuque Formations in its middle slopes, indicating that the Miocene depositional basin once extended farther west than its present fault boundary.

In the Ojo Caliente area, the surrounding steep hills of Precambrian rocks represent the southern end of the Brazos uplift which plunges south beneath the Tertiary section of the Española Basin. These hills are overlapped on their lower flanks by the Los Pinos Formation (Fig. 3-82.3b). Although no faults are exposed in the immediate vicinity, the presence of the mineral springs along the base of the linear ridge we are standing on is suggestive of northeast-trending Laramide(?) faults or fractures beneath the Los Pinos Formation. Minor, northeast-trending, Pliocene–Pleistocene faults under the alluvium of Rio Ojo Caliente are also suspected but not confirmed. These faults are along trend with the Ojo Caliente fault zone, a zone of fracturing, cementation, and anomalous groundwater activity that extends N25°E from the Rio Chama area. Topics to be discussed here include the Precambrian geology, the Miocene stratigraphy (see May, this guidebook), and regional structural control of the geothermal activity.

The Precambrian rocks in this area include metarhyolite, metarhyolite porphyry (Cerro Colorado), dikes and layers of amphibolite (metabasalt?) and pegmatites, all of which have been regionally metamorphosed to middle amphibolite grade. Description of these rocks is summarized mainly from Treiman (1977).

Pink to light-brown metarhyolite and fine-grained granite constitute most of the Precambrian west of Ojo Caliente and south of Cerro Colorado. Quartz and microcline predominate and commonly occur as megacrysts. Plagioclase is common but less abundant than microcline. Small quantities of biotite, muscovite, apatite, chlorite, and magnetite, and traces of zircon, garnet, zenotime, and sillimanite also occur. Treiman (1977) also reported xenolithic inclusions, possible relict glass shards and dropstone structures, implying that these rocks were originally rhyolite tuffs and ash-flow tuffs derived from a nearby vent. A well-developed gneissic foliation strikes N60°E and dips 60–85°SE in most outcrops. Treiman (1977) assigned these rocks to the Arroyo Rancho Formation of the Ortega Group and correlated them with part of the Ortega Group of the Picuris Range.

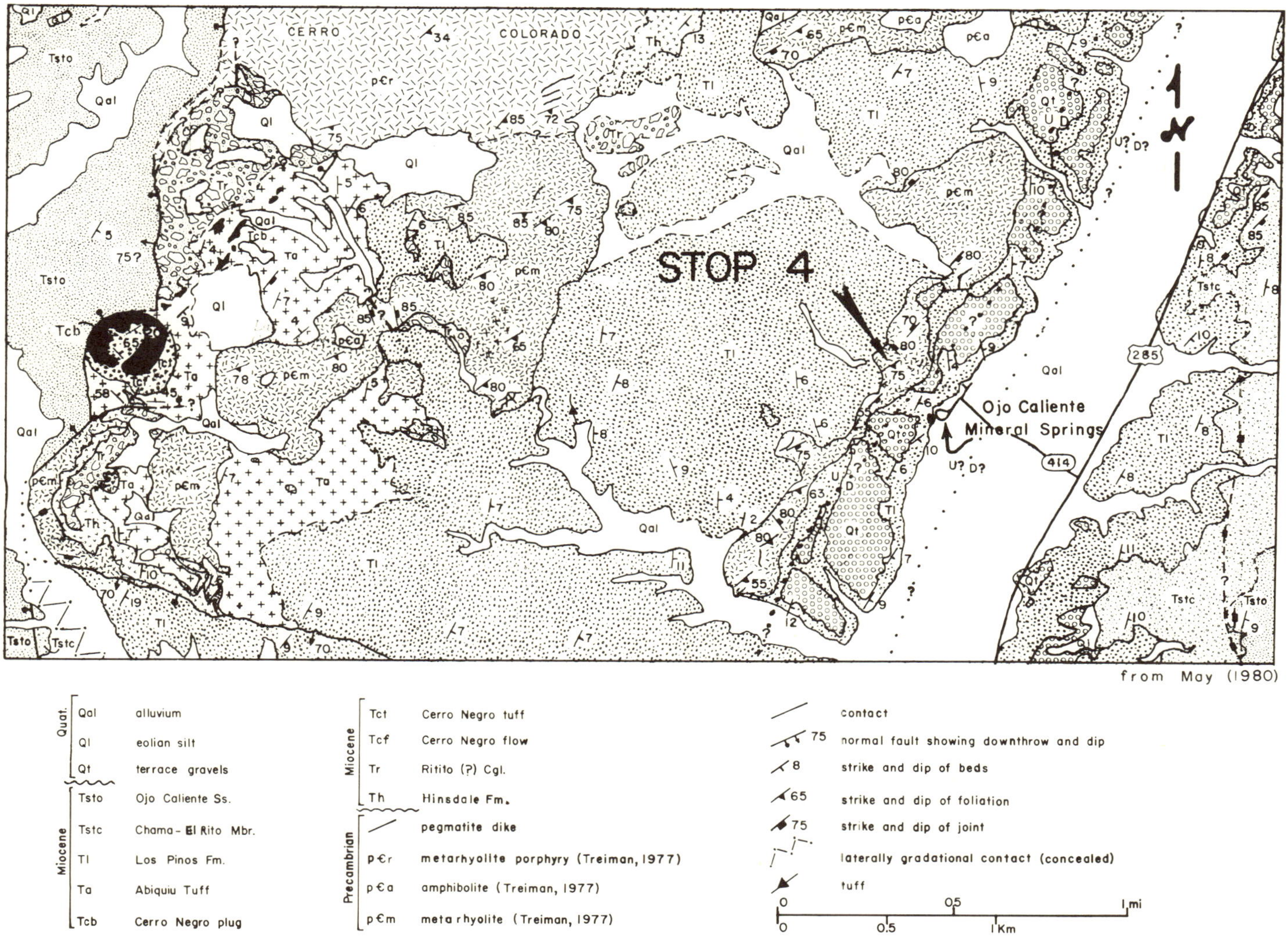

FIGURE 3-82.3b. Geologic map of area around Ojo Caliente. Road log Stop 4. Modified from May (1980).

Small, dark-gray pods and layers of schist, amphibolite, and gneiss occur in the metarhyolite, and contacts of these bodies appear to be discordant with the surrounding rock. The amphibolites may represent metamorphosed basaltic dikes, sills, and flows.

Cerro Colorado is a northeast-trending, elliptical body of resistant, red-brown metarhyolite porphyry, which is discordant with the surrounding layered metamorphic rocks, including the metarhyolite. The rock contains quartz (40%), microcline (40%), plagioclase (15%), and minor biotite, muscovite, chlorite, magnetite, apatite, zircon, fluorite, and garnet. Many quartz and microcline megacrysts occur in the rock. Many relict textures of probable plutonic origin and discordance with the surrounding metamorphic rocks strongly imply that the body formed an intrusion that cooled at very shallow depths (Treiman, 1977, 1978). Possibly the Cerro Colorado pluton was the source of the surrounding metamorphosed rhyolite tuffs and later intruded them.

Pegmatite dikes and a few small quartz veins cut the Precambrian rocks west of Ojo Caliente. Most dikes trend northeast (approximately parallel to the regional metamorphic foliation) and are from 2 to 8 yds wide and 8 to 60 yds long. Jahns (1946) reported that they range from simple quartz and feldspar bodies to complex, zoned pods containing muscovite, biotite, garnet, tourmaline, beryl, fluorite, bismutite, columbite, native copper(?), and samarskite. Muscovite and beryl from some dikes were mined during World War II and later.

The Miocene stratigraphy of the northern Española Basin is moderately complex, with much intertonguing of major units and many lithologic contacts that are time-transgressive. The Los Pinos Formation (Cordito Member) underlies most of the Ojo Caliente area and is a succession of gray rhyolite and andesite pebble conglomerates that grade southward primarily into the arkosic Tesuque and tuffaceous Abiquiu Formations. The Los Pinos here represents a large southwest-sloping fan sequence derived from both local and more northeasterly volcanic areas. This Miocene fan was burying the Laramide Brazos uplift, while to the south it merged with the west-to-southwest-sloping alluvial fans of the Tesuque Formation. The Tesuque was being shed from the east in the Sangre de Cristo Mountains. The resultant intertonguing sequence of volcanic and arkosic beds between the two systems is the Chama–El Rito Member of the Tesuque Formation. The laterally gradational contact between the Los Pinos and Chama–El Rito is exposed in the badlands 3 mi south of Ojo Caliente.

After stop, retrace route to junction with U.S. 285. **0.6**

82.9 **Junction** of U.S. 285 and road to Ojo Caliente Mineral Springs (N.M. 414). **Turn right** on U.S. 285. Our route now takes us south through the Tertiary section of the northern Española Basin. Because of the regional dip of 3–8°ESE into the basin, we will gradually rise in the section through upper Los Pinos Formation at Ojo Caliente, the Chama–El Rito and Ojo Caliente Members of the Tesuque Formation, and finish in the Chamita Formation at the southern end of Black Mesa. **0.5**

83.4 Terrace gravels of Rio Ojo Caliente on right cap Los Pinos Formation in benches above river. Dark-brown ridges behind terraces are southernmost exposures of Precambrian rocks of the Brazos uplift. A Bouguer gravity map by Cordell (1979) shows a gravity high trending southward from these ridges, implying that the uplift continues to plunge gently into the Española Basin beneath the Tertiary section. **0.3**

83.7 Gray volcaniclastic pebble conglomerates of Los Pinos Formation (Cordito Member) in roadcut. Note fluvial channels, crossbeds, and concentration of cobbles in channel axes (Fig. 3-83.7). At 2:00, dark cliffs in distance expose basaltic dikes along northern vent margin of Ojo Caliente tuff ring (see May, this guidebook). Outward-dipping tuff ring facies forms lower, light-gray cliffs to right of vent (best seen in morning light). **0.7**

84.4 Cross dry arroyo. Continuous roadcut for next 0.6 mi exposes Los Pinos Formation grading upward and laterally into orange-brown Chama–El Rito beds as light-gray volcaniclastic conglomerates become less abundant. **0.6**

85.0 Contact between Los Pinos and Chama–El Rito at end of long roadcut. To the right, for the next mile in the hills across the river, note gradual loss of gray Los Pinos conglomerates and increase in orange-brown arkosic beds of Chama–El Rito (best seen in morning light). A few of these volcaniclastic conglomerates extend southward in the Chama–El Rito as far as the Rio Chama and beyond, gradually becoming less common and slightly finer-grained. Age of the Chama–El Rito is about 12–18 m.y. in this part of the basin (May, this guidebook). **0.1**

FIGURE 3-83.7. Volcaniclastic Los Pinos Formation in roadcut at mileage point 83.7.

85.1 Chama–El Rito capped by Rio Ojo Caliente gravels in roadcuts for next 2 mi. **1.5**

86.6 Tongues of light-brown, eolian Ojo Caliente Sandstone interfingering with upper Chama–El Rito in roadcuts. Width of this gradational contact is typically 30–260 ft. **0.5**

87.1 Roadcut exposes contact of Chama–El Rito beds with overlying light-brown, crossbedded Ojo Caliente Sandstone (upper Miocene). Ojo Caliente is in roadcuts for next 12 mi. The sandstone is arkosic to subarkosic and loosely cemented. Clastic grains are typically quartz (50–75%), polygenetic plagioclase and orthoclase (25–35%), lithic fragments of mainly quartzite and rhyolite(?) (3–6%), and opaque minerals (0–2%). The Ojo Caliente Sandstone ranges from about 10 to 12 m.y. in this part of the basin (May, 1980). **0.9**

88.0 Mesa on right is capped with high-level Rio Ojo Caliente terrace gravels. Chama–El Rito underlies gravels in northern one-third of mesa and Ojo Caliente Sandstone underlies gravels in southern two-thirds. **0.2**

88.2 Roadcut exposes tongue of orange-brown Chama–El Rito in Ojo Caliente Sandstone. Also note active sand dunes on alluvium of Rio Ojo Caliente. **0.6**

88.8 Large, festoon crossbeds of Ojo Caliente Sandstone in roadcuts. In most areas these wedge-shaped sets of crossbeds are bounded at the top and base by planar to slightly concave-upward erosion surfaces typical of modern dunes. The wedges are 0–25 ft thick and contain inclined sand laminae 23–72 ft long. The laminae are also concave upward, gradually steepening from 0° to as much as 30°, and are truncated at the top. The lower ends of the crossbeds are tangential to the underlying surface. The steep, downwind slip faces of the crossbeds in this region have a mean dip direction of N80°E with a basically unimodal distribution (Fig. 3-88.8). This indicates that prevailing late Miocene winds were from S80°W, not greatly different from today's west–southwesterly winds (May, 1980). According to Selley (1970), a unimodal dip distribution with only 120° of range is most typical of modern transverse dunes oriented perpendicular to the prevailing wind. **2.6**

91.4 At 7:00, white air-fall tuff in Ojo Caliente Sandstone on lower slopes of Black Mesa is approximately 200 ft above base of the Ojo Caliente. Manley was unable to find sufficient zircons for fission-track dating in samples of this tuff. In this region, a few similar isolated tuffs and sandstones containing reworked tuff occur in the upper 330 ft of the Chama–El Rito and lower 200 ft of the Ojo Caliente. These tuffs gradually become more abundant to the south in the badlands west of Rio Ojo Caliente. The most persistent horizon is approximately 130 ft above the base of the Ojo Caliente Sandstone. Most layers can be traced 250–650 ft laterally and are 0–13 ft thick. Their lenticular shapes imply preservation in the low interdunal areas in the sandstone. Basal contacts are generally sharp, and the upper contacts grade upward into the overlying sandstone because of slight reworking of the tuff after deposition. In thin section they typically contain partly devitrified glass shards and sparse biotite flakes. Pumice and lithic fragments are sparse to absent.

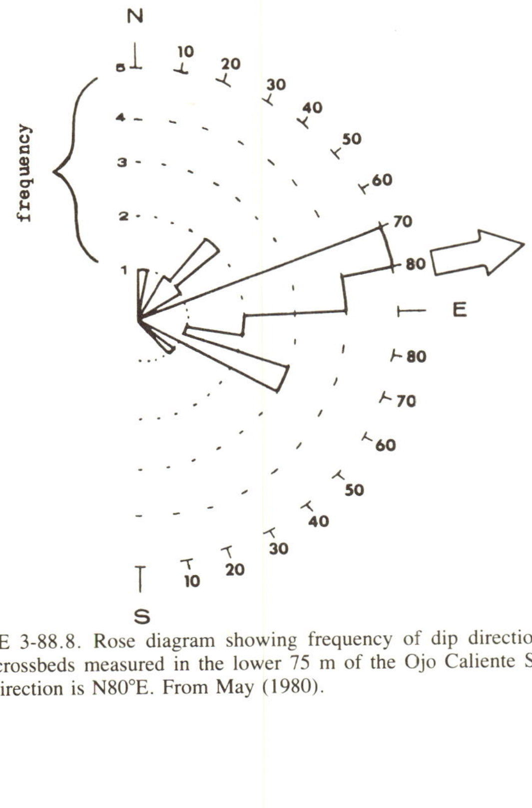

FIGURE 3-88.8. Rose diagram showing frequency of dip directions for 21 eolian crossbeds measured in the lower 75 m of the Ojo Caliente Sandstone. Mean direction is N80°E. From May (1980).

The source of these tuffs may have been the early Jemez volcanic field. Smith and Bailey (1968) state that the Jemez field began erupting about 12 m.y. ago. The tuffs are approximately midway in the section between a 13+-m.y.-old basaltic vent in the middle Chama–El Rito along El Rito Creek (Manley and Mehnert, 1981) and a 9.6-m.y.-old basalt flow in the basal Chamita Formation along U.S. 84 (Baldridge and others, 1980). Izett and Naeser (1981) determined fission-track ages of 14 to 9 m.y. on the abundant tuffs of the Skull Ridge and Pojoaque Members of the Tesuque Formation in the eastern Española Basin. The fact that these tuffs become more abundant to the south and east is consistent with the possibility of predominantly N80°E-moving winds bringing the ash northeastward from early volcanic centers in the Jemez field (May, 1980). **0.1**

91.5 From here south, the Ojo Caliente Sandstone forms the lower half of slopes on Black Mesa and the unconformably overlying Chamita Formation (uppermost Miocene) forms the upper half. Pliocene gravels and a 2.8-m.y.-old basalt flow of Servilleta Basalt cap the mesa (Manley, 1976). The basalt flowed south along the valley floor of the ancestral Rio Grande in the late Pliocene. The present high elevation of the mesa along the structural axis of the basin is therefore a good example of topographic inversion of the central valley by differential erosion of the softer Tertiary "basin-fill" rocks. **0.9**

92.4 On left is a good exposure of Black Mesa section for the next 0.3 mi. Note light-brown color and large crossbeds of Ojo Caliente Sandstone in lower half of slopes. The Chamita Formation above is distinguished by even, parallel fluvial bedding, a slightly darker-brown color, and numerous pale stringers of tuffaceous sandstone. A prominent pale-green tuffaceous zone in the upper middle slopes occurs a few meters above the unconformable con-

FIGURE 3-92.4. Servilleta Basalt capping Black Mesa.

tact and makes a good marker bed along the western slopes of the mesa. Note columnar jointing in Servilleta Basalt. **0.5**

92.9 On right, Rio Ojo Caliente terrace gravels overlie Ojo Caliente Sandstone. **1.0**

93.9 Good exposure of section beneath Black Mesa to left. **0.2**

94.1 At right, badlands west of river expose upper Chama–El Rito beds. Skyline is formed by lighter-brown beds of cemented Ojo Caliente Sandstone downfaulted against Chama–El Rito along main fault of Pliocene Ojo Caliente fault zone. This zone of branching faults, cementation, and anomalous ground-water activity extends N25°E from Rio Chama, past Ojo Caliente mineral springs and the Owl Cliffs travertine benches, to La Madera. For a list of the springs in this area see May (1980, p. 172). The type of cementation appears to be simple mineralogically, with pervasive carbonate cement and local traces of white and black opalite as fracture-filling. No detailed study of this mineralization has been made. Parallelism of the fault zone with the southern end of the Laramide Brazos uplift also suggests pre-Pliocene ancestry. **3.0**

97.1 Ahead is a jagged hogback (Cerrito de la Baca) that is made of Ojo Caliente Sandstone cemented along a northeast-trending fault of the Ojo Caliente fault zone. **0.9**

98.0 On left are spires of Ojo Caliente Sandstone cemented along joints that parallel the Ojo Caliente fault zone. **0.1**

98.1 At 2:00, note well-exposed fault along flank of Cerrito de la Baca. **0.2**

98.3 At 7:00–8:00, the Ojo Caliente Sandstone and Chamita Formation are exposed below the Servilleta Basalt on Black Mesa. **0.7**

99.0 Jagged light-brown cliffs across the river to right are made of fractured and cemented lower Ojo Caliente Sandstone along the Ojo Caliente fault zone. Cliffs end at confluence of Rio Ojo Caliente and Rio Chama. **0.6**

99.6 Cross major northwest-trending fault (poorly exposed) that downdrops Chamita Formation against Ojo Caliente Sandstone. To the northwest, this fault projects under Rio Ojo Caliente alluvium and bends to the northeast as the major fault of the Ojo Caliente fault zone. For the next mile, roadcuts expose downfaulted Chamita Formation overlain by colluvium of Servilleta Basalt and Ojo Caliente Sandstone. Note slightly darker brown fluvial beds of Chamita. **0.6**

100.2 Rio Chama to right of road. In roadcut, light-brown beds of reworked Ojo Caliente Sandstone in the lower Chamita give a pseudo-intertonguing appearance to the Ojo Caliente–Chamita contact.

Major northeast-trending faulting has been recognized to the south of the Rio Chama in Arroyo de la Presa (Dethier and Manley, in press), but is recognized on this side of the river only by steep dips in the Chamita Formation beneath the east edge of Black Mesa at one location. This fault is considered to connect to the north with the Embudo fault and to the south with the Pajarito fault. As such, it forms the west side of the Velarde graben. The structure and stratigraphy of this area are further discussed in Dethier and Martin (this guidebook) and Dethier and Demsey (this guidebook). **0.1**

FIGURE 3-100.2. Fault in Arroyo de la Presa. View southeast.

100.3 On left is same major northwest-trending fault crossed at mile 99.6. It is well exposed 300 ft into amphitheater. Chamita is downdropped against Ojo Caliente with approximately 450 ft of offset. **0.1**

100.4 Roadcut exposes light-brown sandstone bed similar to Ojo Caliente Sandstone in Chamita Formation. **0.5**

100.9 Small dam in Rio Chama on right. Cross major fault zone again. Roadcut exposes Chamita downfaulted against Ojo Caliente Sandstone. Upthrown Ojo Caliente Sandstone now in roadcuts, overlain by Rio Chama terrace gravels. **1.3**

102.2 **Junction** of U.S. 285 and N.M. 74.

End of road log for Third Day.

DAY 3: SUPPLEMENTAL ROAD LOG 3A, FROM TOWN OF TUSAS TO BISCARA CANYON, TUSAS RIDGE, VALLEY OF LITTLE TUSAS CREEK, AND BACK TO TUSAS

SUMMARY

This supplemental road log loops through the area between the Tusas Valley and the Tusas Ridge to provide exposures of two tuffs derived from calderas in the southern San Juan Mountains of Colorado as well as crossing the north–northwest, south–southeast fault zone that forms the western margin of the Rio Grande rift. Route is not recommended for low-clearance vehicles between miles 9.0 and 14.6.

Mileage

0.0 **Junction** of U.S. 64 with N.M. 111 on left. **Continue west** on U.S. 64. **0.8**

0.8 Road on right to Guido Canyon. Near the head of this canyon and along its north wall are several large exposures of the Conejos(?) Formation including both the flow breccias of intermediate composition and volcaniclastic conglomerate derived from them. Overlying the Conejos(?) are exposures of a gray rhyolite-flow breccia (Fig. 3-60.8) that is part of the Cordito Member of the Los Pinos Formation (Manley and Wobus, 1982b). **0.4**

1.2 Roadcut on right in Santa Fe Group rocks with overlying Quaternary terrace gravel. **1.0**

2.2 A fault zone at the base of the hill to the right parallels the Rio Tusas valley. Within the fault zone Precambrian rocks (Tusas Mountain Granite and Maquinita Granodiorite) and the Tertiary El Rito Formation are exposed. Northeast of the zone, the section includes the Tusas Granite overlain by the El Rito, the Treasure Mountain Tuff, and the Esquibel Member of the Los Pinos Formation. Southwest of the zone, both Esquibel and Cordito are exposed. **0.8**

3.0 On right, and in the next several roadcuts, are southwestward-dipping beds of the Cordito Member. **1.5**

4.5 Crossing approximate contact of the Cordito and Esquibel Members. Contact dips steeply to southwest. **0.3**

4.8 **Junction:** Turn right onto Forest Road 421 to Biscara Canyon. **0.1**

4.9 Exposures of Esquibel Member followed by fault-bounded Precambrian Maquinita Granodiorite of Barker (1958). It has a U–Pb age of 1,675–1,725 m.y. (Manley and Wobus, 1982b). Locally, the granodiorite or biotite–quartz diorite contains dark, rounded inclusions of Moppin Series (of Barker, 1958) metavolcanic and metasedimentary rocks. **0.5**

4.8–5.4 Road crosses El Rito Formation. **0.1**

5.5 View ahead of cliffs in Esquibel Member. **0.1**

5.6 Road crosses northwestward-dipping ash-flow tuff from Treasure Mountain Tuff (Figs. 3A-5.6a, b), whose K–Ar age is 29–30 m.y. and was erupted from the Platero caldera 55 mi to the northwest (Lipman, 1975). Thickness is usually about 5 ft. (See Fig. 3-54.6 for distribution map.) **0.6**

6.2 Hillside to west and northwest contains exposures of Conejos(?) and El Rito Formations surrounded by the Esquibel Member. **0.6**

FIGURE 3A-5.6a. Exposure of Treasure Mountain Tuff in lower Biscara Canyon.

6.8 Crossing down-to-the-northeast fault into white tuff of Biscara Canyon, an unwelded ash-flow tuff with lithic fragments of volcanic and Precambrian rocks and numerous, crystal-poor pumice clasts. **0.2**

7.0 Small outcrop on left of El Rito Formation. Exposures in the hillside ahead, across the canyon, include paleohills of Maquinita Granodiorite, and the El Rito and Conejos(?) Formations. **0.1**

7.1 Road passes fault-bounded exposures of El Rito Formation and Esquibel Member on left. **0.7**

7.8 Road crosses contact of white tuff of Biscara Canyon with overlying Esquibel Member. **0.8**

FIGURE 3A-5.6b. Texture of Treasure Mountain Tuff, Biscara Canyon.

8.6 Unnamed Forest Road on left. Continue straight. **0.4**

9.0 **Junction: Bear left** onto Forest Road 167 entering from Tusas Ridge on right. **0.5**

9.5 Crest of Tusas Ridge. **0.8**

10.3 Forest Road 82 to Lamy Peak on right. **0.3**

10.6 **Junction** of Forest Road 167 with Forest Road 133. **Turn left** onto Forest Road 133. (Supplemental road log 3B connects this log to log 3C via Forest Road 133 north to the Rio San Antonio.) **0.7**

11.3 Road crosses same down-to-northeast fault crossed at mile 6.8 and passes into section containing Treasure Mountain Tuff overlain by Masonic Park Tuff. K–Ar age of latter is 28.2 m.y., and it was erupted from the Mt. Hope caldera 70 mi to the northwest (Lipman, 1975). Valley ahead, the Little Tusas Creek valley, is cut by eight additional north–northwest, south–southeast-trending faults that control the lower Tusas Valley and form part of the western boundary of the Rio Grande rift. **0.2**

11.5 Beehive Spring is located on fault crossed at mile 11.3. **0.3**

11.8 Road crosses Masonic Park Tuff (Wobus and Manley, 1982). Hillside to right (west) is dip-slope of the tuff. Tuff overlies El Rito Formation. **0.3**

12.1 Forest Road 80 on right. **0.2**

12.3 Masonic Park Tuff forms cliff on hillside to left (east) (Fig. 3A-12.3). **0.5**

12.8 Crossing down-to-the-southwest fault into the Esquibel Member. **0.5**

13.3 Road crosses another similar fault. Esquibel Member prominently exposed at river level on right. **0.3**

13.6 Road crosses down-to-the-north fault. **0.9**

14.5 Cross down-to-the-northeast fault that is part of the zone paralleling the Rio Tusas. **0.1**

14.6 **Junction:** Forest Road 133 terminates at U.S. 64. **Turn left** (east) onto U.S. 64 toward Tres Piedras. **0.4**

15.0 Road crosses down-to-the-northeast fault into the El Rito Formation. Paleohills of Precambrian rocks also exposed in hillside on left. **0.2**

FIGURE 3A-12.3. View north up valley of Little Tusas Creek. Masonic Park Tuff caps hillside on right.

15.2 Cross down-to-the-southwest fault into the Esquibel Member. **0.1**

15.3 Cross down-to-the-southwest fault into the Cordito Member which is exposed in roadcuts between here and Biscara Canyon. **2.2**

17.5 Forest Road 421 on left to Biscara Canyon.

End of supplemental road log 3A for Third Day.

DAY 3: SUPPLEMENTAL ROAD LOG 3B, FROM JUNCTION OF FOREST ROADS 133 AND 167 THROUGH CISNEROS PARK NORTHWARD TO JUNCTION WITH FOREST ROAD 87 ON THE RIO SAN ANTONIO

SUMMARY

This road log provides a connection between supplemental logs 3A and 3C. It traverses an area dominated by basalts of Tertiary age. The older flows are tilted eastward along a series of rift-bounding faults. The road is rough in places and mileage is approximate.

Mileage

0.0 **Junction** of Forest Roads 133 and 167. **Continue north** on 133. **1.0**

1.0 Mesa ahead is southeast edge of Broke Off Mountain. It is capped with a Hinsdale Formation basalt flow (Manley, 1982b). Similar flows to the north, in Colorado, have K–Ar ages ranging from 25 to 5 m.y. (Lipman and Mehnert, 1975). **1.5**

2.5 Road is between Hinsdale flow on left and Taos Plateau basalt on right. A northwest-trending fault passes through this saddle. **0.8**

3.3 Road climbs onto Taos Plateau basalt. Ahead are the Lucero Lakes (vents?) surrounded by basalt. Relationships here show 3° of eastward tilting of the Hinsdale basalt, whereas the younger flows have surface dips of 2°. **3.8**

7.1 Road crosses Taos Plateau silicic alkali-basalt flow derived from the Cerritos de la Cruz to the southeast. The two cinder cones there retain virtually uneroded morphology and are the youngest volcanic rocks in the San Luis Valley. **0.3**

7.4 **Junction** of Forest Roads 133 and 87.

End of supplemental road log 3B for Third Day.

DAY 3: SUPPLEMENTAL ROAD LOG 3C, FROM TRES PIEDRAS TO NO AGUA, RIO SAN ANTONIO, BEAVER CREEK BASIN, AND SAN MIGUEL

SUMMARY

This log traverses part of the northern Tusas Mountains. The route makes a semicircular path to the west around

San Antonio Mountain to the Rio de los Pinos. Return to U.S. 285 at San Antonio can be made along the river. Route is rough and high-clearance vehicles are recommended. Mileage is approximate.

Mileage

0.0 **Junction** of U.S. 285 with U.S. 64 in Tres Piedras. Drive north on U.S. 285. **1.3**

1.3 View on right is Servilleta shield volcano and cinder cone; at center are rhyolite domes of No Agua; and at left is San Antonio Mountain (Fig. 3C-1.3). **5.6**

6.9 Crossing silicic alkali basalt from Red Hill Cinder Cone to the west. Flow overlies rhyolite clasts from the No Agua domes, and therefore is younger. The rhyolite has K–Ar ages of 3.8 m.y. (Lipman and Mehnert, 1979) and 4.2 m.y. (Breese, this guidebook). **2.1**

RHYOLITE DOMES AND FLOWS AT NO AGUA PEAKS, NEW MEXICO

Richard Breese
Manville Corporation, Denver, Colorado 80217

Two extrusive, endogenous domes and associated coulee lava flows comprise the No Agua Peaks rhyolite-dome complex located within the Taos Plateau subdivision of the Rio Grande rift, approximately 6 mi north of Tres Piedras, New Mexico. Four eroded hills topographically represent the partially dissected domes. Coulee flows, thought to be compositionally related to extrusion of the western dome, constitute the contiguous lower relief that extends to the west and southwest (Fig. 3C-6.9a). The remnants of the exterior obsidian carapace, formed by rapid chilling of flow front margins, have been extensively perlitized by secondary-hydration processes, and are collectively considered to be the world's largest commercial perlite deposit. Open-pit mining at No Agua annually supplies roughly two-thirds of the perlite production in the United States.

Emplacement of both domes, K–Ar dated at 4.2 m.y., is attributed to at least two episodes of Pliocene extrusion. These are considered to represent two compositionally different silica-rich magmas, ranging in SiO_2 content from 73 to 77%. The steep-sided, areally restricted structural features at No Agua are characteristic of highly viscous, silicic-magma extrusion. Surrounding volcanic rocks are compositionally diverse, ranging from volumetrically subordinate silicic rhyolite to basalt (Lipman and Mehnert, 1979).

Major- and minor-element analyses of No Agua obsidian samples collected from partially hydrated portions of the glass carapace present

FIGURE 3C-1.3. View north to No Agua rhyolite domes. Peak to left is the west dome; central and right peaks are part of the east dome. Manville open-pit mine visible in center.

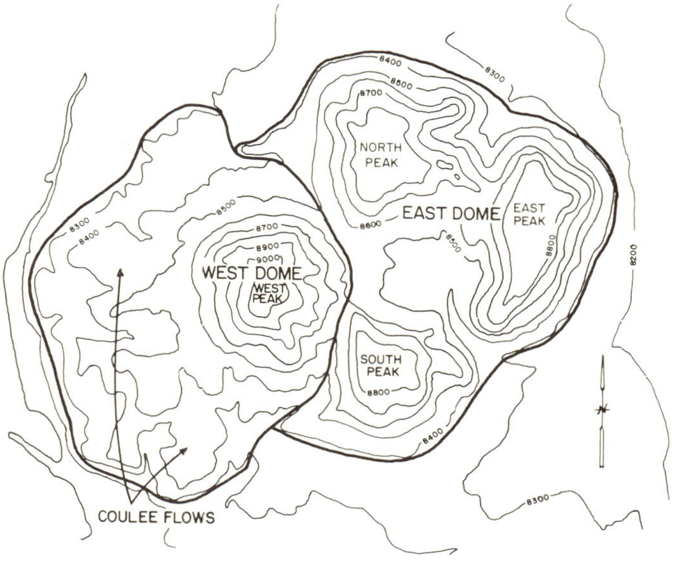

FIGURE 3C-6.9a. No Agua dome complex showing areal extent of compositionally similar rhyolites identified by Ra, Rb, and Sr analyses.

on the four hills are without marked compositional variation (Whitson, 1982). In contrast, analyses of dome and flow perlite have shown compositional variation with respect to Ca and Mg. Trace-element analyses of obsidians and felsites for Ba, Rb, and Sr have shown marked variation (Whitson, 1982), which corresponds geographically with Ca and Mg variation. These data indicate magma sources of contrasting trace-element composition and, combined with geologic mapping, have been useful in outlining the extent of corresponding coulee flows and west and east domal structures.

Dome stratigraphy consists of three distinctive, concentric lithologic units that exhibit gradational contacts. These have been identified on each of the hills and are partially reflected within the coulees. Stratigraphy consists of a devitrified felsitic core peripherally encased by a partially devitrified glass envelope, which is in turn encased by an exterior glass envelope (Fig. 3C-6.9b). Diverse primary textures are readily evident within the exterior glass unit. These are recognized by degree and distortion of vesiculation (controlled by lithostatic and hydrostatic pressure during emplacement) and by the presence of fractures and foliation.

Perlitization has resulted from hydration and the corresponding increase in total water content of the glasses from 0.01–0.5% to 2.8–3.7%, and has been superimposed on the two glassy units as well as on primary texture. Megascopically, perlitic fracture and onionskin texture are only evident in some of the denser glasses; however, in thin section, perlitic fracture is recognizable to a lesser extent within all the textures. To date, deuterium-content analyses show that water extracted from No Agua perlite is related to that of the meteoric water of the region, and that hydration from this water source is the primary mechanism responsible for perlitization of the two exterior glassy units (Friedman and others, 1966; I. I. Friedman, pers. comm. 1984). However, a partially hydrated marekanitic zone, typified by very abundant

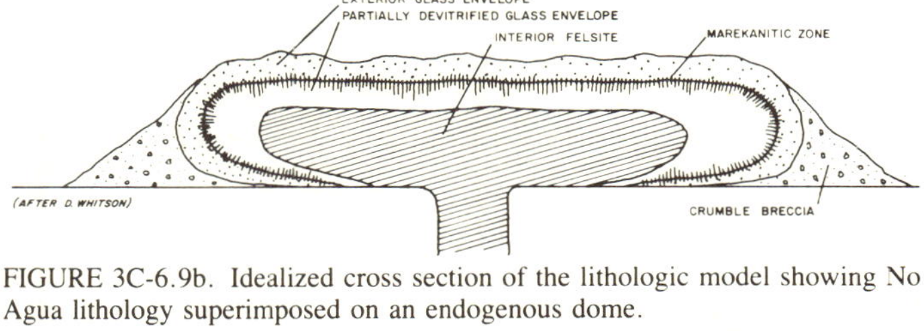

FIGURE 3C-6.9b. Idealized cross section of the lithologic model showing No Agua lithology superimposed on an endogenous dome.

nodules of remnant primary obsidian (Apache tears) within perlitized glass, is characteristically present as a gradational zone at the base of the exterior glass unit, extending into the upper portion of the partially devitified glass envelope. This suggests that a complex meteoric-water hydration mechanism, or more than one source of hydration water, may be responsible for the perlitization of the exterior lithologic units. Friedman is investigating the latter possibility.

9.0 Road cut in Santa Fe Group sedimentary rocks and Servilleta Basalt. **1.1**

10.1 **Junction** of U.S. 285 with Forest Road 87 to Lagunitas. **Turn left** onto Forest Road 87. Road traverses Servilleta Basalt. **1.5**

11.6 San Antonio Mountain (Fig. 3C-11.6), to north of road, is a central volcano of phenocryst-poor rhyodacite. K–Ar age is 3.1 m.y. (Lipman and Mehnert, 1979). To the southwest is Red Hill, a cinder cone composed of silicic alkali basalt. A flow from this cone overlies Servilleta flows and rhyolitic material from the 3.8-m.y.-old No Agua Peaks, and is the youngest flow known within the San Luis Valley (Lipman and Mehnert, 1979). **1.9**

13.5 Crossing contact from Servilleta Basalt onto rhyodacite of San Antonio Mountain. **2.1**

15.6 To left are two cinder cones. The one on right is silicic alkali basalt. The other is xenocrystic basaltic andesite. Their trend, with that of Red Hill, parallels the trend of northwest–southeast faults prominent along the eastern edge of the northern Tusas Mountains (Eppler, 1976). **1.9**

17.5 **Junction** of Forest Road 87 with Forest Road 118 on right. **Bear left** on 87 to Lagunitas. **1.6**

19.1 To right (north) Rio San Antonio has cut a canyon through eastward-dipping Hinsdale Formation basalt. **1.2**

20.3 Forest Road 133 on left to Lucero Lakes and Cisnero Park. (See supplemental road log 3B.) Road traverses Esquibel Member of Los Pinos Formation. **2.3**

22.6 Road crosses Rio San Antonio. **0.1**

22.7 Forest Road 78 to Laguna Larga on right. **0.3**

FIGURE 3C-11.6. Red Hill south of San Antonio Mountain. It is a cinder cone composed of silicic alkali basalt.

23.0 Outcrops in hillside to right are Esquibel Member. **0.9**

23.9 **Junction** of Forest Road 87 with Forest Road 93 on left. **Turn right** on 87 to Lagunitas. **0.6**

24.5 Exposures of Esquibel Member on left. **1.7**

26.2 Road parallels northwest-trending, down-to-the-northeast fault for next 0.3 mi. **2.1**

28.3 Road traverses Hinsdale basalt for next 0.2 mi. **0.4**

28.7 Crossing northwest-trending, down-to-the-northeast fault. This fault, along with two others at mileages 29.9 and 30.1, offsets Hinsdale basalt to the southeast and Masonic Park Tuff to the northwest (Manley, 1982), but is not visible here. **1.8**

30.5 Exposure of Masonic Park Tuff. K–Ar age is 28.2 m.y. (Lipman, 1975). **0.2**

30.7 Road climbs on dip slope of Masonic Park Tuff. **0.6**

31.3 Contact of Masonic Park Tuff with overlying Los Pinos Formation. **0.1**

31.4 Road crosses northwest-trending, down-to-the-northeast fault. **0.1**

31.5 Contact of Los Pinos Formation and Masonic Park Tuff. Road continues mostly on tuff for next 3.4 mi. **1.6**

33.1 **Junction** of Forest Road 87 with Forest Road 87A on right. **Turn right** onto 87A to San Miguel (Figs. 3C-33.1a, b). **1.2**

34.3 Crossing small area of Los Pinos Formation above dip slope of Masonic Park Tuff. View west from this ridge crest is into drainage basin of Diablo Creek which, together with Beaver and Cruces Creeks to the right and far right, respectively, created a steep lower gorge prior to confluence to the north with the Rio de Los Pinos (Fig. 3C-34.3). The erosion in this drainage basin has exposed Precambrian rocks (biotite gneiss, metarhyolite, gneissic granodiorite, and quartz monzonite) and

FIGURE 3C-33.1a. Looking north across Lagunitas Creek to cliff of Masonic Park Tuff.

FIGURE 3C-33.1b. Countryside west of Lagunitas campground. Slabs of weathered Masonic Park Tuff from outcrop on left have been piled into large cairns. Such numerous signal(?) cairns are found on hillstops in the northern Tusas Mountains.

Tertiary rocks including the andesite breccias and volcaniclastic sedimentary rocks of the Conejos Formation, and the tuffs and clastic rocks of the Treasure Mountain Tuff (Manley and Wobus, in press). Many of the ridges are capped by Masonic Park Tuff and Los Pinos Formation. Part of the northwest-trending fault zone crossed repeatedly in the area extends through this basin. Numerous additional faults will be crossed between here and the junction with the Rio de los Pinos; few are observable where they cross the road, but offsets can be measured elsewhere; almost all are normal, steeply dipping, and down to the northeast (Manley, 1982a). **0.6**

34.9 Approximate location of down-to-the-northeast fault. Exposures are now in Los Pinos Formation. **0.4**

35.3 Approximate location of down-to-the-northeast fault. **0.7**

36.0 Approximate location of down-to-the-southwest fault that rotates to opposite offset 1.5 mi to the south. **0.4**

36.4 Approximate location of down-to-the-northeast fault. **0.2**

36.6 Fault crosses the road bringing Masonic Park Tuff up to road level on northeast side. **0.6**

37.2 Crossing fault that drops Masonic Park Tuff below road level. Outcrops are now Los Pinos Formation. **0.1**

37.3 A similar fault here and another at 37.5 drop the Masonic Park Tuff to almost the base of canyon on the right. Five additional faults are crossed between miles 37.5 and 41.3. All trend northwest–southeast. **4.0**

41.3 Route crosses from Los Pinos Formation into Masonic Park Tuff and next into Treasure Mountain Tuff. **0.5**

41.8 Approximate contact with Quaternary deposits of the Rio de los Pinos. **0.3**

42.1 Bridge over Rio de los Pinos. Immediately upstream from bridge, at river level, are exposures of a flow in the Conejos Formation. Flow breccias of the Conejos Formation (Fig. 3C-42.1) are exposed downstream for several miles; the Masonic Park Tuff appears to onlap them and disappear from the section just east of San Miguel.

End of supplemental road log 3C for Third Day.

FIGURE 3C-34.3. View west into Cruces Creek drainage at confluence with Diablo Creek. Hillside on right is composed of Precambrian rocks, Conejos Formation, and Treasure Mountain Tuff. Skyline ridge is capped by eastward-dipping Masonic Park Tuff.

DAY 3: SUPPLEMENTAL ROAD LOG 3D, FROM OJO CALIENTE TO WESTERN EDGE OF TAOS PLATEAU

SUMMARY

This log begins near Ojo Caliente and crosses eastward to the southwestern side of the San Luis Valley.

FIGURE 3C-42.1. Conejos Formation flow breccia on north wall of Rio de los Pinos, upstream from San Miguel.

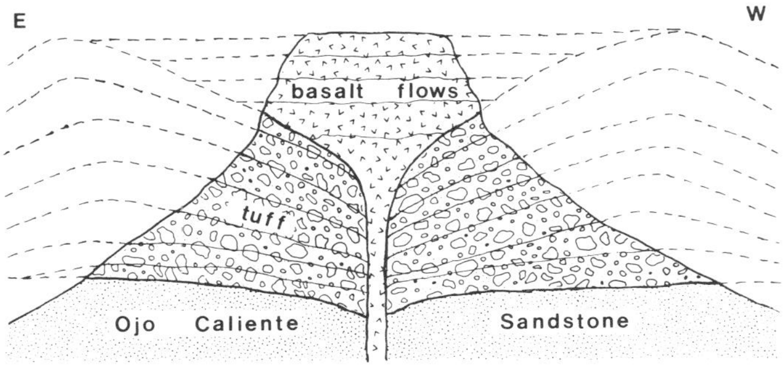

FIGURE 3D-2 4. Schematic cross section of Mesa Vibosa basalt plug.

Mileage

0.0 **Junction** of U.S. 285 and N.M. 96. Proceed northeast on combined U.S. 285 and N.M. 96. **0.1**

0.1 Highway parallels Cañada de los Comanches. La Madera Mountain, at 10:00 exposes Precambrian Ortega Quartzite. Miocene Chama–El Rito Member of the Tesuque Formation exposed on left (north) side of valley for next 4 mi. **0.2**

0.3 Terrace gravels of Rio de los Comanches in roadcut. **0.2**

0.5 Lowermost Ojo Caliente Sandstone exposed in roadcuts for next 7 mi. **0.6**

1.1 Terrace gravels of Rio de los Comanches overlie lowermost Ojo Caliente Sandstone in roadcuts. **0.4**

1.5 At 9:00, cemented ledges expose tongue of Ojo Caliente Sandstone in upper Chama–El Rito. **0.9**

2.4 Butte at 3:00 is Mesa Vibora, a basaltic plug and tuff cone that intrudes the Ojo Caliente Sandstone (Fig. 3D-2.4). Light-gray, inward-dipping basaltic ash forms the middle slopes of the mesa. The ash contains angular blocks of basalt, sparse pebbles of Tertiary rhyolite and latite brought up from the underlying Chama–El Rito, and angular chips of Precambrian quartzite. The upper slopes are flat-lying flows of basalt that overlie the tuff with angular unconformity and probably represent lava ponded in the vent. Three more deeply eroded basalt necks occur within 0.5 mi of the plug. No dates are available on the basalt, but it is probably a part of the Taos Plateau volcanic field. The tuff breccia appears to overlie an erosion surface on the lower Ojo Caliente Sandstone that is about 90–150 ft higher than the present surrounding surface. **0.3**

2.7 Left roadcut exposes light-brown Ojo Caliente Sandstone unconformably overlain by Quaternary wind-blown silt (Española Formation of Galusha and Blick, 1971). For the next 5 mi, this silt is exposed in several of the roadcuts and is distinguished from the underlying Ojo Caliente Sandstone by its darker-brown color and lack of crossbeds. **0.2**

2.9 Enter Carson National Forest. Sangre de Cristo Mountains on skyline. **0.6**

3.5 Road to Cerro Azul to the right (south). Flatlands to the right are eroded lower Ojo Caliente Sandstone discontinuously covered by upper Quaternary Española Formation. **0.2**

3.7 View right (south) into northern Española Basin. At 4:00, Black Mesa is capped with Servilleta Basalt overlying upper Miocene Chamita Formation and Ojo Caliente Sandstone. **0.2**

3.9 At 1:30, forested hill (Cerro Azul) in middle distance is a resistant knob of Precambrian quartzite onlapped by the Ojo Caliente Sandstone. It is part of a basement ridge that separates the San Luis and Española Basins. The ridge is expressed by a gravity high that extends northwest from the Picuris Range to the southern Brazos uplift (Cordell, 1979). **0.4**

4.3 Española Formation in roadcut. **0.3**

4.6 At 10:00–12:00, Comanche Rim on skyline forms the southwestern edge of the Taos Plateau volcanic field. **0.5**

5.1 Road to Cañada de los Comanches on left. **3.2**

8.3 Ascend to top of Comanche Rim. Roadcut exposes Servilleta Basalt overlying unnamed Pliocene sand and gravel. Gravels contain pebbles of Precambrian metamorphic rocks and Tertiary latite and rhyolite probably reworked from the Los Pinos Formation. Vesicular basalt pebbles, possibly derived from the Servilleta Basalt, are also present. These gravels may be approximately age-equivalent with the Puye Conglomerate to the south, but are derived from the north and west. Light-brown fluvial(?) sand, possibly reworked from the Ojo Caliente Sandstone, and light-gray tuffaceous sand of uncertain source are also present below the Servilleta Basalt. Brown eolian silt of the Española Formation overlies the Servilleta Basalt in the roadcut and on the plateau.

End of supplemental road log 3D for Third Day.

ROAD-LOG REFERENCES

Aldrich, L. T., Wetherill, G. W., Davis, G. L., and Tilton, G. R., 1958, Radioactive ages of micas from granitic rocks by Rb–Sr and K–Ar methods: American Geophysical Union, Transactions, v. 39, pp. 1124–1134.

Armstrong, A. K., and Mamet, B. L., 1979, The Mississippian System of north-central New Mexico: New Mexico Geological Society, Guidebook 30, pp. 201–210.

Atwood, W. W., and Mather, K. F., 1932, Physiography and Quaternary geology of the San Juan Mountains, Colorado: U.S. Geological Survey, Professional Paper 166, 176 pp.

Baldridge, W. S., 1979, Petrology and petrogenesis of Plio-Pleistocene basaltic rocks from the central Rio Grande rift, New Mexico, and their relation to rift structure; *in* Riecker, R. E. (ed.), Rio Grande rift—tectonics and magmatism: American Geophysical Union, Washington, D.C., pp. 323–353.

———, Damon, P. E., Shafiqullah, M., and Bridwell, R. J., 1980, Evolution of the Rio Grande rift, New Mexico—new potassium–argon ages: Earth and Planetary Science Letters, v. 51, pp. 309–321.

Barker, F., 1958, Precambrian and Tertiary geology of Las Tablas quadrangle, New Mexico: New Mexico Bureau of Mines and Mineral Resources, Bulletin 45, 104 pp.

———, and Friedman, I., 1974, Precambrian metavolcanic rocks of the Tusas Mountains, New Mexico: major elements and oxygen isotopes: New Mexico Geological Society, Guidebook 25, pp. 115–117.

Baltz, E. H, and Bachman, G. O., 1956, Notes on the geology of the southeastern Sangre de Cristo Mountains, New Mexico: New Mexico Geological Society, Guidebook 7, pp. 96–108.

Bingler, E. C., 1968, Geology and mineral resources of Rio Arriba County, New Mexico: New Mexico Bureau of Mines and Mineral Resources, Bulletin 91, 158 pp.

Bowring, S. A., Reed, J. C., Jr., and Condie, K. C., 1984, U–Pb geochronology of Proterozoic volcanic and plutonic rocks, Sangre de Cristo Mountains, New Mexico (abstract): Geological Society of America, Abstracts with Programs, v. 16.

Boyd, E., 1957, Troubles at Ojo Caliente, a frontier post: El Palacio Magazine, v. 64, November–December, pp. 347–360.

Breese, R., 1984, Rhyolite domes and flows at No Agua Peaks, New Mexico: This guidebook, see supplemental road log 3C.

Brill, K. G., 1952, Stratigraphy in the Permo-Pennsylvanian zeugeosyncline of Colorado and northern New Mexico: Geological Society of America, Bulletin, v. 63, pp. 809–880.

Brown, C. E., 1984, Depositional environments of the Permo-Pennsylvanian Sangre de Cristo Formation, Coyote Creek District, Mora County, New Mexico: This guidebook.

Burns, L. K., and Wobus, R. A., 1983, Correlations and revisions of Precambrian stratigraphy, Needle Mountains, southwest Colorado, and Tusas Mountains, north-central New Mexico (abstract): Geological Society of America, Abstracts with Programs, v. 15, no. 5, p. 424.

Butler, A. P., Jr., 1946, Tertiary and Quaternary geology of the Tusas–Tres Piedras area, New Mexico: Ph.D. thesis, Harvard University, 188 pp.

———, 1971, Tertiary volcanic stratigraphy of the eastern Tusas Mountains, southwest of the San Luis Valley, Colorado–New Mexico: New Mexico Geological Society, Guidebook 22, pp. 289–300.

Cabot, E. C., 1938, Fault border of the Sangre de Cristo Mountains, north of Santa Fe, New Mexico: Journal of Geology, v. 46, no. 1, pp. 88–105.

Carpenter, R. H., 1968, Geology and ore of the Questa molybdenum mine area, Taos County, New Mexico; *in* Ore deposits of the United States 1933–1967: Graton Sales Volume, v. 2, pp. 1328–1350.

Casey, J. M., 1980a, Depositional systems and basin evolution of the late Paleozoic Taos trough, northern New Mexico: Ph.D. thesis, University of Texas at Austin, 236 pp.

———, 1980b, Depositional systems and paleogeographic evolution of the late Paleozoic Taos trough, northern New Mexico; *in* Paleozoic paleogeography of the west-central United States: Rocky Mountain Association of Geologists, Paleogeography Symposium #1, pp. 181–196.

———, and Scott, A. J., 1979, Pennsylvanian coarse-grained fan deltas associated with the Uncompahgre uplift, Talpa, New Mexico: New Mexico Geological Society, Guidebook 30, pp. 211–218.

Chapin, C. E., Chamberlin, R. M., Osburn, G. R., White, D. W., and Sanford, A. R., 1978, Exploration framework of the Socorro geothermal area: New Mexico Geological Society, Special Publication 7, pp. 115–129.

Clark, K. F., and Read, C. B., 1972, Geology and ore deposits of the Eagle Nest area, New Mexico: New Mexico Bureau of Mines and Mineral Resources, Bulletin 94, 152 pp.

———, Johnson, R. B., Lamber, W., and Lisenbee, A. L., 1966, Road log of Sangre de Cristo Mountains and vicinity: New Mexico Geological Society, Guidebook 17, pp. 11–25.

Condie, K. C., 1980, Precambrian rocks of Red River–Wheeler Peak area, New Mexico: New Mexico Bureau of Mines and Mineral Resources, Geologic Map 50, scale 1:48,000.

———, and McCrink, T. P., 1982, Geochemistry of Proterozoic volcanic and granitic rocks from the Gold Hill–Wheeler Peak area, northern New Mexico: Precambrian Research, v. 19, pp. 141–166.

Cordell, L., 1976, Aeromagnetic and gravity studies of the Rio Grande graben in New Mexico between Belen and Pilar: New Mexico Geological Society, Special Publication 6, pp. 62–70.

———, 1978, Complete Bouguer anomaly map of the Taos Basin section of the Rio Grande rift, New Mexico: U.S. Geological Survey, Open-file Map 78-317.

———, 1979, Gravimetric expression of graben faulting in Santa Fe country and the Española Basin, New Mexico: New Mexico Geological Society, Guidebook 30, pp. 59–64.

Corey, A. F., 1960, Kyanite occurrences in the Petaca district, Rio Arriba County, New Mexico: New Mexico Bureau of Mines and Mineral Resources, Bulletin 47, 70 pp.

Cross, W., and Larsen, E. S., 1935, A brief review of the geology of the San Juan region of southwestern Colorado: U.S. Geological Survey, Bulletin 843, 138 pp.

Dane, C. H., and Bachman, G. O., 1965, Geologic map of New Mexico: U.S. Geological Survey, scale 1:500,000.

Dethier, D. P., and Demsey, K. A., 1984, Erosional history and soil development on Quaternary surfaces, northwest Española Basin, New Mexico: This guidebook.

———, and Manley, K. (in press), Geologic map of the Chili quadrangle, New Mexico: U.S. Geological Survey, Miscellaneous Field Studies Map.

———, and Martin, B. A., 1984, Stratigraphy and depositional environment of the Chama–El Rito sandstone northwest of Española, New Mexico: This guidebook.

Dungan, M. A., Muehlberger, W. R., Leininger, L., Peterson, C., McMillan, N. J., Gunn, G., Lindstrom, M., and Haskin, L., 1984, Volcanic and sedimentary stratigraphy of the Rio Grande gorge, and the late Cenozoic geologic evolution of the southern San Luis Valley: This guidebook.

Ekas, L. M., Ingersoll, R. V., Baldridge, W. S., and Shafiqullah, M., 1984, The Chama–El Rito Member of the Tesuque Formation, Española Basin, New Mexico: This guidebook.

Eppler, D., 1976, The geology of the San Antonio Mountain area, Tres Piedras, Taos and Rio Arriba Counties, New Mexico: M.S. thesis, University of New Mexico, 77 pp.

Friedman, I. I., Smith, R. L., Long, W. D., 1966, Hydration of natural glass and the formation of perlite: Geological Society of America, Bulletin, v. 77, pp. 323–328.

Galusha, T., and Blick, J. C., 1971, Stratigraphy of the Santa Fe Group, New Mexico: American Museum of Natural History, Bulletin 144, 127 pp.

Grambling, J. A., 1984, Proterozoic isobaric surfaces and tectonism in northern and central New Mexico (abstract): Geological Society of America, Abstracts with Programs, v. 16, no. 4, p. 222.

———, and Codding, D. B., 1982, Stratigraphic and structural relationships of multiply deformed Precambrian metamorphic rocks in the Rio Mora area, New Mexico: Geological Society of America, Bulletin, v. 93, pp. 127–137.

———, Williams, M. L., and Codding, D. B., 1983, Mn- and Cr-rich marker horizons in multiply deformed Proterozoic rocks, northern New Mexico (abstract): Geological Society of America, Abstracts with Programs, v. 15, no. 5, p. 424.

Gresens, R. L., and Stensrud, H. L., 1974, Recognition of more metarhyolite occurrences in northern New Mexico and a possible Precambrian stratigraphy: Mountain Geologist, v. 1, pp. 109–124.

Gruner, J. W., 1920, Geologic reconnaissance of the southern part of the Taos Range, New Mexico: Journal of Geology, v. 23, no. 8, pp. 731–742.

Hagstrum, J. T., Lipman, P. W., and Elston, D. P., 1982, Paleomagnetic evidence bearing on the structural development of the Latir volcanic field near Questa, New Mexico: Journal of Geophysical Research, v. 87, pp. 7833–7842.

Holcombe, R. J., and Callender, J. F., 1982, Structural analysis and stratigraphic problems of Precambrian rocks of the Picuris Range, New Mexico: Geological Society of America, Bulletin, v. 93, pp. 138–149.

Hovorka, D. S., 1978, Petrology of Precambrian metarhyolites of north-central New Mexico: Unpubl. senior thesis, Williams College, Williamstown, Massachusetts, 67 pp.

Ishihara, S., 1967, Molybdenum mineralization at Questa mine, New Mexico, U.S.A.: Geological Survey of Japan, Report 218, 68 pp.

Izett, G. A., and Naeser, C. W., 1981, Fission-track ages of air-fall tuffs in

Miocene sedimentary rocks of the Española Basin, Santa Fe County, New Mexico: U.S. Geological Survey, Open-file Report 81-0161, 10 pp.

Jahns, R. H., 1946, Mica deposits of the Petaca district, Rio Arriba County, New Mexico: New Mexico Bureau of Mines and Mineral Resources, Bulletin 25, 294 pp.

———, and Ewing, R. C., 1976, The Harding mine, Taos County, New Mexico: New Mexico Geological Society, Guidebook 27, pp. 263–276.

Just, E., 1937, Geology and economic features of the pegmatites of Taos and Rio Arriba Counties, New Mexico: New Mexico Bureau of Mines and Mineral Resources, Bulletin 13, 73 pp.

Kelley, V. C., 1973, Geology of Española Basin, New Mexico: New Mexico Bureau of Mines and Mineral Resources, Geologic Map 48, scale 1:125,000

Kepes, G. J., 1984, Ortega Quartzite at Cerro Azul (abstract):New Mexico Geological Society, 1984 Spring Meeting, Abstracts and Program.

Kues, B. S., 1984, Pennsylvanian stratigraphy and paleontology of the Taos area, north-central New Mexico: This guidebook.

Lambert, P. W., 1966, Notes on the late Cenozoic geology of the Taos–Questa area, New Mexico: New Mexico Geological Society, Guidebook 17, pp. 43–50.

Larsen, E. S., and Cross, W., 1956, Geology and petrology of the San Juan region, southwestern Colorado: U.S. Geological Survey, Professional Paper 258, 303 pp.

Leonardson, R. W., Dunlap, G., Starquist, V. L., Bratton, G. P., Meyer, J. W., Osborn, L. W., Atkin, S. A., Molling, P. A., Moore, R. F., and Olmore, S. D., 1983, Preliminary geology and molybdenum deposits at Questa, New Mexico; *in* The genesis of Rocky Mountain ore deposits—changes with time and tectonics: Proceedings of Denver Region Exploration Geologists Society, pp. 151–155.

Lipman, P. W., 1974, Geologic map of the Platoro caldera area, southeastern San Juan Mountains, southwestern Colorado: U.S. Geological Survey, Map I-828.

———, 1975a, Evolution of the Platoro caldera complex and related volcanic rocks, southeastern San Juan Mountains, Colorado: U.S. Geological Survey, Professional Paper 852, 128 pp.

———, 1975b, Geologic map of the lower Conejos River canyon area, southeastern San Juan Mountains, Colorado: U.S. Geological Survey, Map I-901.

———, 1981, Volcano-tectonic setting of Tertiary ore deposits, southern Rocky Mountains; *in* Dickinson, W. R., and Payne, W. D. (eds.), Relation of tectonics to ore deposits in the southern Cordillera: Arizona Geological Society Digest, v. 14, pp. 199–213.

———, 1983, The Miocene Questa caldera, northern New Mexico—relation to batholith emplacement and associated molybdenum mineralization: Denver Region Exploration Geologists Society, pp. 133–148.

———, and Mehnert, H. H., 1975, Late Cenozoic basaltic volcanism and development of the Rio Grande depression in the southern Rocky Mountains: Geological Society of America, Memoir 144, pp. 119–154.

———, and ———, 1979, The Taos Plateau volcanic field, northern Rio Grande rift, New Mexico; *in* Riecker, R. E. (ed.), Rio Grande rift—tectonics and magmatism: American Geophysical Union, Washington, D.C., pp. 28–311.

———, Steven, T. A., and Mehnert, H. H., 1970, Volcanic history of the San Juan Mountains, Colorado, as indicated by potassium–argon dating: Geological Society of America, Bulletin 81, pp. 2329–2352.

Luddington, S., 1982, Quartz–pyrite–molybdenite stockwork near South Fork Peak, Taos County, New Mexico: U.S. Geological Survey, Open-file Report 82-180, 8 pp.

Machette, M. N., and Personius, S. F., 1983, Quaternary and Pliocene faults in the eastern part of the Aztec and western part of the Raton 1° × 2° quadrangle, New Mexico: U.S. Geological Survey, Miscellaneous Field Studies Map, MF-1465-B.

Mallory, W. W., 1972, Pennsylvanian arkose and the ancestral Rocky Mountains; *in* Geologic atlas of the Rocky Mountain region: Rocky Mountain Association of Geologists, pp. 131–132.

Manley, K., 1976, K–Ar age determination on Pliocene basalts from the Española Basin, New Mexico: Isochron/West, no. 16, pp. 29–30.

———, 1977, Geologic map of the Cejita Member (new name) of the Tesuque Formation, Española Basin, New Mexico: U.S. Geological Survey, Miscellaneous Field Studies Map, MF-877.

———, 1978, Cenozoic geology of the Española Basin; *in* Hawley, J. W. (compiler), Guidebook to the Rio Grande rift in New Mexico and Colorado: New Mexico Bureau of Mines and Mineral Resources, Circular 163, pp. 201–210.

———, 1979a, Stratigraphy and structure of the Española Basin, Rio Grande rift, New Mexico; *in* Riecker, R. E. (ed.), Rio Grande rift—tectonics and magmatism: American Geophysical Union, Washington, D.C., pp. 71–86.

———, 1979b, Tertiary and Quaternary stratigraphy of the northeast plateau, Española Basin, New Mexico: New Mexico Geological Society, Guidebook 30, pp. 231–236.

———, 1981, Redefinition and description of the Los Pinos Formation of north-central New Mexico: Geological Society of America, Bulletin, v. 92, no. 12, pp. 984–989.

———, 1982a, Geologic map of the Bighorn Peak quadrangle, Rio Arriba County, New Mexico, and Conejos County, Colorado: U.S. Geological Survey, Miscellaneous Field Studies Map MF-1451.

———, 1982b, Geologic map of the Broke Off Mountain quadrangle, Rio Arriba County, New Mexico: U.S. Geological Survey, Miscellaneous Field Studies Map MF-1450, scale 1:24,000.

———, and Mehnert, H. H., 1981, New K–Ar ages for Miocene and Pliocene volcanic rocks in the northwestern Española Basin, and their relationships to the history of the Rio Grande rift: Isochron/West, no. 18, pp. 5–8.

———, and Naeser, C. W., 1977, Fission-track ages for tephra layers in upper Cenozoic rocks, Española Basin, New Mexico: Isochron/West, no. 18, pp. 13–14.

———, and Wobus, R. A., 1982a, Reconnaissance geologic map of the Las Tablas quadrangle, Rio Arriba County, New Mexico: U.S. Geological Survey, Miscellaneous Field Studies Map MF-1408, scale 1:24,000.

———, and ———, 1982b, Reconnaisance geologic map of the Mule Canyon quadrangle, New Mexico: U.S. Geological Survey, Miscellaneous Field Studies Map MF-1407, scale 1:24,000.

———, and ——— (in press), Reconnaissance geologic map of the Brazos Peak northeast quadrangle, Rio Arriba County, New Mexico, and Conejos County, Colorado: U.S. Geological Survey, Miscellaneous Field Studies Map MF-1624, scale 1:24,000.

———, Scott, G. R., and Wobus, R. A. (in press), Geologic map of the Aztec 1° × 2° quadrangle, northwestern New Mexico and southern Colorado: U.S. Geological Survey, Miscellaneous Geologic Investigation Map.

May, S. J., 1980, Geology of the Ojo Caliente–Rio Chama area, Rio Arriba and Taos Counties, New Mexico: Ph.D. thesis, University of New Mexico, 198 pp.

Mayo, E. B., 1956, Lineament tectonics and some ore districts in the Southwest: Mining Engineering, v. 10, no. 11, pp. 1169–1175.

McCarty, R. M., 1983, Structural geology and petrology of part of the Vadito Group, Picuris Mountains, New Mexico: M.S. thesis, University of New Mexico, 159 pp.

Miller, J. P., Montgomery, A., and Sutherland, P. K., 1963, Geology of part of the southern Sangre de Cristo Mountains, New Mexico: New Mexico Bureau of Mines and Mineral Resources, Memoir 11, 106 pp.

Montgomery, A., 1953, Precambrian geology of the Picuris Range, north-central New Mexico: New Mexico Bureau of Mines and Mineral Resources, Bulletin 30, 89 pp.

Morgan, T. L., and Broxton, D. E., 1978, Uranium hydrogeochemical and stream sediment reconnaissance of the Raton NTMS quadrangle, New Mexico: Los Alamos Scientific Laboratory, Report LA-7344-MS, 84 pp.

Muehlberger, W. R., 1968, Geology of the Brazos Peak quadrangle, New Mexico: New Mexico Bureau of Mines and Mineral Resources, Geologic Map 22, scale 1:48,000.

———, 1979, The Embudo fault between Pilar and Arroyo Hondo, New Mexico—an active intracontinental transform fault: New Mexico Geological Society, Guidebook 30, pp. 77–82.

Ozima, M., Kono, M., Kaneoka, I., Kinoshita, H., Kobayashi, K., Nagata, T., Larsen, E. E., and Strangway, D. W., 1967, Paleomagnetism and potassium–argon ages of some volcanic rocks from the Rio Grande gorge, New Mexico: Journal of Geophysical Research, v. 72, no. 10, pp. 2615–1621.

Personius, S. F., and Machette, M. N., 1984, Quaternary and Pliocene faulting in the Taos Plateau region, northern New Mexico: This guidebook.

Reed, J. C., Jr., 1984, Precambrian rocks of the Taos Range, Sangre de Cristo Mountains, New Mexico: This guidebook.

———, Lipman, P. W., and Robertson, J. M., 1983, Geologic map of the Latir Peak and Wheeler Peak Wildernesses and Columbine–Hondo Wilderness Study Area, Taos County, New Mexico: U.S. Geological Survey, Miscellaneous Field Studies Map MF-120B.

Roberts, J. W., Barnes, J. J., and Wacker, H. J., 1976, Subsurface Paleozoic stratigraphy of the northeastern New Mexico basin and arch complex: New Mexico Geological Society, Guidebook 27, pp. 141–152.

Sidwell, R., and Warn, G. F., 1953, Pennsylvanian and related sediments of upper Pecos Valley, New Mexico: American Association of Petroleum Geologists, Bulletin 37, pp. 975–1013.

Selley, R. C., 1970, Ancient sedimentary environments: Cornell University Press, 237 pp.

Smith, C. T., Budding, A. J., Pitrat, C. W., 1961, Geology of the southeastern part of the Chama Basin: New Mexico Bureau of Mines and Mineral Resources, Bulletin 75, 57 pp.

Smith, R. L., and Bailey, R. A., 1968, Stratigraphy, structure, and volcanic evolution of the Jemez Mountains, New Mexico (abstract): *in* Epis, R. C. (ed.), Cenozoic volcanism in the southern Rocky Mountains: Colorado School of Mines Quarterly, v. 63, no. 3, pp. 259–260.

Steiger, R. H., and Jager, E., 1977, Subcommission of geochronology—convention on the use of decay constants in geo- and cosmochronology: Earth and Planetary Science Letters, v. 14, p. 74–78.

Steinpress, M. G., 1980, Neogene stratigraphy and structure of the Dixon area, Española Basin, north-central New Mexico: M.S. thesis, University of New Mexico, 127 pp.

———, 1981, Neogene stratigraphy and structure of the Dixon area, Española Basin, north-central New Mexico—summary: Geological Society of America, Bulletin, v. 92, no. 12, pp. 1023–1026.

Stix, J., Pearson, C., Vuataz, F., Goff, F. East, J., and Hoffers, B., 1982, Geology, resistivity, and hydrogeochemistry of the Ojo Caliente hot springs area, northern New Mexico: Geothermal Resources Council, Transactions, v. 6, pp. 55–58.

Sutherland, P. K., 1963, Paleozoic rocks; *in* Geology of the southern Sangre de Cristo Mountains, New Mexico: New Mexico Bureau of Mines and Mineral Resources, Memoir 11, pp. 22–46.

———, and Harlow, F. H., 1973, Pennsylvanian brachiopods and biostratigraphy in southern Sangre de Cristo Mountains, New Mexico: New Mexico Bureau of Mines and Mineral Resources, Memoir 27, 173 pp.

Treiman, A. H., 1977, Precambrian geology of the Ojo Caliente area, New Mexico: M.S. thesis, Stanford University.

———, 1978, A hypabyssal granitic intrusion of Precambrian age, Ojo Caliente quadrangle, New Mexico (abstract): Geological Society of America, Abstracts with Programs, v. 10, no. 5, p. 240.

Tweto, O., 1975, Laramide (Late Cretaceous–early Tertiary) orogeny in the southern Rocky Mountains; *in* Curtis, B. F. (ed.), Cenozoic history of the southern Rocky Mountains: Geological Society of America, Memoir 144, pp. 1–44.

———, 1979, The Rio Grande rift system in Colorado; *in* Riecker, R. E. (ed.), Rio Grande rift—tectonics and magmatism: American Geophysical Union, Washington, D.C., pp. 33–56.

Ulmer, D. S., and Laury, R. L., 1984, Diagenesis of the Mississippian Arroyo Peñasco Group, north-central New Mexico: This guidebook.

Vuataz, F., Stix, J., Goff, F., and Pearson, C. (in press), Low-temperature geothermal potential of the Ojo Caliente warm springs area, northern New Mexico: Los Alamos National Laboratory, Informal Report, 60 pp.

Whitson, D. N., 1982, Geology of the perlite deposit at No Agua Peaks, New Mexico: New Mexico Bureau of Mines and Mineral Resources, Circular 182, pp. 89–95.

Williams, M. L., 1984, Precambrian stratigraphy, structure, and metamorphism in the Tusas Mountains, New Mexico—a progress report (abstract): New Mexico Geological Society, 1984 Spring Meeting, Abstracts and Program.

Wobus, R. A., 1984, An overview of the Precambrian geology of the Tusas Range, north-central New Mexico: This guidebook.

———, and Hedge, C. E., 1980, Rb–Sr isochron age of Precambrian plutons of the San Pedro Mountains, north-central New Mexico: Isochron/West, no. 27, pp. 19–25.

———, and Manley, K., 1982, Reconnaissance geologic map of the Burned Mountain quadrangle, Rio Arriba County, New Mexico: U.S. Geological Survey, Miscellaneous Field Studies Map MF-1409, scale 1:24,000.

Woodward, L. A., Callender, J. F., Seager, W. R., Chapin, C. E., Gries, J. C., Shaffer, W. L., and Zilinski, R. E., Jr., 1978, Tectonic map of Rio Grande rift region in New Mexico, Chihuahua, and Texas: New Mexico Bureau of Mines and Mineral Resources, Circular 163, scale 1:1,000,000.

High sage desert near Monte Blanca, San Luis Valley, Rio Grande rift, Colorado. Photo by Jim Bones.